Biochemical Engineering

Biochemical Engineering

Harvey W. Blanch
Douglas S. Clark

University of California at Berkeley
Berkeley, California

Taylor & Francis
Taylor & Francis Group
Boca Raton London New York

CRC is an imprint of the Taylor & Francis Group,
an informa business

CRC Press
Taylor & Francis Group
6000 Broken Sound Parkway NW, Suite 300
Boca Raton, FL 33487-2742

Printed in the United States of America on acid-free paper
20 19 18 17 16 15
International Standard Book Number-13: 978-0-8247-0099-7 (softcover)

Library of Congress Cataloging-in-Publication Data

Catalog record is available from the Library of Congress

Visit the Taylor & Francis Web site at
http://www.taylorandfrancis.com

and the CRC Press Web site at
http://www.crcpress.com

Preface

Biochemical engineering embraces aspects of biochemistry, cell and molecular biology, bioorganic and bioinorganic chemistry, and has at its core the discipline of chemical engineering. Biochemical engineering may be loosely defined as the use of living organisms, or part of them (for example, enzymes), for the production of chemical or biological materials or the development of new processes. In this sense, the practice of biochemical engineering has a long history, including such early applications as the fermentation of grapes for wine, brewing of beer, cheese manufacture, leavening of bread, and effluent disposal. In more recent times, biochemical engineers have been involved in antibiotic fermentations, production of industrial solvents, organic acids, vaccines, blood and tissue products, animal feedstuffs, commercial enzymes and in the treatment of effluents and wastes.

The advent of recombinant DNA and hybridoma technology in the late 1970s opened new avenues for the production of enzymes and therapeutic proteins, and the modification of metabolic pathways for the production of biochemicals. Today, biochemical engineers work with a variety of cells from microbial, mammalian and plant sources, in the form of single cells and as tissues. Many issues that were vitally important earlier have been resolved in the past decades; these include sterile air filtration, medium and equipment sterilization, aseptic equipment design, and instrumentation. Our knowledge of microbial and cellular physiology has expanded rapidly, together with our understanding of enzyme structure and catalytic function. New biological catalysts, such as ribozymes, catalytic antibodies, and modified enzymes provide new tools for the production of chemicals, biochemicals, materials, and therapeutic agents. Exploitation of advances in recombinant DNA technology relies on early studies of microbial physiology, scale-up and reactor design, heat and mass transfer in fermentations, and on separation and purification processes.

One of the most important tools in biochemical engineering has been continuous culture (the chemostat); it provided a technique to study microbial physiology under well-defined conditions, and a means to develop kinetic models of microbial growth and product formation. The theoretical and experimental basis of chemostat operation was originally proposed by Monod[1]and Novik and Szilard[2] and it was extensively developed by researchers at the Microbiological Research Establishment at Porton Downs (U.K.)[3]and at the Institute of Microbiology of the Czechoslovak Academy of Sciences[4]. The annual review of publications on continuous culture, appearing in Folia Microbiologia[5], included developments in the theory of continuous cultivation, microbial physiology and product formation and applications of the method. In the past twenty years, the number of publications involving chemostats has grown tremendously. Continuous culture has played an important role in understanding the stability of plasmids in recombinant cells, in following events in the cell cycle, and in understanding monoclonal antibody production by hybridoma cells.

Paralleling these advances in fermentation, the area of separation and recovery of biological products has seen the development of techniques such as ion exchange, affinity and gel permeation chromatography, electrophoresis, aqueous two-phase extraction, and membrane separation processes. Most of these techniques rely on newly-developed materials with suitable properties. The high purity requirement of recombinant proteins has been a significant factor in developing highly-efficient small-scale separation processes. The theoretical underpinnings of these separation processes are based on biophysical chemistry, colloidal and interfacial phenomena and mass transfer. Appropriate descriptions of intermolecular interactions are important in understanding processes such as precipitation, crystallization, extraction and interactions of biological materials with surfaces.

There have been several texts that have had a profound influence on the development of biochemical engineering. One of the first texts, written from a chemical engineering perspective, is F.C. Webb's book, "Biochemical Engineering", developed from courses at University College, London[6]. It emphasizes physical chemistry, kinetics, and heat and mass transfer. The first text to integrate microbiology, enzymology and bioprocess engineering was Aiba, Humphrey and Millis' "Biochemical Engineering"[7]. It served as the standard reference for many years and provided a valuable introduction to biochemistry and biology for chemical engineers. Bailey and Ollis' text "Biochemical Engineering

---

(1) Monod, J., Annals Inst. Pasteur **79**, 390 (1950)

(2) Novick, A. and L. Szilard, Science, **112**, 715 (1950)

(3) Herbert, D., Ellsworth, R. and R.C. Telling, J. Gen. Microbiol. **14**, 601 (1956)

(4) Malek, I., Beran K. and Z. Fencl, Theoretical and Methodological Basis of Continuous Culture of Microorganisms, Publishing House of Czech. Academy of Sciences, Prague (1966)

(5) Malek, I. and Z. Fencl, Folia Microbiologia, **6**, 142 (1961) and annual reviews thereafter

(6) Webb, F.C., Biochemical Engineering, van Nostrand (1964)

(7) Aiba, S., Humphrey, A.E. and N.F. Millis, Biochemical Engineering, Academic Press, 1st edition (1964) and 2nd edition (1973)

Fundamentals"[8] incorporated modern biology and included sections on the commercial applications of biochemical engineering. It provided students with sufficient background in the life sciences to enable them to address problems in microbial growth, enzyme kinetics and metabolic control.

This book is a comprehensive textbook of modern biochemical engineering, intended for students in engineering and applied science. With the increasing emphasis on life sciences in engineering curricula, many students interested in biochemical engineering already have backgrounds in biochemistry or cell biology. This text therefore does not include sections on biochemistry, microbiology, and molecular and cell biology. There are a number of excellent texts available that can be used as a supplement to this text, permitting the student to cover biology to any depth required[9]. Material is presented at levels appropriate for both undergraduate and graduate courses in biochemical engineering, and sample course outlines are provided below for courses taken by students with and without backgrounds in the life sciences. Each major topic follows a logical progression from basic principles to more advanced concepts. The major topics include enzyme kinetics and biocatalysis, microbial growth and product formation, bioreactor design, transport in bioreactors, bioproduct recovery, and bioprocess economics and design. Problems are included at the end of each chapter, ranging from straightforward exercises testing knowledge of fundamental concepts to more complex problems dealing with real systems and real bioprocesses. Although the primary aim of the book is to serve as an instructional text for students of biochemical engineering, it should also be useful to practicing biochemical engineers who are concerned with the basic concepts underlying the design and behavior of bioprocesses.

Acknowledgments

We are very grateful to Peter Michels and Jeanne George for reviewing preliminary versions of each chapter. Their critical comments on content and presentation were extremely useful. Suggestions offered by teaching assistants and students as the text progressed to the classroom were also invaluable, especially those of Andrew Shuler, M. Scott Clarke, and Mareia Frost. Thoughtful and useful reviews of selected chapters were furnished by Jonathan Dordick, Charles Goochee, and Timothy Oolman. Thanks are also owed to Larry Erikson, Doug Lauffenburger, and Jay Bailey for generously supplying several homework problems. We are especially grateful to John Martens for his assistance with the illustrations throughout the text.

Harvey W. Blanch
Douglas S. Clark

---

(8) Bailey, J.E. and D.F. Ollis, Biochemical Engineering Fundamentals, McGraw Hill, 1st edition (1977) and 2nd edition (1986)

(9) Stryer, L., Biochemistry, W.H. Freeman, N.Y., 4th edition (1995); Glazer, A.N., and H. Nikaido, *Microbial Biotechnology*, W.H. Freeman & Co. (1995)

Contents

Course Outlines

Undergraduate course on biochemical engineering

Given below is the outline for a course on biochemical engineering, structured for a 15-week semester, assuming three 50-minute lectures per week, and three lecture periods for examinations or review. This outline is intended for students who have had an introductory course in biochemistry or a related biological science, or who are taking such a course concurrently. A course outline for students with no previous experience in the biological sciences follows.

Lecture	**Reading Assignment**
1. Overview of biotechnology	Handouts
2. Principles of enzyme catalysis	pp. 1 - 5; 18 - 27 (Chapter 1)
3. Principles of enzyme catalysis (cont.)	pp. 27 - 39
4. Kinetics of single substrate reactions	pp. 5 - 18
5. Burst kinetics of chymotrypsin. Enzyme inhibition	pp. 50 - 52; 39 - 45
6. Enzyme inhibition (cont.) Enzyme denaturation and inactivation	pp. 39 - 45; 77 - 81
7. Methods of enzyme and cell immobilization	pp. 103 - 116 (Chapter 2)
8. Electrostatic and external mass transfer effects on immobilized enzyme kinetics	pp. 116 - 127
9. Internal mass transfer effects on immobilized enzyme kinetics	pp. 127 - 140
10. Internal mass transfer effects (cont.)	pp. 127 - 140
11. Stoichiometry and energetics of microbial growth	pp. 162 - 181 (Chapter 3)
12. Unstructured models of microbial growth	pp. 181 - 209
13. Unstructured models (cont.)	pp. 181 - 209
14. Structured models of microbial growth	pp. 225 - 230
15. Structured models (cont.)	pp. 230 - 236; 244 - 246
16. Continuous stirred tank bioreactors	pp. 276 - 291 (Chapter 4)
17. Enzyme catalysis in a CSTR Chemostats in series	pp. 292 - 297
18. Plug-flow and packed bed bioreactors Imperfect mixing	pp. 297 - 305
19. Fed batch bioreactors	pp. 305 - 309
20. Gas-liquid mass transfer in bioreactors	pp. 353 - 361 (Chapter 5)
21. Mass balances for two-phase bioreactors	pp. 361 - 364
22. Mass transfer coefficient k_La	pp. 390 - 403
23. Power requirements for bioreactors	pp. 403 - 415
24. Sterilization	pp. 415 - 426

25. Introduction to bioproduct recovery. Centrifugation	pp. 453 - 467 (Chapter 6)
26. Filtration and ultrafiltration	pp. 467 - 470
27. Filtration and ultrafiltration	pp. 482 - 490
28. Precipitation of proteins	pp. 491 - 502
29. Basic concepts of chromatography	pp. 502 - 511
30. Mathematical analysis of chromatography	pp. 512 - 520
31. Fixed-bed adsorption and affinity chromatography. Plate theory of chromatography	pp. 523 - 533
32. Plate theory of chromatography (cont.)	pp. 526 - 533
33. Electrophoresis and crystallization	pp. 533 - 542
34. Review	Chapters 4 - 6
35. Microbial interactions and competition	pp. 578 - 590 (Chapter 7)
36. Competition (cont.)	pp. 581 - 590
37. Predation	pp. 592 - 597
38. Manufacture of biological products	pp. 609 - 643 (Chapter 8)
39. Economic analysis of bioprocesses	pp. 643 - 648
40. Capital costs and manufacturing costs	pp. 648 - 657
41. Case studies	pp. 657 - 677
42. Final Review	Chapters 1 - 8

Additional or alternate topics for a graduate course:

Extended coverage of enzyme kinetics (limiting conditions, multi-substrate reactions, allosteric interactions, pH and temperature effects)	Sections 1.7 - 1.10
Intraparticle diffusion and immobilized enzyme stability	Section 2.5
Growth of eukaryotic cells	Section 3.4
Models of product formation	Section 3.5.3
Gene expression and regulation. Plasmid expression and regulation	Sections 3.6 and 3.7
Further analysis of bioreactors (transient behavior, recycle systems)	Sections 4.6 and 4.7
Expanded survey of transport processes (rheology of fermentation broths, bubble columns, coupled mass transfer and reaction)	Sections 5.1, 5.3, and 5.6
More on heat transfer	Section 5.5.2
Additional separation methods (e.g., extraction)	Section 6.3.1
Microbial interactions in greater depth (chemotaxis, stability)	Sections 7.3, 7.5, and 7.6

For students without background in biological sciences, the following outline is suggested.

Lecture	Reading Assignment
1. Introduction to microorganisms and biological molecules	Supplemental reading
2. Chemistry of amino acids and proteins	Supplemental reading
3. Protein structure and function	Supplemental reading
4. Molecular genetics and protein synthesis	Supplemental reading
5. Recombinant DNA and genetic engineering	Supplemental reading
6. Basic microbiology and cellular structure	Supplemental reading
7. Energetics and metabolism	Supplemental reading
8. Overview of biotechnology	Handouts
9. Principles of enzyme catalysis	pp. 1 - 5, 18-27 (Chapter 1)
10. Kinetics of single substrate reactions	pp. 5 - 18
11. Enzyme inhibition	pp. 39 - 45
12. Enzyme denaturation and inactivation	pp. 77 - 81
13. Methods of enzyme and cell immobilization. External mass transfer and immobilized enzyme kinetics	pp. 103 - 116; 121-125 (Chapter 2)
14. Internal mass transfer effects on immobilized enzyme kinetics	pp. 127 - 140
15. Stoichiometry and energetics of microbial growth	pp. 162 - 181 (Chapter 3)
16. Unstructured models of microbial growth	pp. 181 - 190
17. Structured models of microbial growth	pp. 225 - 230 pp. 244 - 246
18. Continuous stirred tank bioreactors	pp. 276 - 291 (Chapter 4)
19. Enzyme catalysis in a CSTR. Chemostats in series	pp. 292 - 297
20. Plug-flow and packed bed bioreactors. Fed batch bioreactors	pp. 297 - 300 pp. 305 - 308
21. Gas-liquid mass transfer in bioreactors	pp. 353 - 361 (Chapter 5)
22. Mass balances for two-phase bioreactors	pp. 361 - 364
23. Mass transfer coefficient k_La	pp. 390 - 403
24. Power requirements for bioreactors	pp. 403 - 415
25. Sterilization	pp. 415 - 426
26. Introduction to bioproduct recovery; Centrifugation	pp. 453 - 467 (Chapter 6)
27. Filtration and ultrafiltration	pp. 467 - 470
28. Filtration and ultrafiltration	pp. 482 - 490
29. Precipitation of proteins	pp. 491 - 502
30. Basic concepts of chromatography	pp. 502 - 511

31. Mathematical analysis of chromatography	pp. 512 - 520
32. Fixed-bed adsorption and affinity chromatography. Plate theory of chromatography	pp. 523 - 533
34. Plate theory of chromatography (cont.)	pp. 526 - 533
35. Electrophoresis and crystallization	pp. 533 - 542
36. Review	Chapters 3 - 6
37. Microbial interactions and competition	pp. 578 - 590 (Chapter 7)
38. Competition (cont.)	pp. 581 - 590
39. Manufacture of biological products	pp. 609 - 643 (Chapter 8)
40. Economic analysis of bioprocesses	pp. 643 - 648
41. Capital costs and manufacturing costs	pp. 648 - 657
42. Case studies	pp. 657 - 677
43. Final Review	Chapters 1 - 8

Chapter 1. Enzyme Catalysis

Enzymes are one of the essential components of all living systems. These macromolecules have a key role in catalyzing the chemical transformations that occur in all cell metabolism. The nature and specificity of their catalytic activity is primarily due to the three-dimensional structure of the folded protein, which is determined by the sequence of the amino acids that make up the enzyme. The activity of globular proteins may be regulated by one or more small molecules, which cause small conformational changes in the protein structure. Catalytic activity may depend on the action of these non-protein components (known as *cofactors*) associated with the protein. If the cofactor is an organic molecule, it is referred to as a *coenzyme*. The catalytically inactive enzyme (without cofactor) is termed an *apoenzyme;* when coenzyme or metal ion is added, the active enzyme is then termed a *holoenzyme*. Many cofactors are tightly bound to the enzyme and cannot be easily removed; they are then referred to as *prosthetic groups*.

In this chapter we shall examine the nature of enzyme catalysis, first by examining the types of reactions catalyzed and the mechanisms employed by enzymes to effect this catalysis, and then by reviewing the common constitutive rate expressions which describe the kinetics of enzyme action. As we shall see, these can range from simple rate expressions to complex expressions that involve several reactants and account for modification of the enzyme structure.

1.1 Specificity of Enzyme Catalysis

Enzymes have been classified into six main types, depending on the nature of the reaction catalyzed. A numbering scheme for enzymes has been developed, in which the main classes are distinguished by the first of four digits. The second and third digits describe the type of reaction catalyzed, and the fourth digit is employed to distinguish between enzymes of the same function on the basis of the actual substrate in the reaction catalyzed. This scheme has proven useful in clearly delineating many enzymes that have similar activities. It was developed by the Enzyme Commission and the prefix E.C. is generally employed with the numerical scheme.

We shall examine the six main classes of enzymes primarily to illustrate the broad range of reactions which can be catalyzed by enzymes and the types of substrates they can act on.

Class 1. Oxidoreductases

These enzymes catalyze the transfer of hydrogen or oxygen atoms or electrons from one substrate to another. They are often called oxidases or dehydrogenases, and reference is made to the substrate of the reaction, e.g. lactate dehydrogenase. The first digit (1.) of the classification thus indicates the class oxidoreductase; the second digit indicates the donor of the hydrogen atom or electron (i.e., reducing equivalent) involved. The third digit describes the hydrogen atom or electron acceptor.

First digit	*Second digit*	*Third digit*
1.	1 alcohol	1 NAD^+ or $NADP^+$
(Oxidoreductase)	2 aldehyde or ketone	2 Fe^{3+}
	3 alkene -CH=CH-	3 O_2
	4 primary amine	4 otherwise unclassified
	5 secondary amine	
	6 NADH or NADPH	

Class 2. Transferases

Transferases catalyze the group transfer reactions, with a general form given below. However hydrolase and oxidoreductase reactions are excluded, as they are classified above.

$$AX + B \Leftrightarrow BX + A$$

The second digit indicates the general type of group transferred (these are listed below), and the third digit provides details on the exact nature of the group transferred (e.g., 2.1.2 are hydroxymethyl-, formyl and related transferases).

First digit	*Second digit*	*Third digit*
2.	1 1-carbon group	nature of group transferred
(Transferases)	2 aldehyde or ketone	
	3 acyl group (-CO-R)	
	4 glycosyl group	
	7 phosphate group	
	8 sulphur containing groups	

Class 3. Hydrolases

Hydrolases catalyze hydrolytic reactions, with the second digit indicating the type of bond hydrolyzed.

$$A - X + H_2O \Leftrightarrow X - OH + HA$$

First digit	*Second digit*	
3. (Hydrolases)	1	ester
	2	glycosidic (i.e., linking carbohydrate moieties)
	4	peptide
	5	C-N bonds other than peptides
	6	acid anhydrides

Class 4. Lyases

Lyases catalyze the non-hydrolytic removal of groups from substrates. Often the product contains a double bond. The second digit refers to the type of bond broken. The third digit describes the group removed. Also included in this class are enzymes which act in the *reverse* direction to group removal. These are often *synthases* or *hydratases*, e.g. reactions in which groups are added across a double bond.

First digit	*Second digit*		*Third digit*	
4. (Lyases)	1	C-C	1	carboxyl
	2	C-O	2	aldehyde
	3	C-N	3	ketoacid
	4	C-S		

Class 5. Isomerases

The second digit of the classification of isomerases describes the type of reaction involved. The third digit describes the type of molecule undergoing isomerization.

First digit	*Second digit*		*Third digit*	
5. (Isomerases)	1	racemization or epimerization	1	amino acids
	2	cis-trans isomerizations	2	hydroxyacids
	3	intramolecular oxidoreductases	3	carbohydrates
	4	intramolecular transfer reactions		

Class 6. Ligases

Ligases catalyze the synthesis of various types of bonds, where the reactions are coupled with breakdown of energy-containing materials, such as ATP or nucleoside triphosphates. For example;

$$X + Y + ATP \Leftrightarrow X - Y + ADP + P_i$$

or

$$X + Y + ATP \Leftrightarrow X - Y + AMP + PP_i$$

First digit	*Second digit (The second digit indicates the type of bond formed).*	
6. (Ligases)	1	C-O
	2	C-S
	3	C-N
	4	C-C

An example of this classification scheme is given below.

Recommended name (trivial name) : alcohol dehydrogenase
Systematic name: alcohol : NAD^+ oxidoreductase
(an alcohol is the electron donor and NAD^+ is the electon acceptor)
Enzyme number: EC 1. 1. 1. 1.

```
1.      1.      1.      1.
|       |       |       |
|       |       |       number for further
|       |       |       identification
|       |       acceptor is NAD+ or NADP+
|       |
|       donor group is CH-OH
|
number of the primary division
(oxidoreductases)
```

This classification scheme is useful as it unambiguously identifies the enzyme in question. Earlier nomenclature often resulted in one enzyme being identified by several names if its activity was broad.

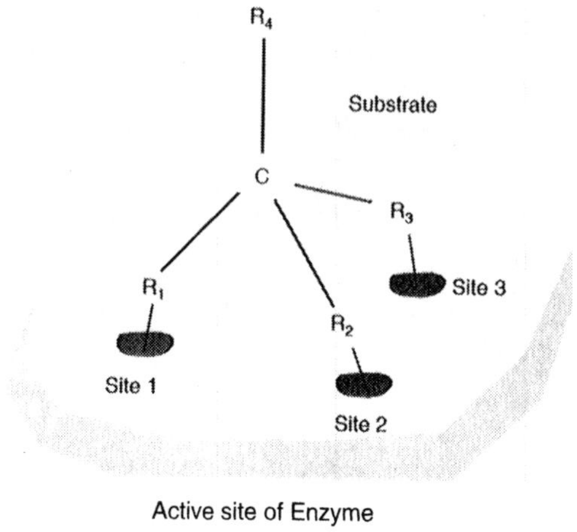

Figure 1.1 *Representation of the three-point interaction of substrate with enzyme.*

While enzymes are specific in function, the degree of specificity varies. Some may act on closely related substrates, and are said to exhibit *group specificity*; others are more exacting in their substrate requirements, and are said to be *absolutely specific*. The product formed from a particular enzyme and substrate is also unique. Enzymes are able to distinguish between stereochemical forms and only one isomer of a particular substrate may undergo reaction. Surprisingly, enzyme reactions may yield stereospecific products from substrates that possess no asymmetric carbon atoms, as long as one carbon is *prochiral*. This chirality is a result of at least three-point interaction between substrate and enzyme in the active site of the enzyme. In Figure 1.1, sites 1, 2 and 3 are binding sites on the enzyme. When two of the R groups on the substrate are identical, the molecule has a *prochiral* center and a chiral center can result from the enzymatic reaction, as the substrate can only "fit" into the active site in one configuration if the site has binding selectivity for three of the R-group substituents. If the substrate has four different R groups, then chirality can be preserved in the reaction as a result of the multipoint attachment.

1.2 Kinetics of Single Substrate Reactions

In this section we shall review various simple constitutive rate expressions which have been applied to describe enzyme-catalyzed reactions. As we saw in Section 1.1, the reaction of an enzyme with a substrate involves the formation of an intermediate which then reacts further with other substrates or decomposes to form products. With recent developments in enzymology, it has been possible to identify many of the elementary reaction steps involved in enzymatic catalysis, and from experimental data on rates of formation of intermediates, the intrinsic kinetics can be found.

Historically, it was found that the kinetics of enzyme catalyzed reactions exhibit features which indicate that a simple single step reaction is not occurring. Invertase, which catalyzes the hydrolysis of sucrose to glucose and fructose, was shown in 1902 by Brown to exhibit kinetics which were first order in the reactant sucrose at low sucrose concentrations, but zeroth order at high concentrations. These results were in contrast to the first order dependence on sucrose concentration found with acid-catalyzed hydrolysis. This hyperbolic dependence of reaction rate on substrate concentration was found to be common for all enzyme-catalyzed reactions. Typically, enzyme rate data are reported as *initial rate* data, where the rate of substrate consumption or product formation is determined over a short period of time following the initiation of the reaction. An example of this is shown in Figure 1.2.

When the initial concentration of substrate is varied, the hyperbolic dependence of the initial reaction rate can be seen (Figure 1.3). This behavior is sometimes referred to as "saturation kinetics". The dependence of the initial rate on enzyme concentration is generally first order.

Specific Activity

Enzyme concentrations are often given in terms of "units" rather than in mole or mass concentration. We rarely know the exact mass of the enzyme in a sample, since it is generally prepared via isolation of the enzyme from microorganisms, or animal or plant tissues and often contains a great deal of non-catalytic protein, the amount of which may vary from sample to sample. Hence a different approach must be adopted, and enzyme concentration is reported in units of *specific activity*. A "unit" is defined as the amount of enzyme (e.g.,

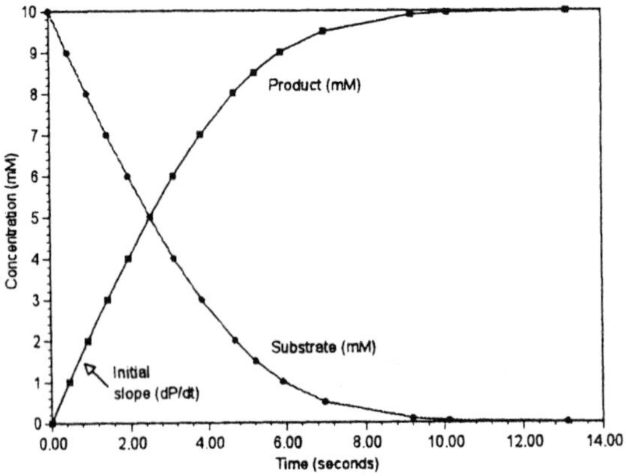

Figure 1.2. *An example of the time course of urea hydrolysis, catalyzed by urease, illustrating the determination of the initial rate of reaction (v_0 in millimoles/liter-sec). The initial rate is the slope of the tangent passing through the origin. The values of the kinetic parameters are k_2 = 30,800 sec^{-1}, K_M = 4.0 mM at pH 8.0 and 20.8°C, $[E_0]$ = 0.1 μM. The reaction catalyzed is*

$$NH_2CONH_2 + H_2O \rightarrow 2NH_3 + CO_2$$

microgram) which gives a certain amount of catalytic activity under specified conditions (e.g., producing 1.0 micromole of product per minute in a solution containing a susbstrate concentration sufficiently high to be in the "saturation" region, as shown in Figure 1.3).

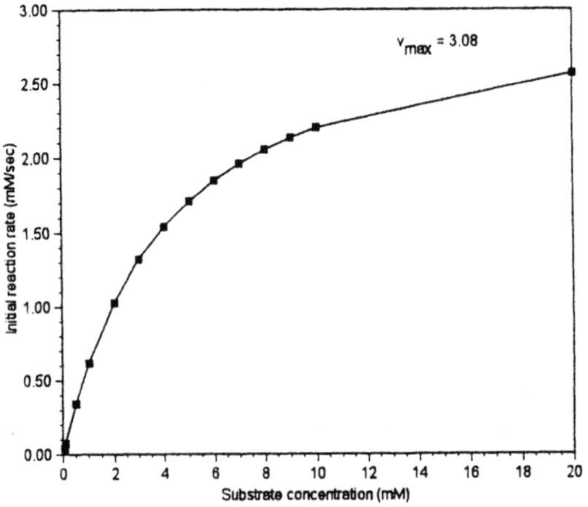

Figure 1.3. *An example of the initial rate of reaction as a function of substrate and enzyme concentrations for urease; reaction parameters are the same as given in Figure 1.2.*

Thus different suppliers of enzymes may have preparations with different units of activity, and care must be taken in analyzing kinetic data. Thus a purified enzyme preparation will have a higher specific activity than a crude preparation; often a protein is considered pure when a constant specific activity is reached during the purification steps.

$$specific\ activity = \frac{activity}{mg\ protein} = \frac{mmole\ product}{(mg\ protein)(ml \cdot min)} \quad (1.1a)$$

The activity is given by the amount of product formed or substrate consumed in the reaction mixture, under the conditions specified (temperature, pH, buffer type and substrate and enzyme concentrations etc.). If the molecular weight of the enzyme is known, the specific activity can also be defined as:

$$specific\ activity = \frac{activity}{mmole\ protein} = \frac{mmole\ product}{(mmole\ protein)(ml \cdot min)} \quad (1.1b)$$

In the case of urease in Figure 1.2, the activity of the enzyme can be found from the slope of the initial rate of product formed.

initial velocity = activity = 3.08 millimoles product (CO_2) released/liter-sec

The reaction mixture contained an enzyme concentration of 0.0001 mM. Thus the specific activity at pH 8.0 and 20.8°C is

$$specific\ activity = \frac{3.08(mM\,CO_2/\sec)}{0.0001(mM\,protein)}$$

$$= 30,800 \quad mM\,CO_2\,(\sec \cdot mM\,protein)^{-1}$$

When only the mass of enzyme is available, the units of specific activity will typically be (mmole product/ml) (μg enzyme/ml)$^{-1}$ sec^{-1}. For reactions following the simple Michaelis-Menten form (described in the following section), the maximum possible specific activity (where the enzyme concentration can be given in molar units) is the first-order rate constant (k_2) for release of product from the [ES] complex. If the enzyme is known to have several active sites or is present in solution as a dimer or tetramer, the specific activity can be divided by the number of active sites, giving the active site specific activity.

1.2.1 Michaelis-Menten Equation

An explanation for the observations of Brown and others was formalized by Henri in 1903 and more completely by L. Michaelis and M. L. Menten in 1913[1]. Michaelis and Menten obtained a great deal of experimental data to support their analysis of dependence

(1) Michaelis, L. and M.L. Menten, *Biochem. Z.*, **49**, 333 (1913).

of the *initial* rate of reaction on substrate and enzyme concentrations. They proposed that the enzyme and substrate first combined to give an enzyme-substrate complex, which was assumed to be a rapid reversible reaction, with no chemical changes occurring to the substrate. The substrate is held by physical forces to the enzyme. This complex then undergoes a chemical change, resulting in the formation of product and the release of product from the enzyme, with a first order dependence on the concentration of the enzyme-substrate complex. This is shown schematically below.

$$K_S \qquad k_2 \rightarrow \text{rate limiting step}$$
$$E + S \Leftrightarrow ES \rightarrow E + P$$

The first step in the Michaelis-Menten analysis is assumed to be an *equilibrium* step, and thus

$$\frac{[E][S]}{[ES]} = K_S \quad \text{dissociation} \tag{1.2}$$

where K_S is the dissociation constant of the enzyme-substrate complex (ES). The rate of product formation is given by:

$$v = \frac{dP}{dt} = k_2 [ES] \tag{1.3}$$

$$v = k_2[ES] \tag{1.4}$$

A conservation equation can be written for the total amount of enzyme in the system, as the enzyme serves a catalytic role and is not consumed.

$$[E_o] = [E] + [ES] \tag{1.5}$$

We can now substitute [E] in terms of the known concentration of the total amount of enzyme initially in the reaction mixture, $[E_o]$, and the enzyme-substrate complex into equation (1.2)

$$[E] = [E_o] - [ES]$$

$$\frac{([E_o] - [ES])[S]}{[ES]} = K_S$$

and thus on rearranging:

$$[ES] = \frac{[E_o][S]}{[S] + K_S}$$

The rate of reaction is then given by

$$v = \frac{k_2[E_o][S]}{K_S + [S]} \tag{1.6}$$

The constants k_2 and E_o are sometimes combined to give a maximum rate of reaction (v_{max}), and Eq. (1.6) is written in the following form

$$v = \frac{v_{max}[S]}{K_M + [S]} \tag{1.7}$$

$K_m = K_S$ in this rxn
but $K_m = \frac{k_{-1} + k_2}{k_1}$
$= K_S + \frac{k_2}{k_1}$

where K_M is known as the Michaelis constant. For the simple reaction mechanism shown above, K_M is equal to K_S, the dissociation constant of (ES). We shall see shortly, however, that in many cases K_M cannot be interpreted as a simple dissociation constant.

The *initial rate* of reaction can be determined from the Michaelis-Menten expression by assuming that the substrate concentration is sufficiently high so that it does not change appreciably from its initial value over the period when the initial rate is determined experimentally (i.e., [S] ~ $[S_o]$)

$$v_o = \frac{k_2[E_o]\,[S_o]}{K_S + [S_o]} = \frac{v_{max}[S_o]}{K_M + [S_o]} \tag{1.8}$$

The form of this equation provides an explanation for the observed hyperbolic dependence of initial rate on substrate concentration and the linear dependence on enzyme concentration. It is the cornerstone of our understanding of enzyme kinetics and is referred to as the Michaelis-Menten equation. However, as we shall see in the following section, the mechanism proposed (equilibrium between enzyme and enzyme-substrate complex) is not always appropriate, although the form of the equation may be applicable.

From the above equation, we see that the maximum reaction rate per enzyme molecule is $v_{max}/[E_o]$, and is an apparent first order rate constant. When the enzyme posesses only one active site, this will also equal k_{cat} (the catalytic center activity, in molecules of substrate converted per active site per time). When the enzyme has n active sites per molecule, k_{cat} will be given by $v_{max}/(n\cdot[E_o])$.

1.2.2 Briggs-Haldane Equation

The equilibrium step in the Michaelis-Menten analysis may not hold when the rate of product formation (k_2) from the (ES) complex is comparable to the rate of breakdown (k_{-1}) of the (ES) complex to enzyme and substrate. Addressing this problem, Briggs and Haldane modified the Michaelis-Menten equation in 1925[2], by introducing a more generally applicable assumption, that the intermediate (ES) complex is at steady state. This is referred to as the "quasi-steady state" assumption. The reaction scheme can be rewritten as

$$E + S \underset{k_{-1}}{\overset{k_1}{\Leftrightarrow}} ES \overset{k_2}{\rightarrow} E + P$$

The rate of formation of (ES) can be written

$$\frac{d[ES]}{dt} = k_1[E]\,[S] - k_2[ES] - k_{-1}[ES] = 0 \quad \textit{at steady state} \tag{1.9}$$

Solving this equation for [ES] and substituting into equation (1.4) gives the reaction rate v in a form similar to the Michaelis-Menten expression

(2) Briggs, G.E. & J.B.S. Haldane, *Biochem. J.*, **19**, 338 (1925)

$$v = \frac{k_2[E_o][S]}{[S]+(k_2+k_{-1})/k_1} = \frac{V_{max}[S]}{K_m+[S]} \tag{1.10}$$

where the Michaelis constant K_M is no longer the dissociation constant for the [ES] complex, but is given by

$$K_M = \frac{k_2+k_{-1}}{k_1} \tag{1.11}$$

K_S, the dissociation constant, is equal to k_{-1}/k_1 and thus

$$K_M = K_S + \frac{k_2}{k_1} \tag{1.12}$$

We recover $K_M = K_S$ when $k_{-1} > k_2$. At high concentrations of enzyme or low substrate concentrations ($[E_o] \sim [S_o]$), the assumption of a steady state value of [ES] is no longer valid. It is then necessary to solve the coupled differential equations for [E], [ES] and [S]. This is discussed in more detail in Section 1.7.1.

In the case when $k_2 >> k_{-1}$, k_{cat}/K_M is equal to k_1, the rate constant for the association of enzyme and substrate. The value of k_1 is typically 10^8 $sec^{-1}M^{-1}$ and this value can be used as a diagnostic for the Briggs-Haldane mechanism.

The Michaelis-Menten equation can also be applied to reaction schemes where other intermediates may occur. In these cases, we define the catalytic center activity rate constant k_{cat} (assuming only one active site per enzyme molecule) by analogy with the Michaelis-Menten form:

$$k_{cat} = \frac{v_{max}}{[E_o]}$$

For simple Michaelis-Menten kinetics, we can see from equation (1.6) that k_{cat} will be equal to k_2. This is also the case in the Briggs-Haldane mechanism. However, in a mechanism involving several intermediates, the constants k_{cat} and K_M (obtained by fitting the experimental reaction rate data to the Michaelis-Menten expression) are no longer equal to the rate constant k_2 and the dissociation constant K_S, but are combinations of other elementary rate constants. This is illustrated in the scheme below.

$$E + S \overset{K_S}{\Leftrightarrow} ES \overset{K}{\Leftrightarrow} ES' \overset{K'}{\Leftrightarrow} ES'' \overset{k_4}{\rightarrow} E + P$$

$$K_M = \frac{K_S}{1+K+K'} \qquad \text{and} \qquad k_{cat} = \frac{k_4 K K'}{1+K+KK'}$$

Thus K_M is less than K_S. If K_S can be separately determined, the observation that K_M is less than K_S can be used as a diagnostic to uncover the presence of intermediates in the reaction pathway.

The Michaelis-Menten Parameters k_{cat}, K_M and the Ratio k_{cat}/K_M

As we have seen from the simple cases above, the Michaelis-Menten form of the reaction velocity as a function of substrate concentration can be used to fit many situations where the actual reaction mechanism differs from that assumed in the derviation of the Michaelis-Menten expression. In these situations, care must be given to interpreting the kinetic parameters. We will briefly describe the significance associated with k_{cat} and K_M.

The Turnover Number (k_{cat})

In the Michaelis-Menten mechanism, k_{cat} is simply the first order rate constant (k_2) for the conversion of the ES complex into enzyme and product. The same is true if an additional intermediate is formed, and the release of P from EP is fast, as shown below:

$$E + S \overset{K_S}{\Leftrightarrow} ES \overset{k_2(slow)}{\rightarrow} EP \overset{k_3(fast)}{\rightarrow} E + P$$

If the dissociation of EP is slow however, it will influence k_{cat}. Hence the rate constant k_{cat}, with units of reciprocal time, is often called the *turnover number* of the enzyme, as it represents the maximum number of substrate molecules that can be converted to product per unit time per active site on the enzyme.

In situations where a number of intermediate steps are involved, we can use the reciprocal of k_{cat} as a characteristic *time constant* for the overall reaction pathway. This leads us to a simple method for examining complex pathways. Consider the pathway where products are interconverted viz:

$$EP_1 \overset{k_1}{\rightarrow} EP_2 \overset{k_2}{\rightarrow} EP_3 \overset{k_3}{\rightarrow} EP_4 \overset{k_4}{\rightarrow} \cdots \rightarrow \cdots \overset{k_{n-1}}{\rightarrow} EP_{n-1}$$

If we assume that the intermediate products are present at steady-state concentrations, the characteristic times for the conversion of one form of the product to another are given by the reciprocal of the first order rate constants (1/k). The overall time constant for the pathway is thus the sum of the individual time constants:

$$\frac{1}{k_{cat}} = \frac{1}{k_1} + \frac{1}{k_2} + \frac{1}{k_3} + \frac{1}{k_4} + \cdots + \frac{1}{k_{n-1}} \tag{1.13}$$

An example of this is the simple case of reversible binding of substrate to enzyme, with the resulting complex being present at a steady-state concentration (i.e., Briggs-Haldane kinetics):

$$E + S \underset{k_{-1}}{\overset{k_1}{\Leftrightarrow}} ES \overset{k_2}{\rightarrow} E + P$$

Rewriting the reaction scheme as

$$E \underset{k_{-1}}{\overset{k_1[S]}{\Leftrightarrow}} ES \overset{k_2}{\rightarrow} E + P$$

we can define a rate constant for the substrate binding reaction as

$$k_1[S] \cdot \frac{k_2}{k_{-1} + k_2}$$

which represents the apparent first order rate constant for the formation of ES from E (i.e., $k_1[S]$) times the probability that ES reacts to P (i.e., $k_2/(k_2+k_{-1})$). By adding the two time constants, k_{cat} is thus

$$\frac{1}{k_{cat}} = \frac{k_{-1} + k_2}{k_1 k_2 [S]} + \frac{1}{k_2} \tag{1.14}$$

Thus k_{cat} will differ from k_2 when $k_1[S]$ is small, i.e., there is a slow rate of binding of enzyme to substrate. At high values of [S] or when $k_1 > k_{-1}$, the first term in the above equation will be negligible and $k_{cat} \sim k_2$.

The Michaelis Constant (K_M)

For a reaction scheme involving one intermediate, K_M is equivalent to the dissociation constant (ie. $K_M = K_S$) when the enzyme-substrate complex is in equilibrium with free enzyme and substrate. K_M can also be considered as an apparent dissociation constant in other cases where the enzyme may be bound in a more complex manner with substrate (e.g., multiple substrate molecules bound, several forms of substrate bound). K_M can then be written in terms of all of the bound enzyme species $\Sigma[ES]$:

$$K_M = \frac{[E][S]}{\Sigma[ES]} \tag{1.15}$$

In the simple Michaelis-Menten rate expression (Eq. 1.7) or in the Briggs-Haldane expression rewritten in terms of v_{max}

$$v = \frac{v_{max}[S]}{K_M + [S]}$$

The Michaelis constant can be determined from the substrate concentration which gives half the maximum reaction rate:

$$\frac{v}{v_{max}} = \frac{1}{1 + \frac{K_M}{S}} \tag{1.16}$$

$$K_M = S \qquad \text{when} \qquad \frac{v}{v_{max}} = \frac{1}{2} \tag{1.17}$$

This is also true even when the reaction mechanism is more complex, as is discussed in the following example.

An Example of Intermediates in the Reaction Pathway: The Mechanism of Chymotrypsin

Chymotrypsin is a serine protease that cleaves the amide linkages in proteins and peptides. It has a binding pocket which is selective for the aromatic residues of amino acids. The reaction occurs by the reversible formation of a Michaelis complex, followed by acylation of Ser-195 to give a tetrahedral acylenzyme intermediate (this is shown in detail later in Figure 1.8). Chymotrypsin will also act as an esterase; we can write the elementary reaction steps in the following form, where RCO-X is an amide or an ester

$$E + RCO - X \overset{K_S}{\Leftrightarrow} RCO - X \cdot E \overset{k_2}{\rightarrow} RCO - E + XH \overset{k_3}{\rightarrow} RCOOH + E$$

where

X = NH-R' (amide) or X = O-R' (ester) and RCO-E is the acyl-enzyme intermediate

This can be written more simply as

$$E + S \underset{k_{-1}}{\overset{k_1}{\Leftrightarrow}} ES \overset{k_2}{\rightarrow} ES' \overset{k_3}{\rightarrow} E + P$$

We can write the mass balances for the intermediates and product as:

$$\frac{d[ES]}{dt} = k_1[E][S] - (k_{-1} + k_2)[ES]$$

$$\frac{d[ES']}{dt} = k_2[ES] - k_3[ES']$$

$$\frac{d[P]}{dt} = k_3[ES']$$

Assuming [ES] and [ES'] are at steady state, we can eliminate [ES'] from the third equation:

$$[ES] = \frac{k_1[E][S]}{k_{-1} + k_2}$$

$$[ES'] = \frac{k_2}{k_3}[ES] = \frac{k_2}{k_3} \cdot \frac{k_1[E][S]}{k_{-1} + k_2}$$

thus

$$v = \frac{dP}{dt} = k_3 \cdot \frac{k_2}{k_3} \cdot \frac{k_1[E][S]}{k_{-1} + k_2}$$

Noting

$$[E_o] = [E] + [ES] + [ES']$$

and substituting for [E] in terms of $[E_o]$ gives

$$v = \frac{\frac{k_2 k_3}{(k_2+k_3)} \cdot [E_o] \cdot [S]}{\frac{k_3}{(k_2+k_3)} \cdot \frac{(k_{-1}+k_2)}{k_1} + [S]}$$

Hence the K_M and v_{max} values are reduced by the ratio $k_3/(k_2+k_3)$, resulting from the presence of the second intermediate ES′. Note that if $k_3 >> k_2$, then k_{cat} reduces to k_2. On the other hand, if $k_2 >> k_3$, then $k_{cat} = k_3$.

The Specificity Constant k_{cat}/K_M

At low substrate concentrations, the ratio k_{cat}/K_M can be seen from equations (1.10) and (1.11) to be an apparent second order rate constant

$$v = (k_{cat}/K_M) \cdot [E_o][S] \tag{1.18}$$

where $[E_o]\sim[E]$ at this low substrate concentration, since most of the enzyme will be unbound. Under these conditions, the ratio (k_{cat}/K_M) represents an effective second order rate constant for the encounter of free substrate with free enzyme, and thus provides an indication of the *specificity* of the enzyme for the substrate. Although it is not a true second order rate constant (except in the case when the enzyme-substrate encounter is the rate limiting step), it does set a lower limit on the association constant for enzyme and substrate. The usefulness of k_{cat}/K_M is that it can be used to describe the specificity of the enzyme for competing substrates. The ratio of reaction rates for low concentrations of substrates A and B is given by

$$\frac{v_A}{v_B} = \frac{(k_{cat}/K_M)_A [A]}{(k_{cat}/K_M)_B [B]} \tag{1.19}$$

It can be shown that equation (1.18) holds at all values of substrate concentrations when K_M is expressed in the form of equation (1.15). Thus equation (1.19) is quite generally applicable.

An Example of the Reversal of Substrate Specificity by Site-Directed Mutagenesis (aspartate transaminase)

Site-directed mutagenesis of the gene encoding a protein permits any amino acid residue in the protein to be changed to any other. By altering the nucleotide sequence of a gene and cloning it into a suitable host, such as *Escherichia coli*, a mutagenized protein can be produced in relatively large quantities. Application of site directed mutagenesis, as well as other mutagenesis techniques, is often referred to as "protein engineering", although this phrase is somewhat misleading in that it connotes an ability to predict results[3]. Site-directed mutagensis has provided significant insights into the nature of molecular recognition and enzymatic catalysis, and recent applications of site-directed mutagenesis have shown notable progress toward the rational alteration of enzyme specificity and reactivity. Ultimately, it

(3) J. Knowles, *Chemical and Engineering News*, July 14, 1986.

may be possible to control in a more predictable fashion a variety of important enzyme properties, including kinetic properties, thermostability, stability and activity in organic solvents, pH optima, and so forth. Indeed, site-directed mutagenesis has enormous potential to increase our understanding of, and to modify in a desirable way, protein activity and stability.

The Reversed Substrate Specificity of Aspartate Transaminase

The enzyme aspartate aminotransferase (AATase) catalyzes the reversible transamination of α-amino to α-keto acids. The preferred substrates are the anionic amino acids L-aspartate and L-glutamate. Crystallographic data indicate that the enzyme's substrate specificity is largely the result of ionic pairing between the side chain of each substrate and an arginine residue at position 292 of the enzyme. In an attempt to verify this hypothesis, Kirsch and co-workers[4] used site-directed mutagenesis to replace Arg292 with aspartate.

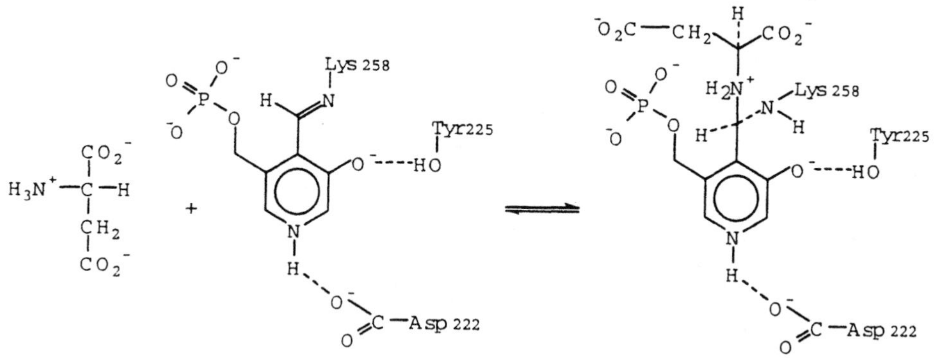

Charge interactions in the active site of AATase. The amino acid substrate (left) combines with the pyridoxal 5'-phosphate (PLP)-enzyme to form the tetrahedral intermediate (right), which is stabilized by ionic pairing of the negative charge on the carboxylate group of the substrate and the positive charge on the guanidinium group of Arg292 [Adapted from Kirsch et al., J. Mol. Biol., **174**, *497 (1984)].*

This amino acid substitution would test the importance of ionic pairing in the substrate specificity of AATase. It was expected that the mutation might reverse the substrate charge preference of the wild-type enzyme, resulting from the replacement of the postiviely charged arginine at the active site with negatively charged aspartate. As the results summarized below indicate, the experiment was highly successful.

(4) C.N. Cronin, B.A. Malcolm, and J.F. Kirsch, *J. Am. Chem. Soc.*, **109**, 2222 (1987).

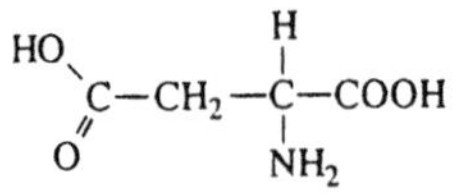

Aspartic acid

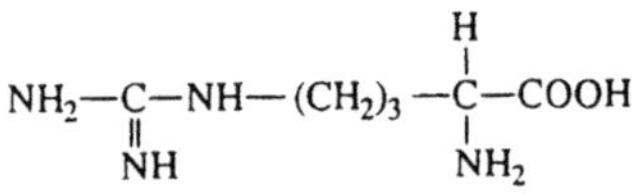

Arginine

Compared to the wild type, the mutant enzyme (R292D) displayed 16-fold and 9-fold higher k_{cat}/K_m values, respectively, toward the *cationic* amino acid substrates arginine and lysine. Although the k_{cat}/K_m value for arginine was low, especially in comparison to k_{cat}/K_m of the wild-type enzyme for aspartate, these results illustrate the usefulness of site-directed mutagenesis for probing the mechanisms of enzymatic reactions and for producing enzymatic activities without precedent in nature, e.g., in this case, a lysine/arginine transaminase.

Substrate Specificities of E. coli Wild Type and R292D Mutant Aspartate Aminotransferases

	(k_{cat}/K_M), $M^{-1}sec^{-1}$		
Substrate	wild type	R292D mutant	selectivity ratio $[(k_{cat}/K_M)_{R292D}]/[(k_{cat}/K_M)_{wild\ type}]$
L-arginine	0.0276 ± 0.0008	0.429 ± 0.026	15.5 ± 1.0
L-lysine	0.0183 ± 0.0002	0.156 ± 0.013	8.5 ± 0.7
L-aspartate	$18{,}500 \pm 500$	0.0695 ± 0.0029	$(3.8 \pm 0.2) \times 10^{-6}$

1.3 Graphical Representation of Kinetic Data

In this section we shall examine some of the various methods for obtaining kinetic constants from rate data. The most general approach involves the direct integration of equation (1.7). By assuming quasi-steady state for [ES] and writing $v = -dS/dt$, the time course of substrate consumption can be obtained, employing the initial condition $S = S_o$:

$$\frac{d[S]}{dt} = \frac{-v_{max}[S]}{K_M + [S]} \tag{1.20}$$

$$v_{max}t = [S_o] - [S] + K_M \ln\frac{[S_o]}{[S]} \tag{1.21}$$

This form does not lend itself to straightforward calculation of the substrate concentration as a function of time; it is more easily used to compute the reaction time which corresponds to a particular substrate concentration. Thus other approaches have evolved.

Most experimental data on enzyme kinetics are initial rate data; that is, substrate or, more commonly, product concentrations are determined as a function of time over rather short initial times. Measurement of product concentration has several advantages. Determination of product concentrations may be relatively straightforward, as the product concentration will be changing significantly over the initial period, and measurement of small changes in substrate concentration may introduce significant errors. By assuming that little change in substrate concentration has occurred, the rate v may be determined by evaluating dP/dt at $[S]\sim[S_o]$. The initial substrate concentration is then varied and a table of $v(S_o)$ versus S_o can be constructed.

Various approaches have been proposed to graphically determine the values of v_{max} and K_M from these data. The most widely employed method is the *Lineweaver-Burk* plot. It is based on a double reciprocal plot of equation (1.11) substituted into equation (1.10)

$$v = v_{max}\frac{[S]}{K_M + [S]} \tag{1.22}$$

$$\frac{1}{v} = \frac{1}{v_{max}} + \frac{K_M}{v_{max}} \cdot \frac{1}{[S]} \tag{1.23}$$

A plot of $1/v$ versus $1/[S]$ (or, more correctly for initial rate data, $1/v_o$ versus $1/[S_o]$) gives $1/v_{max}$ as intercept on the ordinate when $1/[S]$ approaches zero, and an intercept of $-1/K_M$ on the abscissa for v approaching zero. Because the most accurate data are obtained at high values of substrate concentration, the Lineweaver-Burk plot is most suitable for determining v_{max}, but is less accurate for determining K_M.

To overcome this difficulty, the *Eadie-Hofstee* plot is often used. It does not unduly emphasize points at low substrate concentrations. It is derived from equation (1.22) by rearranging the right hand side:

$$v = v_{max} - \frac{K_M v}{[S]} \tag{1.24}$$

v is plotted against v/[S] and K_M is determined from the slope of the resulting line. The ordinate intercept determines the value of v_{max}.

A related plot is the *Hanes* plot, formed by multiplying the Lineweaver-Burk equation by [S]:

$$\frac{[S]}{v} = \frac{[S]}{v_{max}} + \frac{K_M}{v_{max}} \tag{1.25}$$

Because they provide more accurate kinetic parameters, the Hanes and Eadie-Hofstee plots are used by enzyme kineticists, while the more straightforward Lineweaver-Burk plot continues to be most widely employed by enzymologists in general. Several computer programs have been designed to perform curve-fits of data to the Michaelis-Menten equation. The three plots are illustrated in the example below.

An Example of the Determination of Rate Constants: Glycoside Hydrolysis by β-Galactosidase

β-Galactosidase catalyzes the hydrolysis of galactosyl β-linkages in glycoproteins, polysaccharides and disaccharides. This enzyme is one of the most readily available and best characterized enzymes from *E.coli*. Many kinetic experiments have been conducted with this enzyme, as chromogenic substrates, such as ortho- and para-nitrophenyl-β-D-galactosides can be used to follow the course of reaction by the appearance of a yellow color due to the formation of nitrophenylate anions. The reaction is shown below.

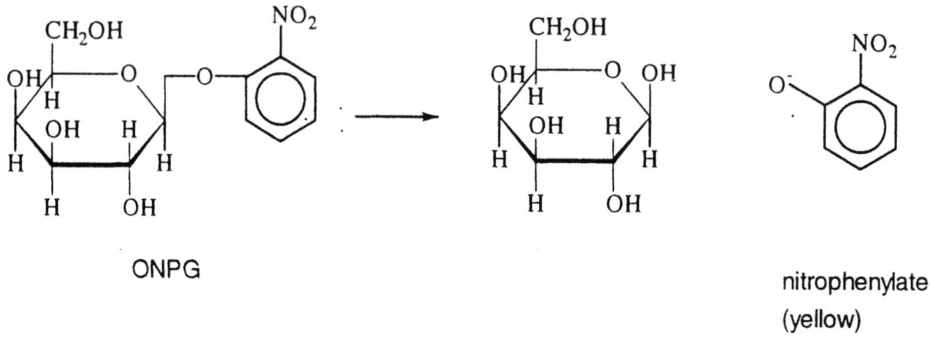

In Figure 1.4, typical plots of data obtained with o-nitrophenyl-β-D-galactosidase are shown. The Michaelis-Menten parameters for β-D-galactosidase are a v_{max} of 178 micromoles substrate hydrolyzed per minute per liter, and a K_M of 0.161 mM. The Eadie-Hofstee plot shows the best separation of the data points.

1.4 Principles of Catalysis

In this section we shall examine the various theoretical approaches that can explain the specificity and the rate enhancements found in enzymatic catalysis. In many cases, an enzyme may enhance the rate of reaction by a combination of mechanisms, making a clear determination of the predominant mechanism difficult. There are several catalytic mechanisms that are important in enzymes. These are:

Approximation — The enzyme brings the reactants together at the active site at higher concentrations than would be found in solution. This is also called the propinquity effect.

Covalent catalysis — A covalent intermediate involving the substrate and the enzyme is formed. Nucleophilic and electrophilic catalysis are two examples of this mechanism.

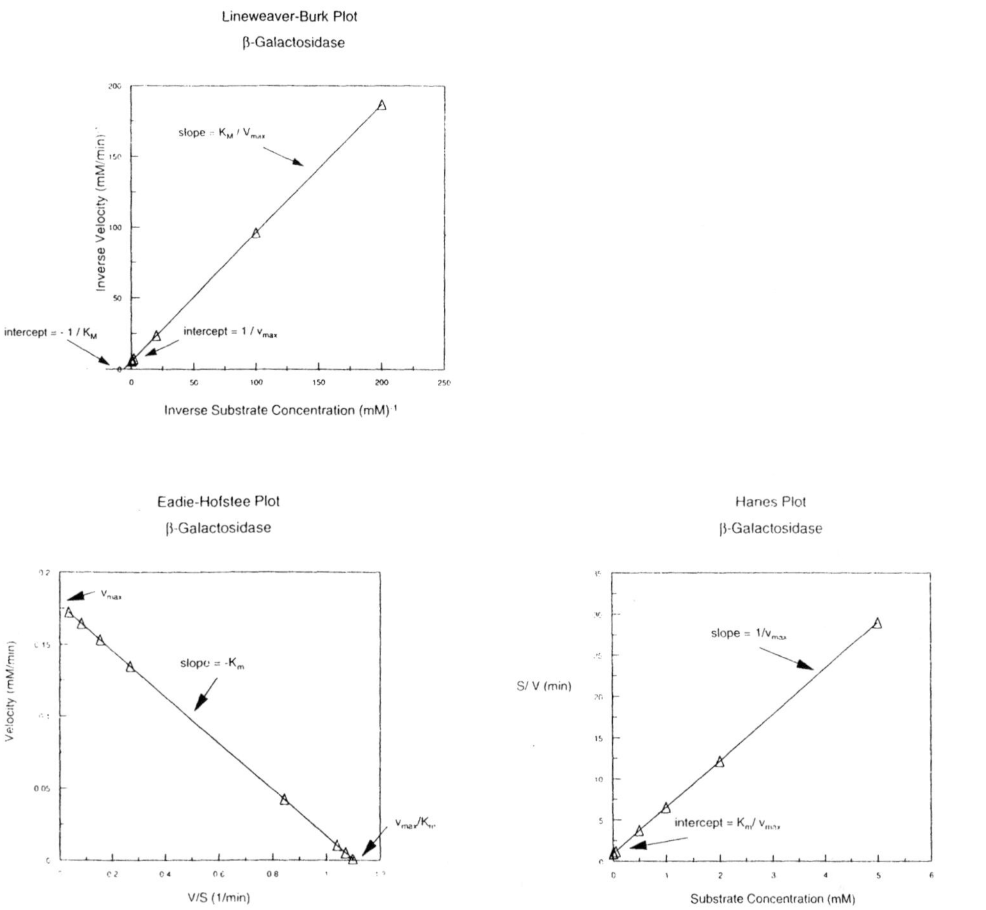

Figure 1.4. *The three common representations of initial rate data as a function of substrate concentration. The Lineweaver-Burk plot illustrates the crowding of data points at high substrate concentrations, while the Eadie-Hofstee plot shows separation of the data points.*

General acid or base catalysis	Amino acids in the enzyme function as general acid or base catalysts in the reaction.

Several other mechanisms may also be involved, including electrostatic catalysis, catalysis involving metal ions and the effects of strain and distortion of substrates induced by the enzyme at the active site. Some further insight into the remarkable specificity that enzymes exhibit for substrates is gained by examining the forces involved in the binding of a substrate to an active site. We shall first examine two approaches from classical chemical kinetics, collision theory and transition state theory, as they provide insights into the upper limits of enzyme catalysis.

Collision Theory

Classical chemical kinetic theories can be applied to enzyme catalysis to determine what mechanisms yield high rates and a high degree of reaction specificity. The simplest approach is that of collision theory, which is based on the assumption that collision of molecules results in reaction. While collision theory does not provide any detailed information on the relationship between enzyme structure and its function, it does establish an upper limit for rates of reaction based on the diffusion of reactants. We consider that the substrate and the enzyme must first collide and that product is then released. For a bimolecular reaction in the gas or liquid phase, the rate of reaction can be expressed as the product of the concentrations of each reactant (assuming ideal behavior) and the bimolecular rate constant k^{coll}. This can be expressed as:

$$k^{coll} = Zpe^{(-E_{act}/RT)} \tag{1.26}$$

where Z is the frequency of collisions, p is the fraction of molecules which have the correct orientation for reaction and E_{act} is the activation energy. The diffusion-limited maximum rate constant will occur when all molecules have the correct orientation (p=1) and the activation energy is zero.

The rate of collision can be related to the radii of the reacting species (r_A and r_B), their diffusion coefficients D_A and D_B and their number concentrations (N_A and N_B, in molecules/cm^3). The flux of A molecules to a single B molecule, $J_{A\rightarrow B}$, can be found by integrating the number of A molecules entering a sphere centered at the B molecule. We intergrate from infinity to contact (at $r_A + r_B$).

$$J_{A\rightarrow B} = 4\pi r^2 \frac{dN_A}{dr}$$

$$J_{A\rightarrow B}\int_{r_A+r_B}^{\infty}\frac{dr}{r^2} = 4\pi D_A \int_0^{N_A} dN_A$$

thus

$$J_{A\rightarrow B} = 4\pi D_A (r_A + r_B) N_A \tag{1.27a}$$

We have neglected the diffusion of B toward A, so the effective diffusion coefficient is increased to ($D_A + D_B$). The total rate of collisions per cm^3 per sec is thus

$$Z' = 4\pi(D_A + D_B)(r_A + r_B)N_A N_B \tag{1.27b}$$

To convert this frequency to units where the concentrations of A and B are expressed in moles/liter, we note $c_A = 1000\ N_A/N_{Avo}$, and the collision frequency in $mole^{-1}\ sec^{-1}$ becomes

$$Z = \frac{4\pi N_{Avo}(D_A + D_B)(r_A + r_B)}{1000} \tag{1.27c}$$

wrong! check notes

Here N_{Avo} is Avogadro's number. More sophisticated treatments of the collision theory include other factors, such as electrostatic interactions and geometric effects, which affect collision rates. → will reduce k^{coll} → decr. rxn rate

The Stokes-Einstein equation can be used to relate the diffusion coefficients of the substrate and the enzyme to their respective radii. Writing for A and B

$$[A/k] \quad k = \frac{R \rightarrow J/mol \cdot K)}{N_A} \qquad D = \frac{k_B T}{6\pi\eta r} \qquad \text{for } \eta = N_s/m^2 = kg/s \cdot m$$

where k_B is the Boltzmann constant, η is the fluid viscosity, and r is the radius of the diffusing species, we obtain

$$Z = \frac{2RT}{3000\eta}\left(\frac{(r_A + r_B)^2}{r_A r_B}\right) \qquad (1.27d)$$

The Collision Theory Rate Constant k^{coll}

If species A and B have equal radii, the collision frequency in water at 25°C, determined from equation (1-27d), is 7.4×10^9 sec^{-1} M^{-1}. The typical range for a small molecule (e.g., an enzyme substrate) interacting with a large one (e.g., an enzyme) is larger, 10^9 to 10^{11} sec^{-1} M^{-1}. For example, the rate of collision of an enzyme of radius r_A =23 Å (a protein of molecular weight ~ 45,000) and diffusion coefficient $D_A = 8 \times 10^{-7}$ cm^{2}sec^{-1} with a substrate molecule of $r_B = 8$ Å (molecular weight ~ 400) and $D_B = 52 \times 10^{-7}$ cm^{2}sec^{-1} would be ~ 1.3×10^{10} sec^{-1}M^{-1}. The increase is due to the fact that the substrate molecule has a higher diffusivity ← due to small size than the enzyme, while the enzyme presents a greater cross-sectional area for collision than another substrate molecule would. └→ due to large enzyme size

Very few enzyme reactions demonstrate these high reaction rates. Most enzyme reaction rates are in the range 10^6 to 10^9 sec^{-1} M^{-1} due to the need for desolvation of the enzyme or for multi-step reactions to occur. However, proton transfers between electronegative atoms approach the diffusional limitation.

One example where a high reaction rate constant is observed occurs with some enzymes ✳ containing histidine. The high pK of the imidazole group present in histidine results in it being a stronger base than water and the rate of proton transfer to imidazole is of order 10^{10} sec^{-1}M^{-1}. As we have seen, the ratio k_{cat}/K_M represents an apparent second order rate constant for enzyme reactions which follow Briggs-Haldane kinetics, and this can be compared with the collision theory value of k^{coll}. A tabulation of some enzymes for which this second order rate constant approaches the diffusional limitation is given in Table 1.1.

Experimental values of the apparent second order rate constant (given by k_{cat}/K_M) provide a means for determining an effective substrate concentration at the catalytic center of the enzyme.

$$\text{observed rate} = \left(\frac{k_{cat}}{K_M}\right) \cdot [\text{enzyme}] \cdot [\text{effective substrate concentration}]$$

$$\text{effective substrate concentration} = \frac{(\text{observed rate}/[\text{enzyme}])}{(k_{cat}/K_M)} \qquad (1.28)$$

Table 1.1. *Enzymes which have apparent second order rate constants approaching the diffusional limitation*[5].

Enzyme	*Substrate*	k_{cat} *(sec^{-1})*	K_M *(M)*	*Apparent second order rate constant* k_{cat}/K_M *(sec^{-1} M^{-1})*
Acetylcholinesterase	Acetylcholine	1.4×10^4	9×10^{-5}	1.6×10^8
Carbonic anhydrase	CO_2	1×10^6	0.012	8.3×10^7
	HCO_3^-	4×10^5	0.026	1.5×10^7
Catalase	H_2O_2	4×10^7	1.1	4×10^7
Crotonase	Crotonyl-CoA	5.7×10^3	2×10^{-5}	2.8×10^8
Fumarase	Fumarate	800	5×10^{-6}	1.6×10^8
	Malate	900	2.5×10^{-5}	3.6×10^7
Triosephosphate isomerase	Glyceraldehyde-3-phosphate	4.3×10^3	4.7×10^{-4}	2.4×10^8
β-Lactamase	Benzylpenicillin	2×10^3	2×10^{-5}	1×10^8

The effective substrate concentration can be quite high. This propinquity effect can be considered as an entropic contribution in enzymatic catalysis; the probability that two substrates will react at the active site of an enzyme is dramatically increased when both are held tightly at the active site in the correct orientation. The reactants thus lose translational and rotational motion, and the subsequent chemical step to form the product will involve very little loss in entropy. Most of the entropy loss thus occurs in the binding step and is supplied by the binding energy of the substrate(s) to the enzyme. We shall consider this in more detail in the following discussion of transition state theory.

Transition State Theory

Transition state theory provides a more detailed picture of enzymatic catalysis. In this theory, the mechanism of interaction of reactants is not considered; the important criterion is that colliding molecules must have sufficient energy to overcome a potential energy barrier (the energy of activation) to react. The reaction is described along a reaction coordinate, against which is plotted the energies of the reactants, transition state species and products. The activation energy may be obtained from values of the reaction rate determined at various

(5) From Fersht, A. "*Enzyme Structure and Mechanism*", Freeman & Co. 2nd Edition, (1985).

temperatures. A rate constant (k) is assumed to be proportional to the concentration of the transition state species. Classical transition state theory[6] leads to the following expression for k:

$$k = \left(\frac{k_B T}{h} \right) K^{\pm} \quad (1.29)$$

where k_B is the Boltzmann constant, h is Planck's constant and $K^{\pm}$ is the concentration equilibrium constant defined as the ratio of the concentration of the activated complex to the concentrations of the reactants:

$$K^{\pm} = \frac{[ES]^{\pm}}{[E][S]}$$

The above definition of $K^{\pm}$ assumes that the enzymatic reaction is bimolecular and is described by the equation

$$E + S \overset{K^{\pm}}{\Leftrightarrow} (ES)^{\pm} \rightarrow E + P$$

where $(ES)^{\pm}$ represents the activated complex. The units of k in equation (1.29) are thus $M^{-1}\ s^{-1}$.

Written in this form, $K^{\pm}$ resembles a classical equilibrium constant, and we can define thermodynamic quantities analogous to the classical Gibbs free energy. Hence the standard Gibbs free energy of activation, $\Delta G^{\pm}$, is defined by the equation

$$\Delta G^{\pm} = -RT \ln[K^{\pm}(c^0)^{-1}]$$

where c^0 is a standard state concentration[7]. Thus $K^{\pm}(c^0)^{-1}$ is unitless. Setting c^0 equal to unit concentration, equation (1.29) can be rewritten as

$$k = \left(\frac{k_B T}{h} \right) e^{(-\Delta G^{\pm}/RT)} \quad (1.29a)$$

where $\Delta G^{\pm}$ is the Gibbs energy of activation.

The Gibbs energy difference between the ground and the transition state thus can be used to predict the rate of reaction. If different reaction conditions or reactants are employed, the Gibbs energy can be used to predict the trends in reactivity. For example, if the ground state is neutral and charge must be transferred from one reactant to another in the transition state, e.g. the base catalyzed hydrolysis of an ester, then esters which have strong electron withdrawing groups will react more rapidly than those with weak electron withdrawing

(6) An excellent introduction to transition state theory can be found in Moore, J.W., and R.G. Pearson, "Kinetics and Mechanism, Third Edition," John Wiley & Sons, N.Y. (1981).

(7) The standard states of reactants and the transition state complex are unit concentrations having the same concentration units as those used to determine the rate constant k. This point is often confusing to instructors of students as well as to students of instructors!

groups. A stronger electron-withdrawing group can more effectively stabilize the negatively-charged transition state.

While entropic factors, acid-base and electrostatic factors (discussed in the following sections) can account for a large part of enzymic catalysis, the binding energy associated with the specific substrate-enzyme interaction is a very significant factor in lowering of the Gibbs free energy change required for reaction. The large binding energies of substrates are due in part to the complementary shape of the active site of the enzyme. The Gibbs energy can be considered as composed of two terms, ΔG_S, the binding energy and $\Delta G_T^{\pm}$, the activation energy involved in the making and breaking of the bonds leading to the transition state ($ES^{\pm}$) from the enzyme-substrate intermediate (ES). These are related viz:

$$\Delta G^{\pm} = \Delta G_T^{\pm} + \Delta G_S \qquad (1.30)$$

This is illustrated schematically in Figure 1.5 . Substituting this expression into equation (1.29a) and writing the rate constant (k) as a second order constant (k_{cat}/K_M) indicates the effect of binding energy on the rate constant.

$$RT\ln\left(\frac{k_{cat}}{K_M}\right) = RT\ln\left(\frac{k_B T}{h}\right) - \Delta G_T^{\pm} - \Delta G_S \qquad (1.29b)$$

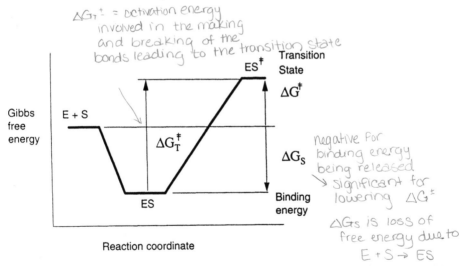

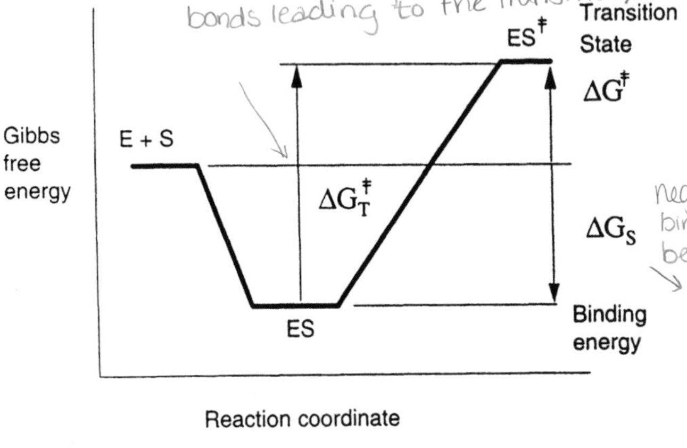

Figure 1.5*. Gibbs free energy for an enzyme catalyzed reaction as a function of the reaction coordinate. Note that ΔG_S is negative as binding energy is released. The uncatalyzed reaction proceeds along a path from the ground state (E + S) to a transition state with a higher activation energy than the catalyzed reaction. The difference between the ground state of the reactants and the products (the standard free energy of reaction ΔG^o) is the same for both the catalyzed and uncatalyzed reactions.*

Transition state theory thus indicates that to increase the binding energy ΔG_S the undistorted enzyme should have an active site which is complementary to the native substrate. As the reaction occurs however, the substrate conformation shifts to that of the transition state. When the enzyme conformation is complementary to the transition state, the binding energy will be less negative; i.e., the initial ES complex will not release all the potentially available binding energy that could be released with an enzyme conformation that is complementary to the native substrate. This is reflected in an increased K_M value, which is related to the formation of the enzyme-substrate intermediate (ES). Complementarity of the enzyme with the transition state will, however, decrease the value of $\Delta G^{\pm}$, so that the reaction rate will increase. This is referred to as *transition state stabilization*. Thus it will be more advantageous for the enzyme to have an active site conformation that matches the transition state substrate conformation ($ES^{\pm}$) than the native substrate (ES).

A final point. In the case of relatively high substrate concentrations ($S_0 >> K_m$) and zero-order kinetics, the enzymatic reaction can be viewed, from the viewpoint of transition-state theory, as a unimolecular reaction. Under these circumstances the rate constant k defined in equation (1.29) reduces to k_{cat} and has units of s^{-1}.

The Role of Entropy in Catalysis

Entropy is composed of translational, rotational and internal entropies. When two molecules react without a catalyst, combining to form an adduct, there is a loss of rotational and translational entropies, as the adduct has three degrees of rotational and translational freedom, whereas the two reactants each had three degrees of rotational and translational freedom. Thus six degrees of freedom are lost. At 25°C and at 1M concentration, this loss corresponds to around 13 to 14 kcal/mole.

An enzyme that can bring together the reactants as an effective intramolecular adduct will not suffer this loss in the reaction step and the intramolecular reaction will be entropically favored. There will only be a small loss of internal entropy. The enzyme-substrate "adduct" reacts without a large entropic loss. This entropic advantage in the reaction, however, is compensated for in the entropy loss which occurs in the binding of the substrate to the enzyme.

We can consider the entropic advantage of the enzyme-substrate adduct to be due to an apparent higher concentration of reactant. For example, the Gibbs energy of activation can be separated into enthalpic and entropic contributions, as seen by rewriting equation (1.29b) as:

$$k = \left(\frac{k_B T}{h}\right)\exp\left(\frac{\Delta S^{\pm}}{R}\right)\exp\left(\frac{-\Delta H^{\pm}}{RT}\right) \tag{1.31}$$

The gain in $\Delta S^{\pm}$ (13-14 kcal/mole) can be substituted into the $\exp(\Delta S^{\pm}/R)$ term, and by combining this with the reactant concentration, we see that the reactant could be considered to be present at a much higher concentration than the corresponding value in free solution, where both the reactants would have to loose entropy to form the transition state. Thus an enzyme can be considered to provide an effective local high concentration of reacting groups by the initial binding to the active site.

While the transition state theory provides more insight into the nature of enzyme catalysis than collision theory, we must consider more specific details of catalysis to fully understand the rate enhancements made possible by enzymes. First, however, there is some interesting experimental evidence detailed in the following example that indicates that the stabilization of the transition state is indeed an important factor in catalysis.

An Example of Transition State Stabilization - Catalytic Antibodies

According to transition state theory discussed above, enzymes act as highly active and selective catalysts by providing a stabilizing environment for the rate-limiting transition state of a given reaction. Among the more remarkable evidence in support of this hypothesis was provided by the advent of *catalytic antibodies*. The concept behind catalytic antibodies is that the highly complementary binding interactions of antibodies can be exploited to catalyze chemical reactions by the overall mechanism of transition-state stabilization. Schultz and coworkers (Science, **234**, 1570 (1986)) translated this idea into practice by showing that an antibody, immunoglobulin A MOPC167, can accelerate the hydrolysis of carbonate **1**.

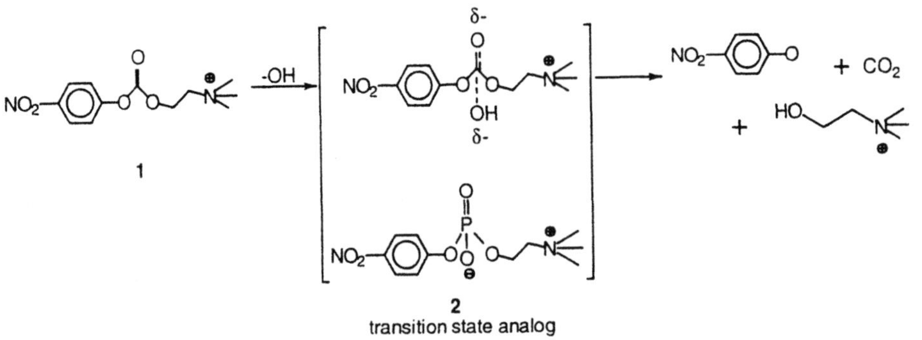

*Aqueous hydrolysis of p-nitrophenyl-N--trimethylammonioethylcarbonate (**1**), and nitrophenylphosphorylcholine (**2**), which is a transition state analog for the reaction.*

MOPC167 binds nitrophenylphosphorylcholine (**2**), which is a transition state analog for the hydrolysis of carbonate **1** (note the structural and electronic similarities between **2** and the bracketed species above). It was expected that binding interactions between the antibody and nitrophenylphosphorylcholine would enable the antibody to lower the free energy of activation by stabilizing the transition state, just as an enzyme would. This expectation was realized, and the antibody was catalytic. In fact, catalysis by the antibody displayed saturation kinetics with a catalytic constant (k_{cat}) of 0.4 min^{-1} and a Michaelis constant (K_M) of 208 μM. The lower limit for the rate of acceleration of hydrolysis by the antibody above the uncatalyzed reaction was 770.

Since the initial work of Schultz and others, antibodies have been shown to catalyze a variety of acyl transfer reactions including lactonization, peptide and ester hydrolysis, and amide bond formation, as well as carbon-carbon bond formation and redox reactions[8]. Some antibodies have produced rate enhancements comparable to those of natural enzymes ($\sim 10^7$), whereas others have displayed a high degree of enantioselectivity. Catalytic antibodies are thus a powerful research tool for studying the mechanisms of enzymatic catalysis. Moreover, catalytic antibodies (or *abzymes*, as they are sometimes called) may also prove useful as unique catalysts in a wide variety of synthetic, medicinal, and analytical applications.

1.4.1 Approximation of Reactants

The simplest mechanism by which an enzyme may enhance the rate of a reaction is to bring the reactants together at the active site. The reactants would then be present at local concentrations that are much higher than those present in solution. This effect is known as approximation of the reactants. How large a rate enhancement might we expect from such approximation? Some insight into this can be obtained by examining *intramolecular* catalysis. Here one group adjacent to a reacting group provides catalytic assistance in the reaction.

An example of intramolecular catalysis is provided by the hydrolysis of tetramethyl succinanilic acid. The reaction is illustrated below.

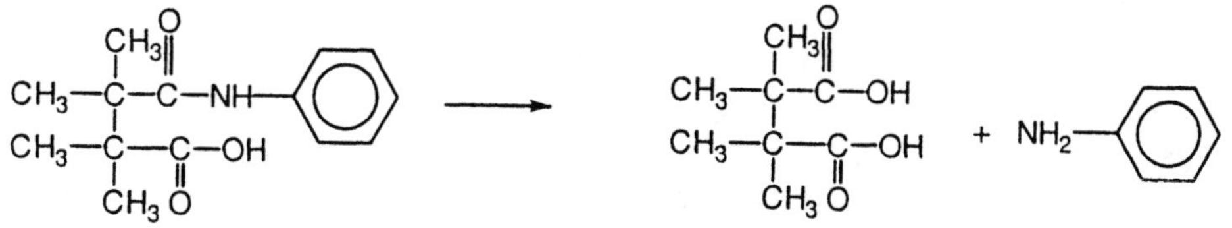

(**8**) K.D. Janda, S.J. Benkovic, and R.A. Lerner, Science, **244**, 437 (1989)

The carboxylic acid moiety provides intramolecular assistance in the hydrolysis of the amide bond. At pH 5 the reaction occurs with a half life of 30 minutes, whereas the corresponding hydrolysis of the unsubstituted acetanilide is some 300 years! This difference in rates corresponds to a rate enhancement of 1.6×10^8. Unsubstituted succinanilic acid is hydrolyzed 1200 times more slowly than the tetramethyl substituted compound; the methyl groups are important in bringing the catalytic group close to the reacting amide bond.

The differences in rates between intramolecular catalysis and the corresponding intermolecular catalysis can be employed to define an apparent concentration of reactant at the reaction site in intramolecular catalysis. In intermolecular catalysis, reactants A and B may react at a rate which is first order in both A and B, the overall rate can be expressed as $k_2[A][B]$, where k_2 is a second-order rate constant (M^{-1} sec^{-1}). In the case of intramolecular catalysis, both A and B are present in the same compound and the rate will now be $k_1[A\text{-}B]$, with k_1 (sec^{-1}) being a first order rate constant. Thus the ratio k_1/k_2, which has units of molarity, represents the effective concentration of reactant A which would need to be present at the catalytic site to cause a smaller concentration of B to react at a pseudo-first order rate equivalent to that of the intramolecularly-catalyzed rate. Hence k_1/k_2 can then be thought of as an "effective" concentration of catalyst A at the reaction site. Such concentrations can be extremely high. When the effective molarities exceed attainable values, then other factors influencing catalysis, in addition to approximation, must be important.

In enzyme-catalyzed reactions, this enhanced local concentration effect can account for some of the rate enhancement, but is generally not sufficiently large. It does provide a lower limit to the rate acceleration that might be expected however. We must turn to other mechanisms to provide an explanation for the catalytic abilities of enzymes.

1.4.2 Covalent Catalysis: Electrophilic and Nucleophilic Catalysis

One of the main routes that an enzyme may employ to enhance rates is to form a covalent bond with one or more reactants and so alter the reaction path from that observed in the uncatalyzed case. The discovery that enzymes may indeed form covalent intermediates relied on early kinetic observations, including "burst" kinetics (Section 1.6.1) and the observation of constant rates of product release from substrates with varying substituents. Today, the crystallographic structures of many enzymes and their substrate-containing intermediates are well known, providing further evidence for the formation of such covalent intermediate compounds.

Covalent catalysis is divided into two types: electrophilic and nucleophilic catalysis. In nucleophilic catalysis, the nucleophilic groups on the enzyme are more electron donating than the normal attacking groups, and a reactive intermediate is formed which breaks down rapidly to form the products. Electrophilic catalysis on the other hand involves catalysts that withdraw electrons from the reaction center of the intermediate. We will first consider nucleophilic catalysis.

The most common nucleophiles in enzymes are the serine hydroxyl (found in serine proteases, esterases, lipases and phosphatases), the cysteine thiol (thiol proteases), the carboxylic group of aspartic acid, the amino group of lysine (aldolase, transaldolase, DNA ligase), the -OH of tyrosine (in topoisomerases) and possibly imidazole (in conjunction with phosphoryl groups in phosphate transfer, otherwise it functions by general base catalysis).

A simple example of nucleophilic catalysis by the serine hydroxyl is afforded by acetylcholine esterase. Acetylcholine is found in the nervous tissue and motor nerve tracts; it is an active neurotransmitter. When a nerve impulse travels along the nerve axon to the synapse, acetylcholine is released and diffuses from the nerve ending to the post-synaptic receptor for acetylcholine on the muscle cell membrane. Acetylcholine esterase functions by breaking down acetylcholine, thus ensuring that the nerve signal is of a short, finite duration. If the enzyme is inhibited, tetanic shock and muscle paralysis follow. The enzyme is thus the target for nerve gases and some insecticides. The enzyme has two sub-sites; one contains the nucleophilic serine which is involved in the formation of an acetyl-enzyme intermediate (called the esteratic site), and the other is negatively charged and provides a salt bridge to enhance recognition and binding of the trimethylammonium region of acetylcholine. The serine acts a nucleophile and attacks the ester linkage, presumably via the formation of a tetrahedral intermediate. Then the choline is released and an acyl enzyme intermediate is formed. Water (OH^-) then releases the acetate from the acyl intermediate. The reaction mechanism is shown in Figure 1.6.

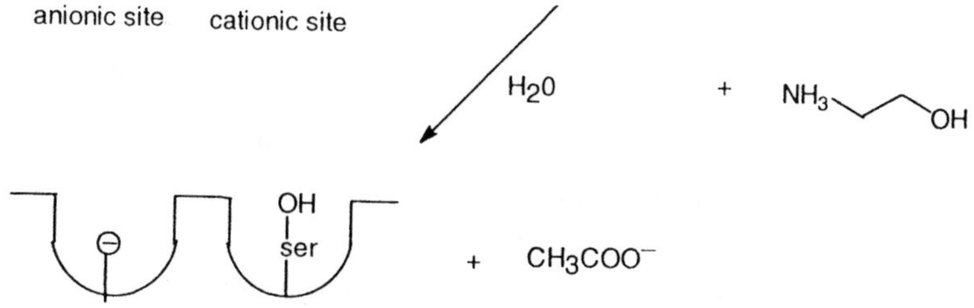

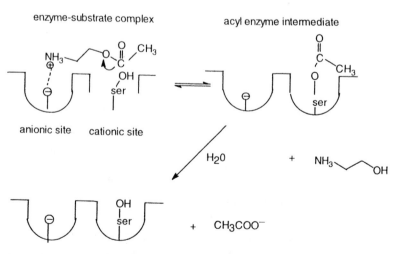

Figure 1.6. *The mechanism of action of acetylcholine esterase.*

The enzyme is remarkably efficient; it has a turnover number (k_{cat}) of 25,000 sec^{-1}, and thus cleaves one substrate molecule every 40 μsec. This rapid rate of cleavage is crucial as nerve impulses can be carried at a rate of 1,000 impulses/sec, necessitating the rapid removal of acetylcholine from the postsynaptic receptor.

Electrophilic catalysts, in contrast to nucleophilic catalysis, act by withdrawing electrons from the reaction center of the intermediate and are thus *electron sinks*. They stabilize a negative charge. Examples of this mechanism involve coenzymes thiamine pyrophosphate and pyridoxal phosphate. In many cases, including these coenzymes, electrophilic catalysis involves the formation of Schiff bases. For example, acetoacetate decarboxylase catalyzes the decarboxylation of acetoacetate to acetone and CO_2. The mechanism involves the formation of a Schiff base involving a lysine residue.

Acetoacetate decarboxylase participates in the production of acetone by fermentation of sugars by anaerobic bacteria, such as *Clostridium acetobutylicum*. This fermentation process was an important route to acetone in World War I, when acetone was employed in the production of the explosive cordite and chemical routes were not available. Other important Schiff base reactions include aldolase and transaldolase reactions.

1.4.3 General Acid-Base Catalysis

Acids and bases can catalyze reactions by either donating or accepting a proton which is transferred in the transition state. When such a charged group develops in the transition state, the resulting positive or negative charge make the transition state unfavorable. The presence of acids or bases result in a stabilization of such a transition state. By providing acid or base groups at the active site, an enzyme is thus able to stabilize the charged transition state. This mechanism is employed by a wide variety of enzymes.

Two types of acid or base catalysis can be distinguished: *general* and *specific*. The distinction between specific and general acids and bases can be best understood by examining experimental observations of catalyzed rates. Consider, for example, the hydrolysis of an ester in buffered solutions. The hydrolysis rate can be determined at a constant pH (by maintaining the ratio of acid and base forms of the buffer at a constant value), at several different total concentrations of buffer. If the rate of reaction increases with increasing buffer concentration, then the buffer must be involved in the reaction and act as a catalyst. This is *general* acid or base catalysis. If the rate is unaffected by buffer concentration, then the reaction involves *specific* acid or base catalysis. The reacting species would be only H^+ or OH^- and the buffer simply serves to maintain these species at constant concentrations.

The linear relationship between the observed rate of reaction and the concentration of buffer can be written as a function of the acid or base form of the buffer. For example for general base catalysis:

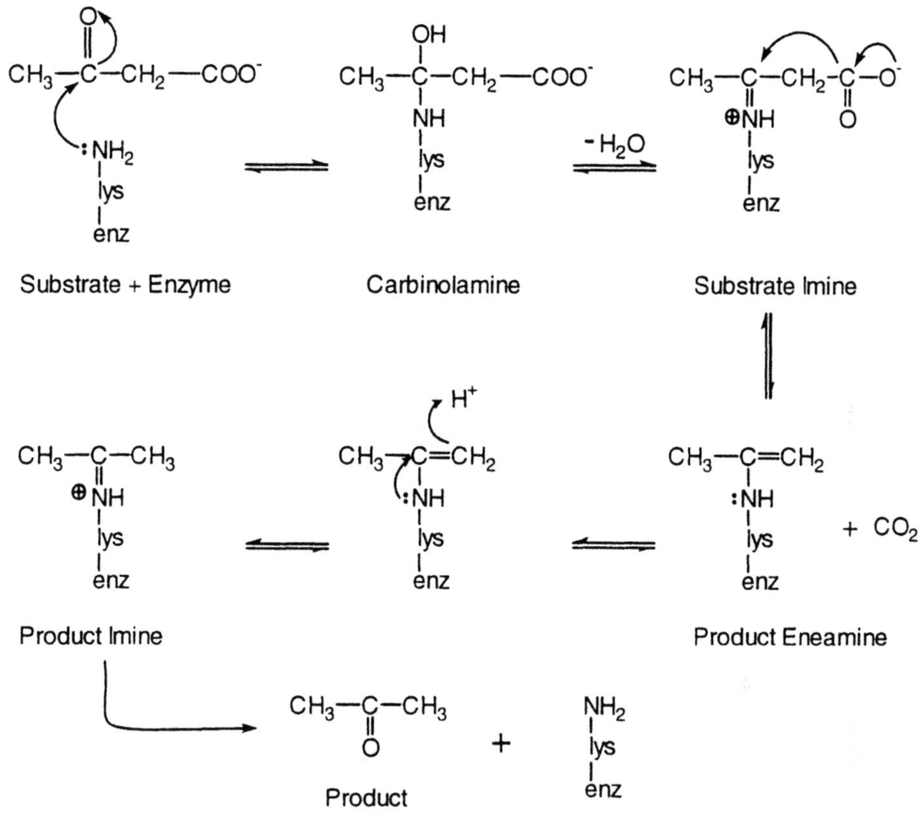

Figure 1.7. *The mechanism of acetoacetate decarboxylase, illustrating the Schiff base intermediate. This is an example of electrophilic catalysis.*

$$k_{obs} = k_0 + k_2[base] \tag{1.32}$$

The constant k_0 is the rate that would be found in the absence of buffer at the specified pH. The second order rate constant k_2 depends on the "base strength" of the base (buffer) employed in the reaction, which in turn depends on its chemical nature. A plot of k_2 against the pKa of the conjugate acid of the base for a variety of bases has been found to yield a linear relationship.

$$\log k_2 = A + \beta pK_a \tag{1.33a}$$

This equation is known as the Brönsted equation, and β is the Brönsted β value. For general acid catalysis a corresponding equation can be written with α being commonly used in place of β:

$$\log k_2 = A - \alpha pK_a \qquad (1.33b)$$

An example of the determination of β is shown in Figure 1.8. The observed first order rate constant for the hydrolysis of ethyl dichloroacetate is shown as a function of base concentration. The zero concentration intercept gives the uncatalyzed rate at the particular pH value employed. Two bases, pyridine and acetate, are shown. The first order rate constant k_{obs} increases more rapidly with pyridine than with acetate anion concentration. Pyridine is thus a more effective catalyst at the same concentration than is the acetate anion. When the same data are replotted in the form of equation (1.33a), where the pK_a of the conjugate acids to these bases is used as the abscissa, and k_2 as the ordinate, we see that the value of β is independent of the chemical nature of the base, depending solely on the strength of the base. If a proton is completely transferred, then the value of α or β is unity; in the absence of transfer these constants have values of zero. Thus stronger acids and bases provide greater rate enhancements than weaker ones; the exact magnitude depends on the values of pK_a and α or β.

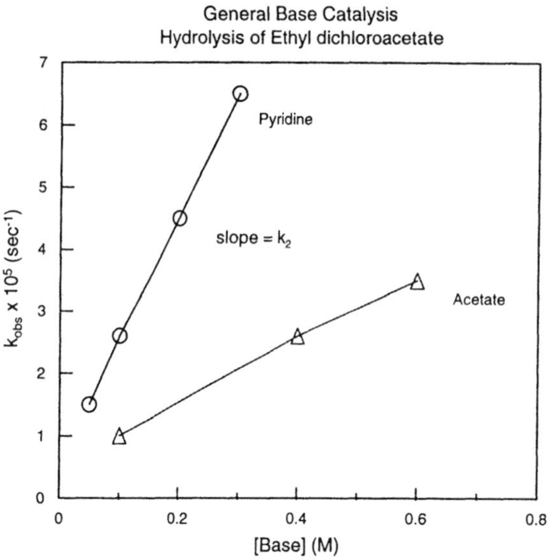

Figure 1.8. The first order rate constant for the base-catalysed hydrolysis of ethyl dichloroacetate, data of W.P. Jenks and J. Carriuolo, J. Am. Chem. Soc. ***83****, 1743 (1961) (from Ferscht, A. "Enzyme Structure and Function", W.H.Freeman, New York). The rate constant is shown for concentrations of two bases, pyridine and acetate. The zero base concentration intercept (k_o) represents the uncatalyzed rate at that particular pH. Pyridine is a more effective catalyst. A Brönsted plot of pK_a versus the logarithms of the slopes (the second order rate constants k_2) for these bases would show a linear relationship, indicating that catalysis depends on the effective base strengths and not on specific chemical factors.*

How does an enzyme obtain the rate enhancements made possible by general acid-base catalysis? The answer lies in the combination of pK_a values for amino acid moieties involved in acid-base catalysis and the typical values of α and β, describing proton transfer. An acid group with a pK_a of 5 is a better general-acid catalyst than one with a pK_a of 7. At pH 7, however, typical of the optimal pH of many enzymes, only 1% of the acid with a pK_a of 5 will be unionized and active in catalysis. The acid with a pK_a of 7 will be 50% unprotonated. By equating values of k_{obs} for each acid group, we see from equations (1.32) and (1.33b) that if α is less than ~0.85, then the acid with the pK_a of 7 will be a more effective catalyst. The same comments apply to base catalysis. If we consider the amino acids, we see that histidine contains an imidazole moiety which has a pK_a value around 6 to 7. Therefore, histidine is widely found in enzymes involved in base catalysis as it is 50% ionized at neutral pH. The imidazole moiety may thus be considered as the most effective amino acid base existing at neutral pH. In fact, all enzymatic acyl transfers catalyzed by proteases (e.g., trypsin, chymotrypsin) involve histidine. The imidazole group of histidine can also function as a nucleophile, and we must be careful to determine whether histidine acts a general base catalyst or a nucleophilic catalyst. In proteases, histidine functions as a base catalyst, but it is typically found closely associated with serine. Histidine is thought to deprotonate this neighbouring serine alcohol moiety by general base catalysis. The serine alkoxide ion ($^-O\text{-}CH_2CH\text{-}enz$) so generated has a pK_a of 13.7 and is thus a stronger base than histidine, but at neutral pH is less reactive. As illustrated below, serine functions primarily as a strong nucleophile in proteases.

The Mechanism of Chymotrypsin Catalysis

An example of both general base and nucleophilic catalysis is provided by the enzyme chymotrypsin, which generates a serine alkoxide at its active site. This nucleophilic serine is used by the enzyme to attack unactivated acyl groups on substrates at very high rates. The structure of the enzyme is represented in Figure 1.9. The mechanism of catalysis by chymotrypsin is shown in Figure 1.10, illustrating the manner in which the histidine residue (His-57[9]) increases the nucleophilic character of the serine (Ser-195). The carboxylate of aspartic acid (Asp-102) is involved in forming a "charge relay system", which stabilizes the serine alkoxide ion formation.

---

(9) The 57 indicates the position of the amino acid histidine in the protein sequence, where the numbering starts from the free amino terminus.

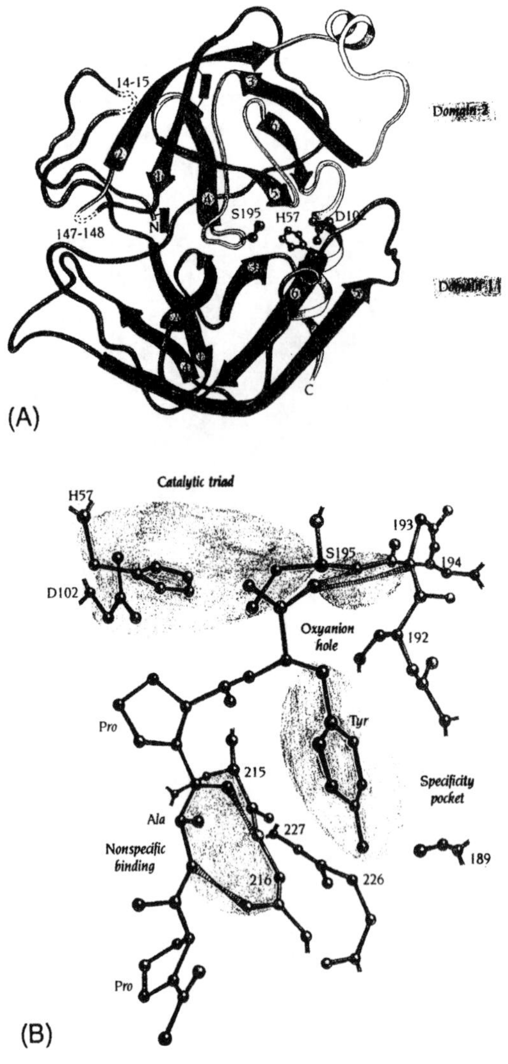

Figure 1.9. (A) Schematic diagram of the structure of α-chymotrypsin, which is folded into two domains of the antiparallel β type. The six β strands of each domain are represented by dark grey arrows. Dotted lines indicate residues 14-15 and 147-148 in the inactive precursor, chymotrypsinogen. These residues are excised during the conversion of chymotrypsinogen to the active enzyme chymotrypsin. The side chains of the catalytic triad, S195, H57, and D102, are also shown. (B) Diagram of the active site of chymotrypsin with a bound tetrapeptide inhibitor, N-acetyl-Pro-Ala-Pro-Tyr-COOH. The diagram illustrates the binding of inhibitor in relation to the catalytic triad (also shown in A, above), the so-called substrate specificity pocket (occupied by the tyrosine side chain of the inhibitor), the oxyanion hole, and the nonspecific substrate binding region. Hydrogen bonds between inhibitor

and enzyme are striped. Note the close alignment between the side chain of S195 and the terminal carboxyl group of the inhibitor. The positioning of this carboxyl group is similar to that of the scissile amide bond of a polypeptide substrate. Adapted from C. Branden and J. Tooze, "Introduction to Protein Structure," Garland Publishing, Inc., N.Y. (1991).

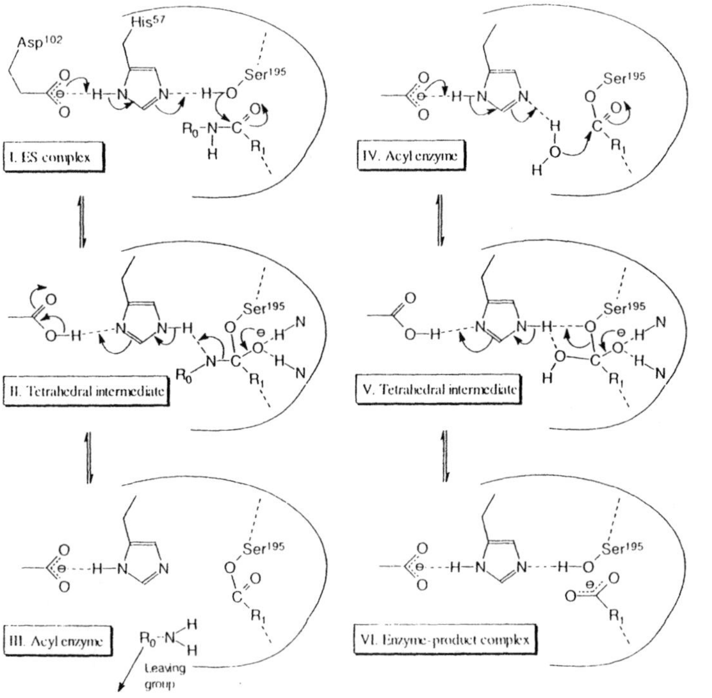

Figure 1.10. *The proposed mechanism of proteolytic action of chymotrypsin involving the charge relay system. Histidine-57 is the ionizing residue.*

1.4.4 Other Mechanisms Involved in Catalysis

In addition to covalent and general acid-base catalysis, enzymes also employ other mechanisms of rate enhancement. One of these is electrostatic catalysis. As we have seen from the transition state analysis described above, electrostatic interactions between substrate and enzyme may stabilize this transition state and thus yield significant rate enhancements. We shall briefly describe this and the other types of catalysis that are found in enzymes.

Electrostatic Catalysis

In water, the large dielectric constant results in a small electrostatic interaction energy between charges, and electrostatic catalysis is not generally important in homogeneous catalysis in aqueous systems. However, the active site of a protein is very heterogeneous and the dielectric constant of the medium between charged groups may be quite different from water. The aromatic and aliphatic amino acid residues present at the active site act to reduce the dielectric constant and charged amino acid residues act as fixed dipoles, thus stabilizing charge quite effectively. The electrostatic interaction energy depends on the charges and is inversely proportional to the dielectric constant and distance between charges. Lowering the dielectric constant can increase this energy considerably. Proteins may thus use parts of their own structure to solvate transition states and induce electrostatic strain. In fact, enzymes may stabilize charged groups in the transition state better than water, as the amino acids which function as dipoles are rigidly positioned and have a direction in relation to the substrate, whereas in water this directionality is lost.

The overall significance of electrostatic catalysis in enzymes is still not clear, as determination of the local dielectric constant is difficult. Electrostatic stabilization of charged transition states certainly plays some role, however.

Catalysis Involving Metal Ions

In metalloenzymes a metal ion is present at the active site and this ion plays an important role in stabilizing negative charges that are formed in electrophilic catalysis. Zinc, copper and cobalt are commonly involved in coordination of oxyanions involved as reaction intermediates. The enzyme carboxypeptidase-A, which is a carboxyl-terminus exopeptidase (i.e., it acts by hydrolyzing the peptide from the carboxylic acid terminus), contains Zn^{2+} which polarizes the carbonyl oxygen of the terminal peptide bond. The terminal carboxylate is charge paired with the guanidinium cation of Arg^{145}, leading to polarization of the terminal carboxylic carbonyl group. This polarization increases the electrophilicity of the carbonyl carbon and facilitates nucleophile-mediated hydrolysis of the amide bond. This is illustrated in the accompanying figure. In addition to stabilizing negative charges, metal ions serve as a source of potent nucleophilic hydroxyl ions. Metal-bound water molecules provide these nucleophilic hydroxyl groups at neutral pH.

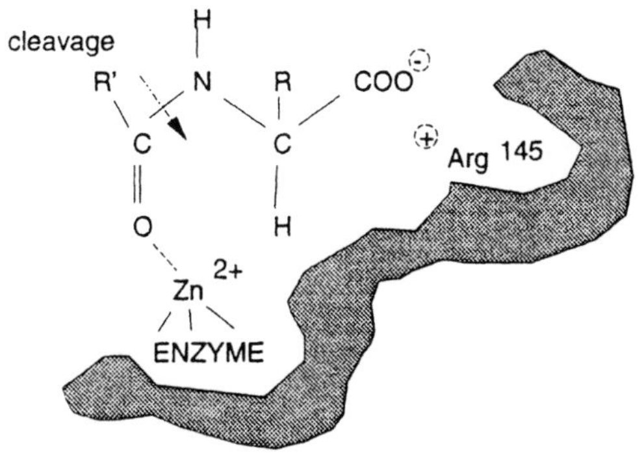

An example is the extremely rapid hydration of CO_2 by carbonic anhydrase to produce bicarbonate. The enzyme contains zinc coordinated to the imidazole groups of three histidines, with the fourth ligand being water, which is ionized. The zinc-bound hydroxyl has a pK_a of 8.7 and the reaction mechanism is thought to be:

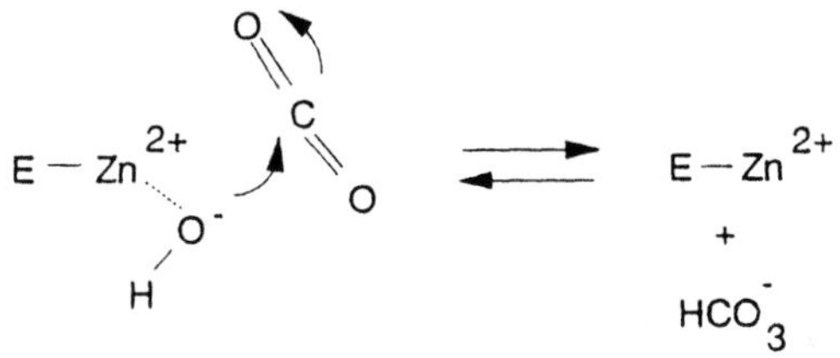

1.4.5 Energetics of Substrate Binding

We now briefly examine the question of how an enzyme binds a substrate during catalysis and why it is able to be so selective in its choice of substrates.

When the substrate enters the active site of an enzyme, it will be held initially by non-covalent forces. These non-covalent forces responsible for binding may be employed to lower the activation energy of the reaction. The types of non-covalent forces that are involved can be summarized as:

(a) *Electrostatic interactions* include charge-charge ($1/r$), dipole-dipole ($1/r^6$), charge-induced dipole ($1/r^4$) and dipole-induced dipole ($1/r^6$) interactions. The magnitude of these forces depends on the distance between molecules, varying with the distance (r) in the manner indicated above. All depend inversely on the dielectric constant of the solvent between the ions or dipoles.

(b) *van der Waals forces* are comprised of electron cloud repulsion ($1/r^{12}$) and attractive dispersion forces (London forces) ($1/r^6$). The sum of these is described by the Lennard-Jones 6-12 potential. Dispersion forces are not large, but in an enzyme the sum of all such forces between substrate and enzyme may be quite significant.

(c) *Hydrogen bonds* are important in biological systems and occur when two electronegative atoms are bound to a common proton. Often oxygen is one of the atoms.
(d) *Hydrophobic forces* reflect the tendency of apolar molecules to partition from an aqueous environment to a hydrophobic one. The driving force for such movement can be thought of as a result of the entropy gain when water molecules, which must be structured around an apolar molecule, are able to assume a more random arrangement when the molecule is transferred. The magnitude of this force is found experimentally to depend on the surface area of the molecule.

When the substrate moves from the external aqueous environment to the active site of the enzyme, its solvation shell is lost and one or more of the above forces are important in determining the strength of its binding. Hydrogen bonding and electrostatic interactions are generally most important. The difference in energies between the solvated state and that of the $E \cdot S$ complex determines the strength of substrate binding. The contributions of various groups to the overall binding energy can be estimated from the dissociation constant for $E \cdot S$ complexes. If we compare K_D values for a substrate S-H and a substrate analog S-X, the difference in the free energies can be interpreted as the binding energy contribution of group X. For example, the binding energy of S-H and S-X are

$$\Delta G_{b(S-H)} = -RT \ln K_{D(S-H)}$$

$$\Delta G_{b(S-X)} = -RT \ln K_{D(S-X)}$$

$$\Delta\Delta G_{bX} = -RT \ln \frac{K_{D(S-X)}}{K_{D(S-H)}}$$

The binding energy ($\Delta\Delta G_b X$) determined in this manner may not reflect all the available binding energy of the X group. The active site of an enzyme may be thought of as being fairly complementary to the substrate. During the reaction, the substrate changes in configuration to the transition state. If the enzyme is complementary to this structure rather than the undistorted substrate, less activation energy is then required to reach the transition state. In this case, some of the binding energy can be thought of as being used to reduce the activation energy barrier. A better measure of the binding energy associated with group X can thus be obtained from the effective second order rate constants for the two reactants, as this includes both binding and activation energies.

$$\Delta\Delta G_{bX} = -RT \ln \frac{(k_{cat}/K_M)_{S-X}}{(k_{cat}/K_M)_{S-H}}$$

Table 1.2 illustrates the binding energies of various X groups for transfer from water to the enzyme aminoacyl-tRNA synthetase and compares them to the corresponding values for tranfer to n-octanol. The higher values for the enzyme active site tranfer indicate that it is more hydrophobic than octanol.

***Table 1.2**. Binding energies of various groups with the enzyme aminoacyl-tRNA synthetase compared to corresponding values with n-octanol. [From Fersht, A., Shindler, J.S., and W.C. Tsui, Biochemistry **19**, 5520 (1980)].*

Group	$\Delta\Delta G_b X$ from enzyme to water (kcal/mol) (relative to H)	$\Delta\Delta G_b X$ from n-octanol to water (kcal/mol)
$-CH_3$	3.2	0.68
$-CH_2CH_3$	6.5	1.36
$-CH(CH_3)_2$	9.6	1.77
-OH	7.0	-1.58

1.5 Enzyme Inhibition

The activity of an enzyme may be reduced by several means. A structural analog of the substrate may bind in the active site, either reversibly or irreversibly, and thereby reduce activity. Alternatively, an inhibitor may bind on some other portion of the enzyme and induce conformational changes that may reduce or eliminate the ability of the enzyme to bind substrate. We thus consider two main types of inhibition - reversible and irreversible inhibition, depending on whether activity can be restored by dilution or removal of the inhibitor.

Irreversible Inhibition

Enzymes can be irreversibly inactivated by the binding of an inhibitor molecule to the active site of the enzyme. The inhibitor typically forms a covalent bond with the enzyme and thus prevents binding of the substrate. It may also act by destroying some part of the active site. Although heat and some chemical agents can denature the enzyme and thereby inhibit it, we generally do not consider these to be irreversible inhibitors. Many irreversible inhibitors attack sulfhydryl groups at the active site (e.g., iodoacetate) or the hydroxyl group of serine (e.g., di-isopropylphosphofluoridate DFP, a nerve gas which inactivates acetylcholinesterase - see Figure 1.6). Irreversible inhibitors thus reduce the effective concentration of the enzyme without affecting the observed kinetics of the enzyme with respect to substrate. Irreversible inhibitors are used in investigations of active site binding and kinetic mechanism, as they can be selected to bind with specific amino acids thought to be present at the active site.

Reversible Inhibition

Unlike irreversible inhibitors, reversible inhibitors can be removed from the enzyme, enabling it to recover its activity. Reversible inhibitors can be removed by dialysis or by addition of another agent which binds with the inhibitor in solution and effectively lowers its concentration. We will consider the four main types of reversible inhibitors in this section; only simple single-substrate reactions which obey Michaelis-Menten kinetics will be considered. More complex, multi-substrate cases are discussed in Section 1.8.

1.5.1 Competitive Inhibition

Competitive inhibitors, as the name implies, compete with the substrate for the active site of the enzyme. They often closely resemble the substrate, but lack the appropriate reactive group or are held at the active site in a manner that prevents the enzyme reaction from occurring. This situation results in the formation of a *dead-end complex* in either case; the inhibitor must diffuse from the active site before a reactive substrate can enter the site. The observed kinetics depend on the concentrations of inhibitor and substrate and their relative affinities for the active site. At high substrate concentrations, the effect of the inhibitor will be reduced and the maximum reaction rate will be unchanged.

Consider the case of a single-substrate reaction where the inhibitor competes directly for the active site:

$$E+S \underset{k_{-1}}{\overset{k_1}{\Leftrightarrow}} ES \overset{k_2}{\rightarrow} E+P \quad \text{and} \quad E+I \overset{K_I}{\Leftrightarrow} EI$$

increases Km → decr. affinity for substrate

The dissociation constant for the binding of the inhibitor to the free enzyme is K_I:

$$K_I = \frac{[E][I]}{[EI]} \tag{1.34}$$

We can derive a rate expression in a similar manner as that obtained previously for the Briggs-Haldane case (Section 1.2.2):

$$v = \frac{dP}{dt} = k_2[ES] \tag{1.35}$$

$$\frac{[E][S]}{[ES]} = \frac{k_{-1}+k_2}{k_1} = K_M \tag{1.36}$$

To substitute for [E] we must now account for the reduction in free enzyme by the action of the inhibitor:

$$[E_o] = [E] + [ES] + [EI] \tag{1.37}$$

Substituting for [EI] from equation (1.34) and rearranging gives:

$$[E_o] = [E]\cdot\left(1+\frac{[I]}{K_I}\right) + [ES] \tag{1.38}$$

and substituting for [E] from equation (1.36)

$$[E_o] = K_M\frac{[ES]}{S}\cdot\left(1+\frac{[I]}{K_I}\right) + [ES] \tag{1.39}$$

We now rearrange (1.39) and substitute [ES] into equation (1.35) to obtain the rate expression in terms of the measurable quantities [E_o], [S] and [I]:

$$v = \frac{k_2[E_o][S]}{S + K_M(1 + [I]/K_I)} \tag{1.40}$$

We note here that the maximum rate of reaction remains unchanged at sufficiently high values of [S] (provided that [I] < [S]). The effect of competitive inhibition is to increase the Michaelis constant K_M by the factor (1+[I]/K_I), i.e., the apparent affinity of the enzyme for the substrate is decreased[10]. This type of binding is generally known as *allosteric* binding, and the inhibition is referred to as *allosteric inhibition*. More often, however, this term is employed in cases where a dead-end complex is not formed and the substrate is still able to bind at the active site and react, although the binding characteristics may now be altered. We will examine this further in Section 1.9.

Equation (1.40) can be rearranged in the double-reciprocal form of the Lineweaver-Burk plot:

$$\frac{1}{v} = \frac{K_M\left(1 + \frac{[I]}{K_I}\right)}{v_{\max}} \cdot \frac{1}{[S]} + \frac{1}{v_{\max}} = \frac{K_M^{app}}{v_{\max}} \cdot \frac{1}{[S]} + \frac{1}{v_{\max}} \tag{1.41}$$

The effect of varying inhibitor concentrations on enzyme activity is shown in Figure 1.11 below, together with the Eadie-Hofstee plot, given in the form:

$$v = v_{\max} - \frac{K_M\left(1 + \frac{[I]}{K_I}\right)v}{[S]} \tag{1.42}$$

The inhibition constant K_I can be determined from either of the above plots by simply graphing the apparent K_M value against inhibitor concentration, provided that the enzyme concentration is held constant. The relationship below can then be employed to determine K_I as the K_M value in the absence of inhibitor is known.

$$K_M^{app} = \frac{K_M}{K_I} \cdot [I] + K_M \tag{1.43}$$

Competitive inhibitors can be used as chemotherapeutic agents in medicine or as insecticides and herbicides. The sulfa drugs, sulfonamides such as sulfanilimide, were employed prior to the discovery of penicillin as antibacterial agents, due to their ability to compete with the substrate p-aminobenzoic acid in the synthesis of the coenzyme tetrahydrofolate (FH_4). This coenzyme is produced in bacteria but not in man, where it is obtained directly from the vitamin folic acid. Thus the sulfanilamides were used with relatively little human risk to treat bacterial infections.

In some cases the competitive inhibitor may be the product of the reaction. An

(10) The same effect could be obtained if the inhibitor was bound to the enzyme but not at the active site, provided that the EI complex so formed would not permit substrate to bind.

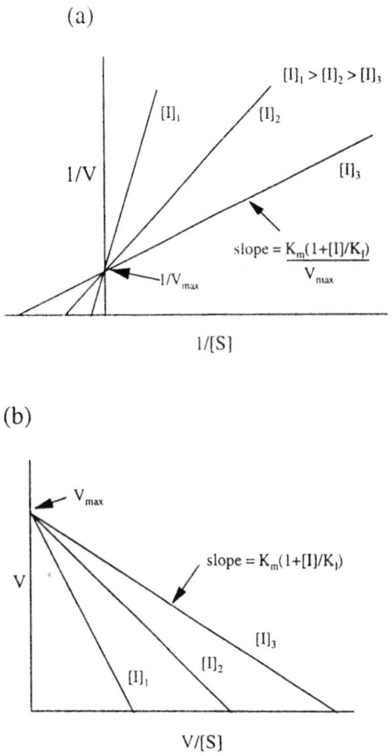

Figure 1.11. *(**a**) Lineweaver-Burke plot, illustrating the change in slope (K_M) and unchanged v_{max} with competitive inhibition. (**b**) Eadie-Hofstee plot with several values of inhibitor concentration [I].*

important case in the glycolytic pathway is the enzyme phosphoglycerate mutase, which transfers the phosphoryl group in 3-phosphoglycerate from the unactivatable C-3 position via a 2,3 diphosphoglycerate intermediate to the activatable C-2 position in 2-phosphoglycerate (2-PGA). It is inhibited by the 2,3 diphosphoglycerate intermediate which prevents excessive carbon flow through this pathway. The 2-PGA is further processed by an enolase to yield the high-energy compound, phosphoenolpyruvate (PEP), with the energy being subsequently captured by ADP in the formation of pyruvic acid.

1.5.2 Uncompetitive, Non-Competitive, and Mixed Inhibition

Uncompetitive inhibitors bind to either a modified form of the enzyme which arises when substrate is bound, or to the enzyme-bound substrate itself. In either case we consider binding to ES to occur, but no binding of the inhibitor to free enzyme. The reaction sequence can be considered to be

$$E + S \Leftrightarrow ES \rightarrow E + P$$

$$ES + I \Leftrightarrow ESI$$

The complex ESI is a dead-end complex. We can develop a rate expression for uncompetitive inhibition following the approach used for competitive inhibition. We assume that ES and ESI rapidly obtain equilibrium concentrations and the inhibitor constant becomes

$$K_I = \frac{[ES][I]}{[ESI]} \tag{1.44}$$

The overall enzyme concentration is now written as

$$[E_0] = [E] + [ES] + [ESI]$$

Eliminating [E] and [ESI] by use of equations (1.36) and (1.44) and substituting in the expression for v yields

$$v = \frac{\frac{v_{max}}{(1+[I]/K_I)} \cdot S}{S + \frac{K_M}{(1+[I]/K_I)}} \tag{1.45}$$

The maximum velocity and the Michaelis constant are now both altered by the factor $(1 + [I]/K_I)$

$$v_{max}^{app} = \frac{v_{max}}{\left(1 + \frac{[I]}{K_I}\right)} \qquad K_M^{app} = \frac{K_M}{\left(1 + \frac{[I]}{K_I}\right)} \tag{1.46}$$

However, the slope of the Lineweaver-Burk plot reflects the actual K_M and v_{max} values, and is independent of the inhibitor concentration.

$$\frac{1}{v} = \frac{1}{v_{max}^{app}} + \frac{K_M^{app}}{v_{max}^{app}} \cdot \frac{1}{[S]} \tag{1.47}$$

$$\frac{K_M^{app}}{v_{max}^{app}} = \frac{K_M}{v_{max}} \tag{1.48}$$

The values of the inhibition constant K_I may be obtained from plots of the reciprocal of the apparent maximum reaction rate or apparent K_M value against inhibitor concentration.

Noncompetitive inhibitors can combine with either the free enzyme or the enzyme-substrate complex to produce a dead-end complex. The reaction pathway is illustrated in the figure opposite.

In the simplest analysis of this reaction scheme, we assume that all the reactions are at equilibrium and that the dissociation constants for binding of substrate and inhibitor are the same (i.e., $K_S = K_S'$ and $K_I = K_I'$). Then

$$K_s$$
$$E \xrightleftharpoons[-S]{+S} ES \longrightarrow E + P$$

$$+I \quad \big\downarrow\big\uparrow \quad -I \qquad\qquad +I \quad \big\downarrow\big\uparrow \quad -I$$

$$K_i \qquad\qquad\qquad\qquad\qquad K_i$$

$$EI \xrightleftharpoons[-S]{+S} EIS$$

$$K_s'$$

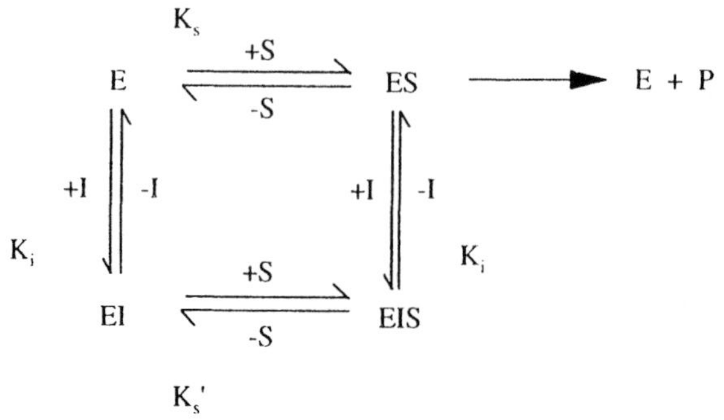

$$K_I = \frac{[E][I]}{[EI]} = \frac{[ES][I]}{[ESI]}$$

with a similar expression for K_S. An overall balance on enzyme and elimination of [EI], [ES] and [EIS] yields

$$v = \frac{v_{max}}{\left(1 + \dfrac{I}{K_I}\right)} \cdot \frac{S}{(S + K_S)} \tag{1.49}$$

Thus the maximum reaction rate is decreased but the Michaelis constant remains the same.

More commonly, the dissociation constant for the [EIS] complex is different from that for [ES] and thus $K_S \neq K_S'$. Both v_{max} and K_S (or K_M in the case where the quasi-steady state assumption is made, rather than the equilibrium assumption employed here) are then altered by the inhibitor concentration, and the inhibition is termed *mixed*.

There are relatively few examples where the inhibition pattern falls clearly into either the uncompetitive or noncompetitive case for single substrate reactions. When the active site includes a proton acceptor (for example chymotrypsin), increasing the hydrogen ion concentration may result in a noncompetitive inhibition pattern over narrow ranges of pH. Figure 1.12 illustrates the general patterns for both uncompetitive and noncompetitive inhibition.

1.5.3 Substrate Inhibition

In some instances, substrate may bind at an alternative site to the active site, and the resulting enzyme-substrate complex may be unreactive. This non-productive binding results in another type of inhibition known as *substrate inhibition*, a special form of uncompetitive inhibition. If we replace [I] by [S] in equation (1.45), we obtain

$$v = \frac{v_{max}[S]}{[S] \cdot \left(1 + \dfrac{[S]}{K_I}\right) + K_M} \tag{1.50}$$

At low values of [S], the term $(1 + [S]/K_I)$ is approximately unity and the resulting rate expression is that of Michaelis-Menten form. At high values of [S],

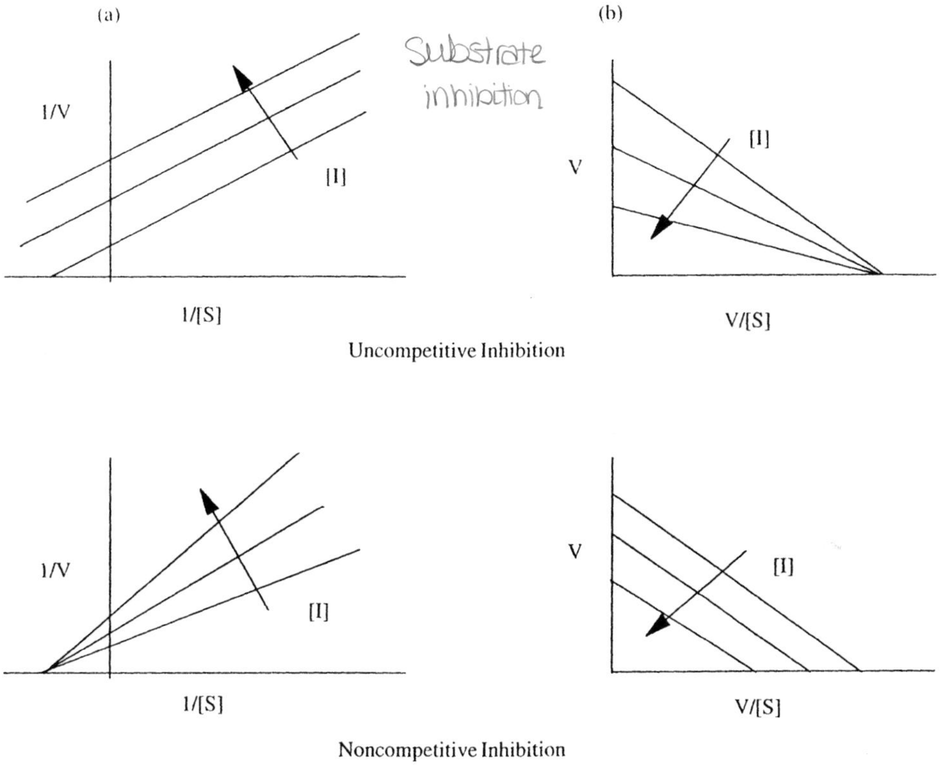

Figure 1.12. ***(a)*** *Lineweaver-Burk and* ***(b)*** *Eadie-Hofstee plots illustrating uncompetitive and noncompetitive inhibition. The inhibitor concentration increases in the direction of the arrows.*

$$1 + \frac{[S]}{K_I} \gg \frac{K_M}{[S]}$$

and the reaction rate becomes inversely dependent on substrate concentration:

$$v = \frac{v_{\max}}{1 + \frac{[S]}{K_I}} \tag{1.51}$$

1.6 Pre-Steady State Kinetics

The steady state assumption that we have made so far in examining the kinetics of enzyme-catalyzed reactions places some limitations on the kinetic constants that can be determined. Only K_M and the first order rate constant k_{cat} can be found. In the case of inhibitors, we may distinguish the type of inhibition and the dissociation constant K_I. To determine whether a particular enzyme follows Michaelis-Menten or Briggs-Haldane type kinetics (indistinguishable on the basis of *initial rate* data), we need to relax the steady-state assumption and examine the time course for the formation of the [ES] complex itself. As typical turnover numbers for enzymes may be of order 100 molecules/sec, rapid methods to measure concentrations of substrate and product must be employed.

1.6.1 Rapid Mixing Methods

One approach to determine substrate or product concentrations over short periods of time involves a technique known as *rapid mixing*. A continuous plug flow reactor is employed with substrate and enzyme introduced at the inlet. As the reaction proceeds, concentration profiles along the reactor can be measured at various distances (for example, spectroscopically) and the time course determined by examining length-concentration profiles.

Considering a first order reaction for simplicity, we can write

$$u \cdot \frac{d[S]}{dz} = -k[S] \tag{1.52}$$

$$\frac{d[S]}{d\tau} = -k[S] \tag{1.53}$$

Solving equation (1.53) for [S] yields:

$$[S] = [S_0]e^{-k\tau} \tag{1.54}$$

where $\tau = z/u$, u is the velocity in the axial direction and z is distance along the reactor. The flow must be turbulent to minimize radial concentration gradients. To obtain a (1/e)-fold reduction in substrate concentration, L must be of order u/k. This may require very long reactors, as u will be large in turbulent flow, and k will be very small.

A more practical approach is the *stopped flow method*. Two syringes, one containing enzyme and the other substrate, are driven at a predetermined rate to deliver their fluids to a mixing chamber (see Figure 1.13.). The mixing of the fluids initiates the reaction and a detector (usually a spectrophotometer) some distance along a tube from the mixing chamber is employed to determine substrate or product concentration. The reaction is stopped by mechanically stopping the flow. By varying the flow rates, the time course of the reaction can be obtained. A variation of this method, the *rapid quenching technique*, involves stopping the reaction by mixing the fluid with a third stream containing a reagent which stops the reaction (e.g. acid which denatures the enzyme). These methods all have dead times associated with delays in mixing and dead spaces in the apparatus.

If this dead time is ignored, we can analyze the product- or substrate-versus-time data to obtain the individual kinetic parameters. For a simple monomolecular first-order reaction

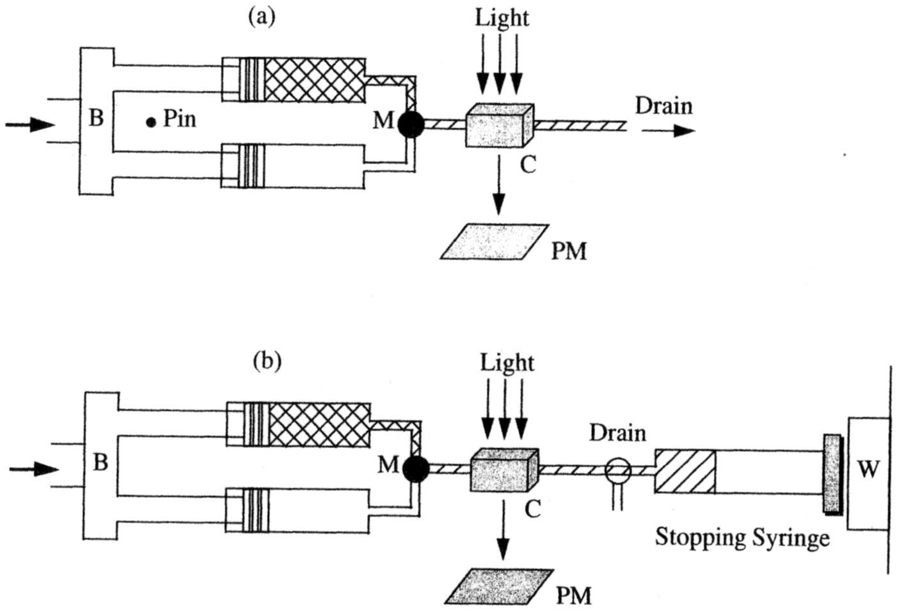

Figure 1.13. *The piston driven stopped flow apparatus, illustrating the method of detection. Two stopping methods are shown (a) front stopping type, where the driving block (B) is arrested by a pin (P) and (b) end-stopping type, where flow is stopped when the stopping syringe plunger hits the wall (W). The static mixer (M), observation cell (C) and photomultiplier (PM) are shown.*

$$A \xrightarrow{k_f} B$$

$$\frac{d[A]}{dt} = -k_f[A] \tag{1.55}$$

Integration of the mass balance on [A] in a batch reactor (noting $[A] = [A]_0$ at t=0) gives:

$$[A] = [A]_0 \cdot e^{-k_f t} \tag{1.56}$$

Now consider a reversible reaction with a reverse first order rate constant k_r, and an equilibrium constant $K_{eq} = [B_{eq}]/[A_{eq}]$. At time t = 0, $[A] = [A]_0$ with $[B]_0 = 0$, and $[B]_t = [A]_0 - [A]_t$; hence we can write

$$\frac{d[A]}{dt} = -k_f[A] + k_r[B]$$

$$= -k_f[A] + k_r([A]_0 - [A]) \tag{1.57}$$

which has the solution:

$$[A]_t = [A]_0 \cdot \frac{k_f}{k_f + k_r} \cdot \left(\frac{1}{K_{eq}} + e^{-(k_f + k_r)t} \right) \quad (1.58a)$$

To determine both k_f and k_r, the value of K_{eq} must be known. The approach to the equilibrium value of [A] is then determined, so that the exponential term (k_f + k_r) can be evaluated. To this end, Equation (1.58a) can be rearranged in a more convenient form for this, noting $[A]_{eq}$ is the equilibrium concentration:

$$\ln\left(\frac{[A] - [A]_{eq}}{[A]_0 - [A]_{eq}}\right) = -(k_r + k_f)t \quad (1.58b)$$

Alternatively, if the association of enzyme and substrate can be considered to be a pseudo-reversible reaction and if [S] >> [E], then

$$E + S \overset{k_f}{\Leftrightarrow} ES$$

becomes

$$E \overset{k_f'}{\Leftrightarrow} ES$$

with a pseudo-first order forward rate constant $k_f' = k_f[S]$. By varying [S] and determining the observed first order rate constant k_{obs} (= $k_f' + k_r$) as a function of substrate concentration, both forward and reverse rate constants may be determined without knowledge of the equilibrium constant. This requires that the stopped flow apparatus be able to operate with residence times in the range of $1/k_r$, the lower limit for k_{obs}. A typical apparatus can only resolve time constants to 1000 sec^{-1}, and many enzyme-substrate dissociation constants are much larger than this. Thus other techniques, for example the relaxation methods described in Section 1.6.2, must be employed.

We shall now examine a single substrate enzyme reaction.

$$E + S \underset{k_{-1}}{\overset{k_1}{\Leftrightarrow}} ES \overset{k_2}{\rightarrow} E + P$$

The time course of the [ES] complex is given by

$$\frac{d[ES]}{dt} = k_1[E][S] - (k_{-1} + k_2)[ES] \quad (1.59)$$

When substrate is present in excess, $[S_0] > [E_0]$, and provided little product is formed, we can write

$$[E]_0 = [E] + [ES]$$
$$[S]_0 = [S] + [ES] + [P]$$
$$\sim [S]$$

Thus equation (1.59) can be simplified by this assumption of constant substrate concentration:

$$\frac{d[ES]}{dt} = k_1([E]_0 - [ES])[S]_0 - (k_{-1} + k_2)[ES] \quad (1.60)$$

which can be integrated with the initial condition $[ES]_{t=0} = 0$ to give

$$[ES] = \frac{k_1[E]_0[S]_0}{k_1[S]_0 + k_{-1} + k_2} \cdot \left(1 - e^{[-(k_1[S]_0 + k_{-1} + k_2)t]}\right)$$

$$[P] = k_2 \int_0^t [ES]dt = \frac{k_1 k_2 [E_0][S_0]}{k_1[S]_0 + k_{-1} + k_2} \cdot t + \frac{k_1 k_2 [E_0][S_0]}{(k_1[S]_0 + k_{-1} + k_2)^2} \cdot e^{[-(k_1[S]_0 + k_{-1} + k_2)t]} \qquad (1.61)$$

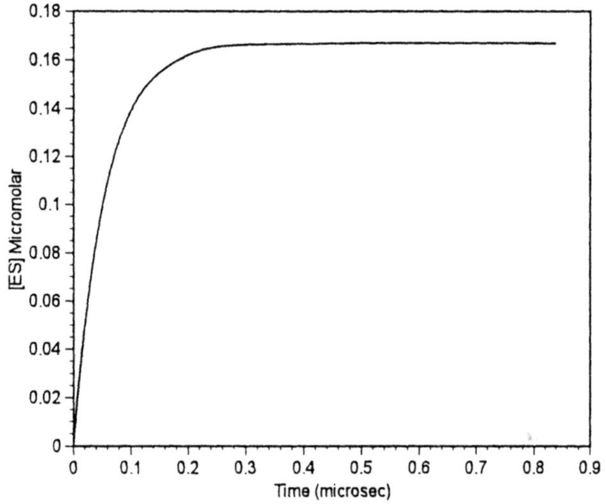

***Figure 1.14**. The initial time course of the ES complex for typical Michaelis-Menten behavior. After a very short period, the ES complex attains an approximately constant concentration, indicating the applicability of the quasi-steady state assumption. The parameter values are* $[E]_o = 10^{-6}$ *M,* $S_o = 10^{-4}$ *M,* $k_1 = 3x10^4$ $M^{-1}sec^{-1}$*,* $k_{-1} = 5$ sec^{-1}*,* $k_2 = 10$ sec^{-1}*.*

The time course of the reaction is shown in Figure 1.14. The pseudo-steady state assumption can thus be seen to be valid after a short time period.

Determining rate constants using the techniques described above requires a knowledge of the concentration of active sites in the enzyme and the enzyme concentration. The enzyme concentration cannot always be determined from its mass, as an isolated enzyme sample is generally not 100% pure. For certain types of enzymes, where there is an accumulation of an intermediate at the active site, determining the initial release of a reaction by-product provides a mechanism for determining the absolute concentration of active enzyme. We shall examine this in the following example.

Determination of the Number of Active Sites - The Initial "Burst" Kinetics of Chymotrypsin

The classical example of the determination of the concentration of active sites in an enzyme preparation is the hydrolysis of *p*-nitrophenylacetate catalyzed by chymotrypsin (see B.S. Hartley and B.A. Kilby, *Biochemistry J.* **56**, 288 (1954)). The reaction proceeds by a fast binding of the substrate to the enzyme and formation of an acyl-enzyme intermediate with the release of *p*-nitrophenol. The second step, hydrolysis by water, releasing acetate, is slow. The reaction pathway is illustrated.

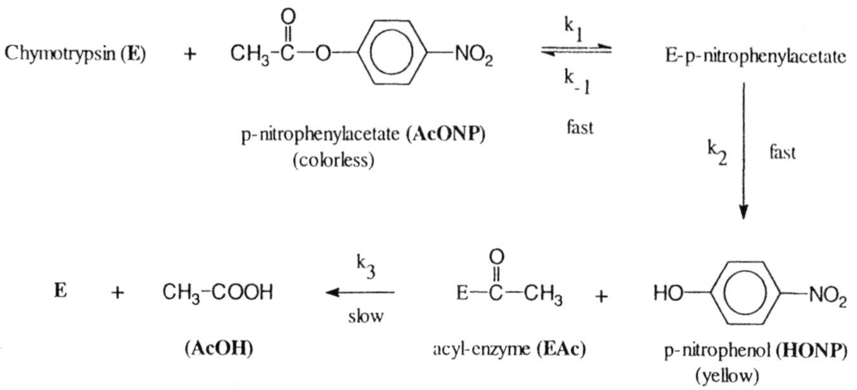

First Order Approximation

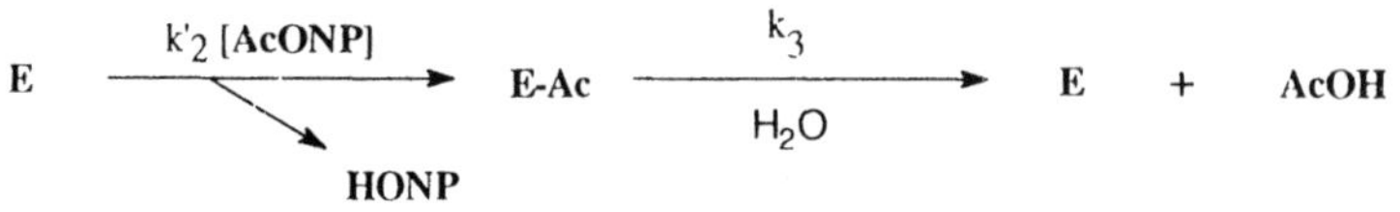

As the first molecule of substrate binds to the enzyme it releases the colored *p*-nitrophenol very rapidly (the "burst"). The subsequent slow rate of production of colored product results from the recycle of free enzyme via deacylation of the acyl-enzyme intermediate EAc. This reflects the rate determining step, the hydrolysis of the acyl-enzyme intermediate to free acetate anion. At this point, the concentration of the acyl-enzyme intermediate form of the enzyme must be greater than that of the chymotrypsin-*p*-nitrophenolacetate intermediate, i.e., k_3 is rate determining. We can analyze the results of Hartley and Kilby by assuming a pseudo-first order reaction for the formation of the acyl-enzyme intermediate (acetyl-chymotrypsin). Following the scheme shown above and noting $k''_2 = k'_2[AcONP]$:

$$\frac{d[HONP]}{dt} = k''_2 \cdot [E] \quad (1.62)$$

$$\frac{d[EAc]}{dt} = k''_2[E] - k_3[EAc] \quad (1.63)$$

$$\frac{d[AcOH]}{dt} = k_3[EAc] \quad (1.64)$$

We note [E] + [EAc] = $[E]_0$, and can substitute for [E] in the equations above. Solving for [EAc] gives

$$[EAc] = \frac{k''_2[E]_0}{k''_2 + k_3} \cdot \left(1 - e^{-(k''_2+k_3)t}\right) \quad (1.65)$$

The concentration profile for *p*-nitrophenol, whose initial concentration is assumed to be zero, is given by

$$[HONP] = \left(\frac{k''_2}{k''_2 + k_3}\right)^2 E_0 + \frac{k''_2 k_3 E_0}{k''_2 + k_3} \cdot t - \left(\frac{k''_2}{k''_2 + k_3}\right)^2 E_0 \cdot e^{-(k''_2+k_3)t} \quad (1.66)$$

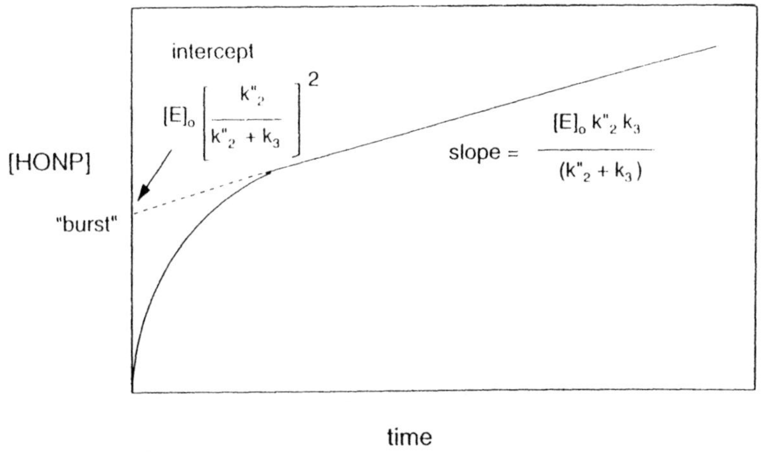

***Figure 1.15**. The time course of p-nitrophenol release from p-nitrophenolacetate at a fixed enzyme concentration.*

At short times, the term which is linear in t is small and the behavior is governed by the exponential term. At longer times (i.e., $t > 1/(k''_2+k_3)$), the influence of the exponential term decreases and the linear term in t predominates. The behavior is illustrated in Figure 1.15.

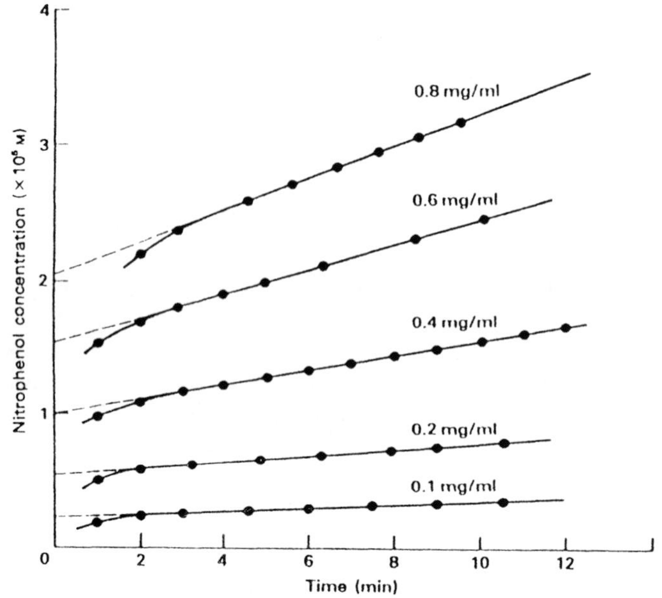

Figure 1.16. The initial burst experiment of Hartley and Kilby. Nitrophenol liberation is shown as a function of time for various enzyme concentrations. Around 0.63 mole of p-nitrophenol is released per mole of enzyme in each burst. The enzyme is either ~63% pure or the rate of formation of the acyl-enzyme intermediate is not sufficiently larger than the rate of deacylation (i.e., k_2'' is not larger than k_3), so that the enzyme does not maintain a fully loaded active site.

The initial burst rate occurs with a time constant of $1/(k''_2+k_3)$. Extrapolation of the linear rate to zero time gives the initial "burst" concentration. If $k''_2 \gg k_3$, the burst concentration of product corresponds to $[E]_0$ and can thus be used to determine the absolute concentration of enzyme. This assumption is often made and the method is referred to as "active site titration". The data of Hartley and Kilby are shown below for various values of enzyme concentration.

1.6.2 Relaxation Techniques

The time scales that can be explored with rapid mixing methods are somewhat restricted to determining rate constants of order 1000 sec^{-1}. To determine larger rate constants, relaxation methods are employed. These methods are able to examine events on a time scale of nanoseconds. In general, the method consists of making a small perturbation in a reaction parameter, such as temperature, pressure or electric field to a reaction at steady state, and continuously monitoring the response of the reaction mixture. The reaction system will move from the initial steady state condition to a new steady state. We shall consider the reversible formation of the [ES] complex as an example of the use of relaxation kinetics for determining the individual rate constants. The reaction is

$$[E]+[S] \underset{k_{-1}}{\overset{k_1}{\Leftrightarrow}} [ES]$$

Suppose that the temperature is suddenly perturbed, so that the system adjusts to a new steady state characterized by $[E]_{eq}$, $[S]_{eq}$, $[ES]_{eq}$. A mass balance in a well-mixed batch reactor can be written as

$$\frac{d[S]}{dt} = k_{-1}[ES] - k_1[E][S] \tag{1.67}$$

Noting as before

$$[E]+[ES]=[E]_0 \tag{1.68a}$$

$$[S]+[ES]=[S]_0 \tag{1.68b}$$

we can write

$$\frac{d[S]}{dt} = k_{-1}\{[S]_0-[S]\} - k_1[S]\{[E]_0+[S]-[S]_0\} \equiv f([S]) \tag{1.69}$$

The value of [S] at the new steady state $[S]_{eq}$ is then found by setting $f([S]_{eq}) = 0$. If the change in concentration from one steady state to the other is small, we introduce a perturbation variable s which is defined by $s = [S] - [S]_{eq}$. Substituting for [S] in equation (1.69) above, we obtain

$$\frac{ds}{dt} = f(s+[S]_{eq}) \tag{1.70}$$

$f(s+[S]_{eq})$ can be expanded in a Taylor series around $[S]_{eq}$, and by retaining only first order terms in s, and noting $f([S]_{eq}) = 0$, we obtain

$$f(s+[S]_{eq}) = f([S]_{eq}) + \frac{df([S])}{ds}\Big|_{[S]=[S]_{eq}} \cdot s + O(s^2) \tag{1.71}$$

$$\frac{ds}{dt} = -(k_{-1}+k_1\{[E]_0-[S]_0+2[S]_{eq}\})\cdot s \tag{1.72}$$

Making use of equations (1.68a) and (1.68b) at the new steady state and integrating yields

$$s(t) = s_0 \exp\{-(k_{-1}+k_1\{[E]_{eq}+[S]_{eq}\})t\} = s_0 \exp\left\{\frac{-t}{\tau}\right\} \tag{1.73}$$

where $s_o = [S]_{t=0} - [S]_{eq}$. Thus the system moves to a new steady state in the first order manner, as we have neglected terms of higher order in s. By obtaining data on the rate at which the system moves to the new steady state, and determining the slope of $\ln(s/s_0)$ versus time $(-1/\tau)$ at various values of $[E]_{eq}$ and $[S]_{eq}$, values for k_{-1} and k_1 can be determined from plots of τ versus $[E]_{eq}$ or $[S]_{eq}$. Different equilibrium conditions can be obtained by changing the temperature or pressure.

1.7 Enzyme Kinetics at Limiting Conditions

1.7.1 Dilute Substrates ($[S]_0$~$[E]_0$)

The Michaelis-Menten (equilibrium) or the Briggs-Haldane (quasi-steady state) treatments for the formation of the (ES) complex both provide models which agree well with experimental data as long as that the concentration of (ES) remains small. If the enzyme concentration is high, or the substrate concentration is low (e.g., an organic reactant which is only sparingly soluble in water), then $[S]_0$ and $[E]_0$ may be comparable. In this case, the transient concentration of (ES) may be quite high as the reaction proceeds. Considerable errors may result in obtaining kinetic parameters from monitoring the substrate concentration throughout the reaction. In these situations, we cannot consider [ES] to be in steady state or in equilibrium with [E] and [S]. The analytical solution of the Briggs-Haldane model for [S(t)] no longer holds, and numerical methods must be employed to solve for substrate and product concentration profiles[11]. We can write material balances and rate equations for the general case viz.

$$[E]+[S] \underset{k_{-1}}{\overset{k_1}{\Leftrightarrow}} [ES] \overset{k_2}{\rightarrow} [E]+[P]$$

$$[E]_0=[E]+[ES]$$

$$[S]_0=[S]+[ES]+[P] \tag{1.74}$$

The rate equations can be written using the above equations to eliminate [E] and [S]:

$$\frac{d[S]}{dt}=-k_1([E]_0-[ES])\cdot([S]_o-[ES]-[P])+k_{-1}[ES] \tag{1.75}$$

$$\frac{d[ES]}{dt}=k_1([E]_0-[ES])\cdot([S]_0-[ES]-[P])-(k_{-1}-k_2)\cdot[ES] \tag{1.76}$$

$$\frac{d[P]}{dt}=k_2[ES] \tag{1.77}$$

We can no longer assume that d[ES]/dt is zero and obtain the quasi-steady state result. We proceed by eliminating [ES] from equations (1.76) and (1.77). The following equation for [P] is obtained

(11) See Lim, H. *"On Kinetic Behavior at High Enzyme Concentrations"*, AIChE Journal, **19**, 659 (1973) for further examples.

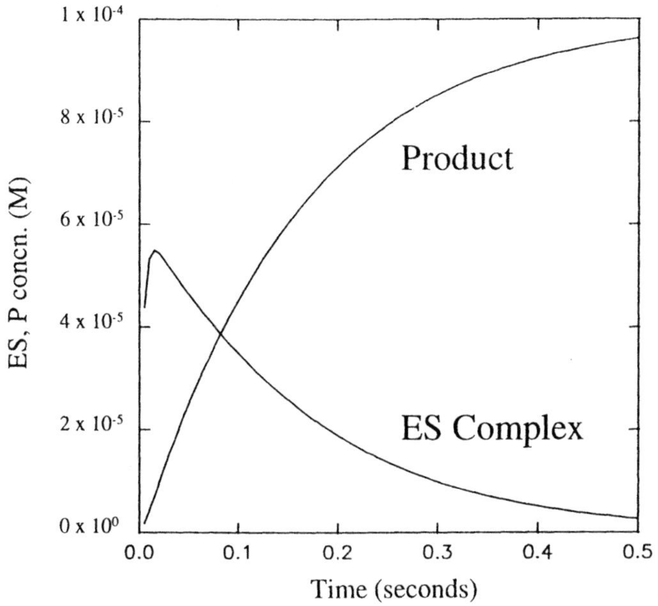

Figure 1.17. *Numerical solution of equations (1.75), (1.76), and (1.77) with the values* $E_0 = 20\ \mu M$, $S_0 = 10\ \mu M$, $k_1 = 10^6\ M^{-1}sec^{-1}$, $k_{-1} = 90\ sec^{-1}$ *and* $k_2 = 10\ sec^{-1}$.

$$\frac{d^2[P]}{dt^2} - \left(\frac{k_1}{k_2}\right)\left(\frac{d[P]}{dt}\right)^2 + \{k_1([E]_0 + [S]_0) + k_{-1} + k_2\}\frac{d[P]}{dt} - k_1[P]\frac{d[P]}{dt}$$
$$+ k_1k_2[E]_0[P] - k_1k_2[E]_0[S]_0 = 0 \qquad (1.78)$$

No analytical solution is available for this equation. The numerical solution is shown in Figure 1.17 for $S_0/E_0 = 4$. Note that the ES complex does not have any region where its concentration remains constant, confirming that the steady-state approximation for ES does not apply.

Single Turnover $[E]_0 >> [S]_0$

When the enzyme is present in large excess over the substrate, only a single turnover of substrate occurs. In this case the equations describing the kinetics (equations 1.74 to 1.77) can be solved analytically, as the simplification [E] ~ $[E]_0$ can be made. Observation of (ES) or product concentration permits the apparent first order rate constant to be obtained. This is illustrated in Problem 10 for the case of two (ES) complexes formed in a three step mechanism, as would be the case, for example, in the formation of an acyl-enzyme intermediate in peptide synthesis.

1.7.2 Solid Substrates

Although we have examined only soluble substrates to this point, there are many examples of enzymes acting on insoluble substrates. Enzymes which act on cell walls (for example, lysozyme breakdown of bacterial cell walls, chitinase action on chitin), on cellulose (endo and exo β-1,4 glucosidases, the "cellulases"), and other insoluble substrates may demonstrate kinetic patterns quite different to those above. If a limited number of sites are available on the surface of the substrate, the enzyme in solution may equilibrate with bound enzyme and exhibit kinetics which are "opposite" to those found for soluble substrates. For example, consider an equilibrium adsorption of enzyme (E) onto substrate (S)

$$E + S \underset{k_{des}}{\overset{k_{ads}}{\Leftrightarrow}} ES$$

If the total concentration of adsorption sites on the substrate surface is S_0, expressed per liquid volume, then we may write

$$[S_0] = [S] + [ES]$$

and the equilibrium rate expression

$$\frac{d[ES]}{dt} = k_{ads}[E][S] - k_{des}[ES] = 0 \tag{1.79}$$

The rate of reaction of substrate to product is now assumed to be first order in [ES], with a rate constant k_2. Thus the reaction velocity is given by

$$v = k_2[ES] = \frac{k_2[S]_0[E]}{\left(\frac{k_{des}}{k_{ads}}\right) + [E]} \tag{1.80}$$

If most of the enzyme is in free form $[E] \sim [E]_0$, and the velocity becomes

$$v = \frac{k_2[S]_0[E]_0}{\left(\frac{k_{des}}{k_{ads}}\right) + [E]_0} \tag{1.81}$$

which resembles the Michaelis-Menten form, but with enzyme and substrate interchanged.

1.7.3 Enzyme Activity at Interfaces

A large number of enzymes are active when adsorbed at interfaces. For example, in eukaryotic cells enzymes are attached to the mitochondrial membrane and their proximity is important in catalyzing multistep reactions. Lipases which split triglycerides into glycerol and fatty acids act at oil-water interfaces. However, some enzymes do not function at solid-liquid or liquid-liquid interfaces as they are subject to interfacial denaturation.

The three-dimensional structure of an enzyme arises from minimization of the Gibbs free energy of all the interatomic interactions within the protein. Those resulting from both apolar and polar interactions of the amino acid residues are important in determining the tertiary structure. Apolar residues tend to be located in the interior of the protein, and are not exposed to the aqueous milieu. When a protein is exposed to a hydrocarbon-water interface, the native conformation of the protein may be altered and it may assume a random coil structure, where all the hydrophobic residues are exposed to the hydrocarbon. The free energy change for the transition to this state is small (typically 0.096 kcal/mole amino acid residue), as the entropy gain from increased rotation of the polypeptide chain compensates for the enthalpy loss in the creation of the large hydrocarbon-water interface. The amphipathic nature of proteins results in their being surface active, and this may sometimes cause the denaturation of proteins at liquid-liquid or gas-liquid interfaces. Some structures remain stable at interfaces, such as the α-helix and disulfide bridges.

When an interface is occupied by a monolayer of protein, the proteins in this layer may lose enzymatic activity which cannot be regained if the interface is removed or reduced, for example by compression of the film. However when additional protein molecules are added to the interface, the denatured proteins act as a protective barrier, and the added enzymes exhibit normal activity. Moreover, near a liquid-liquid or gas-liquid interface the local pH can differ significantly from that of the bulk aqueous solution. The general behavior of proteins at interfaces is thus quite complex, and depends on the enzyme in question[12]. We will examine the behavior of lipases as an example of an enzyme which is activated by interfaces.

Kinetics of Lipolysis[13]

Lipolytic enzymes (lipases) are esterases that catalyze the hydrolysis of fats into glycerol and free fatty acids.

$$
\begin{array}{lll}
CH_2-O-R_1 & CH_2-OH & R_1-COOH \\
CH-O-R_2 \longrightarrow & CH-OH \quad + & R_2-COOH \\
CH_2-O-R_3 & CH_2-OH & R_3-COOH
\end{array}
$$

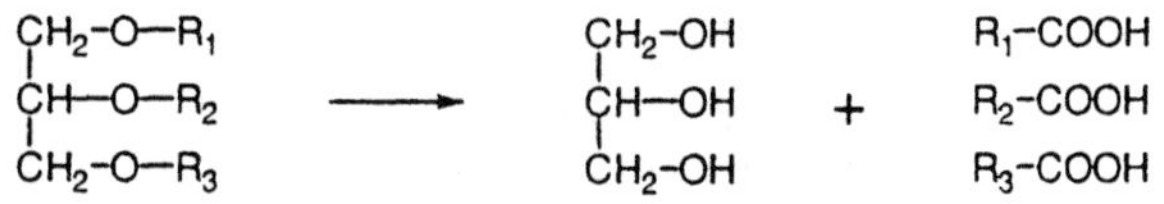

Lipase Catalyzed Triglyceride Hydrolysis

Lipolytic enzymes are typically activated by interfaces, such as membranes, beads, liquid-liquid interfaces or by substrate aggregation in the form of micelles or emulsions. This effect was clearly illustrated in the experiments of Sarda and Desnuelle shown in Figure 1.18. The hydrolysis of triacetin by two enzymes with the same catalytic function is compared. Horse liver esterase is not activated by interfaces and shows a normal Michaelis-

---

(12) The behavior of enzymes at interfaces is reviewed by F. Macritchie, "Interfacial Synthesis", (F. Millich and C. Carraher, eds) **1**, 103 (1977) Marcel Dekker; M.C. Phillips "The Conformation and Properties of Proteins at Liquid Interfaces" (Chemistry and Industry, **5**, 170 (March), 1977; L.K. James and L. Augenstein "Adsorption of Enzymes at Interfaces; Film Formation and the Effects of Activity", Adv. in Enzymology, **28**, 1, 1966; Norde, W. & J. Lyklema, "Why Proteins Prefer Interfaces", J. Biomater. Sci. Polymer Edn. **2** (3) 183 (1991).

(13) The kinetics of lipolysis is discussed in some detail in R. Verger "Enzyme Kinetics of Lipolysis", Methods in Enzymology, **64**, 340 (1980).

Menten dependence on substrate concentration. On the other hand, porcine pancreatic lipase shows very little activity with the same substrate when the substrate is in the monomeric state. When the solubility limit of triacetin in water is exceeded, the substrate exists in an emulsified state and a sharp increase in enzyme activity is seen.

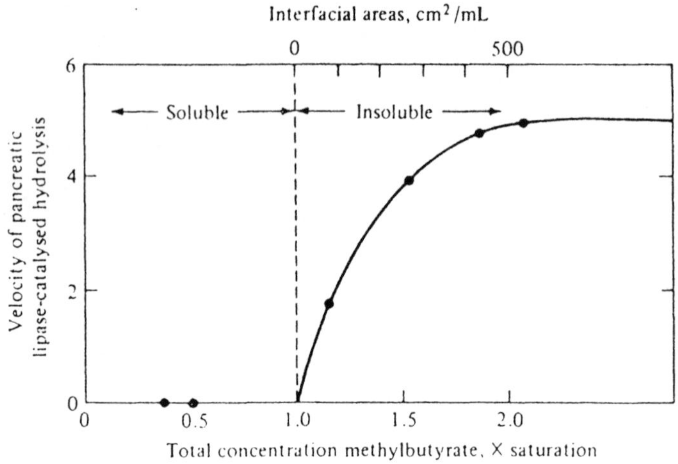

Figure 1.18. *The hydrolysis of triacetin by horse liver esterase and porcine pancreatic lipase. Substrate concentration is expressed in multiples of the saturation concentration. [From L. Sarda and P. Desnuelle, Biochim. Biophys. Acta **30**, 513 (1958)].*

The study of interfacial enzymes is thus complicated by the variable nature of the interface at which activity occurs and by the lack of a readily applicable means for determining interfacial concentrations of substrate and enzyme. The rate enhancement observed in the presence of interfaces varies with the "quality" of the interface. Thus the presence of, for example, bile salts, yields an interface with different properties ("quality") than a monolayer of a surface-active agent.

Models of the kinetics of interfacial enzymes relate bulk substrate and enzyme concentrations to interfacial concentrations through the assumption of some form of adsorption kinetics. We shall develop a simple model based on the assumption that the adsorption steps are all in equilibrium. This assumption may not always be valid, as pancreatic lipase has been shown to slowly penetrate into a monolayer of lecithin at the aqueous-hydrocarbon interface[14]. Thus the $E \Leftrightarrow E_{ads}$ step may be rate limiting. The system is illustrated in the figure.

(14) M.C.Mieras and G.H. de Haas, *J. Biol. Chem.*, **248**, 4023 (1973)

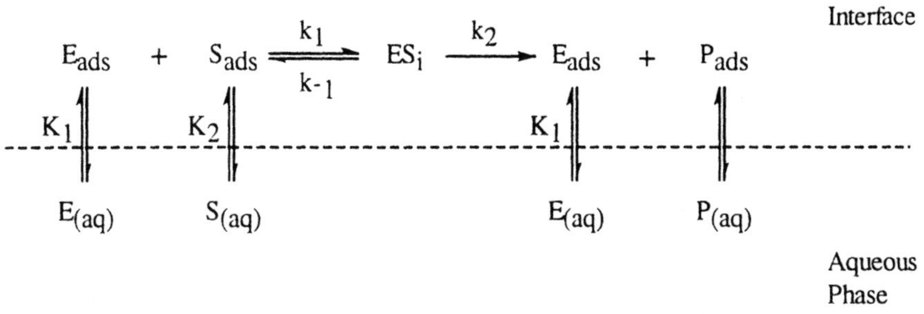

The equilibrium adsorption steps can be written viz:

$$\frac{[E]}{[E_{ads}]} = K_1 \qquad \frac{[S]}{[S_{ads}]} = K_2 \tag{1.82}$$

The adsorbed species concentrations are written in terms of moles per interfacial area. Making the assumption that the ES complex is at steady state:

$$\frac{d[ES_{ads}]}{dt} = k_1[E_{ads}]\,[S_{ads}] - (k_2 + k_{-1})\,[ES_{ads}] = 0 \tag{1.83}$$

The observed rate of reaction is based on the appearance of product P in the bulk aqueous phase. We shall assume that the concentration of product at the interface is at steady state, i.e., the rate of transport of product to the aqueous phase is comparable to the rate of product formation.

$$\begin{aligned} v = \frac{d[P]}{dt} &= k_2 a[ES_{ads}] - a\frac{d[P_{ads}]}{dt} \\ &= k_2 a E_o \cdot \frac{[S_{ads}]}{K_M(a + K_1) + a \cdot [S_{ads}]} \\ &= k_2[E_o] \cdot \frac{[S]}{K_M K_2 \frac{(a+K_1)}{a} + [S]} \end{aligned} \tag{1.84}$$

where

$$K_M = \frac{k_2 + k_{-1}}{k_1} \quad \text{and} \quad a = \frac{\text{interfacial area}}{\text{volume}}$$

$$\begin{aligned} E_o &= E + a \cdot ([E_{ads}] + [ES_{ads}]) \\ &= K_1[E_{ads}] + a \cdot ([E_{ads}] + [ES_{ads}]) \end{aligned}$$

In the case of a small interfacial area (i.e., $a \sim 0.1\ \text{cm}^{-1}$), with most of the substrate residing in the aqueous phase $K_2 >> 1$, and $K_M.K_2.(1+K_1/a) > S$, the equation reduces to the form

$$v = \frac{k_2[E_o][S]}{K_M K_2\left(1 + \frac{K_1}{a}\right)} \tag{1.85}$$

Generally, in a two-phase system containing small droplets, the interfacial area to volume ratio, a, will be ~ 10^3 cm^{-1}, and thus we recover kinetics of the Michaelis-Menten form.

1.8 Kinetics of Multi-Substrate Reactions

The majority of enzymes do not involve just one substrate, they usually involve two. Often the second substrate is a cofactor, such as NAD^+ in oxido-reductases. Many of the concepts we have developed to date can be applied to situations where one of the substrates is held at a constant concentration; the kinetics then follow one of the forms described earlier. Most types of multisubstrate reactions fall into one of two classes, depending on the order of substrate binding and release of products. The description of the reaction in these terms is referred to as the "formal kinetic mechanism". In the first case, when *all* substrates are added to the enzyme prior to catalysis, a complex enzyme-substrate intermediate exists. With two-substrate reactions, the mechanism is referred to as a *ternary complex mechanism.* In the second case, when one product is released before the second substrate is bound (for example, when a group is transferred from one reactant to another), the mechanism is referred to as an *enzyme-substituted mechanism.*

In the case of two-substrate enzyme-substituted systems, there is thus a required order of substrate binding (i.e., a *sequential* mechanism). In the case of ternary complex mechanisms, a number of possible binding orders arise. For example, binding of one substrate may not be possible until a particular substrate is first bound, as this may result in a conformational change in enzyme structure which then permits the binding of the second substrate. This is referred to as *compulsory* order. When such order is not important, the mechanism is known as *random* order. This is summarized below for two-substrate reactions.

Ternary Complex Mechanism (EAB) formed		Enzyme-substituted Mechanism A binds and is released; Group transferred to B
Compulsory order A binds before B	Random order A or B can bind first	Compulsory binding order (Sequential mechanism)

1.8.1 Mechanisms for Two Substrate Reactions

We will consider the two main classes of mechanisms described above using the simplest case, that of two-substrate reactions as an example. The reaction to be considered is

$$A + B \Leftrightarrow P + Q$$

For situations where a ternary-complex mechanism is involved, four possibilities can be considered: either a compulsory-order or random mechanism with either steady state or equilibrium assumptions being used to describe the behavior of the various enzyme-substrate complexes formed. The enzyme-substituted mechanism (also referred to as a double displacement or ping-pong mechanism) is yet another possibility, distinct from the ternary-complex mechanism. In all of these situations, we shall see that the initial rates of reaction can be written in the following form:

$$\frac{[E]_0}{v_0} = \Theta_0 + \frac{\Theta_1}{[A]} + \frac{\Theta_2}{[B]} + \frac{\Theta_{12}}{[A][B]} \tag{1.86}$$

Here Θ_i are constants which contain kinetic rate constants. This form was proposed by Dalziel[15] and has also been generalized to three substrate reactions. We shall examine the various classes of two substrate reactions to determine detailed rate equations which provide expressions for v_0.

Compulsory Order Reactions

Steady State Analysis

In compulsory order, steady state systems, one substrate must be added prior to the addition of the second substrate. The reaction pathway can thus be visualized as:

$$E + A \underset{k_{-1}}{\overset{k_1}{\Leftrightarrow}} EA + B \underset{k_{-2}}{\overset{k_2}{\Leftrightarrow}} EAB \underset{k_{-3}}{\overset{k_3}{\Leftrightarrow}} EPQ \overset{k_4}{\rightarrow} EP + Q \overset{k_5}{\rightarrow} E + P$$

The substrates must be added in the order A and then B, and products are released in compulsory order as well. We can write the equation below for the total amount of enzyme in the system as

$$[E]_0 = [E] + [EA] + [EAB] + [EPQ] + [EP] \tag{1.87}$$

The analysis proceeds by assuming that all enzyme substrate complexes are at steady state and that the initial rate of reaction is given by

$$v_0 = k_5[EP] \tag{1.88}$$

Expressions for each of the enzyme-substrate and enzyme-product complexes can now be written:

$$\frac{d[EA]}{dt} = k_1[E][A] - k_{-1}[EA] - k_2[EA][B] + k_{-2}[EAB] = 0 \tag{1.89}$$

$$\frac{d[EAB]}{dt} = k_2[EA][B] - (k_{-2} + k_3)[EAB] + k_{-3}[EPQ] = 0 \tag{1.90}$$

---

(15) Dalziel, K. (*The Enzymes*, **10**, 2, P. D. Boyer (ed) 3rd Edition (1975)) and Cleland W. W. (*Biochemistry*, **14**, 3220 (1974)) provide short reviews of multisubstrate reactions.

$$\frac{d[EPQ]}{dt} = k_3[EAB] - (k_{-3} + k_4)[EPQ] = 0 \tag{1.91}$$

$$\frac{d[EP]}{dt} = k_4[EPQ] - k_5[EP] = 0 \tag{1.92}$$

Using these equations, we can eliminate the intermediates, and finally express [EA] as a function of [EP] and [B].

$$[EPQ] = \frac{[EP] \cdot k_5}{k_4} \tag{1.93}$$

$$[EAB] = \frac{[EP] \cdot k_5(k_{-3} + k_4)}{k_3 k_4} \tag{1.94}$$

$$[EA] = \frac{k_5(k_{-2}k_{-3} + k_{-2}k_4 + k_3k_4)[EP]}{k_2k_3k_4[B]} \tag{1.95}$$

[E] may now be found in terms of [EP] from the steady state expression for [E]:

$$\frac{d[E]}{dt} = k_5[EP] + k_{-1}[EA] - k_1[E][A] = 0 \tag{1.96}$$

By substituting into equation (1.87) the steady state concentrations of each of the enzyme complexes, we can relate the total enzyme concentration to [EP] viz:

$$[E]_0 = [EP] \cdot (1 + \frac{k_5}{k_4} + \frac{k_5(k_{-3} + k_4)}{k_3k_4} + \frac{k_5 \cdot (k_{-2}k_{-3} + k_{-2}k_4 + k_3k_4)}{k_2k_3k_4 \cdot [B]}$$

$$+ \frac{k_5}{k_1 \cdot [A]} + \frac{k_{-1}k_5 \cdot (k_{-2}k_{-3} + k_{-2}k_4 + k_3k_4)}{k_1k_2k_3k_4 \cdot [A] \cdot [B]} \tag{1.97}$$

Noting that $v_0 = k_5[EP]$, we can express the velocity in Dalziel's form (equation (1.86)):

$$\frac{[E]_0}{v_0} = \frac{\frac{1}{k_5} + \frac{1}{k_4} + (k_{-3} + k_4)}{k_3k_4} + \frac{(k_{-2}k_{-3} + k_{-2}k_4 + k_3k_4)}{k_2k_3k_4 \cdot [B]}$$

$$+ \frac{1}{k_1 \cdot [A]} + \frac{k_{-1}(k_{-2}k_{-3} + k_{-2}k_4 + k_3k_4)}{k_1k_2k_3k_4 \cdot [A] \cdot [B]} \tag{1.98}$$

Under conditions where both substrates are present in excess, the last three terms in equation (1.98) can be neglected and the maximum velocity can be found:

$$\frac{[E]_0}{v_{max}} = \Theta_0 \qquad \frac{v_{max}}{[E]_0} = \frac{1}{\Theta_0} \tag{1.99}$$

If only one of the substrates is present in excess, we retain the terms containing the other substrate and obtain the initial rates as functions of the limiting substrates:

$$\frac{[E]_0}{v_0} = \Theta_0 + \frac{\Theta_1}{[A]} \qquad \text{or} \qquad \frac{[E]_0}{v_0} = \Theta_0 + \frac{\Theta_2}{[B]} \tag{1.100}$$

Thus plots (so called "primary plots") of the inverse velocity against the inverse concentration of one substrate (e.g., [A]) for fixed concentrations of the other (e.g., [B]) should be linear, as seen from equations (1.98). The slopes of these plots will vary with concentration of the fixed substrate [B]. Such plots serve to distinguish the compulsory order mechanism from the enzyme-substituted mechanism, where, as we shall see, the slopes are independent of [B].

Equilibrium Analysis

By assuming that the enzyme-substrate complexes are all in equilibrium, the analysis is considerably simplified. The reaction pathway now becomes

$$E + A \overset{K_1}{\Leftrightarrow} EA + B \overset{K_2}{\Leftrightarrow} EAB \overset{K_3}{\Leftrightarrow} EPQ \overset{k_4}{\rightarrow} EP + Q \overset{k_5}{\rightarrow} E + P$$

where K_i are dissociation constants for the complexes. The release of products from the [EAB] complex can be considered to occur with an overall rate constant k_{obs}. By following an analysis similar to that used previously, we eliminate the concentrations of intermediates by using the equilibrium expressions, and write the reaction velocity as $v_0 = k_{obs}[EAB]$. The resulting expression is

$$\frac{[E]_0}{v_0} = \frac{1}{k_{obs}} + \frac{K_3}{K_1 \cdot [B]} + \frac{K_2 K_3}{k_{obs} \cdot [A] \cdot [B]} \tag{1.101}$$

Note the absence here of the term Θ_1 /[A] found in the case of the steady state analysis.

Random Order Reactions

In the case of random order reactions, we must consider two parallel pathways for the addition of substrates A and B to the enzyme. The resulting EAB complex then undergoes reaction to form EPQ, and these products are then released in any order. The reaction pathway is shown schematically below for the simplest case of rapid equilibrium between all the enzyme-substrate complexes.

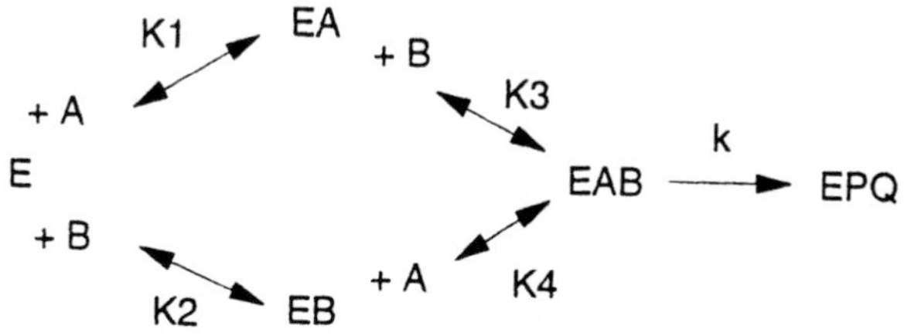

As was the case above, we shall consider the initial rate to be given by $v_0 = k[EAB]$. Using the same method as before to eliminate the concentrations of the enzyme-substrate complexes apart from EAB in the overall enzyme conservation equation, we can derive the following expression for the rate of reaction.

$$\frac{[E]_0}{v_0}=\frac{K_1K_3}{k[A][B]}+\frac{K_3}{k[B]}+\frac{K_4}{k[A]}+\frac{1}{k} \tag{1.102}$$

The case of steady state existing among the enzyme-substrate complexes rather than equilibrium is considerably more complex to analyze. We must consider the two possible routes for release of products P and Q from the EPQ complex. The general solution no longer follows the form proposed by Dalziel, but contains quadratic terms in A and B in both the numerator and the denominator of the resulting expression for $[E]_0/v_0$. Thus, depending on the values of the rate constants, apparent substrate activation or inhibition can be observed in this case.

Enzyme-substituted Reactions (the Ping-Pong Mechanism)

In enzyme-substituted or double displacement mechanisms, both substrates need not be bound together at the active site of the enzyme. No ternary intermediate is formed. This mechanism occurs when, for example, a phosphate-transferring enzyme, such as phosphoglycerate mutase, is phosphorylated. One substrate (A) reacts with the enzyme to give E^* (e.g., a phosphorylenzyme) which then transfers the phosphoryl group to the second substrate B.

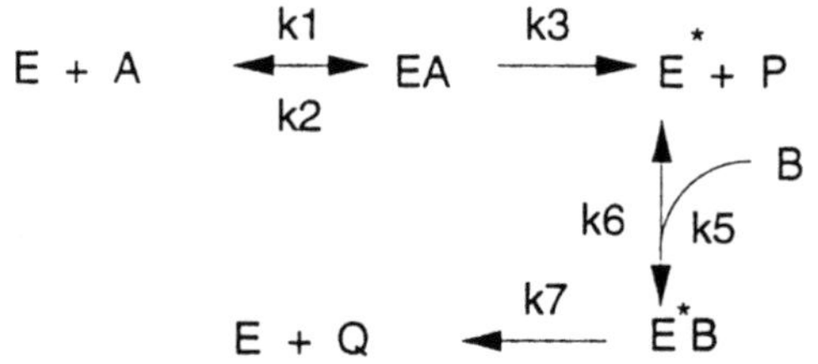

Often these reaction mechanisms are written in abbreviated form as suggested by Cleland

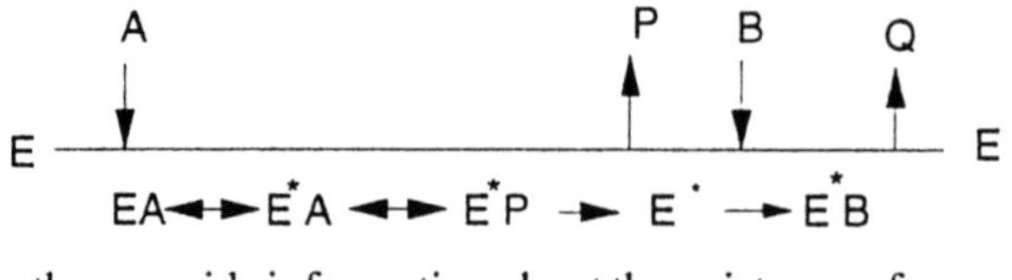

Kinetics of this type thus provide information about the existence of a covalent intermediate. Often the finding of double displacement (or "ping-pong") kinetics is used as evidence for the existence of this intermediate, but other confirming information should be sought. The kinetic pattern of this type of mechanism is unique, as we shall see below. We can analyze the kinetics by writing balance equations for a well-mixed batch reactor in the following manner:

$$\frac{d[E]}{dt}=k_2[EA]+k_7[E^*B]-k_1[E][A]=0 \tag{1.103}$$

$$\frac{d[E^*B]}{dt} = k_5[E^*][B] - (k_6 + k_7)[E^*B] = 0 \tag{1.104}$$

$$\frac{d[EA]}{dt} = k_1[E][A] - (k_2 + k_3)[EA] = 0 \tag{1.105}$$

$$\frac{d[E^*]}{dt} = k_3[EA] + k_6[E^*B] - k_5[E^*][B] = 0 \tag{1.106}$$

solving for E^* from Equation (1.104)

$$[E^*] = (k_6 + k_7)\frac{[E^*B]}{k_5[B]} \tag{1.107}$$

and employing Equation (1.105) to obtain EA yields

$$[EA] = \frac{k_1[E][A]}{(k_2 + k_3)} \tag{1.108}$$

Equations (1.104) and (1.106) can be used to relate EA and E^*B viz:

$$[EA] = \frac{k_7[E^*B]}{k_3} \tag{1.109}$$

From Equations (1.108) and (1.109) we obtain

$$[E] = \frac{k_7(k_2 + k_3)[E^*B]}{k_1k_3[A]} \tag{1.110}$$

The enzyme conservation equation is given by

$$[E]_0 = [E] + [EA] + [E^*] + [E^*B]$$

$$= [E^*B]\left(1 + \frac{(k_6 + k_7)}{k_5[B]} + \frac{k_7}{k_3} + \frac{k_7(k_2 + k_3)}{k_1k_3[A]}\right) \tag{1.111}$$

and the rate of reaction is given by $v = k_7[E^*B]$. Thus equation (1.104) can be rearranged in the Dalziel form, giving

$$\frac{[E]_0}{v} = \frac{1}{k_7} + \frac{1}{k_3} + \frac{(k_2 + k_3)}{k_1k_3[A]} + \frac{(k_6 + k_7)}{k_5k_7[B]} \tag{1.112}$$

or in the general form

$$\frac{[E]}{v} = \Theta_0 + \frac{\Theta_1}{[A]} + \frac{\Theta_2}{[B]}$$

There is no term involving the product AB since Θ_{12} is zero. If rate data are plotted in Lineweaver-Burk form for ping-pong kinetics, straight lines are obtained if one substrate is held constant and the other is varied. The general pattern of behavior is illustrated in Figure (1.19).

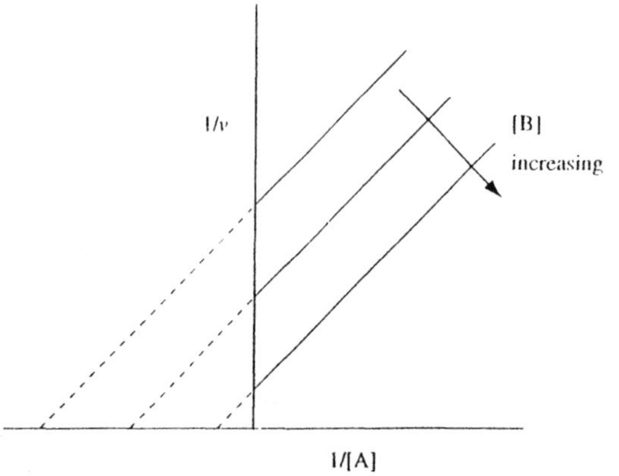

***Figure 1.19**. Lineweaver-Burk plot of ping-pong mechanism. The characteristic parallel lines at varying concentrations of [B] result from the ratio v_{max}/K_M remaining constant; as [B] increases v_{max} increases, as does K_M.*

Determination of Θ values from experimental data

Θ values in the rate expression of Dalziel are determined from initial rate data. Generally, one substrate is maintained at constant concentration and the other is varied. For example, if [B] is held constant and [A] is varied, the general rate expression can be rewritten in the form

$$\frac{[E]_0}{v_0} = \left(\Theta_0 + \frac{\Theta_2}{[B]}\right) + \left(\Theta_1 + \frac{\Theta_{12}}{[B]}\right) \cdot \frac{1}{[A]} \tag{1.113}$$

The slopes and intercepts of the Lineweaver-Burk plots may then be used to determine the kinetic constants and to distinguish between the various two-substrate mechanisms. The ping-pong mechanism has $\Theta_{12} = 0$, and can thus be distinguished from the ternary complex mechanisms which all have positive values of Θ_{12}. In these cases, the slopes of the Lineweaver-Burk plots at increasing values of [B] will vary. In the compulsory order-equilibrium mechanism, Θ_1 is zero, thus if [B] is varied at a constant value of [A] the intercept on the double-reciprocal plot is $1/k_{obs}$ and is independent of the concentration of [A] employed. Conversely, if [A] is varied at constant values of [B], the expected dependence of v_{max} and K_M on [B] is found at different values of the constant concentration [A].

Another approach to determine which of the several possible mechanisms is occurring is to use alternative substrates. Compulsory-order, steady state and random order, rapid equilibrium mechanisms can be distinguished because Θ_1 for the compulsory-order, steady state case is $1/k_1$, where k_1 refers to the binding of the first substrate A to the enzyme. In the case of the random-order mechanism, Θ_1 does not have this same simple physical significance. Thus variations in the chemical nature of substrate B will not affect Θ_1 in the first case, but will in the latter case. If the binding order is not known, then both A and B need to be varied (see Problem 11 for an example of this approach).

1.9 Allosteric Enzymes

Many enzymes are composed of several identical sub-units (or protomers) and thus have multiple sites for binding of substrate molecules. In some cases, the binding sites are well separated and do not interact, so that substrate binding is not influenced by interactions occurring at the other binding sites. However, some enzymes exhibit cooperative binding, i.e., the binding of substrates after the first may occur with increasing affinity. This also is the case with the binding of ligands to some proteins which do not have catalytic function. We may generalize this phenomenon, by considering the broad definition of a ligand as a molecule which binds to a protein; in the case of a protein which is an enzyme, the ligand will be the substrate or an inhibitor.

An example of cooperative ligand binding is afforded by oxygen binding to hemoglobin. Hemoglobin is composed of four peptide chains (each of which resembles myoglobin, which has one oxygen binding site). If the degree of hemoglobin saturation with oxygen is plotted against the partial pressure of oxygen, in the form of an adsorption isotherm, a *sigmoidal* pattern is found. The fourth molecule of oxygen that binds to hemoglobin does so with an affinity that is many hundred times higher than that of the first oxygen bound. Thus there exists an interaction between binding sites, so that conformational changes in the protein must occur once oxygen is bound, changing the affinity of the other binding sites. This is referred to as *positive cooperativity*.

Similar effects have been found with enzymes. When reaction rate v is plotted against [S], sigmoidal kinetics are sometimes observed. This usually occurs with enzymes whose activities are regulated by feedback or feedforward control, as is generally the case with key regulatory enzymes in metabolic pathways (these are often the first enzymes in a particular branch of a pathway). Such enzymes are referred to as *allosteric* enzymes (from the Greek *allos = other, steros = space or solid*). Allosteric enzymes are regulated by molecules which are not usually the natural substrates of the enzyme; these ligands bind at sites distinct from the active site and thereby activate or inhibit the enzyme. This provides, in the case of feedback inhibition, a means for the end product of a pathway to regulate the metabolic flux along the pathway. We shall see examples of this in Chapter 3.

In the following sections, we shall examine the isotherms for the binding of ligands to proteins and then examine the kinetics of allosteric enzymes.

1.9.1 Binding of Ligands to Proteins

We shall consider the simplest case of the binding of a ligand to a protein having only one binding site to illustrate the general procedures that we will employ later. The binding can be described by (P is protein and L is the ligand):

$$P + L = PL$$

the binding constant is K_b and the dissociation constant is K_S, where $K_b = 1/K_S$, and

$$K_b = \frac{[PL]}{[P][L]} \tag{1.114}$$

and the fractional degree of saturation of protein is:

$$\overline{Y} = \frac{[PL]}{[PL]+[P]} = \frac{K_b[L]}{K_b[L]+1} \tag{1.115}$$

Thus the fractional saturation shows a hyperbolic relationship with ligand concentration, similar to the Michaelis-Menten form for v versus [S]. As we saw earlier with the Michaelis-Menten equation, there are several ways in which the ligand-protein binding data can be plotted to obtain linear relationships. One common method is the Scatchard plot:

$$[PL] = [P]_T - \frac{[PL]}{K_b[L]} \tag{1.116}$$

Thus a plot of [PL] versus [PL]/[L] is linear, with $[P]_T$ representing the total protein concentration.

In cases where binding of one ligand affects the binding of subsequent ligands (which may be the same or different from the first), four types of cooperativity must be considered.

Positive Cooperativity occurs when the binding of one ligand increases the binding of subsequent ligands, which may be the same or different from the first ligand bound.
Negative cooperativity occurs when the binding of the first ligand decreases the binding of subsequent ligands.
Homotropic cooperativity refers to the binding of subsequent ligands which are the same as the original ligand bound.
Heterotropic cooperativity refers to binding of ligands which are different to the original ligand bound.

In the case of enzyme regulation by ligands which are not substrates for the reaction, *allosteric inhibition* provides an example of negative heterotropic cooperativity. *Allosteric activation* is positive heterotropic cooperativity. We shall examine some simple cases of cooperativity.

Positive Homotropic Cooperativity: The Adair and Hill Equations

We shall consider the simplest case of two identical ligands binding to a protein, where the second ligand binds more rapidly than the first. The situation can be visualized viz:

$$P_2 + L \rightarrow P_2L$$

$$P_2L + L \rightarrow P_2L_2$$

where P_2 indicates that the protein contains two identical subunits called *protomers*. The binding constants for the two steps are:

$$K_{b1} = \frac{[P_2L]}{[P_2][L]} \qquad K_{b2} = \frac{[P_2L_2]}{[P_2L][L]} \tag{1.117}$$

the fractional degree of saturation in this case is given by

$$\overline{Y} = \frac{\text{concentration of protomers with bound ligand}}{\text{total protomer concentration}}$$

$$= \frac{[P_2L] + 2[P_2L_2]}{2([P_2L] + [P_2L_2] + [P_2])} \tag{1.118}$$

We substitute for $[P_2L]$ and $[P_2L_2]$ from equation (1.117), rewriting $\overline{Y}$ in terms of P_2 and L:

$$\overline{Y} = \frac{K_{b1}[P_2][L] + 2K_{b1}K_{b2}[P_2][L]^2}{2([P_2] + K_{b1}[P_2][L] + K_{b1}K_{b2}[P_2][L]^2)} = \frac{K_{b1}[L] + 2K_{b1}K_{b2}[L]^2}{2(1 + k_{b1}[L] + K_{b1}K_{b2}[L]^2)} \tag{1.119}$$

This equation is the *Adair equation* for the binding of one ligand to a dimeric protein. If the binding demonstrates *complete cooperativity*, then the first step is slow relative to the second, and the binding follows

$$P_2 + 2L \Leftrightarrow P_2L_2$$

and the binding constant K_b is given by

$$K_b = \frac{[P_2L_2]}{[P_2][L]^2} \qquad \text{and thus} \qquad \overline{Y} = \frac{K_b[L]^2}{1 + K_b[L]^2} \tag{1.120}$$

this may be rewritten in the form of the *Hill equation*:

$$\log\left(\frac{\overline{Y}}{1 - \overline{Y}}\right) = 2\log[L] + \log K_b \tag{1.121a}$$

or more generally

$$\log\left(\frac{\overline{Y}}{1 - \overline{Y}}\right) = n\ \log[L] + \log K_b \tag{1.121b}$$

where the second equation is for the case where n ligands are bound. The graph of log $(\overline{Y}/(1 - \overline{Y}))$ versus log [L] is referred to as a Hill plot, and the value of the experimentally determined slope in the region of 50% saturation is the Hill coefficient, h. At very low concentrations of ligand, only one site on the protein will be occupied, and thus the slope of this plot will tend to unity. At high values of [L], very few protein molecules will have more than one binding site available to be filled as most will have all but one site occupied. Thus the slope will again tend to unity. The derivation of the Hill equation assumes that cooperativity is complete, and so the Hill coefficient will assume the value n, the number of binding sites, only as an upper limit. In general, h will be less than n. If $h < 1$, the cooperativity is negative, and for $h = 1$ there is no cooperativity.

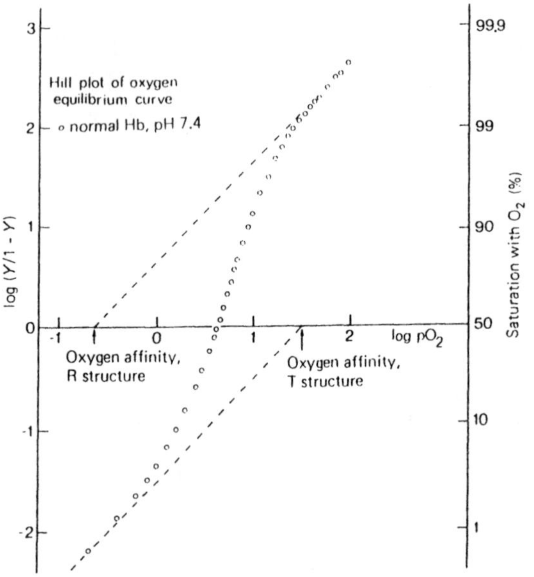

Figure 1.20. The Hill plot (Equation (1.121(b)), illustrated for the case of oxygen binding to hemoglobin. The Hill coefficient h is found from the slope of the binding curve at 50% oxygen saturation. The oxygen affinities of the T-form and the R-form are shown. (From J. Kilmartin, K. Imani & R. Jones, in Erythrocyte Structure and Function, Alan Liss, p21, 1975).

Experimental determination of the Hill coefficient and binding constants can be accomplished by a variety of techniques. Equilibrium dialysis is commonly employed, as it permits the free ligand (if it has a sufficiently small molecular weight) to be separated from the protein-ligand complex, which is retained behind the dialysis membrane. Thus the concentration of free ligand can then be analyzed and the bound ligand concentration can be calculated from a mass balance if the total ligand concentration is known. The ligand concentration can sometimes be determined spectroscopically if there is a difference in the absorption spectrum of bound and free ligand. If an affinity ligand is available for any of the P, L or PL species, affinity chromatography can be employed to separate the components and determine their concentrations.

1.9.2 The Monod-Changeux-Wyman (MCW) Model

Studies on the cooperativity phenomenon in the binding of ligands to proteins led to the observations of rate cooperativity with enzymes and the concept of allosteric regulation. In 1965, Monod, Changeux, and Wyman proposed a concerted symmetry model to explain these rate cooperativity effects. This model and the related model of Koshland, Némethy and Filmer are based on the concept that each protomer can exist in two conformational forms: a T-form (or *tensed* form) that can only exist in the absence of bound ligand and an

R-form (or *relaxed* form) that exists when ligand is bound to the protomer but is also present in the absence of ligand. It is assumed that the most stable states are those where all the protomers exist in the same form (either T or R) and these are consequently the predominant forms of the protein. Thus no hybrid states (i.e proteins containing mixes of T and R forms) will be present. In the absence of the ligand, the two conformational forms of the protein will be in equilibrium, and this equilibrium is disturbed by the binding of the ligand. The MCW model assumes the preexistence of conformational equilibrium and thus differs from others which assume that it is the binding of the ligand which induces the conformational change in the protein.

We will develop the MCW model for the case of a dimeric protein having two identical binding sites for the substrate S. The two forms of the protein are represented as T_2 and R_2 and we ignore the intermediate TR form. An equilibrium exists between the two forms and the equilibrium constant is typically represented by the symbol L (unrelated to that used to represent the ligand concentration [L]). This equilibrium constant is termed the allosteric constant. Ligand (substrate) binding only occurs with the R_2 form. The binding of a second substrate molecule to R_2S is assumed to occur with the same binding affinity. The intrinsic dissociation constant for one molecule of substrate binding is K_R. The situation is shown diagramatically below.

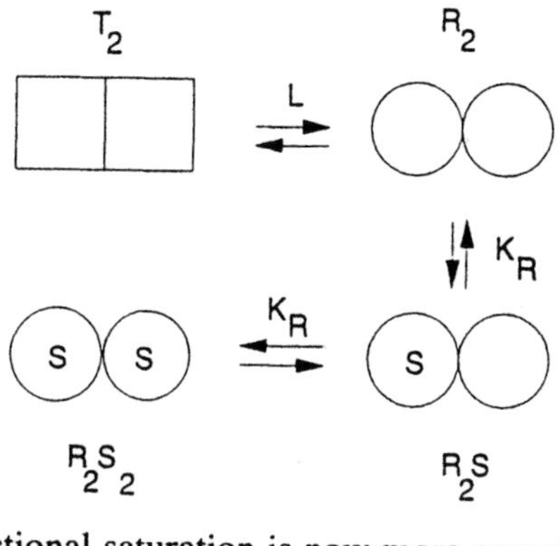

The expression for the fractional saturation is now more complex than was the case with equation (1.118).

$$\overline{Y} = \frac{\text{concentration of protomers with bound ligand}}{\text{total protomer concentration}}$$

$$= \frac{[R_2S]+2[R_2S_2]}{2([T_2]+[R_2]+[R_2S]+[R_2S_2])} = \frac{[R_2S]+2[R_2S_2]}{2(L[R_2]+[R_2]+[R_2S]+[R_2S_2])} \quad (1.122)$$

Since there are two sites available for binding of S to R_2, but only one bound substrate can dissociate in the reverse direction, the equilibrium relationship for the first substrate molecule bound becomes

$$[R_2S] = \frac{2[R_2][S]}{K_R} \tag{1.123}$$

The second substrate molecule has only one binding site available, but the reverse step has two bound substrate molecules which are available to dissociate

$$[R_2S_2] = \frac{[R_2S][S]}{2K_R} = \frac{2}{K_R}\cdot\frac{1}{2K_R}\cdot[R_2][S]^2 = \frac{[R_2][S]^2}{(K_R)^2} \tag{1.124}$$

We can now substitute in the expression for $\overline{Y}$

$$\overline{Y} = \frac{\frac{2[R_2][S]}{K_R} + \frac{2[R_2][S]^2}{(K_R)^2}}{2\left([R_2] + \frac{2[R_2][S]}{K_R} + \frac{[R_2][S]^2}{(K_R)^2} + L[R_2]\right)} = \frac{\frac{[S]}{K_R}\left(1 + \frac{[S]}{K_R}\right)}{L + (1 + \frac{[S]}{K_R})^2} \tag{1.125}$$

In the case of n protomers, a more general form of the MCW model can be derived

$$\overline{Y} = \frac{\frac{[S]}{K_R}\left(1 + \frac{[S]}{K_R}\right)^{n-1}}{L + (1 + \frac{[S]}{K_R})^n} \tag{1.126}$$

Some limiting cases can be considered. When L is zero, only the R form is present (L = [T]/[R]) and a hyperbolic relationship is obtained, as would be expected from the Adair equation with no interaction between the binding sites ($K_{b1} = K_b$, $K_{b2} = (1/2)K_b$ and $K_b = 1/K_R$). If the protein is monomeric, n = 1 and again a hyperbolic relationship is obtained. In addition, we can obtain the Hill equation as a limiting case. If the binding of substrate occurs with a high affinity, then K_R will be small and $[S]/K_R$ will be large, so that $(1 + [S]/K_R) \sim [S]/K_R$ and equation (1.126) reduces to

$$\overline{Y} = \frac{\frac{[S]}{K_R}\left(\frac{[S]}{K_R}\right)^{n-1}}{L + \left(\frac{[S]}{K_R}\right)^n} = \frac{\frac{[S]^n}{(K_R)^n L}}{1 + \frac{[S]^n}{(K_R)^n L}} \tag{1.127}$$

The Hill plot for enzymes generally takes the form of $\log(v_0/(v_{max} - v_0))$ versus log [S] rather than $\log(\overline{Y}/(1 - \overline{Y}))$ versus log [S] for ligand binding.

The MCW can be generalized further by assuming that substrate can bind with the T form of the enzyme, with a dissociation constant given by K_T. Writing the ratio of dissociation constants for the T and R forms as $c = K_R/K_T$, $\overline{Y}$ can be obtained for the case of n sites on the protein from equation (1.122a).

$$\overline{Y} = \frac{\alpha(1+\alpha)^{n-1} + L\alpha c(1 + c\alpha)^{n-1}}{(1+\alpha)^n + Lc(1+c\alpha)^n} \tag{1.122a}$$

where α is given by $[S]/K_R$. Some values of L, c and h for several proteins are given below. It can be seen that c is small and consequently $K_T \gg K_R$, so that most of the substrate binds to the R form, even though the large value of L indicates that the T form predominates at equilibrium.

Allosteric Constants for Selected Proteins[16]

Protein	Ligand	Number of Binding Sites (n)	Hill Constant (h)	L	c
Hemoglobin	O_2	4	2.8	3.10^5	0.01
Pyruvate kinase	PEP	4	2.8	9.10^3	0.01
Glyceraldehyde 3-phosphate dehydrogenase	NAD^+	4	2.3	60	0.04

One of the drawbacks of the MCW model is that it cannot account for negative cooperativity. The binding of the first substrate molecule can only stabilize the high-affinity state (R) and cannot alter the concentration of T forms. Thus evidence of negative cooperativity is often taken to indicate that the KNF model is applicable.

1.9.3 The Koshland-Némethy-Filmer (KNF) Model

The KNF model differs from the MCW model by relaxing the assumption that binding can only occur on the R_2 species. It assumes that the binding of a substrate molecule *induces* the conformational change from T to R form. The binding sequence is shown below.

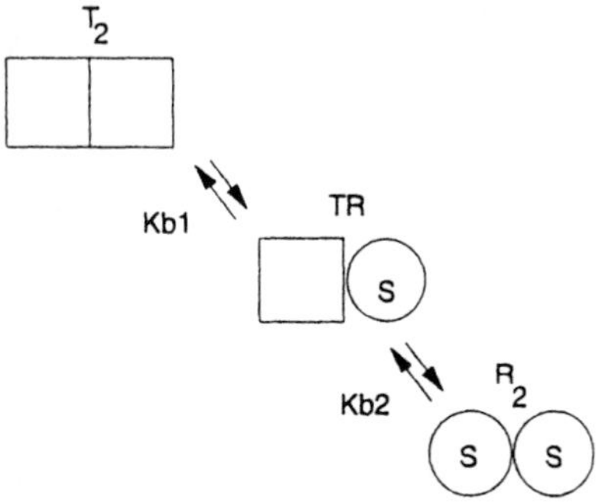

The analysis of this situation is similar to the Adair equation (1.121) discussed earlier, and the binding constants K_{b1} and K_{b2} have the same significance as with ligand-protein binding.

(16) From Fersht, A. *Enzyme Structure and Mechanism*, 2nd Edition, Freeman & Co., 1985.

Both the MCW and KNF models can be visualized as special cases of a general binding scheme, shown below.

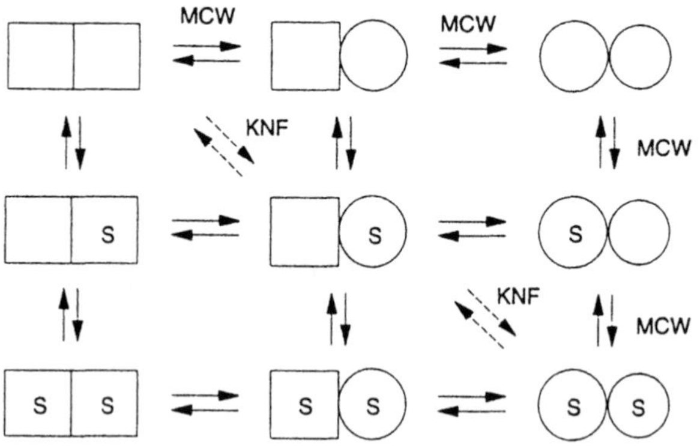

The MCW model can be seen as proceeding along the sides of the rectangle of possible states, with the hybrid TR state not being present to any significant extent. The situation we analyzed earlier did not consider substrate binding to the T form and thus is represented by only one possible pathway. The KNF model moves diagonally along the possible set of states for the enzyme. Obviously a very general model could be constructed which included all possible conformational forms. A further complexity to the modelling is added when the protein is a tetramer, as the number of intermediates increases.

1.9.4 Examples of Cooperative Binding Kinetics

Both the MCW and the KNF models were developed to explain allosteric activation and inhibition of key enzymes in metabolic pathways. Enzymes present early in metabolic pathways serve as rate-limiting enzymes; they regulate the flow of metabolites from starting materials to products. Subsequent enzymes in the pathway operate at reaction rates which are sufficient to prevent the accumulation of significant amounts of intermediates in the pathway. Often the end-products of the pathway feedback inhibit the key enzymes; this serves to conserve raw materials in the cell and prevent overproduction of end-products that are not required.

Phosphofructokinase

One example of feedback control is the enzyme phosphofructokinase (PFK), which regulates the flow of carbon along the glycolytic pathway. To prevent glucose from being "wasted" if energy in the form of ATP is available to the cell or to trigger ATP production under conditions of high energy demand, the enzyme is very tightly regulated by feedback control. Eukaryotic phosphofructokinase is activated by AMP, ADP and 3'-5' cyclic AMP and inhibited by ATP and citrate. Thus glycolysis is regulated, occurring only under conditions of demand. In some prokaryotes, the regulation is somewhat simple in that only phosphoenolpyruvate (PEP) feedback inhibits PFK. PFK from *B. stereothermophilus* is a tetramer that follows the predictions of the MCW model quite well. The inhibitor PEP binds to the T form and stabilizes it. The T and R forms show different binding affinities for the substrate fructose-6-phosphate, the R form binding the substrate more tightly. Both forms have the same k_{cat}. (This is referred to as a *K system* as only the Michaelis-Menten constant is affected, and v_{max} is unaltered.)

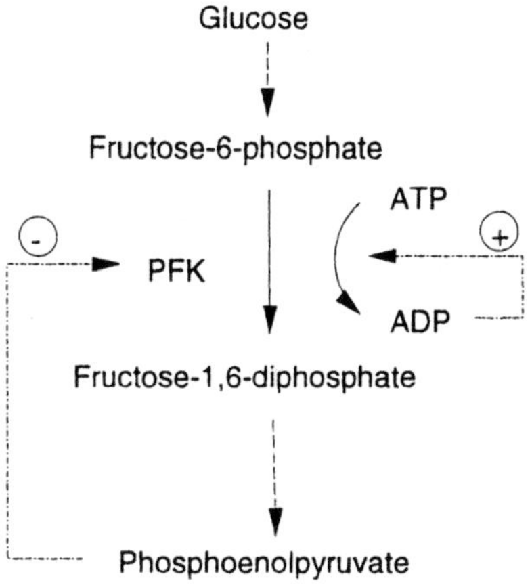

Regulation of PFK in E. coli

***Figure 1.21**. The mechanism of regulation of phosphofructokinase activity in E. coli. PFK is feedback inhibited by PEP [this inhibition is indicated by (-)] and activated by ADP [indicated by (+)].*

In the presence of ADP, the substrate fructose-6-phosphate binds in a highly cooperative manner to PFK, with a Hill coefficient of h = 3.8, very close to the number of binding sites (n = 4) present in the tetrameric enzyme. When ADP is absent, it binds preferentially to the R form and h drops to h = 1.4.

Glyceraldehyde 3-phosphate dehydrogenase

Rabbit muscle glyceraldehyde 3-phosphate dehydrogenase shows negative homotropic cooperativity in the binding of NAD^+, and thus is an example of the KNF model. The reaction catalyzed is:

$$\text{D-glyceraldehyde-3-phosphate} + NAD^+ + P_i \Leftrightarrow \text{3-phospho-D-glyceroylphosphate} + NADH + H^+$$

The enzyme glyceraldehyde-3-phosphate dehydrogenase from yeast, which also binds NAD^+, has been examined by temperature jump methods. Rate constants for the binding of the cofactor to the four binding sites on the enzyme showed support for the MCW model. Thus the same enzyme from different sources bind NAD^+ quite differently.

1.10 Temperature and pH Effects

In this section we shall examine the influence of the external environment on enzyme activity. This is particularly important in employing enzymes in manufacturing processes, as often optima in rates exist for both temperature and pH. In studying the effect of temperature on the rate of an enzyme catalyzed reaction, two factors need to be considered. The first is the influence of temperature on the reaction rate constant; the second is the thermal denaturation of enzymes at elevated temperatures, which counteracts the enhancement in the rate constant at higher temperatures.

1.10.1 Temperature and Rate Enhancement

The Arrhenius equation is the starting point for relating the effect of temperature to the reaction rate constant:

$$k = A \cdot \exp\left(\frac{-E_A}{RT}\right) \tag{1.128}$$

where k is the rate constant, R the gas-law constant, A is a frequency factor and E_A is the activation energy for the reaction. The equation can be rewritten in a convenient form for graphically obtaining the activation energy:

$$\ln k = -\frac{E_A}{R} \cdot \left(\frac{1}{T}\right) + \ln A \tag{1.129}$$

Thus the activation energy can be found by measuring k as a function of temperature. An example is given in Figure 1.22 below.

Studies on the temperature dependence of the rate constant can yield information about the binding and release of substrates and products. To obtain meaningful information however, we must know the formal mechanism of the enzyme. For example, if we consider the simple Michaelis-Menten mechanism, with Michaelis constant K_M (= K_S) for the formation of the ES intermediate and first order rate constant (k_{cat}) for the release of product from ES, the associated free energy changes for each of these reactions are ΔG_S and $\Delta G^{\pm}$. The apparent second order rate constant for the formation of the $ES^{\pm}$ transition state (see Figure 1.5) is k_{cat}/K_M (see section 1.2.2), with $\Delta G_T^{\pm}$ being the corresponding activation energy. We can write $\Delta G_T^{\pm}$ as

$$\Delta G_T^{\pm} = RT \ln\left(\frac{k_B T}{h}\right) - RT \ln\left(\frac{k_{cat}}{K_M}\right) \tag{1.130}$$

$$\Delta G_T^{\pm} = \Delta G^{\pm} + \Delta G_S \tag{1.131}$$

and thus rearranging

$$RT \ln\left(\frac{k_{cat}}{K_M}\right) = RT \ln\left(\frac{k_B T}{h}\right) - \Delta G^{\pm} - \Delta G_S \tag{1.132}$$

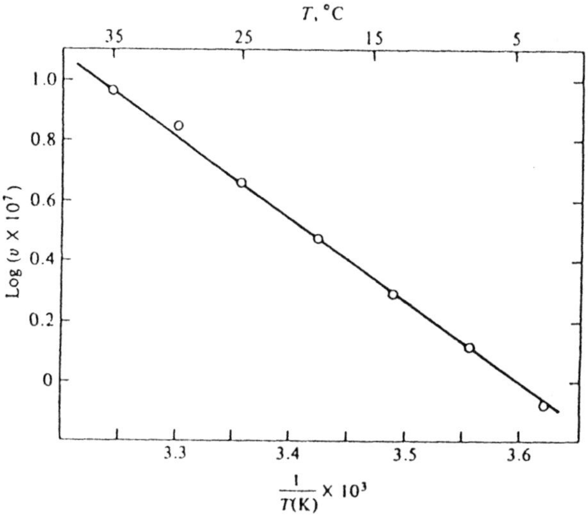

Figure 1.22. *Arrhenius plot of the myosin catalyzed hydrolysis of ATP. Data from Quellet et al., Arch. Biochem. Biophys., **39**, 37 (1952).*

Note that k_B is the Boltzmann constant (1.3807 x 10^{-16} erg.deg^{-1}), h is Planck's constant (6.626 x 10^{-27} erg.sec) and $\ln(k_BT/h)$ has the value 12.793 at 25°C.

The temperature range over which enzymes show activity is rather limited; at temperatures above 100°C relatively few enzymes exhibit activity and at temperatures near freezing, rates are very slow. This last effect is employed in cryoenzymology, the study of enzyme reactions at low temperatures. If a reaction step is too rapid to study using stopped-flow or other techniques, it may be studied at very low temperatures; however, temperatures below 0°C require the use of organic solvents which themselves may affect the reaction of interest. Although enzymes obtained from thermophiles may show enhanced thermal stability and may operate at very high temperatures (over 100°C if the system is under pressure), thermal denaturation results in an optimum temperature for enzyme function.

1.10.2 Thermal Deactivation of Enzymes

Thermal deactivation of enzymes limits their useful lifetime in processing environments and is thus of considerable importance in process design and development. The temperature range over which thermal denaturation occurs varies with the nature of the enzyme considered. For many mammalian enzymes, denaturation may occur at 45 to 55°C, as these enzymes have optimum temperatures around 37°C.

Thermal deactivation may be reversible or irreversible, and depends on the temperature to which the enzyme is raised and the period of time it is held at this temperature. The events leading to thermal denaturation are generally considered to involve an initial *reversible* denaturation, followed by an *irreversible* denaturation step. Often this irreversible step is the result of polymolecular processes, such as aggregation and precipitation. In cases where aggregation is not the cause of irreversible deactivation, conformational or covalent changes may result in deactivation. We shall consider the kinetics of deactivation as following the scheme:

$$E_a \overset{K}{\Leftrightarrow} E_d \overset{k}{\rightarrow} E_I$$

where E_a is native enzyme, E_d is reversibly denatured enzyme and E_I is the irreversibly denatured form.

The reversible formation of E_d is caused by temperature-induced conformational changes in the protein. This change is sensitive to pH. The equilibrium constant K depends on temperature. Typically, irreversible denaturation is examined by raising the solution temperature of the enzyme to some particular value. Aliquots of enzyme are then taken at fixed time intervals to determine the residual enzyme activity at a lower temperature, usually the optimum temperature for the enzyme of interest. If, at the assay temperature, E_d rapidly refolds into the active form E_a, the assay determines the enzyme concentration E_a plus E_d. Therefore we can write the observed rate of deactivation as[17]

$$v = k_d([E_a] + [E_d]) \qquad (1.133)$$

where k_d is the observed first order rate constant for deactivation. The rate at which deactivation occurs is given from the mechanism above as

$$v = k \cdot [E_d] \qquad (1.134)$$

Eliminating E_a by use of the equilibrium relationship $K = E_a/E_d$, we obtain

$$k_d = \frac{k}{1+K} \qquad (1.135)$$

Thus the observed rate constant k_d depends on both the monomolecular rate constant k and the equilibrium constant K. The transition between the E_a and E_d forms is a single step, cooperative process for most enzymes. The fraction of active enzyme E_a is a sigmoidal function of temperature, and depends on the pH as illustrated below.

At high temperatures, $[E_d] >> [E_a]$ and $K << 1$, and thus $k_d \sim k$. If the rate constant for thermal denaturation is determined by incubating the enzyme at a high temperature (T_1) and it is desired to use this value ($k_d(T_1)$) to estimate the thermal denaturation rate at a lower temperature (T_2), where much of the enzyme remains active (i.e., $[E_a]$ has some finite value), use of the the Arrhenius relationship to extrapolate from T_1 to T_2 may result in errors. These arise from the fact that at the lower temperature, where denaturation is slow, $[E_a]$ must be comparable to $[E_d]$ and as a result, $K > 1$. The observed rate of thermal denaturation will be

---

(17) S.E. Zale and A.M. Klibanov, *Biotechnol. Bioeng.*, **25**, 2221 (1983).

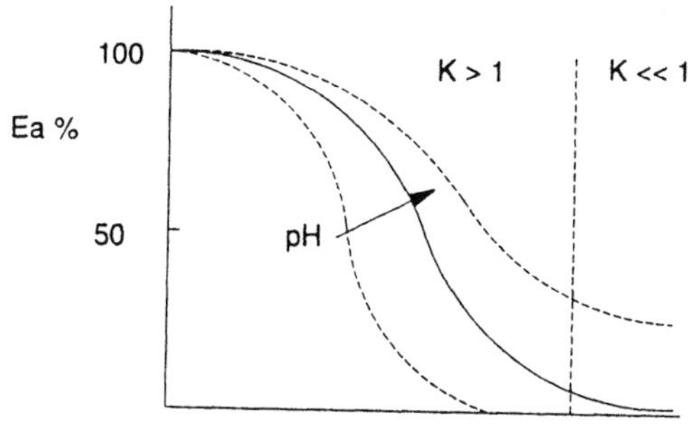

Figure 1.23. *The reversible thermal unfolding of an enzyme in terms of the fraction of the enzyme in its native conformation (E_a) versus temperature. The possible effect of pH is illustrated.*

k/(1+K) rather than k. Thus it is necessary to determine both k and K to satisfactorily account for thermal denaturation in processes operating around or slightly above the optimum for the enzyme of interest.

It is interesting to note that there is not necessarily a correlation between reversible thermal denaturation and irreversible thermal denaturation of the enzyme. Generally, at temperatures where K ~1 or K > 1, the more stable the enzyme is against reversible denaturation, the more stable it will be against irreversible denaturation. However, at temperatures where K << 1, no such correlation is expected[18].

The Mechanism of Irreversible Thermal Denaturation of Lysozyme

Ahern and Klibanov (*Science* **228**, 1280 (1985)) examined the factors leading to the irreversible thermal deactivation of hen egg-white lysozyme. Lysozyme is a small monomeric enzyme (MWt 14,500) which is well characterized; it catalyzes the hydrolysis of bacterial cell walls by breaking the polysaccharide component which consists of N-acetylglucosamine (NAG) and N-acetylmuramic acid (NAM),linked β-1,4. The substrate binds in a cleft which has six sites, designated A,B,C,D,E and F. NAM residues can only bind in the B, D and F sites, whereas NAG can bind at all sites. Bond cleavage occurs between sites D and E.

When lysozyme is heated to 100°C at pH 6.0 it looses its catalytic activity, determined by the rate of lysis of *Micrococcus lysodeikticus* cells. If it is quickly cooled to 25°C, the catalytic acitvity is regained, illustrating the reversible denaturation step described above. If it is held at 100°C for longer periods of time, decreasing fractions of the original activity

---

(18) The mechanisms of reversible denaturation are discussed in more detail in S. Lapanje, "*Physicochemical Aspects of Protein Denaturation*", Wiley, New York (1978); C. Tanford, *Adv. Protein Chem.*, **23**, 121 (1970); and S.E. Zale and A.M. Klibanov, *Biotechnol. Bioeng.* 1983, **25**, 2221.

are recovered. The irreversible inactivation follows first order kinetics, as illustrated in Figure 1.24, and the rate constant is a function of pH. This inactivation was shown to be independent of the initial enzyme concentration, which suggests that enzyme aggregation is not an important mechanism of irreversible deactivation in this case.

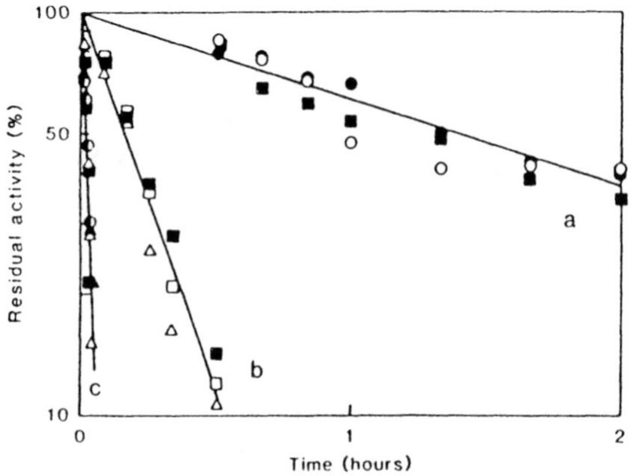

*Figure 1.24. The time course of irreversible deactivation of lysozyme at 100°C as a function of pH. Curves are drawn at various lysozyme concentrations. pH conditions are (a) pH 4.0, (b) pH 6.0 and (c) pH 8.0. [From Ahern and Klibanov, Science, **228**, 1280 (1985)].*

Two possible explanations for irreversible denaturation can be proposed. The first is a conformational process where incorrectly folded enzyme is formed at elevated temperatures; this incorrectly folded enzyme is kinetically or thermodynamically stable at elevated temperatures. When cooled, these "scrambled" structures persist, as there is a high kinetic energy barrier which prevents the enzyme from refolding to its native conformation at lower temperatures. Similar scrambled structures can be induced by the destruction of disulphide bonds with reducing agents. Both these mechanisms can be considered as conformational deactivation. If strong denaturants (such as guanidine hydrochloride) which disrupt non-covalent interactions are added to the enzyme during the thermal treatment, the enzyme should be maintained in a highly unfolded form and upon cooling and removal of denaturant, the active conformation should again result. Thus denaturants should protect the enzyme against this form of thermal denaturation. The data of Ahern and Klibanov show that this unfolding was partially responsible for deactivation at pH 8.0, but not at lower pHs.

The second possibility is that denaturation is a result of covalent processes, such as breakdown of the peptide chain, or deamidation of various amino acid residues. In the case of lysozyme, deamination of asparagine residues was found to be the major cause of irreversible deactivation at pH 4.0 and 6.0. At pH 8.0, destruction of cysteine residues can also be seen to contribute to deactivation. The rate constants for each of the causes of irreversible thermal deactivation are shown below.

	Rate Constants ($hour^{-1}$)		
Irreversible thermoinactivation	*pH 4.0*	*pH 6.0*	*pH 8.0*
Directly measured overall process	0.49	4.1	50
Individual Mechanisms			
deamidation of Asn residues	0.45	4.1	18
hydrolysis of Asp-X peptide bonds	0.12	0	0
destruction of cystine residues	0	0	6
formation of incorrect structures	0	0	32

1.10.3 pH Dependence: Ionization of Acids and Bases

The ionization of α-amino and carboxyl terminal groups in proteins, in addition to the many other ionizable groups on constituent amino acids, could give rise to complex kinetics when pH is varied. We usually find, however, that the rate of reaction depends rather simply on pH, the dependence reflecting the ionization of only one or two groups at most. This results from the fact that the only ionizations that are important are those related to groups at the active site which participate in catalysis, or those which are required to maintain the conformation of the protein. In analyzing these pH effects, we will first briefly review the common methods for representing the data.

When the general reaction for the ionization of a base B is written as

$$B + H^+ \Leftrightarrow BH^+$$

The ionization constant is $K_a = [B][H^+]/[BH^+]$, and the pK_a is then defined as $-\log_{10}K_a$. The definition of the ionization constant can be conveniently rewritten to yield the Henderson-Hasselbach equation:

$$pH = pK_a + \log\left\{\frac{[B]}{[BH^+]}\right\} \tag{1.136}$$

The corresponding equation for an acid HA is thus

$$pH = pK_a + \log\left\{\frac{[A^-]}{[HA]}\right\} \tag{1.137}$$

Denoting $[A]_o$ as the total concentration of acid, the variation of [HA] and $[A^-]$ with pH can be thus expressed as:

$$[HA] = \frac{[A]_o[H^+]}{K_a + [H^+]} \qquad [A^-] = \frac{K_a[A]_o}{K_a + [H^+]} \tag{1.138}$$

If a rate constant for an enzyme k is found to depend on the ionization of some group on the protein, for simplicity assumed to be an acid, such that when the acid is protonated the rate is k_{HA} and when it is unprotonated the rate is k_A, then the rate constant can be expressed at any pH as a function of these rate constants viz.

$$k_H[A_o] = \frac{k_{HA}[A]_o[H^+]}{K_a + [H^+]} + \frac{k_A K_a [A]_o}{K_a + [H^+]} \tag{1.139}$$

$$k_H = \frac{k_{HA}[H^+] + k_A K_a}{K_a + [H^+]} \tag{1.140}$$

If k_H is plotted against pH, a sharp inflection is found at the pK_a value of the acid group. This value occurs at a value of k_H equal to $(k_{HA} + k_A)/2$.

When the system has dibasic sites, analogous to most enzymes

$$H_2A \overset{K_1}{\Leftrightarrow} HA^- + H^+ \overset{K_2}{\Leftrightarrow} A^{2-} + 2H^+$$

and K_1 and K_2 are ionization constants, it can be analyzed in a similar manner with the result

$$k_H = \frac{k_{H_2A}[H^+]^2 + K_1 k_{HA^-}[H^+] + K_1 K_2 k_{A^{2-}}}{K_1 K_2 + K_1[H^+] + [H^+]^2} \tag{1.141}$$

This result is of interest as the activity of many enzymes is found to depend on both an acidic and a basic group, with both groups being required for activity. These groups may be involved in acid-base catalysis or the ionization of the groups may be important in maintaining the active site conformation for substrate binding. If we consider that the catalytic activity of an enzyme depends on these two groups, then the rate depends only on the concentration of HA^-, as we can consider that the fully protonated or fully deprotonated forms are inactive (i.e., $k_{H2A} = 0$ and $k_A = 0$). Equation (1.141) can thus be simplified to give Equation (1.142). A typical plot of k_H versus pH exhibits the bell-shaped character shown in Figure 1.25.

$$k_H = \frac{K_1 k_{HA^-}[H^+]}{K_1 K_2 + K_1[H^+] + [H^+]^2} \tag{1.142}$$

The optimum pH is given by $(pK_1 + pK_2)/2$. pH profiles can thus provide information about which residues may be important at the active site, although care must be exercised when interpreting these pH data, since pK_1 and pK_2 may be altered by the local environment within the protein. For example, lysozyme, which cleaves bacterial cell walls, shows a bell-shaped pH profile, with two pK_a values at pH 4 and 6. The pK of 4 suggests the effect on activity of a carboxylic acid group, which corresponds to aspartic acid-52. However, the other important group is also a carboxylic acid, glutamate-35, which is present in a hydrophobic pocket and thus has an apparent pK around 6. At pH 5, the optimum pH for lysozyme, glu-35 is protonated and thus appears to have basic character.

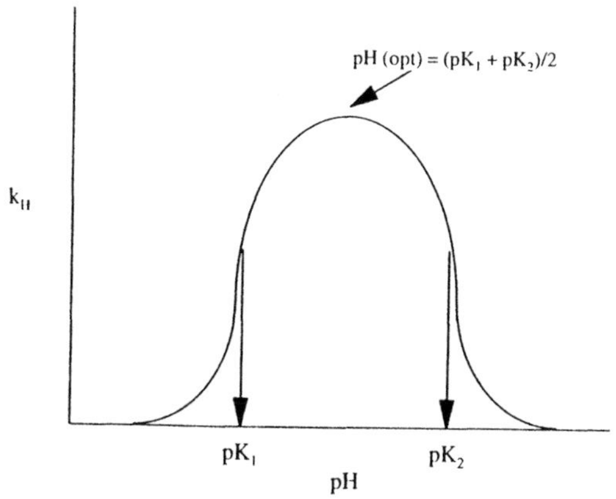

***Figure 1.25**. The pH optimum of a typical enzyme. The pK_a values of the two groups responsible for activity are illustrated.*

pH Effects on the Michaelis-Menten Mechanism

We will now consider the Michaelis-Menten mechanism and obtain expressions for the pH dependence of K_M and k_2 (or k_{cat}, its equivalent in this case). This analysis is slightly more complicated than the scheme considered above, as the rate limiting step is the dissociation of the ES complex (slow k_{cat}), whereas the above system considered the rate to depend only on the concentration of [HA⁻]. The reaction scheme considers binding of substrate to both protonated and unprotonated enzyme.

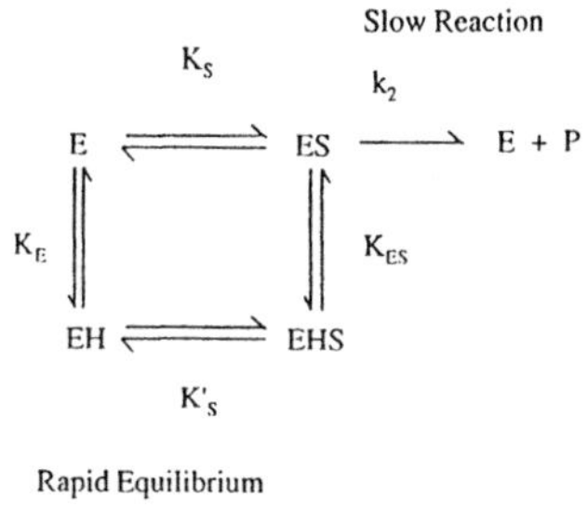

It is assumed that proton transfers are more rapid than chemical steps, and thus a rapid equilibrium exists between E, ES, EH and EHS. The four equilibrium constants (K_E, K_{ES}, K_S and K'_S) are not independent, but are related

$$K_E \cdot K_S' = K_{ES} \cdot K_S$$

Thus if $K_E = K_{ES}$, then $K'_S = K_S$ and there is no difference in substrate binding between protonated and unprotonated species. We can analyze the case when $K_E \neq K_{ES}$ in the same manner as previously, by writing the concentration of species EH and EHS in terms of ES and employing an overall balance on enzyme to relate ES and E_0. The rate expression becomes

$$v = \frac{k_2[E]_o[S]}{K_S\left(1+\frac{[H^+]}{K_E}\right)+[S]\left(1+\frac{[H^+]}{K_{ES}}\right)} \qquad (1.143)$$

Thus when $[S] >> K_S$, the maximum reaction rate can be found from equation (1.143)

$$v_{max} = \frac{k_{cat}[E]_o K_{ES}}{K_{ES}+[H^+]} \qquad (1.144)$$

The pH dependence of K_M can be found by rearranging equation (1.143)

$$K_M = K_S\frac{(1+[H^+]/K_E)}{(1+[H^+]/K_{ES})} \qquad (1.145)$$

The simple Michaelis-Menten analysis breaks down when there are several intermediates in the reaction pathway (e.g., $ES \rightarrow ES' \rightarrow E+P$). If the step from ES' to P is slow, the pH profile will reflect the ionization of ES' as ES' will accumulate and little ES will be present. If both ES and ES' accumulate, the pH profile will be weighted by the ionizations of both intermediates.

A more complex situation arises with the Briggs-Haldane mechanism (Section 1.2.2). The rate constant k_2 is comparable to k_{-1} and thus equilibrium of E and S with ES cannot be assumed. If the formation of product from ES is pH dependent, then k_2 will decrease from its maximum value as the pH is shifted from the optimum. At some point, k_2 will fall below k_{-1}, and the mechanism will change to one where the ES complex will be in equilibrium with its protonated forms (as in the Michaelis-Menten mechanism). Thus pH changes result in a change in the rate determining step, and this is sometimes referred to as a *kinetic effect*, as the change in the apparent pK_a for the reaction is governed by rate constants (k_{-1} and k_2), rather than ionization constants of E or the ES complex.

pH Effects on Protein Structure

The three-dimensional structure of a protein is stabilized by the formation of ion-pairs between various amino acids. For example, lysine, arginine and histidine may all be positively charged at neutral pH; glutamate and aspartate carboxyl side chains may be negatively charged. The exact charge depends on the local environment, as illustrated in the table below. As the pH of the solution containing the protein is altered, the groups involved in the formation of ion-pairs are titrated and the stabilization provided is lost. This may result in local conformational changes, or if the ion-pairs are sufficiently involved in protein stability, unfolding of the protein may occur. At extremes of pH, proteins denature due to disruption of both surface and buried ion-pair interactions. Perturbations in the pK_a of buried charged groups are difficult to predict, as the influence of neighboring residues are difficult to calculate. Some examples of strongly perturbed pK_a values in proteins are indicated below. The ionic strength of the solution is also important, although shielding by ions may prevent the full effects of solutes being felt at a buried site in the protein.

Perturbed pK_a Values in Proteins

Enzyme	*Residue (R)*	pK_a	pK_a *of free amino acid R-group*
Lysozyme	Glu-35	6.5	4.25
Acetoacetate decarboxylase	Lys (ε-NH_2)	5.9	10.53
Chymotrypsin	Ile-16	10.0	9.6
Papain	His-159	3.4	6.0

1.11 Nomenclature

Symbol	*Meaning*
a	interfacial area per unit volume (cm^2/cm^3)
A	constant in the Brönsted equation (Equation 1-8)
D_A	diffusivity of species A in solution
D_{eff}	effective dielectric constant in protein
E_{act}	activation energy for reaction
h	Planck's constant (6.626×10^{-27} erg.sec)
h	Hill coefficient (Equation 1-121b)
I	ionic strength (M)
k_B	Boltzmann's constant (1.3807×10^{-16} erg.deg^{-1})
$K^{\pm}$	quassi-equilibrium constant for the formation of $(ES)^{\pm}$ from E and S
K	equilibrium constant for interconversion of active and thermally deactivated forms of an enzyme (Equation 1.134)
k''_2	apparent first order rate constant (Equation (1.62)
k'_2	apparent first order rate constant (Equation (1.62)
k_{-1}	first order rate constant for the decomposition of the ES complex (Equation 1.17)
k_1	first order rate constant sec^{-1}
k_1	second order rate constant for the formation of ES (Equation 1.17)
K_1, K_2	distribution coefficients (Equation 1.82)
k_2	bimolecular rate constant, sec^{-1}M^{-1} (Equation 1.1)
k_2	second order rate constant for general acid or base catalysis (Equation 1.8)
k_2	first order rate constant for the decomposition of the ES complex (Equation 1.12)

k_{ads}	second order adsorption rate constant (Equation (1.79)
K_{bi}	binding constant for ligand binding to protein (Equation 1.114)
k_{cat}	maximum number of substrate molecules reacting per active site per second
k_d	observed first order thermal deactivation rate constant (Equation 1.133)
k_{des}	first order desorption rate constant (Eqution 1.79)
K_{eq}	equilibrium constant for a monomolecular reversible reaction (Equation (1.58)
k_f	forward first order rate constant (Equation (1.55)
k_i	i=1,2..,n first order rate constants (Equation (1.21))
K_I	dissociation constant for the EI complex (Equation (1.34))
K_M	Michaelis-Menten constant (Equation 1.19)
k_o	first order rate constant for general base catalysis at a fixed pH value (Equation 1.7)
K_M^{app}	apparent Michaelis-Menten constant (Equation (1.43))
k_{obs}	observed first order rate constant (Equation 1.7)
k_r	reverse first order rate constant (Equation (1.55)
K_S	dissociation constant for the ES complex (Equation 1.10)
n	number of ligand binding sites on protein
N_{avo}	Avogadro's number
p	molecular orientation (Equation (1.1)
R	gas constant
r_A	equivalent spherical radius of A
t	time
T	temperature
u	axial fluid velocity (Equation (1.53)
v	reaction rate (velocity) (Equation (1.14)
v_{max}	maximum reaction rate (velocity) (Equation (1.15)
v_{max}^{app}	apparent maximum reation rate (Equation (1.46))
v_o	initial reaction rate (Equation 1.16)
z	reactor length (Equation (1.52)
Z	frequency of molecular collisions, given by Equation (1.2)
[E]	enzyme concentration
$[E_a]$	concentration of active enzyme
$[E_d]$	concentration of thermally deactivated enzyme
[EI]	concentration of enzyme-inhibitor complex
$[E_I]$	concentration of irreversibly thermally deactivated enzyme
[ES]	concentration of the enzyme-substrate complex
[I]	inhibitor concentration
[L]	ligand concentration

[P]	product concentration
[S]	substrate concentration
$[S_o]$	initial substrate concentration
$\Delta G^{\pm}$	Gibbs free energy difference between ground and excited states
$\Delta S^{\pm}$	entropy difference between ground and activated states
$\Delta H^{\pm}$	enthalpy difference between ground and activated states

Greek Symbols

β	Brönsted constant (Equation 1.8a)
α	Brönsted constant (Equation 1.8b)
τ	dimensionless distance (z/u)
Θ_i	i=1,2... constants in the generalized Dalziel rate expression (Equation 1.86)
$\overline{Y}$	fractional degree of saturation of protein with ligand (Equation 1.115)
κ	transmission constant for transition state complex (Problem 1.6.5)
ν	average frequency of energy barrier crossing (Problem 1.6.5)

1.12 Problems

1. Reaction and Diffusion

One of the most active enzymes is catalase (EC 1.11.1.6), which decomposes hydrogen peroxide to oxygen and water. It is a tetramer consisting of four identical subunits, each of which contains one active site. The molecular weight of the tetramer is 240,000. It is found to follow Michaelis-Menten kinetics, with a K_M of 1.1 M at pH 7.0 and 30°C. Estimate the turnover number k_{cat} for catalase, given that the initial rate of reaction has been determined to be 0.57 M.sec^{-1} for an initial substrate concentration of 0.15 M and an enzyme concentration of 30 μg/ml. How does this first order rate constant compare with the diffusion-limited rate of encounter of enzyme and substrate?

2. Enthalpy and Entropy of Reaction

Like other serine proteases, chymotrypsin catalyzes the hydrolysis of specific peptide (amide) and ester susbtrates. Below are values of k_{cat} for the hydrolysis of benzoyl-L-tyrosinamide (BTA) and benzoyl-L-tyrosine ethyl ester (BTEE) measured at various temperatures.

	k_{cat} (sec^{-1})	
Temperature (°C)	BTEE	BTA
30.0	10.2	1.19
21.0	6.66	0.492
13.0	4.32	0.203
5.0	2.81	0.084

From these data, determine the heat of activation (ΔH^*) and the entropy of activation (ΔS^*) for each reaction. Assuming that the rate-limiting steps for amide and ester hydrolysis are described by k_2 and k_3, respectively (as defined in the example on chymotrypsin in section 1.2.2), how might you explain the ΔS^* values?

3. Kinetic Parameters for Trypsin and a Catalytic Antibody

Listed below are initial rate data for the hydrolysis of 4-nitrophenyl methyl carbonate at 30°C, pH 8.1, by a catalytic antibody produced in the laboratory of P.G. Schultz at Berkeley. The concentration of antibody, all of which was assumed to be active, was 1 μM. Included for comparison are data for the hydrolysis of benzoyl-L-tyrosine ethyl ester (BTEE) by trypsin at 30°C, pH 7.5. The enzyme concentration was 3.34 mM, and active-site titration revealed that 80% of the enzyme was active. The initial rates are reported in units of product concentration per minute. Determine k_{cat} and K_M values for the catalytic antibody and trypsin.

Antibody		Trypsin	
Substrate Conc. (μM)	Inital Rate (μM/min)	Substrate Conc. (μM)	Inital Rate (μM/min)
210	14.0	20	330
110	11.3	15	300
85	9.80	10	260
64	7.84	5.0	220
42	6.67	2.5	110

4. On the Dynamics of the Michaelis-Menten Reaction Mechanism

For the general enzyme mecahnism given below, write the mass balances for each of the species E, S, P, and X (the ES complex).

$$E + S \underset{k_{-1}}{\overset{k_1}{\Leftrightarrow}} X \overset{k_2}{\rightarrow} E + P$$

(a) Since the total enzyme and substrate are conserved (assuming that one mole of substrate gives one mole of product), simplify the four differential equations into two; those for substrate and the ES complex.

(b) Apply the quasi-steady state assumption (qss) in the form dX/dt = 0 to obtain an expression for the rate of change of substrate concentration -dS/dt. Show that this is equal to the rate of formation of product dP/dt. Although we have assumed dX/dt = 0, the qss assumption actually only requires that dX/dt << dS/dt. Write your results in terms of the Michaelis constant K_M [defined as $(k_2 + k_{-1})/k_1$].

(c) Apply the equilibrium assumption, that the binding of the exzyme and the substrate is in equilibrium, to determine the equilibrium concentration of the ES complex. Write your expression in terms of K_S, the equilibrium constant. Find the rate of product formation dP/dt, and show that this is the same as that obtained in part (b). Find the rate of disappearance of substrate dS/dt. Show that in this case

$$\frac{dP}{dt} \neq -\frac{dS}{dt}$$

(d) (more difficult) Linearize the two equations for S and X obtained in part (a) about the initial condition $S = S_o$, $X = 0$. Show that the Jacobian for the system is

$$\underline{\underline{J}} = \begin{bmatrix} -k_1E_o & k_{-1} + k_1S_o \\ k_1E_o & -(k_2 + k_1S_o + k_{-1}) \end{bmatrix}$$

The eigenvalues of $\underline{\underline{J}}$ yield two relaxation times τ_1, τ_2. Find the determinant and the trace of the Jacobian matrix. When the ration $det(\underline{\underline{J}})/tr(\underline{\underline{J}})^2$ is small, the relaxation times can be written

$$\tau_1 \rightarrow \frac{-tr(\underline{\underline{J}})}{det(\underline{\underline{J}})} \qquad \tau_2 \rightarrow \frac{-1}{tr(\underline{\underline{J}})}$$

Show that τ_1/τ_2 is small when any of the following conditions apply

(i) $E_o << K_M$
(ii) $S_o >> K_M$
(iii) $E_o >> K_M$
(iv) $k_2 << k_{-1}$

By examining the time scales of motion of X and S in terms of τ_1 and τ_2, show that if condition (i) or (ii) is met, the system can be described by the quasi-steady state approximation. Similarly show that if condition (iii) or (iv) is met, the system is described by the equilibrium assumption. (This is explored further in B.O. Palsson, Chem. Eng. Sci. **42**, (3) 447-458, 1987).

5. *Graphical Representation of Data*

The enzyme alcohol dehydrogenase catalyzes the oxidation of a number of alcohols to their corresponding aldehydes. The reaction involves electron transfer to NAD^+:

$$\text{ethanol} + NAD^+ \Leftrightarrow \text{acetaldehyde} + NADH + H^+$$

The forward reaction was investigated by monitoring the absorbance with time of NADH at 340 nm (NAD^+ does not absorb at this wavelength; both species however absorb at 260 nm). The enzyme concentration was 1.5 mM and the ethanol concentration was 0.5 M in all the experiments conducted. The initial concentration of NAD^+ was varied and the absorbance at 340 nm was recorded as a function of time, as given below.

	Absorbance at 340 nm at times given below (min)					
NAD^+ (Conc.)	0.5	1.5	2.5	3.5	4.5	5.5 (min)
1.5 (mM)	0.009	0.025	0.041	0.055	0.069	0.081
2.0	0.012	0.034	0.055	0.074	0.091	0.108
2.5	0.015	0.042	0.068	0.092	0.114	0.135
3.0	0.017	0.051	0.081	0.110	0.137	0.161
5.0	0.029	0.084	0.135	0.182	0.226	0.267
10.0	0.056	0.164	0.264	0.357	0.445	0.526

The measurements of absorbance were made in a 1 cm cell, and the extinction coefficient for NADH is $\varepsilon_{340} = 6.22 \times 10^3\ cm^{-1}.M^{-1}$. Calculate the value of the Michaelis-Menten constant K_M and the rate constant k_{cat}.

6. Inhibition Kinetics

The initial rate of reaction of an enzyme catalyzed reaction was determined in the presence of several inhibitors, A, B, and C, at various concentrations of the substrate S. The data in the presence and absence of each inhibitor are shown in the table below. Determine the manner of action of each inhibitor and calculate the inhibition constant K_i for A, B and C.

	Initial reaction rate (nmole/min)			
Substrate Concn. (mM)	No inhibitor	A at 5 μM	B at 25 μM	C at 0.1 mM
0.2	8.34	3.15	5.32	4.33
0.33	12.48	5.06	6.26	5.56
0.5	16.67	7.12	7.07	8.75
1.0	25.0	13.3	8.56	14.8
2.5	36.2	26.2	9.45	23.6
4.0	40.0	28.9	9.60	28.5
5.0	42.6	31.8	9.75	30.0

7. Transition State Structures and Mechanism

An important enzyme-catalyzed reaction in the human liver is the deamination of cytidine via the liver enzyme, cytidine deaminase. Cytidine has the structure below:

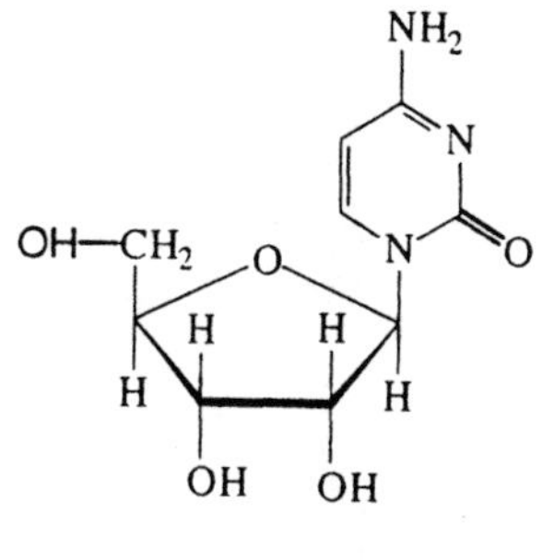

cytidine

The reaction liberates ammonia and produces uridine (the para-amino group is replaced with a keto group).

Transition-state theory is founded on two assumptions; 1) the rate constant k, of any homogeneous chemical reaction is controlled by the decomposition of an activated transition-state complex, and 2) the reaction system can be treated as though the transition-state complex is in equilibrium with the reactants. The resulting equation of the transition-state theory is

$$k = \kappa \nu K^{\pm}$$

where κ is the transmission coefficient, which we shall assume to be unity, and ν is the average frequency of the energy-barrier crossing by the activated complex. $K^{\pm}$ is the equilibrium constant for formation of the transition state complex from the reactants.

(a) For the enzyme catalyzed deamination of cytidine, determine the first-order rate constant for the reaction at 25°C if $\Delta S^{\pm}$ is -1.2 x 10^{-2} kcal/mol.K and $\Delta H^{\pm}$ is 4.7 kcal/mol.

(b) For low substrate concentrations, show the relationship between the Michaelis-Menten parameters k_{cat} and K_M and the rate constant k determined from transition-state theory.

(c) The following three compounds were tested as transition-state analogs for cytidine during the deamination reaction catalyzed by cytidine deaminase [R. Wolfenden, "Transition State Analogues for Enzyme Catalysis," *Nature*, **223**, 704 (1969)]. At 25°C in a Tris buffer at pH 8.0 the molar binding constants ($K_b = k_{-1}/k_1$) for the three compounds were 6.1 x 10^{-3} for analog (I), 1.1 x 10^{-2} for analog (II) and 1.8 x 10^{-6} for analog (III). Discuss the differences between these analogs and propose a reasonable structure for cytidine in its transition state.

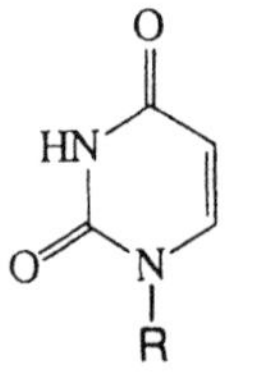

(I)

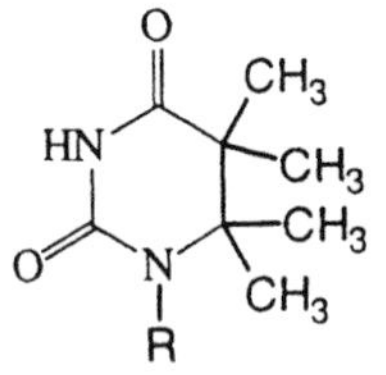

(II)

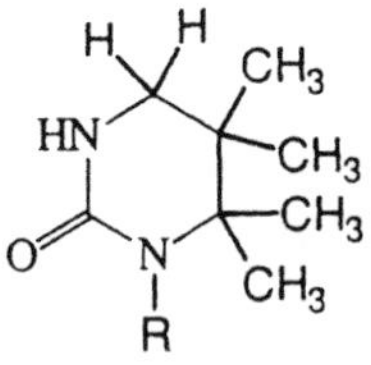

(III)

(d) The deamination of cytidine via enzymatic catalysis occurs at a rate which is roughly 30,000 times faster than the same reaction in aqueous solution without enzyme present. Explain why the dramatic improvement in the rate might occur in the case of the enzyme-catalyzed reaction.

8. Temperature Jump

An enzyme and substrate interact to form an ES complex, with a dissociation constant K_d of 2.0 x 10^3 M at 30°C. The enthalpy change for the formation of the ES complex ΔH is - 5 kcal/mol. What is the concentration change in ES caused by a temperature jump of 3°C (from 27 to 30°C) in a temperature jump apparatus used to study the kinetics of susbtrate binding? The initial enzyme concentration was 0.1 mM and the initial substrate concentration was 1.0 mM. (Hint: use the van't Hoff equation as a starting point).

9. Active Site Titration

We have seen that the hydrolysis of p-nitrophenylacetate by chymotrypsin can be described by the following three step reaction:

$$E + S \underset{k_{-1}}{\overset{k_1}{\Leftrightarrow}} ES \overset{k_2}{\rightarrow} ES' \overset{k_3}{\rightarrow} E + P$$

When the p-nitrophenylacetate-chymotrypsin reaction is carried out at pH 8, an initial burst of p-nitrophenol is observed, followed by the slow (zero-order) production of p-nitrophenol. One way of describing this process was discussed in Section 1.6.1. Below, we will take a slightly different approach.

Assume that the primary adsorptive step E + S $\Leftrightarrow$ ES quickly reaches a quasi-equilibrium state, then show that, if $[S]_0 >> [E]_o$ (and thus $[S] \sim [S]_o$ at small times t), [ES] has the general form

$$[ES] = \frac{[E]_o - (a/b)(1 - e^{-bt})}{1 + \frac{K_S}{[S]_o}}$$

where a and b are constants. Show that $[P_1]$ can be written in the form

$$[P_1] = At + B(1 - e^{-bt})$$

What are the exact expressions for A and B?

Suppose $[P_1]$ is plotted against time t, as would be the case in an active-site titration experiment. What conditions must be met for such a plot to yield the value of the enzyme concentration $[E]_o$?

10. Single Turnover of Substrate when $E_0 >> S_0$

The three step mechanism illustrated below has been proposed to describe the kinetics of several proteolytic enzymes, such as papain and ficin, which form acyl-enzyme intermediates. By using nitrophenyl esters as substrates, the product P_1 released in the reaction is nitrophenol, which can be easily monitored spectrophotometrically. Alternative substrates which show a large change in absorption when the acyl enzyme is formed permit observation of ES' (e.g., N-(2-furyl) acryloyl-L-tyrosine (or tryptophan) methyl ester).

$$E + S \underset{k_{-1}}{\overset{k_1}{\Leftrightarrow}} ES \overset{k_2}{\rightarrow} ES' + P_1 \overset{k_3}{\rightarrow} E + P$$

By writing the mass balances in a well-mixed batch reactor for all the species present, use the simplification that $E_0 >> S_0$ to derive a second order differential equation for the time course of ES in terms of the kinetic constants and the initial enzyme concentration. Show that the assumption that $k_1E_0 >> k_2$ permits the solution of this equation to be simplified to yield the following solutions for ES and P_1

$$ES = \frac{k_1 E_0}{k_1 E_0 + k_{-1} + k_2} \cdot S_0 \cdot \exp\left[-\left(\frac{k_1 E_0}{k_1 E_0 + k_{-1} + k_2}\right) k_2 t\right]$$

$$P_1 = S_0 \left\{1 - \exp\left[-\left(\frac{k_1 E_0}{k_1 E_0 + k_{-1} + k_2}\right) k_2 t\right]\right\}$$

Find the corresponding solutions for ES' and P_2. Show that observation of ES or P_1 permits the apparent first order rate constant to be found. How may k_2 and k_3 be determined? If the first reaction can be considered to be at equilibrium, with K_s being the dissociation constant for the reaction $E + S \Leftrightarrow ES$, show that the apparent first order rate constant is now given by:

$$k_1^{obs} = \frac{k_2 E_0}{K_s + E_0}$$

11. Mechanism of Bisubstrate Reactions (LADH)

The enzyme liver alcohol dehydrogenase (LADH) catalyzes the oxidation of alcohols and the reduction of a variety of aldehydes. NAD or NADH serve as the second substrate (cofactor). The reaction is readily monitored spectrophotometrically at 340 nm or by fluorescence (NADH can be excited at 340 nm and then fluorescences at 450 nm, whereas NAD does not absorb at 340 nm).

$$\text{alcohol} + NAD^+ \Leftrightarrow \text{aldehyde} + NADH$$

In studying both the oxidation and the reduction reaction with a variety of alcohols and aldehydes, Dalziel and Dickinson[19] found that all four Θ parameters in the generalized rate equation for bisubstrate reactions were present for both oxidation and reduction reactions.

$$\frac{[E]_0}{v_0} = \Theta_0 + \frac{\Theta_1}{[A]} + \frac{\Theta_2}{[B]} + \frac{\Theta_{12}}{[AB]}$$

The table below indicates the values of the Θ parameters for various alcohols and aldehydes. Based on these results, what is the kinetic mechanism of the oxidation and of the reduction reactions? Does this depend on the nature of the substrates? Discuss the reasons for your conclusions.

(19) Dalziel, K. and F. Dickinson, *Biochemistry Journal*, **100**, 34-46 (1966)

Substrate	Θ_0 (sec)	Θ_1 (μM-sec)	Θ_2 (μM-sec)	Θ_{12} (mM^2-sec)
Alcohols				
Ethanol	0.37	1.1	66	0.0072
1-Propanol	0.31	1.2	19	0.0012
1-butanol	0.35	1.1	4	0.0004
2-methylpropan-1-ol	0.34	1.7	40	0.0032
Aldehydes				
Acetaldehyde	0.0075	0.1	3.3	-
Propionaldehyde	0.0095	0.14	0.43	-
Butyraldehyde	0.0075	0.10	0.17	-
2-methylpropionaldehyde	0.009	0.12	0.25	-

12. *pH Effect on Transient Kinetics*

A stopped flow experiment is conducted at various pH's to examine the binding constants for an enzyme E and substrate S. Observations of the apparent first order rate constant $k_{1,app}$ are made at various pH values. The enzyme is thought to only show substrate binding activity when deprotonated, with the mechanism proposed viz.

$$E + S \underset{k_{-1}}{\overset{k_1}{\Rightarrow}} ES$$

$$E + H^+ \overset{K_H}{\Leftrightarrow} EH$$

The proton dissociation constant is K_H, and the EH complex is assumed to be in rapid equilibrium with E (proton transfer steps are generally faster than chemical steps). Show that the characteristic time constant for the substrate binding (τ) is given by

$$\frac{1}{\tau} = k_{1,app}([E]_{app} + [S]) + k_{-1}$$

where

$$k_{1,app} = \frac{k_1}{\left(1 + \frac{[H^+]}{K_H}\right)} \quad \text{and} \quad [E]_{app} = [E] + [EH]$$

13. *Kinetic Mechanism from Stopped-Flow Data*

Ohnishi and Hiromi (Carbohydrate Research, **61**, 335 1978) obtained the following data on the binding kinetics of glucoamylase (E) with its substrate maltotriose (S). The method employed was stopped-flow and the absorbance change at 303 nm was monitored.

A single reciprocal relaxation time ($1/\tau$) was found as a function of initial substrate concentration [S_o], at 5°C, pH 4.5 and [E_o] = 11 μM. What conclusions can you draw from this data?

Substrate Concn $[S]_o$ (mM)	$1/\tau$ (sec^{-1})	Substrate Concn $[S]_o$ (mM)	$1/\tau$ (sec^{-1})
0.2	107	4.0	360
0.4	145	5.0	385
0.5	165	6.0	408
0.6	215	7.0	385
1.0	235	8.0	385
1.2	270	9.0	375
2.0	300	10.0	415
3.0	335		

14. Electrostatic Effects

Electrostatic effects are believed to play an important role in enzyme catalysis by stabilizing charged transition states. However, it is difficult to calculate the magnitude of such effects because of variations in the local dielectric constant of the protein. Defining an effective dielectric constant, D_{eff}, is one way to simplify this problem. Accordingly, we will assume that transmission of electrostatic effects from charged surface groups through a protein is governed by D_{eff} of the inhomogeneous protein. The value of D_{eff} between two charged residues can be calculated from the standard formula for electrostatic interactions:

$$D_{eff} = \frac{332 z_1 z_2}{r \Delta\Delta G} \tag{1}$$

where z_1 and z_2 are the two charges, measured in units of the charge of an electron, separated by a distance of $r\,\mathring{A}$ and which interact with an energy of $\Delta\Delta G$ kcal/mol (1 kcal = 4.184 kJ).

One way of measuring D_{eff} involves the use of site directed mutagenesis to modify a charge on the surface of a protein and thereby alter the pKa of an ionizable group at some other position in the molecule. This approach was recently used by Fersht and co-workers to measure D_{eff} in the active site cleft of the enzyme subtilisin [A.J. Russell, P.G. Thomas, and A.R. Fersht, *J. Molec. Biol.*, **193**, 803 (1987)].

In subtilisin, the active site histidine His64 acts as a general base during catalysis, accepting a proton from residue Ser221 as it forms a bond with the substrate carbonyl carbon. The enzyme is active <u>only</u> when His64 is unprotonated; thus, the pKa of His64 can be determined by measuring the dependence of k_{cat}/K_M on pH. Shown in the figure below is the pH dependence of k_{cat}/K_M for the hydrolysis of the tetrapeptide substrate suc-A-A-P-F-pNA ($K_M/S_o \cong 15$) by wild-type subtilisin and the mutant Asp -> Ser99 at ionic strength 0.1 M. The residue Asp99 is in an external loop of the enzyme on the rim of the active site cleft. The distance between the carboxylate oxygen atoms of Asp99 and the His64 imidazole is about 12.4 Å.

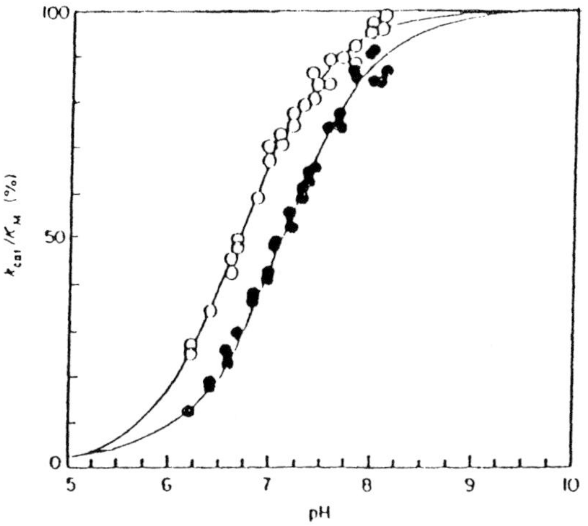

The pH dependence of k_{cat}/K_M *for the hydrolysis of suc-A-A-P-F-pNA by wild-type subtilisin (•) and the mutant Asp -> Ser99 (o) at ionic strength 0.1 M, 25°C. Data are normalized to 100% for the limiting value at high pH for each mutant. From Russell et al., 1987.*

(a) From the pH dependence of k_{cat}/K_M and an equation similar to eqn (1) above (i.e., you must first determine how D_{eff} depends on ΔpKa), estimate D_{eff} between Asp99 and His64 for an ionic strength of 0.1 M.

(b) The dependence of D_{eff} on ionic strength I (in M) can be related to the value at zero ionic stength D_w by

$$\ln D_{eff} = \ln D_w + \alpha\sqrt{I}$$

where α is a constant. Assuming α = 1.71, what is D_w for the region between Asp99 and His64? Compare D_w to the value of 78.5 expected for water, and explain the similarity or dissimilarity of the two values.

15. The MCW Concerted Symmetry Model

Consider a protein with two binding sites, illustrated below. The general concerted--symmetry model permits substrate to bind to both the T and R forms and allows T to R transitions among partially filled forms of the protein. Defining the following equilibrium binding constants

$L = [T_0]/[R_0]$, $L_1 = [TS]/[RS]$ and $L_2 = [TS_2]/[RS_2]$

express L_1 and L_2 in terms of L and c, where $c = K_R/K_T$.

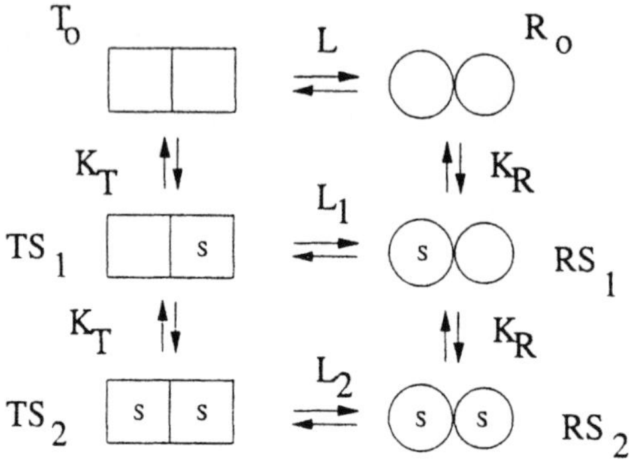

Derive an equation for the fraction of bound protomer $\overline{Y}$ for this general mechanism. Plot $\overline{Y}$ as a function of [S] for the following parameters:

$L = 25$, $K_R = 2$ and $c = 0.2$

16. pH Effects

The active site of an enzyme contains a glutamic acid residue with a pK of approximately 5.0. This residue must be in an anionic form for catalytic function. The substrate for the enzyme is positively charged and has a pK value of 10.0. It remains positively charged over the pH range of interest for catalysis.

(a) What are the reactions which show the effect of pH on the various enzyme species in solution?

(b) Develop the equation describing the initial rate of reaction as a function of pH. (See equation 1.143 for an example of Michaelis-Menten kinetics).

(c) Schematically graph the velocity-pH profile over a wide range of pH and comment on the shape.

17. Thermal Denaturation

The irreversible thermal inactivation of ribonuclease at 70° C is greatly affected by the presence of $CaCl_2$. For example, at 50° C, 2.25 $CaCl_2$ M increases the measured rate constant of thermoinactivation (relative to that in the absence of salt) by a factor of 41, whereas at

75° C, the same concentration of salt *decreases* the rate constant by a factor of 1.3 [S.E. Zale and A.M. Klibanov, *Biotechnol. Bioeng.*, **25**, 2221 (1983)]. How might you explain this behavior?

18. Thermal Stability

Shown below are thermal transition curves of RNase-A followed by observing changes in circular dichroism (a) and in difference spectral intensity (b). θ/θ_N is the relative change in ellipticity at 222 nm, and $\Delta\varepsilon$ was measured at 278 nm[20].

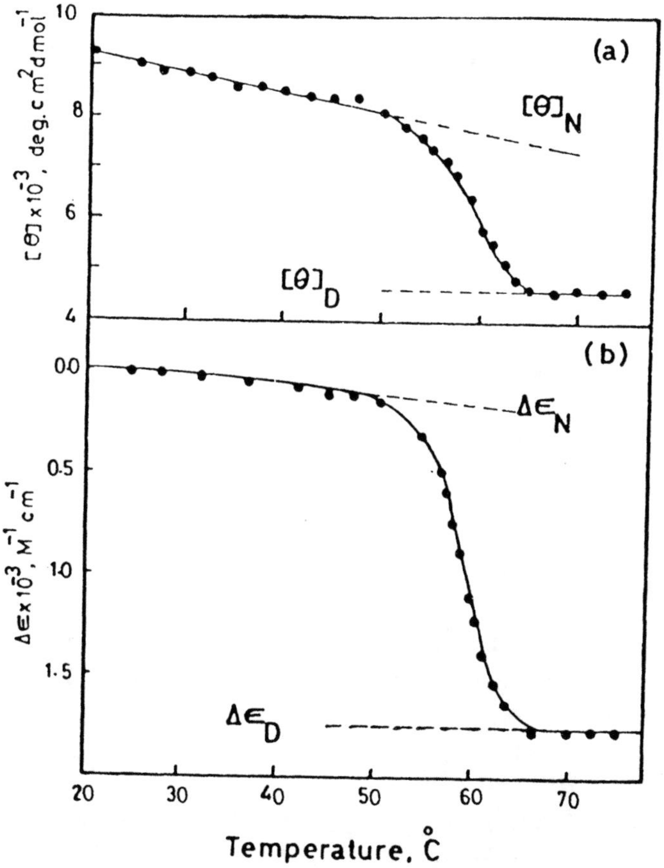

(20) From F. Ahmad, in *Thermostability of Enzymes*, Chap 6., Editor M.N. Gupta, Springer Verlag Narosa Publishing House, (1993)

Based on these plots, how would you expect the following measures to affect the thermal stability of the protein at both 55°C and at 65°C? Why?

1. Chemically cross-link the enzyme to an insoluble polymer.
2. Use site directed mutagenesis to replace an Asp residue with Glu.
3. Replace a Met residue with Ala.
4. Introduce a Cys residue close to an existing Cys residue and oxidize with sodium tetrathionate.
5. Replace an Asn residue with Gln.
6. Replace an Asp residue buried in the protein interior with Met.
7. Replace a Thr residue located about midway between the interior and exterior of the protein with Lys.

19. Probing the Stability of Phage T4 Lysozyme

In an effort to improve the thermal stability of phage T4 lysozyme, Matthews and coworkers [Alber et al., *Nature*, **330**, 41 (1987)] used random mutagenesis to generate mutant proteins. One such mutation had the amino acid substitution threonine 157 -> isoleucine 157 (i.e., threonine was replaced by isoleucine at position 157). The interactions of threonine 157 in the wild-type (natural) lysozyme are shown below, where hydrogen bonds are indicated by broken lines.

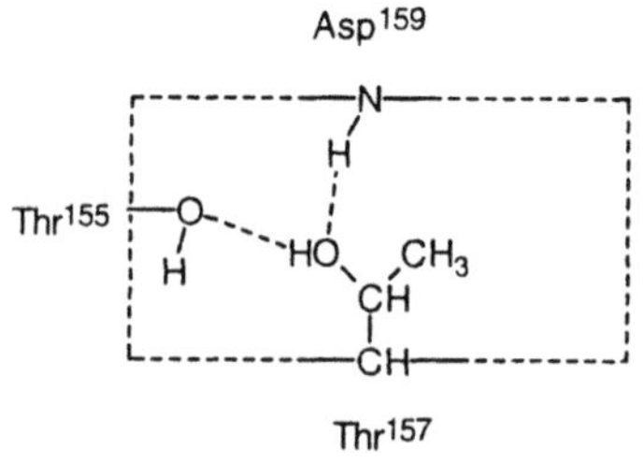

a) Speculate why the isoleucine 157 mutant was <u>less</u> stable than the wild-type enzyme. Explain your answer with the aid of a schematic illustration like the one above.

b) Interestingly, when glycine was substituted at position 157, the resulting mutant was almost as stable as the wild-type enzyme. Explain why the Gly 157 mutant had much better stability than the Ile 157 mutant. (*Hint*: the smaller glycine allows a water molecule to bind at the site previously occupied by the side chain of threonine.)

20. Mechanisms of Thermal Denaturation

The thermal denaturation properties of mutant and wild-type kanamycin nucleotidyltransferase (KNTase) are summarized in the table below.

What are possible mechanisms by which these specific changes stabilized the enzyme? Justify your answers in terms of possible expressions for the deactivation rate constant, k_{obs}.

KNTase	Half-life, min 50° C	55° C	60° C	65° C
Wild type	17	1.9	< 0.3	< 0.3
TK1 (Asp -> Tyr)	Stable	> 60	16.5	< 1
TK101 (Asp -> Tyr); (Thr -> Lys)	Stable	Stable	> 60	15.2

Stable denotes no loss of activity in 60 min.

21. Energetics of Protein Stability

Apolar amino acid side chains in the interiors of proteins are believed to stabilize proteins in two general ways: "hydrophobic effects," through which apolar groups shield the protein interior from external aqueous solvent molecules, and "packing effects," which increase favorable van der Waals interactions. Recently, the gene V protein of bacteriophage fl was modified by site directed mutagenesis in an effort to separate the contributions of packing effects from hydrophobic effects in the protein's stability [W.S. Sandberg and T.C. Terwilliger, "Influence of Interior Packing and Hydrophobicity on the Stability of a Protein," *Science*, **245**, 54 (1989)]. The four proteins studied were the wild type (WT), the single mutants Val^{35} -> Ile (V35I) and Ile^{47} -> Val (I47V), and the reversed double mutant Val^{35} -> Ile and Ile^{47} -> Val (V35I-I47V). Free energies of unfolding for each of these proteins were measured, and the differences between the free energies of unfolding of the mutants and WT ($\Delta\Delta G^o$) are summarized on the next page.

Assume the difference in free energy of transfer from water to octanol between Val and Ile is ~ 0.8 kcal/mol (i.e., $\Delta\Delta G^o_{trans} = (\Delta G_{trans})^{Val} - (\Delta G_{trans})^{Ile}$). Then, from the above results, calculate

a) the free energy of unfavorable packing interactions in V35I compared with WT
b) the energy of destabilization due to packing differences between I47V and WT
c) the energetic effect of packing changes on the reversed mutant.

What do these results suggest about the strategy of adding buried hydrophobic groups to stabilize proteins?

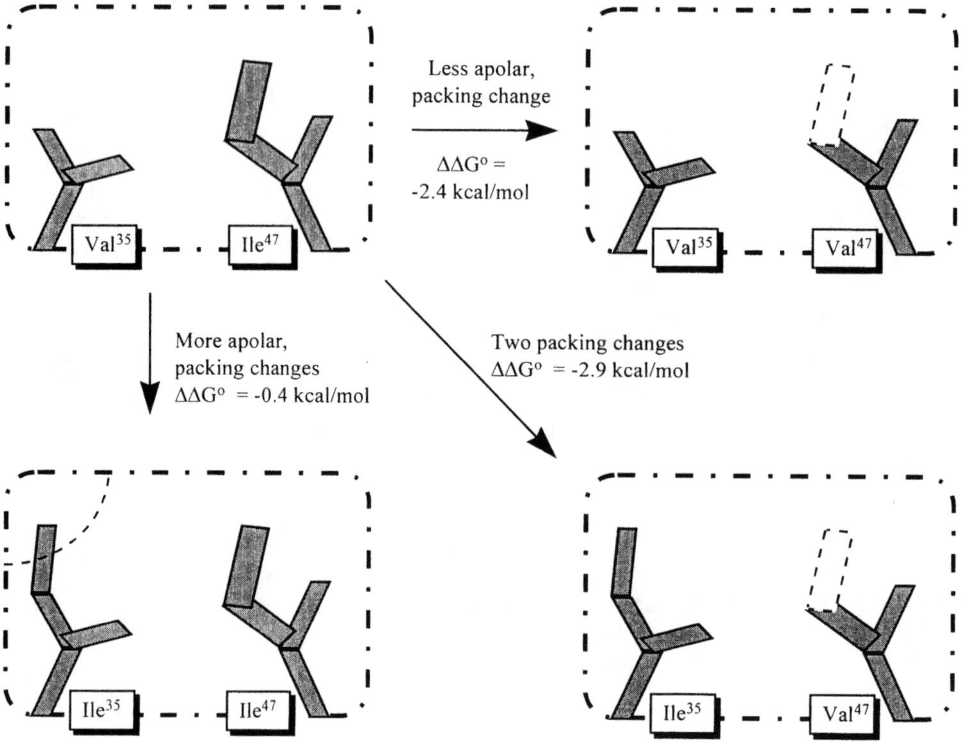

Schematic illustration of stability changes induced by amino acid substitutions.

Chapter 2. Immobilized Biocatalysts

2.1 Introduction

In the previous chapter we considered the general properties and kinetics of enzymes in solution. In many practical applications, however, it is advantageous to employ enzymes in *immobilized* form. In general, immobilized enzymes are enzymes that are attached to, or entrapped within, a macroscopic support matrix so that the resulting catalyst can be reused. Whole cells, which can carry out more complex chemical transformations than individual enzymes, can also be immobilized. Both immobilized enzymes and immobilized whole cells will be considered in this chapter. We will review some of the many methods available for immobilizing enzymes and whole cells, along with different reactor configurations in which immobilized enzymes and cells are often used. Finally, we will see that immobilizing an enzyme can have a substantial effect on the enzyme's kinetic properties and operational stability.

2.2 Rationale and Applications

Immobilized enzymes offer several potential advantages over soluble enzymes. Immobilized enzymes are typically macroscopic catalysts that are retained in the reactor; therefore, continuous replacement of the enzyme is not necessary, and separation of the enzyme from other components in the reaction mixture is simplified (the importance of this point will be underscored in Chapter 6). Immobilized enzymes can be employed in a wide range of different reactor configurations (Figure 2.1) and, because high concentrations of catalyst can be obtained, correspondingly high volumetric productivities are possible. Higher reactor productivities lead to lower capital costs. Moreover, immobilized enzymes are often more stable than enzymes in solution, as discussed in Section 2.5. It is also important to note that the properties of the support, for example, its ionic charge, can in some cases be exploited to modify the behavior of the enzyme, as we will see in Sections 2.4.1 and 2.4.2.2.

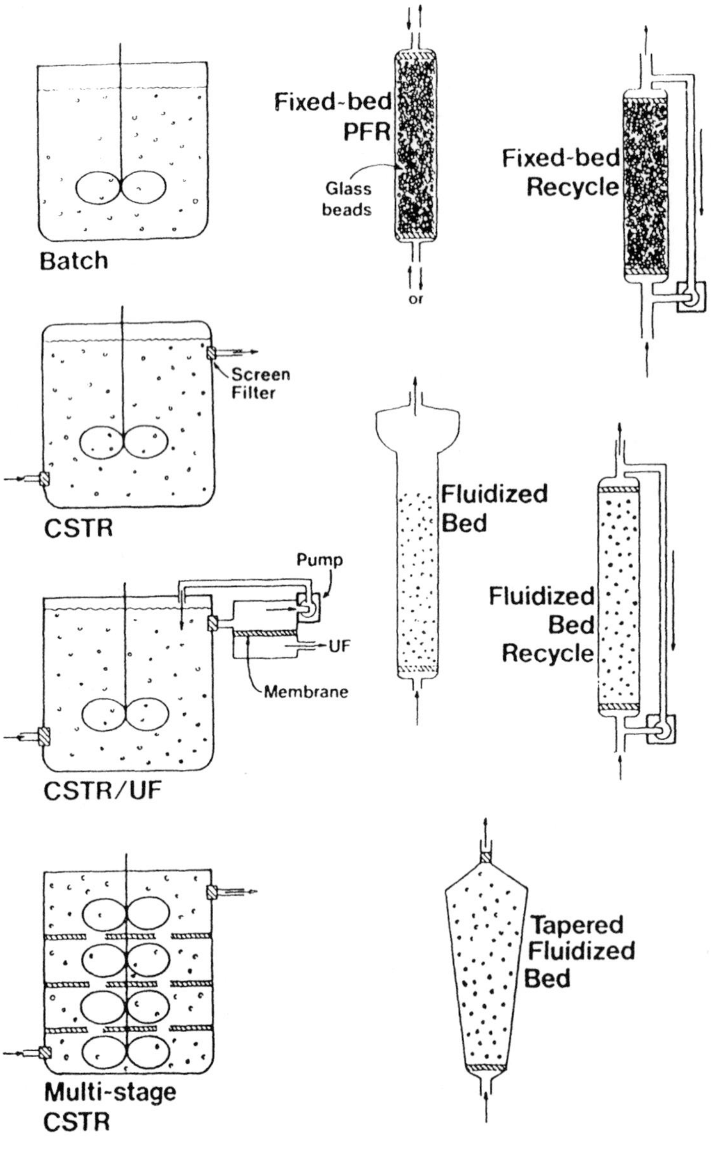

Figure 2.1. *Types of reactors for containment and use of immobilized enzymes (from H.E. Swaisgood, "Immobilization of Enzymes and Some Applications in the Food Industry," in Enzymes and Immobilized Cells in the Food Industry, A.I. Laskin, Ed. (Benjamin/Cummings, Menlo Park, 1985) p. 8)*

Immobilized enzymes are now employed in many analytical devices (for example, in immobilized enzyme electrodes for the detection of various chemical substances), and as catalysts in industrial processes. Although an extensive overview of the applications of immobilized enzymes is not our objective here, some current and potential applications are summarized in Tables 2.1 and 2.2. Table 2.1 also includes a few applications of immobilized microbial cells as industrial catalysts, and Table 2.3 summarizes the characteristics of various microbial sensors based on immobilized *living* cells. In these systems, detection requires assimilation of the analyte by the microorganism followed by changes in respiratory activity or the production of an electroactive metabolite. Thus it is clear that whole-cell sensing devices can be much more complex than single-enzyme electrodes.

Immobilized whole cells represent another type of immobilized biocatalyst. Immobilized cells may be employed in a greater variety of ways than immobilized enzymes. In the simplest case (for example, the immobilized cell systems listed in Table 2.1), dead cells are immobilized for the activity of a single enzyme, thereby avoiding costly purification of the enzyme prior to immobilization. More complicated situations are possible, however, as illustrated by the microbial sensors of Table 2.3. In general, applications of immobilized cells range from single or multistep reactions catalyzed by immobilized nongrowing cells, to complex synthetic transformations catalyzed by actively-metabolizing and dividing immobilized cells. The latter case represents the most elaborate level of immobilized cell catalysis, and enables the *in situ* regeneration of cofactors that may be required for catalysis.

Cofactors are required in approximately one-third of all enzymatic reactions. Cofactors are also quite expensive; thus, it is important that they be recycled, that is, retained and regenerated, in large-scale enzymatic processes. Efficient cofactor regeneration *in vitro* remains a complex challenge, however, and currently represents an obstacle to the industrial application of immobilized enzymes for multi-step reactions.

In single-step reactions catalyzed by immobilized enzymes, cofactor regeneration can be accomplished by several means. One of the more effective strategies involves coupling NAD(H) to a soluble polymer (for example, polyethylene glycol (PEG) or the polysaccharide dextran). The macromolecular polymer-bound cofactor can then be retained, along with the enzymes of interest, in an ultrafiltration apparatus for continuous reuse. Regeneration is achieved through the use of two enzymes: one enzyme utilizes the cofactor in catalyzing the reaction of interest, and the other enzyme regenerates the cofactor in a parallel reaction.

A successful example of polymer-bound coenzyme regeneration in the enzymatic production of amino acids is summarized in Figure 2.2 for the specific case of L-alanine. In this system[1], three enzymes and the PEG-NAD(H) conjugate are retained by an ultrafiltration membrane. (LD)-Lactate is converted to L-alanine via pyruvate by the combined activities of L- and D-lactate dehydrogenases (LDH) and alanine dehydrogenase (ALADH).

(1) C. Wandrey, E. Fiolitakis, U. Wichmann, and M.-R. Kula, "L-Amino acids from a racemic mixture of α-hydroxy acids," Ann. N.Y. Acad. Sci., **434**, 91 (1984).

Table 2.1. *Immobilized enzymes and cells in use or under consideration as industrial catalysts, along with estimated 1992 worldwide production levels of selected products*[*].

Enzyme (and microorganism)	*Application*
Aminoacylase	Optical resolution of DL-amino acids (commercial synthesis of L-amino acids)
Glucose isomerase	Isomerization of glucose to fructose (production of high fructose corn syrup; 8 million tons)
Penicillin amidase	Production of 6-aminopenicilloic acid; 7500 tons (manufacture of semisynthetic antibiotics)
β-Galactosidase	Hydrolysis of lactose to galactose and glucose (treatment of milk and whey)
Lipase	Interesterification of fats
Nitrile hydratase	Production of acrylamide from acrylonitrile; 15,000 tons
L-Aspartate β-decarboxylase (*Pseudomonas dacunhae*)	Production of L-alanine
Aspartase (*Escherichia coli*)	Production of L-aspartic acid
Fumarase (*Brevibacterium ammoniagenes*)	Production of L-malic acid
Aspartic amino transferase (*Escherichia coli*)	Production of L-phenylalanine

[*]Adapted from I. Chibata and T. Tosa, "Immobilized cells: historical background," in Appl. Biochem. Bioeng., **4**, 1 (1983), and E. Katchalski-Katzir, "Immobilized Enzymes: Learning from Past Successes and Failures," in Trends Biotechnol., **11**, 471 (1994).

Feasibility studies achieved continuous operation for more than a month and a space time yield of 134 g L-alanine $(\text{liter-day})^{-1}$. The cycle number (the number of times a cofactor is recycled during its useful lifetime) was estimated at 19,200 mol L-alanine produced/mol NADH deactivated or lost across the membrane. A similar process for L-leucine production resulted in a cycle number of 80,000 for a space-time yield of 211 g L-leucine $(\text{liter-day})^{-1}$.

Table 2.2. *Characteristics of immobilized enzyme electrodes*[*].

Analyte	*Enzyme*	*Sensor*	*Immobilization Procedure*	*Response Time*	*Detection Range*
Galactose	Galactose oxidase	H_2O_2	Covalently bound on collagen membrane	5-6 min (std)[a] 1 min (kin)[b]	5×10^{-7} - 6×10^{-4} M
Ethanol	Alcohol dehydroge-nase	Pt(+350 mV)/ $FeCN_6^{-4}$	Glutaraldehyde/ cellulose triacetate	<5 min	10^{-4} - 5×10^{-3} M
Urea	Urease	O_2	Entrapped; dialysis membrane	NI[c]	10^{-3} - 8×10^{-2} M
Pyruvate	Pyruvate oxidase	O_2	Acetylcellulose	2 min	Up to 8×10^{-4} M
Cholesterol	Cholesterol oxidase	H_2O_2	Covalently bound on collagen membrane	<5 min (std) <1 min (kin)	10^{-7} - 8×10^{-5} M
Lactose	ß-Galactosidase + Glucose oxidase	O_2	Glutaraldehyde and BSA; magnetic film	NI	10^{-3} - 4×10^{-3} M
Aspartame	Chymotrypsin (CT) + alcohol oxidase (AOD)	O_2	Gel of AOD sand-wiched between teflon and retaining membrane; CT in solu-tion	3-5 min	80-800 ppm
Bilirubin	Glucose oxidase + horse radish peroxi-dase	H_2O_2	Entrapped; gelatin dial-ysis membrane	2 min	5×10^{-6} - 5×10^{-5} M

[a]Steady-state response
[b]Kinetic, dynamic response
[c]Not indicated
[*]Adapted from G. Bardeletti et al., "Amperometric Enzyme Electrodes for Substrate and Enzyme Activity Determinations," in Biosensor Principles and Applications, L.J. Blum and P.R. Coulet, eds., Marcel Dekker, Inc., NY, 1991.

Unlike the relatively simple conversion shown in Figure 2.2, many biochemical transformations require several enzymatic reactions with simultaneous cofactor regeneration. For example, the conversion of glucose to ethanol requires twelve enzymatic steps, six of which involve cofactors. The *de novo* synthesis of ajmalacine alkaloids from sucrose by

Catharanthus roseus cells involves thirteen enzymatic steps, the last of which requires NADPH. Clearly, it would be impractical (if not unfeasible) to mimic such a complex system by isolating and immobilizing all of the enzymes involved and devising appropriate cofactor regeneration schemes. The alternative is to employ immobilized whole cells and exploit the inherent coenzyme synthesis and regeneration capacity of the cell. Thus, immobilized cells are presently preferred over immobilized enzymes for reaction sequences that involve several enzymes and coenzymes. Other criteria for choosing one type of biocatalyst or the other are discussed in Section 2.3.2.

Table 2.3. *Characteristics of Microorganism-Based Sensors**.

Analyte	***Immobilized micororganism***	***Device***	***Response time (min)***	***Range (mg dm^{-3})***
Assimilable sugars	*Brevibacterium lactofermentum*	O_2 probe	10	10-200
Glucose	*Pseudomonas fluorescens*	O_2 probe	10	2-20
Acetic acid	*Trichosporon brassicae*	O_2 probe	10	3-60
Ethanol	*Trichosporon brassicae*	O_2 probe	10	2-25
Methanol	unidentified bacteria	O_2 probe	10	5-20
Formic acid	*Citrobacter freundii*	fuel cell	30	10-10^3
Methane	*Methylomonas flagellata*	O_2 probe	2	0-6.6 mmol
Glutamic acid	*Escherichia coli*	CO_2 probe	5	8-800
Cephalosporin	*Citrobacter freundii*	pH electrode	10	100-500
BOD	*Trichosporon cutaneum*	O_2 probe	15	3-60
Lysine	*Escherichia coli*	CO_2 probe	5	10-100
Ammonia	nitrifying bacteria	O_2 probe	10	0.05-1
Nitrogen dioxide	nitrifying bacteria	O_2 probe	3	0.51-255 ppm
Nystatin	*Saccharomyces cerevisiae*	O_2 probe	60	0.5-54 unit/ml
Nicotinic acid	*Lactobacillus arabinosis*	pH electrode	60	10^{-5}-5
Vitamin B_1	*Lactobacillus fermenti*	fuel cell	360	10^{-3}-10^{-2}
Cell numbers	-	fuel cell	15	10^8-10^9 cell/ml
Mutagen	*Bacillus subtilis Rec$^-$*	O_2 probe	60	1.6-2.8x10^3

*From I. Karube, "Microorganism-based Sensors" in Biosensors: Fundamentals and Applications", A. Turner, I. Karube and G.S. Wilson, Eds. Oxford Univ. Press, NY 1987.

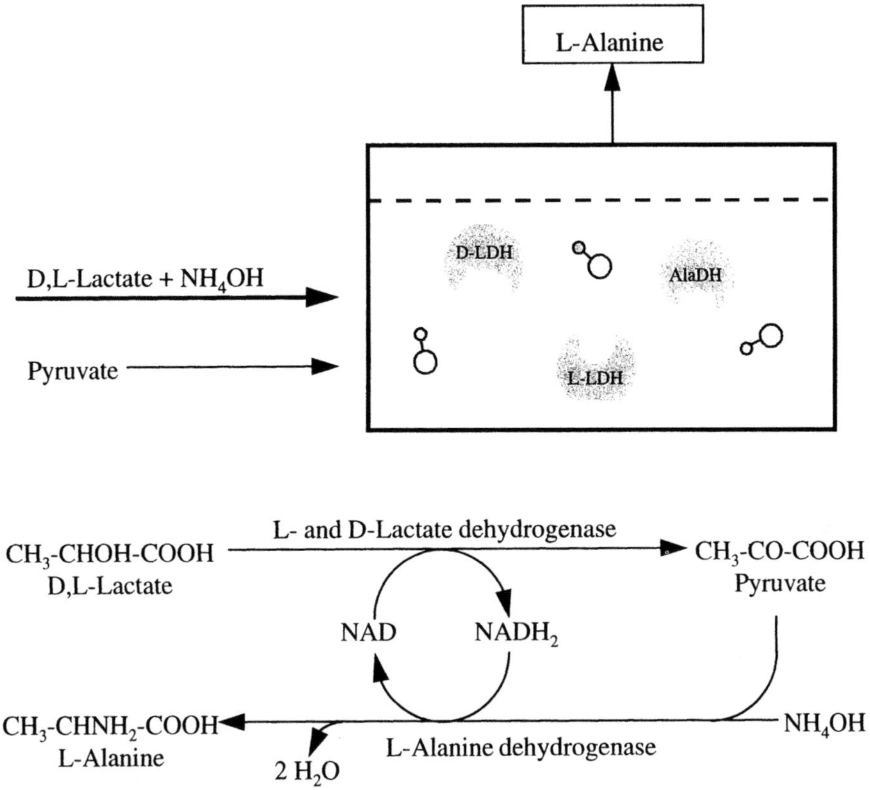

Figure 2.2. *(Top) Simple schematic of membrane reactor for the enzymatic production of L-alanine. Polymer-bound NAD^+ and NADH are represented by the small open and filled circles. (Bottom) More detailed scheme for L-alanine production and cofactor regeneration in the reactor above.*

2.3 Methods of Immobilization

2.3.1 Enzymes

Enzymes can be immobilized to water-insoluble supports by a wide variety of methods, including physical adsorption, covalent bonding or cross-linking, and entrapment. These general methods are depicted in Figure 2.3. The support can be an organic or inorganic material, such as derivatized cellulose or glass, or a membrane. Cellular debris resulting from disruption of the enzyme-producing organism can also serve as a convenient support. In recent years there have been hundreds of papers in the scientific literature describing techniques for immobilizing enzymes (and whole cells as well). We will consider only some of the general characteristics of the major classes of immobilization techniques, and compare their advantages and disadvantages.

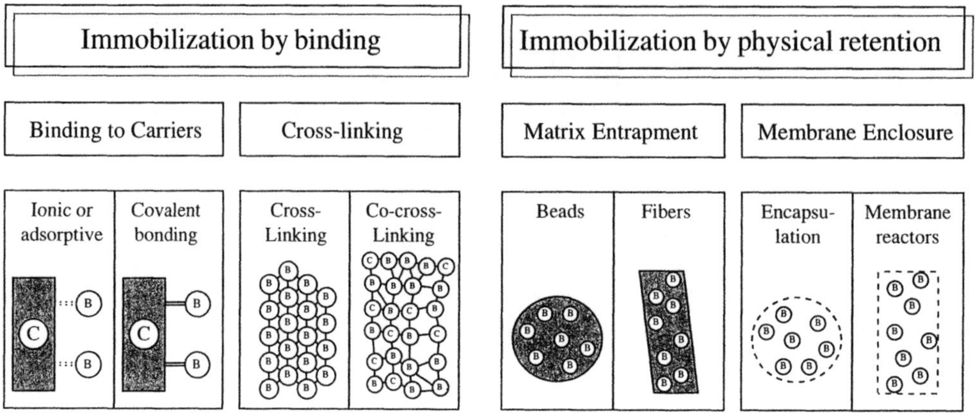

Figure 2.3. *Classification of enzyme immobilization methods;* ***E****: enzyme,* ***C****: carrier.*

Adsorption of enzymes to an insoluble carrier, for example, an ion-exchange resin, is the simplest immobilization technique. The immobilization protocol consists simply of exposing the enzyme in solution to the support under the appropriate conditions. The ease of adsorption, however, can be offset by the ease of desorption, depending on the conditions under which the catalyst is to be used. On the other hand, easy removal of enzyme from the support can be an advantage for regeneration of the catalyst; that is, once the immobilized enzyme has lost its activity, the inactive adsorbed enzyme can often be stripped from the support and subsequently replaced with new enzyme. In the optimal case, this process can be carried out without removing the support itself from the reactor.

Covalently binding an enzyme to a carrier is an alternative to physical adsorption. As polyfunctional macromolecules, enzymes typically contain some combination of hydroxy, mercapto, and amine functionalities that are reactive to a wide variety of common reagents. Thus, enzymes can be covalently coupled through these reactive groups to a suitably activated surface. Some common carriers for covalent immobililzation are summarized in Table 2.4. A variation of direct covalent attachment is crosslinking the enzyme by a multifunctional reagent, for example, glutaraldehyde (Figure 2.4). In this approach, enzyme molecules are extensively crosslinked to each other, either with or without an added support, producing a polyenzyme network (which often resembles gelatin). The main advantage of covalent methods such as these is the strength of the binding; enzyme detachment during use is usually negligible. However, regeneration of the catalyst is not possible, and covalent attachment through multiple sites on the enzyme molecule can lower its activity. In addition, the reagents involved increase the cost of the immobilization procedure.

Table 2.4. *Common carriers for covalent immobilization of enzymes through ε -amino groups of lysine residues.*

Support	Activating or Coupling Agent	Functional Group on Activated Surface
Aminopropyl controlled-pore glass	Glutaraldehyde	-O-Si-$(CH_2)_3$-N=CH-$(CH_2)_3$-CHO
Sepharose (cross-linked agarose)	Cyanogen bromide (CNBr)	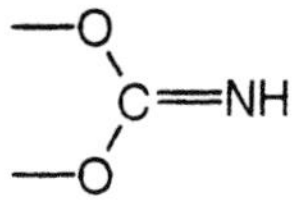
Amberlite IRC-50 (a carboxy resin)	Thionyl chloride ($SOCl_2$)	-COCl
Eupergit (preactivated polyacrylamide)	None	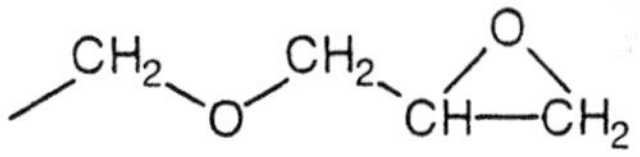

The last type of immobilization method we will consider involves entrapment of the enzyme in a semipermeable membrane or microcapsule. Entrapping an enzyme in a polymeric microcapsule (for example, a microcapsule comprised of polylysine) is often referred to as encapsulation. Entrapment procedures utilize membranes with pores that are small enough to prohibit release of the enzyme but are large enough to permit passage of substrates and products through the membrane. Thus, the enzyme remains in solution and any potentially adverse effects resulting from adsorption or covalent attachment are avoided. Moreover, many variations of this general approach are possible. For example, enzymes can be entrapped within reversed micelles (or inverted micelles). These are formed by adding a small volume of water to a much larger volume of immiscible organic solvent containing a surfactant. The surfactant is oriented at the interface of the two phases so that the hydrophilic head is in contact with the aqueous microphase and the hydrophobic tail is in the bulk organic

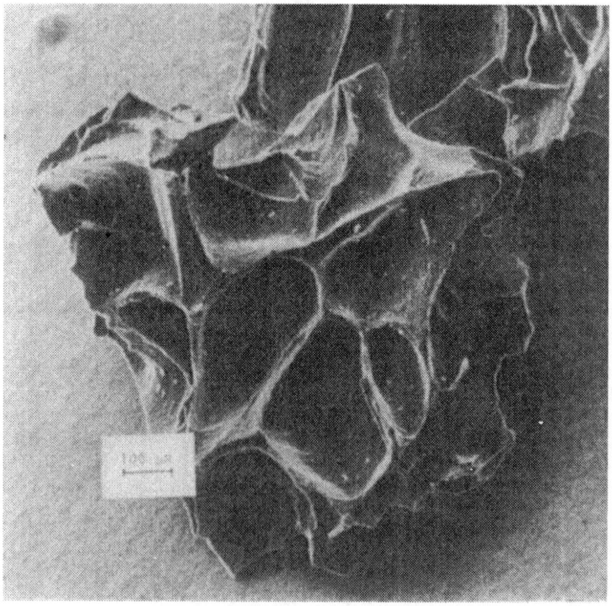

***Figure 2.4**. Scanning electron microscope picture of a Sweetzyme flake, which is a commercial preparation of immobilized glucose isomerase. The flake is manufactured by crosslinking glucose isomerase (along with other proteins) from homogenized cells with glutaraldehyde, followed by freezing, thawing, and drying to shape the catalyst. [From S. H. Hemmingsen, "Development of an immobilized glucose isomerase for industrial application," in Appl. Biochem. Bioeng., **2**, 157 (1979).]*

phase. Reversed micelles allow the enzyme to remain solubilized in the aqueous phase while product and substrate molecules diffuse to and from the surrounding organic phase. This strategy has proven advantageous in cases where the low solubility of a hydrophobic substrate renders the reaction difficult to conduct in a single aqueous phase. Figure 2.5 illustrates the enzymatic production of tryptophan in reverse micelles.

In general, the best immobilization method for a particular enzyme and application will depend on several interrelated parameters. These include the cost of both the enzyme and the support (including necessary reagents), the useful lifetime of the immobilized enzyme (that is, how often the catalyst will need to be regenerated or replaced), the molecular weights of the species involved (for example, are the substrate and product small enough to pass through an entrapment membrane?), and the extent to which the enzyme retains its activity upon immobilization by different means. The potential implications of mass transfer rates are also important, as discussed in Section 2.4. Thus, there is no definitive procedure

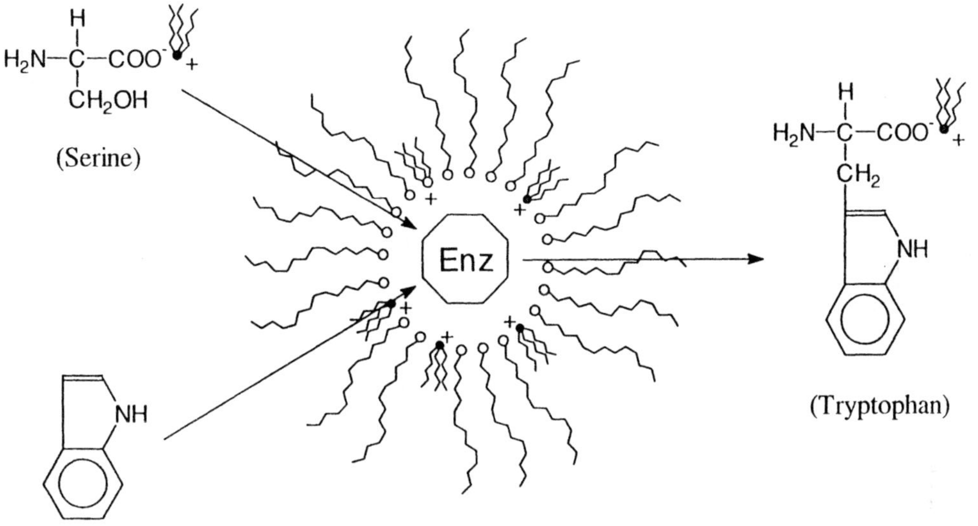

Figure 2.5. *Enzymatic production of tryptophan in reverse micelles (the enzyme tryptophanase is denoted by E). Solubilization and transport of the amino acids to and from the reverse micelle is facilitated by the anion-exchange agent, Aliquat-336, which is represented in the figure as a three-tailed molecule. [From D.K. Eggers and H.W. Blanch, "Enzymatic production of L-tryptophan in a reverse micelle reactor," Bioproc. Eng., **3**, 83 (1988).]*

for choosing the optimal immobilization method. Selection of a suitable immobilization technique usually involves advance consideration of the above factors followed by some amount of trial and error to find an immobilization protocol that gives acceptable results.

2.3.2 Whole Cells

Immobilized cells can carry out more complex reactions than immobilized enzymes due to the far more extensive and sophisticated biochemical capabilities of living cells compared to single enzymes. Whole-cell biocatalysts can also be simpler to prepare. For example, even in cases where dead cells are immobilized for the activity of a single enzyme, isolation of the enzyme from the cell is avoided, simplifying the immobilization procedure and reducing the cost of catalyst preparation. This point is illustrated by Table 2.5, which summarizes the expected properties of immobilized glucose isomerase prepared by different methods. The first method consists of immobilizing glucose isomerase within homogenized bacterial cells; the remaining methods generally involve immobilizing cell-free enzyme following its removal from the cell. Of the three methods in this latter category, only adsorption to an insoluble carrier compares favorably with bacterial-cell immobilization.

Table 2.5. Expected properties of immobilized glucose isomerase prepared by various procedures.

Immobilization procedure	Expressed activity ($\frac{\text{activity observed}}{\text{activity bound}}$)	Potency ($\frac{\text{activity}}{\text{weight}}$)	Carrier cost	Reusability of carrier
Immobilization in bacterial cells	High	Average	Low	No
Adsorption on insoluble carrier	High to Low	High	High	Yes
Entrapped in insoluble matrix	Low	Average to Low	High	No
Covalently bound to insoluble carrier	Low	Low	High	No

(From "Enzyme Technology," Applied Biochemistry and Bioengineering, vol. 2 (L.B. Wingard, Jr., E. Katchalski-Katzir, and L. Goldstein, eds.), Academic Press, New York, 1979, p. 122.)

Immobilized cells are not free of disadvantages, however. For instance, compared to immobilized enzyme systems prepared with purified enzymes, immobilized cells may not have as high a catalytic activity per unit weight. In addition, immobilized cells may contain numerous enzymes that compete for the substrate or react with the desired product. Depending on the type of cell, the nutritional requirements of immobilized living cells can be quite complex, thereby increasing the cost of the feed stream and complicating purification of the product. We should also recognize that recovering a product from an immobilized cell bioreactor generally requires that the product be extracellular, that is, excreted by the cell into the surrounding medium. Nonetheless, immobilized cells represent a very versatile class of biocatalyst, and offer the potential for regenerating both enzymes and cofactors *in situ*. Immobilized living cells are thus well-suited to carry out multi-enzyme, multistep reaction sequences.

Substantial effort has been devoted to developing methods for immobilizing cells. In general, the methods are similar to those for enzyme immobilization and can be categorized as follows:

(1) Carrier-free immobilization (crosslinking or flocculation)
(2) Immobilization onto a pre-formed carrier (adsorption or covalent attachment)
(3) Immobilization in the course of carrier formation (entrapment or encapsulation)

(4) Immobilization within a pre-formed semipermeable membrane (for example, in the extracapillary space of an array of hollow fiber membranes).

Table 2.6 includes some examples of immobilized-cell systems prepared by the above methods, and Figure 2.6 shows pig kidney cells growing on Cytodex microcarriers.

***Table 2.6**. Examples of cells immobilized by various means*[a].

Method	***Support***	***Cells***	***Product***
Floculation	None	*Saccharomyces cerevisiae*	Ethanol
Adsorption	Basic, anionic ion exchanger	*Saccharomyces cerevisiae*	Ethanol
Adsorption	Wood chips	*Saccharomyces cerevisiae*	Ethanol
Entrapment	Polyacrylamide	*Anthrobacter simplex*	Prednisolone
Entrapment	Alginate	*Clostridium acetobutylicum*	Acetone/butanol
Entrapment	Carrageenan	*E. coli*	L-Aspartic acid
Entrapment	Agarose	*Catharanthus roseus*	Ajmalicine
Entrapment	Alginate	*Digitalis lanata*	Digoxin
Entrapment	Alginate	Hybridomas	Monoclonal antibodies
Adhesion to microcarriers	Cytodex	Human fibroblasts	Interferon
Adhesion to microcarriers	Cytodex	Monkey kidney cells	Polio vaccine
Membrane retention	Hollow fibers	*Pseudomonas fluorescens*	L-Histidine

[a]For a more extensive summary of immobilized cell systems and immobilization methods, see *Immobilized Cells and Organelles, Volumes I and II*, B. Mattiasson, ed. (CRC Press, Boca Raton, FL, 1983).

When choosing an immobilization method, the same general considerations apply to whole cells as to single enzymes: the immobilization procedure should be simple and inexpensive, and should yield an immobililzed-cell product with high activity and good operational stability. An important difference, however, is that the preservation of cell viability is often a major goal of cell immobilization. Thus, direct crosslinking of cells to each other by potentially toxic chemicals (for example, with bi- or multifunctional aldehydes via free amino groups in or on the cell walls of bacterial cells) will not be the method of choice for some types of living cells. A gentler and more versatile approach is to entrap cells within the interstitial spaces of a water-insoluble polymeric gel. One of the most popular methods is the immobilization of whole cells by entrapment in calcium alginate, a nontoxic matrix formed by the chaotropic gelation of alginate in the presence of Ca^{2+}. Alginate, which is extracted from seaweed, is a polysaccharide of D-manuronic acid and L-guluronic acid. Other methods of cell entrapment are diagrammed in Figure 2.7.

2.4 Kinetics of Immobilized Enzymes

In this and the following sections we analyze in detail the kinetics of immobilized enzyme-catalyzed reactions. Little more will be said about immobilized whole cells, although many of the concepts described pertain to immobilized biocatalysts in general. The choice to exclude immobilized cells for the time being has been made for the sake of clarity; the discussion will be simplified by referring to single-enzyme reactions only. Further, extension of the following treatment to immobilized living cells involves the use of growth equations introduced in the next chapter. Several problems dealing with the kinetics of immobilized-cell processes are included at the ends of Chapters 3 and 4.

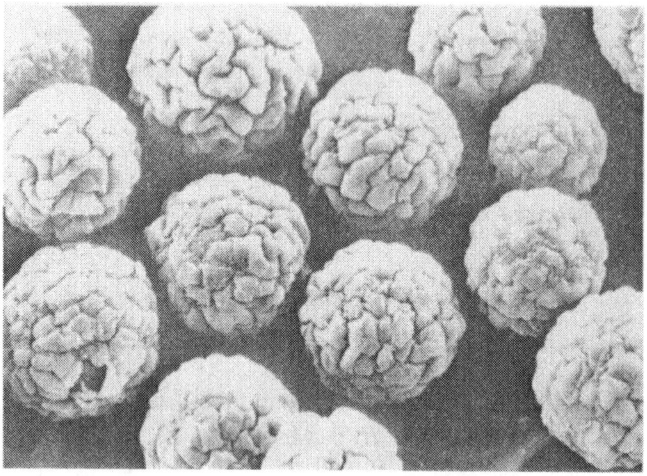

Figure 2.6. *Scanning electron micrograph of pig kidney cells (IBR-S2) growing on Cytodex microcarriers 72 h after inoculation. From M. Hirtenstein and J. Clark, "Microcarrier-bound mammalian cells," in Immobilized Cells and Organelles, Volume I, B. Mattiasson, ed. (CRC Press, Boca Raton, FL, 1983).*

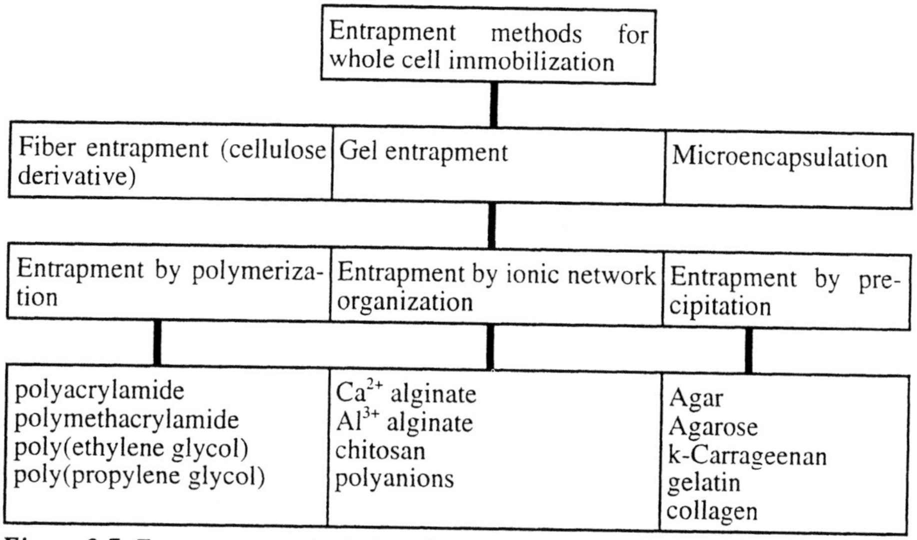

Figure 2.7. *Entrapment methods for whole-cell immobilization.*

Designing a reactor for an immobilized enzyme process requires information on the kinetics of the immobilized enzyme reaction. Knowing the kinetic properties of the immobilized enzyme enables one to predict how changes in the operating conditions will affect reactor performance. Determining such kinetic behavior is an interesting challenge, however, because many factors can cause the kinetic parameters of immobilized enzymes to differ from those of soluble enzymes. These different effects can be classified as follows:

(1) *Conformational effects*: the conformation of the enzyme may be altered by immobilization.

(2) *Electrostatic and partitioning effects*: the concentration of important chemical species (for example, hydrogen ions, substrate molecules, and product molecules) in the immediate environment of the immobilized enzyme may be different from their concentration in the bulk solution due to the physicochemical properties of the support.

(3) *Diffusional, or mass-transfer effects*: the observed kinetics of the immobilized enzyme may not be governed solely by interactions between the enzyme and substrate, but instead may be limited to some extent by the rate of substrate diffusion to the external surface of the support, and/or by the rate of substrate diffusion through the internal pores of the support.

Figure 2.8 depicts the general situation of an enzyme immobilized to a porous, charged support, and illustrates the transport processes involved in immobilized enzyme catalysis.

In cases where effect (1) is significant, the true, or *intrinsic*, kinetic parameters of the immobilized enzyme may differ from those of the soluble enzyme. Thus, when we refer to intrinsic kinetic parameters of an immobilized enzyme (for example, v_{max} or K_m), it must be remembered that these parameters may be different from those of the soluble enzyme, irrespective of partitioning or diffusional effects. Unfortunately, the results of effect (1) cannot be quantitatively predicted, nor is there a theoretical foundation that enables one to calculate catalytic activities for different enzyme conformations. However, effects (2) and (3) can be analyzed by applying basic principles of physical chemistry and chemical reaction engineering, and both of these effects can have a profound influence on the observed activity of immobilized enzymes. Furthermore, once the effects of partitioning and mass tranport have been assessed, the effects of immobilization on the enzyme's intrinsic kinetics can be evaluated.

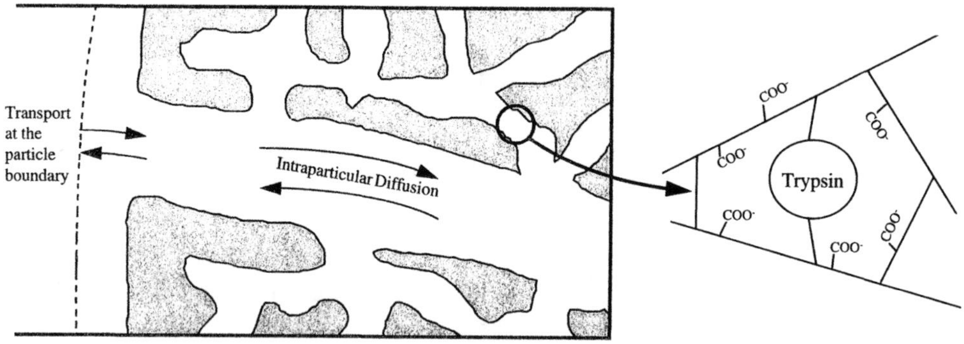

Figure 2.8. Schematic diagram of a porous, charged immobilized trypsin catalyst.

2.4.1 Effects of the Electrostatic Potential of the Microenvironment

pH-Activity Profile[2]

Let us begin our analysis of immobilized enzyme kinetics by considering an enzyme immobilized within a porous support that contains ionized functional groups, for example, the negatively-charged carboxyl groups of a cation exchange resin. For simplicity, neither external nor intraparticle mass transfer effects will be considered. Our aim, then, is to

(2) This analysis was first presented in L. Goldstein, Y. Levin, and E. Katchalski, "A water-insoluble poly-anionic derivative of trypsin. II. Effect of the polyelectrolyte carrier on the kinetic behavior of the bound trypsin," Biochemistry, 3, 1913-1919 (1964).

determine how the electrostatic potential of the charged support affects the local concentration of hydrogen ions, which in turn affects the observed pH-activity profile of the immobilized enzyme.

A single particle of immobilized enzyme suspended in an aqueous medium can be treated as a two-phase system in equilibrium: one phase consisting of the catalyst particle and its immediate microenvironment, and the bulk phase corresponding to the bulk solution (Figure 2.8). An average electrostatic potential, Ψ, is assumed to prevail throughout the particle ($\Psi = 0$ for an uncharged support). The equilibrium condition requires that the electrochemical potential of the hydrogen ions in the particle, $\tilde{\mu}_i^{H^+}$ (Joules/mole) equals that of the hydrogen ions in the bulk phase, $\tilde{\mu}_0^{H^+}$:

$$\tilde{\mu}_i^{H^+} = \tilde{\mu}_0^{H^+} \tag{2.1}$$

Relating the electrochemical potentials to the corresponding chemical potentials, $\mu_i^{H^+}$ and $\mu_0^{H^+}$, gives

$$\mu_i^{H^+} = \mu_0^{H^+} - ZN_{Avo}\varepsilon\Psi \tag{2.2}$$

where ε designates the positive electronic charge of the hydrogen ion (coulombs), Z is the valence (including the sign) and N_{Avo} is Avagadro's number. Rewriting the chemical potentials in terms of the hydrogen-ion activities in the two phases, $a_i^{H^+}$ and $a_0^{H^+}$, we obtain

$$\ln a_0^{H^+} - \ln a_i^{H^+} = \frac{Z\varepsilon N_{Avo}\Psi}{RT} \tag{2.3}$$

(recall that, in general, $\mu = \mu^0 + RT\ln a$). Since $-\log a^{H^+} = \text{pH}$, the difference between the pH of the immobilized-enzyme phase, pH_i, and the pH of the bulk solution, pH_0, is

$$\Delta pH = pH_i - pH_0 = 0.43\frac{Z\varepsilon N_{Avo}\Psi}{RT} \tag{2.4}$$

Thus, depending on the magnitude of Ψ, the pH of the microenvironment of the immobilized enzyme will differ from the pH of the bulk solution. In other words, the hydrogen-ion concentration within a negatively charged support ($\Psi < 0$) will be higher than in the bulk solution, whereas in a positively charged support ($\Psi > 0$) the concentration will be lower than in the bulk solution. An important consequence of this pH difference is that the pH-activity profile of an immobilized enzyme will tend to shift toward more alkaline values when a negatively charged carrier is used, and toward more acidic values when a positively charged carrier is used. The magnitude of this shift is given by Eq. (2.4).

Apparent Michaelis Constant

Now consider a similar situation in which the substrate of the enzymatic reaction bears a charge $Z\varepsilon$ (where Z is a positive or negative integer). Again, diffusional effects are assumed to be negligible. The distribution of the charged substrate between the particle and the bulk phase is then

$$S_i = S_0 e^{-Z\varepsilon\psi/k_B T} \tag{2.5}$$

where S_i and S_0 denote the local (intraparticle) and bulk substrate concentrations, respectively (S_i is the internal concentration expressed per volume of the surrounding *fluid*), and k_B is the Boltzman constant ($k_B = R/N_{Avo}$, erg/degree). Equation (2.5) can be obtained in analogous fashion to Eq. (2.4) by substituting substrate concentrations for the corresponding hydrogen-ion activities and assuming ideal behavior.

Assuming that the reaction catalyzed by the immobilized enzyme obeys Michaelis-Menten kinetics, the rate of reaction v *per unit volume of catalyst* (liquid plus solid) can be expressed in terms of the local substrate concentration S_i by

$$v = \frac{v_{max} S_i}{K_m + S_i} \tag{2.6}$$

In Eq. (2.6), v_{max} is the maximum reaction rate per unit volume of catalyst. Thus, k_{cat} of the immobilized enzyme can be obtained by dividing v_{max} by the loading of immobilized enzyme (enzyme per volume of support):

$$k_{cat} = \frac{v_{max}}{E_{imm}} \quad (\text{sec}^{-1}) \quad = \left[\frac{\left(\dfrac{\mu\text{mol substrate converted}}{\text{volume catalyst . sec}} \right)}{\left(\dfrac{\mu\text{mol active enzyme}}{\text{volume catalyst}} \right)} \right]$$

Determining the true value of k_{cat} therefore requires that we know the concentration of *active* immobilized enzyme. This quantity can be difficult to measure, however, since a fraction of the immobilized enzyme may be inactive (due to conformational changes, for example) or inaccessible. In some cases (as discussed in Chapter 1 for chymotrypsin), active immobilized enzyme can be quantified by *active-site titration* techniques.

It is useful to express the immobilized enzyme reaction rate in terms of S_0 (since S_i is not readily measurable). The Michaelis-Menten equation then becomes

$$v = \frac{v_{max} S_0 e^{-Z\varepsilon\psi/k_B T}}{K_m + S_0 e^{-Z\varepsilon\psi/k_B T}} \tag{2.7}$$

Equation (2.7) indicates that $v = v_{max}/2$ when $S_0 = K_m \exp(-Z\varepsilon\psi/k_B T)$; therefore, the *apparent* Michaelis constant of the immobilized enzyme, $K_{m,app}$, is related to the intrinsic Michaelis constant of the immobilized enzyme, K_m, by

$$K_{m,app} = K_m e^{Z\varepsilon\Psi/k_B T} \tag{2.8}$$

Thus, from Eq. (2.8) it is evident that $K_{m,app} < K_m$ when the substrate and the support are oppositely charged, that is, when the local concentration of substrate is higher than in the bulk solution. On the other hand, when the substrate and the support have the same charge, electrostatic repulsion lowers the local concentration of substrate, and $K_{m,app} > K_m$.

2.4.2 Effects of External Mass Transfer

Uncharged Support

We have seen that partitioning effects can alter the microenvironment of the immobilized enzyme and thus influence the overall reaction rate. In addition, transport of substrate between the bulk fluid and the immobilized enzyme can also affect the reaction rate, and the ramifications of *external* mass transfer will now be considered. Suppose the enzyme is immobilized to the surface of an uncharged, nonporous particle and the entire surface is uniformly accessible. The rate at which substrate molecules can be transported from the bulk solution to the catalyst surface is, of course, finite, and in some cases the limited transport rate will lead to a concentration difference across a stagnant film surrounding the catalyst. In this situation, the average flux N_s of substrate to the fluid-solid interface can be written in terms of the concentration difference and an average mass-transfer coefficient k_s:

$$N_s = k_s(S_0 - S^*) \tag{2.9}$$

where S^* and S_0 are the substrate concentrations (e.g. mmol/cm^3) at the solid-liquid interface and in the bulk fluid, respectively. The mass transfer coefficient k_s (cm/sec) can often be determined from well-established empirical correlations that include the physical properties of the fluid and the flow conditions near the catalyst.

At steady state, the enzymatic reaction rate must be exactly balanced by the rate of substrate transport to the catalyst surface; therefore,

$$v' = \frac{v'_{max} S^*}{K_m + S^*} = k_s(S_0 - S^*) \tag{2.10}$$

where v' and v'_{max} are rates of reaction *per unit surface area of catalyst*. Equation 2.10 can be cast into dimensionless form by introducing the following dimensionless variables

$$x^* = \frac{S^*}{S_0}; \quad \nu = \frac{K_m}{S_0}; \quad Da = \frac{v'_{max}}{k_s S_0} \tag{2.11}$$

Thus we can write

$$v' = \frac{v'_{max} x^*}{x^* + \nu} \tag{2.12}$$

$$\frac{x^*}{x^* + \nu} = \frac{1 - x^*}{Da} \tag{2.13}$$

where Da, an important dimensionless group known as the Damköhler number, can be assigned the following physical interpretation:

$$Da = \frac{\text{maximum reaction rate}}{\text{maximum flux through the diffusion layer}} \tag{2.14}$$

Therefore, if Da << 1, the maximum enzymatic reaction rate is much smaller than the maximum rate of mass transfer to the surface, and the system is known to operate in the *reaction-limited* regime. On the other hand, if Da >> 1, the maximum rate of substrate diffusion to the surface is much smaller than the maximum rate of substrate consumption, and the *observed* reaction rate (also known as the *effective* or *global* reaction rate) is said to be *diffusion-limited.*

Equation 2.12 expresses the observed reaction rate in terms of the dimensionless substrate concentration at the catalyst surface. An analogous expression can be written for the case of no diffusional limitations; that is, when $S^* = S_0$ and hence $x^* = 1$:

$$v'\big|_{S^*=S_0} = \frac{v'_{max} S_0}{K_m + S_0} = \frac{v'_{max}}{1+\nu} \tag{2.15}$$

Therefore, the observed rate can be related to the rate that would be obtained in the absence of external diffusional limitations by

$$v' = \eta_E \frac{v'_{max}}{1+\nu} \tag{2.16}$$

where η_E is known as the *external effectiveness factor,*[3] defined as

$$\eta_E = \frac{\text{observed reaction rate}}{\text{rate evaluated at bulk substrate concentration So}} \tag{2.17}$$

In terms of dimensionless quantities (from Eqs. (2.12) and (2.16)):

$$\eta_E = \frac{(1+\nu)x^*}{x^* + \nu} \tag{2.18}$$

The external effectiveness factor is a numerical measure of the influence of external mass transfer resistance on the observed reaction rate. (We will see shortly that there is an analogous parameter to describe the influence of internal mass transfer resistance.) If η_E << 1, mass-transfer resistance is restricting the supply of substrate to the surface and is thus limiting the catalytic activity of the immobilized enzyme (or in like fashion, an immobilized cell). In contrast, if $\eta_E \cong 1$, the substrate concentration at the surface of the immobilized

(3) The effectiveness factor can be defined with respect to the bulk substrate concentration, S_0, or with respect to the concentration at the catalyst surface, S^*. Both definitions are employed in the chemical and biochemical engineering literature, but they are not always clearly distinguished. We have adopted the nomenclature *external effectiveness factor* and *internal effectiveness factor* to distinguish one definition from the other.

biocatalyst is essentially the same as that in the surrounding bulk, and the reaction rate is not limited by external mass transfer. Note that these two limiting cases are the same as those discussed above in connection with the Damköhler number.

The general dependence of η_E on Da for different values of β, a dimensionless bulk concentration defined as $\beta = 1/\nu = S_o/K_m$, is shown in Figure 2.9. Note that Figure 2.9 plots η_E against the product of β and Da. Here we should point out that $\beta \cdot$ Da, which is equal to $v'_{max}/(k_s K_m)$, can be viewed as an alternative definition of the Damkohler number. In fact, $v'_{max}/(k_s K_m)$ has been assigned the symbol Da in several previous descriptions of external mass transfer resistance in immobilized enzyme catalysis. Thus, one must pay close attention to how Da has been defined when using plots of the kind shown in Figure 2.9.

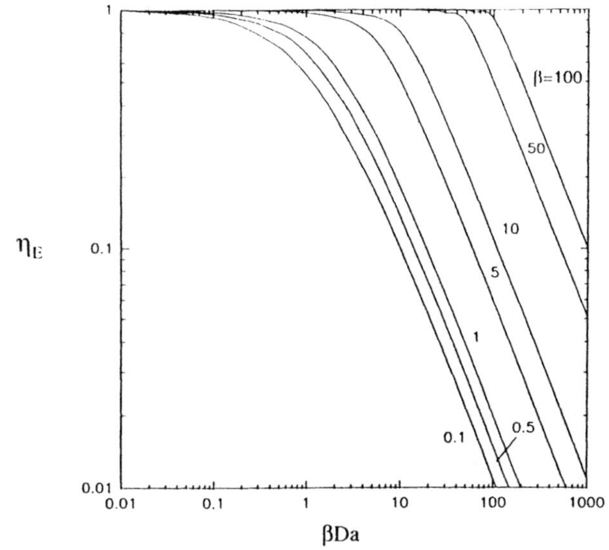

Figure 2.9*. External effectiveness factor, η_E, as a function of $\beta \cdot Da$ for different values of the dimensionless bulk substrate concentration β (S_o/K_m).*

The relationships between η_E and Da for the limiting regimes of kinetic and diffusion control are easily derived. For example, in the diffusion-limited regime ($Da \gg 1$, ν finite), it follows from Eqs. (2.13) and (2.18) that

$$\eta_E = \frac{1+\nu}{Da} \tag{2.19}$$

Therefore, from Eq. (2.16) and the definition of Da [Eq. (2.11)], the observed rate of reaction is

$$v'|_{Da \gg 1} = k_s S_0 \tag{2.20}$$

Physically, Eq. (2.20) implies that the enzymatic reaction rate is so fast the substrate concentration at the catalyst surface approaches zero. The reaction will then appear to follow first-order kinetics, even though the intrinsic rate expression is of Michaelis-Menten form.

At the other extreme, the overall rate is reaction limited when $Da \ll 1$, and Eqs. (2.13), (2.16), and (2.18) indicate that

$$\eta_E|_{Da \ll 1} = 1 \tag{2.21}$$

and

$$v'|_{Da \ll 1} = \frac{v'_{max}}{1 + \nu} \tag{2.22}$$

Therefore, to maximize the effectiveness of a nonporous catalyst, operating conditions should be chosen to minimize Da. One strategy for minimizing Da consists of maximizing k_s by employing a high liquid flow rate past the catalyst particles (a relationship between k_s and the flow rate is given in the example beginning on p. 138). A value of Da much less than unity also ensures that the observed reaction rate reflects the intrinsic kinetics of the immobilized enzyme catalyzed reaction. Consequently, one can determine whether immobilization has altered the intrinsic kinetic properties of the enzyme (an important criterion for evaluating an immobilization method) by performing rate measurements under conditions where Da << 1 and $\eta_E \cong 1$. Otherwise, diffusional limitations will disguise the true catalytic behavior of the immobilized enzyme. Such effects are explored further in problems at the end of this chapter.

Figure 2.9 illustrates that the effectiveness factor of the catalyst can be determined once both Da and β are known. However, evaluating these parameters requires knowledge of the intrinsic-rate parameters v'_{max} and K_m, which are not directly obtainable from measurements of diffusion-limited reaction rates. Thus, at first it appears that the effectiveness factor cannot be determined unless the intrinsic reaction rate is already known. Fortunately, there is an alternative approach for determining η_E that is based on an *observable Damkohler number*, $\overline{Da}$, based on v' and not v_{max}, defined as

$$\overline{Da} = \frac{v'}{k_s S_O} \tag{2.23}$$

Note that $\overline{Da}$ contains the measured overall rate but no intrinsic rate parameters. Thus, $\overline{Da}$ is much more useful than Da for determining effectiveness factors from measured reaction rates. Combining Eqs. (2.23) and (2.16) with the definition of Da from Eq. (2.11) yields

$$\overline{Da} = \frac{\eta_E v'_{max}}{(1+\nu)k_s S_O} = \frac{\eta_E Da}{(1+\nu)} = \frac{\eta_E Da\,\beta}{(1+\beta)} \tag{2.24}$$

Plots of η_E versus $\overline{Da}$ are given in Figure 2.10.

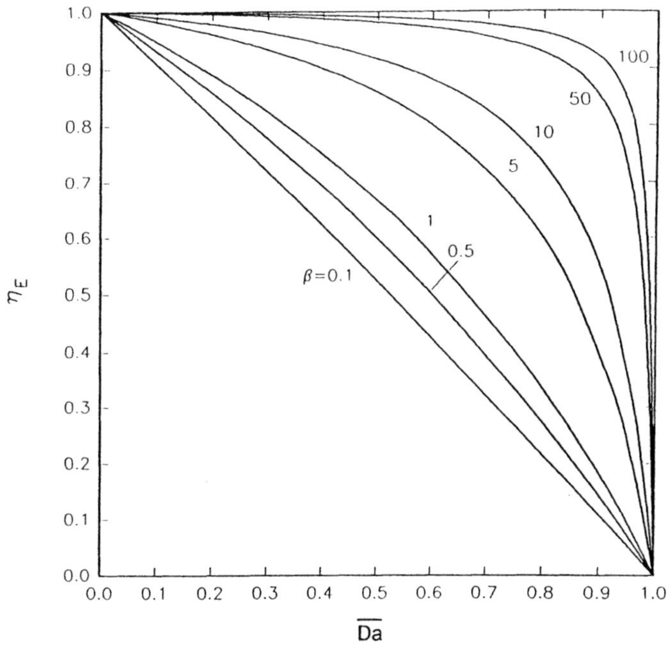

Figure 2.10. *The external effectiveness factor in terms of the observable Damkohler number defined in Eq. (2.23). These curves were generated using Eq. (2.24) and the plots of* η_E *versus* $\beta \cdot Da$ *shown in Figure 2.9.*

Charged Support

We shall now briefly consider transport effects on a reaction involving a charged substrate and enzyme immobilized to the external surface of a planar, charged support (note that here we have specified the geometry of the catalyst). Such a situation is considerably more complex than any of the cases examined so far, and the corresponding equations describing the coupling of mass transfer and chemical reaction are more complicated. Therefore, we will highlight the major results without focusing on all of the intermediate algebraic details. A more complete treatment can be found in an analysis of Shuler et al.[4]

The steady-state molar flux of a charged substrate to a planar surface that has an electrostatic potential of Ψ_0 can be written as

$$N_s = D_s \frac{dS}{dz} + S \frac{ZD_s F}{RT} \frac{d\Psi}{dz} \tag{2.25}$$

(4) *M.L. Shuler, R. Aris, and H.M. Tsuchiya, J. Theor. Biol., 35, 67 (1972).*

where the first term represents Fickian diffusion and the second is the migration due to the gradient of the electrostatic potential in the z direction (the planar surface is at z = 0). Here D_s is the molecular diffusivity of the substrate, $\Psi(z)$ is the electrostatic potential, T is the temperature, Z is the valence of the substrate (including the sign), F is Faraday's constant, and R is the gas constant. Note that at steady state, $dN_s/dz = 0$; thus, N_s is constant. To solve Eq. (2.25), we multiply both sides by the integrating factor $[\exp(\lambda\Psi(z)/\Psi_o)/D_s]$ and integrate from z = 0 to z = δ, the boundary of the diffusional film surrounding the catalyst. Integration and subsequent manipulation yields

$$N_s = \frac{MD}{\delta}(S_0 - S^* e^{\lambda}) \tag{2.26}$$

where λ is the dimensionless potential defined as

$$\lambda = \frac{ZF\Psi_0}{RT} \tag{2.27}$$

and

$$M^{-1} = \frac{1}{\delta}\int_0^{\delta} \exp\left(\frac{\Psi(z)}{\Psi_0}\right) dz \tag{2.28}$$

The integral in Eq. (2.28) can be evaluated by specifying $\Psi(z)$ (for example, if an exponential decline with z is assumed) and noting that M = 1 when $\Psi(z) = 0$. Equating the flux of substrate to the surface (Eq. (2.26)) and the enzymatic reaction rate yields the steady-state relation

$$N_s = Mk_s(S_0 - S^* e^{\lambda}) = \frac{v'_{max} S^*}{K_m + S^*} \tag{2.29}$$

where we have assumed that the mass transfer coefficient, k_s, is equal to (D_s/δ). Using the dimensionless parameters x^* and ν [Eq. (2.11)] and defining a modified Damköhler number, Da_M, as

$$Da_M = \frac{v'_{max}}{Mk_s s_0} \tag{2.30}$$

leads to the dimensionless form of Eq. (2.29):

$$(1 - x^* e^{\lambda}) = \frac{Da_M x^*}{(x^* + \nu)} \tag{2.31}$$

Eqs. (2.31) and (2.18) can be combined to obtain η_E as a function of ν, λ, and Da_M:

$$\eta_E = f(\nu, \lambda, Da_M) \tag{2.32}$$

Substitution of Eqn (2.32) into Eq. (2.16) leads to an apparent Michaelis constant, $K_{m,app}$, that accounts for *both* electrostatic and external mass-transfer effects:

$$K_{m,app} = K_m e^{\lambda}\left\{1 + \frac{v'_{\max}}{Mk_s(S_0 + K_m e^{\lambda})}\right\} \tag{2.33}$$

Thus, the intrinsic Michaelis constant of an enzyme immobilized to a charged planar support will generally be disguised by electrostatic interactions and/or external mass-transfer resistance. Moreover, bear in mind that an apparent Michaelis constant determined under such circumstances is simply a convenient empirical parameter; it has no general validity and is subject to the conditions of the experiment. Consequently, an apparent Michaelis constant measured in one situation will not necessarily apply to another.

2.4.3 *Effects of Intraparticle Diffusion with Uncharged Supports*

Michaelis-Menten Kinetics

So far, our discussion of mass transfer effects has been limited to catalysts consisting of enzyme immobilized to the external surface of a nonporous support. In practice, enzymes are often immobilized to porous materials with large internal surface areas, thereby increasing the amount of immobilized enzyme per unit weight of support. Reactant molecules are converted to product at the interface between the solid (i.e., the support) and fluid phases, and a larger surface area for enzyme immobilization corresponds to a larger interface for reaction. Enzyme immobilized to the internal surface of a porous particle is shown in Figure 2.8.

For reactant molecules to come in contact with immobilized enzyme in the interior of the catalyst, they must first diffuse through the pore structure of the catalyst (Figure 2.7). In most cases, diffusion through the catalyst cannot be characterized by the same difffusion coefficient that applies to *ordinary* or *bulk diffusion* for the following reasons: only a fraction of the catalyst cross section is actually comprised of pores (the remainder is solid); the pore geometry is very complex and individual pores frequently change in size, shape, and direction; the substrate molecules and pores may be of comparable dimensions. All of these effects are typically incorporated into a single diffusion coefficient, the *effective diffusivity*, based on the total cross section of porous material (*void plus solid*) normal to the direction of diffusion. The effective diffusivity can be related to the bulk diffusivity D_s by

$$D_{eff} = D_s \frac{\varepsilon_p}{\tau} H \tag{2.34}$$

where ε_p is the porosity of the particle and τ is the tortuosity factor (an empirical parameter ususally assumed to be in the range of 1.4 to 7). Also included in Eq. (2.34) is H, which is a hindrance factor that accounts for steric interactions between the diffusing solute and the pore wall, and the enhanced drag on the solute resulting from the presence of the wall. This parameter need not be included in Eqn (2.34) if the radius of the pore is much greater than that of the solute molecule; however, in many cases of technological or scientific interest the dimensions of the solute (for example, a protein) are of the same order as those of the pore (for example, the pores of an ultrafiltration membrane). In such cases, the phenomena described by H become important and solute diffusion is said to be *hindered* or *restricted.*

Efforts to derive analytical expressions for the hindrance factor date back more than half a century. One of the earliest results of such efforts is the so-called "centerline approximation"[5]:

$$H = (1-\gamma)^2(1-2.1044\gamma+2.089\gamma^3-0.948\gamma^5) \tag{2.35}$$

where γ is the ratio of the solute radius to the pore radius. Equation (2.35) is often referred to as the Renkin equation. The first term, $(1 - \gamma)^2$, corresponds to the partition coefficient of a hard sphere in a cylindrical pore (that is, the ratio of concentration inside the pore to the concentration outside at equilibrium) arising purely from steric interactions between the sphere and the pore wall. The second quantity enclosed by parentheses accounts for the fractional reduction in diffusivity within the pore that results when the solute and pore size are of comparable magnitude. The centerline approximation is reasonably accurate for $\gamma <$ 0.4. Although slightly more accurate expressions are available[6], they are typically restricted to much narrower ranges of γ. Thus, the centerline approximation is still among the most versatile equations for H and is sufficiently accurate for most design purposes.

Having defined the effective diffusivity for diffusive transport through a porous catalyst, we can proceed to analyze the effect of intraparticle diffusion on the overall rate of reaction that takes place in a porous particle containing immobilized enzyme. Several assumptions will be made throughout our analysis: 1) the immobilized enzyme is distributed uniformly throughout the particle, 2) transport of substrate through the catalyst is described by a Fick's law form relating the diffusive flux to the substrate concentration gradient, and the effective diffusivity of substrate is constant, 3) the reaction is isothermal and involves no change of pH, and 4) electrostatic effects are negligible. These assumptions accurately describe many real systems of practical interest; however, any or all of them may not apply in a particular case. Thus, the appropriateness of each assumption must be carefully considered in any application of the following analysis.

The general differential equation for mass transfer in an immobilized enzyme-catalyst particle is

$$\frac{\partial S_i}{\partial t} + \overline{\nabla N_s} = v(S_i) \tag{2.36}$$

(5) J.R. Pappenheimer, E.M. Renkin, and L.M. Borrero, Am. J. Physiol., **167**, 13 (1951).

(6) For a comprehensive review of liquid-phase diffusion in porous media, see W.M. Deen, Hindered Transport of Large Molecules in Liquid-Filled Pores, *AIChE J.*, **33**, 1409 (1987).

where $\overline{\nabla}$ is a vector differential operator[7], $\overline{N_s}$ is the molar flux vector, and $\overline{\nabla N_s}$ is the net rate of molar efflux of the substrate. We will consider the specific case of enzyme immobilized in a spherical particle of radius R. At steady state, $\partial S / \partial t = 0$, and if we assume that diffusion occurs in the radial direction only, the steady-state material balance becomes

$$\frac{d^2 S_i}{dr^2} + \frac{2}{r}\frac{dS_i}{dr} = \frac{v(S_i)}{D_{eff}} = \frac{v_{max} S_i}{(K_m + S_i) D_{eff}} \tag{2.37}$$

Here v_{max} is the volumetric maximum rate of reaction, in contrast to the surface reaction rate employed in our analysis of external diffusion. In terms of the dimensionless internal concentration $x = S_i/S_o$, the dimenionless radial coordinate $\bar{r} = r/R$, and the dimensionless bulk concentration $\beta = S_o/K_m$, a dimensionless form of Eq. (2.37) is

$$\frac{d^2 x}{d\bar{r}^2} + \frac{2}{\bar{r}}\frac{dx}{d\bar{r}} = \frac{R^2 v(S_i)}{D_{eff} S_0} = \frac{R^2 (v_{max}/K_m D_{eff}) x}{1 + \beta x} \tag{2.38}$$

It is evident from Eq. (2.38) that the concentration profile in the porous particle will depend on the size of the particle, the effective diffusivity of substrate through the particle, and the intrinsic kinetic parameters of the immobilized enzyme. The traditional chemical engineering approach is to combine these three factors in a single dimensionless parameter, the *Thiele modulus* , defined for Michaelis-Menten kinetics by[8]

$$\phi = \frac{R}{3}\left(\frac{v_{max}}{K_m D_{eff}}\right)^{1/2} \tag{2.39}$$

---

(7) This operator, known as *del* or *nabla*, is defined in Cartesian coordinates as

$$\overline{\nabla} = \vec{e}_x \frac{\partial}{\partial x} + \vec{e}_y \frac{\partial}{\partial y} + \vec{e}_z \frac{\partial}{\partial z}$$

where $\vec{e}_x$, $\vec{e}_y$, and $\vec{e}_z$ are unit vectors. For spherical coordinates,

$$\overline{\nabla} = \vec{e}_r \frac{\partial}{\partial r} + \vec{e}_\theta \left(\frac{1}{r}\right)\frac{\partial}{\partial \theta} + \vec{e}_\phi \left(\frac{1}{r\sin\theta}\right)\frac{\partial}{\partial \phi}$$

The molar flux will vary only with radial position, however, so we can neglect the partial derivatives with respect to θ and ϕ .

(8) Alternatively, the Thiele modulus for Michaelis Menten kinetics is sometimes written as

$$\phi = \frac{R}{3}\left(\frac{v_{max}}{D_{eff}(K_m + S_o)}\right)^{1/2}$$

The physical significance of ϕ is essentially the same for both expressions, but one should be aware, as with the Damkohler number, that different definitions do exist. In this chapter and throughout this textbook we will define ϕ by Eq. (2.39).

The above definition of ϕ includes a characteristic length, R/3, which is defined for all geometries as V_p/S_x, where V_p is the gross volume of the catalyst particle and S_x is the gross exterior surface area.

Note that the physical interpretation of ϕ^2 is analogous to that of the Damkohler number, that is, the square of the Thiele modulus is the ratio of an intrinsic chemical-reaction rate in the absence of mass transfer limitations to the rate of diffusion through the particle. This similarity suggests that the Thiele modulus can be related to an effectiveness factor defined for coupled intraparticle diffusion-reaction (likewise, the external effectiveness factor depends on the Damkohler number).

Combining Eqs. (2.38) and (2.39) gives the following dimensionless differential equation for the concentration of substrate throughout the particle:

$$\frac{d^2x}{d\bar{r}^2}+\frac{2}{\bar{r}}\frac{dx}{d\bar{r}}=\frac{R^2v(S_i)}{D_{eff}S_0}=\frac{9\phi^2x}{1+\beta x} \tag{2.40}$$

If the rate of mass transfer through the external film is fast relative to the rate of intraparticle diffusion (low Da), the reactant concentration at the external surface of the catalyst particle will equal that in the bulk fluid. Under these circumstances, the boundary conditions for Eq. (2.40) are

$$x\,|_{\bar{r}=1}=1; \quad \frac{dx}{d\bar{r}}\,|_{\bar{r}=0}=0 \tag{2.41}$$

In writing the first boundary condition, we have assumed that $S_0 = S^* = S_i^*$, where S_i^* is the substrate concentration just within the particle surface. In other words, we have assumed that there is no partitioning of substrate at the solid-liquid interface. This assumption will be modified in Section 2.4.4. The second boundary condition implies that there can be no diffusive flux through the center of the catalyst and applies regardless of the substrate concentration at the catalyst's surface.

Eqs. (2.40) and (2.41) present a nonlinear boundary value problem that can be solved numerically to obtain the radial concentration profile of S as a function of ϕ and β. Figure 2.11 shows the substrate concentration profile, S_i/K_m (= $\beta \cdot x$), within a sphere for different values of ϕ when β is unity.

Once the concentration profile is known, the *observed* overall rate of reaction per unit volume of catalyst can be determined from the integral equation

$$v_{obs}=3\int_0^1 \bar{r}^2\frac{v_{max}x(\bar{r})}{x(\bar{r})+\beta^{-1}}d\bar{r} \tag{2.42}$$

As in our treatment of external mass-transfer effects, an important objective in our analysis of internal mass-transfer resistance is to determine the effectiveness of the immobilized enzyme. An effectiveness factor (note the difference between the definition of η_I is introduced and the definition of η_E given by Eq. (2.16)):

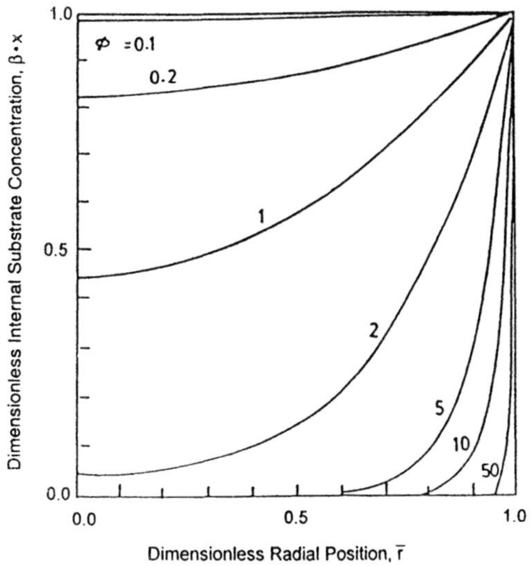

***Figure 2.11**. Concentration profiles within a spherical catalyst particle for various values of ϕ and $\beta = 1$.*

$$\eta_I = \frac{\text{observed reaction rate}}{\text{rate in the absence of intrapellet concentration gradients}} \tag{2.43}$$

As might be expected from the dependence of S(r) on ϕ and β (Eq. 2.40), η_I is also dependent on these two parameters. This dependence is made clear by expressing η_I as the rate of substrate transport into the catalyst (which equals v_{obs}) divided by the reaction rate evaluated at $S = S_o$:

$$\eta_I = \frac{D_{eff}S_x(ds/dr\,|_{r=R})}{V_p[v_{max}S_o/(S_o+K_m)]} = \frac{D_{eff}\left(\frac{3}{R}\right)(ds/dr\,|_{r=R})}{[v_{max}S_o/(S_o+K_m)]} = \frac{3D_{eff}(dx/d\bar{r}\,|_{\bar{r}=1})S_o}{R^2[v_{max}S_o/(S_o+K_m)]} = \frac{3K_mD_{eff}(dx/d\bar{r}\,|_{\bar{r}=1})}{R^2v_{max}/[1+(S_o/K_m)]}$$

Therefore, in general,

$$\eta_I = \frac{3K_mD_{eff}(dx/d\bar{r}\,|_{\bar{r}=1})}{R^2v_{max}[1/(1+\beta)]} = \frac{(dx/d\bar{r}\,|_{\bar{r}=1})}{3\phi^2[1/(1+\beta)]} \tag{2.43a}$$

Plots of η_I versus ϕ for different values of β are shown in Figure 2.12, based on numerical solutions of Eqs. (2.40) and (2.42), and the definition of η_I given by Eq. (2.43a).

As β approaches zero ($S_o << K_m$), η_I converges to the effectiveness factor of the corresponding first-order reaction, η_I^1. In this case, the diffusion-reaction model can be solved analytically, and η_I^1 is given by

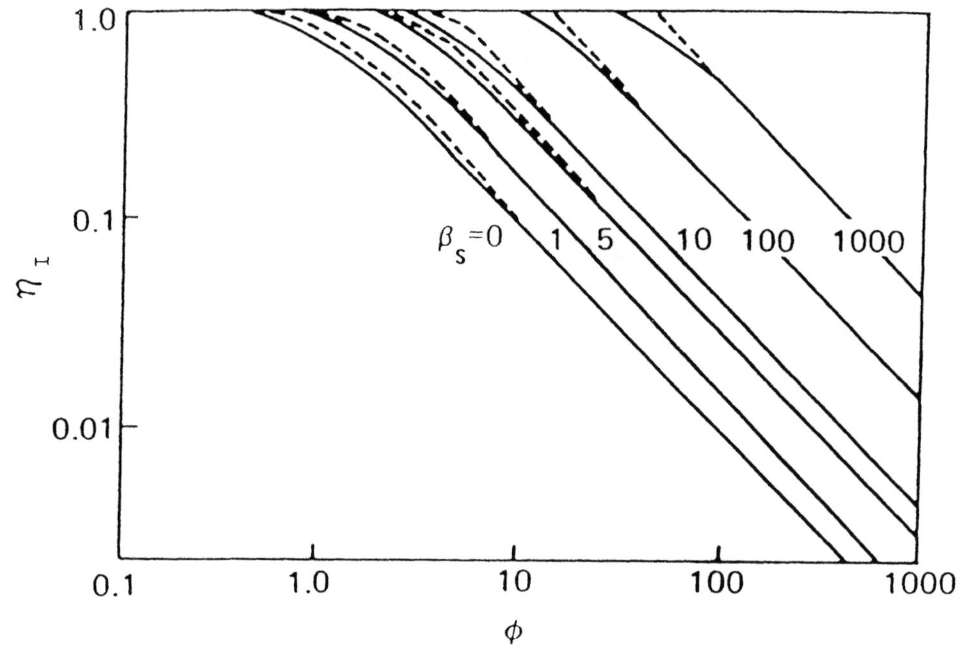

Figure 2.12. *Effectiveness factor, η_I, versus the Thiele modulus, ϕ, for various values of the dimensionless substrate concentration β. The different curves correspond to different values of the dimensionless substrate concentration at the surface of the catalyst, given by β. The effectiveness factors for spherical and slab geometries are represented by the solid and dashed lines, respectively.*

$$\eta_I^1 = \frac{1}{\phi}\left[\frac{1}{\tanh 3\phi} - \frac{1}{3\phi}\right] \tag{2.44}$$

When ϕ is sufficiently small, (slow reaction, rapid diffusion), the effectiveness factor is practically unity for all values of β. In this situation, reactant molecules are able to diffuse far into the pore structure before they are converted to product; thus, the concentration of substrate is nearly uniform throughout the catalyst and the entire accessible surface area effectively promotes reaction. On the other hand, when ϕ is large, reactant molecules are consumed before they diffuse very far, and the reaction is limited to a thin region near the periphery of the particle. The concentration of substrate in the core of the catalyst is very low and η_I is small.

Several other features of Figure 2.12 are worth noting. For example, for each value of ϕ the effectiveness factor increases with increasing β because the enzymatic reaction rate becomes less sensitive to substrate concentration as the kinetics approach zero order (recall the two limiting forms of the Michaelis-Menten equation). Also apparent from Figure 2.12

is that the effectiveness factor of a spherical particle is less than or equal to that of a slab for all values of ϕ . (In the definition of ϕ for a slab, R/3 is replaced by the characteristic dimension of the slab.) The lower effectiveness factor of a sphere relative to a slab is a result of the sphere's lower external surface area to volume ratio.

The effectiveness factor of the catalyst can be determined from Figure 2.12 once both ϕ and β are known. However, as we noted previously in our discussion η_E and its relation to the Damkohler number, this approach requires knowledge of the intrinsic-rate parameters v_{max} and K_m. Since these parameters are not directly obtainable from diffusion-disguised reaction rates, we adopt the same strategy as before and define an *observable modulus*, Φ, which for spherical geometry can be written as[9]

$$\Phi = \frac{v_{obs}}{D_{eff}S_0}\left(\frac{R}{3}\right)^2 \tag{2.45}$$

Like the observable Damkohler number, $\overline{Da}$, Φ contains only the measured overall rate and not intrinsic rate parameters.

Setting $S = S_0$ (i.e., x = 1) in the right-hand side of Eqn (2.40) and rearranging yields

$$\frac{v(S_0)}{D_{eff}S_0}\left(\frac{R}{3}\right)^2 = \frac{\phi^2}{1+\beta} \tag{2.46}$$

Using Eq. (2.46) and the definitions of η_I (Eq. 2.43) and Φ (Eq. 2.45), an independent equation relating η_I, ϕ, β, and Φ is obtained:

$$\eta_I = \frac{\text{observed reaction rate}}{\text{rate in the absence of intrapellet concentration gradients}}$$

$$= \frac{v_{obs}}{[v_{max}S_o/(K_m+S_o)]} = \frac{(R/3)^2(v_{obs}/K_m)\left(\frac{K_m+S_o}{S_o}\right)}{(v_{max}/K_m)(R/3)^2} = \frac{\left(\frac{R}{3}\right)^2\left(\frac{v_{obs}}{D_{eff}K_m}\right)\left(\frac{K_m+S_o}{S_o}\right)}{\left(\frac{R}{3}\right)^2\left(\frac{v_{max}}{D_{eff}K_m}\right)}$$

Therefore,

$$\eta_I(\phi,\beta,\Phi) = \frac{\Phi(1+\beta)}{\phi^2} \tag{2.47}$$

The functional relationship between η_I, ϕ , and β (Figure 2.12) can be combined with Eq. (2.47) to generate plots of η_I versus Φ for different values of β. Plots of the $\eta_I(\Phi)$ relationship for $\beta \to 0$ and $\beta \to \infty$ (first- and zero-order kinetics, respectively) are given in Figure 2.13. Notice that the curves are far less sensitive to β than the analogous $\eta_I(\phi)$ curves of Figure 2.12. In fact, Φ can now be used to establish general criteria for the importance of mass

(9) By convention, the observable modulus is defined analogously to the square of the Thiele modulus [Eq. (2.39)]. This approach was first suggested by Wagner [C. Wagner, *Chem. Tech. (Berlin)*, **18**, 1, 28 (1945)] and later popularized by P. B. Weisz and co-workers.

transfer effects on the overall kinetics. When Φ is less than about 0.3, the internal effectiveness factor approaches unity. On the other hand, for large Φ (that is, $\Phi \geq 3$), η_I is inversely proportional to Φ. To summarize, then,

$$\eta_I \big|_{\Phi \leq 0.3} \cong 1; \quad \eta_I \big|_{\Phi \geq 3} \propto \frac{1}{\Phi} \tag{2.48}$$

Again there is some difference between the function η_I *vs* Φ for a slab and a sphere, with the difference being greater in the zero-order regime than in the first-order regime (Table 2.7).

Table 2.7. *Values of* Φ *corresponding to* η_I *of 0.95.*

Reaction Order	Φ *for Slab*	Φ *for Spheres*
0	2.1	0.66
1	0.15	0.11

Once Φ is known, one can obtain a reasonable estimate of η_I without having the exact value of β . On the other hand, Figure 2.12 illustrates that knowledge of β is critical to determine η_I from ϕ. This point is emphasized further in the example below, which demonstrates a general approach for evaluating the effectiveness factor for immobilized enzyme particles.

Calculating the effectiveness factor from diffusion-limited rate data.

The serine protease α-chymotrypsin (α-CT) has been immobilized to a porous spherical support for use as a catalyst in dipeptide synthesis. In preliminary studies of the immobilized enzyme's activity, reaction rates were measured using N-acetyl-L-tyrosine ethyl ester (ATEE) as the substrate. Chymotrypsin catalyzes the hydrolysis of ATEE to N-acetyl-L-tyrosine and ethanol. Based on the data given below, what is the effectiveness factor of the catalyst for ATEE hydrolysis?

Enzyme loading:	1.27 μmols active α -CT / g support (determined by active-site titration)
Substrate concentration, S_0:	1 mmol ATEE / L
D_{eff} of ATEE in porous support:	3.8×10^{-6} cm^2/s
Radius of particles, R:	60 μm
Density of particles:	0.41 g / cm^3
Measured reaction rate:	$498 \dfrac{\mu\text{mols ATEE}}{\mu\text{mol active } \alpha\text{-CT} \cdot \text{min}}$

Solution

To determine the effectiveness factor, η_I , we first calculate the observable modulus, Φ:

$$\Phi = \frac{\left(\frac{60x10^{-4}cm}{3}\right)^2\left(498\frac{\mu mols\ ATEE}{\mu mol\ active\ \alpha\text{-}CT.min}\right)\left(1.27\frac{\mu mols\ active\ \alpha\text{-}CT}{gm}\right)\left(0.41\frac{gm}{cm^3}\right)}{\left(\frac{3.8x10^{-6}cm^2}{sec}\right)\left(\frac{60\,sec}{min}\right)\left(1\frac{\mu mol\ ATEE}{cm^3}\right)} = 4.5$$

This large value of Φ indicates that the reaction rate is diffusion limited.

To calculate the effectiveness factor we need a relationship between Φ and η_I. The dependence of η_I on Φ is shown in Fig. 2.13; however, Fig. 2.13 includes β, which we shall assume is unknown (recall that K_m in β is the intrinsic Michaelis constant of the immobilized enzyme, which may differ from that of the soluble enzyme). Moreover, since the catalyst's overall activity is diffusion limited, K_m cannot be determined in the conventional manner. Nonetheless, we can assume a value of β and use Figure 2.13 to estimate η_I from Φ. This approach will yield an approximate result which, depending on the required accuracy, may be satisfactory. From Fig. 2.13 we see that η_I will fall between about 0.17 and 0.35.

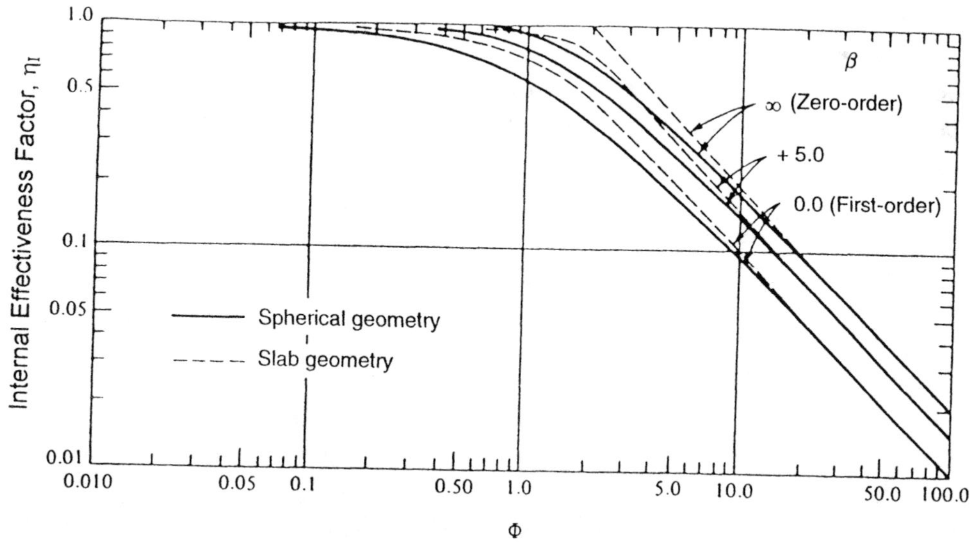

Figure 2.13. Effectiveness factor, η_I, as a function of the observable modulus, Φ, and the dimensionless substrate concentration, β.

An alternative strategy is to combine Eq. (2.47) with an independent equation for η_I and ϕ. Relationships of this kind have been reported for slab geometry by Atkinson and Davies[10]:

$$\eta_I(\phi, \eta_D) \cong 1 - \frac{\tanh\phi}{\phi}\left[\frac{1/\eta_D}{\tanh(1/\eta_D)} - 1\right] \qquad \text{provided} \qquad \frac{1}{\eta_D} \le 1$$

$$\eta_I(\phi, \eta_D) \cong \eta_D - \frac{\tanh\phi}{\phi}\left[\frac{1}{\tanh(1/\eta_D)} - 1\right] \qquad \text{provided} \qquad \frac{1}{\eta_D} > 1 \tag{2.49}$$

where η_D is the asymptotic value of η_I for large ϕ and is given by

$$\eta_D = \sqrt{2}\left(\frac{1+\beta}{\phi\beta}\right)[\beta - \ln(1+\beta)]^{1/2} \tag{2.50}$$

Equations (2.49) and (2.50) can also be used for spherical particles, provided that the appropriate definition of ϕ is employed, that is, Eq. (2.39). Equation (2.47) can now be equated to Eq. (2.49), giving a single equation for ϕ that can be solved numerically by an iterative procedure. Once ϕ is known, η can be calculated from Eq. (2.47). For example, if $1/\eta_D > 1$, we can write

$$\frac{\Phi(1+\beta)}{\phi^2} = \eta_D - \frac{\tanh\phi}{\phi}\left[\frac{1}{\tanh(1/\eta_D)} - 1\right] \tag{2.51}$$

We still need to know the value of β in order to solve Eq. (2.51), so here we assume a reasonable range of β values and compare the results. For example, under the assay conditions the Michaelis constant of soluble α-CT is 0.73 mM (or β = 1.4); thus, we shall calculate η_I for K_m values of 0.73 mM and 1.66 mM (a value of K_m for immobilized α-CT taken from the literature). The results are summarized below.

β	Φ	ϕ	η_I
1.4	4.5	6.1	0.29
0.60	4.5	5.3	0.26

Thus, based on the available information a suitable estimate of η_I would be 0.28.

These results illustrate that the activity of the catalyst is reduced by a factor of about 3.5 due to diffusional limitations. Furthermore, at constant Φ, η_I is not very sensitive to a two-fold difference in β. On the other hand, at constant ϕ the same difference in β corresponds to a larger relative difference in η_I, determined from Eq. (2.49). The comparison is completed below.

(**10**) B. Atkinson and I.J. Davies, The Overall Rate of Substrate Uptake (Reaction) by Microbial Films. Part I--A Biological Rate Equation. *Trans. Instn. Chem. Engrs*, **52**, 248 (1974).

β	ϕ	η_I
1.4	4.5	0.39
0.6	4.5	0.30

Therefore, as mentioned earlier, the dependence of η_I on ϕ is more sensitive to β than is the dependence of η_I on Φ.

2.4.4 Simultaneous External and Internal Mass-Transfer Resistances and Partitioning Effects

So far we have considered the effects of external and internal diffusional resistances separately. However, in practice these two resistances can both be significant and can simultaneously influence the global reaction rate in an immobilized enzyme or cell reactor. Moreover, substrate partitioning, typically expressed by an equilibrium partition coefficient (or distribution coefficient), can also affect the observed kinetics. In this section we shall investigate the combined influence of all of these factors.

We begin by reconsidering the steady-state intraparticle mass balance for substrate in a spherical immobilized enzyme pellet:

$$\frac{d^2x}{d\bar{r}^2}+\frac{2}{\bar{r}}\frac{dx}{d\bar{r}}=\frac{R^2v(S_i)}{D_{eff}S_0}=\frac{R^2(v_{max}/K_mD_{eff})x}{1+\beta x} \tag{2.38}$$

When external film resistance cannot be neglected, the boundary conditions for Eq. (2.38) are

$$\frac{dx}{d\bar{r}}\Big|_{\bar{r}=0}=0 \tag{2.52}$$

and

$$\frac{dx}{d\bar{r}}\Big|_{\bar{r}=1}=Bi(1-x^*) \tag{2.53}$$

where the new parameter appearing here is the *Biot number*, Bi, defined as

$$Bi = \frac{k_sR}{D_{eff}} = \frac{\textit{characteristic film transport rate}}{\textit{characteristic intraparticle diffusion rate}} \tag{2.54}$$

Thus, the Biot number expresses the relative magnitude of external and internal diffusional resistances. The new boundary condition, Eq. (2.53), states that the rate of film transport *to* the pellet equals the rate of transport *into* the pellet by intraparticle diffusion. Note that the previous boundary condition, $x = x^* = 1$ at $\bar{r} = 1$, corresponds to the case when Bi $\rightarrow \infty$, i.e., negligible film mass transfer resistance.

In the presence of partitioning effects, the equilibrium concentration of substrate within the pellet will differ from that in the outside liquid, even in the absence of immobilized enzyme. In such a case, an abrupt concentration change occurs at the solid-liquid interface, described mathematically by

$$S_i^* = K_p S^* \tag{2.55}$$

where S_i^* is the substrate concentration *just within* the particle surface (per intraparticle fluid volume), and K_p is the partition coefficient. Therefore, assuming rapid equilibration at the solid-liquid interface, Eq. (2.53) can be rewritten as

$$\frac{dx}{d\bar{r}}\bigg|_{\bar{r}=1} = \mathrm{Bi}\left(1 - \frac{S_i^*}{K_p S_o}\right) \tag{2.56}$$

K_p is expected to be less than unity for large solutes due to steric exclusion effects. However physicochemical interactions (such as electrostatic, hydrophilic, or hydrophobic interactions) between the substrate and particle matrix can lead to K_p values significantly larger than unity.

The simultaneous effects of both mass-transfer resistances and substrate partitioning can again be accounted for by an effectiveness factor, in this case the *overall effectiveness factor*. The overall effectiveness factor is defined with respect to the reaction rate evaluated at S_0 and accounts for both external and internal mass transfer resistances. The dependence of η_o on ϕ for both $K_p = 0.5$ and $K_p = 3$, determined from numerical solutions of Eqs. (2.38), (2.52), and (2.56), is shown in Figure 2.14. For these plots, η_o was calculated from its definition in terms of dimensionless variables:

$$\eta_o = \frac{(dx/d\bar{r})_{\bar{r}=1}}{3\phi^2 1/(1+\beta)} \tag{2.57}$$

Figure 2.14 illustrates that the influence of Bi on η_o increases as Bi decreases, that is, as external film resistance becomes more significant. Moreover, the dependence of η_o on β decreases as Bi decreases, again owing to the greater importance of external mass transfer. Generally, if Bi is on the order or 100 or more, the effects of external resistance are not significant.

Biot number for glucose isomerization in a packed bed reactor

Listed below are data pertaining to the pilot scale isomerization of glucose to fructose by immobilized glucose isomerase in a fixed bed reactor. Estimate the Biot number for the enzyme-catalyzed reaction in the packed bed reactor.

Reactor operation		*Liquid phase properties*
Bed diameter:	0.2 meter	45% w/w dry solids (≥ 93% glucose, 0%
Bed height:	1.9 meter	fructose, ≤ 7% oligosaccharides)

Bed density:	0.3 gm/cm^3	viscosity $\mu = 2.9$ cP
Superficial fluid		density $\rho = 1.2$ gm/cm^3
velocity:	21 meter/hr	temperature 60°C

Catalyst Properties (Sweetzyme type Q cylinders)

Cylinder diameter (d_p):	0.55 mm
Cylinder length (L):	1.4 mm
Particle density (ρ_p):	1.4 g/cm^3
Particle void fraction (ε_p):	0.45
V_{max} of immobilized enzyme:	0.217 μmol/min/mg catalyst

Solution

To determine the Biot number we must first estimate the mass transfer coefficient, k_s. This coefficient can be calculated from the following empirical correlation for the j_D factor[11]:

$$j_D = \frac{k_s}{v_\infty} Sc^{2/3} = 0.91 Re^{-0.51}\zeta \tag{2.58}$$

Here the Reynolds number is defined by

$$Re = \frac{G_o}{a\mu\zeta} \tag{2.59}$$

where G_o is the superficial mass velocity, a is the external particle surface area per unit bed volume, μ is the solution viscosity, and ζ is an empirical shape factor (= 0.91 for packed beds of cylinders).

The external particle surface are per unit weight of catalyst, a_m, for a single Sweetzyme cylinder is

$$a_m = \frac{\frac{2\pi d_p^2}{4} + \pi d_p L}{\frac{\pi d_p^2 L}{4}\rho_p} = \frac{\frac{2}{L}+\frac{4}{D}}{\rho_p}$$

$$= \frac{\frac{2}{1.4}\frac{1}{mm} + \frac{4}{0.45}\frac{1}{mm}}{1.4\frac{cm^2}{gm}}\left(10\ \frac{mm}{cm}\right) = 73.7\ \frac{cm^2}{gm}$$

Thus the external particle surface area per unit of bed volume is

$$a = a_m\rho_b = (73.7)(0.3) = 22.1\ \text{cm}^{-1}$$

Using this result to calculate the Reynolds number, we obtain

$$Re = \frac{\left(\frac{21m}{h}\right)\left(\frac{1.2g}{cm^3}\right)\left(\frac{100cm}{m}\right)}{\left(\frac{22.1}{cm}\right)\left(\frac{2.9x10^{-2}g}{cm\sec}\right)(0.91)\left(\frac{3600\sec}{h}\right)} = 1.20$$

---

(11) R.B. Bird, W.E. Stewart, and E.N. Lightfoot, *Transport Phenomena* (John Wiley & Sons, N.Y., 1960) p. 411.

Therefore

$$j_D = 0.91Re^{-0.51}(0.91) = 7.55 \times 10^{-1}$$

In similar fashion, we can calculate the Schmidt number:

$$Sc = \frac{\mu}{\rho D_s} = \frac{\left(\frac{2.9\times10^{-2}gm}{cm\ sec}\right)}{\left(\frac{1.2gm}{cm^3}\right)\left(\frac{8.0\times10^{-6}cm^2}{sec}\right)} = 3.02\times10^3 \tag{2.60}$$

In the above calculation we have assumed that the diffusion of glucose in the liquor is described by a simple binary diffusivity.

We can now solve for the mass transfer coefficient by rearranging the j-factor equation:

$$k_s = \frac{0.755}{(3.02\times10^3)^{2/3}}\left(\frac{21m}{h}\right)\left(\frac{1h}{3600s}\right) = 2.11\times10^{-5}m/s$$

Now, to calculate the Biot number we must first estimate the effective diffusivity, D_{eff}, of glucose through the Sweetzyme cylinder. In this case, D_{eff} can be estimated from Eq. (2.34) by assuming H = 1 and $\tau = 4$:

$$D_{eff} = \frac{(8.0\times10^{-6}cm^2/\sec)(0.45)}{4} = 9.0\times10^{-7}cm^2/\sec$$

The Biot number can now be determined from Eq. (2.54). For a cylindrical Sweetzyme particle, the characteristic length is

$$\frac{\frac{\pi d_p^2 L}{4}}{\frac{2\pi d_p^2}{4}+\pi d_p L} = \frac{1}{\frac{2}{L}+\frac{4}{D}} = \frac{1}{\frac{2}{1.4mm}+\frac{4}{0.55mm}} = 0.115 \quad mm$$

Therefore,

$$Bi = \frac{(2.11\times10^{-5}m/\sec)(0.115m)\left(\frac{1m}{1000mm}\right)}{(9.0\times10^{-7}cm^2/\sec)\left(\frac{1m}{100cm}\right)^2} = 27$$

A Biot number of 27 indicates that the influence of external resistance is probably fairly small; however, Figure 2.14 shows that the relative importance of external mass transfer tends to be greater for larger values of K_p.

2.5 Intraparticle Diffusion and Immobilized Enzyme Stability

The stability of an enzyme, that is, its resistance to inactivation, is often a critical factor in determining the enzyme's usefulness in practice. Enzyme denaturation or inactivation occur as a result of heat, chemical denaturants, organic solvents, and the action of proteases. Immobilization often, but not always[12], stabilizes enzymes, where an increase in enzyme

(12) Of 50 cases of enzyme immobilization analyzed by G. J. H. Melrose (*Rev. Pure. Appl. Chem.*, **21**, 83, 1971), 30 involved claims of stabilization, 8 cited decreases in stability, and 12 reported no effect.

stability is manifested by either a decrease in the rate constant of inactivation under denaturing conditions (corresponding to an increase in the enzyme's half-life) or an increase in the degree of denaturing action required to effect a certain extent of inactivation. Some of the factors principles believed to be important in the stabilization of enzymes by immobilization are as follows[13]:

- prevention of either proteolysis or aggregation by spatial fixation of enzyme molecules to the support;
- reduced tendency of the enzyme to unfold due to multipoint covalent or adsorptive attachment to the support, and/or intramolecular crosslinking of the enzyme;
- lower likelihood of multimeric enzymes to dissociate if all subunits are attached to the support;
- exclusion by the support of denaturing agents (e.g., chemical inactivators) from the enzyme's microenvironment, or decomposition of inactivators by the support (e.g., decomposition of hydrogen peroxide, produced during the oxidation of glucose by gluocose oxidase, catalyzed by activated carbon);
- shifting by a charged support of the local pH, thus preventing pH inactivation of the enzyme;
- exclusion by the support (e.g., an encapsulation membrane) of proteases from the enzyme's environment.

It is important to distinguish between an increase in an enzyme's *intrinsic* stability, that is, its true resistance to inactivation, and an increase in its apparent stability. Whereas changes in intrinsic stability must be explained in terms of the physicochemical forces that govern protein structure, changes in apparent stability can be analyzed within the theoretical framework of coupled reaction and diffusion.

Regardless of how immobilization affects the intrinsic stability of an enzyme, mass transfer effects can dramatically influence the *apparent* stability of an immobilized enzyme. In particular, the apparent deactivation rate of an immobilized enzyme will be lower than its intrinsic deactivation rate whenever intraparticle diffusional resistance is significant. Thus, diffusion alone can diminish the sensitivity of an immobilized enzyme to a potential denaturant, for example, temperature. A general method for estimating the intrinsic deactivation rate from diffusion-limited kinetic data will now be developed[14].

We begin by considering immobilized enzyme distributed homogeneously within a carrier particle. To simplify the analysis, we will assume that external mass transfer effects

(**13**) For a more detailed discussion of these mechanisms, see A.M. Klibanov, Enzyme Stabilization by Immobilization, *Anal. Biochem.*, **93**, 1 (1979).

(**14**) This analysis is based on the work of H. Ooshima and Y. Harano, Effect of Intraparticle Diffusion Resistance on Apparent Stability of Immobilized Enzymes, *Biotechnol. Bioeng.*, **23**, 1991 (1981).

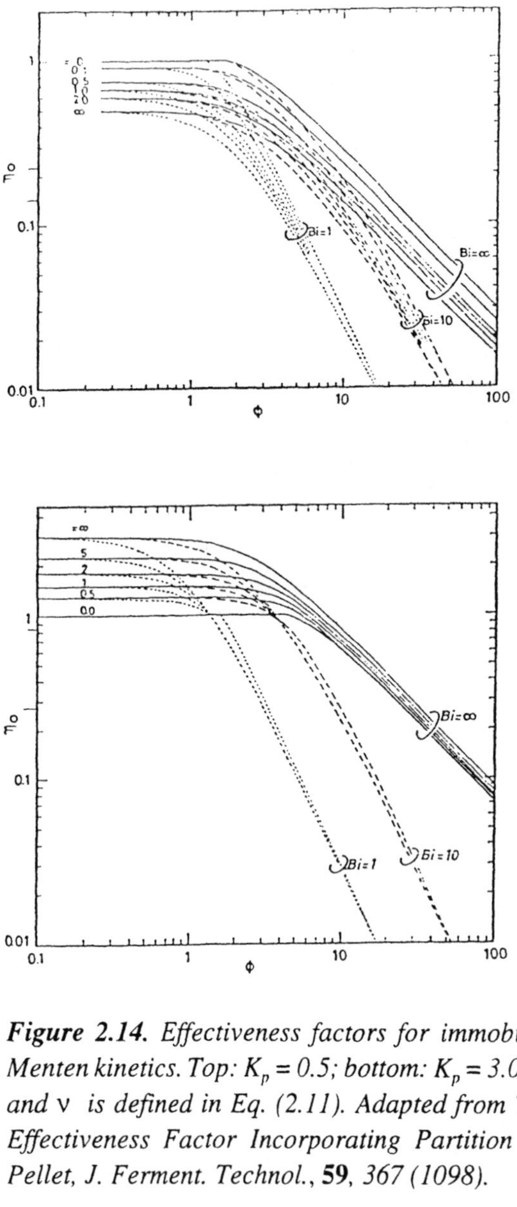

Figure 2.14. *Effectiveness factors for immobilized enzyme catalysts obeying Michaelis Menten kinetics. Top: $K_p = 0.5$; bottom: $K_p = 3.0$. Bi is the Biot number defined in Eq. (2.54), and ν is defined in Eq. (2.11). Adapted from T. Yamane, S. Araki, and E. Sada, Overall Effectiveness Factor Incorporating Partition Coefficient for Gel-Immobilized Enzyme Pellet, J. Ferment. Technol.,* **59**, *367 (1098).*

are negligible and that substrate partitioning between internal and external fluids is uniform ($K_p = 1$). In this case the observed reaction rate per unit volume of catalyst, v_{obs}, is given by

$$v_{obs} = \eta_I v(S_o) = \eta_I \frac{v_{max} S_o}{K_m + S_o} = \eta_I \frac{k_{cat} E_a S_o}{K_m + S_o} \tag{2.61}$$

where E_a is the concentration of *active* immobilized enzyme per particle volume, k_{cat} is the corresponding rate constant, and S_o is the substrate concentration at the surface of the particle (which we have assumed is equal to the concentration in the bulk). At this point it is helpful to take the natural logarithm of each term in Eq. (2.61), which yields

$$\ln(v_{obs}) = \ln\eta_I + \ln[(k_{cat}S_o)/(K_m + S_o)] + \ln E_a \tag{2.62}$$

Now suppose the concentration of active immobilized enzyme decreases with time due to denaturation, but the value of k_{cat} remains the same (that is, the amount of active immobilized enzyme decreases but the specific activity of active immobilized enzyme does not change). In this case, the change with time of $\ln v_{obs}$ is equal to

$$\frac{d\ln v_{obs}}{dt} = \frac{d\ln\eta_I}{dt} + \frac{d\ln E_a}{dt} \tag{2.63}$$

Recall that η_I is a function of ϕ, which will vary with time as the amount of active enzyme decreases (see Eq. [2.39]). Thus, we can write

$$\frac{d\ln v_{obs}}{dt} = \left(\frac{d\ln\eta_I}{d\ln\phi}\right)\left(\frac{\partial\ln\phi}{\partial t}\right) + \frac{d\ln Ea}{dt} \tag{2.64}$$

or, recalling the definition of ϕ given by Eq. (2.39), we have

$$\frac{d\ln v_{obs}}{dt} = \left(\frac{d\ln\eta_I}{d\ln\phi}\right)\left(0.5\frac{d\ln E_a}{dt}\right) + \frac{d\ln E_a}{dt} = \alpha\frac{d\ln E_a}{dt} \tag{2.65}$$

where

$$\alpha = \left(1 + 0.5\frac{d\ln\eta_I}{d\ln\phi}\right), 0.5 \le \alpha \le 1 \tag{2.66}$$

The range of α is obtained from

$$-1 \le d\ln\eta/d\ln\phi \le 0 \tag{2.67}$$

and it should be noted that ϕ and hence α change with the elapsed deactivation time.

As mentioned previously, the intrinsic kinetics of immobilized enzyme deactivation often differ substantially from first-order decay. Nonetheless, first-order deactivation is observed in some denaturing environments and in such cases

$$-\frac{d\ln E_a}{dt} = k_d \tag{2.68}$$

Therefore, from Eqs. (2.65) and (2.68):

$$\frac{d\ln v_{obs}}{dt} = -\alpha k_d \tag{2.69}$$

Equation (2.69) indicates that the deactivation rate constant decreases by a factor of α when diffusional resistance is significant. Thus, whenever $d\ln\eta_I / d\ln\phi < 0$ (that is, whenever $\eta_I < 1$), the apparent stability of the immobilized enzyme will exceed its intrinsic stability. Moreover, the dependence of α on ϕ can be determined once the functional relationship between η_I and ϕ is known. Figure 2.15 contains plots of α vs. ϕ for first-order deactivation of immobilized enzyme that obeys Michaelis-Menten kinetics. In the case of first-order reaction kinetics (the solid line corresponding to $\beta = 0$), Eq. (2.44) was used to obtain $\ln \eta_I$ as a function of $\ln \phi$; in all other cases, Eq. (2.49) was used. Note that for all values of β , α decreases with increasing ϕ and the enzyme appears to become more stable until the limit of $\alpha = 0.5$ is reached. An example of how this information can be used to obtain the intrinsic deactivation constant from diffusion-limited data is given below.

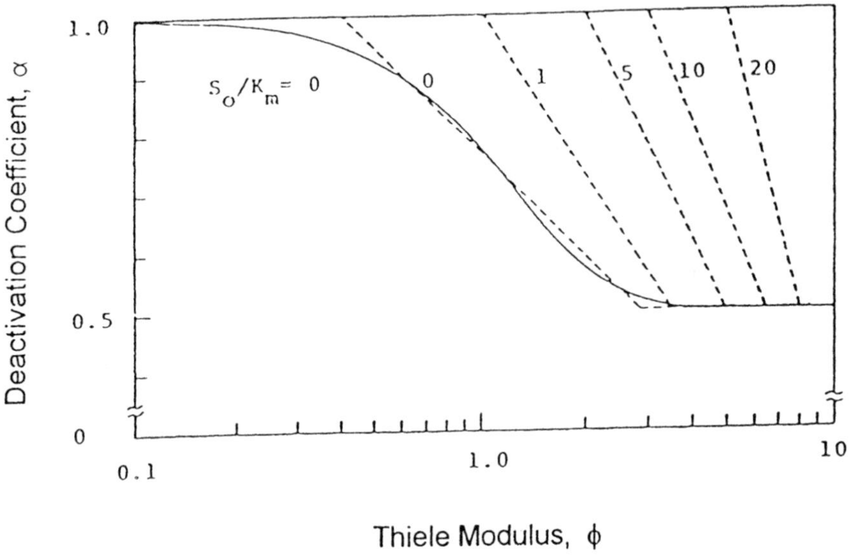

***Figure 2.15**. Dependence of α on ϕ for a flat-plate immobilized enzyme catalyst undergoing first-order deactivation. From H. Ooshima and Y. Harano, Effect of Intraparticle Diffusion Resistance on Apparent Stability of Immobilized Enzymes, Biotechnol. Bioeng.*, **23**, *1991 (1981).*

Determining k_d for the thermal deactivation of immobilized invertase

Four different immobilized invertase catalysts of varying particle size were prepared by adsorbing invertase to the ion exchange resin Amberlite IRA-94 [Ooshima and Harano (1981)]. The catalysts were designated AMI-1, 2, 3, and 4, and the mean particle diameters were 90, 460, 545, and 715 μm, respectively. The deactivation of each catalyst was examined by incubating the immobilized enzyme at 58.5°C and periodically removing samples for activity assays at 40°C. Using each AMI catalyst partially inactivated by the thermal pre-treatment at 58.5°C and pH 4.1, the hydrolysis of sucrose was carried out at 40°C, pH 4.1, with $s_0 = 0.187$ mM. Important characteristics of each catalyst are summarized below.

Catalyst	Mean particle diameter (μm)	Initial Active Enzyme Concentration E_{ao} (μM)	Φ_o	ϕ_o	η_I
AMI-1	90	2.7	0.12	0.50	1.0
AMI-2	460	7.2	1.8	2.4	0.61
AMI-3	545	7.2	2.1	2.8	0.54
AMI-4	715	5.4	2.6	3.3	0.47

Here we have determined values of the observable modulus from the original results of Ooshima and Harano, taking the effective diffusivity of sucrose to be 5.2×10^{-6} cm^2/sec. For each immobilized enzyme preparation it was assumed that $K_m = 0.187$ mM and thus $\beta = 1.0$. In general, once Φ_o is known, ϕ_o can be calculated from Eq. (2.48) or its analog for the case of $1/\eta_D < 1$ (where o denotes the initial value). The effectiveness factor can then be obtained from eqn (2.44). Alternatively, η_I can be obtained from Figure 2.11 and ϕ_o can be determined from Figure 2.10.

Values of $\ln[r_{obs}(t)/r_{obs,o}]$ versus the pretreatment time of deactivation at 59°C are plotted below (Figure 2.16) for AMI-1, AMI-2, and AMI-3. How do the first-order deactivation constants of the different catalysts compare?

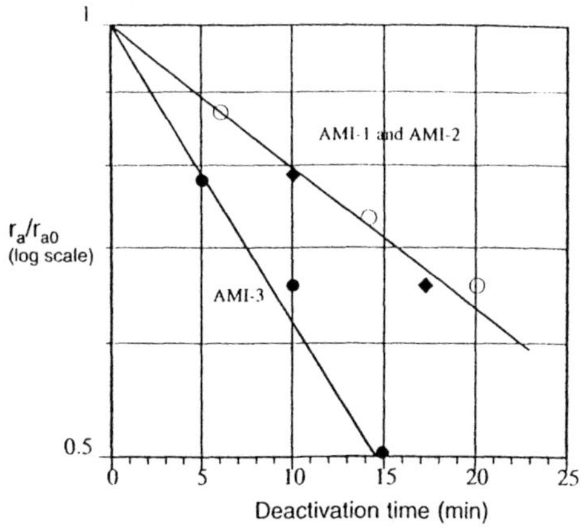

Figure 2.16. *Relative initial rates, $r_{obs}(t)/r_{obs,o}$, versus the pretreatment time t of deactivation at 58.5°C and pH 4.1. From H. Ooshima and Y. Harano (1981).*

Solution

Recall that our analysis in this case presupposes that both k_{cat} and K_m of active enzyme remain constant, and that the loss of activity is due to a decrease in the amount of active enzyme, E_a. This being the case, the slopes of the two lines in the plot above are equal to $-\alpha_o k_d$, where α_o refers to the initial value of α. Moreover, from the ϕ_o values and Figure 2.12, the α_o values of the three catalysts can be determined. Once we know α_o and $\alpha_o k_d$, we can readily obtain k_d. The relevant results are summarized below.

Catalyst	ϕ_o	$\alpha_o k_d$ (min^{-1})	α_o	k_d (min^{-1})
AMI-1	0.50	4.4	1.0	4.4
AMI-2	2.4	2.4	0.60	4.0
AMI-3	2.8	2.4	0.57	4.2

Based on these k_d values, the intrinsic deactivation rate of each immobilized enzyme preparation is roughly the same.

2.6 Nomenclature

Symbol	*Meaning*
a^{H^+}	activity of hydrogen ion
a	external particle surface area per unit bed volume (cm^2/cm^3)
a_M	external particle surface area per unit weight of catalyst (cm^2/g)
Bi	mass-transfer Biot number (2.54)
Da	Damkohler number [$= v'_{max}/(k_s S_0)$]
Da_M	modified Damkohler number [$= v'_{max}/(M k_s S_0)$]
$\overline{Da}$	observable Damkohler number [$= v/(k_s S_0)$]
D_s	molecular diffusivity of substrate (cm^2/sec)
E_a	concentration of active immobilized enzyme per particle volume (mg/cm^3 or µmol/cm^3)
E_{eff}	effective diffusivity of substrate (cm^2/sec)
G_0	superficial mass velocity (cm/sec)
H	hindrance factor [Eq. (2.35)]
k_d	intrinsic first-order deactivation constant of immoblized enzyme (sec^{-1})
K_m	intrinsic Michaelis constant (mM)
$K_{m,app}$	apparent Michaelis constant of immobilized enzyme (mM)
K_p	partition coefficient (S_i^*/S^*)
k_s	mass transfer coefficient (cm/sec)

M	flux modifier [Eq. (2.28)]
N_s	molar flux of substrate to particle surface (mmol cm^{-2} sec^{-1})
$\vec{N}_s$	molar flux vector for substrate (mmol cm^{-2} sec^{-1})
$\bar{r}$	dimensionless radial coordinate (= r/R)
R	gas constant
R	particle radius (μm)
Re	Reynolds number for flow through a packed bed [Eq. (2.59)]
S^*	substrate concentration in the liquid phase at the particle surface (mmol/cm^3)
Sc	Schmidt number [Eq. (2.60)]
S_i	internal substrate concentration (expressed per volume of fluid) (mmol/cm^3)
S_i^*	substrate concentration just within the particle surface (mmol/cm^3)
S_0	bulk substrate concentration (mmol/cm^3)
S_x	external surface area of catalyst pellet (cm^2)
T	temperature
V_p	gross particle volume
v	enzymatic reaction rate per unit volume of catalyst (mmol cm^{-3} sec^{-1})
v'	enzymatic reaction rate per unit surface area of catalyst (mmol cm^{-2} sec^{-1})
v_{max}, v'_{max}	maximum enzymatic reaction rate (per unit volume and per unit surface area, respectively)
v_{obs}	observed reaction rate per unit volume of catalyst (mmol cm^{-3} sec^{-1})
x	dimensionless internal substrate concentration (= S_i/S_0)
x^*	dimensionless surface substrate concentration (= S^*/S_0)
Z	valence, including sign, of charged substrate

Greek Symbols

α	deactivation coefficient [Eq. (2.66)]
β	dimensionless bulk substrate concentration (= S_0/K_m)
δ	thickness of diffusional film [Eq. (2.26)]
γ	ratio of solute radius to pore radius
ε_p	particle porosity
ε	positive electronic charge of hydrogen ion (1.6021 x 10^{-19} coulomb)
η_E	external effectiveness factor [Eq. (2.17)]
η_I	internal effectiveness factor [Eq. (2.43)]
η_D	diffusion-limited effectiveness factor [Eq. (2.50)]
η_o	overall effectiveness factor [Eq. (2.57)]
λ	dimensionless potential [Eq. (2.27)]

μ	solution viscosity (centipoise)
μ^{H^+}	chemical potential of hydrogen ion (Joules/mole)
$\tilde{\mu}_0^{H^+}$	electrochemical potential of hydrogen ions in the bulk phase (Joules/mole)
$\tilde{\mu}_i^{H^+}$	electrochemical potential of the hydrogen ions in a charged matrix
ν	dimensionless Michaelis constant (= K_m/S_o)
ζ	empirical shape factor for packed beds
τ	tortuosity [Eq. (2.34)]
ϕ	Thiele modulus [Eq. (2.39)]
Φ	observable modulus [Eq. (2.45)]
Ψ	electrostatic potential (millivolts)

2.7 Problems

1. Immobilized Enzyme Effectiveness with Reaction-Generated pH Change

An enzyme catalyzing the following reaction:

$$R_1COOR_2 \rightarrow R_1COO^- + H^+ + R_2OH$$

is immobilized throughout an uncharged, flat porous membrane. In solution the enzyme exhibits optimal activity at a pH of 4.6. The following plots of η vs φ were obtained for this system at the indicated *external* pH values [J.E. Bailey and M.T.C. Chow, "Immobilized enzyme catalysis with reaction-generated pH change," Biotechnol. Bioeng., **16**, 1345 (1974)]. Explain this behavior.

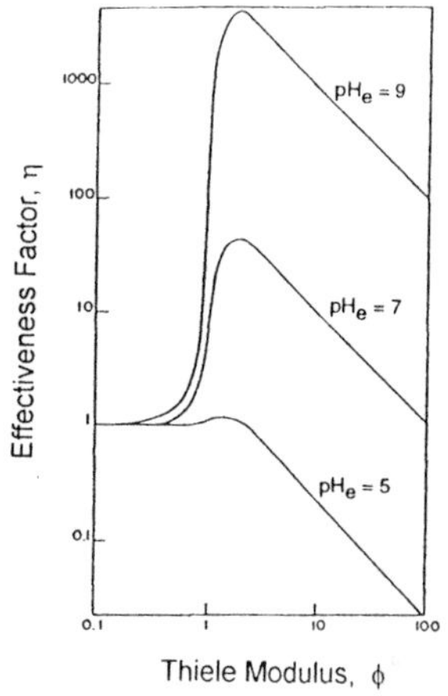

2. Catalytic Properties and Electrostatic Potential of Charged Immobilized Enzyme Derivatives

The following data were measured for the enzyme pyruvate decarboxylase immobilized on different resins of aminomethylpolystyrene, a cationic polymer [J. Beitz, A. Schellenberger, J. Lasch, and J. Fischer, Biochim. Biophys. Acta, **612**, 451 (1980)]. The resins differed in the concentration of charged amino groups, $Z_{m.c}$, throughout the polymer matrix.

Polystyrene -pyruvate decarboxylase	Bound protein (mg/g)	Activity ($\mu mol \cdot min^{-1} \cdot g^{-1}$)	$Z_{m,c}$ (mol-NH_2/liter swollen resin)	$K_{m,app}$ (mM)
P_1-PDC	22	27	0.39	12.3
P_2-PDC	15	15	0.45	11.7
P_3-PDC	11	21	0.65	7.9
P_4-PDC	21	23	1.18	5.1
P_5-PDC	31	35	1.48	4.6

In the absence of diffusional limitations, the dependence of $K_{m,app}$ on the ionic strength, I, and the charge concentration in the matrix volume, $Z_{m,c}$, is given by

$$I/K_{m,app} = I/\gamma K_m + Z_{m,app}/2K_m$$

where γ is the ratio of the mean ion activity coefficients of the matrix and the bulk phase.

a. Assuming I = 0.1 M and $\gamma = 1$, what is K_m of immobilized pyruvate decarboxylase?

b) Combine the above equation with Eq. (2.8) to obtain a single equation for the electrostatic potential. Use this equation to calculate the electrostatic potential of resin P_5.

3. *pH-Activity Curves for Free and Immobilized Trypsin*

Shown below are pH-activity curves for trypsin and IMET-3 (trypsin immobilized within a polyionic gel) at different ionic strengths, ($\Gamma/2$), using benzoyl-L-arginine ethyl ester (BAEE) as substrate (Goldstein et al., 1964). Ionic strengths were adjusted with NaCl. Calculate the electrostatic potential, Ψ, for the ionic strength values of 3.5 x 10^{-3} and 1.0. Is the polyelectrolyte carrier polycationic or polyanionic?

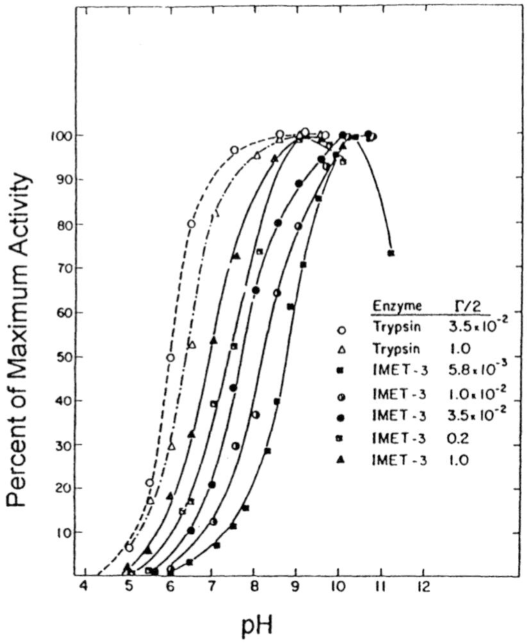

4. *Substrate Consumption by Enzyme Immobilized in a Polyelectrolyte Gel*

Consider an enzyme immobilized in a polyelectrolyte gel. The total volume of the system, V, is given by

$$V = V_i + V_o$$

where Vi and Vo are the volumes of the gel phase and external solution, respectively. In the absence of mass transfer effects, the total amount of product formed per unit time in the whole system, dP/dt, can be written as

$$\frac{dP}{dt} = \frac{-d(S_i V_i + S_o V_o)}{dt} = \frac{k_{cat} e_i S_i V_i}{K_m + S_i}$$

Show that, at time t:

$$\{[S]_o - [S]_t\} + \left(\frac{V_i + V_o e^{-Z\varepsilon\Psi/kT}}{V} \right) K_m \ln \frac{[S]_o}{[S]_t} = k_{cat}[e]t$$

where $[S]_o$ and $[S]_t$ equal the total average concentrations of substrate in the whole system at t = 0 and t, respectively, and [e] is the total average concentration of enzyme in the whole system.

5. Immobilized Enzyme-Catalyzed Reaction in the Presence of Inhibitor

A single enzyme-catalyzed reaction S → P occurs at a planar interface between fluid and a nonporous solid upon which enzyme is immobilized. The substrate and product concentrations in the bulk fluid are s_o and p_o, respectively, and the mass transfer coefficients for transport of S and P between the bulk fluid and the interface are k_s and k_p, respectively. Assuming that P is a noncompetitive inhibitor of the enzyme and that the Michaelis constant K_m is much smaller than the substrate concentration at the interface, determine the steady-state rate at which P appears in the bulk fluid (in units of moles per unit time per unit surface area of solid).

6. Reaction Rate of Immobilized Enzyme Towards a Charged Substrate

The flux of a charged substrate to a solid surface is given by Eq. (2.24):

$$N_s = Mk_s(s_o - se^{\lambda}) \qquad (2.24)$$

An enzyme of known kinetics is attached to the surface. The soluble enzyme's dependence on substrate concentration is shown below:

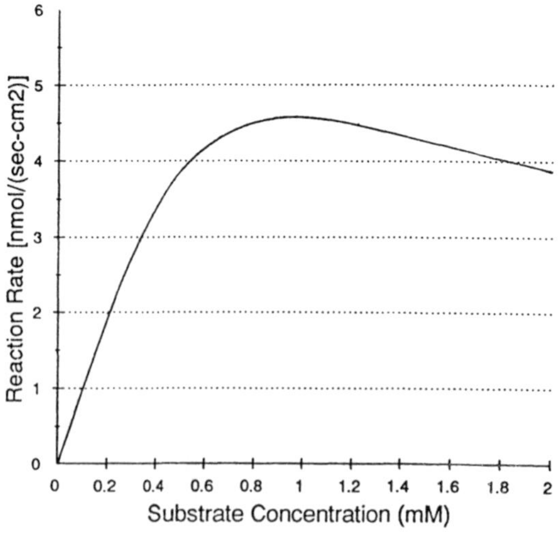

the following parameters are known:

$Mk_s s_o = 5 \times 10^{-9}$ mol/sec-cm^2, $s_o = 10^{-2}$ mol/l, $e^{\lambda} = 30$

a) Estimate the expected rate of reaction at steady state.

b) If the observed reaction rate is 30% less than predicted, what would you conclude?

7. External Mass Transfer Effects on the Observed Kinetics of an Immobilized Enzyme

Consider an enzyme immobilized to an uncharged, nonporous particle. We wish to examine the effects of the dimensionless bulk substrate concentration, β, and the modified Damkohler number, $\beta \cdot Da$, on the dimensionless reaction rate, v/v_{max}. Our approach will be to plot v/v_{max} versus β for 2 different values of $\beta \cdot Da$.

a) First, derive an expression for v/v_{max} in terms of β and $\beta \cdot Da$. Equation (2.16) is a good starting point.

b) Use your expression from part (a) to plot v/v_{max} versus β for $\beta \cdot Da = 100$ and $\beta \cdot Da = 1.0$. Your plot should span β values from zero to 50. What do these 2 plots tell you about the effect of mass transfer limitations on the kinetics of the observed reaction rate?

8. Electrostatic Effects on Immobilized Enzyme Effectiveness

A monolayer of enzyme is immobilized on the surface of a charged membrane for use in an automated blood analyzer. The electrostatic potential at the surface of the membrane is $\Psi(0)$. Below is a contour plot of Da_M versus ve^{λ} for various values of $\eta/(1+v)$.

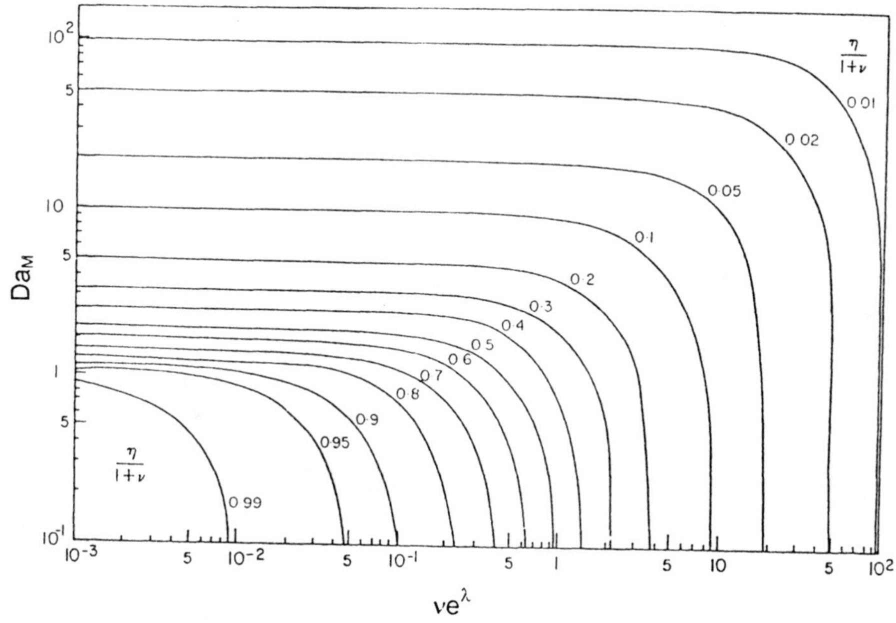

The curves were obtained for the potential distribution of a Gouy diffuse double layer. Recall that the following definitions apply:

$$\nu = \frac{K_m}{S_o}$$

where S_o is the bulk substrate concentration, and

$$\lambda = \frac{zF\Psi(0)}{RT}$$

Determine the effectiveness factor of the immobilized enzyme system for the following parameters: $V_m = 5 \times 10^{-6}$ mol/sec/mg enzyme, $K_m = 10^{-4}$ mol/l, T = 310 K, $k_s = 10^{-3}$ cm/sec, z = 1 equiv/mol, $\Psi(0) = 0.094$ V, and $M \cong 1$.

What would the effectivness factor be if z = -1 equiv/mol?

9. Lineweaver Burk Plot for a Porous Immobilized Enzyme Catalyst

In this problem we will consider the general characteristics of a Lineweaver Burk plot for a porous immobilized enzyme catalyst operating under diffusional limitations. In such a case, the measured rate of reaction per unit volume of porous material will not equal the true, or intrinsic, rate of reaction.

a) What is the general equation for a Lineweaver-Burk plot when internal diffusional limitations are present?

b) Now we shall consider such a plot for a ϕ value of about 10. On a plot of 1/v versus $1/s_o$, draw the line corresponding to very high substrate concentrations, say, $\beta \geq 100$. What is the y-intercept of this line? The slope? Now draw the line corresponding to very low values of β, that is, the line obtained as β approaches zero. What are this line's slope and y-intercept? Finally, fill in what you would expect for the Lineweaver-Burk plot of experimental data for the immobilized enzyme system.

10. Diffusion Through a Porous Mica Membrane

Mica sheets have been made into membranes by bombardment with fission fragments from a U^{235} source and subsequent etching with hydrofluoric acid. The pores formed by this process are essentially straight and extremely uniform. Shown below are measured diffusivities of various solutes through such membranes [R.E. Beck and J.S. Schultz, Biochim. Biophys. Acta, **255**, 273 (1972)]. (The effective diffusivities are based on the total cross-sectional area of membrane.) Based on these data, what is the porosity, ε_m, of the membrane?

Solute	$R_s/R_p(\gamma)$	Diffusivity $cm^2/s \times 10^6$ D_s	D_{eff}
Urea	0.0354	12.35	0.513
Glucose	0.0596	5.55	0.213
Sucrose	0.0745	3.77	0.127
Raffinose	0.0878	2.89	0.0893
α -Dextrin	0.1074	2.15	0.0624
β -Dextrin	0.1138	2.00	0.0576
Ribonuclease	0.289	0.221	0.00192

$R_p = 74.5\ \mathring{A}$

11. Effective Diffusivity of Xylose in Particles of Immobilized Glucose Isomerase

Sweetzyme Q is a commercial preparation of immobilized glucose isomerase manufactured by Novo Industries of Denmark. Plotted below are initial rates of xylose isomerization catalyzed by Sweetzyme Q (μmols xylose isomerized/min/mg Sweetzyme Q) as a function of particle size (Olivier and du Toit, 1986):

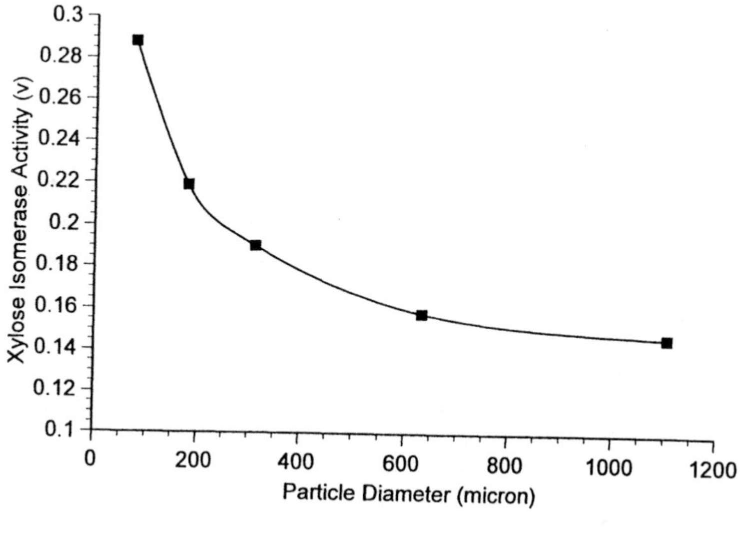

The above data were obtained in a batch reactor at 60°C with 1.5 M xylose, pH 8.0. The density of the catalyst was 1.4 gm/cm^3, and the measured value of K_m was 0.209 M. Using these results and *assuming* the maximum activity plotted above would not increase if the particles were made smaller, estimate the effective diffusivity of xylose in particles of Sweetzyme Q. What is the largest potential source of error in this calculation?

12. Asymtotic Effectiveness Factor for $\phi >> 1$

In this problem we shall derive an expression for the effectiveness factor of immobilized enzyme in the case of large ϕ. The first step is to recognize that if ϕ is sufficiently large, substrate does not penetrate far into the particle and the enzymatic reaction is confined to a thin region near the outer surface of the catalyst. Under these conditions, the effect of curvature can be neglected and a spherical particle can be treated as a slab. Therefore, starting with a dimensionless mass balance for substrate through the particle, combine the solution for $(dx/d\bar{r})|_{\bar{r}=1}$ with Eq. (2.43a) to obtain Eq. (2.50).

13. Effectiveness Factors of Glucose Oxidase Entrapped in Poly(HEMA)

Korus and O'Driscoll (1974) reported the effect of intraparticle diffusion on the kinetics of several gel entrapped enzymes, including glucose oxidase entrapped in poly(hydroxyethyl methacrylate) [poly(HEMA)]. Among the data reported were internal effectiveness factors for different particle diameters. The effectiveness factors were determined by comparing the activities of the larger particles to the activity of very small particles. These results are given below.

d (μ m)	25	57	89	125	200	332
η_I	1.0	0.8	0.65	0.54	0.35	0.27

The experimental results were compared to theoretical plots of η_I versus d generated with the aid of a figure similar to Fig. 2.12. The best agreement between theory and experimental data was obtained for a constant value of $\phi^2/d^2 = 1.11 \times 10^5$ cm^{-2}. Plot the above data and compare it to the theoretical curve corresponding to this value of ϕ^2/d^2. What is a potential source of disagreement between the experimental and theoretical curves? Other useful data: K_m of immobilized glucose oxidase, 4.85 mM; substrate (glucose) concentration, 14 mM.

14. Effectiveness Factor for Large Catalyst Particles

The following data were obtained for porous immobilized enzyme particles of different sizes. Enzyme was immobilized uniformly throughout the particles, and the radius of the particles is given in parentheses.

	Large Particles (1 mm)	Small Particles (0.1 mm)	Smaller Particles (0.01 mm)
V_{obs}	$200 \frac{\mu mols}{cm^3 catalyst\text{-}min}$	$1000 \frac{\mu mols}{cm^3 catalyst\text{-}min}$	$1000 \frac{\mu mols}{cm^3 catalyst\text{-}min}$
S_o	100 mM	100 mM	100 mM

a) Assuming that $\overline{Da} << 0.1$ and $\beta \cong 0$, what is the internal effectiveness factor for the large catalyst particles?

b) *Estimate* the effective diffusivity of the substrate through the catalyst pores.

c) How would you expect the thermal stability of the large catalyst (i.e., the half-life of the catalyst at a high temperature) to compare to that of the soluble enzyme? Why?

15. Eadie-Hofstee Plots for Immobilized Enzymes

Eadie-Hofstee plots are useful for determining kinetic parameters of enzyme-catalyzed reactions. Shown below are two such plots: one for the enzyme chymotrypsin immobilized to small (~ 10 μm) catalyst particles (O), and one for chymotrypsin immobilized to large spherical (~ 120 μm) porous particles (*) [the data are from D.S. Clark and J.E. Bailey, "Structure-Function Relationships in Immobilized Chymotrypsin Catalysis," Biotechnol. Bioeng., **25**, 1027 (1983)]. The reaction velocity, v, is defined per unit volume for soluble enzyme and per unit catalyst volume for immobilized enzyme; s_o is the bulk substrate concentration, e_a is the active enzyme loading per weight of catalyst, and ρ_p is the particle density.

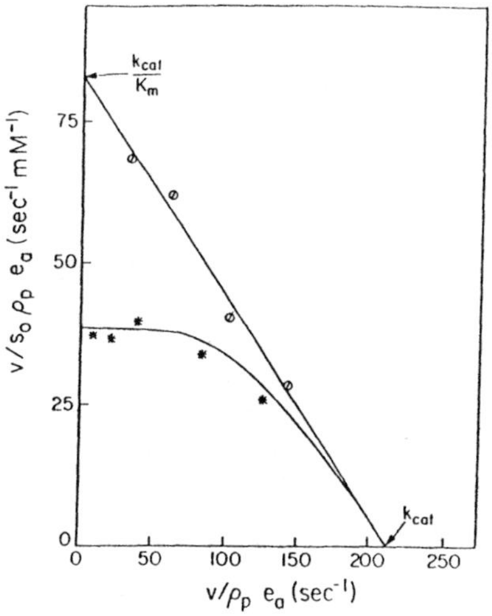

a) Why does the plot for the large catalyst particles deviate from linearity in the manner shown?

b) Now we will derive an expression for the y-intercept of the immobilized enzyme data in terms of the Thiele modulus of the catalyst. If ϕ is sufficiently large, as we will assume it is for the large particles, the effect of curvature can be neglected and the particle can be treated as a slab.

i) Write a steady state material balance on substrate through the particle.

ii) Use the following identity

$$\frac{d^2S}{dr^2} = \frac{1}{2}\frac{d}{dS}\left(\frac{dS}{dr}\right)^2$$

and the material balance from (i) to show that, if $S(r) << K_m$,

$$\frac{dS}{dr}\Big|_{r=R} \cong \left(\frac{v_{max}}{K_m D_{eff}}\right)^{1/2} S(R)$$

where S(R) is the substrate concentration at the particle surface.

(iii) At steady state, the overall reaction rate, v_{obs}, must equal the rate of substrate uptake through the outer surface of the particles, that is,

$$v_{obs} = D_{eff}\left(\frac{3}{R}\right)\frac{dS}{dr}\Big|_{r=R}$$

Combine this expression with the expression in part (ii) to obtain the y-intercept of the Eadie-Hofstee plot of diffusion-limited rate data. In the next problem, we'll see how to apply this result in practice.

16. Kinetics of Chymotrypsin Immobilized to Polymer Microspheres

Note: This problem should be assigned concurrently with Problem 15.

Radiation grafting of polyacrolein to various polymer microspheres, including magnetic carriers (d_p = 10 μ m, ρ_p = 1.4 g/ml), was recently investigated as a technique for improving the capacity of polymers for protein immobilization. Polyacrolein provides a reactive surface layer for covalent immobilization of protein [D.S. Clark , J.E. Bailey, R. Yen, and A. Rembaum, "Enzyme Immobilization on Grafted Polymeric Microspheres," Enzyme and Microbial Technology, **6**, 317 (1984)]. Below are data for the immobilization of the enzyme α-chymotrypsin (α-CT) to polystyrene-magnetite beads with and without polyacrolein grafting.

Sample	Polyacrolein pretreatment (h)	Active enzyme loading (μmol active α-CT g support^{-1})
1	0	0.31
2	2 x 1/2	0.38

For each sample, the rate N-acetyl tyrosine ethyl ester (ATEE) hydrolysis was measured at different substrate concentrations. The results are summarized below.

ATEE Conc. (mM)	Activity (μmol ATEE s^{-1} g catalyst^{-1}) Sample 1	Sample 2
0.25	7.75	6.08
0.49	13.3	11.0
1.00	20.2	22.0
1.25	20.5	23.6
2.50	31.9	33.8

(a) Determine v_{max} and K_m for each preparation of immobilized α-CT.

(b) One set of data reflects diffusion limitations at low reaction rates. From these results, determine the Thiele modulus ϕ for the diffusion-limited immobilized enzyme system, and estimate the effective diffusion coefficient of ATEE through the porous region of the support. For comparison, the diffusivity of ATEE in solution is approximately 5.6 x 10^{-6} cm^2/sec.

(c) In addition, when polystyrene-magnetite beads were treated with polyacrolein for a total of 2 h, the resultant immobilized enzyme loading was only 0.34 μmols active α-CT per gram of support. Provide a possible explanation for why this loading was smaller than that of sample 2.

17. Modeling Enzymatic Production of Tryptophan in Liquid Membranes

The amino acid L-tryptophan possesses market potential as a feed supplement for livestock. At present, L-tryptophan is produced chiefly by chemical synthesis. However, Eggers *et al.* (1988)[15] investigated the tryptophanase-catalyzed synthesis of L-tryptophan (Trp) and the subsequent extraction and stripping of the product from the reaction phase by means of a liquid membrane. The reaction involved was the following:

$$\text{ammonia} + \text{pyruvic acid} + \text{indole} \leftrightarrow \text{tryptophan} + \text{water}$$

The reaction system consisted of a planar organic layer bounded by two aqueous phases of different pH, as shown below. Tryptophan is produced enzymatically in the aqueous reaction phase (pH 9) and partitions into the organic phase as governed by the distribution coefficient. The complex of tryptophan and a carrier molecule, denoted NR_4^+, then diffuses

(15) Eggers, D.F.,& H.W. Blanch, *Bioprocess Engng. 3*, (1988)

across the organic layer to the other interface where tryptophan is exchanged for pyruvate (Pyr^-). Because of the favorable distribution coefficients at the two membrane interfaces, the concentration of typtophan in the aqueous stripping phase may reach values many times larger than the concentration in the reaction phase. Here we will develop a mathematical model for estimating the rates of substrate and product transport and for predicting the sensitivity of the system to changes in specific variables such as the partition coefficients and thickness of the liquid membrane.

a) Why do you suppose it is necessary to maintain the two aqueous phases at pH 9 and 7, respectively?

b) Concentration profiles and nomenclature for the mathematical model are shown below, where C denotes the concentration of tryptophan or the tryptophan/carrier complex.

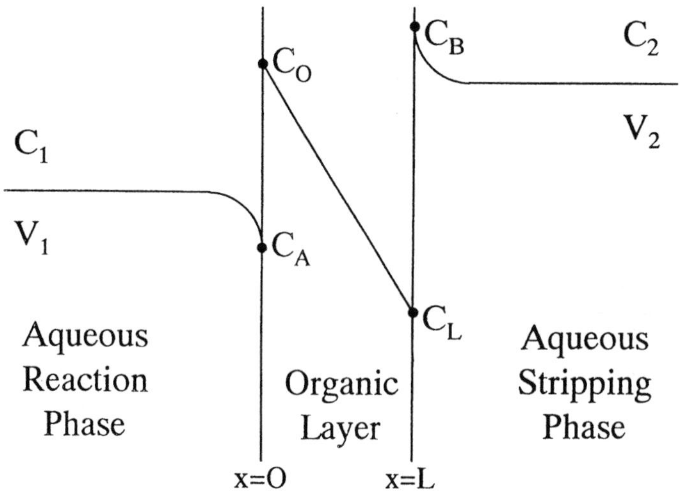

Write an unsteady state mass balance for tryptophan in the organic phase, along with the appropriate boundary conditions. Assume the enzymatic reaction is zero-order. The distribution coefficients between the aqueous phases and organic phase are K_1 and K_2. Note that steady state is not assumed; therefore, C_1, C_2, and C (the concentration in the organic layer) all vary with time. Formulate mass balances for two separate cases: with and without film resistance in each aqueous phase.

The mass balance equations can be solved by means of Laplace transforms. Shown on the next page is a comparison of experimental and calculated results for the transport of tryptophan at 37°C, assuming negligible film resistance. (*Advanced problem: use Laplace transforms or a numerical method to obtain the predictions shown by the solid lines in the plot below*).

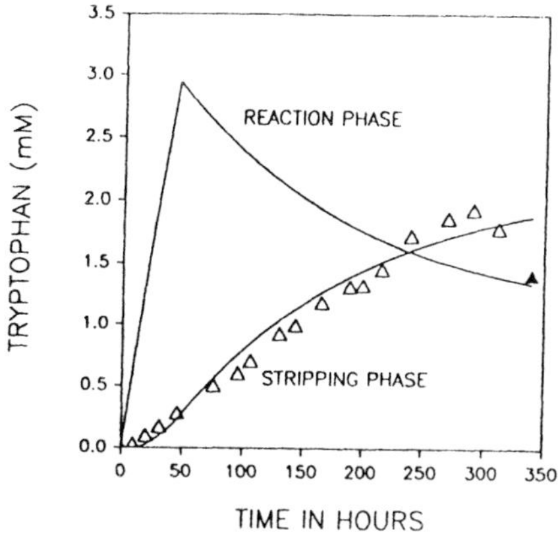

Enzymatic production of tryptophan in a two-phase system employing an organic liquid membrane. Solid lines represent model predictions for a diffusivity $D = 4 \times 10^{-6}$ cm^2/s; an enzyme reaction rate (assumed to be zero-order) = 28.7 $\cdot H(t - \tau)$ nmoles/min, where $H(t - \tau)$ is the Heaviside unit step function, introduced to account for loss of enzyme activity at time τ; $\tau = 45$ h; the volume of each phase $V_1 = V_2 = 23$ cm^3; $K_1 = 1.25$ and $K_2 = 0.6$.

18. Modeling a Cellular Electrode

A biosensor was developed that uses immobilized mammalian cells to evaluate antitumor activities of several known drugs [B.S. Liang, X.-M. Li, and H.Y. Wang, Biotechnol. Prog., **2**, 187 (1986)]. This was accomplished by immobilizing mammalian (tumor) cells to the surface of a dissolved O_2 electrode, inserting the probe into a solution containing the test compound, and monitoring the probe output. An increase in the output current indicated inhibition of the cellular O_2 uptake rate, i.e., antitumor activity. Assuming that the system can be represented schematically as shown below, that the oxygen uptake rate of the cell obeys Monod growth kinetics (Chapter 3), and that drug consumption can be described by nth-order kinetics, formulate (but do not solve) a mathematical model for the biosensor response. In the figure below, C represents the oxygen concentration ($mmol/cm^3$) and G the antitumor drug concentration ($mmol/cm^3$).

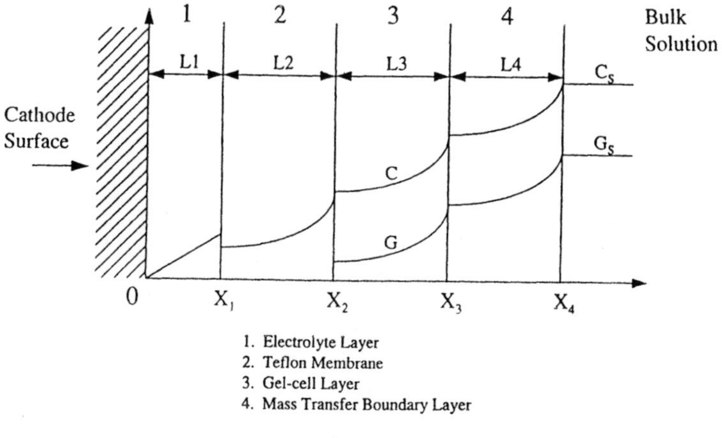

Chapter 3. Microbial Growth

3.1 Introduction

In this chapter we shall consider the processes by which a microbial cell consumes nutrients and generates products in order to reproduce. The set of reactions which are employed by a particular cell to accomplish this objective is referred to as cellular *metabolism*, and details of various reaction pathways are given in biochemistry and microbiology texts. We distinguish microorganisms on the basis of the type of nutrients that the organism employs as carbon and energy sources.

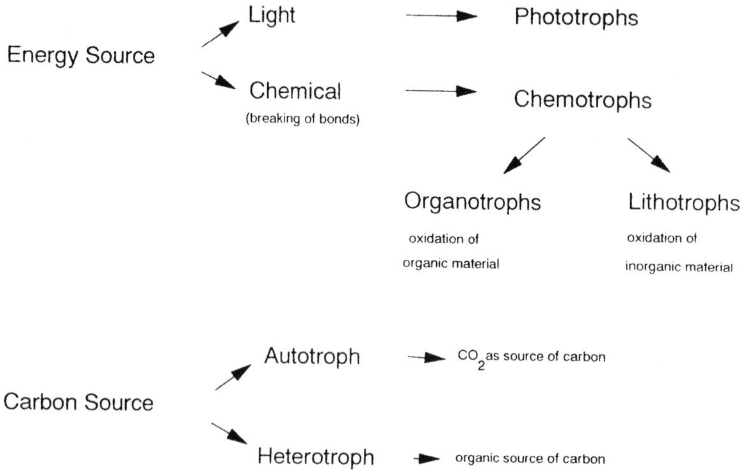

Figure 3.1. *Classification of microorganisms on the basis of their nutritional and energy requirements.*

Most organisms we shall encounter in this text are chemoheterotrophs, i.e., organisms requiring an organic carbon source and a chemical source of energy. In this class we find animal cells, protozoa, fungi (including yeast), and most bacteria. Plant cells fall into the class of photoautotrophs, i.e., they fix CO_2 and employ light as an energy source. Although the efficiency of capture of incident light energy by plants is relatively low (around 1-2%[1]), most organisms are remarkably efficient in retaining and employing the energy in the chemical bonds of their organic energy source.

3.2 Stoichiometery and Energetics of Growth

For many organisms, the energy and carbon requirements for growth and product formation can be met by the same organic compound. This considerably simplifies our analysis of cellular kinetics at both the macroscopic and microscopic levels. We shall examine the growth of microorganisms in the following sections with this simplification; however, the methods employed can also be extended to more complex situations. Our starting point is a macroscopic view of the uptake of nutrients (substrates for cell growth) and subsequent production of new cells and products. In this sense, the cell can be viewed as an "open" or flow system from a thermodynamic viewpoint; energy flows from a low entropy state (chemical energy stored in compounds more reduced than CO_2 or other end products of metabolism) to a high entropy state (heat dissipated at ambient temperature). Thus organisms are not at equilibrium as irreversible processes are taking place. The state of the system is described by extensive quantities, for example, the concentrations of chemical species and the amount of energy.

A general balance equation can be written for each extensive quantity, expressing the rate of accumulation within the system as a function of the flows of the quantity into and out of the system and the rate of generation of that quantity by chemical reaction within the system. For example, for a species S_i we can write the balance equation below, assuming that the volume and surface area of the system are unchanging:

$$\frac{d}{dt}\oint_V S_i dV = \oint_V r_{S_i} dV + \oint_A j_{S_i} dA \tag{3.1}$$

where r_{Si} is the volumetric rate of production of S_i and j_{Si} is the flux (per area) of S_i into the cell. We assume that the system is in steady state (the time derivative in Equation (3.1) is zero). Species can be subdivided into conserved, such as atomic species, and non-conserved species. We shall examine balance equations for atomic species in the following section and then address the more complex issue of energy balances. The atomic species balance equations are useful in the design of nutrient media for microbial growth and in developing

(1) The upper bound for photosynthesis is around 35%, when expressed as the free energy stored as carbohydrate per free energy absorbed at a wavelength of 700nm. However, only ~43% of incident energy is useful for photosynthesis and leaf shading reduces this by ~80%. Around 67% of the energy captured is lost by dark respiration.

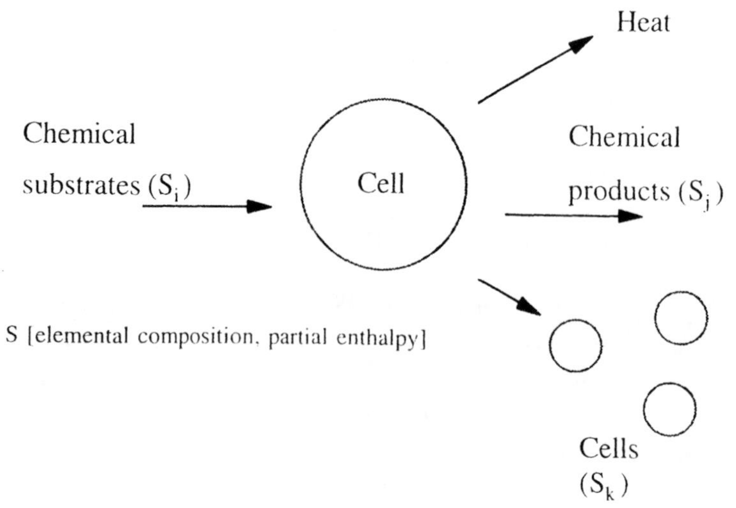

Figure 3.2. *A view of the cell as an open system. Chemical species (S_i) flow into and out of the system. These species are characterized by their elemental compositions and their partial enthalpies. The formation of daughter cells by reproduction is considered as the production of additional chemical species.*

relationships which describe the regulation of carbon and nitrogen metabolism within the cell. By combining measurement of some of the metabolites and these balance equations, we can predict the behavior of other metabolites that may be difficult to measure directly. An example of this approach is given in Problem 8. As we shall see in Chapter 5, the energy balance equations are important in providing information on the rates of heat evolution during growth. Maintenance of constant temperature is important in maximizing fermentation productivity, and design of bioreactor cooling systems requires an understanding of these rates of microbial heat evolution.

3.2.1 Elemental Balance Equations

The total number of nutrients in a medium for growth of an organism may be rather large. While some organisms are able to grow on a *simple* medium generally containing a sugar source, inorganic nitrogen and a mix of salts, more complex organisms, such as eukaryotic cells, require complex and often ill-defined sources of nutrients which may include amino acids, vitamins, growth factors and cofactors for enzymes. Attempting to describe all the components in a growth medium in order to write balance equations would be impractical. Instead, balance equations are written in terms of a growth-limiting component (substrate). This growth-limiting substrate will generally be the first one to be exhausted when the organism is grown batchwise. In principle, a growth-limiting nutrient is usually defined from a kinetic viewpoint, i.e., it is the nutrient which limits the growth *rate* of the organism (we shall see this exemplified in Section 3.3.3). It is also possible that the nutrient can limit the growth of the organism in terms of obtainable cell mass by a stoichiometric limitation. Thus one substrate may limit the growth rate and another the final cell concentration. In what follows, we shall consider a single growth-limiting substrate to limit both kinetically and stoichiometrically.

In order to write atomic species balances for cell growth, we need to obtain an elemental composition for a cell and define a cell formula. In establishing a cell formula and corresponding formula weight, it is assumed by convention that the cell formula is based on one gram-atom of carbon and contains only the major elements. We shall write the formula for a cell as $CH_aO_bN_c$, and ignore for the moment the other minor elements in the cell, such as phosphorus, sulfur and ash.

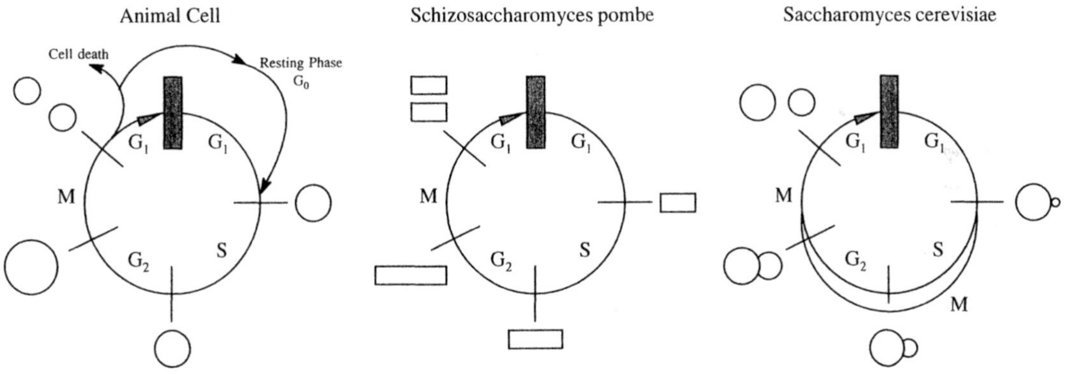

Figure 3.3. *Schematic representation of the mammalian cell cycle, fission yeast cell cycle (S. pombe) and budding yeast cell cycle (S. cerevisiae). The cycle runs clockwise. The incidence of death and resting phases are illustrated only in the mammalian cell cycle, but are similar in all eukaryotic organisms. Homologues of the cell division cycle control genes have been found in yeast and animal cells. For example, the cdc2 gene controls the cycle at the corresponding key points in all three organisms, as illustrated by the black bar. G1 is the gap; S DNA synthesis; G2 gap; M mitosis [From Fiechter, A. and Gmunder, F.K., Metabolic Control of Glucose Degradation in Yeast and Tumor Cells, Advances in Biochemical Engineering,* ***39****:1-28 (1989)].*

Composition of Cells

The various events that occur during the growth of a single cell from its inception till the time of its division into daughter cells are referred to as the *cell cycle*. In the *M phase* of the cell cycle, nuclear division occurs (mitosis), followed by the interphase. During the interphase, daughter cells, formed from the mitosis phase, enter the G_1 *phase*, which is characterized by a high rate of biosynthesis. The *S phase* then begins when DNA synthesis starts and ends when the DNA content of the cell has doubled. The G_2 *phase* follows and ends with the initiation of mitosis. Thus the biochemical content of the cell is constantly changing over the cell cycle. The phases of growth are illustrated below.

We define the rate of cell growth as r_X (volumetric rate of increase of cell concentration X) and the specific rate (μ) as r_X/X. We use the term *specific* to connote a volumetric rate per unit cell concentration, and μ has units of time^{-1}. When cells are grown at different specific growth rates (related to the doubling time of a cell), the duration of the G_1 phase changes. Because cells are usually distributed over all points in the cell cycle, we define an average cellular composition based on all cells in the population. Table 3.1 indicates the typical amounts of cellular components for a bacterium. Figure 3.4 shows the variation of major cellular components with the specific rate of growth of the bacterium *Aerobacter aerogenes*.

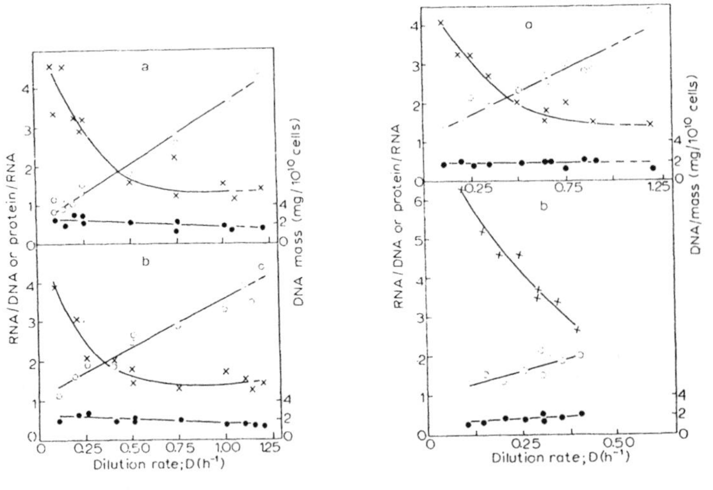

Figure 3.4(a). *The variation of macromolecular composition of Aerobacter aerogenes with dilution rate (equivalent to the specific growth rate) in steady state continuous cultures. Figure (a) contains data from carbon limited growth; figure (b) from nitrogen limited growth. Open circles, RNA/DNA ratio; closed circles, DNA/cell mass; x, protein/RNA. Data from Dean, A.C.R. and P.L. Rogers, Biochim. Biophys. Acta* ***148****, 267 (1967).*

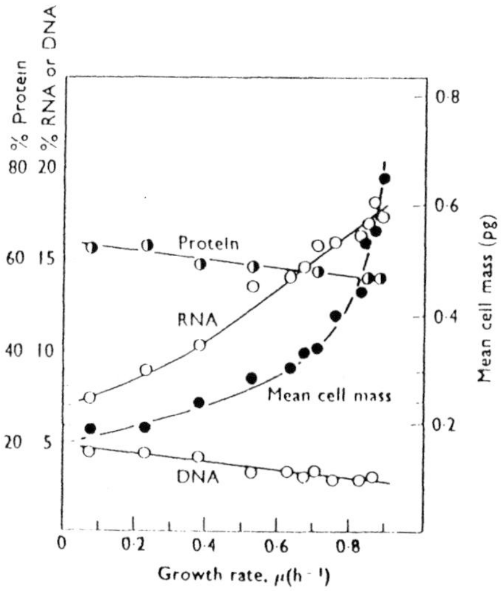

***Figure 3.4 (b)**. Effect of growth rate of A. aerogenes on macromolecular composition and mean cell size for glycerol-limited growth. Data from Herbert, D. Soc. Gen. Microbiol. **11** 391-416 (1961).*

The elemental composition of the cell is shown for various microorganisms in Table 3.2. The effect of growth rate on elemental composition can be seen to be small compared to intraspecies differences. When the rate-limiting nutrient is altered, the composition of the cell changes accordingly.

When cells are grown on ammonia as the only source of nitrogen, and a carbohydrate as sole carbon and energy source, a general form of the reaction resulting in cell growth and formation of products can be written:

$$\alpha CH_lO_m + \beta NH_3 + \gamma O_2 \rightarrow CH_aO_bN_c + \delta CH_pO_qN_r + \varepsilon H_2O + \kappa CO_2 \quad (3.2)$$

where $CH_aO_bN_c$ is the elemental composition of the cell
CH_lO_m is the elemental composition of the carbon source
$CH_pO_qN_r$ is the elemental composition of extracellular products

The stoichiometric coefficient of the cells has been taken as unity for simplicity. If a complex nitrogen source is required by the organism, e.g., an amino acid, then ammonia must be replaced in the above equation. Atomic balance equations can be written for C, H, O and N.

$$C: \quad \alpha = 1 + \delta + \kappa$$

$$H: \quad l\alpha + 3\beta = a + p\delta + 2\varepsilon$$

$$O: \quad m\alpha + 2\gamma = b + q\delta + \varepsilon + 2\kappa$$

$$N: \quad \beta = c + r\delta \quad (3.3)$$

***Table** 3.1. Molecular Composition of a Typical Bacterium [from B. Atkinson and F. Mavintuna, Biochemical Engineering and Biotechnology Handbook, Nature Press, (1983)].*

Component	*Weight (x 10^{13} g)*	*% of total weight*	*Molecular weight*	*Molecules per cell**
Entire cell	15	100		
water	12	80	18	4 x 10^{10}
dry weight	3	20		
Protein				
ribosomal	0.22	1.5	4 x 10^4	3 x 10^5
non-ribosomal	1.5	10	5 x 10^4	1.8 x 10^6
RNA				
ribosomal (16S)	0.15	1.0	6 x 10^5	1.5 x 10^4
ribosomal (23S)	0.30	1.0	1.2 x 10^6	1.5 x 10^4
t-RNA	0.15	1.0	2.5 x 10^4	3.5 x 10^5
m-RNA	0.15	1.0	1 x 10^6	9 x 10^5
DNA	0.15	1.0	4.5 x 10^9	2
	0.15	1.0	1.8 x 10^2	5 x 10^7
Polysaccharides				
Lipids	0.15	1.0	1 x 10^3	9 x 10^6
Small molecules	0.08	0.5	4 x 10^2	1.2 x 10^7

* calculated from 6 x 10^{23} x (weight/cell)/(molecular weight)

A solution for the six unknown stoichiometric coefficients (α, β, γ, δ, ε, and κ) cannot be obtained from these four atomic balance equations alone. Additional empirically-based relationships are required. One of these is provided in aerobic growth by the *respiratory quotient.* This is defined as the rate of carbon dioxide formation divided by the rate of oxygen consumption. It is denoted as RQ.

$$\mathrm{RQ} = \frac{\text{rate of} CO_2 \text{ production (mM/liter-hr)}}{\text{rate of} O_2 \text{ consumption (mM/liter-hr)}} \tag{3.4a}$$

This definition results in the relationship

$$\mathrm{RQ} = \frac{\kappa}{\gamma} \tag{3.4b}$$

Table 3.2. *Elemental composition of selected microorganisms, in percent by weight. [From B. Atkinson and F. Mavintuna, Biochemical Engineering and Biotechnology Handbook, Nature Press, (1983)].*

Microorganism	*Limiting Nutrient*	$\mu(hr^{-1})$	*% C*	*% H*	*% N*	*% O*	*% P*	*% S*	*% ash*	*Empirical chemical formula*	*Formula molecular weight*
Bacteria			53.0	7.3	12.0	19.0			8	$CH_{1.66}N_{0.20}O_{0.27}$	20.7
Bacteria			47.1	7.8	13.7	31.3				$CH_2N_{0.25}O_{0.5}$	25.5
E. aerogenes			48.7	7.3	13.9	21.1			8.9	$CH_{1.78}N_{0.24}O_{0.33}$	22.5
K. aerogenes	glycerol	0.1	50.6	7.3	13.0	29.0				$CH_{1.74}N_{0.22}O_{0.43}$	23.7
K. aerogenes	glycerol	0.85	50.1	7.3	14.0	28.7				$CH_{1.73}N_{0.24}O_{0.43}$	24.0
Yeast			47.0	6.5	7.5	31.0			8	$CH_{1.66}N_{0.13}O_{0.40}$	23.5
Yeast			50.3	7.4	8.8	33.5				$CH_{1.75}N_{0.15}O_{0.5}$	23.9
Yeast			44.7	6.2	8.5	31.2	1.08	0.6		$CH_{1.64}N_{0.16}O_{0.52}P_{0.01}S_{0.005}$	26.9
C. utilis	glucose	0.08	50.0	7.6	11.1	31.3				$CH_{1.826}N_{0.19}O_{0.47}$	24.0
C. utilis	glucose	0.45	46.9	7.2	10.9	35.0				$CH_{1.84}N_{0.20}O_{0.56}$	25.6
C. utilis	ethanol	0.06	50.3	7.7	11.0	30.8				$CH_{1.82}N_{0.19}O_{0.46}$	23.9
C. utilis	ethanol	0.43	47.2	7.3	11.0	34.6				$CH_{1.84}N_{0.20}O_{0.55}$	25.5

where κ and γ are stoichiometric constants defined in Equation (3.2). Another relationship to determine unknown stoichiometric coefficients can be obtained via an electron balance. The *degree of reductance* of an organic compound is defined as the number of moles of electrons available (per g atom carbon) for transfer to oxygen on combustion of a compound to CO_2, H_2O and N_2. In other words, the degree of reductance of a compound is defined as the number of equivalents of available electrons per g atom of carbon of the compound. The number of equivalents of available electrons is taken as 4 for carbon, 1 for hydrogen, -2 for oxygen and -3 for nitrogen. Based on these values, we see that CO_2, H_2O and NH_3 all have zero degrees of reductance. Applying this concept to the species in Equation (3.2), we can define the reductance degree of biomass, substrate and product as γ_b, γ_S and γ_P.

$$\text{Cells:} \quad \gamma_b = 4 + a - 2b - 3c$$

$$\text{Substrate:} \quad \gamma_S = 4 + l - 2m$$

$$\text{Product:} \quad \gamma_P = 4 + p - 2q - 3r \tag{3.5}$$

A balance equation for available electrons can now be written from equation (3.2), noting that CO_2, H_2O and NH_3 all have zero degrees of reductance:

$$\alpha\gamma_S - 4\gamma = \gamma_b + \delta\gamma_P \tag{3.6}$$

Equation (3.6) can be rearranged in the form:

$$1 = \underset{\text{oxygen}}{\frac{4\gamma}{\alpha\gamma_S}} + \underset{\text{cells}}{\frac{\gamma_b}{\alpha\gamma_S}} + \underset{\text{product}}{\frac{\delta\gamma_P}{\alpha\gamma_S}} \tag{3.6a}$$

The reductance degree balance equation is a linear combination of the C, H, O and N atomic balance equations. The reductance degree balance can be derived by elimination of the coefficient for water (ε) from the hydrogen and oxygen balance equations. Thus it does not provide additional information when all the other balance equations are available. However, the production of water is not easily measured in aqueous fermentation systems, and so the reductance balance is quite useful.

Values of γ_b are remarkably constant for various bacteria and yeast. Some typical values are shown in Table 3.3. The substrate and product are typically specified, i.e., values of l, m, p, q and r are known. Eliminating ε as an unknown and assuming that γ_b is constant leads to a set of five equations (the C and N elemental balances, Eq. 3.4b, the definition of γ_b, and Eq. 3.6) that can be solved for the five unknown coefficients (α, β, γ, δ, and κ).

The heat of reaction per oxygen-released electron (Q_o) is approximately constant for most organic molecules, and has the value ~27 kcal/mole of available electrons released. Each term in equation (3.6a) can be multiplied by (Q_o/Q_o) to transform the electron balance

Table 3.3. *Examples of the Degree of Reductance of Biomass (γ_b). The average value of γ_b for a variety of organisms grown on different substrates is 4.291 ± 0.172[2].*

Microorganism	*Carbon Source*	γ_b
Candida tropicalis	n-alkanes	4.385
Bacterium	n-pentane	4.607
Saccharomyces cerevisiae	glucose	4.291
Saccharomyces cerevisiae	acetic acid	4.416
Saccharomyces cerevisiae	ethanol	4.469

into an enthalpy balance. The term $(4\gamma Q_o)/(\alpha\gamma_S Q_o)$ corresponds to the fraction of substrate energy released as heat. Similarly, the second and third terms correspond to the fractions of substrate energy converted to cells and to product.

An empirical correlation between the heat of combustion of biomass ($-\Delta H_c$ kcal/gm) and Q_o (kcal per electron transferred) has been developed[3].

$$Q_o = \frac{12(-\Delta H_c)}{\sigma_b \gamma_b} - 3.328 \tag{3.7}$$

where σ_b is the weight fraction of carbon in biomass (typically 0.46). This equation includes a correction to convert combustion of biomass nitrogen to N_2 instead of NH_3. We need to express the heat of combustion of biomass on a weight basis, since a "mole" of cells is not defined. Equation (3.7) can be used to estimate the ΔH_c in energy balances, if typical values of Q_o, σ_b and γ_b are assumed.

3.2.2 Metabolic Coupling - ATP and NAD^+

Additional equations can be written describing the generation of ATP (the "energy currency" of the cell) and NADH (employed for cellular electron transport) based on the oxidation of the carbon source CH_lO_n. In the process of substrate oxidation, electrons are transferred to NAD^+ in the form of hydrogen atoms. To develop these equations, we need to consider the overall oxidation of the carbon and energy source as three terms; one representing complete substrate combustion to CO_2 for energy production, one representing biosynthesis of cells with partial oxidation of the substrate to CO_2, and a third term in which a product is formed from the carbon source. We will examine each of these terms.

---

(2) Erickson, L.E. Biotechnol. Bioeng., **22**, 451 (1980)

(3) Solomon B.O. and L. E. Erickson, *Process Biochemistry*, ***12***, 44 (1981)

Energy Generation

$$\xi_1 CH_lO_m + \xi_1\gamma_S NAD^+ + \xi_1 H_2O \quad \rightarrow \quad \xi_1 CO_2 + \xi_1\gamma_S(NADH + H^+) \qquad (3.8a)$$

$$\xi_1\eta_S(ADP + P_i) \quad \rightarrow \quad \xi_1\eta_S(ATP + H_2O) \qquad (3.8b)$$

The fraction of substrate which undergoes this reaction is denoted by ξ_1. The stoichiometric coefficient of water in this reaction has been specified as unity for simplicity. The second equation represents the direct formation of ATP from substrate-level phosphorylation reactions that occur as part of oxidation (e.g. pyruvate kinase catalyzed production of pyruvate and ATP from phosphoenolpyruvate in the glycolytic sequence of reactions). η_S denotes the number of such substrate level phosphorylation reactions which occur. These are distinct from ATP formation which occurs via oxidative phosphorylation, which is shown below.

Oxidative phosphorylation

We denote the fraction of NADH that undergoes oxdative phosphorylation as ξ_2. The reaction which occurs is

$$2\xi_2(NADH + H^+) + \xi_2 O_2 \quad \rightarrow \quad 2\xi_2 NAD^+ + 2\xi_2 H_2O \qquad (3.9a)$$

$$2\xi_2(P/O)(ADP + P_i) \quad \rightarrow \quad 2\xi_2(P/O)(ATP + H_2O) \qquad (3.9b)$$

where P/O is the *oxidative phosphorylation ratio* (P/O ratio). This is the ratio of the number of ADP phosphorylations which occur per atom of O_2 consumed. This ratio characterizes the efficiency of oxidative phosphorylation and varies with the carbon source. With glucose, the P/O ratio is generally 3, although this value can vary depending on the organism and the growth conditions. Other substrates have different values, e.g. (P/O) for microbial growth on acetate is typically 2.25.

Biosynthesis of cells

Here the fraction of substrate that is consumed for production of cell matter is denoted by ξ_3. The reaction requires nitrogen, reducing equivalents and energy.

$$\xi_3 CH_lO_m + \xi_3\beta_1 NH_3 + \xi_3(\gamma_B - \gamma_S)(NADH + H^+) \quad \rightarrow \quad CH_aO_bN_c + \xi_3\kappa_1 CO_2$$

$$+ H_2O + \xi_3(\gamma_B - \gamma_S)NAD^+ \qquad (3.10a)$$

$$\xi_3\frac{1}{Y'_{ATP}}(ATP + H_2O) \quad \rightarrow \quad \xi_3\frac{1}{Y'_{ATP}}(ADP + P_i) \qquad (3.10b)$$

ATP is consumed in the macromolecular synthesis reactions involved in cell growth. The moles of C-equivalents in cells produced per mole of ATP consumed is called the ATP yield coefficient, Y'_{ATP}. This is more conveniently expressed in terms of a weight yield, Y_{ATP}, the grams of cells produced per mole of ATP consumed. As we shall see later, this is approximately constant for many anaerobic organisms (10.7 gm cells/mole ATP) grown on a wide variety of substrates. Under aerobic conditions, the cell mass produced per mole of ATP consumed is greater and a constant value is not obtained.

Product Formation

The equation for product formation from substrate requires a knowledge of the reaction pathways involved, in order to determine whether ATP is required or produced by the product synthesis reactions. As we shall see later (Section 3.3.7) product formation may be uncoupled from cell growth. Even in cases where a reaction stoichiometry can be written directly from substrate to product, the actual metabolic pathway employed by the cell may involve intermediates not considered by the simple stoichiometry pathway. Thus this simple reaction can only be considered as a lower limit on the ATP requirement for product synthesis.

Maintenance

A further equation needs to be added to complete our series of ATP and NAD^+ balances. This is an equation which describes the consumption of ATP by the cell for purposes of maintenance of chemical gradients (e.g., regulation of internal pH), transport of nutrients from the surrounding medium through the cell membrane to the cytosol, DNA unwinding and repair, DNA replication, etc. The consumption of ATP for these purposes is referred to as *maintenance*. The amount of ATP consumed by maintenance varies and a constant stoichiometric relationship cannot be written. It is found experimentally that the maintenance stoichiometry varies with the rate of cell growth. We will discuss this later in the following section.

3.2.3 Yield Coefficients

The reaction stoichiometries above provide a convenient means to define various *yield coefficients*. These relate the amount of products and cells formed per unit of substrate consumed by the cells. A knowledge of yield coefficients allows us to design a growth medium that will supply all the required nutrients in balanced amounts, so that a desired nutrient can be made growth-rate limiting. We shall employ the nomenclature that Y' denotes a yield coefficient based on mole C-equivalents, and Y denotes a coefficient based on mass. The following yield coefficients are thus defined:

$$Y'_{X/S} = \frac{\text{moles C-equivalents in cells produced}}{\text{moles C-equivalents in substrate consumed}}$$

$$Y_{X/S} = \frac{\text{mass cells produced}}{\text{mass substrate consumed}} \qquad (3.11)$$

$$Y'_{X/O} = \frac{\text{moles C-equivalents in cells produced}}{\text{moles } O_2 \text{ consumed}}$$

$$Y_{X/O} = \frac{\text{mass cells produced}}{\text{mass } O_2 \text{ consumed}} \tag{3.12}$$

$$Y'_{P/S} = \frac{\text{moles C-equivalents in product produced}}{\text{moles C-equivalents in substrate consumed}}$$

$$Y_{P/S} = \frac{\text{mass product produced}}{\text{mass substrate consumed}} \tag{3.13}$$

Relating Yield Coefficients to Stoichiometry

If cell growth is described by the stoichiometric Equation (3.2), the yield coefficients can be written as

$$Y'_{X/S} = \frac{\text{moles C-equivalents in cells produced}}{\text{moles C-equivalents in substrate consumed}} = \frac{1}{\alpha}$$

$$Y_{X/S} = \frac{\text{mass cells produced}}{\text{mass substrate consumed}} = \frac{\sigma_S}{\sigma_b} \cdot \frac{1}{\alpha}$$

where σ_b is the weight fraction of carbon in the cell mass and σ_S is the weight fraction of carbon in the substrate. σ_b has been found to be constant for many microorganisms and has a value of 0.462 ± 0.023. Oxygen and product yield coefficients can be similarly defined:

$$Y'_{X/O} = \frac{\text{moles C-equivalents in cells produced}}{\text{moles } O_2 \text{ consumed}} = \frac{1}{\gamma}$$

$$Y_{X/O} = \frac{\text{mass cells produced}}{\text{mass } O_2 \text{ consumed}} = \frac{3}{8\sigma_b} \cdot \frac{1}{\gamma}$$

$$Y'_{P/S} = \frac{\text{moles C-equivalents in product produced}}{\text{moles C-equivalents in substrate consumed}} = \frac{\delta}{\alpha}$$

$$Y_{P/S} = \frac{\text{mass product produced}}{\text{mass substrate consumed}} = \frac{\sigma_S}{\sigma_P} \cdot \frac{\delta}{\alpha}$$

where σ_P is the weight fraction of carbon in the product ($CH_pO_qN_r$).

Based on these definitions, it is apparent that $Y'_{X/S} \leq 1$. By substituting the definitions of $Y'_{X/S}$ and $Y'_{X/O}$ into equation (3.5) and examining a limiting case where no product is formed ($\delta = 0$), we obtain:

$$\frac{1}{Y'_{X/O}} = \left(\frac{\gamma_S}{4}\right) \cdot \frac{1}{Y'_{X/S}} - \frac{\gamma_b}{4} \tag{3.14}$$

Since $Y'_{X/S} \leq 1$, we have an upper limit for $Y'_{X/O}$

$$Y'_{X/O} \leq \frac{4}{\gamma_S - \gamma_b} \tag{3.14a}$$

and as $Y'_{X/O} \geq 0$, $Y'_{X/S} \leq \gamma_S/\gamma_b$ (this follows directly from equation (3.6a)). Since γ_b is approximately 4.291, the upper limit for molar cell yields can be plotted as a function of the degree of reductance of the substrate, γ_S. Such a plot is shown in Figure 3.5.

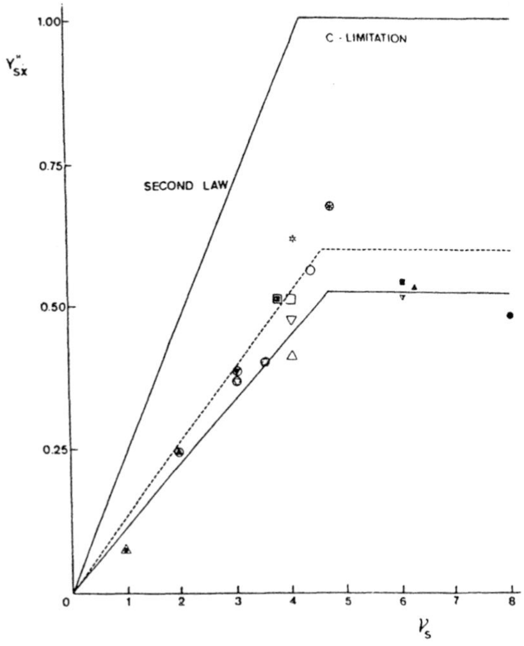

Figure 3.5. *The yield coefficient limits of $Y'_{X/S}$ (mole dry cell weight/C-mole substrate) as a function of the degree of reductance of the substrate for several carbon sourcces. The nitrogen source is NH_3. The theoretical limitation is based on carbon limitation and the second law of thermodynamics. Symbols are for different substrates, including methane, n-alkanes, methanol, ethanol, glycerol, mannitol, acetic acid, lactic acid, glucose, form-aldehyde, gluconic acid, succinic acid, citric acid, malic acid, formic acid, and oxalic acid. [From J.A. Roels, "Energetics and Kinetics in Biotechnology", Elsevier (1983)].*

When product formation is included δ is no longer zero, and Equation (3.6a) can be written as:

$$\frac{1}{Y'_{X/S}} = \left(\frac{4}{\gamma_S}\right) \cdot \frac{1}{Y'_{X/O}} + \frac{1}{Y'_{P/S}} \cdot \frac{\gamma_P}{\gamma_S} - \frac{\gamma_b}{\gamma_S} \tag{3.15}$$

This equation provides a useful test of the consistency of data on cell growth and product formation. The experimentally determined values of the cell and product yield coefficients can be checked by their insertion in the above equation. Inconsistencies may point to other unidentified products being formed or to errors in the data.

Other Yield Coefficients; Y_{ATP}, $Y_{ave\ e}$ *and* Y_{kcal}

The ATP yield coefficient Y'_{ATP} (moles C-equivalents in cells per mole ATP consumed) has been introduced in Equation (3.10b). It is, however, typically used in the form of the weight yield Y_{ATP} (grams of cells produced per gram mole ATP consumed). In aerobic organisms, the fraction of carbon source which is employed for energy production can be quite large. For *E. coli* grown aerobically only on glucose and inorganic salts, a maximum value of Y_{ATP} has been estimated as 28.8 gm cells/gmol ATP, based on the known biosynthetic pathways[4]. Anaerobic growth shows an average Y_{ATP} of 10.7 gm cells/mol ATP, with a typical range of 8.3 to 12.6 gm cells/mol ATP. These lower values are a result of less efficient energy production by cells in the absence of oxidative phosphorylation in mitochondria. In growth media containing metabolic precursors ("complex" media), there are generally sufficient amounts of monomer units (e.g., amino acids) which can be directly incorporated into polymers (e.g. proteins) and the growth-limiting substrate can be completely catabolized. Thus an average value for Y_{ATP} is found for a variety of anaerobically grown organisms with different carbon sources.

The growth yield coefficient $Y_{ave\ e}$ was introduced to overcome some inadequacies associated with the yield coefficient $Y_{X/S}$, which is based solely on the carbon content of the substrate without regard to its degree of oxidation. To generalize the cell yield to an *energy* content of the substrate, $Y_{ave\ e}$ is defined as the weight of dry cell mass produced per electron equivalent available from the substrate. Thus we write:

$$Y_{ave\ e} = \frac{Y_{X/S}}{Y_{ave\ e/S}} \tag{3.16}$$

where $Y_{ave\ e/S}$ is the number of electrons available from the substrate. This can be calculated by multiplying the number of moles of oxygen required to completely combust one mole of the substrate by four (the number of electrons required to reduce one molecule of oxygen). For example, $Y_{ave\ e/S}$ for glucose is 24, as 6 moles of O_2 are required to combust one mole of glucose and one mole of O_2 corresponds to 4 equivalents of electrons. The average value for $Y_{ave\ e}$ is 3.07 gm cell/gm ave e for C_4 to C_6 carbon substrates. When lower carbon number substrates are examined, much lower $Y_{ave\ e}$ values are obtained. This can be explained by the weight basis employed in the definition of $Y_{ave\ e}$. When one mole of glucose (180 gm) is transported into the cell one mole of ATP is expended. When one mole of acetate (60 gm)

---

(4) Stouthamer, A.H. "A Theoretical Study on the Amount of ATP Required for the Synthesis of Microbial Cell Material", *Anton. Van Leeuwenhoek*, **39**, 545 (1973).

is transported, one mole of ATP is similarly expended. On a weight basis, transport of 180 gram of each substrate requires 3 times more ATP in the case of acetate than glucose. Thus the value of $Y_{ave\,e}$ will be lower. In addition, more energy must be expended with low carbon number substrates rather than high (5 or 6) carbon number substrates to produce macromolecules such as DNA and proteins.

Following a similar approach, a second type of growth yield Y_{kcal} relates cell production to microbial heat evolution. It relates growth to an enthalpy balance and is useful in determining the heat released during fermentation:

$$Y_{kcal} = \frac{\Delta X}{\Delta Q} \tag{3.17}$$

Here ΔQ is the total heat evolved during microbial growth (kcal) and can be found from an enthalpy balance viz:

$$\Delta Q = (-\Delta H_S)(-\Delta S) + (-\Delta H_N)(-\Delta N) - (-\Delta H_C)(\Delta X) - \sum_j(-\Delta H_{Pj})(\Delta P_j) \tag{3.18}$$

where ΔH_S, ΔH_N, ΔH_C, ΔH_{Pj} (kcal/gm) are the heats of combustion of carbon substrate, nitrogen source, microbial cells and jth product respectively. ΔS, ΔN, ΔX[5] and ΔP_j refer to the amounts of the respective materials utilized or produced. Heats of combustion of carbon substrate and nitrogen source can be obtained from standard chemistry reference tables. The value of the heat of combustion of cells, ΔH_C, can be estimated from a modification of the Dulong equation given below, using the experimentally-determined weight fractions of carbon, hydrogen and oxygen (C, H, and O, respectively) in a typical cell:

$$(-\Delta H_C) = \frac{kcal}{gm - cells} = 8.076C + 34.462\left(H - \frac{O}{8}\right) \tag{3.19}$$

The caloric content of most microorganisms (ΔH_C on dry weight, ash-free basis) ranges from 5 to 6 kcal/gm, with a mean of 5.41 kcal/gm cells.

Several correlations for (ΔQ) as a function of the amount of oxygen consumed during aerobic growth have been experimentally determined. For example, Mayberry et al[6] proposed that ΔQ (in kcal) could be related to the heat of combustion per available electron equivalent transferred to oxygen

$$\begin{aligned}\Delta Q &= \left(\frac{kcal}{\text{mole available electrons transferred}}\right)(-\Delta O_2) \\ &= \left(27\frac{kcal}{\text{ave e}}\right)\left(\frac{\text{4 avail e}}{\text{mole } O_2}\right)(-\Delta O_2) \\ &= 108(-\Delta O_2) \end{aligned} \tag{3.20}$$

(5) We shall use X rather than C to denote cell concentration.

(6) Mayberry, W.R. et al. *Appl. Microbiol.*, **15**, 1332 (1967).

where $(-\Delta O_2)$ is the molar amount of O_2 consumed during aerobic growth. Other expressions include those of Cooney et al.[7] (Equation 3.21a), and Luong and Volesky[8] (Equation (3.21b).

$$\Delta Q = 124.1\ (kcal/mole)(-\Delta O_2) \tag{3.21a}$$

$$\Delta Q = 110.1\ (kcal/mole)(-\Delta O_2) \tag{3.21b}$$

Thus Y_{kcal} can be determined from oxygen consumption data using the above correlations. We can also determine Y_{kcal} for growth on a variety of substrates by examining the components of ΔQ more closely. We can lump terms in Equation (3.18) together and insert terms into (3.17) to obtain

$$Y_{kcal} = \frac{\Delta X}{(-\Delta H_C)(\Delta X) + \Delta Q_{cat}} \tag{3.17a}$$

where ΔQ_{cat} is the heat generated by catabolism. The denominator thus represents the sum of the energy incorporated into cellular material and the energy expended by the cells in catabolism. When cells are grown on a complex medium, most of the carbon source is consumed for energy and little of it is incorporated into biomass, as metabolic precursors are available from the medium. When cells are grown on a simple salts medium, the carbon source is only partly metabolized by catabolic pathways, and some of it must be employed for biosynthesis. Thus the term ΔQ_{cat} must be modified to account for that part of the carbon that is incorporated into cells. The amount of carbon source that is consumed by energy producing processes is

$$(-\Delta S)\left(1 - \frac{\sigma_b}{\sigma_S} Y_{X/S}\right)$$

and ΔQ_{cat} can be written as

$$\Delta Q_{cat} = (-\Delta H_S)(-\Delta S)\left(1 - \frac{\sigma_b}{\sigma_S} Y_{X/S}\right) - \sum_j (-\Delta H_{Pj})(\Delta P) \tag{3.22}$$

We can divide equation (3.17a) by ΔS and substitute for ΔQ_{cat} to obtain

$$Y_{kcal} = \frac{Y_{X/S}}{(-\Delta H_C) Y_{X/S} + (-\Delta H_S)\left(1 - \frac{\sigma b}{\sigma_S} Y_{X/S}\right) - \sum_j (-\Delta H_{Pj})(Y_{Pj})} \tag{3.23}$$

Values of Y_{kcal} for cases where no products other than cells are produced are shown in Table 3.4. Values for growth on glucose are higher than those obtained when lower carbon number substrates are consumed, due to the higher ATP requirements for transport of an eqúivalent mass of substrate molecules.

(7) Cooney, C.L., Wang, D.I.C. and R.I. Mateles, *Biotech. Bioeng.*, **11**, 269 (1968).
(8) Luong, J.H. and B. Volesky, *Can. J. Chem. Eng.*, **58**, 497 (1980).

Table 3.4. *Calculated values of Y_{kcal} from Equation (3.23) for heterotrophs growing on minimal medium without producing products. Values based on observed values of $Y_{X/S}$ and ΔH_S, assuming σ_b = 0.5 gm/gm and ΔH_C = 5.3 kcal/gm. [From Nagai (1979)][9].*

Organism	*Substrate*	*Y_{kcal} (gm/kcal)*
Aerobacter aerogenes	maltose	0.104
Candida utilis	glucose	0.126
Penicillium chrysogenum	glucose	0.107
Pseudomonas fluorescens	glucose	0.096
Rhodopseudomonas spheroides	glucose	0.112
Saccharomyces cerevisiae	glucose	0.123
Aerobacter aerogenes	glucose	0.101
Average:		***0.116***

Maintenance and Endogeneous Respiration

As was described in section 3.2.2, ATP is consumed by the cell for biosynthesis, for DNA repair, and for maintenance of various chemical potentials in the cell. The energy for maintenance is generally supplied by the energy released by consumption of substrate, resulting in the formation of CO_2 and any intermediate products, denoted as CH_xO_y . We can represent this by the maintenance reaction

$$CH_lO_m + \lambda_1 O_2 \rightarrow \lambda_2 CO_2 + \lambda_3 CH_xO_y + \lambda_4 H_2O \qquad (3.24)$$

The ATP that is generated from this reaction provides the maintenance energy. The rate of substrate consumption for maintenance is denoted by m, with units of (gm substrate consumed for maintenance)/(gm dry cell-hr). We can write a general equation for substrate consumption which includes biosynthesis, the maintenance contribution and product formation:

$$r_{S_i} = \frac{r_X}{Y_{X/S_i}} + m_i X + \sum_j \frac{r_{P_j}}{Y_{P_j/S_i}} \qquad (3.25)$$

where r_{Si} is the rate of consumption of substrate S_i (gm/liter-hr)
r_X is the rate of increase of dry cell weight (gm/liter-hr)

(9) S.Nagai, *Mass and Energy Balances for Microbial Growth Kinetics*, Adv. Biochem. Eng. **11**, 49 (1979)

r_{Pj} is the rate of product P_j formation (gm P_j/liter-hr)
$Y_{X/Si}$ is the yield of cells on substrate S_i (gm cells/gm S_i)
$Y_{Pj/Si}$ is the yield of product P_j on substrate S_i (gm P_j/gm S_i)
X is the dry cell weight (gm/liter)
m_i is the maintenance coefficient, i.e., the rate of consumption of S_i due to maintenance processes (gm S_i/gm dry cell-hr)

The value of m_i depends on the type of substrate consumed and on environmental conditions such as pH and temperature. We can write the above equation for a carbon substrate, a nitrogen-containing substrate, for oxygen or for available electrons. The values of the corresponding maintenance coefficients can then be related from the stoichiometry of the maintenance reaction (Equation (3.24)). In practice, only maintenance coefficients for substrate and oxygen are used. Most of the maintenance requirement is involved in maintaining osmotic gradients. Thus if a cell is placed in a medium of high salinity or one which has an extreme pH value, the maintenance coefficient increases significantly.

Table 3.5*. Values of the maintenance coefficient for various organisms*[10]

Organism	*Substrate*	*Maintenance coefficient for substrate* m_S	*Maintenance coefficient for oxygen* m_O
Aerobacter aerogenes	Glucose	5.4	5.4
Aerobacter aerogenes	Glycerol	7.6	10.9
Saccharomyces cerevisiae	Glucose	1.8	1.92
Escherichia coli	Glucose	5.4	1.6
Methane bacteria	Methane	2.0	8.0
Penicillium chrysogenum	Glucose	2.2	2.4
Aerobacter aerogenes	Citrate	5.8	4.8

m_s (x 10^2)(gm/gm dry cell weight-hr)
m_O (x 10^2)(gm/gm dry cell weight-hr)

When the cell is in an environment where no substrate is available to supply its maintenance requirements, it may degrade some of its own cell mass for maintenance. This is called *endogenous respiration* or *endogenous metabolism*. This degradation results in a decrease in cell mass with time. Endogeneous metabolism may also occur even if sufficient substrate for maintenance is available. The rate of decrease of cell mass for endogenous

(10) From J.A. Roels and N.W.F. Kossen, *On the Modelling of Microbial Metabolism*, Progress in Industrial Microbiology, **14**, 95 (1978)

metabolism is given by k_eX (gm cell/hr), where the coefficient k_e is the endogenous rate constant (gm cell mass/gm dry cell-hr). While the maintenance term m_iX appears in the substrate mass balance equation (Eq. 3.25), endogenous metabolism appears in the cell mass balance. When the substrate concentration falls below that required for maintenance, the substrate balance equation will predict a cessation of growth, whereas the endogenous metabolism term in the cell mass balance will predict a decline in cell mass with time. This decline can be observed experimentally. Despite this, the maintenance term in the substrate balance is more commonly used. Some experimental values of the maintenance coefficient for substrate and oxygen are given in Table 3.5.

3.3 Unstructured Models of Microbial Growth

The growth of microbial cells can be viewed from various perspectives and with varying degrees of complexity, depending whether we distinguish between individual cells in a reactor and whether we examine the individual metabolic reactions occurring within the cell. While the most realistic model of the growth of a microbial population would consider all reactions occurring within each cell and the variations from cell to cell in a population, such a model would be very unwieldy. We must therefore make some simplifications; the extent of these will depend on the proposed use of the model. We shall make the following distinctions in models describing cell growth. When the population is segregated into individual cells that are different from one another in terms of some distinguishable characteristics, the model is *segregated*. On the other hand, *nonsegregated* models consider the population as lumped into one "biophase" which interacts with the external environment, and can be viewed as one "species" in solution; the cell concentration can be described by one variable alone. Nonsegregated models have the advantage that they are mathematically simple. The usefulness of segregated models depends on our experimental ability to distinguish between cells in a population. Often this is difficult.

Consideration of the details of the reactions occurring within a cell gives rise to the concept of *structure*. Structured models consider individual reactions or groups of reactions occurring within the cell. The biomass is subdivided into a number of components, e.g. DNA, RNA, proteins etc., and the reactions occurring among these components are described. The external environment can influence the response of the cell in a variety of ways, e.g., lactose may induce the production of the enzyme β-galactosidase, which breaks lactose into its component saccharides, glucose and galactose. A structured models would attempt to describe this behavior, e.g., the mechanism of how the new enzyme is induced (its transcription and translation), the kinetics of breakdown of lactose to its component saccharides, and their subsequent utilization. *Unstructured* models simply view the cell as an entity in solution which interacts with the environment.

An additional level of complexity results when we consider that models may be *stochastic* or *deterministic* in nature. A stochastic model considers the distribution of cell characteristics of interest. For example, the generation time of a cell population may be found from sampling the generation times of individual cells in that population and calculating a distribution of generation times based on the sample, e.g., a normal or Gaussian distribution. As the number of cells in the sample increases, the uncertainty in the mean generation time decreases (even if the standard deviation is large) and the population can be adequately described by a lumped parameter having the average generation time of the sample. Stochastic models are therefore only useful for situations where the cell number is very small and the distribution of cellular characteristics becomes important. For example, when sterilization kinetics are considered and the number of viable cells falls below 10^4 per ml, the probability that a cell will become non-viable becomes significant in determining the effectiveness of the sterilization operation. On the other hand, deterministic models have outputs that are completely determined by the inputs to the models; i.e., random variations in system properties are not considered. Deterministic models are the most common (and useful) and we shall only consider such models in the following sections.

3.3.1 Phases of the Batch Growth-Cycle

When microbial cells are inoculated into a batch reactor containing fresh culture medium and their increase in concentration is monitored, several distinct phases of growth can be observed. There is an initial *lag phase*, which is of variable duration. This is then followed by the *exponential growth phase*, where cell number (and dry weight) increases exponentially. This is also referred to as the *logarithmic phase*, the name arising from the common method of plotting the logarithm of cell number against time. Following this is a short phase of *declining growth*, and then the *stationary phase*. Here the cell numbers are highest. Finally the cell numbers decline during the *death phase*. These phases are illustrated in Figure 3.6 below.

The *lag phase* results from several factors. When cells are placed in fresh medium, intracellular levels of cofactors (e.g., vitamins), amino acids and ions (e.g., Mg^{2+}, Ca^{2+} etc.) may be transported across the cell membrane and thus their concentration may decrease appreciably. If intermediates in metabolic pathways are required for enzyme activity, this dilution may reduce the rate at which various pathways operate. Cells must then metabolize the available carbon sources to replenish the intracellular pools prior to initiating cell division. Similarly, if the inoculum is grown in a medium containing a different carbon source from that of the new medium, new enzymes may need to be induced to catabolize the new substrate and this will also contribute to a lag. The point in the growth cycle from which the inoculum was derived is also important. Cells taken from the exponential phase and used as an inoculum generally show a shorter lag phase than those taken from later phases. These exponentially-derived inocula will have adequate concentrations of intermediates and will not suffer from the dilution effect. If an inoculum is placed in a *rich*

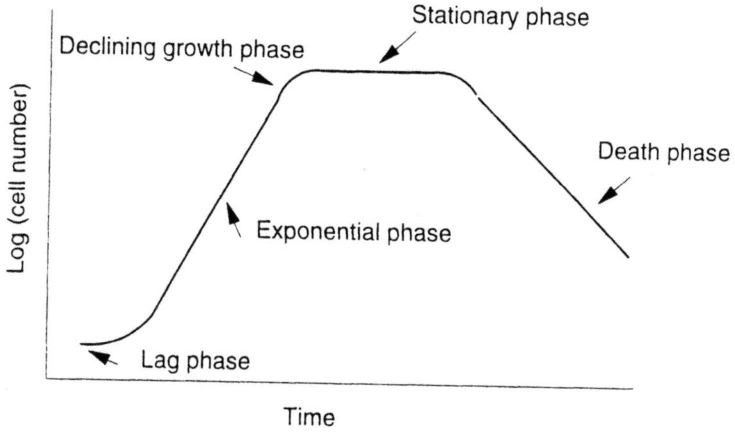

Figure 3.6. *Typical growth characteristics of a microorganism in a batch reactor.*

medium, one containing amino acids and other complex carbon and nitrogen sources, a shorter lag phase results as the intermediates of metabolism are already provided.

When cells are placed in a medium which contains several carbon sources, several lag phases may result. This is known as *diauxic growth*. Cells preferentially use one carbon source prior to consuming the second, due to catabolite repression of the enzymes required to metabolize the second carbon source. For example, when *E. coli* is placed in a medium containing both glucose and lactose, glucose is consumed first and a lag phase follows as cells synthesize the enzyme β-galactosidase, which is required for lactose utilization. During growth on glucose, the formation of this enzyme is under catabolite repression by cyclic AMP, and thus indirectly by glucose. Its concentration in the cell is thus very low. The cells consume glucose (with no additional metabolic energy expenditure) prior to lactose, which requires the synthesis of this enzyme. We shall examine some models of diauxic growth in Section 3.3.4.

Cell division occurs in the *exponential phase*. The rate of increase of cell number (N) is proportional to the number of cells. Cells increase in a geometric progression $2^0, 2^1, 2^2, .. 2^m$ after m divisions. For example, if the initial cell number was N_0, the number after m generations is $2^m N_0$.

Instead of cell number, it is often more convenient to use dry cell weight per volume X as a measure of cell concentration. During the exponential phase in a batch reactor we can write

$$\frac{dX}{dt} = \mu X \tag{3.26}$$

where μ is the *specific growth rate* of the cells. The above equation can be integrated from the end of the lag phase ($X = X_0$, $t = t_{lag}$) to any point in the exponential phase (X,t)

$$X = X_o e^{\mu(t - t_{lag})} \qquad \text{or} \qquad \ln\left(\frac{X}{X_o}\right) = \mu(t - t_{lag}) \tag{3.27}$$

The time required for the cell numbers or dry weight to double, the *doubling time* t_d, is related to the specific growth rate by

$$t_d = \frac{\ln 2}{\mu} \tag{3.28}$$

Occasionally it is found that the doubling times for cell number and cell dry weight may differ, as a result of a non-constant cell mass per cell during the exponential phase. Therefore, we define the cell number specific growth rate (ν hr^{-1}) separately from the specific growth rate (μ hr^{-1}) as follows:

$$\nu = \frac{1}{N}\frac{dN}{dt} \qquad \mu = \frac{1}{X}\frac{dX}{dt} \tag{3.29}$$

When μ and ν are equal, growth is referred to as *balanced.* In balanced growth, there is an adequate supply of all non-limiting nutrients, so that the composition of the cell is constant even though the concentrations of all other nutrients may be decreasing. On the other hand, when growth is unbalanced, variations in cell composition (e.g., protein content) may occur. Although the cell number growth rate may be constant, the cell mass growth rate will vary.

At low nutrient concentrations, it is found that the specific growth rate depends on the nutrient concentration. At high concentrations, the specific growth rate reaches a maximum value, set by the intrinsic kinetics of intracellular reactions, which are related to DNA transcription and translation. The end of the exponential phase arises when some essential nutrient, for example the carbon or nitrogen source, is depleted or when some toxic metabolite accumulates to a sufficient level. Even if very high concentrations of nutrients are employed, the accumulation of toxic metabolites (e.g., acetic acid in the case of *E. coli* growing on glucose) will limit the concentration of cells that can be attained in the exponential phase in a batch reactor. This limitation can be overcome by retaining cells by filtration, while supplying a continuous flow of nutrients and removing products.

Following the exponential growth phase, the rate of exponential growth decreases (*declining growth phase*) and is followed by the stationary phase. The duration of the stationary phase may vary with cell type, previous growth conditions etc. Some cells may lyse, releasing nutrients that can be consumed by other cells, and thus maintain the cell population. Following this is the *death phase.* During the death phase, it is thought that cell lysis occurs and the population decreases. Intracellular metabolites are scavenged by different enzyme systems within the cell and toxic metabolites may accumulate. The rate of decline is also exponential, and is represented during the death phase as

$$\frac{dX}{dt} = -k_d X \tag{3.30}$$

We shall now turn our attention to models which relate the specific growth rate to substrate concentrations and other external variables. The simplest of these do not consider the various phases of the growth cycle, but predict only the rate of growth in the exponential phase. We shall then consider more complex models.

3.3.2 Unstructured Growth Models

The simplest relationships describing exponential growth are unstructured models. These models view the cell as a single species in solution and attempt to describe the kinetics of cell growth based on cell and nutrient concentration profiles. The models that were first developed for cell growth did not account for the dependency of the exponential growth rate on nutrient concentration; they were devised to have a maximum achievable population built into the constitutive expressions employed. Such models find applicability today when the growth-limiting substrate cannot be identified. The simplest model is that of Malthus:

$$r_X = \mu X \tag{3.31}$$

where r_X is the volumetric rate of increase in dry cell weight (which we shall abbreviate as DCW) (e.g., gm DCW/liter-hr) and μ (hr^{-1}) is constant. This model predicts unlimited growth with time ("Mathusian" growth). To provide a means to limit growth, Verlhulst (1844) and later Pearl and Reed (1920) proposed the addition of an inhibition term which was cell concentration dependent:

$$r_X = kX(1 - \beta X) \tag{3.32a}$$

which for a batch system becomes

$$\frac{dX}{dt} = r_X \qquad \text{and thus} \qquad X = \frac{X_o e^{kt}}{1 - \beta X_o (1 - e^{kt})} \tag{3.32b}$$

where $X = X_o$ at $t = 0$. This result is known as the *logistic equation*. The maximum cell concentration attained at large times is $1/\beta$, and the initial rate of growth is approximately exponential, as β is usually considerably less than unity.

3.3.3 The Monod Model

One of the simplest models which includes the effect of nutrient concentration is the model developed by Jacques Monod[11] based on observations of the growth of *E. coli* at various glucose concentrations. It is assumed that only one substrate (the *growth-limiting* substrate, S) is important in determining the rate of cell proliferation. The form of the Monod equation is similar to that of Michaelis-Menten enzyme kinetics; in fact if substrate transport to the cell is limited by the activity of a permease, cell growth might well be expected to follow the Michaelis-Menten form given below:

---

(11) J. Monod, Ann. Inst. Pasteur, Paris, **79**, 390 (1950).

$$\mu = \frac{\mu_{max} S}{K_S + S} \tag{3.33}$$

Thus for batch growth at constant volume:

$$\frac{dX}{dt} = \frac{\mu_{max} SX}{K_S + S} \tag{3.34}$$

where μ_{max} is the maximum specific growth rate of the cells, and K_S is the value of the limiting nutrient concentration which results in a growth rate of half the maximum value. This equation has two limiting forms. At high substrate concentrations, $S >> K_S$, and Eq. (3.34) reduces to a zeroth order dependence on substrate concentration. At low substrate concentrations, $S << K_S$ and a first order dependence results.

$$\mu = \mu_{max} \quad \text{for} \quad S \gg K_S \quad \text{and} \quad \mu = \frac{\mu_{max}}{K_S} \cdot S \quad \text{for} \quad S \ll K_S \tag{3.35}$$

Typical results for the dependence of specific growth rate on substrate concentration are shown in Figure 3.7.

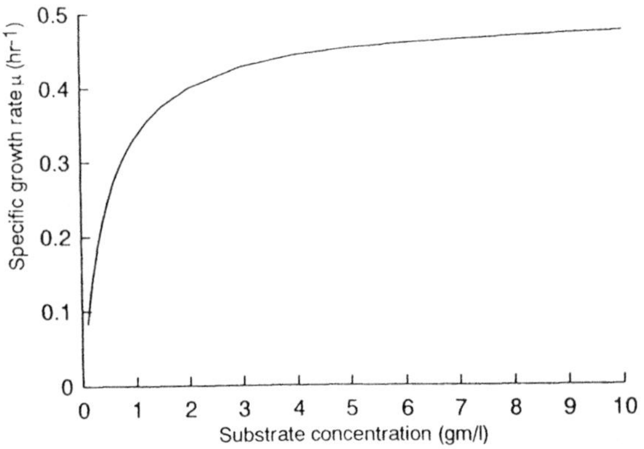

***Figure 3.7**. The dependence of the specific growth rate μ on the growth-limiting substrate concentration S. Here the maximum growth rate is 0.5 hr^{-1} and K_S is 0.5 gm/l.*

Values of μ_{max} vary with the type of organism and the value of K_S depends on the nature of the substrate. Some typical values for various organisms are shown in Table 3.6. As shown, values of K_S are generally quite small, implying that the specific growth rate is near its maximum value for much of the period of batch growth. This apparent zeroth order dependence on substrate concentrations justifies the term "exponential growth".

Table 3.6. *Typical values of* μ_{max} *and* K_S *for various organisms and substrates (at the optimum growth temperatures).*

Organism and growth temperature	*Limiting nutrient*	μ_{max} *(hr⁻¹)*	K_S *(mg/liter)*
Escherichia coli (37°C)	glucose	0.8-1.4	2-4
Escherichia coli (37°C)	glycerol	0.87	2
Escherichia coli (37°C)	lactose	0.8	20
Saccharomyces cerevisiae (30°C)	glucose	0.5-0.6	25
Candida tropicalis (30°C)	glucose	0.5	25-75
Candida sp.	oxygen	0.5	0.045-0.45
	hexadecane	0.5	
Penicillium chrysogenum	glucose		
Klebsiella aerogenes	glycerol	0.85	9
Aerobacter aerogenes	glucose	1.22	1-10

Temperature and pH Effects

The specific growth rate μ_{max} is temperature and pH dependent. The value of the optimum growth temperature is used to classify microorganisms. In general, prokaryotes show a wide range of temperature optima, and eukaryotes show a much narrower range. A typical classification is presented below:

	Growth Temperature (°C)		
Classification	*Minimum*	*Optimum*	*Maximum*
Psychrophiles	-5 to 5	15 to 18	19 to 22
Mesophiles	10 to 15	30 to 45	35 to 45
Thermophiles	25	45 to 75	60 to 80
Extreme thermophiles	50	60 to 75	75 to 95
Hyperthermophiles			≥100

The temperature dependence of μ_{max} and the yield coefficient ($Y_{X/S}$) for *Aerobacter aerogenes* are shown in Figure 3.8.

The Arrhenius relationship generally holds for the temperature dependence of the growth rate. Values of the activation energy (E_A) for microbial growth are of order 5 to 25

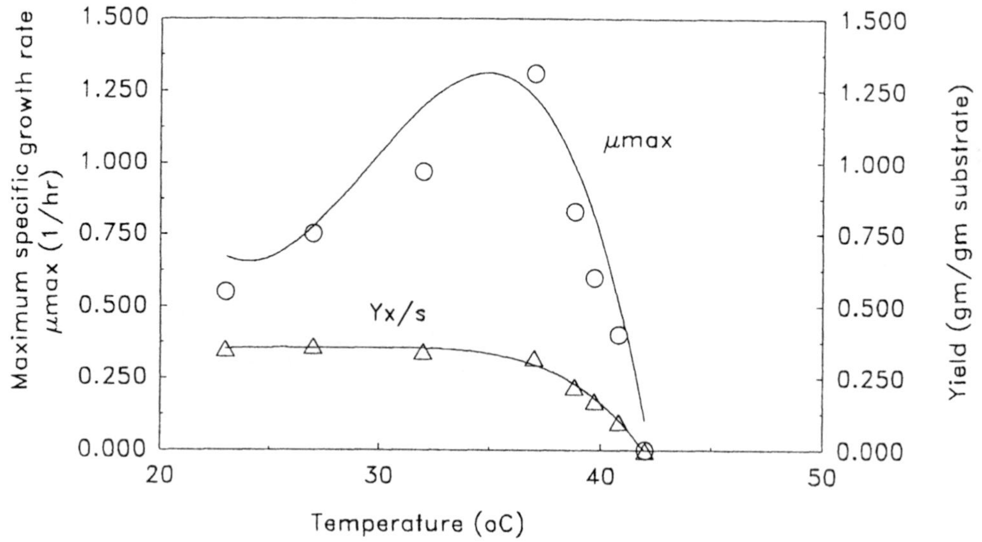

Figure 3.8. *The effect of temperature on the growth rate (μ_{max}) of Aerobacter aerogenes; aerobic growth on glucose. [From W.J. Payne, Ann. Rev. Microbiol.* ***24****, 17 (1970)].*

kcal/mole. The cell yield coefficient is relatively constant until, at a sufficiently high temperature, the enzymes within the cell denature and cell death occurs. Some typical values of E_A are given in Table 3.7.

The effect of pH on microbial growth parallels that observed for enzymes. Most bacteria grow in the range pH 6.5 to 7.5. Yeast and fungi can tolerate lower pHs and often have optima in the range pH 4 to 5. Acidophilic bacteria such as *Ferrobacillus* and *Thiobacillus* grow at pH 2.0. At the other extreme, the urea-decomposing bacteria grow poorly below pH 8.0 and alkali-resistant algae are often found in natural waters above pH 10.0; these organisms often have cytoplasmic membranes which are not permeable to H^+ or OH^-. Organisms that have a pH optimum for growth between pH 10 and 12 are called *alkalophiles*.

We will consider the influence of other factors, such as dissolved oxygen concentration and other nutrients later in Section 3.3.6.

3.3.4 Other Constitutive Models of Growth

A relationship between the growth of cells and substrate concentration is too complex to describe in terms of fundamental physical laws. Thus *constitutive equations*, which are based on experimental observation, are used to relate the response of a cell to its environment. A variety of constitutive equations, in addition to the Monod equation (Eq. 3.33), have been proposed to describe microbial growth. None of these has the simplicity of the Monod model nor its justification by analogy with Michaelis-Menten enzyme kinetics. In this section we shall examine some of the more common models.

Table 3.7. *Values of the activation energy for various microorganisms*[12].

Organism	*Temperature Range (°C)*	*Activation energy E_A (kcal/mole)*
Aspergillus nidulans	20-37	14.0
E. coli	23-37	13.1
K. aerogenes	20-40	14.2
Psychrophilic pseudomonad	2-12	23.8

A simple modification to the Monod model is often made for high substrate concentrations. It is found experimentally that the rate of growth decreases at high values of the initial substrate concentration (S_0), due to the influence of ionic strength, osmotic pressure, or overloading of membrane transport systems. In addition, metabolites that may not be toxic at normal substrate levels may accumulate. These effects can be seen in the modified Monod expression below, where a term $K_{S_0}S_0$ is added to reduce to growth rate:

$$\mu = \mu_{max}\frac{S}{K_S + K_{S_0}S_0 + S} \tag{3.36}$$

At high initial substrate concentrations (S_0), a decrease in μ will result.

A rather general empirical form for the $\mu(S)$ relationship has been proposed by Konak[13]. This is not based on any biological mechanism but is a useful relationship and describes a variety of growth curves.

$$\frac{d\mu}{dS} = k(\mu_{max} - \mu)^p \tag{3.37}$$

Both k and p are adjustable parameters. When p = 1 a solution known as Tiessier's equation results.

$$\mu = \mu_{max}(1 - e^{kS}) \tag{3.38}$$

For other values of p, the general solution is

$$\mu_{max}^{1-p} - (\mu_{max} - \mu)^{1-p} = (1-p)kS \tag{3.39}$$

(12) from S.J. Pirt *Principles of Microbial and Cell Cultivation*, Halsted Press (1975).

(13) A.R. Konak, J., Appl. Chem. Biotechnol. **24**, 453 (1974)

which reduces to the Monod equation for the case p = 2

$$\mu = \frac{\mu_{max} S}{1/k\mu_{max} + S}$$

Another empirical modification to the Monod equation has been proposed by Moser [14]:

$$\mu = \mu_{max} \frac{S^{\lambda}}{K_S + S^{\lambda}} \tag{3.40}$$

where λ is an adjustable parameter. This provides a degree of flexibility in fitting data and can predict interesting dynamic behavior in continuous stirred tank reactors.

Another model of cell growth has been developed by analogy with enzymatic reactions[15]. Cell growth is considered as occurring via a series of reversible enzymatic reactions, with substrate transport into the cell being the first reaction. Assuming maximum rates of forward and reverse reactions and employing the pseudo-steady state hypothesis to the intermediates, a two-step sequence leads to the following equation for the relationship between growth rate and substrate concentration:

$$S = \lambda_1 \mu + \frac{\mu \lambda_2}{\mu_{max} - \mu} \tag{3.41}$$

This equation is similar to the Blackman equation, which was postulated in 1905 to describe the CO_2 and light dependence on the rate of photosynthesis[16]. It has the following limiting behavior:

$$\mu = \frac{S}{\lambda_1} \quad for \quad S < \mu_{max}\lambda_1$$

$$\mu = \mu_{max} \quad for \quad S \geq \mu_{max}\lambda_1 \tag{3.42}$$

The Monod equation however remains the most widely employed relationship describing microbial growth. Variations of this equation will be developed in subsequent sections, describing growth on multiple substrates, and the effects of inhibitors and activators. We shall also examine its dynamic behavior in Chapter 4.

The Coupling of Mass Transfer and Monod Kinetics.

A simple model illustrating the coupling of substrate transport and microbial growth will be considered[17]. The situation is illustrated below.

---

(14) H. Moser, *The Dynamics of Bacterial Populations Maintained in the Chemostat"*, The Carnegie Institute, Washington, DC (1958).

(15) Dabes, J.N., Finn R.K. And C.R. Wilke, Biotech. Bioeng. **15**, 1159 (1973).

(16) Blackman F.F., Ann. Bot. **19**, 281 (1905).

(17) A related approach, considering diffusion of substrate into the cell, has been developed by E.O. Powell, "Microbial Physiology and Continuous Culture", p34, HMSO, London (1969). It involves solution of the Laplace equation describing the concentration of substrate in the external liquid film and subsequently evaluating the substrate flux at the cell surface.

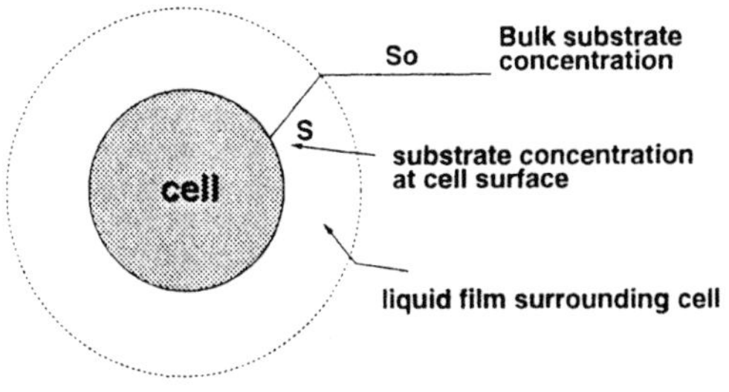

We consider that diffusion may limit substrate transport from the surrounding fluid to the surface of the cell; this may be a result of a stagnant fluid layer surroundng the cell or from a resistance offered by the cell membrane. We represent either of these resistances by a mass transfer coefficient k_L. A steady state rate expression for substrate transport to the cell can be written in terms of the external concentration of substrate S_o and that at the cell surface, S.

$$\text{rate of substrate transport} = k_L a'\left(\frac{X}{\rho_{cell}}\right)(S_o - S)$$

where k_L is the mass transfer coefficient through the liquid film surrounding the cell (e.g. cm/sec) and a' is the surface area to volume ratio of the cell (cm^{-1}), (X/ρ_{cell}) is the volume ratio of cells to solution. The steady state rate of substrate consumption is given by

$$\text{rate of substrate uptake} = qX = \frac{q_{max} S X}{K_S + S}$$

Equating these rates, we obtain

$$\frac{q_{max} S X}{K_s + S} = k_L a'\left(\frac{X}{\rho_{cell}}\right)(S_o - S)$$

Noting that

$$q = \frac{q_{max} S}{K_s + S} \qquad S = \frac{q K_s}{q_{max} - q}$$

the above equation can be solved for the external substrate concentration S_o.

$$S_o = \frac{K_s q}{q_{max} - q} + \frac{\rho_{cell} q}{k_L a'}$$

Solving for q, and noting that $q = \mu / Y_{X/S}$ we can convert the equation into one describing the specific growth rate

$$S_o = \frac{K_s\mu}{\mu_{max} - \mu} + \frac{\rho_{cell}\mu}{k_L a' Y_{X/S}}$$

$$\mu = \frac{S_o + K_s + \left(\frac{\mu_{max}\rho_{cell}}{k_L a' Y_{X/S}}\right)}{2\left(\frac{\rho_{cell}}{k_L a' Y_{X/S}}\right)}\left[1 - \left\{1 - \left(\frac{4\frac{\rho_{cell}}{k_L a' Y_{X/S}}\mu_{max}S_o}{\left[S_o + K_s + \left(\frac{\mu_{max}\rho_{cell}}{k_L a' Y_{X/S}}\right)\right]^2}\right)\right\}^{1/2}\right]$$

An approximate solution to the above equation can be found by noting that all the quantities are positive and

$$\left(S_o + K_s + \left(\frac{\rho_{cell}\mu_{max}}{k_L a' Y_{X/S}}\right)\right)^2 > \left(S_o + \left(\frac{\rho_{cell}\mu_{max}}{k_L a' Y_{X/S}}\right)\right)^2 = \left(\left(\frac{\rho_{cell}\mu_{max}}{k_L a' Y_{X/S}}\right) - S_o\right)^2 + 4S_o\left(\frac{\rho_{cell}\mu_{max}}{k_L a' Y_{X/S}}\right)$$

hence

$$\frac{4S_o\left(\frac{\rho_{cell}\mu_{max}}{k_L a' Y_{X/S}}\right)}{\left(S_o + K_s + \left(\frac{\rho_{cell}\mu_{max}}{k_L a' Y_{X/S}}\right)\right)^2} < 1$$

Thus we can expaapnd the terms inside the square root, and by retaining only the first term obtain

$$\mu = \mu_{max}\frac{S_o}{S_o + K_s + \frac{\mu_{max}\rho_{cell}}{k_L a' Y_{X/S}}} = \mu_{max}\frac{S_o}{S_o + K_{app}} \quad \text{where} \quad K_{app} = K_s + \frac{\mu_{max}\rho_{cell}}{k_L a' Y_{X/S}}$$

Thus the observed value of K_S will depend on the mass transfer coefficient k_L and the surface area of the cell per unit cell volume a', which is equal to $6/d_{cell}$ if the cell is spherical. This suggests that the observed value of K_s should increase linearly with increasing cell size if the rate of substrate transport through an external film or through the membrane is limiting. This may indeed be the case for the oxygen limited growth. Data on the apparent Monod constant for oxygen uptake as a function of the size of the cell are illustrated below, for different types of cell. The organisms and their K_S values are listed. As can be seen, larger cells have higher K_S values for oxygen uptake, as predicted from the relationship:

$$K_S, app = K_s + \frac{\mu_{max}\rho_{cell}}{k_L a' Y_{X/S}} = K_s + \frac{\mu_{max}d_{cell}\rho_{cell}}{6k_L Y_{X/S}}$$

3.3.5 Multiple Substrate Models and Models of Inhibition

Multiple Substrates

The Monod model and its variants are all able to describe the experimental observations of growth as a function of a single substrate. There are many situations however when multiple substrates may simultaneously limit the growth rate. This may be the case when several metabolites enter a metabolic pathway and each metabolite may be limited by a different enzyme. For example, glucose and oxygen may simultaneously limit growth by regulation of hexokinase activity (glucose) and TCA cycle activity (oxygen). In mammalian cell cultures, glucose and glutamine are generally both required for growth, glucose serving as a source of pentose sugars (for nucleic acids) and supplying energy via lactate formation, while glutamine supplies organic nitrogen and a large amount of energy via the TCA cycle.

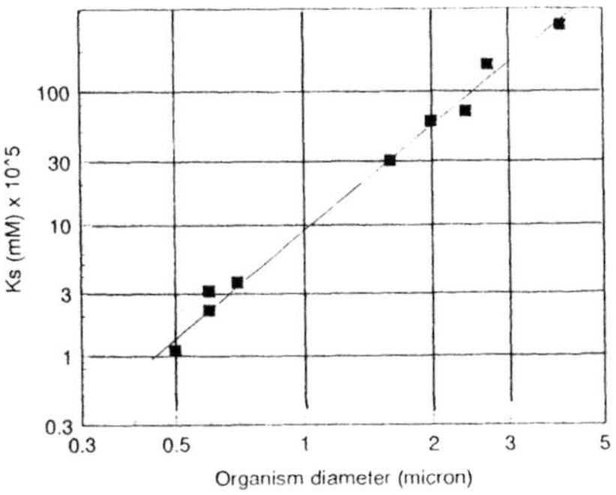

The effect of organism size on the apparent Monod constant for oxygen uptake. Data from J. Longmuir, Biochem. J., **57** *81 (1954). The organisms from which the data were obtained are listed below.*

Values of the Monod constant for oxygen limited growth of several types of cell.

Organism	*Temp (°C)*	*Size (micron)*	K_s *(mM O_2)*
Micrococcus candicans	20.2	0.5	1.1×10^{-5}
Aerobacter aerogenes	19.0	0.6	3.1×10^{-5}
Escherischia coli	19.2	0.6	2.22×10^{-5}
Serratia marescens	18.8	0.7	3.60×10^{-5}
Azotobacter indicum	19.6	1.6	3.00×10^{-4}
Bacillus megatherium	19.2	2.0	5.97×10^{-4}
Bacillus megatherium (LiCl)	20.0	2.4	7.07×10^{-4}
Acetobacter suboxydans	19.2	2.7	1.57×10^{-3}
Bacillus megatherium (glycine)	20.6	4.0	3.12×10^{-3}

Two types of models have been proposed to describe growth on two limiting substrates. The first considers cell growth to follow a scheme analogous to Michaelis-Menten kinetics[18]

(18) Yoon, H.Y., Klinzing, G. and H.W. Blanch, Biotech. Bioeng., **19**, 1193 (1977).

$$X + a_1 S_1 \underset{k_{-1}}{\overset{k_1}{\Leftrightarrow}} X' \overset{k_3}{\rightarrow} 2X$$

$$X + a_2 S_2 \underset{k_{-2}}{\overset{k_2}{\Leftrightarrow}} X'' \overset{k_4}{\rightarrow} 2X$$

where a_1 and a_2 can be related to yield coefficients for each substrate. The psuedo-steady state hypothesis can be applied to the two intermediate microbial "species" X' and X''.

$$\frac{dX'}{dt} = k_1 X S_1 - k_{-1} X' - k_3 X' = 0 \tag{3.43a}$$

$$\frac{dX''}{dt} = k_2 X S_2 - k_{-2} X'' - k_4 X'' = 0 \tag{3.43b}$$

$$X_T = X + X' + X'' \tag{3.43c}$$

From these equations (3.43 a,b,c), X' and X'' can be found as functions of X_T, by eliminating X. The spescific growth rate is found by substituting for X' and X'' in the equation

$$\frac{dX_T}{dt} = k_3 X' + k_4 X'' \tag{3.43d}$$

and hence

$$\mu = \frac{\mu_{max1} S_1}{K_1 + S_1 + \alpha_2 S_2} + \frac{\mu_{max2} S_2}{K_2 + S_2 + \alpha_1 S_1} \tag{3.43e}$$

where

$$\alpha_2 = \left(\frac{k_2}{k_1}\right) \frac{k_{-1} + k_3}{k_{-2} + k_4} = \frac{K_1}{K_2} \quad \text{and} \quad \alpha_1 = \frac{1}{\alpha_2}$$

$$K_1 = \frac{k_{-1} + k_3}{k_1} \quad \text{and} \quad K_2 = \frac{k_{-2} + k_4}{k_2}$$

$$\mu_{max1} = k_3 \quad \text{and} \quad \mu_{max2} = k_4$$

Each substrate thus exhibits a competitive inhibition effect on the utilization of the other. This model can also account for the phenomenon of *diauxic* growth. When two metabolizable substrates are supplied to a microorganism, one is generally consumed prior to the other. If $S_1 \gg \alpha_1/\alpha_2$ then substrate S_1 will be consumed in preference to substrate S_2. Once the concentration of S_1 is reduced, then S_2 will be subsequently consumed. If $\alpha_1 \sim \alpha_2$ both substrates will be consumed at rates which depend on their relative concentrations. Although the development of this model results in the relationship $\alpha_1 = 1/\alpha_2$, the model may be generalized by permitting α_1 and α_2 to be independent parameters.

A second approach is to consider that substrate consumption is non-competitive and that the manner in which they influence the growth rate, by analogy with enzyme kinetics, is additive or multiplicative. This results in equations of the form

Additive kinetics

$$\mu = \frac{\mu_{max}S_1}{K_1 + S_1} + \frac{\mu_{max_2}S_2}{K_2 + S_2} \tag{3.44a}$$

Multiplicative

$$\mu = \mu_{max}\left(\frac{S_1}{K_1 + S_1} \cdot \frac{S_2}{K_2 + S_2}\right) \tag{3.44b}$$

which may be generalized to n substrates viz.

$$\mu = \sum_{i=1}^{n}\left(\frac{\mu_{max_i}S_i}{K_{S_i} + S_i}\right) \tag{3.45a}$$

$$\mu = \mu_{max}\prod_{i=1}^{n}\left(\frac{S_i}{K_{S_i} + S_i}\right) \tag{3.45b}$$

If some substrates are considered to be essential and others simply enhance the rate of growth, a modification to the above equation can be made. Denoting essential substrates as S_{ess} and those which simply enhance the rate of growth by S_{enh}, Tsao and Hanson[19] proposed the following expression:

$$\mu = \left(\mu_{max,o} + \sum_{i=1}^{n}\frac{k_i S_{enh,i}}{K_{enh,i} + S_{enh,i}}\right)\left(\prod_{j=1}^{m}\frac{S_{ess,i}}{K_{ess,j} + S_{ess,j}}\right) \tag{3.45c}$$

The first term in the above equation shows the stimulatory effect of the growth -enhancing substrates, while the second term is analogous to that in Eq. (3.45b). The term $\mu_{max,0}$ represents the maximum specific growth rate in the absence of growth-enhancing substrates.

The multiplicative approach (Equation (3.45b)) has the disadvantage that when several substrates are growth-rate determining, the predicted overall growth rate is considerably reduced. For example, if four substrate are all present at concentrations that would result in 90% of their individual maximum growth rates, the resultant net growth rate would be 65.6% of the maximum possible. This large reduction in the growth rate is not generally experimentally observed. Despite this, Equation (3.45b) has been widely applied to the case of multiple substrate growth and in many models of cellular metabolism. The differences between these types of models is illustrated in Problem 9. Figure (3.9) illustrates the growth of *Pseudomonas vulgaris* on two carbon sources, glucose and citrate. Here citrate is the preferred substrate and glucose is not consumed until the citrate is alomost consumed.

Inhibition

(19) G.T. Tsao and T.P. Hanson, Biotech. Bioeng. **17**, 1591 (1975).

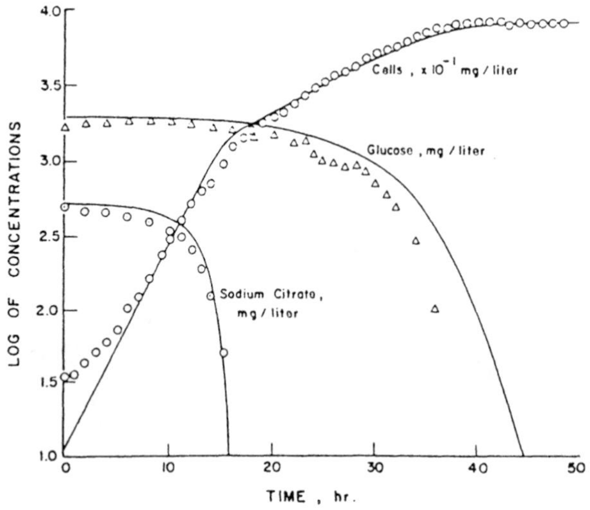

Figure 3.9. Growth of Pseudomonas vulgaris on citrate and glucose. Data from M.M. Tseng, Ph.D. Thesis, Univ. of Toronto, Ontario, Canada (1974). Lines are fitted to the data using the additive model (Equation 3.44a) by G. Tsao and C.M. Yang [Biotech. Bioeng. ***23*** *1827 (1976)].*

There are a number of external variables that may influence microbial growth in addition to pH, temperature and nutrient composition. Most substrates, if present at sufficiently high concentrations, can inhibit microbial growth by one or more mechanisms. Any changes in the physicochemical properties of the medium by high substrate concentrations might be expected to have some effect on cell growth. These properties include the osmotic pressure, ionic strength, dielectric constant and activity of solutes present, any one of which might result in an alteration of membrane fluidity. In addition, overproduction of a metabolite by one pathway may result in inhibition of another pathway. The Pasteur effect provides an example where high levels of oxygen inhibit glycolysis and cause respiration to be the predominant form of metabolism. Table 3.8 summarizes some of the effects an inhibitor may have on microbial growth.

Table 3.8*. Possible types of inhibitor action*

(a) modify chemical potential of substrates, intermediates or products
(b) alter cell permeability
(c) alter enzyme activity
(d) dissociate enzyme aggregates
(e) affect enzyme synthesis
(f) influence functional activity of the cell

One way to model the effect of inhibitory substrates is by analogy with enzyme kinetics. We consider growth to follow the equation

$$X + S \overset{K_S}{\Leftrightarrow} XS \overset{k}{\rightarrow} \text{ new cells}$$

where K_S is [X][S]/[XS]. The effect of an inhibitory substrate can be considered to result from the reversible formation of a complex XS_2, which can no longer form new cells

$$XS + S \overset{K_i}{\Leftrightarrow} XS_2$$

Here K_i is $[XS][S]/[XS_2]$. The overall rate of growth is given by $\mu = k[XS]$, and the resulting expression for the specific growth rate has the form of uncompetitive inhibition

$$\mu = \frac{\mu_{max} S / \left(1 + \frac{S}{K_i}\right)}{S + K_S / \left(1 + \frac{S}{K_i}\right)} = \mu_{max} \frac{S}{K_S + S + \frac{S^2}{K_i}} \tag{3.46}$$

The rate of growth increases with substrate concentrations up to a maximum value (S_{crit}) beyond which it declines. From the equation above, we see that the maximum growth rate occurs when

$$\frac{d\mu}{dS} = 0 \qquad S_{crit} = \sqrt{(K_S K_i)} \tag{3.47}$$

The form of the $\mu(S)$ relationship allows two different substrate concentrations to give the same specific growth rate. As we shall see later, this can lead to two possible steady states in a continuous stirred tank reactor. The values of K_S and K_i are difficult to determine accurately from batch data. However, if $S >> K_S$, a plot of $1/\mu$ against S has an intercept of $1/\mu_{max}$ and a slope of $(K_i \mu_{max})^{-1}$, and K_i can thus be found. This plot does not, however, provide information on K_S, which is best found from values of the specific growth rate obtained at very low substrate concentrations.

The analogy with enzyme inhibition can be extended further, where multiple complexes of the form XS_i can be formed. This results in the following expression for μ.

$$\mu = \mu_{max} \frac{S}{K_S + S + S \sum_{i=1}^{n} (S/K_i)^i} \tag{3.48}$$

If allosteric inhibition (see Chapter 1, Section 1.5.2) is considered, and both substrate "binding sites" are identical, cell growth can be represented by the following reactions

$$X + S \underset{k_{-1}}{\overset{k_1}{\Leftrightarrow}} XS \overset{k_2}{\rightarrow} X + \textit{daughter cells}$$

$$XS + S \underset{k_{-3}}{\overset{k_3}{\Leftrightarrow}} SXS \overset{\beta k_2}{\rightarrow} XS + \textit{daughter cells}$$

Using the steady state assumption for XS and SXS, an equation results which contains the constant β:

$$\mu = \mu_{max}{}' \frac{S(1+\beta S/K_M{}')}{K_M + S + \frac{S^2}{K_M{}'}} \tag{3.49}$$

where $K_M = (k_{-1} + k_2)/k_1$ and $K_M{}' = (k_{-3} + \beta k_2)/k_3$

When $\beta > 1$ both sites are occupied and an increase in the overall growth rate occurs. When $0 < \beta < 1$, daughter cells are formed from the SXS complex and a maximum in rate occurs with increasing substrate concentration. For $\beta = 0$, no cells are formed from the SXS complex and we obtain a growth rate expression of the same form as non-competitive inhibition in enzyme kinetics

$$\mu = \mu_{max} \frac{S}{(K_S + S)\left(1 + \frac{S}{K_i}\right)} \tag{3.50}$$

Here the constants are related to K_M and $K_M{}'$ viz:

$$K_M{}' = K_i + K_S \qquad \frac{1}{K_M} = \frac{1}{K_S} + \frac{1}{K_i} \qquad \mu_{max}{}' = \frac{\mu_{max} K_i}{K_S + K_i)}$$

A plot of $1/\mu$ against S for $S >> K_S$ has an intercept of $1/\mu_{max}$ and a slope of $(K_i \mu_{max})^{-1}$. It cannot therefore be distinguished from a model based on uncompetitive inhibition[20]. Although more complex models could be constructed, discrimination amongst them becomes difficult due to the lack of adequate data on substrate inhibition over a range of substrate concentrations. Figure 3.10 illustrates the inhibition of bacterial growth by the substrate n-pentane.

Substrate Inhibition in Batch Growth

The growth of a microorganism on an inhibitory substrate can be simply modeled in a batch, well-stirred tank reactor. The specific rate of growth of the organism is given by equation (3.50):

---

(20) See G. Hill and C.W. Robinson [Biotech. Bioeng. **17** 1599 (1975)] for a comparison of models used to fit data on the inhibition of *Ps. putida* with phenol as the limiting substrate, and A. Mulchandani and J.H.T. Luong [Enzyme & Microbiol. Technol. **11** 66 (1989)] for a review of inhibition models of microbial growth.

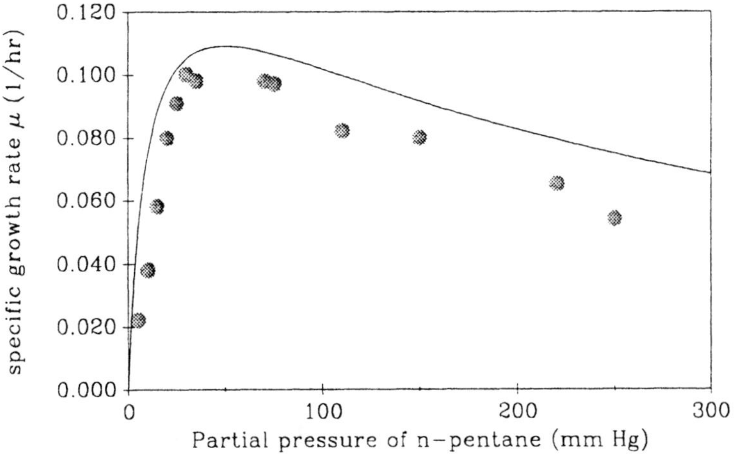

Figure 3.10. *Data of Uemura et al. [J. Ferm. Technol. (Japan)* ***47****, 220 (1969)] for the the inhibition of bacterial growth by n-pentane. The solid line is the prediction based on Equation (3.46) with* K_i*=250 mm Hg,* K_S*=10 mm Hg and* μ_{max}*=0.20* hr^{-1}.

$$\frac{dX}{dt} = \mu_{max} \frac{SX}{(S + K_S)\left(1 + \frac{S}{K_i}\right)} \tag{3.51}$$

The initial conditions are (X_0, S_0). The cell concentration can be eliminated using an overall mass balance and the definition of the yield coefficient $Y_{X/S}$

$$S = S_0 - \frac{X - X_0}{Y_{X/S}} \tag{3.52}$$

We can define dimensionless variables and constants as follows:

$$X_0^* = \frac{X_0}{Y_{X/S} S_0} \qquad K^* = \frac{K_S}{S_0} \qquad S^* = \frac{S}{S_0} \qquad K_i^* = \frac{K_i}{S_0} \tag{3.53}$$

These variables are inserted into Equation (3.51) and we eliminate X. Integration of the resulting equation gives the time course of substrate concentration. As can be seen from Figure 3.11, competitive inhibition has an immediate effect on the growth of cells. The period of growth is considerably extended.

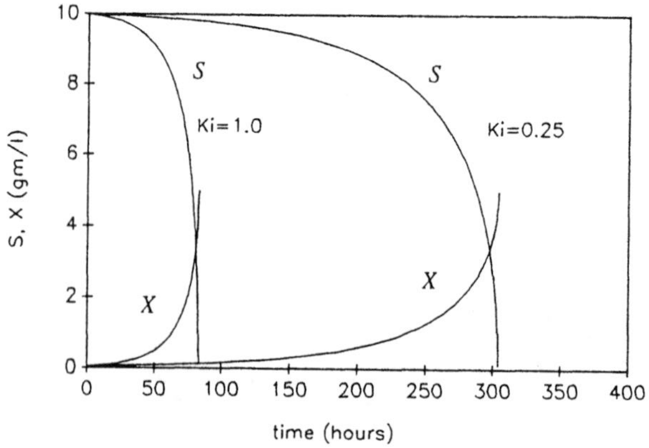

***Figure 3.11**. The effect of competitive inhibition on batch growth. The constants in the equations above are μ_{max} = 0.5 hr^{-1}, $Y_{X/S}$ = 0.5 gm/l, S_0 = 10.0 gm/l, X_0 = 0.05 gm/l, K_S = 0.05 gm/l, K_i values are indicated.*

$$\mu_{max}t = \left(\frac{K_S^*}{1+X_0^*}\right)\ln\left(\frac{1+X_0^*-S^*}{S^*X_0^*}\right)+\left(\frac{K_S^*+X_0^*+1}{K_i^*}+1\right)\ln\left(\frac{1+X_0^*-S^*}{X_0^*}\right)+\frac{S^*-1}{K_i^*} \quad (3.54)$$

Growth Activation

An activator is defined as a substance which is not metabolized but acts to enhance the rate of growth of a culture. Activators may include eukaryotic growth factors, hormones, and insulin, for example, in the case of HeLa cells. Activators increase the maximum specific growth rate but the K_S value remains the same. The effect can be modelled by analogy with enzyme activators, and the model development is similar to that we have seen for substrate inhibitors.

3.3.6 Specific Nutrient Uptake Rates

In section 3.2.3 we saw a general expression for the volumetric rate of substrate consumption which includes substrate conversion to biomass and products, and substrate consumed by maintenance [Equations (3.24), (3.25)]. This expression can be rewritten in terms of a *specific* substrate uptake rate (r_{Si}) by dividing by the cell concentration, generally expressed on a dry cell weight basis;

$$q_{S_i} = \frac{r_{S_i}}{X} = \frac{\mu}{Y_{X/S_i}} + m_i + \sum_j \frac{r_{P_j}}{XY_{P_j/S_i}} \quad (3.55)$$

where q_{Si} is the specific uptake rate (e.g., gm S_i/gm dry cell weight-hr) of substrate i. In the absence of maintenance and significant product formation, the specific substrate uptake rate is simply the specific growth rate divided by the yield coefficient $Y_{X/Si}$. Thus q_{Si} will generally show the same Monod-type behavior with substrate concentration as does μ. If the amount of product formed is small during the phase of rapid growth, the last term in Equation (3.55) may be neglected and we can rewrite the equation for one substrate as

$$q_S = \frac{\mu}{Y_{X/S}} + m = \frac{q_{Smax} S}{K_S + S} \tag{3.56}$$

where we have written a Monod-type expression for q_S as a function of S. By rearranging this equation we can express the growth rate μ as a function of S

$$\mu = \frac{q_{Smax} Y_{X/S} S}{K_S + S} - m Y_{X/S} = \frac{\mu_{max} S}{K_S + S} - m Y_{X/S} \tag{3.57}$$

The growth rate of the organism can thus decrease to zero (due to the effect of maintenance) at a non-zero substrate concentration ($mK_S Y_{X/S}/(\mu_{max} - mY_S)$). This is illustrated in Figure 3.12. If m is large, deviations from the usual Monod behavior of $\mu(S)$ may thus be observed.

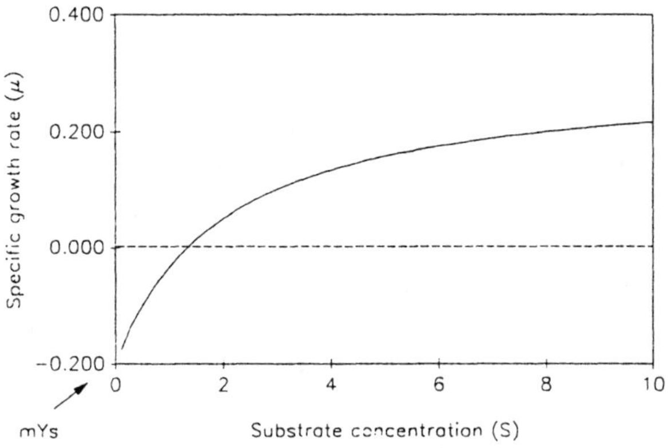

Figure 3.12. *The dependence of the specific growth rate on substrate concentration for a constant maintenance coefficient, m. We see that the growth rate is zero at a finite substrate concentration (1.33 g/l). Extending the calculated* μ *below this concentration provides mYs on the y-intercept. The values of the constants are* $\mu_{max} = 0.5\ hr^{-1}$, $mY_{X/S} = 0.2\ hr^{-1}$, $K_S = 2.0$ *gm/l.*

Values of q_{Smax} for yeast growth on glucose are ~1 gm/gm DCW-hr; for oxygen $q_{O2,max}$ is ~ 10 mmole O_2/gm DCW-hr.

Maintenance effects must be considered when calculating the yield coefficients based on plots of q_S versus μ. An observed yield coefficient, $Y_{X/S,obs}$, can be obtained from the ratio of μ/q_S. This is related to the actual yield coefficient $Y_{X/S}$ from equation (3.56) viz:

$$q_S = \frac{\mu}{Y_{X/S}} + m \tag{3.56}$$

By substituting for $Y_{X/S,obs}$

$$Y_{X/S,obs} = \frac{\mu}{q_S} \qquad \frac{\mu}{Y_{X/S,obs}} = \frac{\mu}{Y_{X/S}} + m \qquad \text{hence} \qquad \frac{1}{Y_{X/S,obs}} = \frac{1}{Y_{X/S}} + \frac{m}{\mu} \tag{3.57}$$

Example: Aerobic Yeast Growth on Ethanol

Consider the aerobic growth of a yeast, such as *C. utilis*, on ethanol, with no other products apart from cell mass and CO_2 produced. The true yield coefficient $Y_{X/S}$ is 0.6 gm DW/gm ethanol and the maintenance coefficient m is 0.02 gm/gm.hr. The maximum specific growth rate of the yeast is found to be 0.4 hr^{-1}. At a specific growth rate of 10% of μ_{max}, the observed yield coefficient will be

$$\frac{1}{Y_{X/S,obs}} = \frac{1}{Y_{X/S}} + \frac{m}{\mu} = \frac{1}{0.6} + \frac{0.02}{0.04} = \frac{1}{0.46} \qquad \text{i.e.,} \qquad Y_{X/S,obs} = 0.46$$

Thus the concentration of biomass at low specific growth rates will be considerably less than that found at values closer to μ_{max}.

3.3.7 Models of Growth and Non-Growth Associated Product Formation

A wide variety of products can be synthesized by the cell. Several schemes have been proposed to classify these products, and kinetic models of product formation have been developed for some of the simpler schemes. One of the most widely applied schemes is based on whether the product is formed as a result of the primary metabolic functions of the cell or whether it is formed from secondary metabolism. End products of energy and carbon metabolism are primary metabolites. Examples include ethanol, produced by the anaerobic fermentation of glucose by yeast, and the production of gluconic acid from glucose by *Gluconobacter*. These products are referred to as *growth-associated* products, as their rate of production parallels the growth of the cell population.

On the other hand, products such as antibiotics and vitamins are generally produced in batch cultures at the end of the exponential phase. Such secondary metabolites are termed *non-growth associated* products, and their kinetics do not depend on the rate of growth of the culture. An intermediate class of products can also be identified, where product formation kinetics lie between the two classes above. Such products are *partially growth associated* and include amino acids, lactic acid, intermediates from the citric acid cycle (including citric acid), extracellular polysaccharides (such as xanthan and pullulan) and solvents such as acetone and butanol.

The development of constitutive rate expressions for these classes of product formation arise from the studies of Luedeking and Piret on the formation of lactic acid by *Lactobacillus delbrueckii*[21]. Lactic acid production was found to depend on both the concentration of cells as well as their growth rate. The resulting expression was proposed:

---

(21) Leudeking, R. And E.L. Piret, Biotech. Bioeng. 1, 393 (1959).

$$r_P = \alpha r_X + \beta X = \alpha\mu X + \beta X \tag{3.58}$$

We see that this expression can be divided into growth ($\alpha\mu X$) and non-growth (βX) associated terms. Based on this approach, the three classes of products can be characterized by the relationship between production kinetics and cell growth:

growth associated	$r_P = \alpha\mu X$	(3.59*a*)
non-growth associated	$r_P = \beta X$	(3.59*b*)
partially growth associated	$r_P = \alpha\mu X + \beta X$	(3.59*c*)

Examples of these kinetics are given in the following sections.

Growth Associated Kinetics. The Production of Gluconic Acid

Gluconic acid is used primarily in dishwashing detergents, where its ability to chelate metal ions is important in reducing "streaking" of glassware. The production of gluconic acid by *Gluconobacter* has been examined by Koga et al.[22] and serves to illustrate the coupling of growth and product formation kinetics. Gluconic acid is formed by the microbial oxidation of glucose to glucono-δ-lactone and the subsequent hydrolysis of this intermedate to gluconic acid.

glucose → glucono-δ-lactone → gluconic acid

The rate of growth of the cells is governed by Monod kinetics:

$$r_X = \frac{\mu_{max} S X}{K_S + S} \tag{3.60}$$

The formation of the intermediate glucono-δ-lactone (L) occurs via a growth associated mechanism, although the value of the Monod constant is assumed to be different to the constant K_S describing cell growth:

$$r_L = \frac{\alpha\mu_{max} S X}{K_S' + S} \tag{3.61}$$

The product gluconic acid is formed by the first-order hydrolysis of the lactone L.

$$r_P = k_L L \tag{3.62}$$

Glucose is consumed for growth and for the production of lactic acid.

---

(22) Koga S., Berg, C.R. And A.E. Humphrey, Applied Microbiol. **15**, 683 (1967).

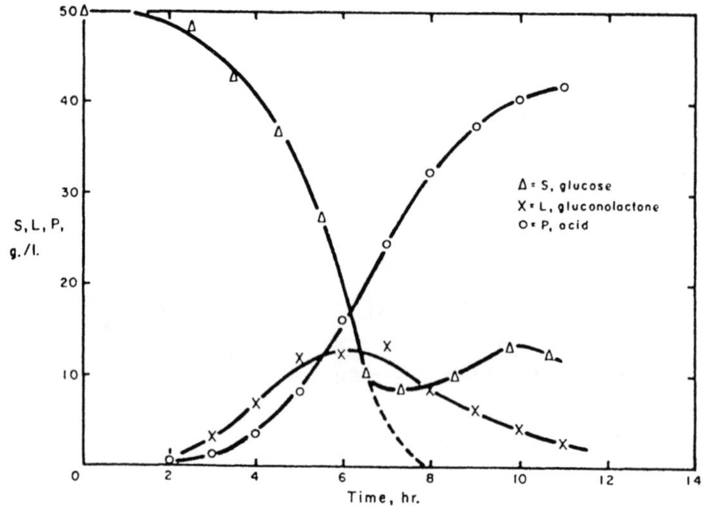

Figure 3.13. The time course of glucono-δ-lactone formation by Gluconobacter. From Humphrey, A.E. and P.J. Reilly, Biotech. Bioeng. 7, 229 (1965)

$$r_S = -\frac{1}{Y_{X/S}} r_X - \frac{1}{Y_{L/S}} r_L \qquad (3.63)$$

Here the Y's are the yield coefficients for cell growth and product formation from substrate. Thus we may write in a batch reactor:

$$\frac{dX}{dt} = \frac{\mu_{max} SX}{K_S + X} \qquad \frac{dL}{dt} = \frac{\alpha SX}{K_S' + S} - k_L L$$

$$\frac{dS}{dt} = -\frac{\mu_{max} SX}{K_S + X}\left(\frac{1}{Y_{X/S}}\right) - \frac{\alpha SX}{K_S' + S}\left(\frac{1}{Y_{L/S}}\right) \qquad (3.64)$$

The concentration of the lactone thus shows an initial increase then reaches a maximum and decreases as gluconic acid is formed. The rate of lactone formation is initially faster than the hydrolysis step. This is illustrated in Figure 3.13.

Non-Growth Associated Kinetics. The Penicillin Fermentation.

Penicillin was the first antibiotic to be produced in submerged fermentation and since its initial large-scale manufacture during the second World War, it has continued to be one of major bulk antibiotics. Penicillin production can be considered to be non-growth associated as a first approximation. The production of penicillin occurs after the phase of rapid growth (the trophophase) of the fungus *Penicillium chrysogenum*. This second phase is referred to as the idiophase. At the end of the growth phase, two key enzymes are found to increase in activity. One is an enzyme which activates phenylacetic acid, which forms the side chain of penicillin, and the other is penicillin acyl-transferase, which attaches activated phenylacetate to 6-aminopenicillanic acid, which is the penicillin nucleus. Penicillin production is subject to catabolite repression. A slow supply of the carbon source is necessary to ensure that intermediate metabolites do not accumulate and reduce production by end-product inhibition.

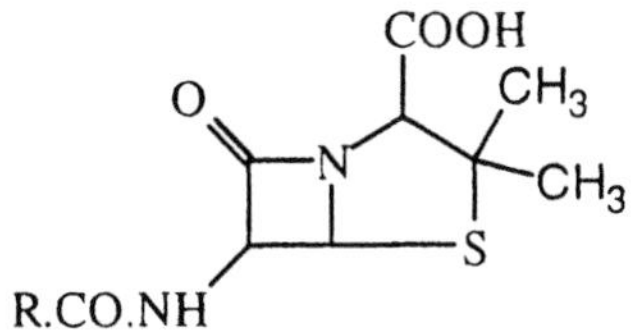

R = Ph.CH_2-
Penicillin G

General Structure of Penicillins

A typical time course of penicillin production is shown in Figure 3.14(a). After the period of exponential growth on lactose (0-24 hours), glucose is slowly fed to the culture and penicillin production increases from 24 to 140 hours. The specific rate of penicillin biosynthesis as a function of specific growth rate is shown in Figure 3.14(b). This illustrates the non-growth rate associated behavior at specific growth rates greater than 0.015 hr^{-1}. Below this growth rate, growth-associated product formation is seen.

Partial Growth and Non-growth Associated Kinetics. Lactic Acid Production

The kinetics of lactic acid production by *Lactobacillus* provide an example of product kinetics that are intermediate between growth and non-growth associated. Lactic acid is an important product arising from the anaerobic fermentation of sugars such as glucose and lactose. It is mainly used as stearyl-2-lactolate, a dough conditioner, and in polymer form as polylactic acid, for biodegradable sutures. Although the anaerobic conversion of glucose to lactic acid is a direct route for energy production, the production kinetics are not soley growth associated. The relationship between the rate of lactic acid production r_p and specific growth rate μ is shown in Figure 3.14 (c).

3.3.8 Models of Product Inhibition

In section 3.3.5 we saw that substrates may inhibit the growth of cells. In addition, the products of cellular metabolism may also inhibit growth and hence slow the rate of their own production. The mechanism for inhibition of growth by end-products of metabolism is not well understood and may result from several effects. As an example, some possible target sites of ethanol inhibition in yeast are shown in Figure 3.15.

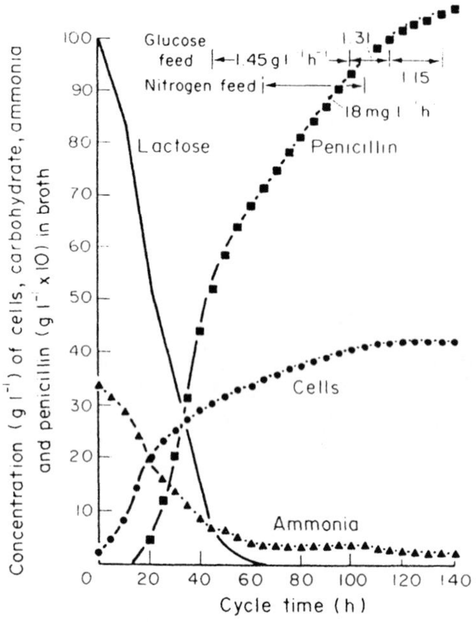

Figure 3.14(a). *Time course of penicillin production by Penicillium chrysogenum. [From Queener, S. and R. Swartz, Economic Microbiology, Ed. A.H. Rose, Vol 3, pp 35-123 (1979)].*

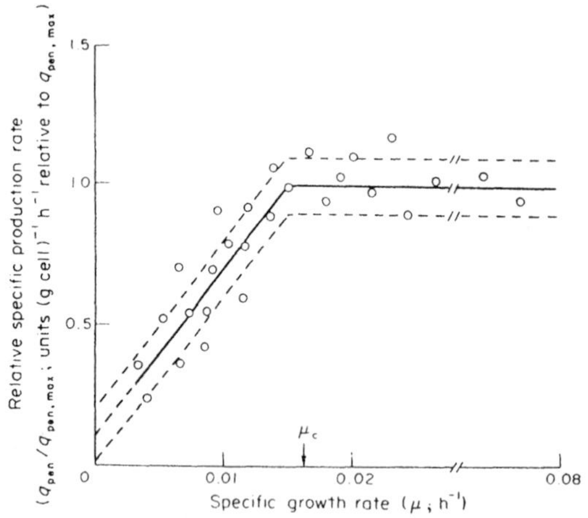

Figure 3.14(b). *Relationship between specific growth rate and specific rate of penicillin synthesis, with glucose as the limiting substrate. [From Ryu, D. and J. Hospodka, 174th ACS Meeting, Chicago, Abstract 8, MBT Division (1977)].*

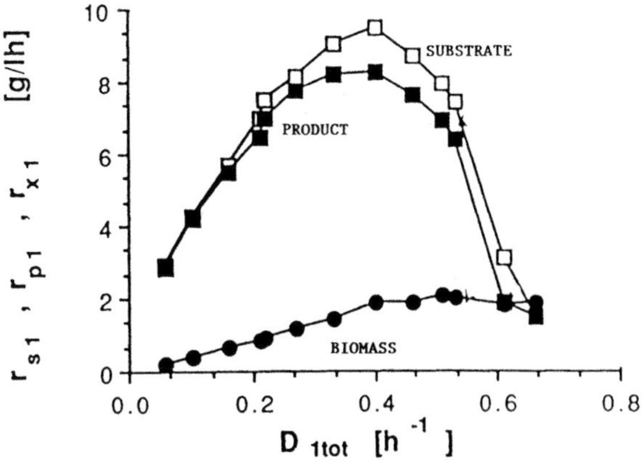

***Figure 3.14 (c)**. The relationship between the rate of lactic acid production (r_P), substrate (whey) consumption (r_S), biomass production (r_X) and the specific growth rate of Lactobacillus helveticus. Data from A. Aeshlimann, L. Di Stasi and von Stockar (Enzyme & Microbiol. Technol. **12** 926 (1990)).*

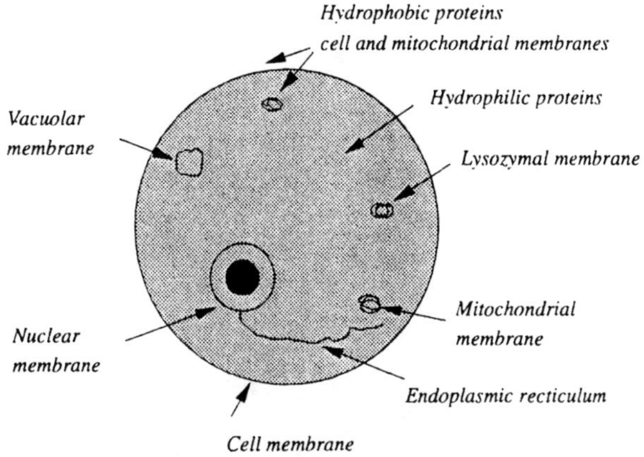

***Figure 3.15**. Possible target sites of ethanol inhibition in yeast. The transport of sugars and amino acids is inhibited by alcohols such as ethanol, butanol, isopropanol and propanol. These inhibitors act in a non-competitive manner on the hydrophobic region of the plasma membrane. Alcohols also affect the membrane potential and proton gradient by inhibiting proton exclusion from the cell. In addition, alcohols may inhibit some of the enzymes within the cell. [From Mulchandani, A. and J.H.T. Luong, Enz. Microb. Technol. **11**, 66 (1989)].*

Using the same types of inhibition models as earlier, product inhibition effects on cell growth can be represented in a variety of ways. One common approach is analogous to that of non-competitive inhibition

$$\mu = \frac{\mu_{max} S}{K_S + S} \cdot \frac{K_P}{K_P + P} \tag{3.65}$$

A second approach is the use of an exponential expression

$$\mu = \frac{\mu_{max} S}{K_S + S} \cdot e^{-k_p P} \tag{3.66}$$

The exponential term can be expanded in a Taylor series and truncated after the first term, provided that the product concentration is not too high:

$$\mu = \frac{\mu_{max} S}{K_S + S} \cdot (1 - k_p P) \tag{3.67}$$

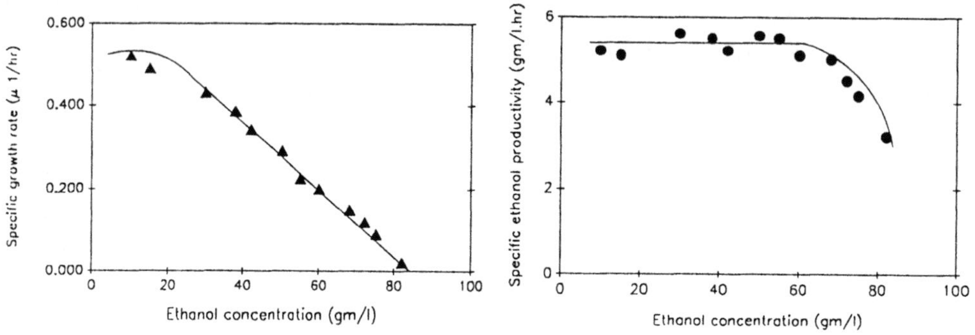

Figure 3.16. *The effect of ethanol on the specific growth rate and specific ethanol production rate for Z. mobilis (ZM4). The data are derived from the continuous culture experiments by Rogers et al. [Adv. Biochem. Engng.* ***23****, 37 (1982)]. The linear portion of the* μ *versus P curve has been fitted with the model* $\mu = \mu_{max}(1 - P/P_m)$, *where* $\mu_{max} = 0.66\ hr^{-1}$, $P_m = 82.5$ *gm/l.*

This gives a result that is of the same form as a similar expansion of the non-competitive inhibition equation (3.65), and the two expressions cannot be distinguished at low product concentrations. Experimentally, a product concentration is often found which completely inhibits the growth of the cell. This concentration is denoted by P_m. Several models describing the effect of the product on the rate of growth μ have been employed to represent experimental data. They include

$$\mu = \mu_{max}\left(1 - \frac{P}{P_m}\right)\left(\frac{S}{K_S + S}\right) \qquad \mu = \mu_{max}\left(1 - \frac{P}{P_m}\right)^{\beta}\left(\frac{S}{K_S + S}\right)$$

$$\mu = \mu_{max}\left(1 - \left(\frac{P}{P_m}\right)^{\gamma}\right)\left(\frac{S}{K_S + S}\right) \tag{3.68}$$

The constants β and γ need to be empirically determined, but they provide a wide range of possible behavior, as they can show a slow initial dependence of growth rate on P and a rapid decline at higher concentrations. The inhibition of fermentative cell growth by ethanol follows this pattern, as shown in Figure 3.16 for the bacterium *Zymomonas mobilis*. A linear dependence of μ on product concentration is apparent, described by the first of the inhibition equations above. The specific rate of ethanol production, however, is relatively unaffected at ethanol concentrations up to 60 gm/l, beyond which there is a rapid decrease in ethanol productivity.

3.4 Growth of Eukaryotic Cells

Eukaryotic cells exhibit a more complex variety of metabolic reactions than those of prokaryotic cells. While simple eukaryotes such as yeast can be grown on well-defined media at growth rates comparable to bacteria, animal and insect cells require a complex set of nutrients, which may include growth hormones and autocrine factors produced by the cells. Animal and plant cells do not have a cell wall to protect them from exposure to the shearing motion of surrounding fluid and thus require carefully-defined hydrodynamic conditions for cultivation. Most primary animal cells require a solid surface on which to attach and grow. Tumorigenic animal cells lose this surface dependence and can be grown in suspension cultures. Other animal cell lines can be grown on microcarrier supports, on hollow fibers in hollow-fiber reactors or on the surfaces of flasks.

In general, the rates of growth of fungi, plant and animal cells are considerably less than those of bacteria, and the range of conditions, such as pH and temperature, over which the cells will grow is considerably more restrictive. Mammalian cells, for example, show a narrow optimum growth temperature of 37°C and a pH optimum of 7.2 to 7.4, corresponding to physiological conditions. At pH 6.5, for example, growth is considerably reduced. In the following sections we shall examine the growth of fungal, animal and plant cells.

3.4.1 Growth of Fungi

Filamentous fungi are an important class of organisms as they produce a wide variety of intermediate metabolites, such as citric, gluconic and itaconic acids, and antibiotics such as penicillins, cephalosporins, erythromycin and tetracyclines (produced by the filamentous bacteria *Streptomyces*, which show fungal-like growth behavior). Fungi generally do not require a complex medium for growth, but in industrial use are often grown with inexpensive complex carbon and nitrogen sources, such as soy-bean meal, fish meal and corn-steep liquor. These nutrients are broken down by the fungi and provide a slowly available source of sugars, amino acids, phosphorus etc. This slow supply of nutrients avoids the effects of catabolite repression on the formation of the desired product. In many instances, the kinetics of fungal growth in stirred tanks resembles those observed for bacteria. The growth rate can be described by Monod kinetics. There are other situations, described below, where fungal growth shows deviations from exponential growth.

Fungi may be cultivated on solid or liquid surfaces, where the two dimensional nature of the surface results in growth rates that are quite different from those found in submerged culture of bacterial cells. Molds, which are higher forms of fungi that exhibit mycelial growth (a highly branched system of tubes containing many nuclei and cytoplasm), may also grow on surfaces or in suspension, where they form pellets under certain growth conditions[23].

When grown on a solid surface, such as an agar plate, molds often show a constant rate of increase of the radius of the mold colony. We can express this as

$$\frac{dr}{dt} = k \tag{3.69}$$

where k is a constant. For a circular colony of constant height, the total biomass (X) can be written in terms of the height of the colony (h), the radius r and the colony density ρ as

$$X = \pi r^2 h \rho \tag{3.70}$$

The rate of change of the total biomass is then

$$\frac{dX}{dt} = 2\pi r h \rho \frac{dr}{dt} \tag{3.71}$$

and substituting for r from equation (3.70) in terms of X we obtain

$$\frac{dX}{dt} = 2\pi h k \rho \left(\frac{X}{\pi h \rho} \right)^{1/2} \tag{3.71}$$

integrating this equation, with X_0 as the initial biomass concentration gives

$$X = (\lambda t + X_o^{1/2})^2 \qquad \text{where} \qquad \lambda = k(\pi h \rho)^{1/2} \tag{3.72}$$

Thus the colony grows with a quadratic dependence on time on the agar surface. A similar analysis for the case of fungal pellet growth can be developed. In this case, the rate of growth of the radius of the pellet is assumed to be constant and the biomass is given by $X = 4\pi r^3 \rho/3$. The resultant expression for the time dependence of the biomass concentration is

$$X = \left(X_o^{1/3} + \frac{\gamma t}{3} \right)^3 \qquad \textit{where} \qquad \gamma = k(36\pi\rho)^{1/3} \tag{3.73}$$

Here the biomass grows with a cubic dependence on time. This has been experimentally observed in the case of pellets of *Aspergillus nidulans*[24]. Once the size of a pellet reaches a critical value however, transport of oxygen to the center of the pellet becomes rate limiting

---

(23) The various factors which influence pellet formation are reviewed in B. Metz and N. Kossen, *The Growth of Molds in the Form of Pellets - A Literature Review*, Biotech. Bioeng., **19** 781 (1977).

(24) Trinci, A.P.J., *Arch. Mikrobiol.*, **73** 353 (1970).

and the hyphae at the center autolyse and lose their cytoplasm. For *Penicillium chyrsogenum* the critical radius is about 0.1 mm, while that for *Aspergillus nidulans* is 2.5 mm, indicative of the more open structure of these fungal pellets.

The analysis of the diffusion of oxygen into a fungal pellet follows that developed for immobilized enzymes in Chapter 2, Section 2.4.3. The intrapellet concentration of oxygen can be obtained from the solution of equation (2.38):

$$\frac{d^2x}{d\bar{r}^2}+\frac{2}{\bar{r}}\frac{dx}{\bar{r}}=\frac{R^2(v_{max}/K_mD_{eff})x}{1+\beta x} \tag{2.38}$$

where x is now the dimensionless oxygen concentration $O_{2,int}/O_{2,ext}$, and D_{eff} is the effective diffusion coefficient for oxygen within the pellet. We are interested in predicting the pellet radius R at which the concentration of oxygen at the center of the pellet, $O_{2,int}(r=0)$ becomes zero. An upper limit can be obtained by simplifying the problem. We consider the oxygen uptake to be zeroth order in oxygen concentration (i.e., $O_{2,int} >> K_m$), so that equation (2.38) becomes

$$\frac{d^2x}{d\bar{r}^2}+\frac{2}{\bar{r}}\frac{dx}{\bar{r}}=\frac{R^2v_{max}}{D_{eff}S_o} \tag{2.38a}$$

which may be solved with the boundary conditions

$$x\,|_{\bar{r}=1}=1 \qquad \frac{dx}{d\bar{r}}\,|_{\bar{r}=0}=0 \tag{2.41}$$

to give

$$x=1-\frac{R^2v_{max}}{6D_{eff}S_o}(1-\bar{r}^2) \tag{3.74}$$

At the critical radius (R_{crit}) when the center of the pellet is depleted of oxygen, x = 0 at $\bar{r}$ = 0

$$\frac{R_{crit}^2v_{max}}{6D_{eff}S_o}=1 \qquad R_{crit}=\sqrt{\left(\frac{6D_{eff}S_o}{v_{max}}\right)} \tag{3.75}$$

Within a pellet, the maximum volumetric rate of oxygen consumption v_{max} can be expressed as $q_{o2,max}$·X, where X is the concentration of hyphae in the pellet. For a typical fungus v_{max} is of order 10~20 mmole O_2/liter-hr. Taking a value of the diffusivity of oxygen within the pellet as slightly less than that in water (1 x 10^{-5} cm^2/sec) and S_o ($O_{2,ext}$) as approximately the solubility of oxygen in water at 30°C (0.20 mmoles/liter), we see

$$R_{crit}=\sqrt{\left(\frac{6D_{eff}S_o}{v_{max}}\right)}=\sqrt{\left(\frac{6x\,1x\,10^{-5}(cm^2/\sec)0.2(mmoles/liter)}{10/3600(mmole/liter-\sec)}\right)}=0.066\ cm$$

which lies between the experimental values obtained for *Aspergillus* and *Penicillium*.

3.4.2 Growth of Animal Cells

Many therapeutic protein products require post-translational modification, such as glycosylation, to be biologically active. This is generally best accomplished in eukaryote hosts, particularly mammalian cells, where correct protein folding patterns can also be achieved. Prokaryotes do not generally glycosylate proteins and thus commonly-used recombinant host bacteria, such as *E. coli*, are not ideal for production of high value therapeutics. The cultivation of animal cells is, however, considerably more difficult than fermentations involving bacterial systems. Animal cell cultivation is much more susceptible to contamination and the medium employed is more complex that the simple salts and sugar solutions that can be employed with simpler organisms.

Two types of animal cells may be distinguished: transformed cells, which are able to be cultivated for extended periods without undergoing "crisis" resulting in cell death, and normal cells which show limited life-spans depending on the tissue of origin. Some transformed cells are able to grow in *suspension* culture, while others require a surface for attachment and growth (*anchorage-dependent* cells). Cells such as those in lymph tissue, the blood stream and in tumors can be typically adapted to grow in suspension, while others grow in monolayers attached to a support matrix.

If a cell population can be grown in an organized manner which simulates the architecture from which it was derived, it may often retain specialized functions from its tissue of origin. Such growth is referred to as *tissue culture*. If cells are grown as discrete entities, with loss of histological organization, the culture is referred to as *cell culture*. These cells no longer communicate by a direct flow of ions and small molecules as do cells in tissues. We shall now examine some of the properties of animal cells that are relevant to their cultivation.

Cell Structure

The composition of animal cells differs from bacteria. There is a variation in the range of compositions which depends on the type of cell. Typical values are given in Table 3.9.

Animal cells contain a nucleus, which is contained within a double membrane structure. The DNA within the nucleus is surrounded by negatively charged nuclear proteins, called histones. The DNA contains about 3×10^9 nucleotides, only a small fraction of which are thought to code for essential proteins. This complex of DNA and histones within the nucleus is known as *chromatin*. Different regions of DNA pack differently in the chromatin; the structural units of chromatin correspond to banding in the mitotic chromosomes. Mitochondria size within animal cells ranges from 0.5 to 5.0 micron and in liver cells, for example, there are an estimated 1000 mitochondria per cell. An important distinguishing feature of animal cells is the presence of a cytoskeleton. The cytoskeleton consists of two types of cytoplasmic fibers: microtubules (tubulin protein) which are 25 nm in diameter and microfilaments (actin protein) 10nm diameter. These fibers are involved in mechanical support of the cell, cell division and membrane transport. Microtubules are associated with

Table 3.9. *Average values for the chemical consitituents of an animal cell*[25].

Constituent	*picogram/cell*	*Range*	*% dry weight*
wet weight	3500	3000-6000	-
dry weight	600	300-1200	-
protein	250	200-300	10-20
carbohydrate	150	40-200	1-5
lipid	120	100-200	1-2
DNA	10	8-17	0.3
RNA	25	20-40	0.7
water	-	-	80-85
volume	4×10^{-9} cm^3	-	-

cilia and flagellae. The plasma membrane contains glycoproteins, glycolipids and lipids. The membrane proteins are mobile and surface receptors can form cross-linked complexes. This is important in the adhesion of cells to surfaces.

The site of adhesion of cells to surfaces has a morphological pattern that is known as a tight junction. When several cells attach around a locus, points of contact may develop between cells; these are known as gap junctions. Both types of junctions (around 50 Å apart) play an important role in allowing coordination and metabolic exchange to occur, resulting in metabolic cooperativity. The type of attachment material, known as the "substrate" or "substratum" in the cell culture literature, is important. As cells are negatively charged, the best substrata are positively charged, holding cells by electrostatic forces. An excellent material is fibronectin, a cold-insoluble globular glycoprotein. Fibronectin promotes cell-cell and cell-substrate adhesion and cell spreading. The type of substrata used for cultivation of anchorage-dependant cells is thus quite important and we shall briefly discuss both natural and artificial substrata.

Substrata for Anchorage-Dependant Cells

When cells attach to other cells or structures, the structures are generally termed the extracellular matrix, which is condensed into microscopically distinguishable layers called "basement membranes". These natural substrata show a wide variety of compositions and structures. Typical natural substrata include collagens, which are proteins in the form of triple helices, consequently containing a large amount of glycine, proline and hydroxy-proline. Various types of collagen exist, one type is found in skin, bones and tendons, another

(25) From Griffiths, J.B. And P.A. Riley, "Animal Cell Biotechnology", Volume 1, p17 Academic Press, (1985).

in blood vessels, yet another in smooth muscles. Proteoglycans are also natural substrata. These contain glucosamines attached to a core protein. Other proteins in the extracellular matrix include fibronectin (as we have seen earlier), laminin, elastin, chondronectin and other glycoproteins.

Artificial substrata are important in cultivation of anchorage-dependant cells. Over the years, various types of substrata have been employed, the first used being glass. The term *in vitro*, meaning "in glass", refers to the initial use of glass containers for cell cultivation. Since both the cells and glass have a negative charge, divalent cations are often required in the medium to shield the electrostatic repulsion that otherwise occurs. Polystyrene is the most common plastic used in tissue culture flasks. It may be modified by sulfonation to give a net negative surface charge. Positively charged polymers are also used, including DEAE-dextran, polylysine, polyornithine, polyhistidine and polyacrylamides. The number of surface charges is important in cell attachment with either negatively or positively charged substrata. This charge density dependence may be related to the attachment of proteins to the substrata at a required density. In some cases the type of substratum used can affect the metabolism and even the differentiation of attached cells.

Cell Growth

The growth of animal cells is considerably more complex than the bacterial and yeast systems we have encountered so far. The human body contains around 10^{14} cells, and while most are non-dividing, around 20 million divisions per second occur for maintenance functions. Cell growth is regulated by a complex series of signals, including hormones, metabolic products, neural signals and cell-cell contact. When cells are transformed by the action of viruses, genetic mutation or other means, the properties of the cells change. In particular, the transport of nutrients across the cell membrane is significantly altered. Some of the changes are illustrated in Table 3.10.

Cell type	G1 phase (hrs)	S phase (hrs)	G2 + M (hrs)
HeLa 53	8.0	9.5	3.5
WI-38	6.0	7.5	4.0
Mouse fibroblasts	8.0	6.0	4.0
Chinese hamster ovary (CHO) cells	5.5	4.5	4.0
Human diploid fibroblasts	6.5	7.5	4.0

The animal cell growth cycle contains the same four main phases we have seen earlier: cell division or mitosis (M), a gap period G1, the period (S) in which DNA is synthesized, and a second gap phase G2. When the cells cannot grow due to an insufficient supply of nutrients they are arrested in the G1 phase. For convenience, cells which can re-enter the proliferative cycle are said to be in a G0 phase. Cells enter G0 from G1, giving rise to the concept that malignant cells are unable to enter G0 and thus must continue through the cell cycle or die. This concept of the cell cycle is referred to as *restriction point* theory. The

***Table 3.10.** Altered properties of transformed cells*[26]

1.	Increased cell population density
2.	Increased life-span
3.	Loss of anchorage-dependance; cells can grow in suspension
4.	Decreased requirements for serum and growth factors
5.	Increased glucose transport and glycolysis
6.	Changes in intracellular metabolic concentrations
7.	Increased lectin agglutination
8.	Increased mobility of surface proteins
9.	Absence of 250,000 Dalton surface proteins
10.	Decrease in ganglioside content of lipids
11.	Alterations in cytoskeletal elements
12.	Changes in surface antigens

other concept of growth regulation is a continuum model, in which an initiator of DNA synthesis is continuously formed and that a threshold amount of this initiator is required for DNA synthesis to occur. If synthesis of this initiator is impaired, then cells become arrested in the cell cycle after mitosis. Typical times to traverse each period of the cell cycle are illustrated.

Carbon Metabolism in Animal Cells

For most animal cells, glucose and glutamine are the major carbon and energy sources. Both nutrients are required; glucose provides pentose sugars via the pentose phosphate pathway, glucosamine-6-phosphate and the widely used precursor glyceraldehyde-3-phosphate. While other sugars such as fructose or galactose can be used in place of glucose, glucose is more rapidly utilized. Glutamine is required for the synthesis of purines and the formation of guanine nucleotides. Glutamine is also the primary source of nitrogen in the cell, via transamidation and transamination reactions. Both carbon sources are required, although asparagine may replace glutamine in some cells. The pathways of glucose and glutamine utilization are illustrated in Figure 3.17. Glutamine enters the cell and may be deamidated to glutamate in the cytosol or in the mitochondria.

As can be seen on Figure 3.17, the metabolism of both glucose and glutamine are interrelated; however, glutamine typically provides most of the energy required by the cell through respiration. In all mammalian cells, glucose is metabolized to pyruvate. In normal (i.e., non-tumor) cells, pyruvate is converted to acetyl-CoA and oxidized via the TCA cycle. The ATP produced by mitochondrial respiration regulates glycolysis as a result of its inhibition of phosphofructokinase (PFK). Glucose-6-phosphate then accumulates and regulates the phosphorylation of glucose via its action on hexokinase. When oxygen is less

(26) From Griffiths and Riley, "Animal Cell Biotechnology", Vol. 1, p17, Academic Press (1985).

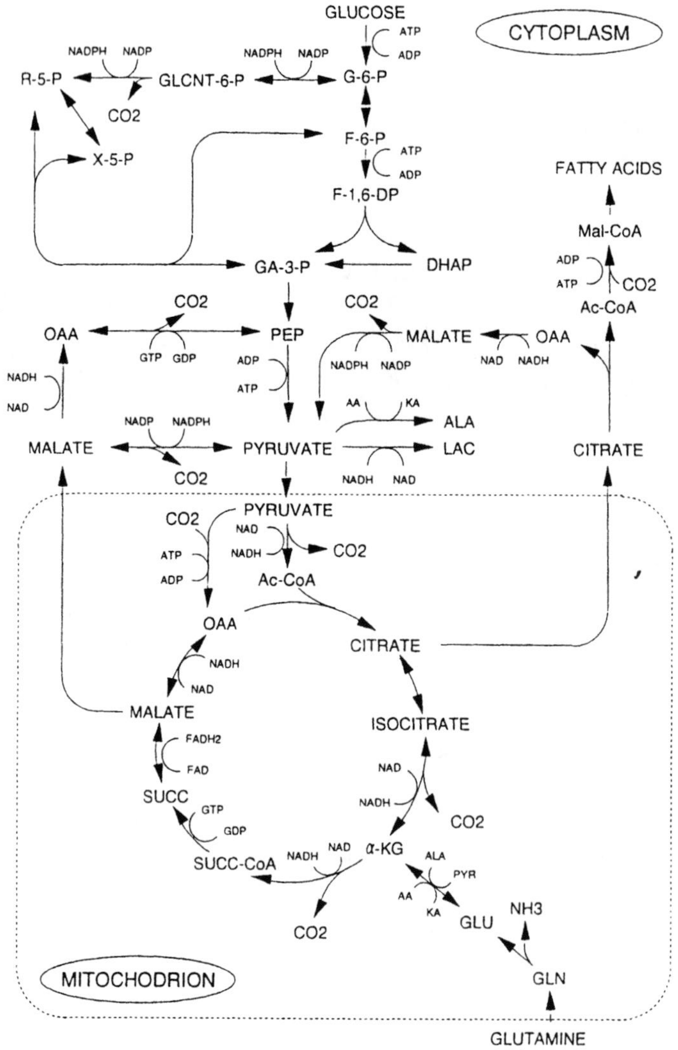

Figure 3.17. The major metabolic pathways in mammalian cells. The cytosolic and mitochondrial reactions are shown together with the compounds which can cross from the mitochondria to the cytosol. [From Mancuso et al., Biotech. Bioeng. ***44****, 563-585 (1994)].*

available, the production of ATP in the mitochondria is reduced and PFK is deregulated. More glucose-6-phosphate is consumed, increasing the glucose flux into the cell. As oxidative phosphorylation is restricted at low oxygen concentrations, pyruvate is converted to lactic acid as a means of regenerating NAD^+ from the NADH generated by glycolysis.

In contrast to normal mammalian cells, cultured cells, tumor cells and proliferating cells exhibit high rates of aerobic glycolysis. High rates of lactate production, similar to normal cells under oxygen limitation, are found. Transformed cells have glycolytic enzymes which exhibit altered regulation as a result of the action of protein kinases (resulting from proto-oncogene expression). The number of glucose transporters, responsible for movement of glucose into the cell, is increased when the cell is transformed. Thus there is a greater potential for glucose uptake. A hexokinase isozyme is found bound to the mitochondria and exhibits reduced inhibition by glucose-6-phosphate and ATP. This decreased inhibition results in less control of glucose entry into the cell. Pyruvate kinase (PK) also shows a reduced affinity for phosphoenolpyruvate. Thus the increased flux of glucose into the cell results in higher levels of glucose -6-phosphate (G-6-P) and fructose-6-phosphate (F-6-P). Higher concentrations of fructose-1,6-diphosphate and fructose-2,6-diphosphate consequently occur, overcoming the normal regulation of PFK. As PK is also less tightly regulated, there is a higher flux of carbon to pyruvate. The respiratory capacity of the cell is limited, and the excess pyruvate is metabolized to lactate as a means of regenerating NAD^+.

As the glucose concentration is increased from about 5 μM to about 5 mM, the specific rate of glucose consumption increases significantly. Below about 0.5 mM, over half the glucose consumed by rat hepatomas, for example, is incorporated into nucleotides, but beyond 5 mM over 90% of the glucose is converted to lactate.[27]

The metabolism of glutamine is altered in tumor cells and proliferating normal cells. Many normal cells produce glutamine, but cells which grow in culture show high rates of glutamine consumption. Glutamine metabolism occurs primarily in the TCA cycle, where glutamine enters as α-ketoglutarate. More than half the CO_2 production by normal diploid fibroblasts is derived from glutamine[28].

The Effect of Environmental Factors on Cell Behavior

(i) pH Effects

Different cell types exhibit widely differing pH optima. A plateau of about 0.5 to 1.0 pH units for growth is common. Table 3.11 illustrates this spread in pH optima with the source of cell.

The pH of a cell culture medium is often controlled by bicarbonate buffering. The gas in the headspace above tissue flasks or in bioreactors generally contains CO_2, oxygen and

---

(27) More detailed descriptions of carbon metabolism in animal cells may be found in Eigenbrodt, E., Fister, P., and Reinacher, M., *Regulation of Carbohydrate Metabolism*, Vol II, p.141, Beitner, R. ed. CRC Press, Boca Raton (1985) and in Pedersen, P.L., Prog. Exp. Tumor Research, **22**, 190 (1978) and in Miller, W.M. and H.W. Blanch, *Regulation of Animal Cell Metabolism in Bioreactors*, p.119 in "Animal Cell Bioreactors" editors D. Wang, and C.S. Ho, Butterworth-Heinemann Press (1991).

(28) Zeilke, H.R., Ozand, P.T., Tildon, J.T., Sevdalian, D.A., and M. Cornblath, J. Cell Physiol. **95** 41 (1978).

Table 3.11. pH Optima for Growth of Mammalian Cells[29].

Species	*Cell Type*	*Specific Strain*	*Optimum pH*
Human	Normal		
	embryonic lung fibroblast	WI 38	7.7
	skin fibroblast	MS2	7.6, 7.7
	Cancer	HeLa	6.9 - 7.4
Monkey	Normal		
	green monkey kidney	primary culture	6.65 - 7.5
Mouse	Normal		
	skin fibroblast	929	7.1 - 7.3
	skin fibroblast	3T3	7.4 - 7.7
Rat	Normal		
	lung fibroblast	BL	7.35
	liver epithelium	B1	7.5 - 7.9
	Cancer		
	glial tumor	C6	7.1, 7.15
Hybrids	Mouse-human	C1 1D x 18VA	7.0 - 7.65
		RAG x WI 38	7.2

nitrogen, reflecting conditions found *in vivo*. In the physiological range of pH, most dissolved CO_2 is in the form of bicarbonate, whereas below pH 5 it is present as dissolved CO_2, and above pH 11 it is present as carbonate. The set of equilibrium reactions is

$$CO_2 + H_2O \Leftrightarrow H_2CO_3$$
$$H_2CO_3 \Leftrightarrow HCO_3^- + H^+$$
$$HCO_3^- \Leftrightarrow H^+ + CO_3^{2-}$$

The total concentration of dissolved CO_2 is taken to include the carbonic acid (H_2CO_3). The pK's of the second and third reactions based on this total of ($[CO_2] + [H_2CO_3]$) are 6.3 (pK_1) and 10.25 (pK_2). As the concentration of H_2CO_3 is small at pH ~ 7, we can neglect its contribution and write

$$pH = pK_1 + \log\left(\frac{[HCO_3^-]}{[CO_2]}\right) \tag{3.76}$$

(29) From Eagle, H. J. Cell Physiol. 82, 1 (1973))

Thus the pH of a medium in a bicarbonate-buffered solution is inversely proportional to the logarithm of the CO_2 concentration in the liquid phase and hence in the gas phase. CO_2 is very soluble relative to oxygen (the Henry's law coefficient for CO_2 is 30 times that for oxygen, and is 0.030 mol/atm at 30°C)[30]. Variation of the gas phase CO_2 concentration can be used for pH control. However, organic buffer systems, such as HEPES and TES, are more widely employed in tissue flask cultivation, and in bioreactor systems NaOH or NH_4OH addition can be employed to maintain pH.

(ii) Effect of Dissolved Oxygen

Oxygen is required for energy production via oxidative phosphorylation and is also used in synthesis of cellular components such as tyrosine and cholesterol. In commercial bioreactors, the low solubility of oxygen may make it the limiting nutrient. The specific oxygen uptake rate varies with the cell type, but is generally between 0.05 and 0.5×10^{-9} mmol $O_2.cell^{-1}hr^{-1}$ for cultured animal cells. In tissue, uptake rates are higher, for example, 0.5 to 2.5×10^{-9} mmol $O_2.cell^{-1}hr^{-1}$ have been reported for rat liver slices. For many cells, the optimum dissolved concentration corresponds to 50% of air saturation. For hybridoma cells, the optimum is considerably lower. Figure 3.18 illustrates the effect of dissolved oxygen on the behavior of hybridoma cells in continuous culture. Similar trends have been observed with hepatocytes and WI-38 cells.

At oxygen concentrations above the optimum, toxicity is observed. This is a result of oxidative damage, causing DNA degradation, lipid peroxidation, and polysaccharide depolymerization. Less chromosomal damage occurs at lower oxygen concentrations, presumably as a result of the decreased concentration of excited oxygen species. The optimum oxygen concentration for growth may not correspond to the optimum for product formation. as illustrated in Figure 3.18 for antibody production.

3.4.3 Growth of Plant Cells

The plant kingdom is a rich chemical resource for pharmaceuticals and food additives. About 75% of the world's population relies on plants and plant extracts for medical use. For example, of the 3500 new chemical structures that were discovered in 1985, 2619 were isolated from plants. The cultivation of plant cells provides an alternative to the growth of whole plants. The advantage that cultured cells have over whole plants is their susceptibility to modifications of the growth medium. In whole plants, the desired products are generally synthesized only for short periods during the growth of the plant and often products are found only in scarce tissue types. By manipulating the growth medium, it may be possible to induce product formation by each plant cell in a culture. Some examples of natural products arising from plants are given in Table 3.12.

---

(30) A complete discussion of the bicarbonate buffering systems is given in Umbriet W.W., "Manometric and Biochemical Techniques", 5th Ed., Burris, R.H. and Stauffer, J.F., eds. p20 Burgess, Minneapolis (1972).

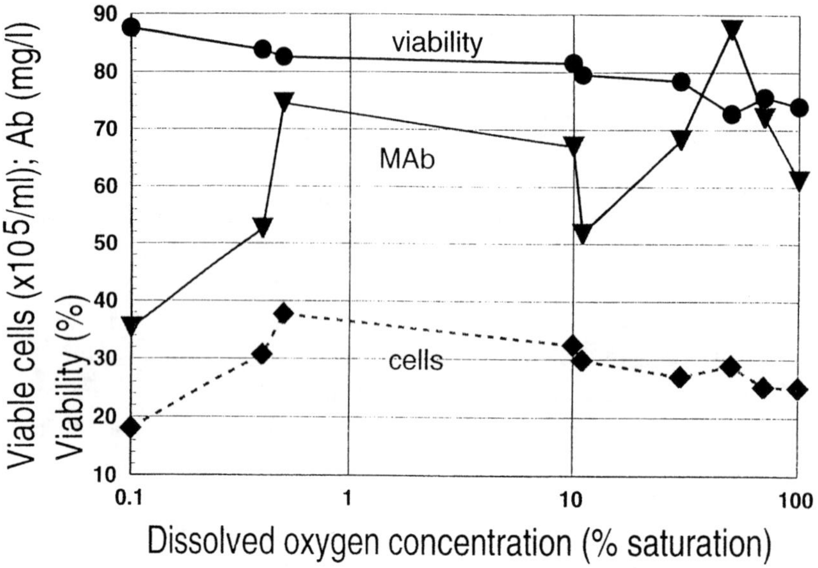

Figure 3.18. *The effect of dissolved oxygen concentration on the viable cell concentration, percent viability and antibody concentration of an IgG_{2a} producing hybridoma cell (Sp2/0 hybrid). [Data from Miller, W.M., Wilke, C.R. and H.W. Blanch, J. Cell Physiol.* ***132*** *524 (1987)].*

Table 3.12. *Commercial products derived from plants.*

	Plant Product	*Plant Species*	*Industrial Use*
Pharmaceuticals	codeine (alkaloid)	*Papaver somniferum*	analgesic
	diosgenin (steroid)	*Diascorea deltoidea*	anti-fertility agent
	quinine (alkaloid)	*Cinchona ledgeriara*	antimalerial
	digoxin (cardiac glycoside)	*Digitalis lanata*	cardiatonic
	scopolomine (alkaloid)	*Datura stramonium*	antihypertensive
	vincristine (alkaloid)	*Catharanthus roseus*	antileukaemic
Agrochemicals	pyrethrin	*Chrysanthemum cinerariaefolium*	insecticide
Food and Drink	quinine (alkaloid)	*Chinchona ledgeriana*	bittering agent
	thaumatin (chalcone)	*Thaumatococcus danielli*	non-nutritive sweetener
Cosmetics	jasmine	*Jasminum sp.*	perfume

From *Plant Biotechnology*, Soc. Exptl. Biology Seminar Series **18**, eds S.H. Mantell & H. Smith, Cambridge University Press)

We can conveniently distinguish five classes of plant tissue culture, depending on the type of plant tissue used. These are:

(i) *Callus culture*. The culture of cell mass on solid media (such as agar). The culture is derived from an explant of a seedling or other plant source.

(ii) *Cell culture*. The culture of cells in liquid media in aerated and agitated vessels.

(iii) *Organ culture*. The aseptic culture on nutrient media of embryos, anthers (microspores), ovaries, root shoots or other plant organs. This is employed primarily for crop improvement.

(iv) *Meristem culture*. The aseptic culture of shoot meristems or other explant tissues on nutrient media for the purpose of growing whole plants. The meristem is usually the dome of tissue located at the extreme tip of a shoot and is generally 0.1 mm in diameter and 0.3 mm in length. Apical meristems are formed during embryo development and remain in an active state of division throughout the vegetative phase of the plant growth. Meristem cells remain totipotent (able to regenerate as a whole plant) and genetically stable; thus they have been widely used in the horticultural industry.

(v) *Protoplast culture*. The aseptic isolation and culture of plant protoplasts from cultured cells or plant tissues. This is generally employed for somatic hybridization by protoplast fusion, for producing plant hybrids between closely related as well as unrelated plants. Cell fusion permits desirable crosses in plant breeding to be obtained when they are not feasible by other methods. The study of protoplast cultures, which lack the cellulosic wall, has been important in understanding the mechanism of nutrient transport in plants as well as viral infection.

The choice of culture method depends on the purpose of plant growth. Plant cell cultures are started by planting a section of sterile tissue on an agar medium. Over a period of around 2 to 4 weeks, a callus or mass of unorganized cells forms. This can be subcultured by transferring a small piece to fresh agar medium. The callus mass can also be transferred to a liquid medium and incubated on a shaker. Over a period of several weeks, depending on the plant species and the composition of the culture medium, a suspension culture can be obtained.

The nutritional requirements of plant cell and tissue cultures involve essential and optional components. Essential nutrients include carbon and energy sources, inorganic salts, vitamins and growth regulators (phytohormones). Optional nutrients include organic nitrogen, organic acids and other complex materials. The usual carbon and energy sources are glucose or sucrose. Fructose and other carbohydrates can be used but are generally less effective. The usual concentration ranges around 2 to 3 wt. %. Sources of inorganic ions are essential for plant growth. These include N, K, P, Ca, S and Mg, which need to be present in millimolar quantities. Micromolar levels of Fe, Mn, Zn, B, Cu and Mo are also required.

Normally plants synthesize the vitamins required for growth and development; however in culture some vitamins become limiting. Thiamine is absolutely required. Growth is usually improved by the addition of nicotinic acid and pyridoxine. Other vitamins which may also improve growth rates include pantothenate and biotin. Phytohormones are required for induction of cell division in culture. There are two classes of phytohormones that are important: cytokinins and auxins. Hormones which occur naturally include indoleacetic acid (IAA), an auxin, and zeatin, a cytokinin. The most important synthetic growth regulator is the auxin 2,4-dichlorophenoxyacetic acid (2,4-D). 2,4-D is most commonly used in suspension cultures. The cytokinins are all derivatives of adenine. Those most commonly used include benzyladenine, kinetin and isopentyladenine.

Many of the important products of plant cell cultures are secondary metabolites, without any vital role in the primary growth of the cells. They include alkaloids, isoprenoids, phenolics, volatile oils and specialized proteins. These are produced in the *iodiophase* of growth, which follows the period of rapid cell division. The production kinetics of these materials resembles that of non-growth associated microbial product formation.

***Table 3.13**. Natural Product Yields from Cell Cultures and Whole Plants.*

		Yield	
Natural Product	Species	Cell culture	Whole Plant
anthraquinones	*Morinda citrifolia*	900 nmol/gm dry wt	110 nmol/gm dry root wt
anthraquinones	*Cassia tora*	0.334% fresh wt	0.2% dry seed wt
ajmalicine and serpentine	*Catharanthus roseus*	1.3% dry wt	0.26% dry wt
diosgenin	*Diascorea deltoidia*	26 mg/gm dry wt	20 mg/gm dry tuber
ginseng saponins	*Panax ginseng*	0.38% fresh wt	0.3-3.3% fresh wt
nicotine	*Nicotiana tabacum*	3.4% dry wt	2-5% dry wt
thebaine	*Papaver bracteatum*	130 mg/gm dry wt	1400 mg/gm dry leaf wt
ubiquinone	*Nicotiana tabacum*	0.5 mg/gm dry wt	16 mg/gm dry leaf wt

From *Plant Biotechnology*, Soc. Exptl. Biology Seminar Series **18**, eds S.H. Mantell & H. Smith, Cambridge University Press.

The growth of plant cells in suspension culture is sensitive to a number of factors, including the level of agitation and light. Although cultures do not grow photoautrophically (except in a few special cases), the activities of enzymes involved in the biosynthesis of cinnamic acids, coumarins, lignins, flavones, flavanols, chalcones and anthocyanins are influenced significantly by light. It has been shown that activity of the enzyme phenylalanine ammonia lyase is increased by exposure to "cool white" fluorescent light in parsley cell cultures and a variety of flavone and flavanol glycosides are subsequently produced.

The degree of agitation can also have a significant influence on secondary metabolite production. At very high levels, the shear-sensitivity of the cell membrane may result in rupture of the cell. Clumping of cells is common and a minimum degree of agitation is required to ensure adequate supply of nutrients to the cells. Below this, low levels of accumulation of secondary metabolites may result.

Kinetics of Plant Cell Growth and Product Formation

Plant cell growth in suspension cultures follow microbial growth patterns to a large extent, however the maximum rates of growth are considerably slower. Typical batch growth profiles of *D. deltoidea* and *C. roseus* are shown in Figure 3.20 below. *D. deltoidea* shows a maximum specific growth rate of 0.25 day^{-1}, while *C. roseus* grows considerably faster (0.45 day^{-1}). The specific growth rates are quite temperature sensitive; for example *D. deltoidea* has specific growth rates of 0.18, 0.24, and 0.27 day^{-1} at 25, 26, and 28°C, respectively.

The growth of *D. deltoidea* can be adequately represented by Monod kinetics. Figure 3.19 shows a double reciprocal plot of μ versus sucrose concentration.

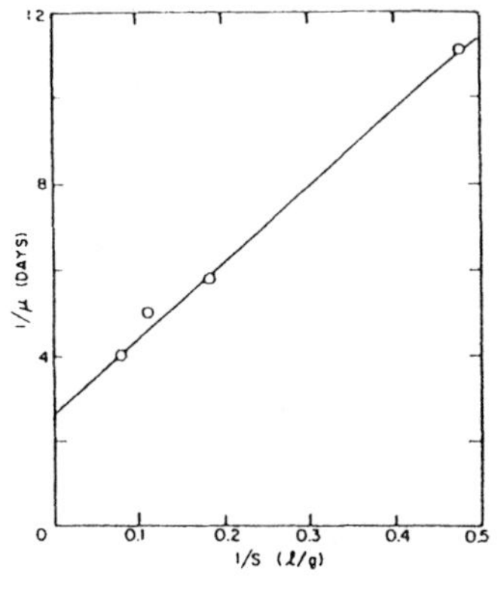

***Figure 3.19**. Lineweaver-Burk plot for the dry weight of D. deltoidea as a function of sucrose concentration. The K_S value is 7 gm/l [from D. Drapeau, H. Blanch and C. Wilke, Biotech. Bioeng., **23**, 1555 (1986)].*

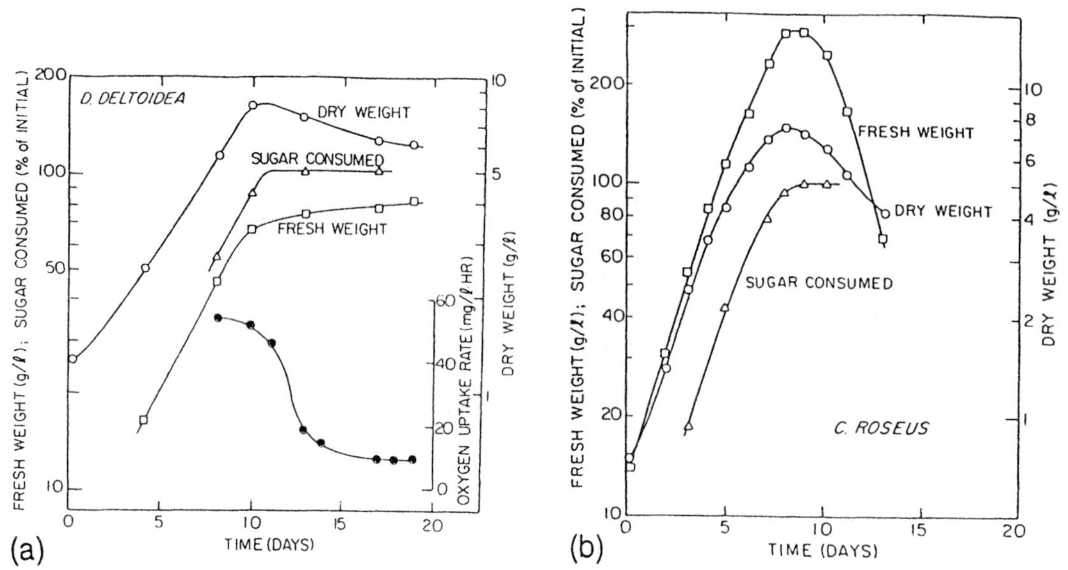

Figure 3.20. *Batch growth of (a) D. deltoidea and (b) C. roseus on 19 gm/l sucrose in a stirred tank fermenter with air sparging [from D. Drapeau, H. Blanch and C. Wilke, Biotech. Bioeng.,* ***23****, 1555 (1986)].*

Specific growth rates based on dry weight are consistently lower than those based on fresh (wet) weight. Plant cells in their natural environment are exposed to a phloem fluid that can contain sucrose at concentrations up to 250 gm/l. When exposed to high external sugar concentrations, the ratio of dry weight to fresh weight increases, as a result of either an osmotic effect, which may govern cell expansion, or reglation of the internal starch content by the external sugar concentration.

The consumption of oxygen by plant cell cultures can be described by the equation

$$q_{O_2} = A + B\mu \tag{3.77}$$

Typical data for *C. roseus* and *D. deltoidea* are given in Figure 3.21. Both cultures show high maintenance requirements for oxygen (12 mg/gm fresh weight.day for *D. deltoidea* and 3.5 mg/gm fresh weight.day for *C. roseus*.

The biosynthesis of many plant cell products follows non-growth associated kinetics. For example, diosgenin, exclusively an intracellular product, accumulates rapidly in batch cultures once sugar has been depleted. The rate of diosgenin biosynthesis is independent of growth rates for sucrose limited cultures (see Figure 3.22), however under nitrogen limitation there are indications from continuous cultures that some growth-associated component to the specific productivity may be present.

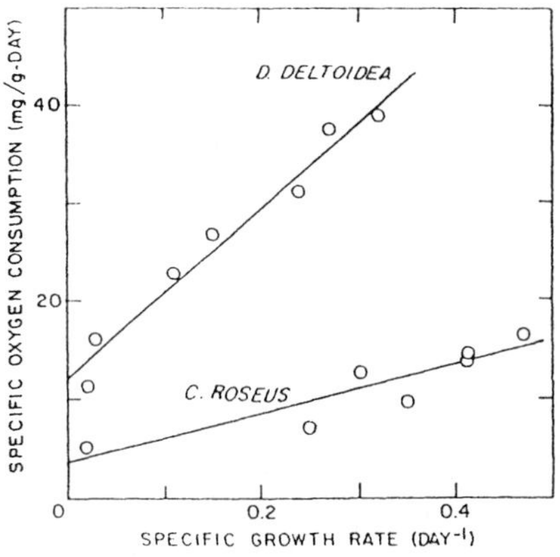

Figure 3.21. Relationship between specific oxygen uptake rate relative to fresh weight and the specific growth rate of the fresh weight [from D. Drapeau, H. Blanch and C. Wilke, Biotech. Bioeng., ***23****, 1555 (1986)].*

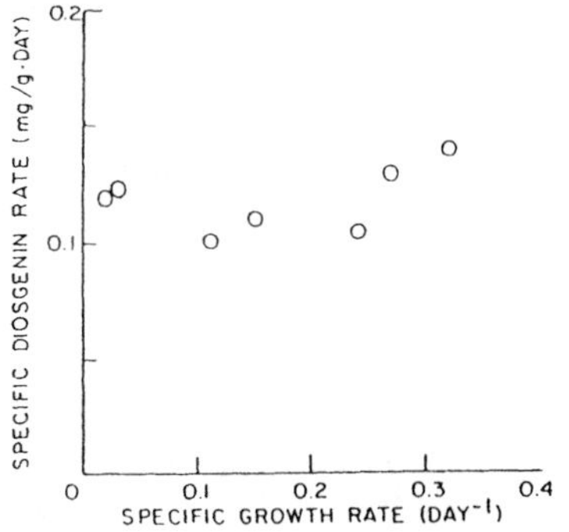

Figure 3.22. The specific diosgenin biosynthesis rate relative to fresh weight as a function of the specific growth rate of fresh weight (from D. Drapeau, H. Blanch and C. Wilke, Biotech. Bioeng., ***23****, 1555 (1986)).*

3.5 Structured Models of Metabolism and Growth

Unstructured models do not recognize the complex set of metabolic reactions occurring within the cell. Therefore we should not expect these simple models to be able to predict the dynamic behavior of cells subject to changing external conditions. In addition, unstructured models can predict intracellular concentrations only if we assume there is a constant fraction of the particular metabolite in the cell; for example, that the fraction of

RNA or DNA within a cell is constant. They thus have limited utility in guiding research aimed at understanding cellular regulation and dynamics. Such studies and the models derived from them are also required for implementation of sophisticated control of biological reactors or biological processes.

A number of approaches have been developed to tackle the problem of developing more sophisticated models of cell metabolism and growth. We shall examine some simple approaches in the following sections. It is clear that attempting to model all the reactions within a cell would be an extremely complex undertaking, particularly since we know relatively little about the kinetics and regulation of many of the enzyme-catalyzed steps involved. We can, however, group many of these reactions based on their characteristic time constants or relaxation times. The temporal organization within the cell can be described in terms of the following systems:

Genetic system: the longest timescale, associated with changes in DNA composition due to random mutagenesis, mutagenesis arising in response to changing environmental parameters, or natural genetic transfer of new characteristics via a vector such as a virus. This timescale is in multiples of the organism's generation time.

Epigenetic system: the intermediate timescale and is related to DNA transcription and translation. Enzyme induction (synthesis) and repression occur over this timescale.

Metabolic system: the short timescale of individual enzyme reactions. Usually much shorter than the epigenetic system. The behavior of the genetic and epigenetic systems may be considered constant over the period of observation of metabolic events.

In most situations, there is sufficient difference between these timescales that the behavior of other systems can be considered constant. For example, if we are examining the substrate and product concentration changes of a particular enzyme-catalyzed reaction within the cell, the enzyme concentration can be considered constant over the time period of our observations. Changing enzyme concentrations involves transcription and translation of DNA on a timescale of order tens of minutes or hours, while the enzyme-catalyzed reaction may have a relaxation time of order seconds or minutes. On the other hand, if changes occurring in the external environment are on the timescale of the epigenetic system, the behavior of the metabolic system will be very fast and we can consider it to be in a quasi-steady state. This approach greatly simplifies model development.

Models which incorporate the details of intracellular metabolism are referred to as *structured* models. Such models attempt to account for *unbalanced* growth of microorganisms, i.e., when the composition of the major cellular constituents, such as RNA, enzyme concentrations etc. vary as a result of changing external conditions. Such conditions apply in batch growth, in fed-batch growth, and in transient situations in well-stirred tank reactors.

The transient responses of cells to these changing external conditions can be modelled by analogy with classical reactor modelling using the transfer function approach. By using an appropriate forcing function and determining the transient response of the cells, the behavior of various cellular constituents can be modelled as first or higher order[31]. This approach has some advantages in developing and analyzing strategies for process control, but does not provide much insight into the factors that regulate metabolism. We shall now turn our attention to models which incorporate more of the features of microbial metabolism.

3.5.1 Compartmental Models

The earliest attempts to include structure in models of cell growth and metabolism generally subdivided the cell mass into various components on the basis of the function of parts of the cell's internal machinery[32]. We shall examine a simple two-compartment model to illustrate the features of this approach.

The Model of Williams[33]

The model of Williams divides the cell into two compartments, a synthetic one (k-compartment) that we can consider as consisting of RNA and pools of small metabolites, and a genetic one (g-compartment) consisting of DNA and protein. The third component is the external substrate concentration. A simple model based on these compartments can be developed as follows. If K and G are the concentrations of the components in the k- and g-compartments, written as mass per unit *cell* volume (V_c), we can write mass balances for a constant *reactor* volume (V_R) as follows.

$$\frac{dXV_R}{dt} = k_1 SXV_R \qquad i.e. \qquad \frac{dX}{dt} = k_1 SX \tag{3.78a}$$

$$\frac{dSV_R}{dt} = -\frac{1}{Y_S} k_1 SXV_R \qquad i.e. \qquad \frac{dS}{dt} = -\frac{1}{Y_S} k_1 SX \tag{3.78b}$$

The rate of substrate uptake is assumed to be first order in substrate concentration S and in total cell concentration X (both S and X being expressed as mass/reactor volume). We assume that the structural-genetic compartment material is produced from the synthetic compartment at a rate that depends directly on the concentration of species in each compartment. The mass balances are based on the reactor volume, thus the concentration per cell must be multiplied by the total cell volume per unit reactor volume, X/ρ_c, where ρ_c is the cell density (cell mass per unit cell volume).

---

(31) See for example Young, T.B., Bruley, D.F. and H.R. Bungay, Biotech. Bioeng. **12**, 747 (1970).

(32) A review of some of the many compartment models is provided in Harder, A. and J.A. Roels, Adv. Biochem. Engng., **21**, 55 (1982).

(33) Williams, F.M., J. Theoret. Biol. **15**, 190 (1967)

$$\frac{dG(X/\rho_c)}{dt} = k_2 GK(X/\rho_c)$$

Expanding the left hand side yields

$$(X/\rho_c)\frac{dG}{dt} + G\frac{d(X/\rho_c)}{dt} = k_2 GK(X/\rho_c)$$

$$\frac{dG}{dt} = k_2 GK - G\frac{1}{X}\frac{dX}{dt}$$

and combining this result with Eq. (3.78a)

$$\frac{dG}{dt} = k_2 GK - (k_1 S)G \tag{3.79}$$

We have assumed that the density of the cell, ρ_c , is constant. The synthetic portion of the biomass is produced at a rate that is first order in substrate concentration and depends on the cell density ρ_c (equal to the sum of K and G):

$$\frac{dKX/\rho_c}{dt} = k_1 S(G+K)(X/\rho_c) - k_2 GK(X/\rho_c)$$

$$\frac{dK}{dt} = k_1 S(G+K) - k_2 GK - K\frac{1}{X}\frac{dX}{dt} \qquad \text{and noting that} \qquad \frac{1}{X}\frac{dX}{dt} = k_1 S$$

$$\frac{dK}{dt} = k_1 S\rho_c - k_2 GK - k_1 KS \tag{3.80}$$

We note that equations (3.79) and (3.80) can be added to show that the relationship below holds:

$$\frac{dK}{dt} + \frac{dG}{dt} = k_1 S\rho_c - k_1 S(K+G) = 0 \qquad (\text{as} \quad K+G=\rho_c) \tag{3.81}$$

Equations (3.78) and (3.79) can be added to eliminate X

$$\frac{d(X+Y_S S)}{dt} = 0 \qquad \text{thus} \qquad X + Y_S S = X_0 + Y_S S_0 \tag{3.82}$$

Equation (3.78) can now be solved for S

$$S = \frac{S_0\left(1+\frac{X_0}{Y_S S_0}\right)}{1+\frac{X_0}{Y_S S_0}\exp\{k_1(X_0+Y_S S_0)/Y_S\}t} \tag{3.83}$$

If we assume that the cell number depends only on the amount of genetic component present (i.e., G), then the cell number will be proportional to GX/ρ_c cells/reactor volume. The cell volume will change as a reflection of the changing amounts in the genetic and synthetic compartments, hence the cell size will be proportional to (K+G)/G, ie. ρ_c/G. The behavior of the model is shown in Figure 3.23.

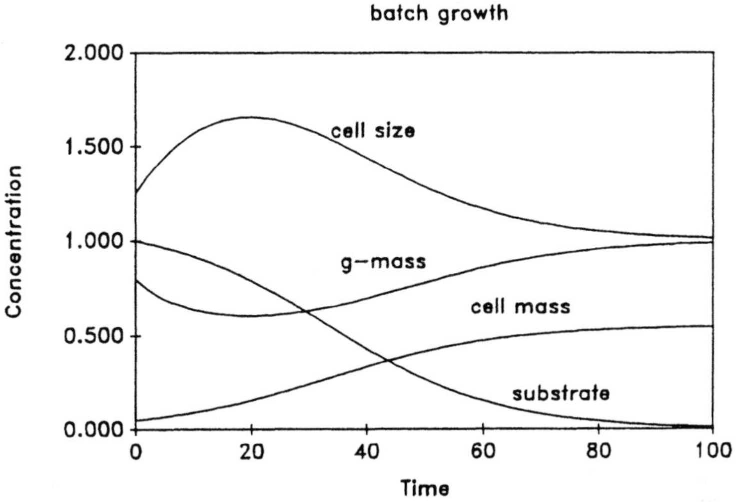

***Figure 3.23**. Simulation of the two-compartment model of Williams for batch growth with the following initial conditions: X_0= 0.05 mg/ml, S_0= 1.0 mg/ml, K_0= 0.6 mg/ml, G_0= 0.4 mg/ml. Time is expressed in hours, concentrations in mg/ml of cell (for g-mass) or reactor volume (for cell mass and substrate). Values of the constants are k_1= 0.0125 $(mg/ml)^{-1}.hr^{-1}$, k_2= 0.025 $(mg/ml)^{-1}.hr^{-1}$, Y=0.5 mg/mg, ρ_c = 1.0 gm/ml. Cell number is proportional to the g-mass.*

The compartment model of Williams illustrates some important properties of cell growth. It predicts the existence of a initial lag phase (see the cell mass curve in Figure (3.23) and, if the inoculum is not fully adapted (ie. K_0 and G_0 do not correspond to values found in exponential growth), cell mass will increase (due to an increase in the K compartment), while the cell number will not change initially (the cell density ρ_c remains constant). During the growth phase, both cell mass and cell numbers increase exponentially. The stationary phase is attained with respect to cell mass prior to cell number (as illustrated in Figure (3.23) by the curves of g-mass and cell mass). These phenomena are typically observed in batch culture experiments with bacteria. The model can be refined by changing the linear dependence of the rate of substrate uptake and growth from first order to a Monod type in equation (3.78). As can be seen from the batch simulations above, the concentration of the g-compartment approaches the cell density $\rho_c$at long times, while that of the k-compartment approaches zero. Neither of these predictions is consistent with experimental observation. The inclusion of maintenance in the formulation of the model would change this inconsistent result and improve the model.

The Model of Ramkrishna et al.[34]

An analogous compartment model to that of Williams has been developed by Ramkrishna et al. The cell is divided into two compartments, G-mass, comprising RNA and DNA, and D-mass, which mostly consists of proteins. An inhibitor (T) is produced during growth which converts both G- and D-mass to inactive forms of biomass. The reactions assumed are the following:

$$G + a_S S + D \rightarrow 2G + D + a_T T$$

$$G + a_S' S + D \rightarrow 2D + G + a_T' T$$

$$G + T \rightarrow N_G + (1 + a_{TI})T$$

$$D + T \rightarrow N_D + (1 + a_{TI}')T$$

In the first reaction, D-mass catalyzes the formation of G-mass, consuming a_S units of substrate and producing a_T units of an unidentified inhibitor T. In the second reaction G-mass catalyzes the formation of D-mass. The last two equations show that both G- and D-mass are deactivated by inhibitor (producing inactive forms N_G and N_D). The rate expressions assumed in the model for production of G- and D-mass are of the double substrate form of equation (3.44), while those for the deactivation reactions are assumed to be first order in each reactant. Mass balance equations for batch and continuous stirred tank reactors can be written and the model predictions can be compared with experimental data. This model can predict oscillatory behavior about a steady state.

3.5.2 Models of Cellular Energetics and Metabolism

As we move beyond the compartmental models of the type above, the level of metabolic detail required increases. At this point, it will be useful to consider the types of reactions that occur within the cell. Metabolic pathways can be distinguished as *catabolic* and *anabolic*. In catabolism, energy-containing molecules, such as carbohydrates, hydrocarbons and other reduced carbon-containing compounds are degraded to CO_2 or other oxidized end-products and the energy is stored in ATP, GTP, and other energy-rich compounds. In anabolism, intermediates and end-products formed from catabolism are incorporated into cell constituents (such as DNA, RNA, lipids, carbohydrates, etc.) and their intermediate precursors (amino acids, purines and pyrimidines, simple sugars etc.). Anabolic reactions generally require energy, which is supplied via ATP and other high-energy phosphates generated during catabolism. As the concentration of these high-energy intermediates within the cell is rather small, anabolism is linked to catabolism and ATP is rapidly turned over. This implies that energy producing and energy consuming processes must be tightly regu-

---

(34) Ramkrishna, D., Fredrickson, A.G. and H. Tsuchiya, Biotech. Bioeng., 9, 129 (1967).

lated within the cell. It is thus necessary to consider both carbon and energy flows within the cell in developing these more complex models. An example of such a model is given in the following section.

A Model for Aerobic Growth of the Yeast Saccharomyces cerevisiae

The yeast *S. cerevisiae* (bakers' yeast) has been well studied and considerable information on its cell cycle, regulation and metabolism is known. Hall and co-workers[35] have formulated a model of the rather complex metabolism exhibited by *S. cerevisiae* when grown on glucose. This yeast can use either the *respiratory pathway*, in which glucose is converted to CO_2 and cell mass, or the *fermentative pathway*, resulting in the formation of ethanol, CO_2 and cell mass. At low growth rates, metabolism is fully oxidative, i.e., the respiratory quotient (RQ), defined as the ratio of the rates of CO_2 production to O_2 consumption, is unity and $Y_{X/S}$ is 0.50 gm cells/gm glucose. This situation is maintained up to a critical growth rate, beyond which the metabolism becomes increasingly fermentative. In the fermentative pathway, the yield coefficient decreases and there is an increase in the specific carbon dioxide production rate (q_{CO2}) and ethanol production. This critical growth rate is slightly higher than the value of μ_{max} on ethanol. There is a change in the enzyme pattern that reflects this switch from respiration to fermentation: typical respiratory enzymes, such as isocitrate lyase, malate dehydrogenase and the cytochromes are repressed at high growth rates, and glycolysis provides the main source of energy. At low growth rates, reduced levels of glycolytic enzymes are found.

As the growth rate increases, the percentage of budding yeast cells increases almost linearly. Using this linear relationship and the mean generation time ($\ln 2/\mu$), the length of the budding period can be calculated. There is little variation in the duration of the budding period at different growth rates. At low growth rates, the generation time increases due to lengthening of the gap-phase following cell division (the post-mitotic gap, or G_1 phase). Thus, referring to the cell cycle, the time periods for DNA replication (S), mitosis (G_2) and cell division (M) phases are all constant. The duration of the G_1 phase appears to be variable. During the single-cell G_1 phase, substrate is accumulated and there is a buildup of reserve carbohydrates (trehalose and glycogen) within the cell that are then depleted for energy and carbon during the period of budding.

The model we shall examine is based on this two-stage breakdown of the cell cycle. The length of the G_1 phase depends on the availability of the limiting substrate; the length of the division phase (the sum of G_2, M and S phases) is assumed to be independent of substrate. The cell mass is considered to be comprised of two parts: A mass, which carries

---

(35) Pamment, N.B., Hall, R.J. and J.P. Barford, Biotech. Bioeng. **20**, 345 (1978) and Bijkerk, A.H. and R.J. Hall, Biotech. Bioeng. **19**, 267 (1977).

out substrate uptake and energy production; and B mass, which carries out reproduction and division. B mass is converted to A mass at a constant rate, whereas A mass consumes substrate and produces B mass at a variable rate.

We now turn to the regulation of respiration and fermentation. The repression of respiratory enzymes by glucose (the "Crabtree" effect) was initially thought to result from glucose acting as a catabolite repressor. More recent evidence suggests that a high catabolic flux is the direct cause of respiratory inhibition and that glucose concentration plays a secondary role. Thus the model proposes that both glycolysis and respiration are carried out by A mass and both provide energy for growth. The following reactions were proposed to describe respiration and glycolysis:

$$A + a_i S \xrightarrow{r_A} 2B + a_2 E + CO_2$$

$$O_2 + A + a_3 E \xrightarrow{r_B} 2B + CO_2$$

$$B \xrightarrow{r_C} A$$

(A + B) is the total cell mass (X) and E is the ethanol concentration.
The following rate expressions are assumed:

$$r_A = k_1 A \frac{S}{K_S + S} \tag{3.84}$$

$$r_B = k_2 A \frac{E}{K_E + E} \tag{3.85}$$

$$r_C = KB \tag{3.86}$$

The budding process (B → A) is assumed to occur at a constant rate whereas respiration and fermentation follow Monod type kinetics. Mass balances can now be written for each of the species (assuming constant cell volume).

$$\frac{dB}{dt} = 2r_A + 2r_B - r_C \tag{3.87}$$

$$\frac{dA}{dt} = -r_A - r_B + r_C \tag{3.88}$$

$$\frac{dE}{dt} = a_3 r_A - a_4 r_B \tag{3.88a}$$

$$\frac{dS}{dt} = -a_1 r_A \tag{3.89}$$

Estimates of the yield coefficients can now be made to evaluate the constants a_1 and a_3. For fermentation, $Y_{X/S}$ (equal to $1/a_1$) is 0.15 gm/gm, and for respiration $Y_{X/S}$ (equal to $1/a_3$) is 0.45 gm/gm. The constant a_2, estimated from the stoichiometry of ethanol production, is 2.80. The specific uptake and production rates can be calculated from

$$q_S = -\frac{1}{X}\frac{dS}{dt} = -\frac{a_1 r_A}{X} \tag{3.90}$$

$$q_E = \frac{1}{X}\frac{dE}{dt} = \frac{a_2 r_A - a_3 r_B}{X} \tag{3.91}$$

$$\mu = \frac{1}{X}\frac{dX}{dt} \tag{3.92}$$

Uptake of oxygen is proportional to the respiratory growth rate, thus the specific oxygen uptake rate $q_{02} = \alpha\, r_B/X$. The constant α can be found from stoichiometry. In respiration, complete oxidation of ethanol to CO_2 and water requires 3 moles of oxygen for each 2 moles of CO_2 produced (assuming that the oxidation state of the cell is the same as that of the carbon source). From the elemental composition of yeast, 1.53 gm of oxygen are required for each gram of dry cell weight produced. Thus q_{O2} can be estimated as

$$q_{O_2} \quad (\text{mol/gm.hr}) = (1.53/32)\frac{r_B}{X} = 0.048\frac{r_B}{X} \tag{3.93}$$

The specific CO_2 production rate q_{CO2} can be found from a carbon balance:

carbon in CO_2 = *carbon in glucose - carbon in ethanol - carbon in cells*

$$q_{CO2}\,(\text{mole/gm.hr}) = (0.40\, q_S - 0.52\, q_E - 0.48\, \mu)/12 \tag{3.94}$$

The other constants in the model were obtained from experimental data on the specific growth rates in each of the phases of growth (details are provided in Bijkerk and Hall *op. cit.*). Their values are

$k_1 = 5.0\ \text{hr}^{-1}$	$a_1 = 6.67$
$K = 0.51\ \text{hr}^{-1}$	$a_2 = 2.80$
$k_2 = 0.30\ \text{hr}^{-1}$	$a_3 = 2.22$
$K_S = 0.5$ gm/l	$K_E = 0.02$ gm/l (3.95)

The predictions of this model are illustrated in Figure 3.24 and 3.25.

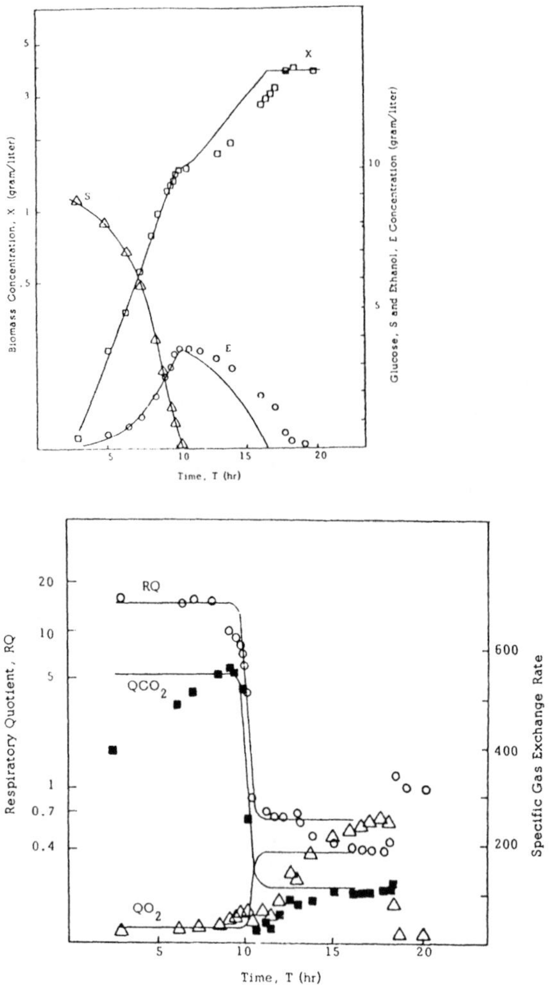

Figure 3.24. *The predictions of the model compared with the data of von Meyenburg. [From H.K. von Meyenburg, "Katabolit Repression und der Sprossungzyklus von* S. cerevisiae*", Ph.D. Dissertation, ETH, Zürich (1969)). The dry weight (X), glucose (S), and ethanol (E) concentrations for batch growth of* S. cerevisiae *are shown. The model parameters are given in the text. The respiratory quotient (RQ) and the specific gas exchange rates Q_{O2} and Q_{CO2} (ml/(gm dry weight.hr) are also illustrated.*

This relatively simple model describes the carbon and energy metabolism of *S. cerevisiae* quite well. The two-stage structure of this model has the consequence that the fraction of B mass increases linearly and the fraction of A mass decreases with specific growth rate. At low specific growth rates, the total specific substrate uptake rate is low and the fraction of A mass available to carry out this activity is large. The substrate flux into A mass is low and the culture can thus make use of the slower, but more efficient, respiratory pathway. At

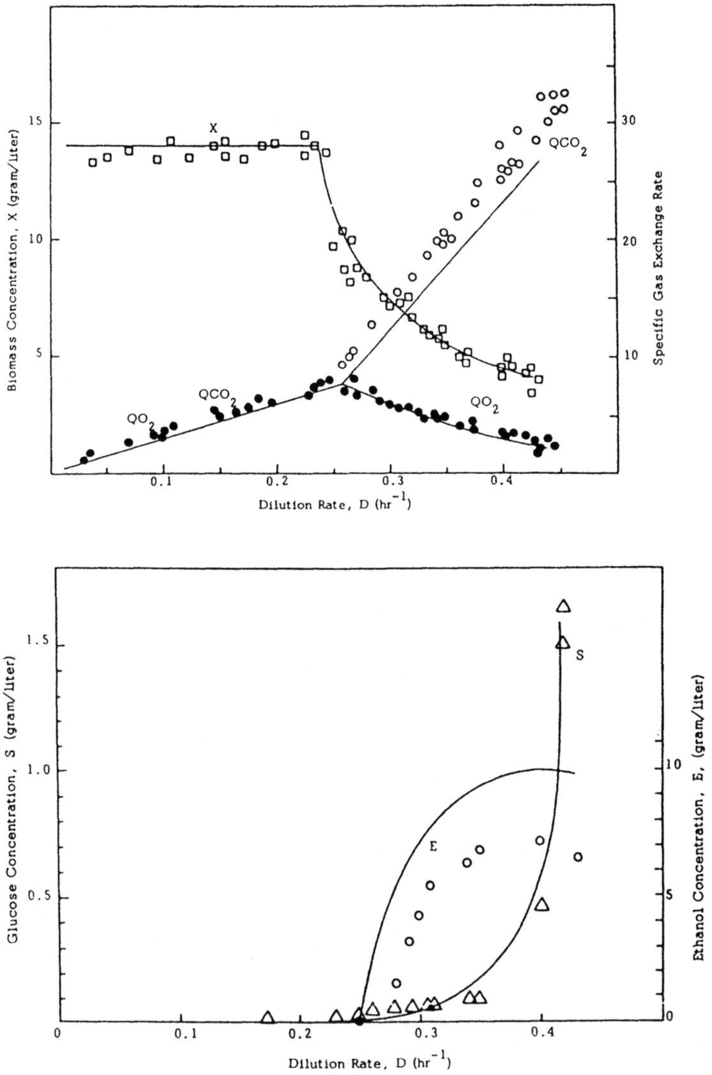

Figure 3.25. *Model predictions and experimental data for* S. cerevisiae *in continuous culture. The specific growth rate is equal to the dilution rate D. The gas exchange rates are in ml/(gm dry weight-hr). Parameter values are the same as in Figure 3.24, except that* a_3 *is 1.65 and* k_2 *is 0.5* hr^{-1}.

higher growth rates, the substrate flux through A mass increases for two main reasons. The rate of catabolism per cell increases and the fraction of A mass decreases. As a result of this increased flux, the respiratory pathway becomes saturated at a growth rate determined by

k_2. The glycolytic pathway is able to support a much higher energy flux than the respiratory pathway and the model demonstrates a higher specific glucose uptake rate through the glycolytic pathway, required for growth to occur at these higher growth rates.

When the glycolytic pathway becomes saturated, the maximum specific growth rate for glucose is attained. Thus in batch culture we see that the rapid catabolic flux produced by initial glycolysis supports a high specific growth rate and a low proportion of A mass, with reduced respiratory activity. The respiratory pathway is operating at its maximum flux; however, since the A mass concentration is small, the respiratory activity per total cell mass is quite small. Ethanol accumulates, but once glucose is exhausted, the culture converts to respiration, consuming the accumulated ethanol. The reduced catabolism required by respiration results in an increase in A mass and thus there is a period of adaptation and a second phase of growth with a higher proportion of A mass, as required by the lower catabolic flux.

3.5.3 Models of Product Formation

The formation of microbial products can be considered to result from either growth-related activities of the organism or from secondary metabolism, or a combination of both. The simple unstructured models for growth and non-growth associated product formation kinetics we examined earlier were attempts to simplify the complex set of metabolic reactions involved in product formation. In many cases, these simple models are not adequate to guide experiments aimed at increasing product yields and concentrations. In antibiotic fermentations, pilot-scale experimentation is costly and a well-formulated model is necessary for process development. Such models need to reflect more details about the metabolic reactions involved than can be obtained with simple Monod-type relationships between substrates and growth rates. Early attempts to model complex fermentations, such as fungal fermentations producing antibiotics, were hampered by a lack of knowledge about metabolism. As more information on the metabolic routes to various products has been obtained, structured models can now include information on rate-determining pathways and nutrient transport kinetics.

With structured models of product formation available, it is possible to examine the effects of genetic amplification of various enzymes in a pathway. Such studies, for example, examination of the effect of removing feedback regulation in a key reaction step, can guide strain development. These studies also highlight the sometimes unforeseen consequences of genetic manipulation, in that other reactions may quickly become rate-limiting and the expected gain in product yield may not be obtained. They may also show that several enzymes must be altered to achieve the desired production levels. This approach, the alteration of more than a single gene to enhance the flow of metabolites through an entire pathway, is often referred to as "metabolic engineering". In this section we shall examine a structured and a segregated model for product formation.

Example: **The Production of α-Galactosidase by Monascus (A Structured Model)**

The production of the enzyme α-galactosidase is subject to catabolite repression; when the mold *Monascus* is grown on glucose as a carbon and energy source, no α-galactosidase enzyme is made. However, if both glucose and galactose are present, glucose is consumed first. When the glucose concentration is reduced to a low level, α-galactosidase is induced and it is produced about 80 minutes later in batch culture. Similarly, when glucose is added to a system producing α-galactosidase, enzyme production is repressed over a 40 minute period. A structured model for the production of the enzyme α-galactosidase by *Monascus* has been formulated by Imanaka et al.[36] and is described here as it illustrates the coupling of substrate transport and regulation of gene expression.

We shall denote extracellular glucose and galactose concentrations as S_A and S_B; the internal galactose concentration is S_{Bi}. By assuming that galactose inhibits the consumption of glucose in a competitive manner, the specific growth rate of the mold is assumed to depend on both substrates in the following manner:

$$\mu = \frac{\mu_{mA} S_A}{K_{SA} + S_A + \frac{K_{SA}}{K_i} S_B} + \frac{\mu_{mB} S_B}{K_{SB} + S_B} \tag{3.96}$$

Transport of galactose into the cell is assumed to be governed by an active transport mechanism. The maximum concentration of galactose-transporter binding sites is G_B. External galactose binds to these sites by an adsorption isotherm that follows a Langmuir dependence on S_B. The rate of galactose transport is described by the following expression.

$$\text{rate of galactose transport} = U\left(\frac{G_B S_B}{K_m + S_B} - S_{Bi}\right) \tag{3.97}$$

Here U is a transport rate coefficient in hr^{-1}, G_B is in μg galactose per mg cell mass and the intracellular concentration of galactose is in μg/mg cells. Intracellular galactose is consumed at a first order rate given by $k_1 S_{Bi}$. Thus a mass balance on intracellular galactose gives

$$\frac{d(S_{Bi}X)}{dt} = U\left(\frac{G_B S_{Bi}}{K_m + S_B} - S_{Bi}\right)X - k_1 S_{Bi} X \tag{3.98}$$

The effect of glucose repression is modelled by considering that when the external glucose concentration exceeds a critical value S_{Ac}, the rate of galactose transport into the cell immediately ceases, i.e., when $S_A \geq S_{Ac}$, then $U = 0$. As a consequence, the specific growth rate μ_{mB} decreases to zero. Experimental data, shown in Figure 3.26 indicates that in the presence of 5 gm/liter glucose, galactose uptake is negligible. Galactose transport can also be seen to be constitutive, as there is no time delay in uptake.

---

(36) Imanaka, T., Kaieda T., and H. Taguchi, J. Ferment. Technol. **51** (6) 423 (1973).

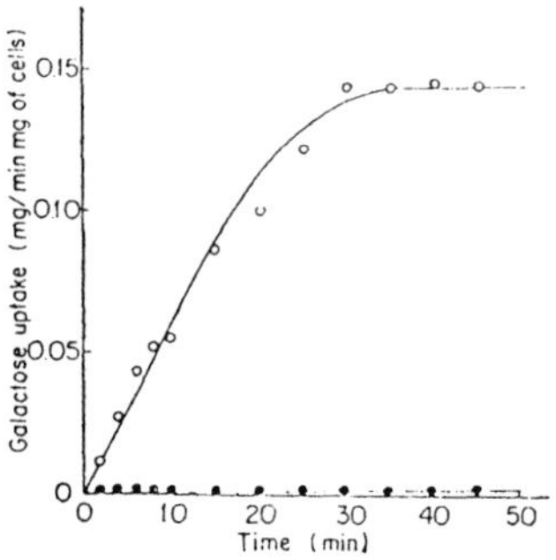

Figure 3.26. Galactose uptake by Monascus. Uptake of galactose at 35°C, with a galactose concentration of 5 gm/liter containing 1 μci ^{14}C-galactose. Open circles represent cells grown on glucose, washed and incubated on 5 gm/liter galactose medium. Closed circles indicate incubation in the presence of 5 gm/liter glucose. No galactose uptake is evident under glucose repression. (From Imanaka et. al.)

A simple model for enzyme induction by galactose, based on the Monod-Jacob operon model, can be developed. In the absence of galactose, a repressor (R) acts to inhibit the synthesis of mRNA for α-galactosidase. When galactose is present, the repressor and galactose combine to form $[RS_{Bi}]$ which can no longer bind to the DNA and enzyme synthesis can occur. The amount of the (unidentified) intracellular repressor is found by considering it to be produced at a constant rate k_2, degraded by a first order process (k_3R) and reacting reversibly with intracellular galactose to form a complex $[RS_{Bi}]$ by mass action kinetics. The mass balance on repressor R yields

$$\frac{dRX}{dt} = (k_2 - k_3R - k_4RS_{Bi} + k_5[RS_{Bi}])X \tag{3.99}$$

The action of the repressor on the transcription of DNA and production of mRNA is as follows. The synthesis of mRNA depends on the concentration of free repressor; when galactose is present and repressor is bound to it, the concentration of free repressor is reduced and it is no longer as effective in blocking mRNA synthesis. The rate of mRNA synthesis is assumed to be proportional to the reduction in free repressor from some maximum value R_c. The mRNA is assumed to decay with a first order rate constant k_7. The synthesis of mRNA (M) is thus described by

$$\frac{dMX}{dt} = \{k_6(R_c - R) - k_7M\}X \tag{3.100}$$

This scheme for regulation of enzyme production is illustrated in Figure 3.27. The rate of α-galactosidase formed by the mold depends on the concentration of mRNA:

$$\frac{dEX}{dt} = k_8MX \tag{3.101}$$

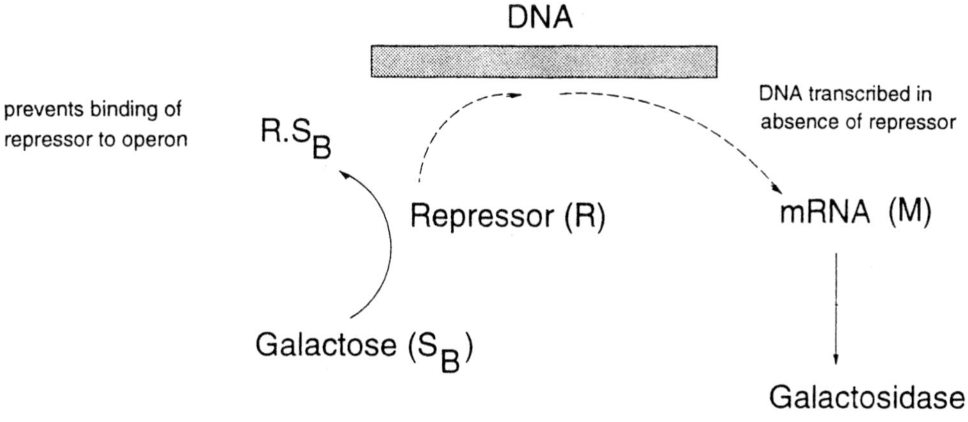

Figure 3.27. *Schematic of the gene level regulation of α-galactosidase by galactose. The repressor R is unknown, but is probably related to glucose or its metabolites.*

A complete set of mass balances can now be written, including a balance for repressor-galactose complex $[RS_{Bi}]$. We introduce the individual yield coefficients Y_A and Y_B, these being the cell mass yields on glucose and galactose substrates, respectively. In writing the mass balances, the terms describing the intracellular concentrations have been expanded viz.

$$\frac{dS_{Bi}X}{dt} = X\frac{dS_{Bi}}{dt} + S_{Bi}\frac{dX}{dt} = X\frac{dS_{Bi}}{dt} + \mu S_{Bi} \tag{3.102}$$

Thus the intracellular concentrations are effectively diluted by the expanding volume of the cell.

$$\frac{dX}{dt} = \mu X = \left\{ \frac{\mu_{mA}S_A}{K_{SA} + S_A + \frac{K_{SA}}{K_i}S_B} + \frac{\mu_{mB}S_B}{K_{SB} + S_B} \right\} X \tag{3.103}$$

$$\frac{dS_A}{dt} = -\frac{X}{Y_A}\frac{\mu_{mA}S_A}{K_{SA} + S_A + \frac{K_{SA}}{K_i}S_B} \tag{3.104}$$

$$\frac{dS_B}{dt} = -\frac{X}{Y_B}\left\{ \frac{\mu_{mB}S_B}{K_{SB} + S_B} \right\} \qquad \text{where} \qquad S_A \geq S_{Ac} \quad \mu_{mB} = 0 \tag{3.105}$$

$$\frac{dS_{Bi}}{dt} = U\left(\frac{G_B S_{Bi}}{K_m + S_B} - S_{Bi} \right) - k_1 S_{Bi} - \mu S_{Bi} \qquad \text{where} \qquad S_A \geq S_{AC} \quad U = 0 \tag{3.106}$$

$$\frac{dR}{dt} = (k_2 - k_3 R - k_4 R S_{Bi} + k_5[RS_{Bi}]) - \mu R \tag{3.107}$$

$$\frac{dM}{dt} = \{k_6(R_c - R) - k_7 M\} - \mu M \qquad where \qquad R \geq R_C \quad R = R_C \tag{3.108}$$

$$\frac{dE}{dt} = k_8 M - \mu E \tag{3.109}$$

$$\frac{d[RS_{Bi}]}{dt} = k_4 RS_{Bi} - k_5[RS_{Bi}] - \mu[RS_{Bi}] \tag{3.110}$$

The kinetic and microbial parameters in the above set of equations were determined by Imanaka et al. from batch and steady state continuous culture data. The intracellular parameters were estimated, whereas the yields and specific growth rates were obtained from the experimental data. The values of the constants are listed in Table 3.14.

***Table 3.14**. Values of constants in the model of Imanaka et al.*

Estimated			Experimental values		
k_1	= 40	hr^{-1}	$\mu_{mA,g}$	= 0.215	hr^{-1}
k_2	= 1	μg/mg cells-hr	$\mu_{mB,g}$	= 0.208	hr^{-1}
k_3	= 1	hr^{-1}	$\mu_{mA,p}$	= 0.190	hr^{-1}
k_4	= 0.1	mg cells/μg-hr	$\mu_{mB,p}$	= 0.162	hr^{-1}
k_5	= 1 x 10^{-4}	hr^{-1}	$K_{sA,g}$	= 1.54 x 10^{-4}	gm/ml
k_6	= 1	hr^{-1}	$K_{sB,g}$	= 2.58 x 10^{-4}	gm/ml
k_7	= 8	hr^{-1}	$K_{sA,p}$	= 1.54 x 10^{-4}	gm/ml
k_8	= 4.0	units/μg M-hr	$K_{sB,p}$	= 3.07 x 10^{-4}	gm/ml
$k_{8,p}$	= 6.67	units/μg M-hr	K_i	= 1.39 x 10^{-4}	gm/ml
S_{Ac}	= 2.25 x 10^{-4}	gm/ml			
U	= 100	hr^{-1}	$Y_{A,g}$	= 0.530	gm/gm
G_B	= 3.5	μg/mg cells	$Y_{B,g}$	= 0.516	gm/gm
K_m	= 1 x 10^{-8}	μg/mg cells	$Y_{A,p}$	= 0.377	gm/gm
R_c	= 0.803	μg/mg cells	$Y_{B,p}$	= 0.361	gm/gm

The subscripts g and p denote values during the growth phase and enzyme production phase repectively.

The predictions of this model compared to the experimental batch culture data are illustrated in Figure 3.28. The model can succesfully predict the catabolite repression of galactose uptake by glucose and the slow rise in the concentration of α-galactosidase once glucose is consumed and galactose is transported into the cell. The key feature of this model of enzyme production is the incorporation of a simple model of enzyme repression according to the scheme proposed by Jacob and Monod. It is able to effectively describe the observed production kinetics. We shall examine some more complex models of gene expression in Section 3.6.

The ability of this model to predict the dynamics of α-galactosidase repression by glucose is considered in Problem 16. The half life of mRNA is a key constant in determining the strength of glucose repression.

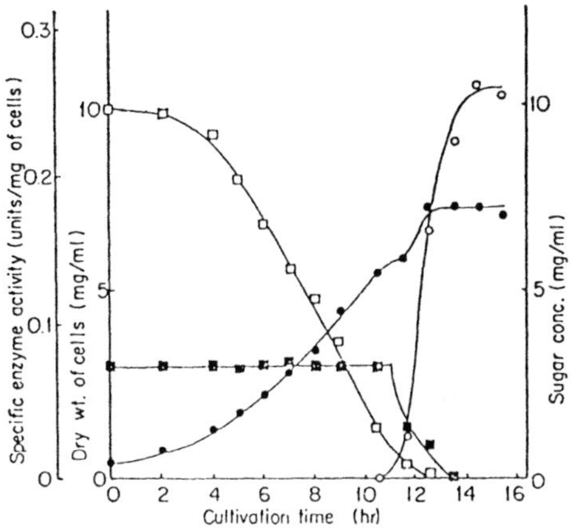

Figure 3.28. Time course of cell growth and α-galactosidase production by Monsacus. The inoculum contained glucose grown cells. The medium composition was 10 gm/l glucose, 3 gm/l galactose, 5 gm/l NH_4NO_3, KH_2PO_4 5 gm/l, $MgSO_4.7H_2O$ 1 gm/l, yeast extract 0.1 gm/l. Open squares, glucose; closed squares, galactose; open circles, α-galactosidase; closed circles, cell mass. The initial conditions for the model were: $X = 0.5$ gm/l; $S_A = 10$ gm/l; $S_B = 3$ gm/l; $S_{Bi} = 0$ μg/gm cells; $R = 0.910$ μg/gm cells; $M = 0$ μg/mg cells; $E = 0$ units/mg cells; $[RS_{Bi}] = 0$ μg/gm cells.

Example: **An Age Distribution Model for the Production of Antibiotics (A Segregated Model)**

Secondary metabolites, including most fungal antibiotics, are either produced late in the growth phase (the idiophase) or in the stationary phase and are often classified as non-growth associated products. Antibiotics produced by unicellular microorganisms usually only appear when the cell ceases dividing. When the organism is filamentous, secondary metabolites are formed even though the cell dry weight may still be increasing. This occurs because cells located in filaments may no longer be dividing, have entered the iodiophase and produce the desired metabolite, while cells at the tips of the hyphae continue to grow and are do not produce product. Examples of this behavior include production of antibiotics such as rifamycin, the polyene anitbiotics, and chloramphenicol.

One simple approach to modelling the production of such secondary metabolites, which does not require detailed information of the complex metabolic pathways involved, is to incorporate the concept of cell age into the model. We can consider two age groups of cells: "mature" cells capable of product synthesis, and "immature" cells unable to form product[37]. The growth rate of mature cells is described by Monod-type kinetics (with constants α, β, whereas immature cells age at a first-order rate (a_{12}) to form mature cells. A schematic of the growth and maturation process is given in Figure 3.29.

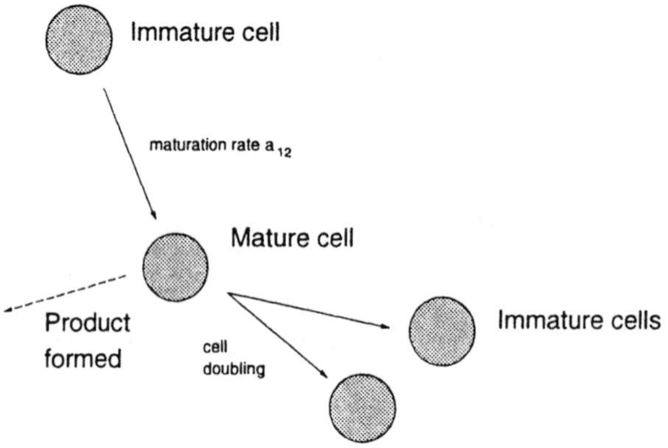

Figure 3.29. *The formation of mature cells by a first-order aging process. Mature cells undergo fission to produce immature cells.*

Denoting the immature cell concentration as X_1, the concentration of mature cells by X_2, and the product by P, the following set of mass balances for batch culture can be written:

$$\frac{dX_1}{dt} = -a_{12}X_1 + \frac{2\alpha S X_2}{S+\beta} \tag{3.111}$$

$$\frac{dX_2}{dt} = a_{12}X_1 - \frac{\alpha S X_2}{S+\beta} \tag{3.112}$$

$$\frac{dS}{dt} = -\gamma \frac{\alpha S X_2}{S+\beta} \tag{3.113}$$

$$\frac{dP}{dt} = \left(\frac{1}{k}\right)\frac{dX_2}{dt} \tag{3.114}$$

The rate of growth of the cell population ($X_1 + X_2$) follows a Monod-type dependence on substrate (S), while the rate of product formation is assumed to be proportional to the number of mature cells. The cell yield coefficient is $1/\gamma$, and the rate of product formation is proportional to the rate of formation of mature cells. This model can simulate the production kinetics of the cyclic decapeptide gramicidin S, a bacterial antibiotic produced by *Bacillus brevis*. Gramicidin S synthetase activity increases dramatically in the late logarithmic phase of growth, resulting in gramicidin S production in the late logarithmic and stationary phases of growth. The kinetics of growth and product formation are illustrated in Figure 3.30.

(37) The model described is that of Blanch, H.W. and P.L. Rogers, Biotech. Bioeng. **13**, 843 (1971).

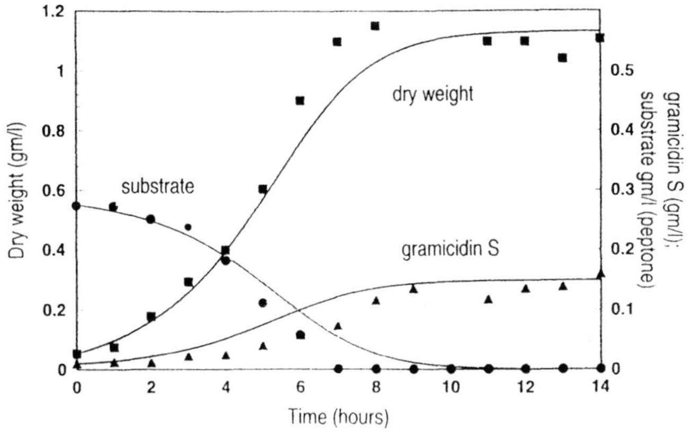

***Figure 3.30**. The predictions of the age-dependent model for gramicidin S production by Bacillus brevis. The parameters were fitted to the experimental data. Values of the constants were a_{12} = 0.905 hr^{-1}, α = 1.190 hr^{-1}. β = 0.75 g/l, γ = 0.51 gm/gm, k = 3.79 gm cells/gm gramicidin S.*

Evaluation of the constants in this model from batch culture data can be simplified by introducing a "maturation time" (t_m)[38]. The rate at which cells mature or enter the idiophase depends on the formation rate of immature cells at a period t_m earlier. We can express this relationship as

$$\frac{dX_2}{dt}\Big|_t = \frac{dX}{dt}\Big|_{t-t_m} \tag{3.115}$$

The rate of formation of product can now be written in terms of the cell concentration

$$\frac{dP}{dt}\Big|_t = \left(\frac{1}{k}\right)\frac{dX}{dt}\Big|_{t-t_m} \tag{3.116}$$

Equation (3.116) can be integrated, assuming no product is initially present, and the inoculum contains a cell concentration X_0:

$$P(t) = \left(\frac{1}{k}\right)\{X(t-t_m) - X_o\} \tag{3.117}$$

A graphical trial and error procedure can be employed to estimate k and t_m. By plotting P(t) against X at times (t - t_m) earlier, the value of the maturation time is established when a linear relationship is found for some value of t_m. The slope of the linear relationship yields the constant k. Table 3.15 illustrates the values of maturation times determined for various antibiotic fermentations this way.

(**38**) D.E. Brown and R.C. Vass, Biotech. Bioeng. **15**, 321 (1973).

Table 3.15. Cell maturation times and values of (1/k) for several antibiotic fermentations.[39]

Antibiotic Produced	Microorganism	Maturation time (hr)	Constant (1/k)
Candidin	*Streptomyces viridoflavus*	14	7.5 (μg polyene/μg DNA)
Candicidin	*Streptomyces griseus*	17.5	21.3 (μg polyene/μg DNA)
Chloramphenicol	*Streptomyces venezuelae*	12	0.2 g chloramphenicol/gm cells
Rifamycin	*Streptomyces mediteranei*	55	1×10^5 units rifamycin/gm cells
Gramicidin S	*Bacillus brevis*	2	0.237 mg gramicidin/mg cells
Penicillin	*P. chrysogenum*	40	2.2×10^5 units penicillin/gm cells

3.5.4 Single Cell Models

By considering reactions occurring in a single cell as being representative of the behavior of the whole microbial population, more sophisticated models of cell behavior can be developed. Such models are certainly less complex than models which consider both the chemical structure of the cell and variations from cell to cell (i.e., segregation). Single cell models have the advantages that they can incorporate cell geometry (surface to volume ratios) and its influence on metabolite transport; they can predict temporal events during the cell cycle (e.g., changes in cell size); they can incorporate details of the spatial arrangements within the cell (e.g., mitochondrial concentrations may be distinct from those in the cytosol); and they can include details of the metabolic pathways. The price for this increasing sophistication is that determination of rate expressions for the large number of reactions is difficult and estimates must be made for many of the constants involved. An example of this approach is provided by the model of *E. coli* growth and cell division formulated by Shuler and coworkers[40].

---

(39) Data from J.F. Martin and L. McDaniel, Biotech. Bioeng. **17** 925 (1975); D.E. Brown and R.C. Vass, Biotech. Bioeng. **15** 321 (1973); and H.W. Blanch and P.L. Rogers, Biotech. Bioeng. **13** 843 (1971).

(40) Shuler, M.L. and M.M. Domach, *Mathematical Models of the Growth of Individual Cells. Tools for Testing of Biochemical Mechanisms*, in *Foundations of Biochemical Engineering*, eds. Blanch, Papoutsakis & Stephanopolous, ACS Symp. Series **207**, 93 (1983). See also references therein. Extension of this model to encompass transcription and translation is given in Peretti, S.W and J.E. Bailey, *Biotech. Bioeng.* ***28***, 1672 (1986)

In this approach, the cell is treated as an expanding reactor, i.e., mass balances are written which include the effect of the changing cell volume resulting in a dilution of intracellular concentrations. A representation of the model by Shuler and co-workers is shown in the Figure 3.31.

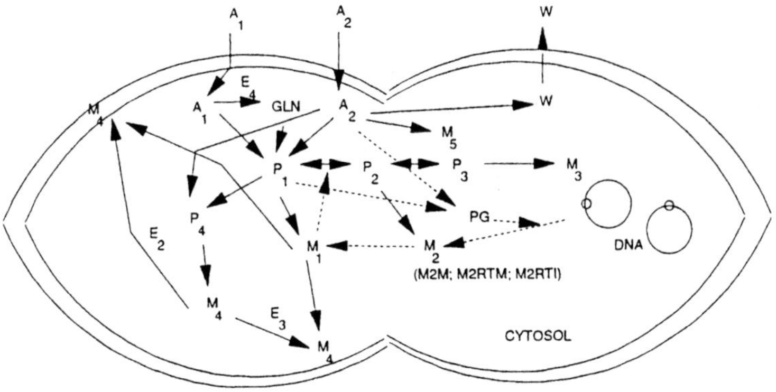

***Figure 3.31**. A representation of the key metabolic reactions of E. coli growing on glucose and ammonium salts, on which the model of Shuler and coworkers is based. In the figure above, the cell has completed a round of DNA replication and initiated cross-wall formation. The solid lines indicate reaction pathways, while the dashed lines represent regulatory steps.*

The metabolic components indicated above are:

A_1 = ammonium ion
A_2 = glucose
W = waste products (e.g., CO_2, acetate and water)
P_1 = amino acids
P_2 = ribonucleotides
P_3 = deoxyribonucleotides
P_4 = cell envelope precursors
M_1 = protein (cytosolic and envelope)
$M_{2,RTI}$ = immature, stable RNA
$M_{2,RTM}$ =mature, stable RNA (t-RNA and r-RNA)
$M_{2,M}$ = messenger RNA
M_3 = DNA
M_4 = non-protein part of cell envelope
M_5 = glycogen
PD = ppGpp
E_1 = enzymes in conversion of P_2 to P_3
E_2,E_3 = enzymes involved in directing cross wall formation and cell envelope synthesis
GLN = glutamine
E_4 = glutamine synthetase

Equations can be developed for each of the species listed above in terms of total mass of each metabolite (rather than in terms of concentration). In the figure above, the dashed lines indicate the structure of the metabolic regulatory processes. In addition, stoichiometric relations are required for the lumped energy, mass and reductant consumption processes in the cell. In the case of anaerobic growth, electron balances must be added so that the amount of ATP and reducing power generated meet the demands of energy consumption.

As an example of the model formulation, consider the mass balance for DNA synthesis:

$$\frac{dM_3}{dt} = \mu_3 \left(\frac{P_3/V}{K_{M_3P_3} + P_3/V} \right) \left(\frac{A_2/V}{K_{M_3A_2} + A_2/V} \right) F \tag{3.118}$$

where M_3 is the mass of DNA, P_3 is the mass of deoxynucleotides, etc. The constitutive rate expression is *ad hoc*; DNA formation is assumed to depend on the intracellular concentration of nucleotide precursors and on the intracellular glucose concentration, which we might consider to reflect the availability of energy to the cell. The rate expressions are formulated in concentrations expressed as mass per cell volume, noting that the cell volume (V(t)) changes with time. F is the number of replication forks; μ_3 is a rate constant for the maximum rate of DNA formation per fork, in units of DNA mass per fork per time; the K's are saturation constants. μ_3 can be determined from data on the size of the *E. coli* chromosome, the number of replication forks and the time required for a fork to traverse the chromosome under conditions of maximum growth. To determine the number of replication forks, F, a separate set of equations describing the control of chromosome replication must be solved.

Clearly an enormous amount of metabolic information is required in formulating single cell models. However, these models can provide information on the transient response of cells to environmental changes and are capable of predicting measureable quantities, such as cell size and nucleic acid content. These can be used to test the assumptions inherent in the rate expressions. Models such as these involve a very large number of equations and parameters; thus they are not described in detail here. It may be interesting however to examine the wide range of predictive responses such models can generate[41].

3.6 Models of Gene Expression and Regulation

In this section we shall examine some more complex models which describe an important aspect of both prokaryotic and eukaryotic systems, namely the regulation of gene expression. We have seen a rather simple model of regulation earlier in the example of α-galactosidase production by *Monascus*. In the examples we shall review, the rates of DNA transcription and translation are considered for both chromosomal DNA and for foreign plasmid-encoded DNA. Models of this type are important in understanding the processes involved in enzyme induction and repression and provide insights into the factors that are important in the stability of plasmids and the expression of plasmid-encoded proteins.

Example: **The Dynamics of the Epigenetic System**

A model for the control of macromolecule synthesis within cells has been developed by Goodwin[42]. A simple description of metabolic feedback control is incorporated in this model, and since it illustrates many of the key features of the real system, we shall examine

---

(41) A well-developed model of the red blood cell is instructive in this regard; A. Joshi and B. Palsson, "Metabolic Dynamics in the Human Red Cell: Parts I-IV", J. Theoret. Biol. **141** 515-528; 529-545 (1989); **142** 41-68; 69-85 (1990).

(42) B.C. Goodwin, *Temporal Organization in Cells*, Academic Press (1963).

it in some detail. The control scheme is shown in Figure 3.32. L represents the genetic locus which synthesizes mRNA in amounts described by the variable X, with units of molecules/cell. The mRNA then moves to the ribosome (R), where synthesis of protein Y occurs. This protein then exerts metabolic control at a cellular locus C. This influence may be a result of enzyme action (if we consider Y to be an enzyme), which results in the formation of a metabolite M, which in turn acts on the genetic locus as a repressor or co-repressor (by combining with another metabolite) of mRNA synthesis. Thus the feedback control loop is established as shown in Figure 3.32.

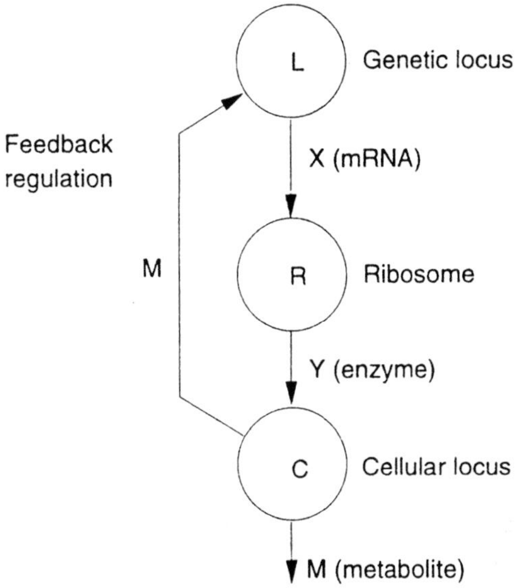

Figure 3.32. *Schematic representation of the regulation of gene expression by feedback of a metabolite M to the operator region of a particular structural gene.*

Based on this rather simple scheme, we can derive equations describing protein synthesis. Assuming that the rate of protein synthesis within the cell (which is assumed to have a constant volume) is proportional to the amount of mRNA present, and that protein is degraded at a constant rate, we can write

$$\frac{dY}{dt} = \alpha X - \beta \tag{3.119}$$

Protein synthesis is considered to be an irreversible process for this step to be rate controlling.

Messenger RNA synthesis can now be modelled. The general form of of this equation will be

$$\frac{dX}{dt} = \Phi(X, Y, M) - \Psi(X, Y, M) \tag{3.120}$$

where the first term represents synthesis; the second, degradation. Regulation of gene expression is a result of the reversible binding of repressor metabolite M at a template site on the operator. This same template combines reversibly with precursors (A) for mRNA. These precursors can be considered as an average concentration of nucleotides. Templates which are free of repressor molecules can function in mRNA synthesis. The reversible binding of repressor to the template T can be written as

$$T + R \underset{k_2}{\overset{k_1}{\leftrightarrow}} TR$$

where the equilibrium association constant K_R is given by [TR]/[T][R]. The reaction between templates T and the precursors for mRNA synthesis (A) can be written in the same manner, where K_A is [TA]/[T][A]. A conservation equation for templates gives

$$[T_0] = [T] + [TA] + [TR]$$

We can substitute for [T] and [TR] in terms of [TA]:

$$[T_0] = \frac{[TA]}{K_A[A]} + [TA] + \frac{K_R[R][TA]}{K_A[A]} \tag{3.121}$$

and solving for [TA], we find that

$$[TA] = \frac{K_A[A][T_0]}{1 + K_A[A] + K_R[R]} \tag{3.122}$$

A relationship between the repressor R and the metabolite M produced by the control locus needs to be developed. The metabolite concentration may have a typical concentration of [S]. When [M] exceeds this value, the difference ([M]-[S]) reflects the strength of the feedback signal. It is assumed that $[R] = \sigma([M]-[S])$ and the rate of mRNA synthesis depends on the amount of precursor-template complex, [TA], with a first order rate constant of λ. Thus

$$\Phi(X, Y, M) = \frac{\lambda K_A[A][T_0]}{1 + K_A[A] + \sigma K_R([M] - [S])} \tag{3.123}$$

The rate of degradation of mRNA will be assumed to be a constant, given by b. Thus the expression for mRNA synthesis becomes

$$\frac{dX}{dt} = \frac{\lambda K_A[A][T_0]}{1 + K_A[A] + \sigma K_R([M] - [S])} - b \tag{3.124}$$

Let us now turn to the metabolite M. This variable belongs to the metabolic pool, which has a different relaxation time to that of the epigenetic system. We can consider that M is in steady-state with respect to variations in X and Y. The mass balance which describes its synthesis is first order in enzyme concentration (Y) and its consumption by cellular reactions is first order in M:

$$\frac{dM}{dt} = \alpha_1 Y - \beta_1 M \tag{3.125}$$

This differential equation reduces to a quasi-steady state algebraic equation with solution

$$M = \frac{\alpha_1 Y}{\beta_1} \tag{3.126}$$

Substituting for M and simplifying, the mRNA mass balance becomes

$$\frac{dX}{dt} = \frac{a}{B + kY} - b \tag{3.127}$$

where

$a = \lambda\ K_A[A].[T_0]$

$B = 1 + K_A[A] - \sigma\ K_R[S]$

$k = \sigma\ K_R\ \alpha_1\ /\ \beta_1$

We now have two coupled differential equations (Eq. 3.119 and 3.127) describing the time course of X and Y. These can be combined into a single equation

$$(\alpha X - \beta)\frac{dX}{dt} + \left(b - \frac{a}{B + kY}\right)\frac{dY}{dt} = 0 \tag{3.128}$$

which can be integrated to give the phase-plane behavior of X and Y:

$$\alpha\frac{X^2}{2} - \beta X + bY - \frac{a}{k}\ln(B + kY) = \text{constant} \tag{3.129}$$

According to this relationship, X and Y cannot both decay to zero, nor can they increase without bound, as the constant in the equation is determined by the initial conditions (X_0,Y_0). Hence X and Y must oscillate around these steady state values (provided they did not initially start at the steady state values). Some estimates of the constants in this model can now be made for a typical cell.

In bacteria, many messenger RNAs have rather small concentrations, of order 10 per cell, too small to be represented by a continuous variable. However if we examine a situation where perhaps 100 mRNA molecules might be present in a cell, then the fluctuations in number will be smaller and representation of concentration by a continuous variable may be appropriate. In higher organisms, the population numbers of mRNA are considerably higher, and the assumption is much more realistic. If we consider that a typical protein population of 24,000 molecules per cell has a mean half-life of 20 hours, then β, the decay constant in Eq. (3.119) is 20 molecules/min. A typical protein synthesis time in higher organisms is of order 5 minutes, giving a value of $\alpha = 0.2\ \text{min}^{-1}$. With these values, the protein mass balance equation becomes

$$\frac{dY}{dt} = \alpha X - \beta = 0.2X - 20 \tag{3.130}$$

This gives a steady state mRNA population (X_{ss}) of 100 molecules/cell. Estimating the constants in the mRNA synthesis equation is not straightforward and it is necessary to make some assumptions about pool size of metabolites. Returning to the original equation for mRNA:

$$\frac{dX}{dt}=\frac{\lambda[T_0]K_A[A]}{1+K_A[A]+K_R[R]}-b \tag{3.131}$$

λ is the rate constant for mRNA synthesis, taken to be 1 molecule/min. $[T]_0$ can be taken to be 2 if we assume that only two DNA templates are available for mRNA synthesis (i.e., two gene copies for a protein of interest). [A] is the size of the nucleotide pool, which might be estimated as equivalent to 100 mRNA molecules, i.e., ~ 30,000 nucleotides (300 nucleotides per messenger of molecular weight ~ 10^5, coding a polypeptide unit of 100 amino acids). The equilibrium constant K_A for the reaction between activated nucleotides and template can be defined from Equation (3.132):

$$K_A[A]=\frac{[TA]}{[T]} \tag{3.132}$$

where [TA]/[T] is the ratio of templates engaged in mRNA synthesis to free templates. We can assume that this ratio is heavily in favor of mRNA synthesis and has a value of 100. Since [A] is 100, K_A is 1 cell/molecule. If mRNA has an approximate half life of 4 hours in higher organisms, and the steady state value of [A] is 100 molecules, then b is 100/(4x60) molecules/minute. At steady state

$$\frac{dX}{dt}=\frac{2\cdot 100}{1+100+K_R[R]}-\frac{5}{12}=0 \tag{3.133}$$

Solving Eq. (3.133) yields $K_R[R]_{ss} = 379$. To evaluate K_R, some estimate of the steady state number of repressor molecules $[R]_{ss}$ must be made. We select a value of 100, giving K_R = 3.79, i.e., the affinity of the repressor for the DNA template is greater than that of nucleotides (K_A was assumed to have a value of 1). This seems to be a reasonable assumption. We now need to turn to the metabolite equation to evaluate the remaining parameters. The ratio $\sigma\alpha_1/\beta_1$ is difficult to estimate, as is the size of the metabolite pool [S]. We shall simply assume that $\sigma\alpha_1/\beta_1 = 25$ and $\sigma[S]$ has a value of 20 molecules/cell.

Summarizing, we have

K_A = 1 cells/molecule	$[T]_0$= 2 molecules/cell
K_R = 3.79 cells/molecule	b= 0.417 molecules/minute
α = 0.2 minute^{-1}	$\sigma\alpha_1/\beta_1$= 25
β = 20 molecules/min	σ [S]= 20 molecules/cell
λ = 1 minute^{-1}	[A]= 100 molecules/cell

We can also evaluate the constants in the mRNA mass balance:

$a = \lambda.K_A[A].[T_0] = (1)(1)(100)(2) = 200$

$B = 1 + K_A[A] - \sigma.K_R[S] = 1 + (1)(100) - (3.79)(20) = 25.2$

$k = \sigma K_R \alpha/\beta_1 = 94.75$

Estimating initial conditions for X and Y as $[X]_0$ = 95 molecules/cell and $[Y]_0$ = 5 molecules/cell, the two dynamic equations for [X] and [Y] thus become

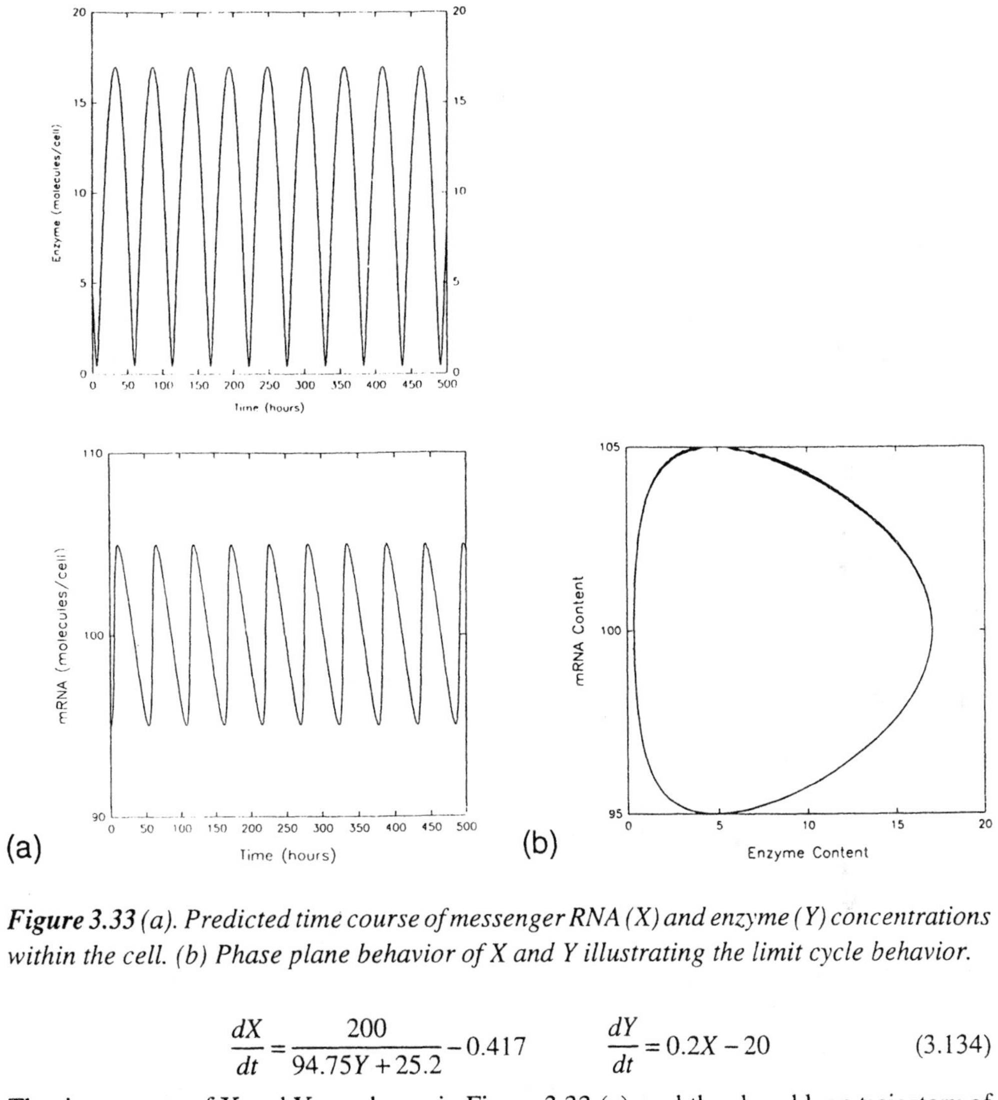

***Figure 3.33** (a). Predicted time course of messenger RNA (X) and enzyme (Y) concentrations within the cell. (b) Phase plane behavior of X and Y illustrating the limit cycle behavior.*

$$\frac{dX}{dt} = \frac{200}{94.75Y + 25.2} - 0.417 \qquad \frac{dY}{dt} = 0.2X - 20 \tag{3.134}$$

The time course of X and Y are shown in Figure 3.33 (a), and the closed loop trajectory of X(t) versus Y(t) is given in Figure 3.33 (b). The sustained oscillations predicted by this model (limit cycle behavior) parallel observations of variations in the levels of various proteins within the cell. By coupling several systems of this type together, for example by assuming that metabolite M_1 regulates mRNA synthesis of a second protein at genetic locus L_2, the period of the oscillations can be lengthened. This period can then approach that of circadian rhythms, the "natural clocks" that can be found in many higher organisms.

3.7 Plasmid Expression and Replication

Two of the difficulties associated with the use of recombinant organisms for production of plasmid-encoded proteins are their more complex growth patterns and the stability of the plasmid within the host cell, particularly for high copy-number plasmids. In this section, we shall examine models describing the replication of plasmids within the cell and more complex models describing the expression of the encoded plasmid product.

The number of plasmids within a cell may vary depending on the nature of the plasmid and the growth rate of the host. The amount of plasmid DNA in the cell is an important determinant of the host-plasmid system. When plasmid expression occurs, an additional metabolic burden is imposed on the cell and a deterioration in cellular growth occurs. When there is a large amount of plasmid DNA present, this metabolic burden may become quite high. Plasmids may be lost from the host by several mechanisms. These are a result of segregational effects, where plasmids may partition unevenly between mother and daughter cells at the point of cell division, and structural effects, where loss occurs due to a reduction in the rate of growth of plasmid-containing cells. Partitioning of plasmids at cell division from the mother cell to the daughter cell is generally regulated in low and intermediate copy number plasmids (e.g., RP1 plasmids) by genetic information contained on the plasmid at the *par* locus (from *par*tition). These plamids are thus desirable for their stability characteristics. High copy number plasmids (typically used for their high levels of expression of encoded protein) do not contain a *par* locus. Segregational instability in the absence of this type of genetic regulation can be related to the number of plasmids in the cell.

The probability (Θ) that a plasmid-free daughter cell may arise from a plasmid-containing mother cell in the absence of specific partitioning effects described above is

$$\Theta = 2^{1-N} \tag{3.135}$$

where N is the number of plasmids in the mother cell. When there are relatively few non-par-containing plasmids in the host cell, the probability of appearance of a plasmid-free segregant is high. On this basis, high copy number plasmids might not be expected to show significant segregational instability. However, plasmids may form multimers within the cell and reduce the apparent copy number. Thus, even a high copy number plasmid may show segregational instability.

We shall now examine an unstructured model for plasmid replication which describes the interplay of plasmid properties and the growth characteristics of the host cell.

***Example:* A Generalized Model of Plasmid Replication**[43]

We consider that plasmid replication, resulting in a doubling of plasmid number within the cell, is governed by two separable factors: the host cell and the plasmid itself. Thus for the reaction

$$p \rightarrow 2p$$

a rate expression for plasmid replication $r_p(p,h)$ can be written

(43) From Satyagal, V.N. and P. Agrawal, Biotech. Bioeng. **33**, 1135 (1989).

$$r_p(p,h) = r_p(p) \cdot r_p(h) \tag{3.136}$$

where $r_p(p)$ and $r_p(h)$ are the plasmid- and host-cell regulated reaction rates, respectively. The host cell regulates the host cell rate factor $r_p(h)$ through the availability of enzymes for plasmid synthesis and through components involved in the reactions of synthesis. The plasmid-regulated component of the above rate expression $r_p(p)$ is governed by the amount of plasmid present. Because it is an enzyme-regulated replication, we expect this rate expression to follow Michaelis-Menten kinetics:

$$r_p(p) = \frac{V_p^{\max} p}{K_p + p} \tag{3.137}$$

where p is the plasmid number, $V_p^{\max}$ is the maximum rate of plasmid synthesis, and K_p is a saturation constant. Both constants are characteristic of the host-plasmid system, and $V_p^{\max}$ can be thought of as the maximum rate in the presence of a surplus of all host-required components for plasmid synthesis.

We now turn to the expression for $r_p(h)$. The host cell, and the conditions under which it is growing, influence the plasmid synthesis rate. It is assumed that these conditions limit synthesis when growth activity is low and that host functions saturate at high levels of cellular activity. The general metabolic activities that influence $r_p(h)$ can be assumed to be linearly proportional to the specific growth rate of the cell, μ. An expression that shows the appropriate limiting behavior is

$$r_p(h) = \frac{V_h^{\max} \mu}{\mu + K_h} \tag{3.138}$$

Equation (3.138) shows that at high rates of cellular activity (and thus growth rate), plasmid synthesis reaches a saturation rate. At low cellular growth rates, plasmid synthesis depends on the cellular growth rate. K_h can be thought of as a measure of the dependence of the plasmid on the host for replication. The equations for $r_p(h)$ and $r_p(p)$ can be combined as follows:

$$r_p(p,h) = \frac{V_h^{\max} V_p^{\max} p \mu}{(K_p + p)(K_h + \mu)} = \frac{v^{\max} p \mu}{(K_p + p)(K_h + \mu)} \tag{3.139}$$

Thus the rate of plasmid synthesis has the same form as that for double-substrate limiting kinetics. A mass balance over the cell (noting that the volume may change during growth) gives the following expression for the plasmid number:

$$\frac{dpV_c}{dt} = p\frac{dV_c}{dt} + V_c\frac{dp}{dt} = r_p(p,h) \cdot V_c \tag{3.140}$$

$$\frac{dp}{dt} = r_p(h,p) - \mu p \tag{3.141}$$

When the cell is in a state of balanced growth, (e.g. cells grown in a continuous well-mixed reactor or in the exponential growth phase), the value of the intracellular components will tend to a constant value. Thus we can set dp/dt to zero and calculate the steady-state plasmid number (p_s) from

$$p_s = \frac{v^{\max}}{K_h + \mu} - K_p \qquad \mu \neq 0 \tag{3.142}$$

An estimate of the steady-state concentration of plasmid p_{so} can be made from

$$p_{so} = \lim_{\mu \to 0}\left(\frac{v^{\max}}{\mu + K_h} - K_p \right) = \frac{v^{\max}}{K_h} - K_p \tag{3.143}$$

Equation (3.143) implies that at low growth rates, the host cell, through K_h, influences the plasmid number. A low value of K_h would give the case of *runaway replication*, where extremely high copy numbers are found. If K_h is large the plasmid number remains small.

A specific growth rate where the steady-state number of plasmids falls to zero can be found by setting $p_s(\mu)$ equal to zero. This defines a plasmid "washout" growth rate, μ_{pwo} :

$$\mu_{pwo} = \frac{v^{\max}}{K_p} - K_h \tag{3.144}$$

Using the definition of p_{so} and the expression for $p_s(\mu)$, we can eliminate K_p and rearrange the resulting equation to provide a linear relationship for determining the parameters K_h and v^{max}:

$$\frac{1}{(p_{so} - p_s)} = \frac{K_h}{v^{\max}}\left(1 + \frac{K_h}{\mu} \right) \tag{3.145}$$

Alternatively, we can use the definition of μ_{pwo} and $p_s(\mu)$ to eliminate K_h and obtain

$$\frac{1}{(\mu_{pwo} - \mu)} = \frac{K_p}{v^{\max}}\left(1 + \frac{K_p}{p_s} \right) \tag{3.146}$$

Predictions from this model can now be compared with the experimental data of Seo and Bailey[44] for *E. coli* HB101 containing pDM247 plasmids. This is a low molecular weight plasmid which is present in high copy number, but the plasmid number decreases with increasing growth rates. The experimental data is shown in Figure 3.34. The curve through data has been extrapolated to determine μ_{pwo}, and a value of 2.0 hr^{-1} is obtained. This is clearly greater than μ_{max} for *E. coli* (usually around 1.0 hr^{-1}). This value of μ_{pwo} is used to transform the data and $1/(\mu_{pwo} - \mu)$ is then plotted against $1/p_s$. As can be seen in Figure 3.35, a linear relationship results. The values of v^{max} and K_p can be determined from this graph to be 1.08 (mg/gm cell-hr) and 0.53 (mg/gm cell), respectively.

(44) Seo, J.H. and J.E. Bailey, *Biotech. Bioeng.*, **27**, 1668 (1985).

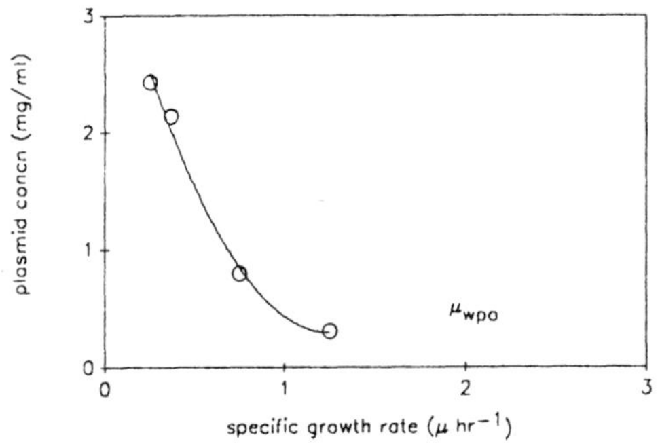

Figure 3.34. *Plasmid concentration within E. coli as a function of the specific growth rate (μ_{pwo} is estimated as 2.0 hr^{-1}).*

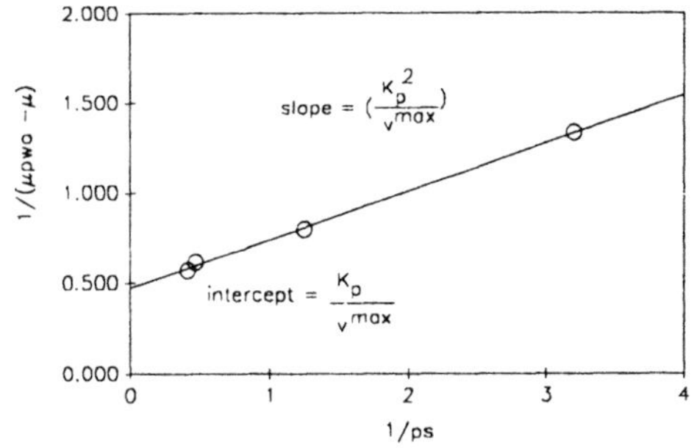

Figure 3.35. *Linearized representation of the data according to the model equations.*

Thus this model provides a simple representation of the essential features of plasmid replication. Like the Monod model for microbial growth, it is a simplification that cannot be expected to be valid under transient conditions. In the next section, we will examine a structured model that is based on the approach described in this section that might be expected to be more generally applicable.

Example: **A Simple Structured Model for Plasmid Replication**

The equation employed in the preceding model describing the effect of the plasmid itself on its rate of replication ($r_p(p)$) was a purely constitutive one. We shall now develop a mechanistic model which incorporates our understanding of the nature of Col E1 plasmid replication and show that the simplification employed in the above constitutive model is reasonable. The model is that of Satyagal and Agrawal[45].

Replication of Col E1 plasmids is controlled by a *replicon*, which consists of an origin of replication, a gene for initiator synthesis and a gene for repressor synthesis. The initiator and the repressor are assumed to be produced constitutively. The repressor controls the replication rate by complexing with and inactivating the initiator. The formation of this complex is a second order reaction. A schematic of replication control is shown below.
The initiator and repressor molecules are RNA in Col E1 plasmids. We can now write mass balances around the cell, denoting the intracellular concentrations of initiator and repressor molecules as I and R respectively. The plasmid concentration is given by p. We need to note that the cell volume V_c will change with the growth rate of the cell and this must be included in our mass balances. For both I and R formation (assumed in both cases to be first order in plasmid concentration), degradation (first order) and reaction terms are included:

(45) Satyagal, V.N. and P. Agrawal, *Biotech. Bioeng*, **33** 1135 (1989).

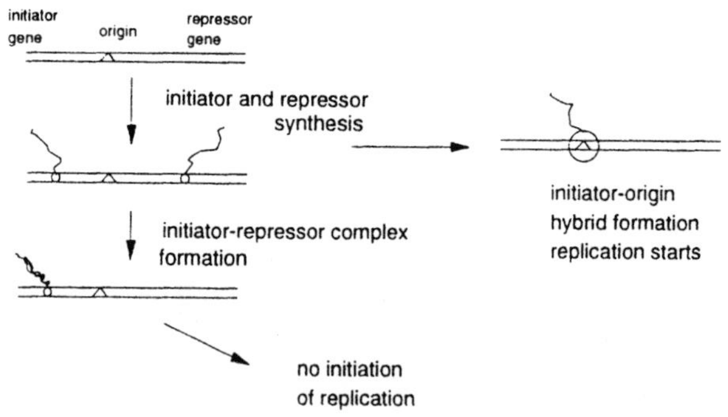

Replication control of colicinogenic plasmids

$$\frac{dRV_c}{dt} = k_2 p V_c - k_3 R V_c - kIRV_c \tag{3.147}$$

If the density of the cell is constant, then

$$\mu = \frac{1}{V_c}\frac{dV_c}{dt} \tag{3.148}$$

Equation (3.147) can now be simplified:

$$\frac{dR}{dt} = k_2 p - k_3 R - kIR - \mu R \tag{3.149}$$

Similarly, the mass balance for I becomes

$$\frac{dI}{dt} = k_4 p - k_5 I - kIR - \mu I \tag{3.150}$$

and that for plasmid concentration is

$$\frac{dp}{dt} = k_1 I \frac{V_h^o \mu}{(\mu + K_h) - \mu p} \tag{3.151}$$

where the same form for $r_p(h)$ as used in the simplified model above has been retained and the rate of plasmid replication is assumed to be first order in initiator I. The case of balanced growth can now be considered. The time derivatives are equated to zero and the following assumptions made: (a) the rate of deactivation of R is much greater than its rate of dilution due to cell growth i.e., $k_3 \gg \mu$; and (b) $(k_5 + \mu)\mu \ll k_1 k_4$.
The concentrations under balanced growth then become

$$p = \frac{k_1 V^o k_3 k_4}{k(\mu + K_h)(k_2 - k_4)} \tag{3.152}$$

$$I = \frac{k_3 k_4}{k(k_2 - k_4)} \tag{3.153}$$

$$R = \frac{k_1 V^o k_4}{k(\mu + K_h)} \quad (3.154)$$

The initiator concentration is a constant, independent of the cell growth rate, whereas the repressor and plasmid concentrations decline with increasing cell growth rates. Equation (3.153) shows that a positive I requires $k_2 > k_4$. This implies that the rate of repressor synthesis must be greater than the rate of initiator synthesis on a unit plasmid basis. We can further examine the model equations by considering that the changes in repressor and initiator concentrations are rapid with repect to changes in the plasmid concentration, i.e., the quasi-steady state assumption that dR/dt = 0 and dI/dt = 0. Further, let us assume that the dilution terms due to cell growth are negligible for I and R (i.e., μI and μR) and that the initiator degradation rate is small. The equations for I and R then become

$$k_2 p - k_3 R - kIR = 0 \quad (3.155)$$

$$k_4 p - kIR = 0 \quad (3.156)$$

Solving for I we obtain

$$I = \frac{k_3 k_4}{k(k_2 - k_4)} \quad (3.157)$$

The dynamic behavior of plasmid concentration can now be described by employing this expression for I:

$$\frac{dp}{dt} = \frac{k_1 k_3 k_4}{k(k_2 - k_4)} \frac{\mu V^o}{(\mu + K_h)} - \mu p \quad (3.158)$$

This expression is analogous to that for $r_p(p,h)$ employed in the simple model examined earlier with K_p equated to zero. Thus this more complex model shows the validity of the earlier simple constitutive model under these limiting conditions.

3.8 Nomenclature

Symbol	Definition	Typical units
a'	interfacial area per volume	cm^{-1}
D_{eff}	effective diffusion coefficient	cm^2/sec
D	dilution rate or inverse residence time	hr^{-1}
E_A	activation energy for reaction	kcal/mole
E	ethanol concentration	gm/liter
G_B	galactose concentration	mg/gm cell mass
G	concentration of species in g-compartment	gm/liter
h	colony height (Eq. 3.70)	cm
I	initiator concentration (Eq. 3.147)	initiator/cm^3 cell
K_1, K_2	apparent Monod constants (Eq. 3.43)	gm substrate/liter
K_A	equilibrium binding constant	cells/molecule

k_c	endogenous rate constant (cells-hr)	gm cell mass/gm
k_d	cell death rate constant (first order)	hr^{-1}
K_i	substrate or product inhibition constant	gm/liter
k_L	mass transfer coefficient	cm/sec
K_p	saturation constant for plsmid production	plasmids/cm^3 cell
K_S	Monod coefficient	gm substrate/liter
K	concentration of species in k-compartment	gm/liter
k	constant in Konak Eq. (3.37)	$(hr)^{1-p}$/(gm S/liter)
K	first order reaction rate/mass transfer rate $v_{max}/(k_L a' K_S)$	[-]
k	rate constant (Eq. 3.32)	hr^{-1}
$k_{\pm j}$	first (or second order) rate constants	hr^{-1} (liter/gm-hr)
m_i	maintenance coefficient (rate of substrate S_i consumed by maintenance processes)	gm S_i/gm cells-hr
n_T	total cell concentration	cells/ml
n_v	viable cell concentration	cells/ml
N	cell number	cells/liter
p	constant in Konak Eq. (3.37)	[-]
p	plasmid concentration	plasmids/cm^3 cell
P	product concentration	gm/liter
q_{CO2}	specific rate of CO_2 production	mM/gm cells-hr
q_{O2}	specific rate of oxygen consumption	mM O_2/gm cells-hr
Q_o	heat of reaction released per electron transferred	~27 kcal/mole e
q_{Pi}	specific rate of product P_i formation	gm P_i/gm cells-hr
q_{Si}	specific rate of nutrient S_i consumption	gm S_i/gm cells-hr
r_p	rate of plasmid replication	plasmids/cm^3 cell-hr
RQ	respiratory coefficient (Eq. 3.4)	
r_{Si}	rate of consumption of ith substrate	gm/liter-hr
r_X	rate of production of cells	gm cells/liter-hr
r	colony radius (eqn. 3.69)	cm
R	fungal pellet radius (Eq. 3.74)	cm
R	repressor concentration (Eq. 3.147)	repressor/cm^3 cell
S_i	concentration of ith substrate	gm/liter
S_o	intital substrate concentration	gm/liter
t_d	doubling time	hr
t	time	hr
U	transport rate coefficient (Eq. 3.97)	hr^{-1}
V_c	cell volume	cm^3
V^{max}_p	maximum rate of plasmid synthesis	plasmids/cm^3 cell.hr
V_R	reactor volume	liters
V	reactor volume	liters
X_1	concentration of immature cells	gm cells/liter
X_2	concentration of mature cells	gm cells/liter
X	cell concentration (dry weight basis)	gm/liter
X	messenger RNA concentration (Eq. 3.120)	molecules/cell
$Y_{ave\ e/S}$	number of electrons available from substrate	mole electron/mole S
$Y_{ave\ e}$	yield coefficient (cell dry mass/electron available in substrate)	gm/mole electron

Y_{cal}	yield coefficient (mass cells/kcal heat evolved during growth)	gm/kcal
$Y_{P/Si}$	yield coefficient (mass product formed/mass substrate i consumed)	gm/gm
Y_{S} S	cell yield coefficient	gm cells/gm substrate
$Y_{X/O}$	yield coefficient (mass cells produced/mass oxygen consumed)	gm/gm
$Y_{X/S.obs}$	observed cell yield coefficient	gm cells/gm S_i
$Y_{X/Si}$	yield coefficient (mass cells produced/mass substrate i consumed)	gm/gm
y	dimensionless substrate concentration (S/K_S)	[-]
Y	enzyme concentration (Eq. 3.119)	molecules/cell
α	growth associated product formation constant	gm product/gm cells
α	stoichiometric coefficient	
β	non-growth associated product formation constant	gm product/gm cells.hr
β	stoichiometric coefficient	
β	rate contant (Eq. 3.32)	liters/gm cells
δ	stoichiometric coefficient	
ΔQ	heat evolved in microbial growth	kcal
ΔQ_{cat}	heat evolved by catabolic reactions	kcal
ΔH_c	heat of combustion of biomass	kcal/gm
λ_i	constants (Eq. 3.40-3.42)	[-]
γ	stoichiometric coefficient	
γ_b	degree of reductance of cells	
γ_S	degree of reductance of substrate	
γ_P	degreee of reductance of product	
μ	specific growth rate of cells	hr^{-1}
μ_{max}	maximum specific growth rate of cells	hr^{-1}
ν	cell number specific growth rate	hr^{-1}
ρ_c	density of cells	gm/liter
ρ	density of colony on surface (Eq. 3.70)	gm cells/cm^3
ε	stoichiometric coefficient	
σ_b	weight fraction of carbon in biomass	[-]
σ_P	weight fraction of carbon in product	[-]
σ_S	weight fraction of carbon in substrate	[-]

3.9 Problems

1. Stoichiometry

The growth of an organism on hexadecane can be described by the following stoichiometric equation:

$$C_{16}H_{34} + 12.4O_2 + 2.09NH_3 \dashrightarrow 2.42(C_{4.4}H_{7.3}N_{0.86}O_{1.2}) + \varepsilon H_2O + 5.33CO_2$$

Calculate the following:

a) the coefficient ε

b) the respiratory coefficient

c) the yield coefficients $Y_{X/S}$, $Y'_{X/S}$, $Y_{X/O}$, $Y'_{X/O}$

d) the heat of combustion of the biomass

2. Reductance

Photosynthetic purple bacteria have been found to have the following composition in weight percent: 56% carbon; 7% hydrogen;12% nitrogen; 16% oxygen; 9% ash. For organisms with this elemental composition, calculate the degree of reductance of biomass, γ_b.

3. Yields and Reductance

Consider the following equation describing glucose consumption and biomass formation for butyric acid bacteria (E.T. Papoutsakis, Biotech. Bioeng. **26** 174 (1984):

$$2(\text{glucose}) + a\text{NADH}_2 + b\text{ATP} \dashrightarrow 3C_4H_{4p}O_{4n}N_{4q}$$

The values of a and b can be calculated using the additional information that the weight fraction of carbon in the biomass, σ_b , and the degree of reductance of biomass, γ_b , are relatively constant:

$$\sigma_b = 0.462 \pm 0.023$$

$$\gamma_b = 4.291 \pm 0.172$$

a) The ATP needs for biomass synthesis can be estimated from the ATP weight yield, Y_{ATP}, which is defined as the amount of dry biomass produced (in grams) per mole ATP used in biosynthesis. Assuming Y_{ATP} is 10.5 g/mol, calculate how much ATP is required for biomass synthesis for every 2 mol of glucose incorporated into biomass.

b) Calculate how much $NADH_2$ is needed for every 2 mol of glucose incorporated into biomass to bring glucose to the oxidation level of $C_4H_{4p}O_{4n}N_{4q}$. Recall that the degree of reductance of a "compound" is defined as the number of equivalents of available electrons per g atom of carbon of the "compound."

4. Electron Balances

An available electron balance (Eq. 3.5) can be used to analyze experiments in which biomass and lysine were produced in batch culture. *Brevibacterium* was cultured in a medium containing molasses, corn extract, and other nutrients [L.E. Erickson et al., *Biotechnol. Bioeng.*, **20**, 1623 (1978)]. Summarized below are measured values of biomass productivity, lysine productivity, and oxygen consumption rate measured over a 48 hour period.

Biomass productivity, $\Delta X/\Delta t = 0.260$ g/(liter hr)
Lysine·HCl productivity, $\Delta P/\Delta t = 0.375$ g/(liter hr)
O_2 consumption rate, $Q_{O_2} = 1.40$ g/(liter hr)

From these data, calculate the fraction of chemical energy in the organic substrate converted to heat, the fraction of chemical energy in the substrate converted to biomass, and the fraction of chemical energy in the substrate incorporated into extracellular product. In these calculations, assume that $\gamma_b = 4.25$ and $\sigma_b = 0.480$, where σ_b is the weight fraction of carbon in dried biomass.

5. Yield Coefficients

Consider the aerobic growth of *S. cerevisiae* on glucose, described approximately by

$$CH_2O + aO_2 + bNH_3 \rightarrow cCH_{1.8}N_{0.2}O_{0.5} + dH_2O + eCO_2$$

Show that the yield coefficient for biomass, $Y'_{X/S}$, can be related to the degree of reductance of glucose, γ_s, and the respiratory quotient, RQ, by the following equation:

$$Y'_{X/S} = \left[\frac{1 - 0.25\gamma_s RQ}{1 - (1.05)RQ}\right]$$

6. ATP Requirements for Growth

Consider a cell which has the macromolecular composition given below. The ATP requirement for synthesis of these macromolecules is also provided, based on the energy requirements of the biosynthetic parthways. If the bacterium has a generation time of 30 minutes, determine the specific rate of ATP formation and the efficiency of ATP utilization based on a value of Y_{ATP} of 10.5 gm cells/(mole ATP generated).

Macromolecule	ATP requirement (mole ATP/(gm macromolecule) x 10^4
DNA	330
RNA	373.2
Protein	391.1
Lipid	123.6
Polysaccharides	114.8

7. Redox and Electron Balances

Clostridium acetobutylicum anaerobically converts glucose to acetone , butanol and smaller concentration of butyrate , acetate etc. In a fermentation, the following products were obtained from 100 moles of glucose and 11.2 moles of NH_3 as nitrogen source.

Products formed	moles
cells	13
butanol	56
acetone	22
butyric acid	0.4
acetic acid	14
CO_2	221
H_2	135
ethanol	0.7

(a) By performing a carbon, nitrogen, hydrogen and oxygen balance, determine the elemental composition of the cells.
(b) Determine the redox status of the fermentation using available electron balance equations.
(c) If the fermentation product composition were determined as below, with the same glucose and NH_3 feeds, calculate the excess or shortfall of available electrons in the products.

8. Yields and Heat Production

Corynebacterium with an empirical formula $C_8H_{17}O_7N$ is grown in a batch reactor for the production of the amino acid threonine, with glucose as the sole carbon source. The rate of cell growth and the rate of threonine production are 0.3 gm cells/liter.hr and 0.4 gm threonine/liter.hr, respectively. Approximately 50% of the energy in the carbon source is released as heat. Estimate (a) the fractions of glucose used for cell mass production and for threonine production; (b) the yields of cell mass based on glucose and oxygen ($Y_{X/S}$ and $Y_{X/O}$); and (c) the rate of heat evolution.

9. Stoichiometry and Energetics of Growth

Microbial growth on hydrocarbons (hydrocarbon fermentation) has been analyzed theoretically by Erickson [*Biotechnol. Bioeng.*, **23**, 793 (1981)]. In order to determine the maximum expected yield of biomass from n-alkanes such as n-hexadecane, the maximum yields with respect to both carbon and available energy need to be examined. A useful relationship for the maximum energetic yield (fraction of chemical energy in the substrate converted to biomass), ξ_{max} , is given below:

$$\xi_{max} = \frac{\left(\frac{\sigma_b \gamma_b}{12} Y_{ATP}\right)}{\left(\frac{\sigma_b \gamma_b}{12} Y_{ATP} + \frac{2}{\eta}\right)}$$

where η is the moles of ATP produced/g atom oxygen incorporated. The maximum value of η associated with the tricarboxylic acid cycle and oxidative phosphorylation is 3.

In this problem we shall consider microbial growth on n-hexadecane. The n-hexadecane is first converted to acetate, and most of the cellular compounds are then formed from acetate, and 82% of the carbon in acetate is incorporated into biomass. The biochemical pathways by which n-hexadecane is converted to acetate and acetate to cellular compounds also produce CO_2. The CO_2 can also be converted to biomass by additional pathways. Malic acid is produced as extracellular product. The fraction of carbon incorporated into extracellular product is fixed at 0.2. Additional parameters are given below (notation from Section 3.1).

$\sigma_b = 0.462$	$Y_{ATP} = 10$ for acetate; 6.5 for CO_2
$\gamma_b = 4.291$	$\gamma_P = 3$ for malic acid
$\alpha^{-1} = 0.82$	$\gamma_S = 6.125$ for n-$C_{16}H_{34}$

For the conversion of $C_{16}H_{34}$ to acetate, 2.125 equivalents of available electrons/(g atom carbon) are transferred to oxygen, and 0.91 mol ATP is produced/(equivalent of available electrons transferred to oxygen)

a) For every g atom of carbon in hexadecane, how much cell mass is produced, and what is the ATP requirement?

b) Based on an ATP balance, how much additional biomass can be formed from CO_2?

c) How many equivalents of electrons remain after biomass and product formation, as determined in parts (a) and (b)? How much *additional* biomass can be produced from these equivalents and CO_2, according to the above equation for ξ_{max} ?

d) Based on your answers above, is biomass production carbon- or energy-limited?

10. Balanced Growth

Balanced growth of a microorganism in batch culture occurs when the intracellular levels of all metabolites attain constant values. It follows that under balanced growth conditions, $d(q_s)/dt = 0$, where q_s is the specific rate of substrate consumption in batch culture.

Show that when cells are grown in balanced growth, the shape of the substrate concentration versus time curve is concave downward, and the shape of a growth-associated product concentration profile must be concave upward. When the yeast *S. cerevisiae* is grown on glucose, the ethanol concentration profile is concave upward, but concave downward when the yeast is grown on ethanol. Is balanced growth occurring in both cases?

11. Stoichiometry of the Butryic Acid Bacteria

This example illustrates the use of the reaction pathway stoichiometric method developed by Papoutsakis and coworkers (Biotech. Bioeng. **26** 174 (1984) and Biotech. Bioeng. **25** 50 (1985)) for the butryic acid fermentation. The metabolic pathways involved are illustrated in the figure below.

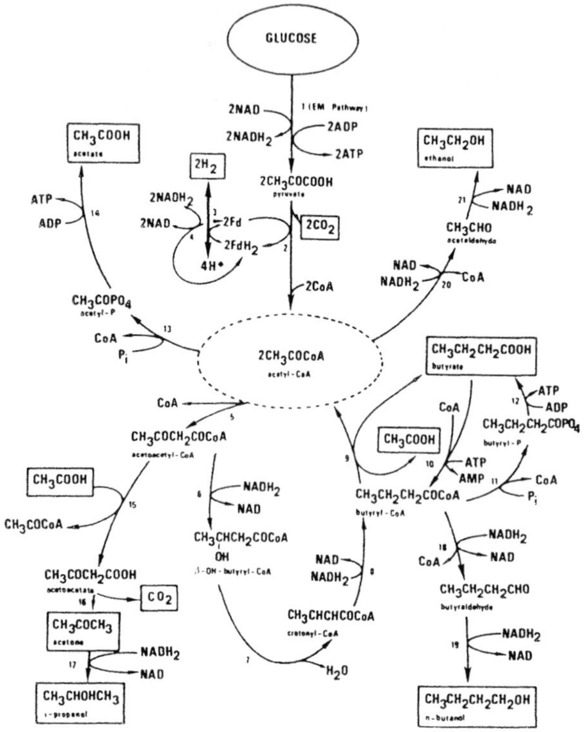

Figure 3.36. Metabolic pathways involved in the butryic acid bacteria, from C.L. Meyer and E.T. Papoutsakis, "Handbook of Anaerobic Fermentations", p83-118, Eds. ERickson, L.E. and Fung, D. Marcel Dekker (1988). The pathways or enzyme systems for the reactions shown are (1) PEP phophotransferase system and the EMP pathway; (2) pyruvate-ferredoxin oxidoreductase; (3) hydrogenase; (4) NADH-ferredoxin oxidoreductase; (5) acetyl-CoA acetyltransferase; (6) L-β-hydroxybutrylCoA dehydrogenase; (7) L-3-hydroxyacylCoA hydrolase (8) butrylCoA dehydrogenase; (9) CoA transferase; (10) butryl CoA synthetase; (11) phophotransbutyrylase; (12) butryate kinase; (13) phopho-transacetylase; (14) acetate kinase; (15) CoA transferase; (16) acetoacetate decarboxylase; (17) i-propanol dehydrogenase; (18) butyraldehyde dehydrogenase; (19) butanol dehydrogenase; (20) acetaldehyde dehydrogenase; (21) ethanol dehydrogenase.

(a) Each reaction or linear pathway in the figure can be represented by a single reaction. For example, the production of pyruvate from glucose through the EMP pathway and the phosphoenolpyruvate phosphotransferase system can be represented by the net reaction

$$\text{glucose} \rightarrow 2 \text{ pyruvate} + 2 \text{ ATP} + 2 \text{ NADH}$$

Develop similar reaction schemes for the following metabolites

pyruvate
hydrogen production from the electron carrier ferredoxin (Fd)
reduction of NAD by FdH_2
production of acetate from from acetyl-CoA (via phosphotransacetylase and acetate kinase
formation of butyryl-CoA from acetyl-CoA
formation of butyrate from butyryl-CoA (note there are two possible routes that are equivalent due to the acetate reaction from acetyl-CoA)
butanol production from butyryl-CoA
acetone production
ethanol production
isopropanol production from acetone

Although it is not illustrated on the figure, acetoin can be considered to be produced via acetolactate by the reaction

$$2 \text{ pyruvate} \rightarrow \text{acetoin} + 2 \, CO_2$$

(b) Consider the biomass to be produced by a single chemical reaction. The empirical formula for biomass may be taken as $CH_aO_bN_c$. Noting from Section 3.2.3 that the weight fraction σ_b is approximately constant (0.462) and the degree of reductance of biomass γ_b is 4.291 (Table 3.3), show

$$\gamma_b = 4 + a - 2b - 3c = 4.291$$

Use an available electron balance and the above equation to show that 0.873 mol of NADH is required for 1 mol of glucose converted to biomass.

(c) The amount of ATP required for biomass synthesis can now be estimated. If Y_{ATP} is the gm dry weight biomass per mole ATP required, show

$$\frac{\text{mol ATP}}{\text{mol glucose}} = \frac{(12)(6)}{Y_{ATP}\sigma_b}$$

Write a reaction that describes the formation of biomass ($CH_aO_bN_c$) from glucose, ATP and NADH.

(d) There are 13 independent chemical equations describing the metabolism and 17 biochemical species (ATP, acetyl-CoA, butyryl-CoA, NADH, FdH_2, pyruvate, ethanol, acetone, acetate, butanol, acetoin, butyrate, isopropanol, H_2, CO_2, biomass and glucose). Write a balance for each chemical species in the form

$$x_i = \sum_{j=1}^{13} r_j$$

where x_i is the net rate of production of metabolite i, and r_j is the rate of the jth reaction. Sum these equations to obtain an overall stoichiometric reaction describing the conversion of glucose to biomass and other products.

(e) Metabolic intermediates can be assumed not to accumulate, and their net rates of production can be set to zero. Identify these five metabolites and set their net rates of production to zero (equivalent to the pseudo-steady state hypothesis). All the remaining metabolites except ATP can be experimentally measured. If the ATP equation is neglected, there are 16 equations and 13 independent reactions; five of which are for intermediate metabolites. Based on the data for the amounts of metabolites produced given below, use a least squares approach to determine the best-fit values of r_j which give x_i values that best match the experimental data. Show the x_i values that you determine.

By setting x_{ATP} to zero, use the values you have obtained to determine Y_{ATP}.

Metabolite	*Amount produced in batch (mol per 100 mol glucose fermented)*
ethanol	2.8
acetone	9.0
acetate	43.7
butanol	18.8
acetoin	1.8
butyrate	36.4
CO_2	180
H_2	183
glucose	100
biomass	66.0

12. Growth of a Facultative Anaerobe based on ATP use

In aerobic or obligate anaerobic growth of microorgansims only one metabolic pathway is generally used for energy production. Facultative anaerobes, on the other hand may use two parallel pathways. With facultative anaerobes, the observed cell yield will vary, depending on the extent to which each pathway is used. The oxygen limited growth of *Klebsiella pneumoniae* is an example of a facultative anaerobe that uses the TCA cycle under aerobic conditions, but when the electron transport chain is limited by oxygen availability, inusfficient ATP is produced to meet metabolic demands. Then fermentative production of 2,3 butanediol occurs. This metabolic pathway produces far less ATP than aerobic growth and the cell yield consequently decreases.

A model can be formulated for the growth of *Klebsiella pneumoniae* by assuming that the ATP required for growth and maintenance of cells is equal to the ATP synthesized by the two energy producing pathways. We shall assume that all the metabolic activity can be represented by the following five reactions:

$$\frac{120}{Y_{ATP}} ATP + xylose + 0.2 \cdot NADH_2 \rightarrow 120(gm\ new\ cells) \tag{1}$$

$$m_e ATP + 1(gm\ cells) \rightarrow 1(gm\ cells) \tag{2}$$

$$Xylose \rightarrow 5CO_2 + 10NADH_2 + \frac{10}{3} ATP \tag{3}$$

$$\frac{1}{2} O_2 + NADH_2 \rightarrow 2ATP \tag{4}$$

$$Xylose \rightarrow \frac{5}{6} CO_2 + \frac{5}{6} NADH_2 + \frac{5}{6} ATP + \frac{5}{6}(butanediol) \tag{5}$$

Where m_e is the maintenance requirement of the cells for ATP (moles ATP/gm cells.hr).

The rates at which each reaction occur can be describes as follows:

$(Q_S)_{aerobic}$ (or $(Q_S)_A$) is the rate at which reaction (1) occurs in moles xylose respired/liter.hr

Q_{ETS} is the rate at which the electron transport chain operates (reaction (4)), and is limited to a maximum value by the oxygen supply rate, given by $k_L a.C_{O2}^*$ (moles/liter.hr)

Reactions (3) and (4) thus represent the respiratory pathway.

$(Q_S)_{fermentation}$ (or $(Q_S)_F$) is the rate at which reaction (5) proceeds in moles xylose/liter.hr.

During growth at low oxygen concentration, $(Q_S)_R$ accounts for all the $NADH_2$ sent to the electron transport chain. As time proceeds, however, $NADH_2$ is produced in greater amounts by the fermentation pathway. We will not concern ourselves with the mechanism of regulation of these two pathways, but simply assume that $(Q_S)_R$ decreases to zero to satisfy the $NADH_2$ balance, at which point the formation of butanediol is sufficient to saturate the electron transport chain capacity. If more $NADH_2$ is produced thatn the electron transport chain can accommodate, the excess $NADH_2$ is metabolized by other means at a rate Q_{H2} (moles/liter.hr). this may occur via hydrogen production or reaction with a hydrogen acceptor.

(a) Using the above five equations, write steady state material balances on the following compounds; xylose, ATP, and $NADH_2$. For example, the total xylose utilization rate is the sum of xylose respired to CO_2, xylose fermented to 2,3 butanediol and xylose assimilated into cell mass.

(b) The volumetric ATP uptake rate can be written as a function of the specific growth rate of the cells, the concentration of cells and the maintenance requirement:

$$Q_{ATP} = \frac{1}{Y_{ATP}} \mu X + m_e \cdot X \tag{6}$$

For aerobic growth, when $(Q_S)_F = 0$, show that

$$(Q_S)_R = \frac{1}{23.3} Q_{ATP} \tag{7}$$

$$(Q_S)_A = \frac{1}{120} \frac{dX}{dt} \tag{8}$$

with Q_{ATP} being defined in equation (6) and $(Q_S)_A$ is the rate of xylose assimilation into biomass (moles/liter.hr).

(c) During growth of the cells, aerobic growth continues until the electron transport chain becomes limited by oxygen transfer. At this point $Q_{ETS} = 2.k_L a.C_{O2}^*$, where C_{O2}^* is the concentration of oxygen in the fermentation liquid in equilibrium with that in the air. During this period of oxygen limitation, show that

$$(Q_S)_R = \frac{1}{5} k_L a C_{O2}^* - \frac{25}{3} (Q_S)_F$$

$$(Q_S)_F = \frac{18}{25} \left\{ Q_{ATP} - \frac{14}{3} k_L a C_{O2}^* \right\}$$

Write the corresponding equations when the electron transport chain is saturated and the growth becomes oxygen limited.

(d) Using the following parameter values, numerically integrate the differential mass balances for X (cell concentration), dissolved oxygen concentration and butanediol. You will need to include the switch in metabolism from aerobic to oxygen limited growth in your numerical scheme. The following parameters have been obtained for *Klebsiella pneumoniae* and the reactor:

$\mu_{max} = 0.62\ hr^{-1}$	$k_L a.C_{O2} = 0.025$ moles/liter.hr
$Y_{ATP} = 12$ gm cells/moles ATP	$m_e = 0.1$ moles/gm.hr

Assume initial concentrations of xylose and cells of 50 gm/liter and 0.5 gm/liter

13. Models of Growth on Two Limiting Substrates

The following data were obtained for the growth of a strain *Lactobacillus* with glucose as growth-limiting nutrient in the presence of excess nitrogen, and for growth with nitrogen as limiting nutrient in the presence of excess glucose.

Glucose limitation

	Expt A		*Expt B*		*Expt C*	
Time (hrs)	*cell conc. (gm/l)*	*glucose (gm/l)*	*cell conc. (gm/l)*	*glucose (gm/l)*	*cell conc. (gm/l)*	*glucose (gm/l)*
0	0.1	0.25	0.1	1.00	0.500	5.00
0.5	0.106	0.225	0.11	0.958	0.526	4.942
1.5	0.113	0.200	0.122	0.912	0.554	4.88
2.0	0.119	0.175	0.134	0.862	0.583	4.816
2.5	0.125	0.151	0.148	0.800	0.613	4.748
3.0	0.131	0.128	0.163	0.750	0.645	4.677
3.5	0.136	0.106	0.178	0.686	0.679	4.60
4.0	0.141	0.088	0.195	0.619	0.714	4.524
4.5	0.145	0.071	0.213	0.547	0.75	4.442
5.0	0.149	0.056	0.232	0.474	0.79	4.356

Nitrogen limitation

	Expt D		*Expt E*		*Expt F*	
Time (hrs)	*cell conc. (gm/l)*	*glucose (gm/l)*	*cell conc. (gm/l)*	*glucose (gm/l)*	*cell conc. (gm/l)*	*glucose (gm/l)*
0	0.1	0.25	0.1	1.00	0.500	5.00
0.5	0.106	0.225	0.11	0.958	0.526	4.942
1.5	0.113	0.200	0.122	0.912	0.554	4.88
2.0	0.119	0.175	0.134	0.862	0.583	4.816
2.5	0.125	0.151	0.148	0.800	0.613	4.748
3.0	0.131	0.128	0.163	0.750	0.645	4.677
3.5	0.136	0.106	0.178	0.686	0.679	4.60
4.0	0.141	0.088	0.195	0.619	0.714	4.524
4.5	0.145	0.071	0.213	0.547	0.75	4.442
5.0	0.149	0.056	0.232	0.474	0.79	4.356

(i) Determine the values of μ_{max} ,the yield coefficient ($Y_{X/S}$) and K_S for each substrate. Based on these values, use Equation (3.43) to write mass balances for cell mass and each substrate in a batch reactor. Note that the uptake rates of each substrate contains only one of the terms in Equation (3.43).

(ii) Using a numerical method, solve these equations for batch growth of the *Lactobacillus* on 10gm/l glucose and 5 gm/l nitrogen source.

(iii) Compare the predictions of this model with those based on the additive (Equation 3.44 (a)) and the multiplicative (Equation (3.44 (b)) models, using a numerical scheme with the same initial conditions. What is the appropriate value of μ_{max} to use in these equations? Comment on the differences in the patterns of substrate uptake of the three models. Can these models adequately represent diauxic growth?

14. Anaerobic Digestion

Anaerobic digestion is the degradation of complex organic matter to gaseous products, CO_2 and CH_4. Although the complete process involves complex interacting microbial species, the majority of methane formed in an anaerobic digestion is produced by acetate-utilizing methanogens. Yang and Okos (S.T. Yang and M.R. Okos, paper No. 39d presented at the AIChE 1984 Annual Meeting) have studied the kinetics of methanogenesis from acetate for *Methanococcus mazei* and two strains of *Methanosarcina barkeri* in batch culture. Significant substrate inhibition was observed for all three methanogens grown on acetate. Rate constants estimated from growth data for two of the methanogens are given below.

Strain	μ_{max} (h^{-1})	K_s (g acetate/L)	K_i (g acetate/L)
M. mazei S6	0.029	1.00	48.7
M. barkeri MS	0.63	100	0.46

Using these data, calculate for each methanogen the optimal acetate concentration and the maximum observable specific growth rate.

15. Growth Constants

Listed below are the specific growth rates of *M. barkeri* strain 227 measured for different acetate concentrations. Use these data to calculate K_s (g acetate/L), K_i (g acetate/L), and μ_{max} .

Acetate conc. (g/L)	0.20	0.60	1.6	2.9	6.0	7.1
Growth rate (h^{-1})	0.005	0.10	0.017	0.019	0.0176	0.0156

16. Metabolic Fluxes

Clostridium acetobutylicum is an anaerobic bacterium capable of producing organic acids and solvents from a variety of substrates. A highly branched metabolic pathway enables *Clostridium acetobutylicum* to produce acetic acid, butyric acid, acetone, butanol and several other compounds in varying amounts, depending on cultivation conditions. A simplified schematic of the pathway is showm below [K.F. Reardon, T.-H. Scheper, and J.E. Bailey, *Biotech. Prog.*, **3**, 153 (1987)]. The v_i indicate the specific rates (mole i-unit cell mass^{-1}-time^{-1}) along the pathway branch i, and the boxed components denote extracellular compounds.

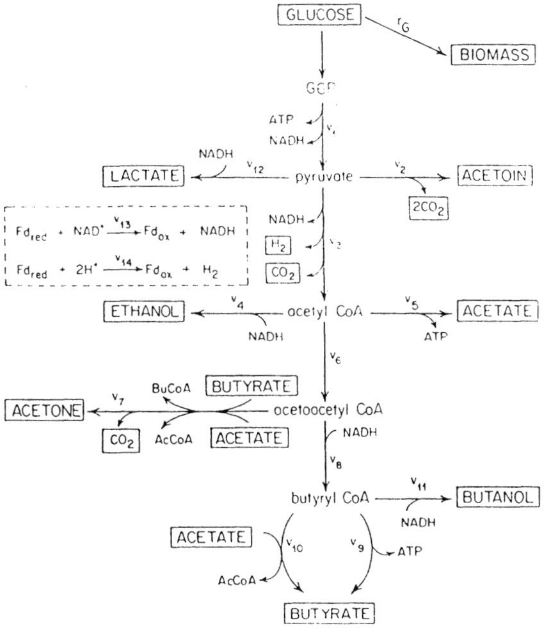

A material balance for an intracellular metabolite i (mol i·unit cell mass^{-1}) can be written as (summed over j)S[46]

$$\frac{dc_i}{dt} = \sum_j v_{ij} v_j$$

where v_{ij} are stoichiometric coefficients for each of the branch reactions. In addition, a material balance on each fermentation product C_i (mole i per unit culture volume) in a constant volume batch reactor can be related to metabolic rates by

$$\frac{dC_i}{dt} = X \cdot \sum_j \delta_{ij} v_j$$

where X is the biomass concentration in the reactor and δ_{ij} is the appropriate stoichiometric coefficient.

It can be assumed that the overall growth equation can be written as

$$C_6H_{12}O_6 + 1.5\ NH_3 + 0.75\ (NAD(P)H + H) + 14.9\ ATP \rightarrow$$
$$6\ CH_2O_{0.5}N_{0.25} + 0.75\ NAD(P)^+ + 14.9(ADP + P_i) + k\ H_2O$$

(46) This balance applies to resting, non-growing cells that use glucose exclusively for the formation of metabolites and end-products (ie. no glucose is used for biomass synthesis). Alternatively, we could assume, as did the original authors, that the rate of formation of metabolites due to biomass synthesis is balanced by the dilution of metabolites due to cell growth.

(a) Write material balances (assuming a constant volume batch fermenter) for all of the compounds shown in the figure. For example, assume glucose can can be described by $dC_{glu}/dt = -X \cdot v_0$ and proceed from there. Note that pathway 7 actually consists of 3 reactions. Make sure the material balances of compounds involved in the overall growth equation include terms that account for their overall consumption for cell growth.
(b) Assume that the quasi-steady state hypothesis applies to each metabolic intermediate at a node in the pathway, i.e., $dc_j/dt = 0$, where c_j is the concentration of species j at a node (e.g., G6P). Write equations (6 total) for the above material balances.
(c) Assume that the extracellular concentration profiles can be determined as a function of time for acetic acid, butryic acid, acetoin, ethanol, acetone, butanol, glucose and biomass. How can you use this information to determine the pathway-branch reaction rates? How many of them can you determine?

17. Interacting Reactions within the Cell

(a) This problem is based on the approach of A.C.R. Dean and Sir Cyril Hinshelwood [*Growth, Function and Regulation in Bacterial Cells*, Clarendon Press (1966)] for modelling networks of interacting reactions within the cell. Consider the simplest case where the cell mass is divided into two compartments, one containing X and the other Y. The rate of formation of X shows a first order dependence on Y, while that of Y depends on X in the same manner, i.e.,

$$\frac{dX}{dt} = \alpha Y \qquad \text{and} \qquad \frac{dY}{dt} = \beta X$$

where X and Y represent concentrations on a per cell volume basis (assumed to be constant). We might consider that X represents small molecular weight components (e.g., amino acids) produced by catabolism which is regulated by macromolecules (e.g., enzymes) and Y represents these macromolecules whose synthesis (anabolism) depends on the concentration of the amino acids present in the cell. Show that such an interacting system results in exponential growth of both X and Y with a growth rate of $\sqrt{\alpha\beta}$. If X_0 and Y_0 are the initial concentrations of X and Y, show that after a sufficient period growth becomes *balanced*, i.e., the ratio of X/Y within the cell is constant. Will balanced growth occur if cells are taken from a balanced growth phase and used as an inoculum for a second stage of growth?
(b) Bacterial cells show a large variation in the proportions of key components as the external growth conditions are changed. They also exhibit adaptation and improve their growth rates after exposure to sub-optimal growth conditions. A more complex model of a branched metabolic pathway which illustrates this phenomenon is shown below.

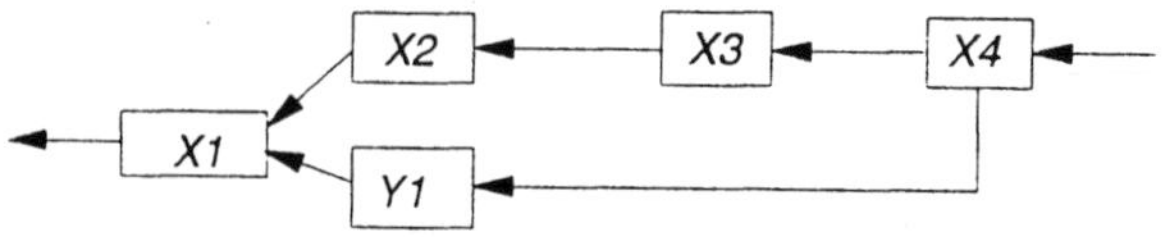

The mass balance equations for species X_1, X_2, X_3 and Y_1 are given by

$$\frac{dX_1}{dt} = \alpha_1 X_2 + \beta_1 Y_1 \qquad \frac{dX_2}{dt} = \alpha_2 X_3 \qquad \frac{dX_3}{dt} = \alpha_3 X_4 \qquad \frac{dY_1}{dt} = b_1 X_4$$

Show that the rate of balanced growth of the cells (after a rather complex transient period) can be represented by the expression

$$\mu = \left\{ (\alpha_1 \alpha_2 + \beta_1 b_1) \left(\prod_{i=3}^{4} \alpha_i \right) \right\}^{1/4}$$

and that during balanced growth $(1/X_i)(dX_i/dt) = \mu$, where

$$\mu = \alpha_1 \frac{X_2}{X_1} + \beta_1 \frac{Y_1}{X_1}$$

During normal growth we can assume that $\alpha_2 > \beta_1$ which results in most of the carbon flux proceeding along the X branch of the pathway, with relatively little carbon flowing along the Y branch. Suppose now that the cells are placed in a medium in which the flux of carbon through the pathway $X_2 \rightarrow X_1$ (i.e., α_2) becomes small. Show that a new steady state is established in which the flux through the Y branch of the pathway is increased and that the resultant growth rate is not significantly less than the previous growth rate μ, i.e., an increase in the amount of Y_1 has occurred which compensates for the reduced flux through the X-branch of the pathway. This illustrates that a rather simple model may show the selection by an organism of a less efficient pathway in a time of "stress" to compensate for environmental changes.

18. Structured Models

Develop a simple structured model for bacterial growth on two nutrients, A and B, where the bacteria prefer to utilize A. Use the following assumptions:

(i) The rate of growth on A is proportional to the concentration of a constitutive intracellular enzyme, which is produced at a constant rate per unit biomass and degraded by a first-order process.

(ii) The rate of growth on B is proportional to the concentration of an inducible intracellular enzyme, which is produced at a rate that decreases with increasing concentration of A and approaches a constant as the concentration of A goes to zero, per unit biomass. This enzyme is also degraded by a first-order process.

Write the appropriate equations for growth in a well-mixed batch reactor. For balanced growth, develop an expression for the dependence of specific growth rate on the concentrations of A and B, using the additional assumption that the enzyme degradation rates are rapid compared to cell growth rates. Discuss the implications of your result for transient batch growth. *(Courtesy of D.A. Lauffenburger)*

19. Structured Models

Consider a structured model that assumes the cell is comprised of four components: P, the concentration of intracellular precursors; M_1, the concentration of enzymes and RNA involved in cell synthesis; M_2, the concentration of structural macromolecules; and G, storage polymers. The total biomass is thus equal to the sum of the four components:

$$X_T = P + M_1 + M_2 + G$$

where all concentrations are in g/l of reactor volume. Develop expressions for dG/dt, dM_1/dt, dM_2/dt, and dP/dt by assuming the following:

(i) growth of the cell is limited by the concentration of glucose, S;

(ii) dG/dt is equal to the rate of G formation minus the rate of G degradation. The rate of G formation per unit reactor volume obeys Michaelis Menten kinetics in the precursor concentration per cell mass and the concentration of M_1 per unit reactor volume. The rate of G degradation is first-order in G.

(iii) dM_2/dt obeys Michaelis Menten kinetics in the precursor concentration per cell mass and the concentration of M_1 per unit reactor volume.

(iv) dM_1/dt is first order in both M_1 and M_2; the second-order rate constant exhibits Michaelis-Menten dependence on the precursor concentration per cell mass.

(v) dP/dt is proportional to the rate of S uptake per cell mass minus the time derivatives of M_1, M_2, and G.

Use the following notation:

k_1 = gm G/gM_1-hr　　k_{1D} = gm P/gG-hr

k_2 =gm M_2/gM_1-hr　　k_3 = gm M_1/gm M_2-gm M_1-hr

v_s = gm S transported/gm X_T-hr　　K_i = saturation parameters

γ_i = stoichiometric coefficients

20. Models of Enzyme Repression

The model of Imanaka et al. for the production of α-galactosidase was described in Section 3.5.3. We shall use it to predict the effect of catabolite repression on α-galactosidase production by *Monascus*. Consider an inoculum of 5 gm/l cells growing on a medium containing 3 gm/l galactose in the absence of glucose. After 90 minutes, 5 gm/l glucose is added to the batch culture. Set up the series of mass balance equations describing the model of Imanaka et al. using the constants given in Table 3.14. During this enzyme production phase, use the values subscripted "p" for production. The initial values of the intracellular components are:

M = 0.1 μg/mg cells

S_{Bi} = 0.15μg/mg cells

[RS_{Bi}] = 0.2 μg/mg cells

R = 0.5 μg/mg cells

E = 0.0 units/mg cells

Using a numerical scheme, solve the series of batch mass balances for all extra and intracellular components for the time period from initial growth to 250 minutes later. Use a value of mRNA decay of k_7 of 8 hr^{-1}. Experimentally, Imanaka et al. observed that 70 minutes after glucose addition, enzyme production ceased as a result of glucose repression. Vary k_7 from 4 to 10 hr^{-1} and plot the time at which enzyme production ceases against k_7. Vary the repressor-galactose binding constants k_4 and k_5 and plot their influence on the time for enzyme production to cease. Why is enzyme production more sensitive to the decay rate of mRNA than to the repressor-galactose binding kinetics?

21. Plasmid Synthesis

The expression for steady-state plasmid concentration, Eq. 3.143, developed in Section 3.6 does not account for possible nutrient limitations on the plasmid synthesis rate. However, in continuous culture and in post-exponential batch growth, the nutritional condition of the environment is an important determinant of cellular behavior and hence may affect the plasmid concentration. To account for such effects, we introduce a new factor, $r_p(s)$, into the overall rate expression, where $r_p(s)$ is the ratio of the specfic growth rate under nutrient limitation to its value in the absence of any limiting nutrients:

$$r_p(s) = \left(\frac{\mu}{\mu_{max}}\right)^n$$

Incorporate this term into the overall rate expression for plasmid replication and develop the corresponding expression for plasmid concentration at balanced growth.

22. Plasmid Replication

Koizumi and coworkers [*Biotechnol. Bioeng.*, **27**, 721 (1985)] cultivated *B. stearothermophilus* harboring plasmid pLP11 in continuous culture at different dilution rates and at different temperatures. Their data for T = 47.5°C are given below.

Dilution rate (h^{-1})	0.22	0.43	0.47	0.66	0.69	0.92	0.94
Plasmid concentration (mg/g cell)	0.14	0.14	0.20	0.28	0.25	0.35	0.38

Are these results consistent with the expanded model of plasmid replication developed in Problem 21? State any assumptions needed to justify your answer.

Chapter 4. Bioreactor Design and Analysis

The bioreactor is the center of all biochemical processing. By combining our knowledge of the kinetics of biological reactions with material and energy balances, we can, at least in principle, design and analyze the behavior of the bioreactor. In practice, this procedure is made more complex because of the nature of the biological catalyst, which may have time-varying properties and may show complex kinetic patterns, and the nature of the fermentation broth, which may exhibit striking deviations from ideality, resulting in complex flow patterns, and mass and heat transfer characteristics. In this chapter we shall initially consider various simplifications that permit the essential features of bioreactor design to become apparent. This involves the use of simplified kinetic models and the assumption of ideal behavior for both liquid and gas phases in the bioreactor. By considering several idealized bioreactors, mathematically tractable design solutions can be obtained. The deviations from ideality can then be incorporated by modifying these solutions. In the next chapter, we shall consider the role of heat and mass transfer in bioreactors. We generally assume that the cell or enzyme in an idealized bioreactor is exposed to a spatially uniform environment; thus the reaction rate does not vary locally. These assumptions are often adequate for describing the behavior of large-scale industrial reactors, even though there may be deviations from ideal mixing and uniform distributions of cells or enzymes in the vessel.

Most biological reactors are multiphase systems. The biocatalyst may be present as a solid phase, for example as an immobilized enzyme or as an individual cell. Typically, gas is sparged into microbial reactors to supply oxygen and remove carbon dioxide. The hydrodynamic behavior of each phase needs to be considered. There are several types of bioreactor operation we shall consider. To date, we have only considered *batch* reactors, with no flows in or out of the system. In the idealized case, batch reactors have a homogeneous continuous phase; the liquid phase is well-mixed and of uniform temperature and composition. There are no spatial variations in reactant or product concentrations. The gas phase in aerobic batch reactors may be well-mixed or may be modelled by plug-flow. The kinetics of the reaction considered are key in determining the bioreactor behavior. These features will be illustrated in the following sections.

We shall consider *continuous* bioreactors of two types, *well-mixed continuous stirred tank reactors* (CSTRs) and *plug-flow* or *tubular reactors*. Semi-continuous operation is also common for industrial bioreactors and is referred to as *fed-batch* operation. In this situation, the reactor is initially operated as a batch reactor and when a nutrient (usually the carbon source) has been consumed, it is fed to the reactor following a predetermined protocol (hence "fed-batch").

4.1 Batch Reactors

An ideal well-mixed batch reactor is spatially homogeneous as a result of intensive agitation. With appropriate temperature and pH control, these parameters can also be maintained at uniform values throughout the reactor. As we have seen earlier, mass and energy balance equations can be written around the batch reactor for the species of interest. The most general form of these equations is

$$\frac{dc_i V}{dt} = V \cdot r(c_i, c_j) \tag{4.1}$$

where V is the reactor volume; c_i is the concentration of component i and c_j represents the concentrations of other species which may influence the rate of formation or consumption of i in the reaction. $r(c_i,c_j)$ is the volumetric rate of formation of c_i by reaction. We may also write an overall mass balance for the system

$$\frac{d\rho V}{dt} = 0 \tag{4.2}$$

Provided that the total density of the reacting liquid phase is constant, the liquid volume remains constant, and we may remove V from the differential and thus obtain

$$\frac{dc_i}{dt} = r(c_i, c_j) \tag{4.3}$$

With ρ and heat capacities taken as constant, the conservation of energy equation may be written as

$$\rho V c_p \frac{dT}{dt} = [-\Delta H_R] r_X V + Q - W_S \tag{4.4}$$

where r_x is the volumetric rate of increase of dry cell weight (Section 3.2.2) and c_p is the heat capacity. Although the bioreactor may be operated in a batch mode, the energy balance equation contains terms for heat flow from the reactor associated with temperature control (Q) and a term for the generation of heat by agitation ($-W_S$, shaft work).

Some simple models for enzyme reactions and microbial growth have been presented in earlier chapters. For example, microbial growth can be represented by the two equations for cell mass and substrate, using the Monod model:

growth
cell mass = *growth rate* × *cell conc.*
over time

$$\frac{dX}{dt} = \mu X = \frac{\mu_{max} S X}{K_S + S} \tag{4.5}$$

consumption
of subs /time = $\frac{1}{yield\ coefficient}$ · *cell conc.*

$$\frac{dS}{dt} = -\frac{1}{Y_{X/S}} \frac{\mu_{max} S X}{K_S + S} \tag{4.6}$$

with initial conditions $X(0) = X_o$, $S(0) = S_o$.

These equations can be added after multiplication of equation (4.6) by $Y_{X/S}$ to yield

$$\frac{d(X + Y_{X/S}S)}{dt} = 0$$

and thus

$$X + Y_{X/S}S = X_o + Y_{X/S}S_o \tag{4.7}$$

Expressing X in terms of S from the above equation, and substituting into Equation (4.6) gives

$$\frac{dS}{dt} = -\frac{\mu_{max}}{Y_{X/S}} \frac{(X_o + Y_{X/S}\{S_o - S\})S}{(K_S + S)} \tag{4.8}$$

which can be analytically integrated to give

$$[X_o + Y_{X/S}(S_o + K_S)]\ln\left(\frac{X_o + Y_{X/S}\{S_o - S\}}{X_o}\right) - K_S Y_{X/S} \ln\frac{S}{S_o} = \mu_{max}(X_o + Y_{X/S}S_o)t \tag{4.9}$$

This equation is implicit in S and the time course of substrate consumption can best be found by calculating t values which correspond to a specified value of substrate concentration. Substitution of values of S into equation (4.9) provides the time course of biomass concentration.

If the culture forms a product which is partially growth and non-growth associated (Section 3.2.7), such that

$$\frac{dP}{dt} = \alpha \frac{\mu_{max} S X}{K_S + S} + \beta X \tag{4.10}$$

The mass balance for substrate must be rewritten to account for substrate conversion to product, (Eq. 4.11) and the resulting three mass balances, equations (4.5), (4.10) and (4.11) can be numerically solved to yield typical concentration profiles as shown in Figure 4.1.

$$\frac{dS}{dt} = -\frac{1}{Y_{X/S}} \frac{\mu_{max} S X}{K_S + S} - \frac{1}{Y_{P/S}}\left(\alpha \frac{\mu_{max} S X}{K_S + S} + \beta X\right) \tag{4.11}$$

4.1.1 Death of Cells in Batch Culture

In a growing culture, some cells may become dormant or die as a result of mistakes in autosynthesis (misreading of DNA, for example). Although the fraction of viable cells in bacterial cultures may be quite high in batch growth, eukaryotic cells may die at appreciable rates, and the viability of such cultures will be less than 100%. If we consider that only viable cells (X_v) generate non-viable cells (X_d) at a rate which is first order in viable cell concentration with a rate constant k, we can write for batch growth

$$\frac{dX_v}{dt} = \mu X_v - kX_v \tag{4.12}$$

$$\frac{dX_d}{dt} = kX_v \tag{4.13}$$

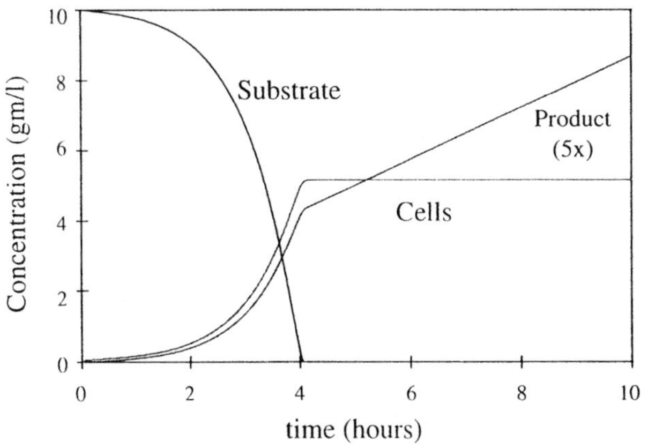

Figure 4.1. *Batch reactor profile of biomass, substrate and product concentrations. Initial conditions* $X_o = 0.05$ *gm/l,* $S_o = 10.0$ *gm/l,* $P_o = 0.0$ *gm/l; parameter values* $Y_{X/S} = 0.5$ *gm/gm,* $Y_{P/S} = 0.1$ *gm/gm,* $K_S = 0.5$ *gm/l,* $\alpha = 0.5$ *(-),* $\beta = 0.1\ hr^{-1}$, $\mu_{max} = 0.2\ hr^{-1}$.

The growth of the total cell population is given by

$$\frac{dX_T}{dt} = \frac{d(X_v + X_d)}{dt} = \mu X_v \tag{4.14}$$

If we consider that μ is approximately constant for most of the exponential phase of batch growth, Equation (4.12) can be integrated for $X_v(t)$ and this result substituted into Equation (4.14) for X_T. Integrating the resultant equation gives the concentration of viable cells and the viability (X_v/X_T):

$$X_v = X_v(0)e^{(\mu - k)t} \tag{4.15}$$

$$\frac{X_v}{X_T} = \frac{X_v(0)e^{(\mu-k)t}}{X_v(0)\frac{X_T(0)}{X_v(0)}\left(1+\frac{\mu}{(\mu-k)}e^{(\mu-k)t}\right)} \tag{4.16}$$

If the inoculum consists mainly of viable cells, $X_v(0)/X_T(0) \sim 1$, and the expression for viability reduces for sufficiently large times ($\exp(\mu - k)t > 1$) to

$$\frac{X_v}{X_T} \sim \frac{\mu-k}{\mu} \tag{4.17}$$

i.e., the fractional viability remains constant for most of the exponential growth period. If the death rate k is equal to the growth rate μ, then linear growth of the total biomass results:

$$X_T = X_T(0) + X_v(0)\mu t \tag{4.18}$$

4.2 Continuous Stirred Tank Bioreactors

4.2.1 The Chemostat: the Ideal CSTR

The use of a continuous stirred tank reactor to extend the duration of culture of microbes was developed in the 1950s by Novick and Szilard[1] and Monod[2]. The realization that a CSTR could be used to maintain microbial growth at a steady state value, which could be varied from any growth rate up to the maximum μ_{max}, was an important advance, as it broke the traditional thinking at the time that stable microbial growth was only possible at the maximum rate, corresponding to the minimum doubling time found in batch cultures. Subsequently, the use of a well-mixed continuous microbial reactor to study microbial physiology led to important advances in understanding the cell cycle, metabolic regulation and microbial product formation.

The configuration of a typical well-mixed continuous reactor is shown in Figure 4.2. Agitation may be provided by an impeller or by the motion imparted to the liquid phase by rising gas bubbles. In aerobic systems, supply of oxygen to the organism generally occurs via air sparging. In the ideal case, the liquid phase is completely mixed, i.e., the liquid phase composition is uniform throughout the vessel. Similarly, temperature is maintained constant and uniform by circulation of cooling water through coils in the vessel or in a jacket surrounding the vessel. Typically the pH of the culture medium is controlled by the addition of acid or base.

We may write material balance equations for each of the important variables in the CSTR. We shall first consider the case where only one substrate (S) limits the growth rate of the organism, and that the volumetric rate of growth is given by μX. The balance equations are:

(**1**) Novick, A. and L. Szilard, *Science*, **112**, 715 (1950).

(**2**) Monod, J. *Ann. Inst. Past.*, **79**, 390 (1950).

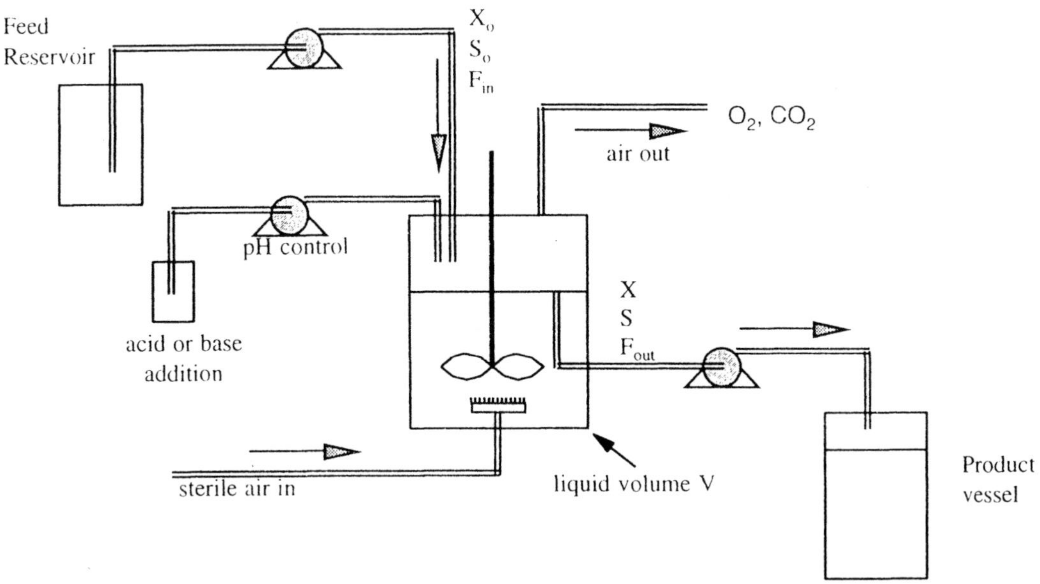

Figure 4.2. *A schematic of a continuous stirred tank bioreactor. Typically the pH and temperature are controlled, as are the flow rates of nutrients into the vessel. The notation used to model CSTR systems is indicated; X is the dry cell weight, S is the substrate concentration and F is the flow rate of nutrients into the vessel.*

$$\frac{dXV}{dt} = F_{in}X_o - F_{out}X + \mu XV \tag{4.19a}$$

$$\frac{dSV}{dt} = F_{in}S_o - F_{out}S - \frac{1}{Y_{X/S}}\mu XV \tag{4.19b}$$

$$\frac{dV}{dt} = F_{in} - F_{out} \tag{4.19c}$$

When the volumetric feed rates, F_{in}, F_{out}, into and from the vessel are maintained constant and equal (F), the equations simplify to (note that dX/dt no longer equals μX, as it does during batch growth)

$$\frac{dX}{dt} = \frac{F}{V}(X_o - X) + \mu X \tag{4.20a}$$

$$\frac{dS}{dt} = \frac{F}{V}(S_o - S) - \frac{1}{Y_{X/S}}\mu X \tag{4.20b}$$

The ratio F/V is generally referred to as the *dilution rate*, denoted as D, with units of reciprocal time. It is the inverse of the average residence time τ. It equals the number of reactor volumes that pass through the reactor per unit time. At steady state, the time derivatives are set to zero and the equation for cell concentration has the solution

$$DX_o = (D - \mu)X \tag{4.21}$$

When the feed stream is sterile (generally the case), X_0 is zero and two solutions to the above equation are possible:

$$X_{ss} = 0 \qquad \text{or} \qquad \mu = D \tag{4.22}$$

In the unusual case that the specific growth rate of the culture ($\mu(S)$) is independent of substrate concentration and is constant, the steady-state concentration of cells that results when the dilution rate is set equal to μ is indeterminate. Solution of the second mass balance shows that the steady state substrate concentration is also indeterminate, although both X_{ss} and S_{ss} must satisfy

$$S_{ss} = S_o - \frac{1}{Y_{X/S}} X_{ss} \tag{4.23}$$

Thus a range of values of cell and substrate concentrations is possible. Experimentally, this can be occasionally seen at very low inlet substrate concentrations. Time-varying cell mass and substrate concentrations are observed.

Generally however, the specific growth rate is a function of substrate concentration. When the Monod relationship between μ and S is employed, the mass balance equations are no longer indeterminate and we find

$$D = \mu = \frac{\mu_{max} S}{K_S + S} \tag{4.24}$$

which can be solved for S:

$$S_{ss} = \frac{DK_S}{\mu_{max} - D} \qquad \text{provided} \qquad X_{ss} \neq 0 \tag{4.25}$$

and from the substrate balance equation we see

$$D(S_o - S_{ss}) - \frac{1}{Y_{X/S}} \mu X_{ss} = 0 \tag{4.26}$$

and noting $D = \mu$, we obtain

$$X_{ss} = Y_{X/S}\left(S_o - \frac{DK_S}{\mu_{max} - D}\right) \tag{4.27}$$

The second steady state solution occurs when $X_{ss} = 0$. The corresponding value of the substrate concentration is $S_{ss} = S_0$. This steady state is referred to as *washout*, as cells are no longer present in the reactor. The dilution rate at which washout occurs can be found by examining equation (4.24). When S_{ss} equals the feed concentration S_0, the corresponding dilution rate is

$$D_{max} = \frac{\mu_{max} S_o}{K_S + S_o} \tag{4.28}$$

The maximum dilution rate is thus slightly smaller than the maximum specific growth rate. If the dilution rate is greater than this value, the system moves to the second steady state solution $X_{ss} = 0$. This can be seen from the behavior of S_{ss}; as $D \rightarrow \mu_{max}$, S_{ss} becomes indeterminate.

Summary of Steady State Solutions

$D < \mu_{max}$	$D > \mu_{max}$
$S_{ss} = \frac{DK_S}{\mu_{max} - D}$ provided $S_0 > \frac{DK_S}{\mu_{max} - D}$	$S_{ss} = S_o$
S_{ss} is indeterminant if $S_0 < \frac{DK_S}{\mu_{max} - D}$	
$X_{ss} = Y_{X/S}\left(S_o - \frac{DK_S}{\mu_{max} - D}\right)$	$X_{ss} = 0$

As we saw in Chapter 3, values of K_S are usually small, particularly when compared to the inlet substrate concentration S_o. Thus the steady state substrate concentration is quite small, and, remarkably, is *independent of the inlet substrate concentration*. The non-trivial solutions can only apply in the case where the inlet substrate concentration is greater than S_{ss}. If $S_o < S_{ss}$, then the specific growth rate μ is constant, the equations are indeterminate, and a variety of steady states could be observed.

The cell concentration is approximately $Y_{X/S}S_o$ for dilution rates up to values approaching μ_{max}. The behavior of the steady state solutions X_{ss} and S_{ss} as a function of the dilution rate is shown in Figure 4.3. The operation of a CSTR under conditions where only one substrate is growth-limiting gives rise to an almost constant value of the substrate concentration over a wide range of dilution rates. Other substrates which are consumed at rates proportional to the specific growth rate of the cells will also have steady state concentrations that are independent of D and are constant. For this reason, this type of bioreactor operation is referred to as *chemostat* operation (i.e., the *chem*ical environment is *stat*ic).

As the dilution rate approaches μ_{max}, the cell concentration decreases very rapidly. Operating the chemostat at dilution rates close to μ_{max} is experimentally difficult due to this sensitivity. Small variations in inlet substrate concentration, feed flow rate or small variations in the growth rate of the organism may result in washout of the cells.

The volumetric productivity of cells in the chemostat is given by DX_{ss} (typically expressed as gm cells per liter reactor volume per hr). The dilution rate at which the maximum productivity occurs can be found from:

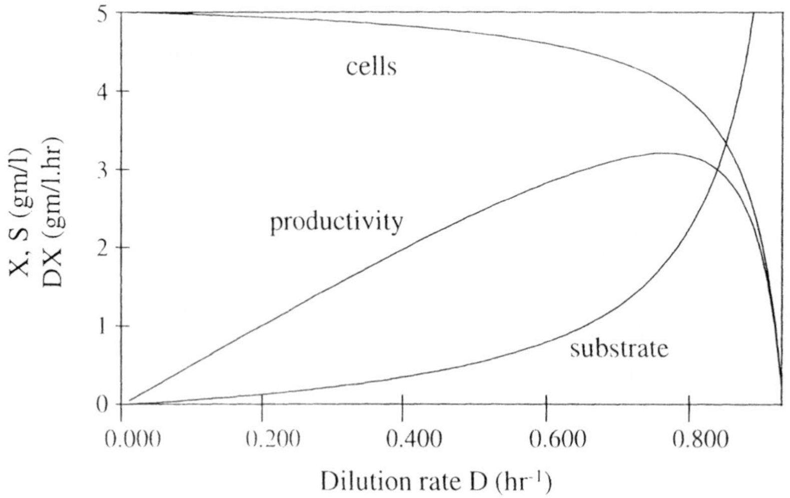

***Figure 4.3**. The dependence of the steady state cell and substrate concentrations on the dilution rate D. The values of the constants in the Monod model are μ_{max} = 1.0 hr^{-1}, K_S = 0.5 gm/l, $Y_{X/S}$ = 0.5 gm cells/gm substrate with S_o = 10 gm/l. Also shown is the cell productivity, DX_{ss}, in gm/(liter-hr).*

$$\frac{dDX_{ss}}{dD}=0 \qquad \text{thus} \qquad D_{max}=\mu_{max}\left(1-\sqrt{\frac{K_S}{K_S+S_o}}\right) \tag{4.29}$$

The corresponding cell concentration is

$$X_{ss,max}=Y_{X/S}(S_o+K_S-\sqrt{K_S(S_o+K_S)}) \tag{4.30}$$

If $K_S << S_o$, then the volumetric productivity becomes $Y_{X/S}\mu_{max}S_0$.

In contrast to the behavior of steady state bacterial and yeast cultures, continuous suspension cultures of animal cells show viable and total cell concentrations that decrease with increasing dilution rates, although the percentage of the cells that are viable may increase with increasing dilution rate. Hybridoma cells, which are fusion products of myeloma and spleen cells used for the production of monoclonal antibodies, become slightly larger at higher dilution rates, which may explain some of these trends. The behavior of these cells in continuous suspension cultures is illustrated in Figure 4.4. The decrease in cell viability at low dilution rates illustrates a high maintenance requirement. At low viabilities, the specific growth rate is no longer equal to the dilution rate, and we must distinguish between viable and non-viable cells. The mass balance equations thus become

$$\frac{dn_v}{dt}=\mu n_v-k_d n_v-Dn_v \tag{3.76}$$

$$\frac{dn_T}{dt} = \mu n_v - D n_T \tag{4.31}$$

where n_v and n_T are the number concentrations of viable and total cells. The viability (n_v/n_T) at steady state thus depends on the death rate k_d and the specific growth rate μ , and thus varies in an indirect manner with D.

$$\left(\frac{n_v}{n_T}\right)_{ss} = \frac{\mu - k_d}{\mu} \tag{4.32}$$

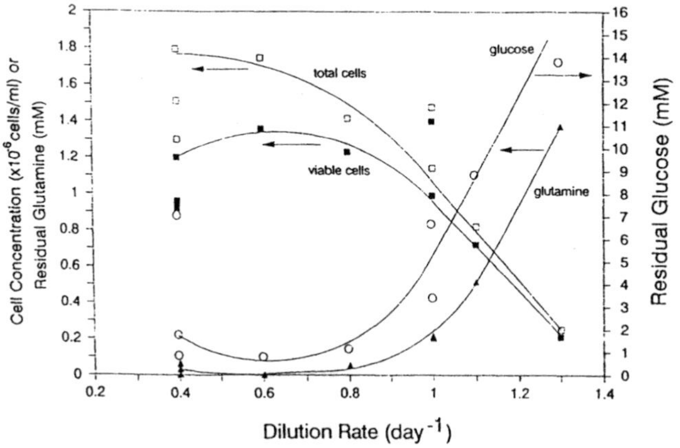

Figure 4.4*. The steady state behavior of hybridoma cells in continuous culture. The cell line is a mouse hybridoma (Ab2-143.2). Data from Hiller, G.W., Aeshlimann, A., Clark, D.S. and H.W. Blanch, Biotech. Bioeng.* ***38*** *733-741 (1991)*

At large dilution rates, the viability is high and the specific growth rate approaches the dilution rate. The residual glucose (substrate) concentration increases with increasing dilution rate more gradually than would be predicted by the Monod model, consistent with the gradual decrease in cell concentration with increasing dilution rate. At low dilution rates, the lactate production from glucose is reduced, as more glucose is required for maintenance and production of nucleotide sugars. The relationships between specific metabolite production or consumption rates are typically not linear, as is usually the case with bacterial systems. The more complex carbon and energy metabolism of mammalian cells, particularly the requirements for two substrates, glucose and glutamine, mean that the Monod assumption that one limiting nutrient regulates the specific growth rate no longer applies.

The formation of products from animal cells in suspension cultures has been most extensively examined with antibody production by hybridoma cells. While many cells show some genetic instability, others have been found to be relatively genetically stable with respect to antibody production over long periods in culture (more than 100 generations). Generally, antibody production has been found to be non-growth associated, following the trends illustrated in Figure 4.5.

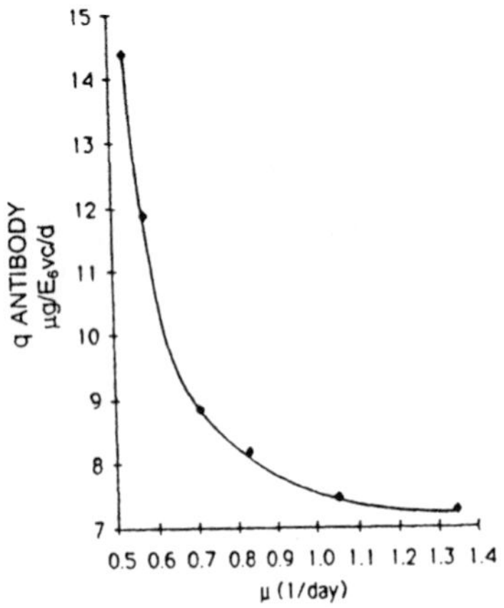

***Figure 4.5**. The specific rate of antibody formation (q_{Ab}) as a function of the specific growth rate of hybridoma cells. [Data from Miller, W.M., Blanch, H.W. and C.R. Wilke, Biotech. Bioeng. **32**, 947 (1988)].*

Comparison of the Productivity of Batch and Continous Cultures

The biomass productivity of batch and continuous cultures can be compared under the same broadly applicable conditions where $S_o >> K_S$. The solution of the batch mass balance equation for biomass can be simplified if we assume $\mu \sim \mu_{max}$ for the duration of the exponential phase. We can then obtain the duration of the exponential phase t_{exp}:

$$X = X_o e^{\mu_{max} t_{exp}}$$

which can be rearranged

$$t_{exp} = \frac{1}{\mu_{max}} \ln\left(\frac{X}{X_o}\right) \tag{4.33}$$

The total time required for a batch fermentation is the sum of the lag and exponential phases and the time required to empty the vessel and resterilize and inoculate it again. Let us lump these other times into one constant, called the turnaround time, t_{turn}. The total batch time then becomes the sum of the time for exponential growth plus the turnaround time:

$$t_{batch} = \frac{1}{\mu_{max}} \ln\left(\frac{X}{X_o}\right) + t_{turn} \tag{4.34}$$

If the initial substrate concentration (S_o) is the same as that in the feed to a chemostat, the biomass produced in this time will be approximately $Y_{X/S}S_o$. The volumetric productivity of the batch reactor is thus $Y_{X/S}S_o/t_{batch}$. Comparing this to the maximum productivity of a chemostat, which we see from Equation (4.29) occurs at a dilution rate of ~ μ_{max}, we find that

$$\frac{Chemostat}{Batch} = \frac{\mu_{max}Y_{X/S}S_o}{\left(\dfrac{\mu_{max}Y_{X/S}S_o}{\ln\left(\frac{X}{X_o}\right) + \mu_{max}t_{turn}}\right)} = \ln(\frac{X}{X_o}) + \mu_{max}t_{turn} \tag{4.35}$$

Typically, the ratio of final cell concentration to inoculum concentration (X/X_o) is approximately 10. Thus, continuous culture offers a minimum of 2.3 [ln(10)] times the productivity of a batch fermentation.

The Effect of Endogeneous Metabolism and Maintenance on Chemostat Behavior

As we saw in Chapter 3, some substrate may be consumed by the cell for DNA repair, maintenance of chemical gradients within the cell, etc. When maintenance is included, the substrate mass balance becomes

$$\frac{dS}{dt} = D(S_o - S) - \frac{1}{Y_{X/S}}\mu X - mX \tag{4.36}$$

Values of m are given in Table 3.5. At steady state and assuming Monod kinetics, the substrate and cell concentrations are given by

$$S_{ss} = \frac{DK_S}{\mu_{max} - D} \qquad X_{ss} = \frac{D(S_o - S_{ss})}{\left(m + \frac{D}{Y_{X/S}}\right)} \tag{4.37}$$

An observed yield coefficient, based on the ratio of biomass produced per unit substrate consumed, can thus be written as

$$Y_{X/S,obs} = \frac{X_{ss}}{(S_o - S_{ss})} = \frac{DY_{X/S}}{D + mY_{X/S}} \tag{4.38}$$

Figure 4.6. shows the effect of substrate consumption for maintenance on the steady state cell concentration and observed yield coefficient, using values typical of *E. coli* grown on glucose. AT low dilution rates, the observed yield coefficient decreases rapidly, as a larger fraction of substrate is consumed for maintenance.

Cells may also consume part of their internal carbohydrate and protein reserves to supply the maintenance energy. The cell mass balance then reflects this endogenous metabolic requirement, and cell mass is consumed at a rate given by k_eX.

$$\frac{dX}{dt} = \mu X - DX - k_eX \qquad \frac{dX}{dt} = 0 \quad \text{at steady state} \tag{4.39}$$

The substrate mass balance is unaffected by endogenous metabolism. With Monod kinetics, the cell mass balance shows the effect of endogenous metabolism on the steady state substrate concentration.

$$S_{ss} = \frac{(D + k_e)K_s}{(\mu_{max} - D - k_e)} \tag{4.40}$$

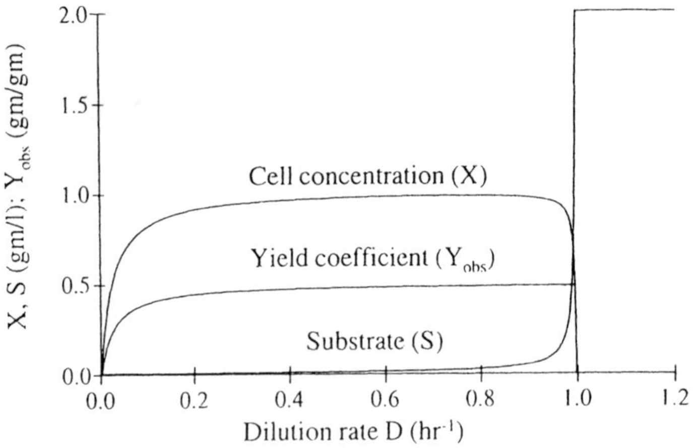

Figure 4.6. *The effect of maintenance on the steady state cell concentration (X_{ss}) and observed yield coefficient. Values of the constants are μ_{max} = 1.0 hr^{-1}, $Y_{X/S}$ = 0.5, K_S = 0.005 gm/l, S_o = 2 gm/l and m = 0.05 gm/gm-hr.*

Here the specific growth rate μ is no longer equal to the dilution rate (in fact, $\mu = D + k_e$). The observed yield coefficient in this case can therefore be written as

$$Y_{X/S,obs} = \frac{X_{ss}}{(S_o - S_{ss})} = \frac{Y_{X/S}D}{\mu} = \frac{Y_{X/S}D}{(D + k_e)} \tag{4.41}$$

This result is equivalent to that obtained in the case of a substrate maintenance requirement, with $k_e = mY_{X/S}$. It is thus not possible to distinguish between the two types of maintenance requirements on the basis of the *observed* yield versus dilution rate curves.

As we saw in Chapter 3 (Section 3.4), it is often convenient to express substrate uptake rates in terms of specific uptake rates, $q_{nutrient}$. These were defined as

$$q_{Si} = \frac{r_{Si}}{X} = \frac{\mu}{Y_{X/Si}} + m_i + \frac{r_{P_i}}{XY_{P_i/S}} \tag{4.41}$$

where the effect of maintenance is included. Ignoring product formation for the moment, and noting that in the absence of endogenous metabolism (i.e., $k_e = 0$) the growth rate is equal to the dilution rate ($\mu = D$), a linear relationship between q_{Si} and the dilution rate exists.

$$q_{S_i} = \frac{D}{Y_{X/S_i}} + m_i \tag{4.42}$$

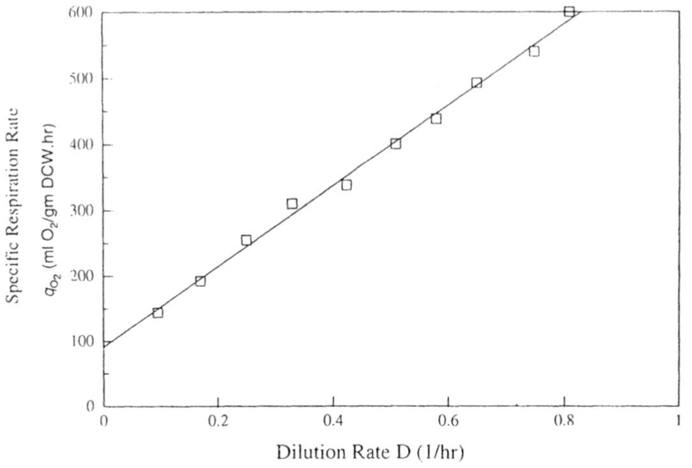

Figure 4.7. *The specific respiration rate of A. aerogenes (q_{O2}, mL O_2/gm DCW-hr) as a function of the dilution rate. The maintenance requirement is 95 mL/gm DCW-hr). [Data from D. Herbert, Continuous Culture of Microorganisms: A Symposium, p.49, Czech. Acad. Sciences (1958)].*

In the case of endogenous metabolism, the corresponding expression is given by Eq. (4.43), which again shows the same functional form as the expression for maintenance. Figure 4.7 shows the effect of maintenance/endogenous metabolism on the respiration rate of *Aerobacter aerogenes*. The non-zero y-axis intercept reflects this maintenance requirement.

$$q_{S_i} = \frac{D + k_e}{Y_{X/S_i}} = \frac{D}{Y_{X/S}} + \frac{k_e}{Y_{X/S}} \tag{4.43}$$

4.2.2 Product Formation

The simple unstructured models developed earlier which describe the kinetics of product formation can be incorporated into mass balances for the chemostat. The general growth and non-growth associated product formation model, applicable to a wide variety of microbial products, will be examined. The steady state mass balances for biomass, substrate and product are

$$\frac{dX}{dt} = \mu X - DX = 0 \tag{4.44a}$$

$$\frac{dS}{dt} = D(S_o - S) - \frac{1}{Y_{X/S}}\mu X - \frac{1}{Y_{P/S}}(\alpha\mu X + \beta X) = 0 \tag{4.44b}$$

$$\frac{dP}{dt} = \alpha\mu X + \beta X - DP = 0 \tag{4.44c}$$

with corresponding steady state solutions

$$S_{ss} = \frac{DK_S}{\mu_{max} - D} \tag{4.45a}$$

$$X_{ss} = \frac{Y_{X/S}Y_{P/S}(S_o - S_{ss})}{\left(Y_{P/S} + Y_{X/S}\left(\alpha + \frac{\beta}{D}\right)\right)} \tag{4.45b}$$

$$P_{ss} = \alpha X_{ss} + \frac{\beta X_{ss}}{D} \tag{4.45c}$$

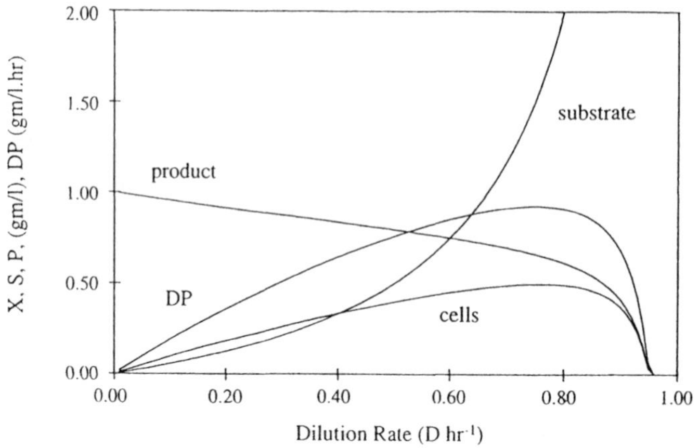

***Figure 4.8**. The steady state concentrations of a product that is growth- and non-growth associated. The values of the constants are the same as those in Figure 4.3, with $Y_{P/S} = 0.1$, $\alpha = 0.05$ and $\beta = 0.5$. Note that at low dilution rates the steady state cell concentration decreases as most substrate is converted to product rather than biomass. DP is the volumetric productivity of the product.*

Low dilution rates favor production of non-growth associated products, while growth associated products parallel the behavior of cell mass. The volumetric productivity is given by

$$DP = \alpha DX_{ss} + \beta X_{ss} \tag{4.46}$$

As before, the dilution rate corresponding to maximum productivity can be found by setting d(DP)/dD equal to zero. Figure 4.8 illustrates the steady state concentrations and productivity of a chemostat with a partially growth and non-growth associated product.

4.2.3 Substrate Inhibition and Multiple Steady States

As we saw in Chapter 3, compounds such as phenolics may serve as substrates but also may inhibit the growth of microorganisms. This situation often occurs in the biological treatment of toxic wastes, where xenobiotics (compounds not found in nature) must be removed from waters or soils. These compounds are toxic to cells at high concentrations but can be degraded at low concentrations. The model generally used to represent the kinetics of this inhibition follows that of noncompetitive inhibition in enzyme kinetics, viz:

$$\mu = \frac{\mu_{max} S}{K_S + S + \frac{S^2}{K_i}} \tag{4.47}$$

The steady state mass balances for cells and substrate are

$$\frac{dX}{dt} = -DX + \frac{\mu_{max} SX}{K_S + S + \frac{S^2}{K_i}} = 0 \tag{4.48a}$$

$$\frac{dS}{dt} = D(S_o - S) - \frac{1}{Y_{X/S}}\left(\frac{\mu_{max} SX}{K_S + S + \frac{S^2}{K_i}}\right) = 0 \tag{4.48b}$$

These can be solved for the two non-washout steady states given by X_{ss} and S_{ss}, and the steady state $X_{ss} = 0$, $S_{ss} = S_o$.

$$S_{ss} = \frac{K_i}{2D}\left((\mu_{max} - D) \pm \left((\mu_{max} - D)^2 - \frac{4D^2 K_S}{K_i}\right)^{\frac{1}{2}}\right) \tag{4.49a}$$

$$X_{ss} = Y_{X/S}(S_o - S_{ss}) \tag{4.49b}$$

A total of three steady state solutions are possible, with two non-trivial steady states occurring for all dilution rates except at D_{max}. This is the dilution rate which equals the maximum growth rate μ_{max}, occurring at a substrate concentration of $(K_S K_i)^{1/2}$. Equation (4.49) shows that both non-washout steady states have positive S_{ss} values, provided that the inhibition constant K_i is not so small that

$$(\mu_{max} - D) < 2\left(\frac{D^2 K_S}{K_i}\right)^{\frac{1}{2}} \tag{4.50}$$

If the above relationship were to hold, the extent of inhibition would be such that, over the range of dilution rates $0 < D < \mu_{max}$, the growth rate μ would vary with the reciprocal of the substrate concentration and only one steady state would occur at a growth rate of $\mu \sim (\mu_{max} K_i)/S$. This steady state is not stable to small perturbations of the system. We shall examine the stability of the steady states in the chemostat in Section 4.5.

4.2.4 Enzyme Catalysis in the CSTR

There are many configurations of CSTRs employing enzymes in single or occasionally multistep reactions. Generally, the high cost of even a crude enzyme preparation prevents its continuous addition to the feed of a CSTR. The enzyme thus must be retained in the reactor, by either preventing its removal by use of an ultrafiltration membrane at the exit of the reactor or by immobilizing the enzyme in a pellet which is held in the reactor or prevented from exiting the reactor by a screen. If the kinetics of the reaction being catalyzed are sufficiently slow, recirculation of the liquid phase through a packed bed enzyme reactor at a high rate will approximate the behavior of a CSTR system. Some examples of CSTRs for enzyme-catalyzed reactions are shown in Figure 4.9.

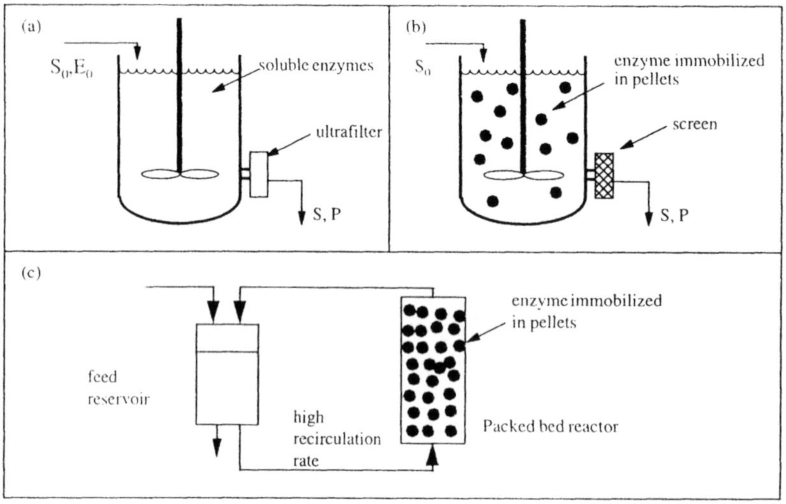

Figure 4.9. Some methods for retaining enzymes in a CSTR system; (a) retention by means of an ultrafilter, (b) immobilized enzyme is in a pellet form, prevented from leaving by a screen in the exit line, (c) rapid recirculation through a packed bed reactor yields kinetic behavior approximating that of a CSTR.

The analysis of enzyme-catalyzed reactions in a CSTR is somewhat simpler than that of microbial growth and product formation. Consider a reaction involving soluble enzyme following Michaelis-Menten kinetics, with an overall stoichiometry of one mole of product formed per mole of substrate reacted ($S \rightarrow P$). The steady state substrate and product balances are:

$$D(S_o - S) = r_S(S) = \frac{V_{max} S}{K_M + S} \tag{4.51}$$

$$S_o - S = P - P_o \tag{4.52}$$

These can be solved for the steady state substrate concentrations as a function of the dilution rate. Then the conversion can be determined.

$$S = \frac{1}{2}\left(S_o - K_M - \frac{V_{max}}{D}\right) - \frac{1}{2}\sqrt{\left(S_o - K_M - \frac{V_{max}}{D}\right)^2 + 4S_o K_M} \tag{4.53}$$

A simpler approach for reactor design is to initially specify a desired fractional conversion of substrate

$$\delta = \frac{S_o - S}{S_o} \qquad S = S_o(1-\delta) \tag{4.54}$$

With this value of S, equation (4.51) can be solved for the residence time (1/D) required for the corresponding conversion:

$$\frac{1}{D} = \frac{\delta(K_M + S_o(1-\delta))}{V_{max}(1-\delta)} \tag{4.55}$$

Other expressions for $r_S(S,P)$ can be inserted into equation (4.51) and the same approach employed to solve for the residence time required for a specified degree of substrate conversion. Table 4.1 indicates the solutions for some common enzyme kinetic forms.

***Table 4.1**. Kinetic forms and corresponding residence times for specified degrees of substrate conversion ($\delta = (S_o - S)/S_o$).*

Rate expression $r_S(S,P)$	Residence time (1/D) required for specified degree of conversion δ.
Michaelis-Menten $\dfrac{V_{max}S}{K_M + S}$	$\dfrac{\delta(K_M + S_o(1-\delta))}{V_{max}(1-\delta)}$
Competitive substrate inhibition $\dfrac{V_{max}S}{K_M + S + \frac{S^2}{K_i}}$	$\delta S_o\left(1 + \dfrac{K_M}{S_o(1-\delta)} + \dfrac{(1-\delta)S_o}{K_i}\right)$
Competitive product inhibition $\dfrac{V_{max}S}{K_M(1+PK_i)+S}$	$\dfrac{\delta}{(1-\delta)}\left(K_M + S_o(1-\delta) + K_M\dfrac{(P_o + S_o\delta)}{K_i}\right)$

4.2.5 Chemostats in Series

In batch cultures, the individual phases of growth result from the changing physiology of the cell; each phase is dependent on the one preceding it. On the other hand, in continuous culture, the physiological state is no longer determined by the one preceding it; rather it is controlled by the dilution rate. Thus the culture has no "history." In situations where the culture must pass through some physiological state in order to produce the desired product, a combination of reactor types will provide the optimal reactor configuration, defined as the smallest reactor volume for a specified degree of product conversion. In situations where the optimal rate of product formation may occur at a different temperature or pH from that of growth, two or more stirred tank reactors operated in series permit growth and product formation conditions to be optimized in each reactor. When mixed substrates are employed (such as two saccharides or a substrate such as lactose, which is hydrolyzed into two saccharides), the preferential substrate is consumed first (diauxic effect). The use of two CSTRs in series enables the preferred substrate to be completely consumed in the first reactor and the second substrate can then be consumed in the second reactor, minimizing the required reactor volume. We shall examine the steady state behavior of chemostats in series with multiple feed streams.

Let us examine a two stage system, with a second feed stream into the second stage. The configuration is shown in Figure 4.10.

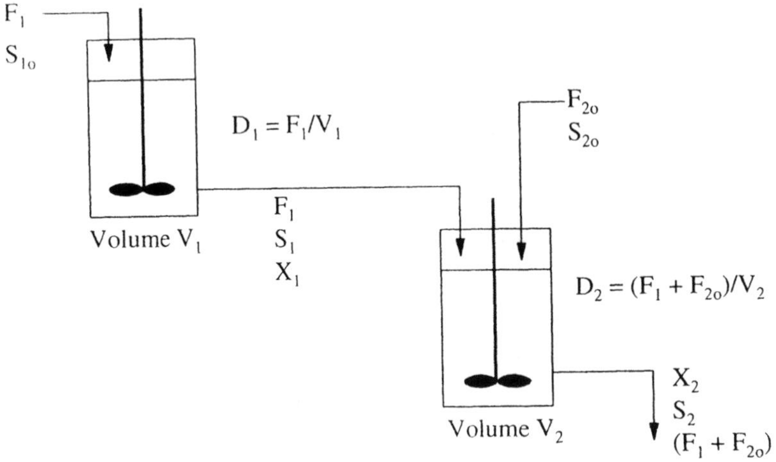

Figure 4.10. Two chemostats in series, with a separate feed stream entering the second reactor. The steady state cell and substrate concentrations are indicated.

We can write steady-state mass balances for cells and substrate for each stage, noting that the specific growth rates in each stage will be different due to the different substrate concentrations and environmental conditions (such as pH) in each vessel. For simplicity we will assume that the values of μ_{max} and K_S are the same in each stage.
For the first stage:

$$F_1 X_1 = \mu_1 X_1 V_1 \tag{4.56a}$$

$$F_1(S_{1o} - S_1) = \frac{1}{Y_{X/S}} \mu_1 X_1 V_1 \tag{4.56b}$$

For the second stage:

$$(F_1 + F_{2o})X_2 - F_1 X_1 = \mu_2 X_2 V_2 \tag{4.57a}$$

$$F_1 S_1 + F_{2o} S_{2o} - (F_1 + F_{2o})S_2 = \frac{1}{Y_{X/S}} \mu_2 X_2 V_2 \tag{4.57b}$$

Denoting $D_1 = F_1/V_1$ and $D_2 = (F_1 + F_{2o})/V_2$, equation (4.57a) can be rearranged to show that

$$\mu_2 = D_2 - D_1 \frac{X_1}{X_2} \cdot \frac{V_1}{V_2} \tag{4.58}$$

i.e., the growth rate in the second stage is always less than the overall dilution rate (D_2) in the second stage. Combining Eqs. (4.57a) and (4.57b) and solving for X_2 gives

$$X_2 = Y_{X/S}\left(\frac{F_1}{V_2 D_2} S_{1o} + \frac{F_{2o}}{V_2 D_2} S_{2o} - S_2\right) \tag{4.59}$$

If we employ the Monod equation to describe the specific growth rate in the second stage, we see

$$\mu_2 = \frac{\mu_{max} S_2}{K_S + S_2} \tag{4.60}$$

then the steady state substrate concentration S_2 is found by inserting the expression for X_2 into (4.57b) and solving. Only the positive root has significance.

$$(\mu_{max} - D_2)S_2^2 - \left\{\frac{\mu_{max} F_1 S_{1o}}{V_2 D_2} + \frac{\mu_{max} F_{2o} S_{2o}}{V_2 D_2} - \frac{F_1 S_1}{V_2} - \frac{F_{2o} S_{2o}}{V_2} + K_S D_2\right\} S_2$$
$$+\left(\frac{K_S F_1 S_1}{V_2} + \frac{F_{2o} S_{2o} K_S}{V_2}\right) = 0 \tag{4.61}$$

Figure 4.11 shows the effect of varying the dilution rate D_2 in the second vessel on the behavior of the steady state values of X_{2ss} and S_{2ss} (F_1, the volumetric feed rate to the first vessel, is constant).

If there is no separate flow into the second stage, the steady state solutions can be found from

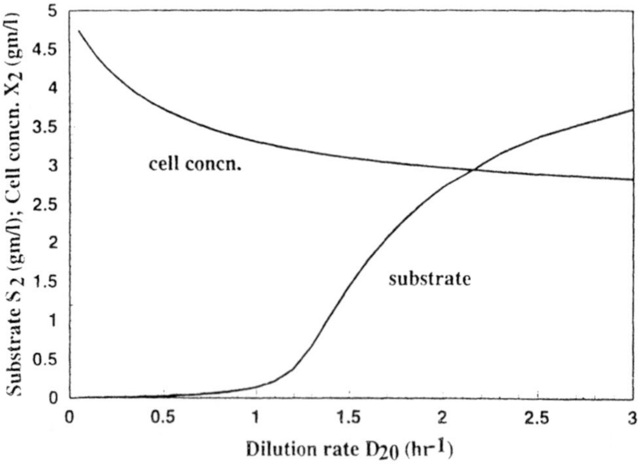

***Figure 4.11**. Steady state values of substrate and cell concentrations in the second stage of two chemostats in series. The dilution rate D_{2o} in the second stage is varied while the feed rate into the first stage is maintained constant at a dilution rate of $D_1 = 0.5\ hr^{-1}$. The other parameter values are $\mu_{max} = 1.0\ hr^{-1}$, $K_S = 0.05$ gm/l, $S_{1o} = 10.0$ gm/l, $S_{2o} = 5.0$ gm/l, $Y_{X/S} =$ 0.5 gm/gm. The steady-state values are obtained from equations (4.59) and (4.61).*

$$(\mu_{max} - D)S_{2,ss}^2 - \{\mu_{max}S_o + D_2(K_S - S_{1,ss})\}S_{2,ss} + K_S D_2 S_{1,ss} = 0 \qquad (4.62a)$$

$$X_{2,ss} = Y_{X/S}\{S_o - S_{2,ss}\} \qquad (4.62b)$$

$$\mu_2 = \frac{D\{X_{2,ss} - X_{1,ss}\}}{X_{2,ss}} \qquad (4.62c)$$

Generally $X_{1,ss}$ and $X_{2,ss}$ are not significantly different and the specific growth rate in the second stage will be approximately zero. If product inhibition is present, however, the specific growth rates may vary significantly (see Problem 1 for an example).

4.2.6 Graphical Design Procedures

When no constitutive model describing the dependence of the growth rate on substrate or product concentrations is available, a graphical procedure based on batch data for cell, substrate or product concentrations can be employed to predict the steady-state behavior of one or more chemostats in series. Consider batch culture data for cell concentration as a function of time. By graphically differentiating the data, dX/dt may be obtained as a function of time. We see from the batch mass balance

$$\frac{dX}{dt} = \mu X \qquad (4.5)$$

that a cross-plot of dX/dt versus X will have a slope of μ. In a chemostat operating at a dilution rate D, μ will equal D at steady state. Thus a line with slope D drawn from the inlet cell concentration X_0 (which will generally be zero in a single stage chemostat) will intercept the dX/dt versus X curve at the point where $\mu = D$. The corresponding X value (X_1) will be the steady state cell concentration in the reactor.

If we extend this now to a multistage system, we can draw a line with slope D_2, the dilution rate in the second stage, from the point (X_1,0) and it will intercept the batch data at the second stage steady state cell concentration X_2. Figure 4.12 extends this procedure to a three-stage CSTR. This approach can be extended to determine the number of stages required to achieve a specific cell mass at any desired dilution rate in each stage.

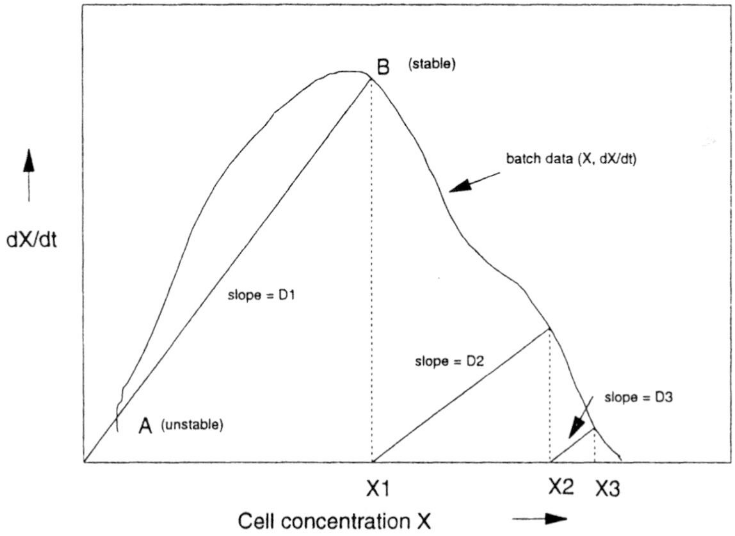

Figure 4.12. *Graphical determination of the steady states in a multistage chemostat. The batch data (X, dX/dt) are plotted. Two possible steady state values can be seen to exist for the first stage. If the slope of the growth curve (dX/dt) is steeper than a line with slope D_1, the steady state is unstable (point A on the figure). When the slope is less, the steady state is stable (point B). Steady state cell concentrations in subsequent stages are given by X_2 and X_3.*

4.3 Plug Flow and Packed Bed Bioreactors

The opposite extreme of the well-mixed reactor is the *plug flow* reactor, where fluid moves along a pipe or channel such that there is negligible mixing in the direction of flow (the axial direction), but the fluid is well-mixed in the radial direction. Such reactors are common in the chemical industry, but, for reasons we shall see shortly, are infrequently employed for cell growth. However, they are used frequently for immobilized cells and enzyme reactions. Glucose isomerization to fructose, a key step in the manufacture of high fructose corn syrup, is an important example of a commercial process carried out in a packed bed immobilized enzyme reactor. A schematic of a plug-flow and a packed bed reactor is shown below.

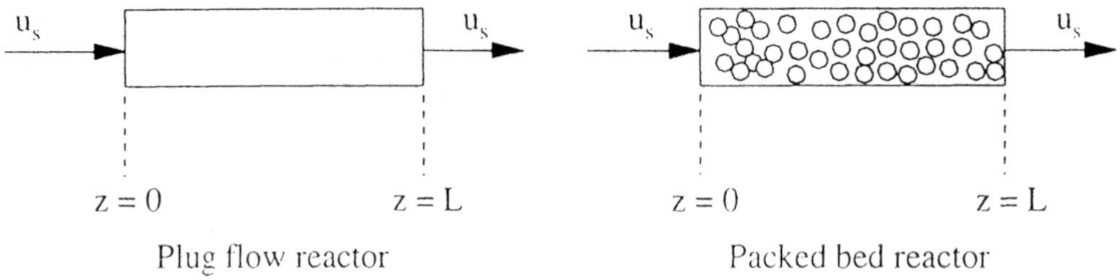

***Figure 4.13**. Schematic of a plug flow and a packed bed reactor.*

Consider first a plug flow reactor with no catalyst containing cells or enzyme. If the axial (z-direction) superficial velocity is u_s[3], we may write steady state mass balances for cells and substrate as follows:

$$\frac{d(u_s X)}{dz} = r_X \qquad \frac{d(u_s S)}{dz} = -r_S \tag{4.63}$$

where r_X and r_S are the reaction rates based on the liquid volumes. If the axial velocity is constant, as would be the case if the fluid density remained constant, these equations become

$$u_s \frac{dX}{dz} = r_X \qquad u_s \frac{dS}{dz} = -r_S \tag{4.64}$$

which can be rewritten in terms of a characteristic time constant θ (= z/u_s)

$$\frac{dX}{d\theta} = r_X \qquad \frac{dS}{d\theta} = -r_S \tag{4.65}$$

Therefore, the equations reduce to those we have seen before for a batch reactor, except that θ replaces time t. The initial and final conditions over which this equation is integrated are as follows:

$z = 0$	$\theta = 0$	$X = X_o$	$S = S_o$
$z = L$	$\theta = L/u$	$X = X_f$	$S = S_f$

(3) The superficial velocity is the average linear velocity the fluid would have if no packing were present in the reactor, and is equal to F/A where A is the reactor cross-sectional area and F is the volumetric flow rate into the reactor.

Thus the solutions to the above equations for various constitutive rate expressions r_X and r_S (e.g., the Monod equation) can be directly taken from the batch results (Section 4.1). The plug flow reactor offers a greater degree of substrate conversion than does the CSTR of equal volume (and thus equal residence time) because the reaction rate in a CSTR will be lower than in a plug flow reactor. In a CSTR, the substrate concentration immediately drops to the exit value and the reaction rate is determined by this value. With Monod-type kinetics, the rate decreases with decreasing substrate concentration; thus, the plug flow reactor offers the advantage of higher reaction rates than a CSTR because the substrate concentration decreases from its initial value along the axial direction.

The main difficulties in operating a plug flow reactor with suspended cells is the need to provide a continuous inoculum of cells (X_0 must be supplied at the reactor inlet) and the difficulties in maintaining constant pH and oxygen supply along the length of the reactor. These problems are generally so extreme that there are few examples of pure plug-flow biological reactors. Some reactor types may approach plug flow behavior as a limit; for example, an air-lift reactor with a strong upward liquid circulation or large wastewater treatment tanks. Analysis of plug-flow reactors, however, does provide an indication of one extreme of continuous operation.

In contrast to microbial growth, plug flow and packed bed reactors are commonly employed with free and immobilized enzymes. Here the constraints of oxygen supply and pH control are not usually applicable, as many enzyme reactions do not require O_2 or generate acid or base. As mentioned, one of the largest industrial applications of this type is the isomerization of glucose to fructose for production of high fructose corn sweetener. This reaction is catalyzed by glucose isomerase immobilized on a support, such as alumina, in packed bed reactors.

In a packed bed reactor, the axial fluid velocity will be larger than in an open plug flow reactor and depends on the void fraction ε, defined as

$$\varepsilon = \frac{\text{free reactor volume}}{\text{total reactor volume}} = 1 - \frac{\text{total particle volume}}{\text{total reactor volume}} \tag{4.66}$$

and the *interstitial* fluid velocity, v_i, is

$$v_i = \frac{F}{\varepsilon(\text{cross-sectional area})} = \frac{F}{\varepsilon V/L} \tag{4.67}$$

The interstitial velocity is then employed to calculate the liquid residence time τ. This has the value $\tau = \varepsilon(V/F)$. In the case of enzymes in solution, the void fraction ε will be unity. A typical value of ε for immobilized enzymes on spherical supports is ~ 0.4. In the case of free or immobilized enzymes, the substrate mass balance has the form

***Table 4.2.** Residence time τ as a function of substrate conversion $\{\delta = (S_o\text{-}S)/S_o\}$ for enzyme reactions in plug flow and packed bed reactors.*

Constitutive rate expression	*Residence time* $(\tau = \varepsilon L/u_s = L/u_i)$
Michaelis-Menten $\frac{V_{max}S}{K_S + S}$	$\frac{S_o\delta}{V_{max}} - \frac{K_M}{V_{max}}\ln(1-\delta)$
Substrate inhibition $\frac{V_{max}S}{K_M + S + \frac{S^2}{K_i}}$	$\frac{S_o\delta}{V_{max}} - \frac{K_M}{V_{max}}\ln(1-\delta) + \frac{S_o^2}{K_iV_{max}}\left(\delta - \frac{\delta^2}{2}\right)$
Competitive product inhibition $\frac{V_{max}S}{S + K_M\left(1 + \frac{P}{K_i}\right)}$	$\frac{S_o}{V_{max}}\left(1 - \frac{K_M}{K_i}\right)(\delta + \ln(1-\delta))$ $-\frac{\ln(1-\delta)}{V_{max}}\left\{K_M + S_o + \frac{K_M}{K_i}P_o\right\}$

$$\frac{dS}{d\tau} = -r_S \tag{4.68}$$

and analytical integration of this equation for several forms of the rate expression is possible. Table 4.2 gives examples of the residence time expression in terms of the parameters in the rate expression. Here it should be noted that V_{max} is the product of v_{max} (maximum reaction rate per catalyst volume [eqn. 2.6]) and the volume of catalyst per liquid volume, i.e., $(1\text{-}\varepsilon)/\varepsilon$. Thus, $V_{max} = v_{max}\,(1\text{-}\varepsilon)/\varepsilon$.

4.4 Imperfect Mixing

The models we have developed describing the behavior of a stirred tank bioreactor to date have assumed that the liquid phase in the reactor is "well-mixed", without carefully specifying what is meant by this expression. The equations we have employed describing the chemostat assume that the inlet substrate concentration instantly falls to the average value in the bioreactor, corresponding to that of effluent stream. In a large bioreactor this is clearly not the case; there will be a finite time required for the inlet substrate to be uniformly mixed throughout the vessel. A completely mixed reactor is one in which there is a uniform concentration of material throughout the vessel. Mixing is characterized by its *scale* and its *intensity*.

The scale of mixing is a length scale that depends on the process under consideration. Below this length scale, gradients in concentrations of materials can be found. Above this scale, the system is well-mixed. The intensity of mixing describes how far the fluid in the reactor deviates from complete mixing. Mixing is accomplished by a combination of mechanisms. At the molecular scale, diffusion results in a reduction of concentration gradients over a very short length scale. In laminar flows, shearing of adjacent fluid layers results in exchange of material and mixing occurs between layers. In turbulent flows, mixing is enhanced by the random motion of small fluid packets (eddies) that move rapidly throughout the liquid phase. In contrast to this, in a *segregated* system, such as the plug-flow reactors we considered in Section 4.3, the packets of fluid do not interact and move in an isolated manner throughout the reactor.

We are most often interested in the behavior of the bioreactor operated at a specified scale and intensity of mixing. Its behavior is conveniently described by the *mixing time*, which is the time required to attain the specified mixing intensity at the specified mixing scale. This micromixing depends on the power imparted to the liquid by the agitator and the geometry of the system. Methods for predicting the mixing time are examined in Chapter 5. On the other hand, macromixing describes the overall behavior of fluid elements in the vessel. The residence time distribution is used to characterize macromixing. Some fluid elements may circulate more times than others and result in a range of liquid residence times. Other elements may short-circuit the vessel and leave very quickly. This distribution of residence times may have a significant effect on the extent of any microbial or enzymatic reaction occurring in the vessel. In the case of microbial growth, a broad residence time distribution will mean that some cells may spend very long times in the vessel, while others are quickly washed out. The age distribution of the population thus depends on both the variation in doubling times and the mixing in the vessel.

We illustrate the concept of the residence time distribution by examining first a well-mixed stirred tank reactor. The steady-state mass balance on a non-reacting material fed to the vessel at a concentration C_0 is

$$\frac{dC}{dt} = \frac{F}{V}(C_o - C) \tag{4.69}$$

If at time t=0 a tracer is injected into the feed at concentration C^*, the boundary conditions become

$$C(0) = 0 \text{ at } t<0 \quad \text{and} \quad C_0(t) = C^* \text{ at } t\geq 0.$$

and Eq. (4.69) has a solution

$$\frac{C(t)}{C^*} = 1 - e^{-(F/V)t} \tag{4.70}$$

The residence time distribution function ($\xi(t)$) is defined as the fraction of liquid in the exit stream that has been in the vessel from a time t to t+dt, divided by the time interval dt. This gives

$$\xi(t) = \frac{d\left(\frac{C(t)}{C^{*}}\right)}{dt} = \frac{F}{V} e^{-(F/V)t} \tag{4.71}$$

Hence $\xi(t)$ is found by differentiating the experimental C(t) data. It is often useful to consider the reactor as equivalent to N well-mixed vessels in series, each containing 1/N of the total volume of the vessel. Each vessel has a volume of $V_i = V/N$. The mean residence time in each of the N vessels is $\bar{t}_i$, equal to V_i/F or V/NF. The residence time in the N vessel system is thus $N\bar{t}_i$, while that of the series of tanks is $\bar{t}$ (equal to V/F). The residence time distribution for the N vessel system can be determined by Laplace transforms of the material balance equation to give

$$\xi(t) = \frac{Nt^{N-1}}{(N-1)!}\left(\frac{NF}{V}\right)^{N-1} e^{-N(F/V)t} \tag{4.72}$$

Modeling the behavior of a single imperfectly-mixed vessel as a series of well-mixed stirred tanks in series provides a useful means to account for the effects of mixing on substrate conversion or biomass production. If we know how many tanks in series are required to model the mixing behavior, sets of substrate and cell mass balance equations can then be written for each tank and solved either numerically or analytically. It remains to determine N from the experimental data. This is accomplished by first obtaining C(t) data from either a step change or pulse input of tracer to the vessel[4]. We then determine the mean and variance of the C(t). These are found from discrete or continuous data as follows. The first moment or mean and the second moment of the exit concentration data C(t) are found from

$$\bar{t} = \frac{\sum_{j}^{n} t_j C_j(t)}{\sum_{j}^{n} C_j(t)} = \frac{\int_0^\infty tC(t)dt}{\int_0^\infty C(t)dt} \tag{4.73}$$

$$m_2 = \frac{\sum_{j}^{n} t_j^2 C_j(t)}{\sum_{j}^{n} C_j(t)} = \frac{\int_0^\infty t^2C(t)dt}{\int_0^\infty C(t)dt} - \tag{4.74}$$

(4) Modelling non-ideal flow behavior in CSTR and plug flow systems is described in detail in O. Levenspiel, *Chemical Reaction Engineering*, 2nd Edition, John Wiley, (1972).

The second central moment, or the variance is obtained from these quantities

$$\sigma^2 = m_2 - \bar{t}^2 \tag{4.75}$$

The variance is additive for flow through all N vessels, so that the overall variance of the system is N times the vaiance for each vessel. The number of equivalent tanks in series can thus be evaluated from the ratio of the square of the mean residence time and the variance.

$$N = \frac{(\bar{t})^2}{\sigma^2} \tag{4.76}$$

As the mixing behavior of the vessel tends toward plug flow, the number of tanks in series increases; in the limit of an infinite number of tanks, pure plug flow behavior is achieved. The residence time distribution function is illustrated in Figure 4.14 as a function of the number of tanks in series.

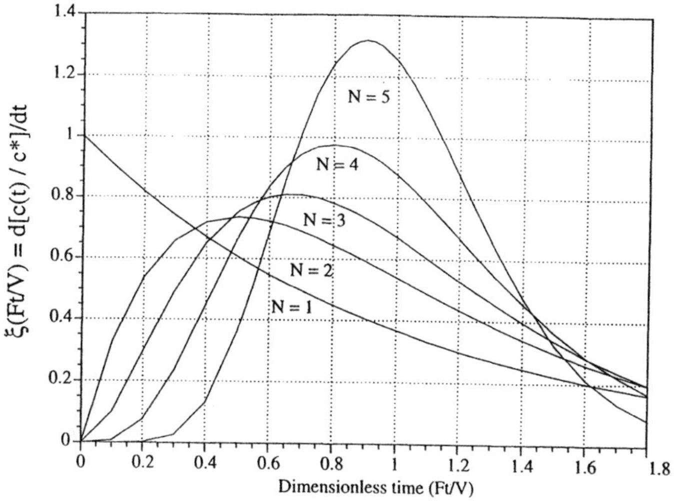

***Figure 4.14**. The normalized residence time distribution function ($\xi(t/\bar{t})$) for N well-mixed tanks in series as a function of normalized time $t/\bar{t}$ (tF/V). The function is plotted for 1, 2, 3, 4 and 5 tanks. As N increases, the distribution function approaches that of a plug flow reactor.*

This approach is very useful in interpreting data from bubble columns and air-lift reactors with internal draft tubes, which aid liquid circulation (these are discussed in Chapter 5). The liquid flow patterns may indeed resemble a number of tanks in series.

Often the residence time distribution data can be best interpreted by models of the bioreactor which combine tanks in series, plug flow reactors, stagnant regions and by-passing effects. The early appearance of fluid tracer, for example, in the C(t) response, would be indicative of by-passing in the vessel. Some of these flow patterns are illustrated in Figure 4.15.

For simple models such as shown in Figure 4.15, the residence time distribution can be determined. We can write the mass balances for a step change in inlet concentration of a tracer material and determine the output. or example, for the reactor liquid mixing pattern shown in Figure 4.15 (b), with a stagnant region, we can write

$$\frac{d\alpha VC}{dt} = FC^{*} - F_1C + F_1C_1 - FC$$

$$\frac{d(1-\alpha)VC_1}{dt} = F_1C - F_1C_1 \tag{4.77}$$

where F_1 is the flow rate of liquid to and from from the tank and the stagnant region, and C_1 is the concentration of tracer in this region. It is assumed that both regions are well-mixed, with the stagnant region comprising a fraction (1-α) of the total vessel volume. These equations can be solved with the initial conditions C(t) = 0, C_1(t) = 0 at t = 0, corresponding to the case of a step change in tracer input. The solution is shown in Figure 4.16.

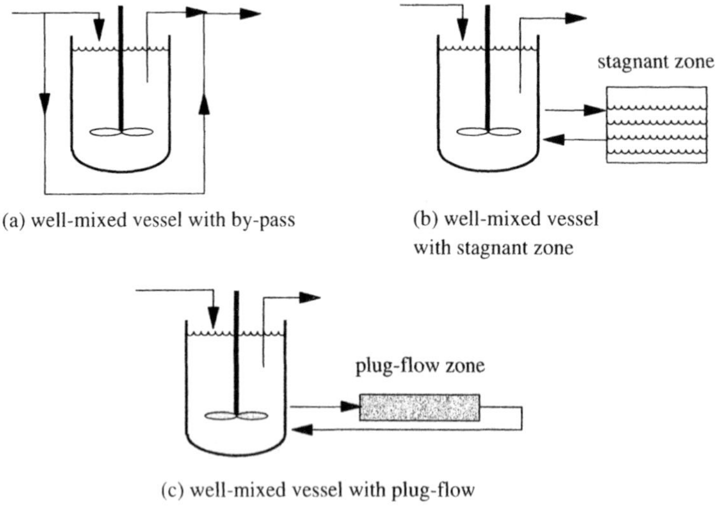

***Figure 4.15**. Models of the liquid flow patterns in bioreactors. (a) a stirred tank with by-passing, (b) a stirred tank with a stagnant fluid region, (c) a stirred tank with a plug flow reactor by-pass.*

4.4.1 Wall Growth

In small bioreactors and in laboratory equipment, microbial growth on the walls of the vessel is often encountered. Even if such growth is not visible, it may be important if the microbial concentration in the vessel is not high. In small bioreactors the surface to volume ratio is much larger than in industrial-scale equipment; thus wall growth is primarily a laboratory artifact. A significant fraction of the substrate may be consumed by microorganisms attached to the wall. At dilution rates approaching the maximum, microbial attachment to the wall prevents complete washout. The extent of wall growth may depend on the level of agitation in the vessel, and the microbial film that develops may reach a concentration that is constant, as excess cells are sloughed from the walls into the bulk liquid.

The difficulty in obtaining reliable kinetic data from a bioreactor with significant wall growth is illustrated in Problem 10.

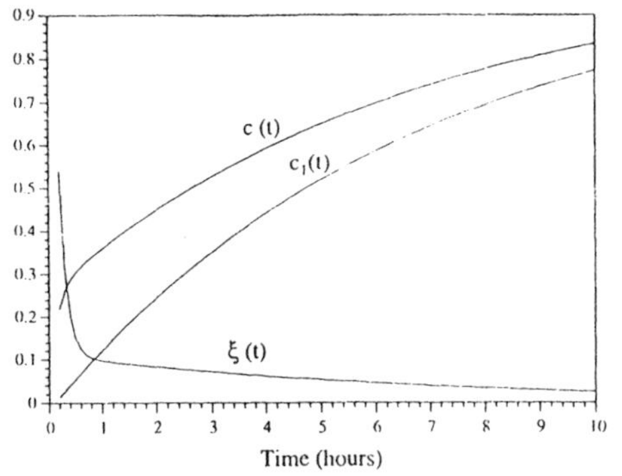

Figure 4.16. *The time course of tracer concentration in the liquid, C(t), and that in the stagnant region, C1(t) following a step input to the vessel. The values for the simulation are* $F/V = 0.2\ hr^{-1}$, $F_1/V = 0.5\ hr^{-1}$, $\alpha = 0.1$, $C^* = 1.0$. *The residence time distribution* $\xi(t)$ *is also shown.*

4.5 Fed Batch Reactors

In many industrial situations, such as the production of bakers yeast and antibiotics, reactors are operated in a semi-continuous fashion, with a substrate feed stream and no exit stream. Fed batch operation permits the substrate concentration to be maintained at some predetermined level. In bakers yeast production, for example, an excess of glucose results in the onset of fermentation and ethanol production, which reduces the cell yield. By slowly supplying glucose at a rate such that it can be completely consumed by the yeast, the residual glucose concentration is maintained at approximately zero and maximum conversion of substrate to cells is obtained. Similarly, in penicillin production, it is common to first obtain a high cell concentration in the reactor during the exponential growth phase, during which little penicillin is produced, and then supply carbon and nitrogen to the culture for antibiotic production at a rate which matches the biosynthetic and maintenance requirements of the organism (the maintenance requirements for penicillin production are quite high). During this substrate feeding, the reactor volume increases. At some point, part of the reactor volume is withdrawn (typically 10 to 25%) and the process is repeated. The broth withdrawn contains product at a high concentration, which has advantages in product recovery. This type of operation is an intermediate mode between batch and continuous operation, increasing the duration of a batch fermentation and the overall reactor productivity. We shall analyze the fed batch process with constant feed rates and the repeated fed batch operation.

Our starting point for the analysis is the set of mass balances for cells, substrate and product, with an additional total mass balance being required because the reactor volume changes with time. Under the assumption of constant liquid density, these equations are

$$\frac{dXV}{dt} = \mu X V \tag{4.78a}$$

$$\frac{dSV}{dt} = FS_f - \frac{1}{Y_{X/S}}\mu X V - \frac{1}{Y_{P/S}} q_P X V \tag{4.78b}$$

$$\frac{dPV}{dt} = q_P X V \tag{4.78c}$$

$$\frac{dV}{dt} = F \tag{4.78d}$$

We shall use the Monod expression for the specific growth rate μ and examine different expressions for the specific rate of product formation q_P, including growth and non-growth associated forms. If we neglect the formation of product for the moment, the left-hand side of the mass balances may be expanded and the equations rewritten as

$$\frac{dX}{dt} = \mu X - \left(\frac{F}{V}\right) X \tag{4.79a}$$

$$\frac{dS}{dt} = \left(\frac{F}{V}\right)(S_f - S) - \frac{1}{Y_{X/S}}\mu X \tag{4.79b}$$

$$\frac{dP}{dt} = q_P X - \frac{F}{V} P \tag{4.79c}$$

While these equations resemble those of the chemostat, there is an important difference. The term F/V, analogous to the dilution rate D, changes with time as V increases, whereas D is constant for steady state chemostat operation.

The cell and substrate mass balances can be combined and integrated to obtain

$$X(t) + Y_{X/S}S(t) = Y_{X/S}S_f - \{Y_{X/S}S_f - (Y_{X/S}S(0) + X(0)\}e^{-Ft/V} \tag{4.80}$$

If the feed concentration S_f satisfies

$$Y_{X/S}(S_f - S(0)) = X(0) \tag{4.81}$$

then

$$X(t) + Y_{X/S}S(t) = Y_{X/S}S_f \tag{4.82}$$

which will also be true at large times (i.e., t >> V/F) where the exponential term is small. Let us now examine a special case. If the feed rate F is constant, by analogy with the behavior of the chemostat, we might expect a quasi-steady state to exist where the cell concentration is constant in time i.e., dX/dt ~ 0. This requires that the specific growth rate μ, equal to F/V, should decrease as V is increasing in time. If μ is continuously decreasing, the substrate concentration must also decrease in time (hence dS/dt cannot also be zero). Substituting the Monod relationship for μ, we see

$$S = \frac{F}{V}\frac{K_S}{\mu_{max} - \frac{F}{V}} \tag{4.83}$$

$= \frac{DK_s}{Um - D}$ D changing

The total cell mass (XV) under these conditions is given by

$$\frac{dXV}{dt} = \mu XV = \frac{F}{V}XV = FX = constant \tag{4.84}$$

Therefore the total cell mass increases linearly with time. We shall now examine how the specific growth rate must vary under quasi-steady state conditions. Since F is constant, V increases linearly with time ($V = V_o + Ft$) and

$$\frac{d\mu}{dt} = \frac{d(F/V)}{dt} = \frac{d}{dt}\left(\frac{F}{V_o + Ft}\right) = \frac{-F^2}{(V_o + Ft)^2} \tag{4.85}$$

Initially, when $V \sim V_o$

$$\frac{d\mu}{dt} \sim -\frac{F^2}{V_o^2} \tag{4.86}$$

and at longer times the volume increases and $V >> V_o$ and hence

$$\frac{d\mu}{dt} \sim -\frac{1}{t^2} \tag{4.87}$$

The specific growth rate will thus rapidly decrease initially and then decrease at a much slower rate at longer times. This can be seen by numerically solving the substrate and cell mass balance equations. To simplify these, we shall introduce the following dimensionless variables:

$$\tilde{X} = \frac{X}{YS_f} \qquad \tilde{V} = \frac{V}{V_o} \qquad \tilde{S} = \frac{S}{S_f} \qquad \tilde{F} = \frac{F}{\mu_{max}V_o}$$

$$\tilde{D} = \frac{F}{\mu_{max} V} \qquad \tilde{K}_S = \frac{K_S}{S_o} \qquad \tilde{t} = t \cdot \mu_{max} \qquad \tilde{\mu} = \frac{\tilde{S}}{\tilde{K}_S + \tilde{S}}$$

The mass balances now become

$$\frac{d\tilde{X}}{d\tilde{t}} = -\tilde{D}\tilde{X} + \tilde{\mu}\tilde{X} \qquad \frac{d\tilde{S}}{d\tilde{t}} = \tilde{D}(1 - \tilde{S}) - \tilde{\mu}\tilde{X} \qquad \tilde{D} = \frac{\tilde{F}}{1 + \tilde{F}\tilde{t}} \tag{4.88}$$

As an example, consider the following initial conditions and parameter values

$$\tilde{X}(0) = 1 \qquad \tilde{S}(0) = 0.1 \qquad \tilde{F} = 2 \qquad \tilde{K}_S = 0.01 \tag{4.89}$$

we see from Figure (4.17) that indeed there is a quasi-steady state in which the cell concentration becomes constant and that after a start up period μ and D approach each other. During the quasi-steady state, the substrate concentration is constantly decreasing as the reactor volume increases.

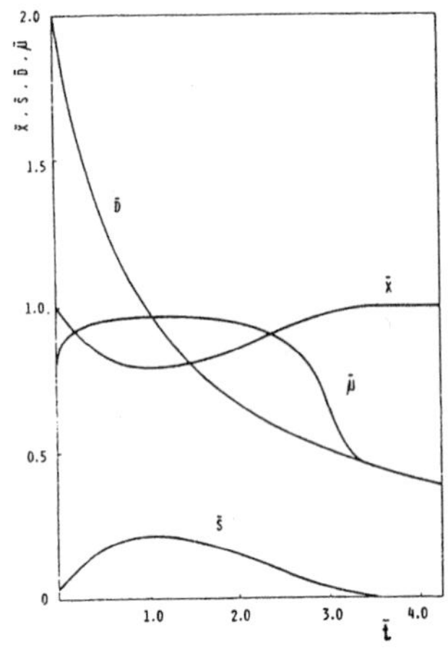

***Figure 4.17**. Approach to and the attainment of the quasi-steady state in a fed-batch culture. [From Dunn, I.J. and J.-R. Mor, Biotech. Bioeng. **17**, 1805 (1975)].*

If we now consider the case where F is no longer constant but varies with time, we may select F such that F/V is constant, denoted by λ. This implies

$$\frac{dV}{dt} = F = \lambda V \qquad \text{hence} \qquad V = V_o e^{\lambda t} \qquad F = \lambda V_o e^{\lambda t} \tag{4.90}$$

Hence, with exponential feeding the system will achieve steady state concentrations corresponding to those of the chemostat. Both X and S can be maintained at constant values.

4.6 The Transient Behavior of Bioreactors

4.6.1 Stability Analysis

To date we have examined the steady state solutions for various constitutive models describing the growth of the cells in the reactor. Here we shall examine whether such steady states can be attained in practice, by examining whether these steady states are stable to perturbations in the system parameters. The method used is a linearized stability analysis, based on the method of Lyapunov[5]. The stability of the non-linear system is examined by considering the behavior of the eigenvalues of the linearized system around the steady state point. The general approach is as follows.

Consider the set of unsteady state mass balances around a stirred tank reactor. These may be written in vector notation as

$$\frac{d\underline{C}}{dt} = \underline{f}(\underline{C}, \underline{p}) \tag{4.91}$$

where $\underline{C}$ is the vector of concentrations and $\underline{p}$ is a vector of parameters, including the kinetic parameters, feed concentrations, etc. The steady state solutions may be found from

$$\underline{f}(\underline{C}_{ss}, \underline{p}) = 0 \tag{4.92}$$

If the vector $\underline{x}$ is introduced as representing the deviations from the steady state:

$$\underline{x}(t) = \underline{C}(t) - \underline{C}_{ss} \tag{4.93}$$

the mass balances may be rewritten in terms of deviations from the steady state

$$\frac{d\underline{x}}{dt} = \underline{f}(\underline{C}_{ss} + \underline{x}, \underline{p}) \tag{4.94}$$

We may now expand $f(\underline{C}_{ss}+\underline{x}, \underline{p})$ in a Taylor series around $\underline{C}_{ss}$. Noting that $f(\underline{C}_{ss}, \underline{p})$ is zero at steady state, we obtain

$$\frac{d\underline{x}}{dt} = \underline{f}(\underline{C}_{ss}, \underline{p}) + \underline{\underline{A}}\,\underline{x} + (\text{ higher order terms}) \cong \underline{\underline{A}}\,\underline{x} \tag{4.95}$$

where the elements of the matrix $\underline{\underline{A}}$ are evaluated at each steady state viz.

$$a_{i,j} = \frac{\partial f_i(\underline{C}_{ss}, \underline{p})}{\partial C_j} \tag{4.96}$$

The solutions of the above equation have the form

$$\underline{x}(t) = \sum_{i=1}^{n} \alpha_i \underline{\beta}_i e^{\lambda_i t} \tag{4.97}$$

(5) See O. Bilous and N. Amundson, AIChE Journal **1**, 513 (1955) and R. Aris, *Elementary Chemical Reactor Analysis*, Prentice Hall, N.J. (1969) for examples of stability analyses applied to chemical reactors.

where n is the number of species for which mass balances have been written, and $\underline{\beta}_i$ and λ_i are corresponding pairs of eigenvectors and eigenvalues of the matrix $\underline{\underline{A}}$. The eigenvalues are found from the n roots of

$$det(\underline{\underline{A}} - \lambda \underline{\underline{I}}) = 0 \tag{4.98}$$

where $\underline{\underline{I}}$ is the n x n identity matrix. The eigenvectors satisfy

$$(\underline{\underline{A}} - \lambda_i \underline{\underline{I}})\underline{\beta}_i = 0 \qquad i = 1, \ldots, n \tag{4.99}$$

and the values of α_i are chosen to satisfy the initial conditions $\underline{x}(0)$

$$\sum_{i=1}^{n} \alpha_i \underline{\beta}_i = \underline{x}(0) \tag{4.100}$$

The eigenvalues can be seen to have, in general, both real and imaginary parts of the form $\lambda_i = a_i \pm b_i i$. If all the eigenvalues have negative real parts and no imaginary parts ($b_i = 0$), the steady state at which the eigenvalues are calculated is locally stable. This is seen from equation (4.97), where the vector $\underline{x}(t)$ decays to zero at sufficiently long times if the λ_i values are all negative. Conversely, if even one of the eigenvalues is positive, then the steady state is unstable, as the deviation from the steady state value (represented by $\underline{x}(t)$) will grow in time. If λ_i has only imaginary parts and zero real parts, the stability is determined by the higher order terms that we have neglected in this analysis.

The behavior of the system around a steady state is given below. It should be noted that the system of equations has been linearized and that a locally unsteady state solution may not be globally unstable. Our expansion is only valid in the neighborhood of the steady state.

Eigenvalues			
Real Part	Imaginary Part	Behavior	Type
Negative	$= 0$	exponential decline to zero	stable node
Positive	$= 0$	exponential growth	unstable node
Negative	$<> 0$	damped oscillations	stable focus
Positive	$<> 0$	undamped oscillations	unstable focus
Zero	$<> 0$	sustained oscillations	vortex point

The behavior of one of the components of $\underline{x}$ for each of these five types of steady state is shown in Figure 4.18.

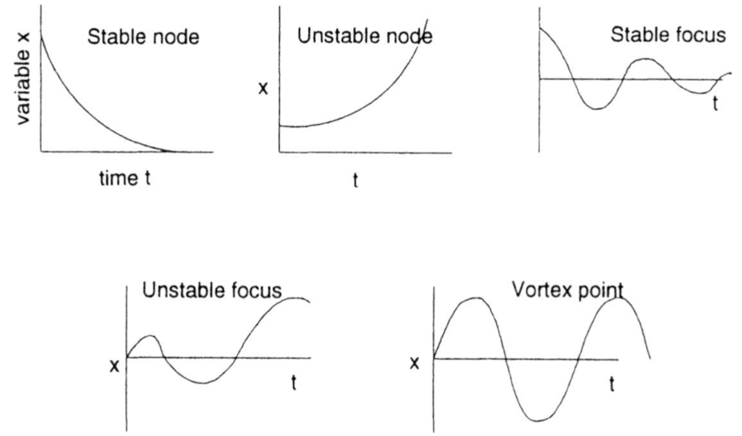

***Figure 4.18**. The time course of deviations from the steady state values, illustrating the possible types of behavior.*

We need not compute all the eigenvalues from Equation (4.98) to determine if the real parts are positive or negative. If the nth order equation for λ is written in the following form

$$\lambda^n + B_1\lambda^{n-1} + \ldots + B_{n-1}\lambda + B_n = 0 \tag{4.101}$$

the Hurwitz criterion states that all the roots of the above equation for λ will have negative real parts, if and only if the following conditions are met:

$$B_1 > 0$$

$$det\begin{pmatrix} B_1 & B_3 \\ 1 & B_2 \end{pmatrix} > 0$$

$$det\begin{pmatrix} B_1 & B_3 & B_5 \\ 1 & B_2 & B_4 \\ 0 & B_1 & B_3 \end{pmatrix} > 0 \tag{4.102}$$

This test is considerably easier to apply than calculation of the eigenvalues for systems with three or more components. We shall examine the simplest case of the chemostat to illustrate the technique.

Stability of the Chemostat

The governing mass balances for cells and substrate are given by

$$\frac{dX}{dt} = -DX + \mu(S)X \qquad \frac{dS}{dt} = D(S_o - S) - \frac{\mu(S)}{Y_{X/S}}X \tag{4.103}$$

The signs of the eigenvalues λ_1 and λ_2 can be found as follows. The matrix $(\underline{\underline{A}} - \lambda\underline{\underline{I}})$ has elements

$$\begin{bmatrix} a_{11} - \lambda & a_{12} \\ a_{12} & a_{22} - \lambda \end{bmatrix}$$

Solving $det(\underline{\underline{A}} - \lambda\underline{\underline{I}}) = 0$ we obtain

$$\lambda^2 - (a_{11} + a_{22})\lambda + (a_{11}a_{22} - a_{12}a_{21}) = 0 \tag{4.104}$$

For both eigenvalues to have negative real parts

$$(a_{11} + a_{22}) < 0 \qquad \text{and} \qquad (a_{11}a_{22} - a_{12}a_{21}) > 0 \tag{4.105}$$

From the mass balances, the matrix $\underline{\underline{A}}$ has components

$$a_{11} = \frac{\partial}{\partial X}\left(\frac{dX}{dt}\right)_{ss} = \mu(S_{ss}) - D \qquad a_{12} = \frac{\partial}{\partial S}\left(\frac{dX}{dt}\right)_{ss} = X_{ss}\left(\frac{d\mu}{dS}\right)_{ss}$$

$$a_{21} = \frac{\partial}{\partial X}\left(\frac{dS}{dt}\right)_{ss} = -\frac{\mu(S_{ss})}{Y_{X/S}} \qquad a_{22} = \frac{\partial}{\partial S}\left(\frac{dS}{dt}\right)_{ss} = -\left[\frac{X_{ss}}{Y_{X/S}}\left(\frac{d\mu}{dS}\right)_{ss} + D\right] \tag{4.106}$$

which satisfies the criterion given by Equation (4.105). We can now introduce the various constitutive models for the specific growth rate. For the Monod model

$$\left(\frac{d\mu}{dS}\right)_{ss} = \frac{\mu_{max}K_S}{(K_S + S_{ss})^2} > 0 \qquad \text{for all} \quad S_{ss} \tag{4.107}$$

The steady states are:

Dilution rate	$S_o > \dfrac{DK_S}{\mu_{max} - D}$	$S_o < \dfrac{DK_S}{\mu_{max} - D}$
$D < \mu_{max}$	$S_{ss} = \dfrac{DK_S}{\mu_{max} - D}$ $X_{ss} = Y_{X/S}(S_o - S_{ss})$	$S_{ss} = S_o$ $X_{ss} = 0$
$D > \mu_{max}$	$S_{ss} = S_o$ $X_{ss} = 0$	$S_{ss} = S_o$ $X_{ss} = 0$

Generally, $S_o > \frac{DK_S}{\mu_{max} - D}$ and the matrix $\underline{\underline{A}}$ has elements

for $D < \mu_{max}$ for $D > \mu_{max}$

$$\begin{bmatrix} 0 & X_{ss}\left(\frac{d\mu}{dS}\right)_{ss} \\ -\frac{D}{Y_{X/S}} & -\left(\frac{X_{ss}}{Y_{X/S}}\left(\frac{d\mu}{dS}\right)_{ss}+D\right) \end{bmatrix} \qquad \begin{bmatrix} -D+\frac{\mu_{max}S_o}{(K_S+S_o)} & 0 \\ -\frac{\mu_{max}S_o}{(K_S+S_o)} & -D \end{bmatrix}$$

We see that in both cases the criteria for negative real parts of the eigenvalues are satisfied. We also note that both the eigenvalues only contain real parts; there are no complex parts. When complex parts exist and the real parts are negative, the solution for $\underline{x}(t)$ (Equation (4.97)) can be rewritten in terms of pre-exponential sine and cosine terms, indicating the presence of damped oscillations around the steady state. In the cases where $S_o < DK_S/(\mu_{max} - D)$ the real parts of the eigenvalues are positive and both steady states are unstable.

The eigenvalues yield further information about the *characteristic times* of the chemostat. For the non-washout steady state, we have

$$\lambda_1 = -D \qquad \lambda_2 = -\frac{(\mu_{max}-D)\,[S_o(\mu_{max}-D)-DK_S]}{\mu_{max}K_S} \tag{4.108}$$

The characteristic times are given by

$$t_i = |\lambda_i|^{-1} \qquad i = 1, \ldots n \tag{4.109}$$

Thus the response time of the chemostat to perturbations will depend on both t_1 (D^{-1}) and t_2, which is a function of D and the set of microbial (K_S and μ_{max}) and environmental (S_o) parameters. We shall examine this in Section (4.6.2).

Stability of the Chemostat with Substrate Inhibition

The case of an inhibitory substrate gives rise to the possibility of two steady states existing at the same dilution rate (Section 4.2.3). Here we shall examine the stability of these steady states. The chemostat mass balances may be written as above, noting that now

$$\mu = \frac{\mu_{max}S}{K_S + S + \frac{S^2}{K_i}} \tag{4.110}$$

In this case we see that $(d\mu/dS)_{ss}$ may assume positive or negative values, if $S_{ss}^2 < K_SK_i$ or if $S_{ss}^2 > K_SK_i$, respectively.

$$\frac{d\mu}{dS} = \frac{\mu_{max}(K_SK_i - S_{ss}^2)}{\left(K_S + S_{ss} + \frac{S_{ss}^2}{K_i}\right)^2} \tag{4.111}$$

Let us now examine the elements of the matrix $\underline{\underline{A}}$, considering only the two non-washout steady states, given by the two roots of the equation below for S_{ss}. We note that for these two steady state values, $\mu = D$.

$$S_{ss} = \frac{K_i}{2D}\left((\mu_{max} - D) \pm \sqrt{\left((\mu_{max} - D)^2 - \frac{4D^2K_S}{K_i}\right)}\right)$$

$$X_{ss} = Y_{X/S}(S_o - S_{ss}) \tag{4.112}$$

$$\underline{\underline{A}} = \begin{bmatrix} 0 & X_{ss}\left(\frac{d\mu}{dS}\right)_{ss} \\ -\frac{D}{Y_{X/S}} & -\left(\frac{X_{ss}}{Y_{X/S}}\left(\frac{d\mu}{dS}\right)_{ss} + D\right) \end{bmatrix} \tag{4.113}$$

To meet the requirement $(a_{11} + a_{22}) < 0$, only the steady state where $S_{ss}^2 < K_S K_i$ will have $(d\mu/dS)_{ss} > 0$ and thus satisfy this requirement. If this condition is met, the second stability condition $(a_{11}a_{22} - a_{12}a_{21}) > 0$ will also hold. Thus both eigenvalues will have negative real parts and no imaginary parts. The steady state will be stable. The second steady state has $(d\mu/dS)_{ss} < 0$ and has positive real eigenvalues with complex parts. It is thus an unstable node. If we attempt to operate the chemostat at this steady state, any perturbations, such as an increase in substrate concentration, will result in the specific growth decreasing and cells will wash out of the system. In the following section (Section 4.6.2) we will examine the dynamic behavior of the chemostat.

Operating Diagrams

In operating the chemostat, the only parameters that can be manipulated by the investigator are the inlet substrate concentration and the dilution rate. It is useful therefore to construct diagrams in (S_o,D) space which show the boundaries between the different steady states. For a given (S_o,D) we can see which steady state the system will reach. These diagrams are called *operating diagrams*. The boundaries between regions corresponding to each steady state can be found by using the stability criteria we have developed. When the real part of one of the eigenvalues is positive, the system is unstable. If, by changing S_o or D, the real parts of the eigenvalues can be made negative, the system will be stable. Thus the stable/unstable boundary must occur at (S_o,D) values which make the real parts of the eigenvalues zero. For a limited number of models of growth kinetics, we can determine these boundaries analytically. In more complex cases we must resort to numerical methods.

We shall consider substrate inhibition kinetics to illustrate the method. Consider the Jacobian matrix $\underline{\underline{A}}$ (Equation 4.96) resulting from the linearization. The eigenvalues are found from

$$\lambda^2 - (a_{11} + a_{22})\lambda + (a_{11}a_{22} - a_{12}a_{21}) = 0 \tag{4.114}$$

The eigenvalues will have no real parts if $(a_{11} + a_{22}) = 0$ and $(a_{11}a_{22} - a_{12}a_{21}) < 0$. Inserting the appropriate values of a_{ij} from Equation (4.113) for substrate inhibition, we have

$$\frac{X_{ss}}{Y_{X/S}}\left(\frac{d\mu}{dS}\right)_{ss} + D = 0 \quad \text{and} \quad \frac{DX_{ss}}{Y_{X/S}}\left(\frac{d\mu}{dS}\right)_{ss} < 0 \tag{4.115}$$

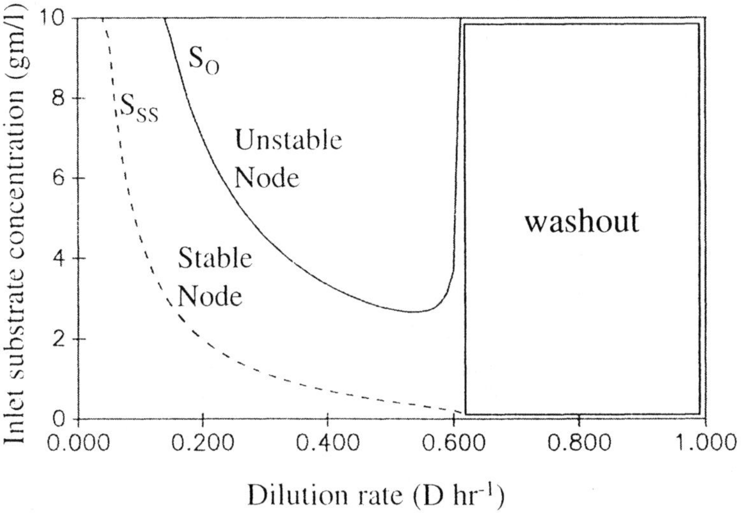

***Figure 4.19**. The operating region for substrate inhibited growth. Parameter values are μ_{max} = 1.0 hr^{-1}, K_S = 0.05 gm/l, K_i = 0.5 gm/l. At dilution rates greater than 0.6125 hr^{-1} the culture washes out, as D > μ. The unstable node region is indicated. The dashed line represents the unstable steady state values of the substrate concentration (S_{ss}), showing it is less than the inlet substrate concentration (S_o).*

Inserting the general expression for $(d\mu/dS)_{ss}$ we see that the first condition reduces to (for the non-washout case)

$$\frac{X_{ss}}{Y_{X/S}}\frac{\mu_{max}(K_S K_i - S_{ss}^2)}{\left(K_S + S_{ss} + \frac{S_{ss}^2}{K_i}\right)^2} + D = 0 \tag{4.116}$$

Substituting for X_{ss} from Equation (4.112) and noting

$$D = \frac{\mu_{max} S_{ss}}{K_S + S_{ss} + \frac{S_{ss}^2}{K_i}} \tag{4.117}$$

$$(S_o - S_{ss})(K_S K_i - S_{ss}^2) + \left(K_S + S_{ss} + \frac{S_{ss}^2}{K_i}\right) S_{ss} = 0 \tag{4.118}$$

rearranging

$$S_o = S_{ss} + \frac{\left(K_S + S_{ss} + \frac{S_{ss}^2}{K_i}\right) S_{ss}}{(S_{ss}^2 - K_S K_i)} \tag{4.119}$$

The relationship between S_o and D can be found by substituting for S_{ss} as a function of D and the microbial kinetic parameters (μ_{max}, K_S and K_i) from Equation (4.117). Values of S_o which are greater than the steady state values (S_{ss}) can only be found when the unsteady state solution is substituted into Equation (4.119). In this case, $S_{ss}^2 > K_S K_i$. This defines the extent of the unstable steady state region, indicated by the solid line on Figure 4.19. The operating diagram is shown in Figure 4.19.

4.6.2 Transient Responses of the Chemostat

In this section we shall examine the time course of the solutions to the mass balance equations describing the chemostat behavior. Although the chemostat with Monod kinetics has a stable, non-trivial steady state, if perturbations in substrate or cell concentrations are made, the system may exhibit underswings or overshoots during its return to the original steady state. This situation could arise from a transient change in the inlet substrate concentration or the feed rate (D), or a sudden increase in cell concentration arising from cells growing around the walls of the vessel suddenly falling into the reactor.

Transient patterns are most easily seen by removing time as an explicit variable in the mass balance equations and examining the dynamic behavior in (X,S) space. This approach is called *phase-plane* analysis. On an (X,S) diagram, the trajectory from an initial point (X_o,S_o) to the steady state value describes the system dynamics, e.g., S may increase and then decrease as X moves from X_o to X_{ss}. This then corresponds to an overshoot in S, while X responds monotonically. The transient patterns are found by dividing the cell mass balances by the substrate balance equation, thereby eliminating time.

$$\frac{dX}{dS} = \frac{\left(\frac{\mu_{max} S}{K_S + S} - D\right)X}{D(S_o - S) - \frac{X}{Y_{X/S}}\left(\frac{\mu_{max} S}{K_S + S}\right)} \tag{4.120}$$

The sign of dX/dS determines the dynamic response. This is found by subdividing the (X,S) plane into *sepatrices*, formulated viz.

Line A

$$S = \frac{K_S D}{\mu_{max} - D} \tag{4.121a}$$

Line B

$$S = S_o - \frac{X}{Y_{X/S}} \tag{4.121b}$$

Line C

$$D(S_o - S) = \frac{X}{Y_{X/S}} \mu_{max}\left(\frac{S}{K_S + S}\right) \tag{4.121c}$$

Six domains result (indicated as I to VI in Figure 4.20). The steady state occurs at the intersection of lines A, B and C. Consider the point (X_o,S_o) in Figure 4.20. Both X and S values are higher than the steady state values for this set of (D,S_o) conditions. The dynamic response is indicated by the dashed line. Substrate decreases below the steady state value and then increases, i.e., it undershoots, while the cell concentration overshoots and then returns to the value X_{ss}. The characteristic behavior of the system for initial sets of (X,S) in each of the six domains is given in Table 4.3.

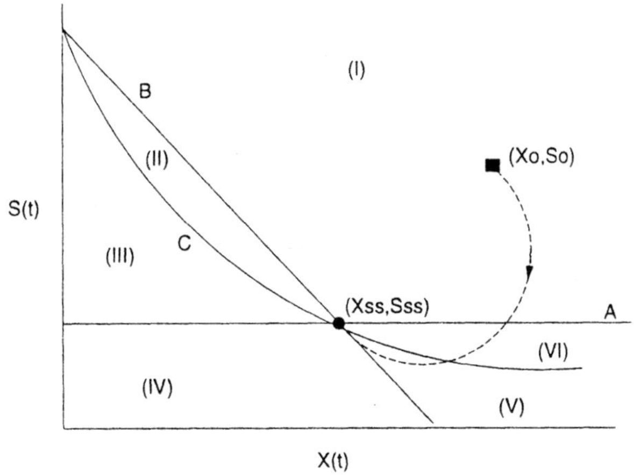

***Figure 4.20**. The phase plane behavior of a chemostat with the Monod model. The lines A, B and C delineate the six domains (I through VI). The transient behavior of point (X_0,S_0) is indicated.*

***Table 4.3**. Classification of the transient response of (X,S) in each of the six domains in the (X,S) phase plane.*

(X_0,S_0) in domain	Behavior of X(t)	Behavior of S(t)
I	overshoot	underswing
II	monotonic increase	monotonic decrease
III	monotonic increase	overshoot
IV	underswing	overshoot
V	monotonic decrease	monotonic increase
VI	monotonic decrease	underswing

If we modify the Monod model to include the effects of maintenance on substrate concentration, then the dynamic response of the system may show damped oscillations for certain values of the maintenance coefficient, m, instead of the underswings or overshoots above. Maintenance is likely to be most important when the dilution rate is low. The effect of maintenance is explored further in Problem 19.

4.6.3 Control of the CSTR

When a chemostat is operated in situations where fluctuations in incoming substrate concentration or flow rate are expected, as is the case in operation of sewage treatment plants, a means of preventing large fluctuations of residual substrates or products in the effluent stream is highly desirable. Control of large scale fermentations for cell mass and primary metabolite production require effective control strategies. We shall explore several of these schemes, but first need to introduce the concepts of *controllability* and *stabilizability*[6]. With chemostat operation, the cell and substrate mass balance equations

$$\frac{dX}{dt} = (\mu(S) - D)X \qquad \frac{dS}{dt} = D(S_o - S) - \frac{\mu(S)X}{Y_{X/S}} \tag{4.122}$$

can be compactly written in vector form

$$\frac{d\underline{x}}{dt} = f(\underline{x}, D) \tag{4.123}$$

where

$$\underline{x} = \begin{bmatrix} X \\ S \end{bmatrix} \qquad f(\underline{x}, D) = \begin{bmatrix} (\mu - D)X \\ D(S_o - S) - \dfrac{\mu X}{Y_{X/S}} \end{bmatrix}$$

The concept of a controllable system is one where a control policy $\underline{U}(t)$ can be found such that it results in the system moving from state $\underline{x}_o$ to $\underline{x}_f$ (the final desired state) in a finite time. If we linearize the chemostat model such that

$$\frac{d\underline{x}}{dt} = \underline{\underline{A}}\underline{x} + \underline{\underline{B}}\underline{U} \qquad \underline{Y} = \underline{\underline{C}}\underline{x} \tag{4.124}$$

where $\underline{x}$ is the state vector of order n, $\underline{U}$ is the control vector of order m, $\underline{Y}$ is an output vector of order l, and $\underline{\underline{A}}$ (the Jacobian), $\underline{\underline{B}}$ and $\underline{\underline{C}}$ are matrices of dimension (n,n), (m,n) and (l,n). The system is said to be controllable if and only if the rank of the (n,nm) controllability matrix $\underline{\underline{L}}$ is n, where $\underline{\underline{L}}$ is

$$\underline{\underline{L}} = [\underline{\underline{B}} \mid \underline{\underline{A}}\underline{\underline{B}} \mid \underline{\underline{A}}^2\underline{\underline{B}} \mid \cdots \mid \underline{\underline{A}}^{n-1}\underline{\underline{B}}] \tag{4.125}$$

Rigorous controllability can only be derived for a linear system, but we can examine the local controllability by linearizing the mass balances for the chemostat around the non-trivial steady state. We consider the dilution rate D as the control variable.

$$\underline{\underline{A}} = \frac{\partial f}{\partial \underline{x}}\Big|_{SS} = \begin{bmatrix} 0 & \dfrac{d\mu}{dS} \\ -\dfrac{\mu}{Y} & -\mu - \dfrac{X}{Y}\dfrac{d\mu}{dS} \end{bmatrix}_{SS} \tag{4.126}$$

---

(6) Further details and references may be found in Agrawal, P. and H. Lim, "Analysis of Various Control Schemes for Continuous Bioreactors" in Adv. Biochem. Eng. **30**, 61 (1984).

$$\underline{\underline{B}} = \frac{\partial f}{\partial D}\Big|_{SS} = \begin{bmatrix} -X \\ S_o - S \end{bmatrix}_{SS} = \begin{bmatrix} -X \\ \frac{X}{Y} \end{bmatrix}_{SS} \tag{4.127}$$

the local controllability matrix is

$$\underline{\underline{L}} = [\underline{\underline{B}} \mid \underline{\underline{AB}}]_{SS} = \begin{bmatrix} -X & \frac{X^2}{Y}\frac{d\mu}{dS} \\ \frac{X}{Y} & \frac{-X^2}{Y^2}\frac{d\mu}{dS} \end{bmatrix} \tag{4.128}$$

For local controllability the rank of $\underline{\underline{L}}$ must be two, or $\det(\underline{\underline{L}}) \neq 0$. We see that $\det(\underline{\underline{L}}) = 0$ and thus the system is not locally controllable. A weaker criterion is the stabilizability of the system. If the unstable steady states of the system can be made stable by the action of a controller, the system is stabilizable. If a system is controllable, it is automatically stabilizable. The linearized chemostat system is stabilizable if the real parts of each eigenvalue of the Jacobian $\underline{\underline{A}}$ can be made negative by controller action. If the real parts are negative. the system is automatically stabilizable, even without controller action.

If we consider a proportional feedback control action on the state variables (eg. cell mass or substrate concentrations)

$$\underline{U}(t) = -\underline{\underline{K}}\underline{x}(t) \tag{4.129}$$

$\underline{\underline{K}}$ is an (m,n) gain matrix. The linearized chemostat equations become

$$\frac{d\underline{x}}{dt} = (\underline{\underline{A}} - \underline{\underline{BK}})\underline{x} \tag{4.130}$$

Provided the elements of $\underline{\underline{K}}$ can be selected such that the eigenvalues of ($\underline{\underline{A}}$-$\underline{\underline{BK}}$) are negative, the system can be stabilized by the proportional controller. For example, a proportional controller which is fed-back to the state variables

$$\underline{U}(t) = -\underline{\underline{KC}}\underline{Y}(t) \tag{4.131}$$

$$\frac{d\underline{x}}{dt} = (\underline{\underline{A}} - \underline{\underline{BKC}})\underline{x} \tag{4.132}$$

Stabilizability of the output requires that the eigenvalues of the matrix ($\underline{\underline{A}}$ - $\underline{\underline{BKC}}$) can be made negative by some choice of the feedback gain matrix $\underline{\underline{K}}$. The chemostat, although not controllable, is stabilizable. We can see this by reducing the cell and substrate mass balance equations to one equation by introducing the variable Z, where

$$Z = X - Y_{X/S}(S_o - S) \tag{4.133}$$

Summing the mass balances for X and S shows

$$\frac{dZ}{dt} = -DZ \tag{4.134}$$

which has the solution

$$Z = Z_o \exp\left(-\int_0^t D(t)dt\right) \tag{4.135}$$

The cell concentration can be controlled by manipulating the dilution rate and the eigenvalue associated with Z is always negative, thus the system is stabilizable. It is thus possible to have a feedback controller that can maintain the cell concentration X at some desired value, with the substrate concentration reaching a steady state value given by (S_o - $X/Y_{X/S}$). This type of operation is termed the *turbidostat*, as the turbidity (optical density) is constant. In a similar manner, it would be possible to control the substrate concentration at some selected value by manipulating the dilution rate and allowing the cell concentration to achieve a corresponding steady state value. This operation is termed *nutristat*. We shall examine both of these control schemes.

Turbidostat Operation

Measurement of the optical density directly in a stirred tank reactor is relatively simple, using either an *in situ* probe or using a separate sample loop from the reactor. The dilution rate can be adjusted to maintain a constant value of the cell mass. In turbidostat operation, cells can be grown with all nutrients present in excess. It also permits cells to be grown at near their maximum specific growth rates. The metabolism of a microorganism can thus be examined under conditions removed from those found in chemostat operation[7]. If the dilution rate is altered to maintain a constant cell concentration, a proportional controller action has the following result

$$D(t) = D_{ss} + K_t(X_{set} - X) \geq 0 \tag{4.136}$$

where K_t is the proportional gain constant and X_{set} is the desired cell concentration. The set of cell and substrate balance equations now become

$$\frac{d\underline{x}}{dt} = (\underline{\underline{A}} - \underline{\underline{B}}\underline{K}_t^T)\underline{x} \tag{4.137}$$

where $\underline{\underline{A}}$ and $\underline{\underline{B}}$ are given by equations (4.126) and (4.127), and $\underline{K}_t^T$ is the transpose of $\underline{K}_t$ ($[K_t, 0]$). The matrix ($\underline{\underline{A}}$ - $\underline{\underline{B}}\underline{K}^T$) is given by

$$(\underline{\underline{A}} - \underline{\underline{B}}\underline{K}_t^T) = \begin{bmatrix} K_t X & \dfrac{d\mu}{dS} \\ \left(-K_t X/Y_{X/S} - \dfrac{\mu}{Y_{X/S}}\right) & \left(-\mu - \dfrac{X}{Y_{X/S}}\dfrac{d\mu}{dS}\right) \end{bmatrix}_{SS} \tag{4.138}$$

For the system to be stable, K_t must be chosen such that the real parts of the eigenvalues of the above matrix have negative values. The determinant and trace are

(7) See D. Zines and P.L. Rogers, [Biotech. Bioeng. **12**, 561 (1970)] for an example of the use of the turbidostat to study the growth of *A. aerogenes* subject to ethanol inhibition.

$$det(\underline{\underline{A}} - \underline{\underline{B}}\underline{K}_t^T) = \left[\left\{\frac{d\mu}{dS} - K_t Y_{X/S}\right\}\frac{\mu X}{Y_{X/S}}\right]_{SS} \qquad trace(\underline{\underline{A}} - \underline{\underline{B}}\underline{K}_t^T) = \left[K_t - \frac{X}{Y_{X/S}}\frac{d\mu}{dS} - \mu\right]_{SS} \quad (4.139)$$

Both det ($\underline{\underline{A}}$ - $\underline{\underline{B}}\underline{K}^T$) will be > 0 and trace ($\underline{\underline{A}}$ - $\underline{\underline{B}}\underline{K}^T$) will be < 0 provided

$$\left(\frac{d\mu}{dS} - K_t Y_{X/S}\right)_{SS} > 0 \quad (4.140)$$

In the case where $(d\mu/dS)_{SS} > 0$, the system without the controller action will be stable, but the stability characteristics of the closed loop system with the controller can be improved by selecting a proportional gain such that $K_t < 0$. When $(d\mu/dS)_{SS} < 0$, such as would be the case with substrate inhibition at the steady state with the higher substrate concentration, the open loop system would be unstable, but by choosing K_t such that

$$K_t < \left[\frac{1}{Y_{X/S}}\frac{d\mu}{dS}\right]_{SS} < 0 \quad (4.141)$$

the steady state can be stabilized. The steady state gain of the system is given by dX/dD, which can be found from

$$X_{SS} = Y_{X/S}\left(S_o - \frac{DK_S}{\mu_{max} - D}\right) \quad (4.141a)$$

$$K_t = \frac{dX}{dD} = \frac{-\mu_{max} K_S Y_{X/S}}{(\mu_{max} - D)^2} \quad (4.141b)$$

At low dilution rates, the gain will be small, i.e., a large controller action is required to effect a change in the cell concentration. This can be seen from the small changes in X as the dilution rate D is varied in chemostat operation. On the other hand, at values of the dilution rate close to μ_{max} the gain will be high and the cell concentration will change rapidly with dilution rate. Substrate will be present at high concentration. In this region the turbidostat will be most effective, and this indeed is the experimental observation.

Nutristat Operation

It is generally difficult to determine the concentration of a growth limiting substrate directly in the reactor, due to lack of appropriate monitoring devices. Nevertheless, nutristats have potentially attractive features in terms of their control. We shall consider the dilution rate to be proportionally controlled in response to changes in effluent substrate concentration viz.

$$D(t) = D_{SS} + K_N(S_{set} - S) \quad (4.142)$$

Analyzing the system as for the turbidostat, we have

$$\frac{d\underline{x}}{dt} = (\underline{\underline{A}} - \underline{\underline{B}}\underline{K}_N^T)\underline{x} \quad (4.143)$$

Here, $\underline{K}_N^T$ is $[0, K_N]$ and

$$(\underline{\underline{A}} - \underline{\underline{B}}K_N^T) = \begin{bmatrix} 0 & X\frac{d\mu}{dS} + K_N X \\ \left(-\frac{\mu}{Y_{X/S}}\right) & \left(-\mu - \frac{X}{Y_{X/S}}\frac{d\mu}{dS} - \frac{K_N X}{Y_{X/S}}\right) \end{bmatrix}_{SS} \tag{4.144}$$

In this case, the determinant and trace are give by

$$det(\underline{\underline{A}} - \underline{\underline{B}}K_N^T) = \left[\left\{\frac{d\mu}{dS} - K_N\right\}\frac{\mu X}{Y_{X/S}}\right]_{SS} \qquad trace(\underline{\underline{A}} - \underline{\underline{B}}K_N^T) = [\frac{-K_N X}{Y_{X/S}} - \mu - \frac{X}{Y_{X/S}}\frac{d\mu}{dS}]_{SS} \tag{4.145}$$

The nutristat will be stable at all steady states proved the determinant is positive and the trace is negative. For both these conditions to be met, we require

$$\left(K_N + \frac{d\mu}{dS}\right)_{SS} > 0 \tag{4.146}$$

Thus open loop steady states, which are normally stable, have $(d\mu/dS)_{SS} > 0$ and any positive values of K_N will stabilize the corresponding closed loop system. For cases where $(d\mu/dS)_{SS} < 0$ we require

$$K_N > -\left(\frac{d\mu}{dS}\right)_{SS} > 0 \tag{4.147}$$

The steady state gain is

$$K_N = \frac{dS}{dD} = \frac{1}{\left(\frac{d\mu}{dS}\right)_{SS}} \tag{4.148}$$

Thus $K_N = K_t/Y_{X/S}$ and as $Y_{X/S}$ is less than unity for carbohydrate substrates, the gain of the nutristat will be larger than that of the turbidostat, hence the nutristat should have a better response to process upsets. This has been confirmed by extensive simulations[8].

4.7 Recycle Systems

4.7.1 Biological Waste Water Treatment

A variety of materials are found in wastewater, including organic and inorganic compounds, colloids and solids. Some of these can be separated from the waste stream by settling, while the organic material is usually removed by microbial oxidation. Domestic sewage contains about 300 ppm suspended solids and 500 ppm organic material (eg. cellulosics). Industrial wastewater may contain considerably higher concentrations of organics, but the volume to be treated is generally much smaller than domestic wastewater. The concentration of organic material in wastewater is usually denoted by the *biological oxygen demand (BOD)*, which is determined by incubating the waste for a certain period of time (5

---

(8) Edwards, V.H., Ko, R.C., and S. Balogh, *Dynamics and Control of Continuous Microbial Propagators Subject to Substrate Inhibition*, Biotech. Bioeng. **14**, 939 (1972).

days being common), and determining the amount of dissolved oxygen consumed at 20°C. This is referred to as the BOD_5. Not all the organic material in the waste is biodegradable and the total is determined as the *chemical oxygen demand (COD)*, equal to the milligrams of oxygen that a liter of sample will require to be completely oxidized by potassium dichromate. The effluent from fermentation plants has a high BOD_5 (4,000 - 5,000 mg/l); domestic sewage has around 300 mg/l; chemical plants 500 - 1,000 mg/l; and food processing plants may range from 500 to 10,000 mg/l.

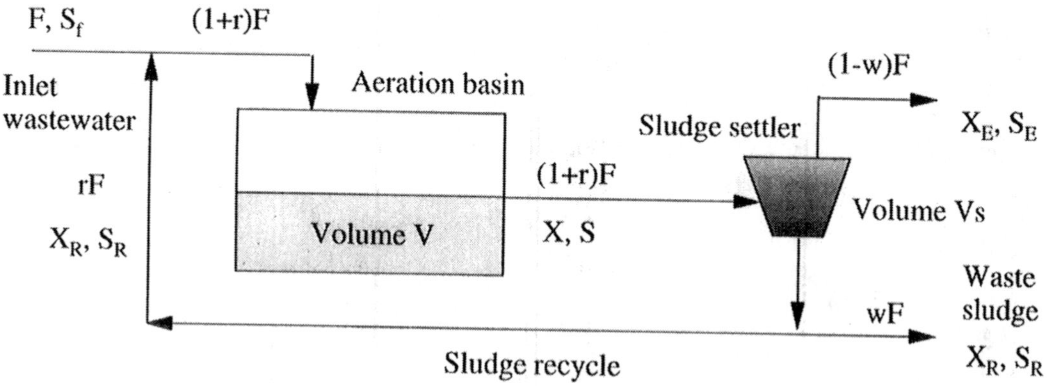

Figure 4.21. *The activated sludge process for wastewater treatment. The sludge is recycled from the settling tank to the aeration basin. The recycle stream flows are indicated as fractions of the input flow rate F.*

Domestic and many industrial wastes are treated by a process known as the activated sludge process. An aerated reactor is continuously fed with wastewater and the organic material is oxidized by the action of a mixed microbial population which develops over time. The effluent contains the cells and any unconsumed organic material, and is known as activated sludge. It is separated from the effluent in a large settling tank, whereas the water with reduced BOD is discharged. Part of the sludge is recycled to the aerated reactor, known as the aeration basin. The microbial population that comprises the sludge is a mixed culture, with *Zoogloea ramigera* as the predominant species. A variety of methods for the activated sludge process have emerged, including those in which the wastewater is introduced at several points in the aeration basin (the step aeration process) and those where the sludge is re-aerated prior to being recycled to the aeration basin (the contact stabilization method).

We shall examine the activated sludge process and develop the material balances describing its steady-state and transient behavior. The process is illustrated in Figure 4.21. A fraction r of the inlet flow rate F is recycled to the aeration basin from the settler; some of the sludge is removed and the effluent from the settler is discharged. Mass balances on the cells and substrate around the reactor can be written, assuming constant volume, as follows:

$$V\frac{dX}{dt} = rFX_R + \mu XV - k_d XV - F(1+r)X \tag{4.149}$$

$$V\frac{dS}{dt} = FS_f + rFS_R - \frac{1}{Y}\mu XV - F(1+r)S \tag{4.150}$$

Here k_d accounts for the death of cells, and a constant cell yield coefficient will be assumed. A steady-state mass balance around the settler gives

$$F(1+r)X = (1-w)FX_E + (r+w)FX_R \tag{4.151}$$

The substrate concentration can be assumed to be constant, i.e., no additional BOD is removed in the settler. Thus

$$S = S_R = S_E \tag{4.152}$$

In a well-designed settler, the effluent will be free of cells and X_E will be zero. This permits X_R to be determined from X, r, and w:

$$X_R = \frac{1+r}{r+w}X \tag{4.153}$$

The cell and substrate mass balances around the aeration basin can be rewritten as

$$\frac{dX}{dt} = \mu X - (1+r)\left(\frac{w}{r+w}\right)\frac{F}{V}X - k_d X \tag{4.154a}$$

$$\frac{dS}{dt} = \frac{F}{V}(S_f - S) - \frac{1}{Y}\mu X \tag{4.154b}$$

In a conventional activated sludge process r ranges from 0.1 to 0.3 and w is approximately 0.015. The ratio $FS_f/(XV + X_R V_S)$ is 0.2 to 0.5 lb BOD/lb VSS-day. VSS refers to volatile suspended solids, and is a measure of the total concentration of viable and non-viable cells in the system. The cell concentration in the recycle stream is often expressed as the sludge volume index (SVI), in units of ml per gram of activated sludge. This is based on the volume of settled sludge in one liter of aeration basin effluent after a period of 30 minutes. Hence it may be expressed in more familiar terms as

$$X_R(mg/liter) \sim \frac{10^6}{SVI} \tag{4.155}$$

We shall examine the effect of the sludge recycle on the steady state behavior of the system. For simplicity we will assume that the death term k_d can be neglected. The steady state balances for S and X then reduce to

$$\mu = (1+r)\left(\frac{w}{r+w}\right)\frac{F}{V} \tag{4.156}$$

$$X = Y\frac{F}{V}(S_f - S) \tag{4.157}$$

If we assume that the specific growth rate of the sludge can be represented by the Monod expression, the steady state effluent substrate concentration can be determined:

$$\frac{\mu_{\max}S}{K_S+S} = \frac{F}{V}(1+r)\left(\frac{w}{r+w}\right) \tag{4.158a}$$

$$S = \frac{\frac{F}{V}(1+r)\left(\frac{w}{r+w}\right)K_S}{\mu_{\max} - \frac{F}{V}(1+r)\left(\frac{w}{r+w}\right)} \tag{4.158b}$$

The effect of cell recycle is to extend the point at which washout of cells will occur. Consider washout occurring when the apparent dilution rate approaches $\mu_{\max}$:

$$\frac{F}{V} = \left(\frac{r+w}{w(1+r)}\right)\mu_{\max} \tag{4.159}$$

By inserting typical values for r (0.25) and w (0.015) above, we see that washout will not occur until approximately 14.13 $\mu_{\max}$, far in excess of the chemostat value. With these values, the cell concentration in the recycle stream is also greatly increased as a result of the settler action.

4.7.2 Feed-Forward Control of the Activated Sludge Process

The flows and BOD content of the stream entering a wastewater treatment plant may vary considerably. To minimize the effect of these process changes on the quality of the effluent (i.e., S), various control schemes are implemented. An effective means of maintaining S at a constant value is to monitor F(t) and S_f(t) and use the deviations of these values from the steady state to control the recycle rate of sludge to the aeration basin. This feedforward control scheme is illustrated in Figure 4.22. The equations governing cell and substrate concentrations are given below, noting that F and S_f are now functions of time.

$$\frac{dX}{dt} = \mu X - \frac{F(t)}{V}(1+r)\left(\frac{w}{r+w}\right) \tag{4.160a}$$

$$\frac{dS}{dt} = \frac{F(t)}{V}(S_f(t) - S) - \frac{1}{Y}\mu X \tag{4.160b}$$

F and S_f are measured and r will be manipulated to control S at the desired steady state value. The above equations can be linearized about the steady state values, noting

$$x = X - X_{SS} \qquad s = S - S_{SS} \qquad f = F - F_{SS} \qquad R = r - r_{SS} \qquad s_o = S_f - S_{f,SS} \quad (4.161)$$

The linearized cell and substrate balance equations become

$$\begin{bmatrix} \frac{dx}{dt} \\ \frac{dy}{dt} \end{bmatrix} = [A] \begin{bmatrix} x \\ s \\ f \\ R \\ s_f \end{bmatrix} \qquad (4.162)$$

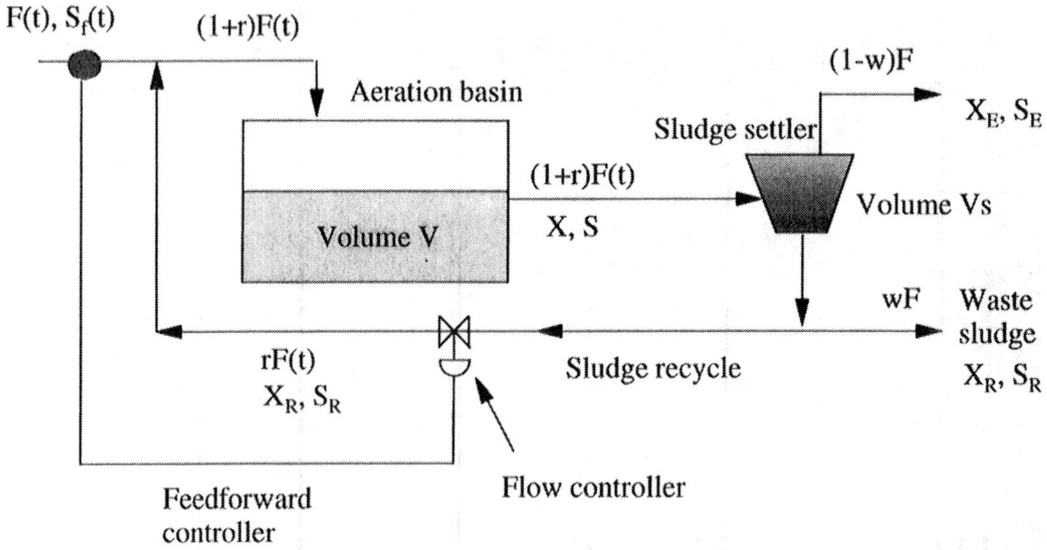

***Figure 4.22**. Feedforward control of the activated sludge process. The inlet parameters F(t) and $S_f(t)$ are monitored and used to determine the recycle factor r of the activated sludge returned to the aeration basin.*

The elements of the (2x5) matrix **A** are determined from the partial derivatives of the cell and substrate balance equations with respect to each of the variables, X, S, F, r and S_f, evaluated at the steady state.

$$a_{11} = \mu - \frac{F}{V}(1+r)\left(\frac{w}{r+w}\right) = 0 \qquad a_{21} = -\frac{\mu}{Y} = -\frac{F}{VX}(S_f - S)$$

$$a_{12} = X\frac{\partial \mu}{\partial S} \qquad a_{22} = -\frac{F}{V} - \frac{X}{Y}\frac{\partial \mu}{\partial S}$$

$$a_{13} = \frac{-X}{V}(1+r)\left(\frac{w}{r+w}\right) \qquad a_{23} = \frac{(S_f - S)}{V}$$

$$a_{14} = \frac{FX}{V}\left(\frac{w(1-w)}{(r+w)^2}\right) \qquad a_{24} = 0$$

$$a_{15} = 0 \qquad a_{25} = \frac{F}{V}$$

Since we wish to examine the behavior of s(t), we eliminate x(t) from the above equations. This is accomplished by taking Laplace transforms of equations (4.162). Denoting the Laplace variable by p, we have

$$\hat{x} = \frac{1}{(p - a_{11})}(a_{12}\hat{s} + a_{13}\hat{f} + a_{14}\hat{R} + a_{15}\hat{s}_o) \tag{4.163}$$

$$\hat{s} = \frac{1}{(p - a_{21})}(a_{21}\hat{x} + a_{23}\hat{f} + a_{24}\hat{R} + a_{25}\hat{s}_o) \tag{4.164}$$

$$\left(p - a_{22} + \frac{a_{21}a_{12}}{p - a_{11}}\right)\hat{s} = \left(a_{23} + \frac{a_{21}a_{13}}{p - a_{11}}\right)\hat{f} + \left(a_{24} + \frac{a_{21}a_{14}}{p - a_{11}}\right)\hat{R} + \left(a_{25} + \frac{a_{21}a_{15}}{p - a_{11}}\right)\hat{s}_o \tag{4.165}$$

Thus the effluent substrate concentration response can be expressed in terms of the changes in inlet flow rate ($\hat{f}$), substrate concentration ($\hat{s}$) and the changes made in the control variable ($\hat{R}$). We shall consider the case of constant inlet BOD and examine the control action in response to changes in inlet flow rate. In this case, we want the value of s(t) to be zero for perfect control action. Thus $\hat{s}$ will be zero and r(t) can be found by inverting the Laplace transform expression

$$\hat{R} = -\frac{\left(a_{23} + \frac{a_{21}a_{12}}{p - a_{11}}\right)}{\left(a_{24} - \frac{a_{21}a_{14}}{p - a_{11}}\right)}\hat{f} \tag{4.166}$$

The steady state values of the elements of $\underline{\underline{A}}$ can now be inserted into the above expression, giving

$$\hat{R} = \left[-\frac{V}{F^2}\frac{(r+w)^2}{w(1-w)}p + \frac{V(r+w)^2}{Fw(1-w)}\left(\frac{\partial \mu}{\partial S}\right)_{SS}\right]\hat{f} \tag{4.167}$$

If the form of the disturbance f(t) is known, the response of the recycle ratio r(t) (equal to [R(t) - r_{SS}]) can be found by substituting $\hat{f}$ in the above equation and inverting the Laplace transform. The form of equation (4.167) suggests that a proportional-derviative (PD) controller action will successfully maintain the effluent substrate concentration constant in response to any inlet disturbances in flow rate. Note that equation (4.167) only contains the derivative of the specific growth rate with respect to substrate concentration. It is thus independent of the specific details of the growth rate expression.

A similar analysis of the recycle response to changes in the inlet substrate concentration can also be performed. In this case we obtain

$$\frac{\hat{R}}{\hat{s}_f} = -\frac{\left(a_{25} + \frac{a_{21}a_{15}}{p - a_{11}}\right)}{\left(a_{24} + \frac{a_{21}a_{14}}{p - a_{11}}\right)} \tag{4.168}$$

Substituting for the a_{ij} and simplifying yields

$$\hat{R} = \left[-\frac{V}{F} \frac{(r+w)^2 p}{w(1-w)(S_f - S_{SS})} \right] \hat{s}_f \tag{4.169}$$

Here a derivative action controller will maintain the effluent substrate concentration at its set point value. The above expression illustrates that the derivative action depends only on the steady state values of S_f and S_{ss}, and the steady state recycle flow ratios, and not on the details of the kinetic expression for growth and substrate consumption. This feed-forward type of control is easily implemented. Given the changes of the microbial population in activated sludge and the variations in the type of waste being treated, this result is highly attractive.

4.8 Nomenclature

Symbol	Definition	Typical units
a_{ij}	elements of Jacobian matrix $\underline{\underline{A}}$ [Eq. (4.96)]	
c_i	concentration of component i	gm/liter
c_p	heat capacity	cal/gm-°C
D	dilution rate (F/V)	hr^{-1}
D_i	dilution rate of ith vessel (F_i/V_i)	hr^{-1}
D_{max}	maximum or optimum dilution rate	hr^{-1}
F_i	inlet flow rate to i^{th} vessel	liters/hr
F_{in}	inlet medium flow rate	liters/hr
F_{out}	exit medium flow rate	liters/hr
k	cellular death rate	hr^{-1}
k_d	first order death rate constant	hr^{-1}
k_e	endogenous metabolism rate constant	gm substrate/gm cells-hr
K_i	inhibition constant	gm/liter
K_M	Michaelis constant	gm/liter
K_N	nutristat gain constant	
K_S	Monod saturation constant	gm/liter
K_t	turbidostat gain constant	
m	maintenance coefficient	gm substrate/gm cells-hr
m_i	i^{th} moment of residence time distribution function	

N	number of vessels	
n_T	total cell number concentration	numbers/ml
n_v	viable cell number concentration	numbers/ml
P	product concentration	gm/liter
q_P	specific rate of product formation	gm product/gm cells-hr
q_S	specific substrate uptake rate	gm substrate/gm cells-hr
r	recycle ratio (Fig. 4.21)	(-)
S	substrate concentration	gm/liter
S_i	substrate concentration in i^{th} vessel	gm/liter
S_o	initial substrate concentration	gm/liter
S_R	substrate concentration in recycle stream	gm/liter
S_{ss}	steady state substrate concentration	gm/liter
t	time	hours
u_i	interstitial liquid velocity [Eq.(4.67)]	cm/sec
u_s	superficial liquid velocity (PFR)	cm/sec
V	reactor volume	liters
V_i	volume of ith vessel	liters
V_{max}	maximum rate of enzymatic reaction	gm S/liter-hr
w	sludge recycle ratio (Fig. 4.21)	
X	biomass concentration (DCW)	gm/liter
X_d	concentration of dead cells	gm/liter
X_i	cell concentration in i^{th} vessel	gm/liter
X_R	cell concentration in recycle stream	gm/liter
X_{ss}	steady state cell concentration	gm/liter
X_T	total concentration of cells	gm/liter
X_v	viable cell concentration	gm/liter
$Y_{P/S}$	product yield coefficient	gm product/gm substrate
$Y_{X/S}$	yield coefficient for biomass	gm cells/gm substrate
Z	variable defined in Eq. (4.133)	gm/liter
α	volume ratio of dead zone mixing model	(-)
α	growth associated product formation constant	gm product/gm cells
β	non-growth associated product formation constant	gm product/gm cells-hr
δ	fractional conversion of substrate	(-)
ε	void fraction [Eq. (4.66)]	(-)
λ	eigenvalues [Eq. (4.98)	
μ	specific growth rate	hr^{-1}

μ_{max}	maximum specific growth rate	hr^{-1}
ρ	density of liquid phase	gm/liter
τ	residence time [$\varepsilon V/F$	(-)
ξ	residence time distribution function	hr^{-1}

4.9 Problems

1. Product Inhibition in a Two-Stage Chemostat

A non-growth associated product is to be produced in two CSTRs operated in series, with the effluent from the first reactor being fed to the second, which has no other feed streams entering. The first reactor volume is half that of the second. What dilution rate in the first stage will maximize the rate of production of the non-growth associated product? Prepare a graph of productivity (DP) for each stage. Compare the maximum productivity obtainable in the two stage reactor with that obtainable in a one stage CSTR which has the same total volume as the sum of both the reactors in the two stage system. What reactor configuration would you recommend for this type of product? The values of the Monod parameters for the organism employed are given below.

Parameters:

$\mu_{max} = 0.4\ hr^{-1}$	$S_o = 10$ gm/l
$Y_{X/S} = 0.5$ gm/gm	$Y_{P/S} = 0.1$ gm/gm
$K_S = 0.05$ gm/l	$\beta = 0.1\ hr^{-1}$

2. Michaelis-Menten Kinetics in a CSTR

Consider an enzymatic reaction occurring in a well-mixed, isothermal, continuous stirred tank reactor of volume V_R. The volumetric flow rate of the feed and effluent streams is F, and the concentrations of S, E, (ES), and P in the feed streams are s_f, e_f, zero, and zero, respectively. We wish to determine the rate (moles/time) at which product P is produced by this reactor in steady-state.

You are to do this calculation in two different ways. First, assume that the overall reaction $S \rightarrow P$ is described by Michaelis-Menten kinetics, with v_{max} and K_M defined in the manner derived by Briggs and Haldane.

Second, calculate the steady-state rate of P production by analyzing this reactor taking full account of the entire reaction sequence as presented in Chapter 1, Section 1.2.2.

Compare the results of these two calculations. Under what conditions does the first approach give approximately correct results.

(Based on a problem provided by J.E. Bailey.)

3. Differential Reactor for Determining Enzyme Kinetics

A differential recirculation reactor enables straightforward analysis of immobilized enzyme kinetic data by differential rate equations, without having to resort to the integrated rate expressions typically required by packed bed reactors. Furthermore, such a reactor prevents spurious kinetic data due to concentration and temperature gradients within the catalyst bed. Consider a reactor consisting of a small column connected to a recirculation loop. The column contains enzyme immobilized to a porous carrier, e.g., porous glass beads, referred to as the matrix. Important parameters include the following:

Q: volumetric flow (recirculation) rate
S_b: the substrate concentration in the bulk solution
S_m: the substrate concentration in the matrix
V_b: volume of the bulk solution
V_m: internal volume of the matrix within the reactor
K_s: partition coefficient of the substrate for the enzyme matrix
K_p: partition coefficient of the product for the enzyme matrix
R_b: observed initial rate of product formation in the bulk solution

The only theoretical requirement for the establishment of a differential reactor is that the conversion per pass of substrate across the catalyst bed, $\Delta S_m/S_m$, be 2% or less based on the initial substrate concentration in the matrix phase.

Derive an expression for $\Delta S_m/S_m$ in terms of R_b, S_b, Q, and any other necessary, but measurable, parameters. Using this equation, the conversion per pass can be calculated to determine whether the system is functioning as a differential reactor.

4. *Multiple Steady States in a CSTR*

Consider an enzymatic reaction carried out in a CSTR. The enzyme is subject to substrate inhibition, and the rate equation is given by Eqn. (1.50):

$$v = \frac{v_{max}[S]}{[S]\cdot\left(1+\frac{[S]}{K_I}\right)+K_M} \tag{1.50}$$

a) Sketch a plot of reaction rate versus substrate concentration for this enzyme reaction. Use this plot to show that for a sufficiently high inlet substrate concentration, S_o, there is a range of dilution rates for which multiple steady states are possible in a CSTR.
b) Explain how you would calculate the highest and lowest dilution rates for which multiple steady states are possible. Your answer should include an expression(s) for the substrate concentrations from which you could calculate these dilution rates.

5. *CSTR versus PFR: Comparison of Biocatalyst Requirements*

Consider a plug flow reactor and a CSTR, each of which contains the same immobilized enzyme catalyst. If each reactor is to have the same average liquid residence time, compare the amounts of immobilized enzyme needed to achieve the same conversion.

6. *Michaelis-Menten Parameters in a PFR*

Shown below are fractional conversions of substrate, X, obtained for the hydrolysis of benzoyl-arginine ethyl ester (BAEE) by the enzyme ficin immobilized to CM-cellulose [Lilly et al., Biochem. J., 100, 718 (1966)]. The immobilized enzyme was packed in a column (plug-flow) reactor, and the data were obtained for different flow rates through the column, F (ml/hr), and substrate concentrations, S_o (mM).

	Substrate Concentration (mM)				
	2.0	6.0	20.0	33.0	55.0
F = 40	0.36	0.33	0.26	0.22	0.16
F = 80	0.30	0.27	0.17	0.13	0.095
F = 140	0.27	0.22	0.12	0.087	0.061

Estimate the Michaelis constant for the hydrolysis of BAEE. State any assumptions you must make, and explain the trend shown by the three sets of data.

7. Intrinsic Kinetics from Transport-Limited Data

Consider a reaction catalyzed by immobilized enzyme in a packed bed reactor. We shall assume that intraparticle diffusional resistance is negligible but that external film resistance may be significant. In this problem we shall consider how to determine the ratio of intrinsic kinetic parameters, V_{max}/K_m, from transport-limited rate data. [Patwardhan and Karanth, Biotechnol. Bioeng., 24, 763 (1982)].

a) Begin by writing a steady-state mass balance for substrate in the packed bed reactor (assume plug flow). This mass balance should contain S^*, the substrate concentration at the catalyst surface. Next, write the equation relating S^* to the bulk substrate concentration, S_b, the external mass transfer coefficient, k_s, and a, the interfacial area per liquid volume.

b) Recast these equations (including the boundary condition) into dimensionless form by introducing the following dimensionless parameters:

$X = (S_o - S_b)/S_o$

$\alpha = S^*/S_b$

$\tau = V_{max}z/u_sS_o$

$P = K_m/S_o$

$Q = V_{max}/k_sS_o$

where S_o is the inlet substrate concentration.

Simplify your dimensionless equations for the case of a very low S_o, and show that under this condition, we can solve for X to obtain:

$$\ln(1-X) = -\frac{V_{max}(z/u_s)}{K_m + V_{max}/k_s a}$$

c) Now consider a plot of S_oX versus ln(1-X). Since S_oX approaches zero in the limit of low S_o, Eqn. (4.8.1) corresponds to the intercept on the ln(1-X) axis. Rearrange Eqn. (4.8.1) to develop a graphical procedure for calculating K_m/V_{max}.

d) Calculate K_m/V_{max} for the immobilized ficin system described in Problem 4.5. The diameter of the immobilized ficin reactor was 1.0 cm, and the reactor length was 4.0 cm. Hint: note that as $1/F \rightarrow 0$, $k_s \rightarrow \infty$.

8. Balance Equations in Continuous Culture

Algae and photosynthetic bacteria have the potential to produce renewable chemicals and to convert light energy into useful forms of chemical energy. Microorganisms have a higher photosynthetic efficiency than agricultural plants and growth conditions can be more easily controlled. A chemical balance equation for light energy conversion by photoautotrophic microorganisms can be written as follows:

$$CO_2 + \beta NH_3 + \varepsilon H_2O + h\nu \rightarrow CH_aO_bN_c + \gamma O_2$$

a) Show that for continuous culture with dilution rate, D (h^{-1}), a carbon balance and available electron balance can be written as

$$\frac{12Q_{CO_2}}{D\sigma_b} = 1 \qquad \text{carbon balance}$$

$$\frac{D\sigma_b\gamma_b}{48Q_{O_2}} = 1 \qquad \text{electron balance}$$

where Q_{CO_2} is the specific consumption rate of CO_2 (g mol CO_2/g cell h) and Q_{O_2} is the specific production rate of O_2 (g mol O_2/g cell h).

b) Starting with a steady state energy balance, show that the biomass energetic yield (η_{kcal}) can be written as

$$\eta_{kcal} = \frac{FX\sigma_b\gamma_b Q_o}{12I_aA}$$

where F is the volumetric flow rate, Q_o is the energy of biomass per equivalent of available electrons, I_a is the intensity of light (kcal/cm^2 h), and A is the surface area (cm^2).

9. Biomass and Substrate Concentrations in a Chemostat

Consider a chemostat operating under steady state conditions with biomass production according to Monod kinetics. Assume the maximum specific growth rate is 1 hr^{-1} and the saturation constant is 500 mg/liter. The feed is sterile and contains 10 g/liter of the carbon substrate; the feed flow rate is 300 liters/hour. The volume of liquid in the fermenter is 1000 liters. The biomass yield is 0.5 g dry biomass/g substrate consumed.

a) Determine the steady state concentration of biomass and substrate for the given conditions.

b) Determine the washout dilution rate for the given kinetics and feed concentration.

c) Your supervisor wants to know if the production of biomass can be improved by adding a divider so that you have two completely mixed tanks in series with 500 liters of liquid in each tank. Determine the biomass and substrate concentrations in each tank for this case. *(Courtesy L.E. Erickson)*

10. Wall Growth

A small laboratory reactor is used to obtain data on the growth of *E. coli* on glucose. The vessel is cylindrical with a liquid volume of 1.0 liters and is 10 cm in diameter. The walls and the base of the vessel are covered with a film of *E. coli*, which grows to a concentration of 10^9 cells/ml. Under the agitation conditions employed, the film is uniform with an average thickness of 0.1 mm. The reactor is operated as a chemostat, with an inlet substrate concentration of 2.0 gm/liter glucose. The true yield coefficient for *E. coli* may be taken to be 10^7 cells/gm glucose, and the maximum specific growth rate is 0.8 hr^{-1}. The Monod constant K_S is 100 mg glucose/liter.

The vessel is to be used to collect data on the kinetics of *E. coli* growth under glucose limitation. How would the observed yield coefficient vary with dilution rate under these conditions of wall growth? Plot Y_{obs} as a function of D. What fraction of the substrate is consumed by cells attached to the wall? In determining the kinetics parameters, only the bulk cell and substrate concentrations are measured, and the contribution by cells attached to the wall is ignored. The cells on the wall consume substrate at a rate described by the Monod equation with the true kinetic constants given above. Their number remains constant due to the combined effects of cell growth and cell lysis.

If a washout experiment is performed (D is set at a value greater than the maximum growth rate and the cell concentration in the exit stream is monitored with time) what is the behavior you would expect in the presence and absence of wall growth? Plot the cell concentration as a function of time for both cases.

11. Optimal Design of CSTR's in Series

In this problem we shall derive a simple analytical expression for the optimal design of CSTR's in series, assuming Michaelis-Menten kinetics and a constant activity of biocatalyst in the reactors [K. Ch. A.M. Luyben and J. Tramper, Biotechnol. Bioeng., **24**, 1217 (1981)]. We define the optimum as the smallest total reactor size (residence time) to obtain a specific conversion. This problem was originally solved by Luyben and Tramper.

a) Consider N CSTR's in series with an inlet substrate concentration of S_o. Derive an expression for the average residence time of the ith reactor, τ_i, in terms of S_{i-1}, the substrate concentration entering the reactor, S_i, the substrate concentration in the reactor, and the enzyme's kinetic parameters. Rewrite this equation in dimensionless form by introducing the following parameters:

$$\alpha = S/S_o \qquad \kappa = K_m/S_o \qquad \theta = (\tau V_{max})/S_o$$

b) Now we must find the intermediate α_i values which correspond to the maximum total residence time. Mathematically, these values must satisfy the following equation:

$$\frac{d\left[\sum_{j=1}^{N} \theta_j\right]}{d\alpha_i} = 0 \qquad i = 1, 2, \ldots, N-1$$

Show that this equation reduces to the following simple result:

$$\alpha_i = \alpha_{i+1}^{i/i+1} \qquad i = 1, 2, \ldots, N-1$$

c) The above equation predicts that using multiple reactors in series can have a dramatic effect on the total residence time (reactor volume), but the effect decreases as N increases. To illustrate this result, calculate the optimal dimensionless total residence time for a final conversion of 99%, $\kappa = 0.1$, and N = 1, 2, 3, and 4.

12. Stability of Col E1 Plasmids in Continuous Culture

Plasmid instability can result in the plasmid-bearing cells in a population being displaced by plasmid-free cells, which typically have the advantage of faster growth. An approach to provide an advantage to plasmid-containing cells is to have them produce a bacteriocin that kills cells without the plasmid. This problem involves modeling the dynamics of a mixed cell population containing both plasmid-containing cells (which produce the bacteriocin Col E1) and plasmid-free cells.

(a) Using the following notation:

p	number density of plasmid-containing cells
b	number density of plasmid-free cells
s	glucose concentration
c	colicin concentration
μ_{max}	maximum specific growth rate of cells
K_p	Monod constant for plasmid-containing cells
K_b	Monod constant for plasmid-free cells
f	frequency of birth of plasmid-free daughter cells from plasmid-containing mother cells
g	first-order rate constant for production of colicin by plasmid-containing cells
q	first-order rate constant for killing of plasmid-free cells by colicin

formulate material balances for plasmid-containing cells, plasmid-free cells, glucose, and colicin in a chemostat operating at a dilution rate of D. Do not assume steady state operation.

(b) Assume that the pseudo-steady state hypothesis apples to glucose and colicin. Assume that the glucose concentration is low relative to the saturation constants for both cell types. Using these assumptions, and scaling the parameters and variables, formulate dimensionless equations for plasmid-containing and plasmid-free cells. Use the following ratios in the scaling: qg/D^2, K_b/K_p, and μ_{max}/DK_b.

If this pair of differential equations is abnalyzed by phase-plane methods, two types of phase-plane behavior are observed. These are shown in the accompanying figures. In the top phase plane (I), the only possible steady state is the one representing plasmid instability. In the bottom phase plane plot (II), two stable steady states exist: one representing

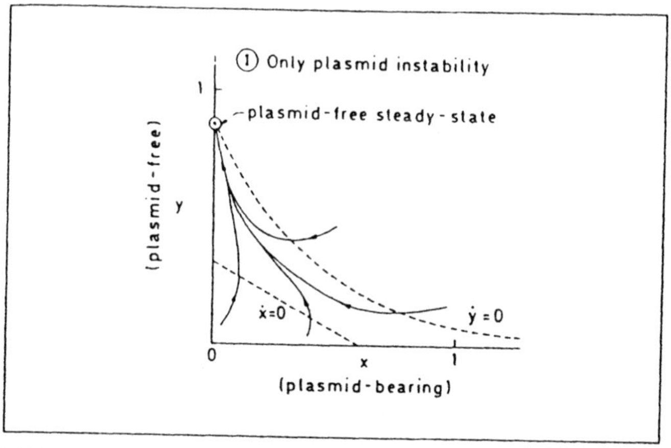

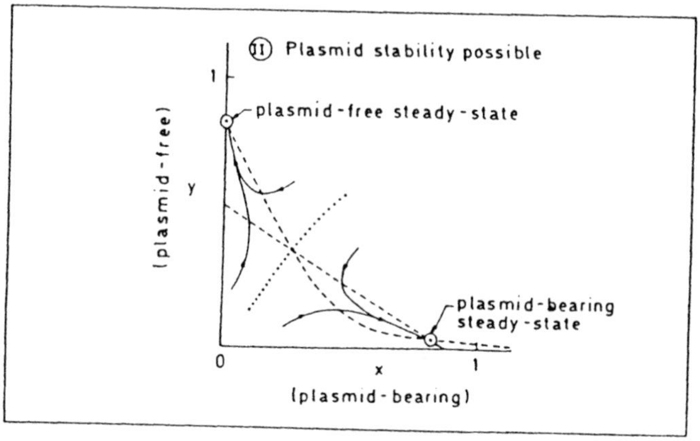

plasmid instability and one representing plasmid stability. In (II) the steady state that will result depends on the initial conditions. Further analysis indicates that a type (II) phase plane will be obtained if

$$\theta > \frac{1-\alpha(1-f)}{\left[\alpha(1-f)-\frac{1}{K}\right](1-f)}$$

where θ, α and K are dimensionless parameters containing the groups indicated earlier. What are they in terms of these groups? The model predicts that the range of of initial conditions leading to a plasmid-containing steady state will be very sensitive to the dilution rate. Is this what you would expect?

13. Methane Production Kinetics in a Chemostat

Methanococcus jannaschii is an extremely thermophilic methanogen isolated from a deep-sea hydrothermal vent at a depth of about 2500 m. This organism was recently grown in a chemostat at 80°C using different flow rates of gaseous substrate into the fermenter. The substrate was a mixture of 80% H_2 and 20% CO_2. Listed below are specific methane production rates (mol CH_4/g dry cell weight-h) for different steady-state dilution rates and gas flow rates (2, 9, and 14 L/hr). For each flow rate, calculate the growth associated methane formation constant (mol of CH_4 evolved (g dry weight of cells)$^{-1}$) and the non-growth associated methane formation constant, β.

Dilution Rate (days^{-1})	q_p [mol CH_4 (g cell-hr)$^{-1}$] 2 (L/hr)	9 (L/hr)	14 (L/hr)
0.058		0.41	0.52
0.080	0.054		
0.10	0.086	0.44	
0.16			0.71
0.20	0.15		
0.30		0.58	0.77
0.40			0.87
0.50			0.84
0.56			0.95

14. *Growth Parameters of a Murine Hybridoma*

The steady-state metabolic parameters for a murine hybridoma cell line have been determined in continuous suspension culture over a wide range of dilution rates [Hiller et al., *Biotechnol. Bioeng.*, **38**, 733 (1991)]. Listed below are the glucose and glutamine specific uptake rates, q_{gluc} and q_{gln}, at different dilution rates. Using these data, estimate the yield coefficients and maintenance requirements for glucose and glutamine.

Dilution Rate (days^{-1})	q_{gln} (mmol/10^9 cells/day)	q_{gluc} (mmol/10^9 cells/day)
0.54	1.00	4.20
0.60	0.81	5.44
0.66	1.03	6.69
0.77	1.07	7.21
0.92	1.55	10.4
1.05	1.59	9.81
1.15	2.26	10.6
1.25	3.03	12.7

15. Design Equations for a Perfusion-Recycle System

A simplified diagram of a perfusion-recycle system for growing mammalian cells is shown below. A continuous flow of feed into the fermenter determines the perfusion, or volumetric feed rate. Cell-containing broth is continuously circulated through the lumen side of a hollow fiber filtration unit. A cell free product (or permeate) is removed from the shell side of the hollow fiber unit, and a product stream containing cells is removed directly from the reactor (cell bleed). With such an experimental configuration, the reactor may operate as a cell retention, cell recycling, or perfusion reactor.

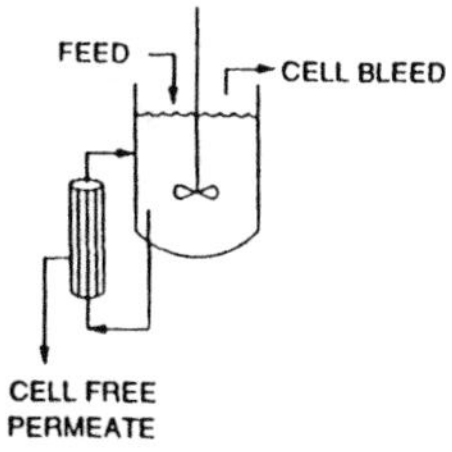

a) Write material balances for the total concentration of cells, n, and the concentration of viable cells, n_v. Important parameters include the *apparent* specific growth rate, μ_{app}, the cell bleed rate, CB (equal to the flow rate of cell-containing broth divided by the reactor volume), and the specific cell death rate, k_d.
b) Show that at steady state, the actual specific growth rate, μ, can be written in terms of the cell bleed rate and the specific cell death rate. This differs from the steady-state result for a conventional chemostat, in which the specific growth rate is equal to the dilution rate.
c) Derive expressions for q_s and q_{O_2}, the specific substrate and oxygen consumption rates.

16. Fermentation With Gas Recycle

Consider a fermentation in which the major gaseous nutrient, e.g., oxygen, is a minor component of the gas phase; thus, the total gas flow rates into and out of the fermenter are approximately equal. We wish to quantify the improvement in conversion of the gaseous nutrient that would be achieved by introducing gaseous-phase recycle, as shown diagramatically below [G. Hamer, *Biotechnol. Bioeng.*, **24**, 511 (1982)]. In the figure below, C_{in} is the gas phase concentration in the inlet stream; C_{out} and C'_{out}, the gas phase concentration in the outlet stream with and without recycle, respectively; G, the gas flow rate through the dispersion; G', the recycle gas flow rate; and Q and Q', the total oxygen uptake by the microorganisms with and without recycle, respectively.

Derive expressions for the ratio of fractional conversion possible with and without recycle, (conversion with recycle)/(conversion without recycle) for the case where the gaseous-phase nutrient under consideration is not limiting growth (assume identical liquid-phase conditions in the fermenter and equal gas flow rates), and the case where growth

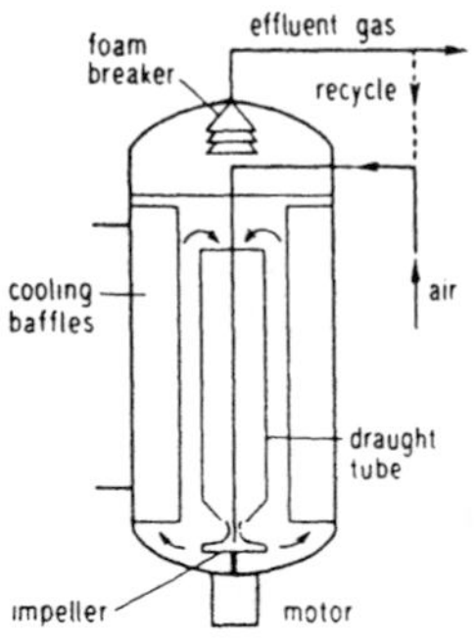

Gas-inducing impeller-agitated fermentor with gas-phase recycle.

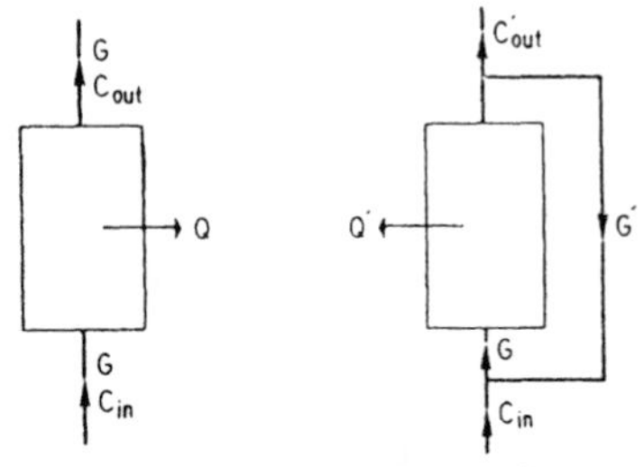

is limited by the gaseous-phase nutrient. In this latter situation, let *a* equal the interfacial area per unit liquid volume; K_L, the overall liquid-side mass transfer coefficient; and H', a constant.

17. Maximum Dilution Rate in a Recycle Reactor

We have seen in Section 4.2.1 for Monod growth kinetics that the maximum possible dilution rate in a conventional chemostat cannot exceed μ_{max}. However, Toda and Dunn [*Biotechnol. Bioeng.*, **24**, 651 91982)] found theoretically and verified experimentally that the maximum dilution rate, above which microbial cells were washed out from the fermenter, could be elevated well beyond the maximum specific growth rate if a particular fermenter combination was used. One such combination, shown below, consists of a CSTR (or backmix flow fermenter) and a tubular-loop fermenter in which liquid is mixed incompletely.

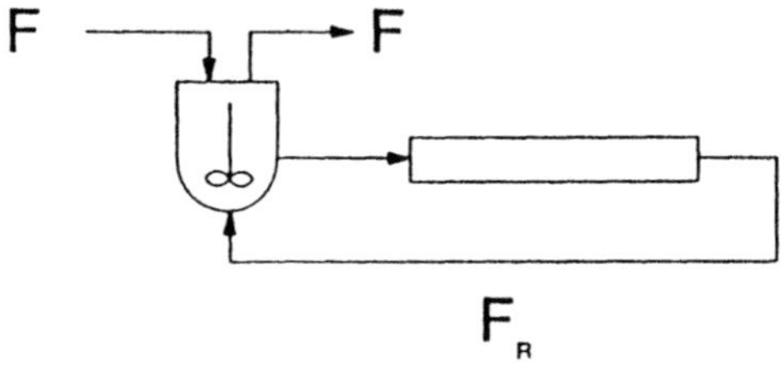

Fresh medium is fed to the inlet at a feed rate F and the exit stream is withdrawn at the same flow rate. The flowrate of recycled culture is denoted by F_R and the dimensionless recycle rate is given by

$$R = F_R/(F + F_R)$$

The volumes of the backmix and tubular-loop fermentors are V_B and V_T, respectively, and the dilution rate of the combined fermenter configuration is defined as

$$d = F/(V_B + V_T)$$

a) Show that a steady-state biomass balance for the backmix section can be written as

$$0 = \left[\frac{D}{(1-R)(1-\varepsilon_T)} \right] (RX_{Z=1} - X_B) + U(X_B)X_B$$

where D is the dimensionless dilution rate defined as $D = d/\mu_{max}$

U(X) is a dimensionless function of μ,

$$U(X) = (1 - X)/(K_S + 1 - X)$$

ε_T is the volume fraction of the tubular-loop section,

$$\varepsilon_T = V_T/(V_b + V_T)$$

and $X = x/(Ys_f)$, $K_S = k_s/s_f$. Additional nomenclature: x, cell mass concentration; Y, yield coefficient of cell growth; s_f, substrate concentration in the feed; k_s, saturation constant.
b) Show that the maximum dimensionless dilution rate, D_{max}, can exceed unity. Experimentally, D_{max} values as high as 4.5 have been obtained.

18. Ethanol Production in a Packed Bed Reactor

A packed bed bioreactor system is loaded with calcium alginate spheres of 3 mm diameter which contain 20 gm of immobilized enzyme (wet basis) per liter of alginate gel. The glucose diffuses into the beads, is converted by the enzyme and the products ethanol and CO_2 diffuse out of the beads. The packed column has an internal diameter of 1.75 cm, a length of 80 cm and is operated isothermally at 33°C and pH 6.0 at steady state. A peristaltic pump provides a constant volumetric flow rate of glucose solution at a concentration of 50 gm/L. The packed bed bioreactor is operated at a steady-state dilution rate of 0.9 hr^{-1}. Assume that the ethanol concentration within the beads is low enough that it does not inhibit ethanol production. The following data are available:

D_{glu}	effective glucose diffusivity in the alginate beads (1.42 10^{-6} cm^2/sec)
$D_{ethanol}$	effective ethanol diffusivity in the fbeads (2.59 10^{-6} cm^2/sec)
D_z/uL	dimensionless dispersion group (1.1)
D_z	axial dispersion coefficient (cm^2/sec)
u	fluid velocity (cm/sec)
L	length of the bioreactor
ε	void fraction in the column (0.4)

The following intrinsic kinetic parameters are available for free enzyme in the absence of mass transfer limitations at 33°C and pH 6.0:

v_{max}	maximum reaction rate (0.016 mol glucose/(sec-liter of wet beads)
K_M	Michaelis constant (1.7 gm glucose/liter)
$Y_{P/S}$	yield coefficient for ethanol (0.48 gm ethanol/gm glucose)

(a) On the basis of the given operation conditions, make a simple mass balance for glucose and ethanol in the packed bed reactor, and derive expresssions for the concentration profiles of both glucose and ethanol along the dimensionless reactor length, at the following points, Z = 0.1, 0.3, 0.5, 0.8 and 1.0, where Z=z/L.

(b) Compare the calculated concentration profiles found in part (a) with the following experimental data from a packed bed reactor operated at the same conditions. All concentrations were measured in the bulk liquid along the reactor. Do you think the experimental data are reliable?

Dimensionless length Z	*Glucose concn. (gm/L)*	*Ethanol concn. (gm/L)*
0.0	50.0	0.0
0.08	38.0	5.8
0.39	17.5	15.3
0.71	8.1	19.6
1.00	3.1	22.2

(c) Calculate the effectiveness factor for glucose conversion to ethanol as a function of bioreactor dimensionless length Z.

(d) Use quantitative arguments to show the effect of using smalle Ca-alginate beads on 0.3 mm diameter on the concentration profiles of both glucose and ethanol if all other operating conditions reamin the same.

19. Effect of Maintenance on the Transient Response of the Chemostat

Consider chemostat operation where the Monod relationship describes the growth of cells, but where substrate is consumed for maintenance at a rate given by mX (gm substrate/liter-hr). The mass balance equations describing the system are given in Section [XREF]. Linearize the equations about the non-trivial steady state and find the conditions where the maintenance coefficient m will result in eigenvalues of the Jacobian matrix which are complex numbers with negative real parts. Can the system ever show undamped oscillations?

Chapter 5. Transport Processes

Biochemical reactors are widely used in the food industries, in microbial fermentations, in waste treatment systems and in some biomedical devices. In virtually all of these reactors, several phases are involved and substrates and nutrients must be transferred from one phase to another. To be effective in achieving the desired degree of conversion of reactants to products, or supplying sufficient nutrients for maintenance of cell viability, interphase heat and mass transfer must occur to a sufficient extent.

One of the key nutrients for all aerobic cells is oxygen, which is sparingly soluble in water. Supply of oxygen from the gas phase to the liquid phase is critical in most aerobic fermentations. Oxygen transport determines the level of aerobic activity in lakes and in the soil. Another important gas in biological systems is carbon dioxide. Carbon dioxide regulates the pH in cultures of mammalian cells. Its transport and interconversion between its various aqueous forms (CO_3^{2-}, HCO_3^-, H_2CO_3) in solution must often be considered in anaerobic waste treatment tanks (anaerobic digesters) and in lake ecosystems. The transport of O_2, CO_2 and other sparingly soluble gases is one of the mass transfer processes we shall consider in some detail in this chapter. We shall focus particularly on the supply of oxygen in aerobic fermentations.

We shall also examine heat transfer, as it is important to control temperature in almost all commercial biological reactors. Liquid-liquid mass transfer is encountered in aqueous-organic extraction systems for the recovery of biological products, and in aqueous two-phase systems for enzyme and whole cell reactions. These are described in Chapter 6.

In examining both theoretical approaches and empirical correlations to predicting transport rates in biological systems, the behavior of the aqueous phase containing cells or enzymes, nutrients and products is clearly important. We shall start by examining the rheological behavior of biological systems and then turn to the problems associated with heat and mass transfer and mixing.

5.1 The Rheology of Fermentation Broths

The rheological properties of the bulk phase (generally aqueous) in biological reactors influence the extent of mixing of that phase and consequently the rates of mass and heat transfer. In bioreactors containing fungi, for example in antibiotic fermentations, filamentous growth may lead to high degrees of entanglement and result in high broth viscosity. In many fermentations, inexpensive substrates containing particulates are employed, and these too can lead to broths with highly non-ideal characteristics. All fermentations containing microorganisms have the potential to exhibit complex rheology. The size of most microorganisms is small (1 to 20 microns), but at high cell concentrations deviations from Newtonian behavior are apparent. In addition, many fermentations result in the production of extracellular polymers, such as polysaccharides, which may result in strongly non-Newtonian rheological behavior. We shall review some fundamentals of the flow behavior of fluids and then examine some biological systems.

5.1.1 Rheological Models

Constitutive rheological models must be included in the set of dynamic equations that define flow behavior. These equations include the rheological model, a thermodynamic equation of state and mass, momentum and energy balances. The simplest constitutive relationship is Newton's law of viscosity. More complex models show the dependence of flow behavior on the flow field and even on the previous flow history of the fluid.

Newtonian Fluids

For unidirectional flow, Newton's law defines the viscosity μ (gm cm^{-1} sec^{-1}, or poise) as the ratio of the shear stress (τ_{xy} gm cm^{-1} sec^{-2}, or dyne cm^{-2}) to the shear rate dv_x/dy (sec^{-1}), which in rectangular coordinates is

$$\tau_{xy} = -\mu \frac{dv_x}{dy} \tag{5.1}$$

Most inviscid, incompressible fluids exhibit Newtonian behavior. Fermentation broths which contain simple salts and cells at low concentrations often follow this behavior, with viscosities not significantly different from that of water. Equation (5.1) can be generalized to other coordinate systems:

$$\underline{\underline{\tau}} = -\eta \underline{\underline{\Delta}} \tag{5.2}$$

where the shear stress tensor $\underline{\underline{\tau}}$ and the rate of shear tensor $\underline{\underline{\Delta}}$ are related by an apparent viscosity η, which is a constant in the case of Newtonian fluids, but generally depends on both the rate and duration of the shear. The case of the time dependent (history) behavior of viscosity with shear has not been examined in much detail for fermentation systems, but many non-ideal broths follow the non-Newtonian pseudoplastic behavior described below.

Non-Newtonian Fluids

When the apparent viscosity η varies with the shear rate, we can distinguish two types of behavior; *shear thinning* (or *pseudoplastic*) behavior, where the viscosity decreases with increasing shear rate; and *shear thickening* (or *dilatant*) behavior, where the viscosity increases with increasing shear rates. Fermentation broths are typically pseudoplastic and often exhibit a *yield stress*, i.e., a stress below which shear is not observed. The behavior of both types of fluids can be described by several models. For the case of no yield stress, the *power law model* is commonly employed. For unidirectional flows the model reduces to:

$$\tau_{xy} = -K\left(\frac{dv_x}{dy}\right)^{n-1}\frac{dv_x}{dy} \tag{5.3}$$

n is the power law index and K is the consistency index or rigidity. For pseudoplastic fluids, n<1 and for dilatant fluids n>1. Newtonian fluids have n=1. In the general case, the power law model is more complex, and can be written as

$$\underline{\underline{\tau}} = -\left\{K\left(\sqrt{\frac{1}{2}\underline{\underline{\Delta}}:\underline{\underline{\Delta}}}\right)^{n-1}\right\}\underline{\underline{\Delta}} \tag{5.4}$$

Here $\underline{\underline{\Delta}}$ is the rate of deformation tensor and $1/2\underline{\underline{\Delta}}:\underline{\underline{\Delta}}$ is the second invariant, which may be found in various coordinate systems in any of the standard fluid mechanics texts[1]. As we shall see later, much of the data on the behavior of mycelial broths has been correlated with the pseudoplastic power law equation. It describes the behavior of the system over a wide range of shear rates. However, at low shear rates a yield stress is often observed, i.e., a certain stress must be applied before the fluid will move. This is generally modelled by modifying the Newtonian equation, the result is known as the Bingham equation:

$$\underline{\underline{\tau}} = -\left(\mu_o + \frac{\tau_o}{\sqrt{\frac{1}{2}\underline{\underline{\Delta}}:\underline{\underline{\Delta}}}}\right)\underline{\underline{\Delta}} \qquad \text{for} \qquad \frac{1}{2}(\underline{\underline{\tau}}:\underline{\underline{\tau}}) > \tau_o$$

$$\underline{\underline{\Delta}} = 0 \qquad \text{for} \qquad \frac{1}{2}(\underline{\underline{\tau}}:\underline{\underline{\tau}}) < \tau_o \tag{5.5}$$

or for unidirectional flows

$$\tau_{xy} - \tau_o = -\mu_o\left(\frac{dv_x}{dy}\right) \qquad \text{for} \qquad |\tau_{xy}| > \tau_o$$

$$\frac{dv_x}{dy} = 0 \qquad \text{for} \qquad |\tau_{xy}| < \tau_o \tag{5.6}$$

(1) See for example *Transport Phenomena*, Bird, R.B., Stewart W. and E. Lightfoot, John Wiley.

A Bingham fluid which is at rest will thus not flow until a shear stress is imposed upon the system that exceeds the yield stress τ_o . A second model which includes a yield stress is the Casson equation

$$\sqrt{\tau_{xy}} = \sqrt{\tau_o} + k_c \sqrt{\frac{dv_x}{dy}} \tag{5.7}$$

The Casson equation fits the experimental data for mycelial broths in the low shear region better than the power law equation. Other models which have been used to describe non-Newtonian behavior, such as the Power-Eyring model, have not found much application in describing fermentation broth rheology. Although some broths (especially polysaccharide containing broths) exhibit viscoelastic behavior, viscoelastic models have rarely been used to describe their behavior, in part because the concentration of the polysaccharide changes during the course of the fermentation and the rheological parameters would require continuous estimation.

The rheological behavior of the main types of fluids is illustrated in Figure 5.1.

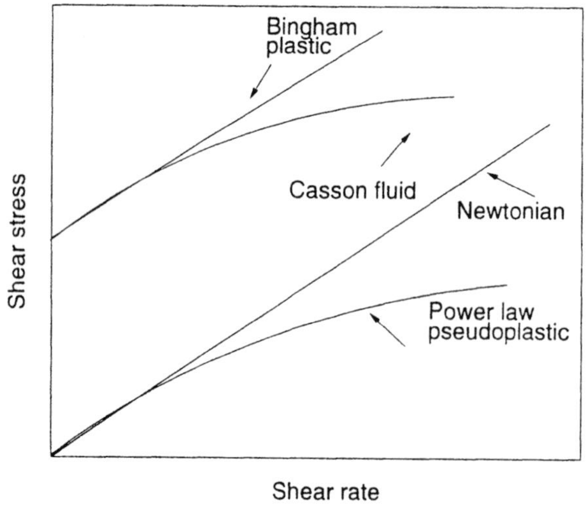

Figure 5.1. Stress-strain relationships for various rheological models commonly applied to fermentation broths.

Rheological models for non-Newtonian fluids are only empirical fits of experimental data. Generally, the constants so obtained are valid only over a limited range of shear rates. It is thus necessary to determine the constants over the range of shear that will be encountered in the actual process. The choice of device for measurement of viscosity will often be dictated by this shear range.

5.1.2 Measurement of Rheological Parameters

There are three general types of instruments commonly used for measuring rheological properties: 1) falling sphere viscometers, 2) capillary viscometers and 3) rotational viscometers. The first two types are seldom used with fermentation broths, while rotational viscometers are quite frequently employed.

The falling sphere is the simplest of all viscometers. The viscosity of a Newtonian fluid can be directly related to the settling velocity of a sphere by Stokes law:

$$\mu = \frac{2gR^2(\rho_s - \rho)}{v_t} \tag{5.8}$$

where v_t and R are the terminal settling velocity and radius of the sphere of density ρ_s. This method is limited to Newtonian fluids with Reynolds numbers Re < 2.

Capillary viscometers make use of the time for a specified volume of fluid to flow through a fine bore tube of known diameter. A momentum balance gives the shear stress at the wall

$$\tau_w = \frac{\Delta P R}{2L} \tag{5.9}$$

where ΔP is the pressure drop along the tube of length L and radius R. In the general case, the shear at the wall is given by the Rabinowitsch-Mooney equation

$$\gamma_w = \left(\frac{4Q}{\pi R^3}\right)\left[\frac{3}{4} + \frac{1}{4}\frac{d\log\left(\frac{Q}{2\pi R^3}\right)}{d\log\left(\frac{\Delta P R}{2L}\right)}\right] \tag{5.10}$$

where Q is the volumetric flow rate of the liquid. The derivative in Equation (5.10) can be evaluated by plotting $\log(Q/2\pi R^3)$ versus $\log(R\Delta P/2L)$ over a range of applied pressure drops or tube sizes and measuring the slope at arbitrary points. For Newtonian fluids (5.10) simplifies to

$$\gamma_w = \frac{4Q}{\pi R^3}$$

$$\mu = \frac{\tau_w}{\gamma_w} \tag{5.11}$$

There are a wide variety of rotational viscometers which operate by rotating a spindle of defined geometry within the fluid. The rotational rate of the spindle is related to the rate of shear and the torque required to turn the spindle is related to the shear stress within the fluid. In the coaxial cylinder viscometer (Couette viscometer) the fluid is placed in the annulus and the inner cylinder is rotated. For a Newtonian fluid, the shear rate within the fluid is related to the angular velocity of the cylinder:

$$\gamma = \frac{2\Omega}{r^2}\frac{R_1^2 R_2^2}{(R_2^2 - R_1^2)} \tag{5.12}$$

r is the radial position in the fluid, R_1 and R_2 are the radii of the inner and outer cylinders, respectively, Ω is the angular velocity of the cylinder. If the fluid gap is small relative to the cylinder radius, the shear rate can be assumed constant throughout the fluid and

$$\gamma = \frac{\Omega R_1}{(R_2 - R_1)} \tag{5.13}$$

The shear stress can be expressed as a function of radial position $\tau = M/2\pi r^2 h$, where M is the applied torque and h is the height of liquid in contact with the inner cylinder. Thus the Newtonian viscosity can be determined from

$$\mu = \frac{\tau}{\gamma} = \frac{M}{4\pi h\Omega}\left(\frac{1}{R_1^2} - \frac{1}{R_2^2}\right) \tag{5.14}$$

For non-Newtonian fluids, the assumption of constant shear greatly simplifies the analysis. Further details can be found in handbooks and reviews[2].

One of the difficulties in the use of coaxial viscometer with suspensions, such as mycelial broths, is the separation of the mycelium from the wall, leading to falsely low viscosities. This is usually apparent when the viscosity decreases with time as the microbial phase settles and separates from the wall. In addition, microbial coagulation can occur, resulting in the formation of loose aggregates or densely packed pellets. If particles near the size of the gap are formed, the shear profile is significantly distorted and erroneous rheological measurements result.

Impeller viscometers been employed to overcome some of the difficulties in determining the behavior of fungal suspensions. In an impeller viscometer, the cylindrical cylinder of a Couette viscometer is replaced with a turbine impeller. Due to the mixing induced by the impeller, phase separation is minimized. The geometric similarity between this type of viscometer and the stirred tank suggest that the hydrodynamic conditions within the viscometer will resemble those within a fermenter. Details of the original impeller viscometers may be found in Bongenaar et al. and Roels et al.[3]

The fluid dynamics around a turbine impeller are too complex to allow for direct theoretical analysis. Thus the rheological properties measured by an impeller viscometer

---

(2) See Mavituna and B. Atkinson, *Biochemical Engineering Handbook*, and in Oolman, T.O. and H. W. Blanch, *Non-Newtonian Fermentation Systems*, CRC Critical Reviews in Biotechnology, **4**, 133 (1986).

(3) Bongenaar, J., Kossen, N.W., Metz, B., and F. Meiboom, *Biotech. Bioeng*, **15**, 201 (1973) and Roels, J., van den Berg, J. and R. Voncken, *Biotech. Bioeng.*, **16**, 181 (1974).

are not necessarily those defined by Equation (5.1) and (5.2). However, by using power number versus Reynolds number correlations (see Section 5.4.4), a simple relation between shear stress and measured torque in an impeller viscometer can be derived[4]:

$$M = C_1\tau \tag{5.15}$$

where C_1 is an instrument constant dependent on the geometric configuration. This above expression is limited to laminar flow. The shear rate at the impeller can be related to the impeller tip speed[5]:

$$\gamma = kN \tag{5.16}$$

where k depends on the impeller geometry. Thus the rheology of the broth can be determined by measuring the torque at various impeller speeds.

5.1.3 Examples of Microbial Broth Rheology

The rheological properties of microbial broths are primarily determined by cell concentration and morphology. High cell concentrations tend to produce viscous broths. Filamentous morphology and intracellular structuring often result in non-Newtonian behavior. Metabolic products, such as polysaccharides and extracellular proteins, and solid substrates may also result in non-Newtonian behavior. Thus the age of the culture is significant in determining its rheological behavior.

Newtonian Broths

Broths containing cells which are approximately spherical and in low concentration are usually Newtonian and have a viscosity which varies with cell concentration. Yeast and bacterial cultures exhibit this behavior, which can be represented in the following form

$$\mu_s = \mu_L(1 + f(J, \Phi)) \tag{5.17}$$

where μ_s is the viscosity of the suspension, μ_L is the viscosity of the suspending liquid, Φ is the volume fraction of the particles and J is a geometric ratio for the particles. At low volume fractions, yeast suspensions can be represented by Einstein's equation

$$\mu_s = \mu_L(1 + 2.5\Phi) \tag{5.18}$$

and at higher concentrations (up to 14% volume fractions) the Vand equation represents the data more closely (see Figure 5.2).

$$\mu_s = \mu_L(1 + 2.5\Phi + 7.25\Phi^2) \tag{5.19}$$

---

(4) Roels, J., van den Berg, J. and R. Voncken, *Biotech. Bioeng.*, **16**, 181 (1974).

(5) Metzner, A. B. and R.E. Otto, *AIChE*, **3**, 3 (1957)

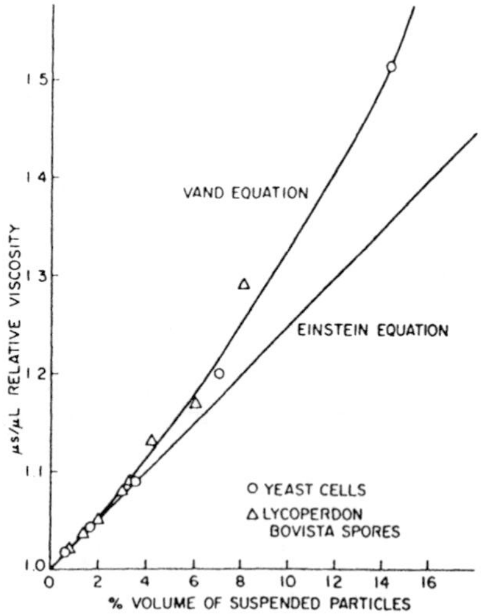

Figure 5.2. *The viscosity of suspensions of yeast and L. bovista spores at various volume fractions. At higher volume fractions the Vand equation can be seen to represent the experimental data more closely.*

As the osmotic pressure of the suspending fluid changes, the diameter of yeast changes and consequently so does the the broth viscosity. Reuss et al[6] have proposed the following relation:

$$\frac{\mu_s}{\mu_L} = \left(1 - \frac{1}{2}h_s\Phi\right)^{-1} \qquad \text{where} \qquad h_s = 0.0487\Pi_{osm} + 1.59 \tag{5.20}$$

Here h_s can be interpreted as the maximum packing volume of the suspension as a function of the osmotic pressure in bar.

Non-Newtonian Broths

A broth containing filamentous fungi usually gives rise to pseudoplastic rheological behavior, often showing a yield stress. Fungal broths are "structured" as a result of the intertwining of the hyphae and hydrogen bonding of the electronegative surfaces of the cells. At low shear stress, this structure is maintained. If the cells also adhere to solid surfaces, a yield stress is observed. However, as the shear stress increases, the hyphal mass breaks into large fragments and circulates throughout the reactor. With increasing rates of shear, the broth structure gradually disintegrates and results in a decrease in the apparent viscosity of the fluid; i.e., shear thinning or pseudoplastic behavior. At sufficiently high rates of shear,

(6) Reuss, M., Debus, D. and G. Zoll, *Chem. Eng.*, p233 June, 1982.

individual cells will be formed and a return to Newtonian behavior can be expected. Thus fungal morphology must be considered in describing filamentous broth rheology by a power law model, as the power law constant will vary.

As the cell concentration increases, the apparent viscosity increases in a non-linear manner. Various correlations have been proposed, typically of the form

$$\eta \sim X^{m} \tag{5.21}$$

where X is the biomass concentration, and we dentoe the apparent viscosity by η. The exponent m has been reported to vary from 1.0 to 2.65 with filamentous molds, and from 0.3 to 3.0 when pellets are formed. Variations in stress with biomass concentration are generally correlated viz:

$$\tau \sim X^{2.3-2.5} \tag{5.22}$$

The Casson equation has been applied to penicillin broth rheology with some success. Roels et al.[7] incorporated the influence of cell concentration and morphology into the Casson equation by assuming the organism to be shaped like a coiled filament, analogous to a polymer chain. Using the excluded volume concept, the Casson equation was modified;

$$\sqrt{\tau} = \sqrt{\delta_1} X \cdot [1 + f(X)\sqrt{\gamma}] \tag{5.23}$$

Here δ_1 is a geometric parameter for the hyphae and f(x) is a complex function of δ_1 and X. Thus both parameters of the Casson equation, the yield stress (τ_o) and the consistency index (k_c) in Equation (5.7), are dependent on the cell concentration and the cell morphology. At low cell concentrations, the above equation reduces to Newtonian behavior. Figure 5.3 shows the behavior of penicillin broths at various times, plotted with curves based on the Casson model and the Bingham model for comparison.

Unlike other fermentation broths, the rheology of microbial polysaccharide broths is generally controlled by the extracellular product concentration rather than by the cell concentration. The rheology of *Xanthamonas campestris* broth, shown in Figure 5.4, exhibits pseudoplastic, power-law behavior, with the consistency index depending on xanthan concentration. Very high viscosities can be attained with these polysaccharides and they find use as thickening agents in foods and related products (eg. salad dressings and toothpastes) and in secondary and tertiary oil recovery. There are several reviews of the rheology of fermentation broths that may be consulted for further examples[8].

---

(7) Roels, J.A., van den Berg, J. and R.M. Vonken, *Biotech. Bioeng.*, **16**, 181 (1974).

(8) See the reviews by Banks, G.T., *Topics in Enzyme and Fermentation Technology*, ed. A. Wiseman, **1**, 72 (1977); Blanch H.W. and S. M. Bhavaraju, *Biotech. Bioeng.*, **18**, 745 (1976); Charles, M. *Adv. Biochem. Eng.*, **8**, 1 (1978); and Oolman, T.O. and H.W. Blanch, *CRC Critical Reviews in Biotechnology*, **4** (2), 133 (1986).

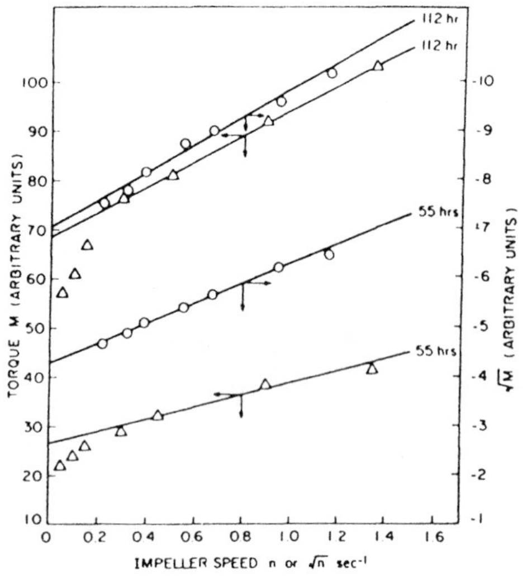

***Figure 5.3**. The rheological behavior of penicillin broths at various times (indicated on the figure). The data have been correlated with the Casson equation (indicated by circles) (as $\sqrt{\tau}$ versus $\sqrt{n}$) and by the Bingham model (triangles) (as τ versus n).*

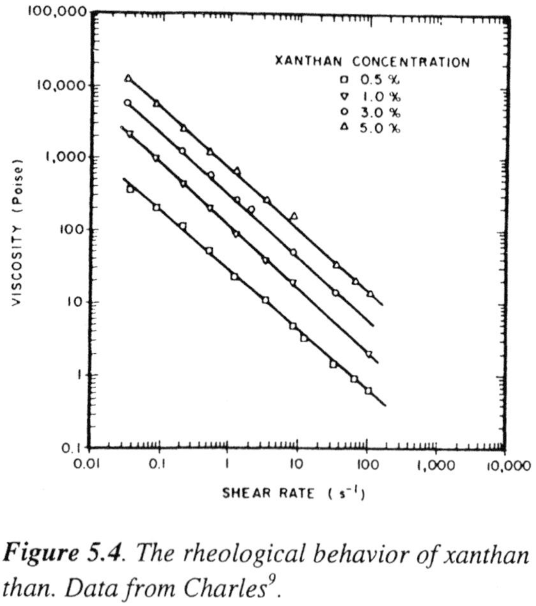

***Figure 5.4**. The rheological behavior of xanthan broths at various concentrations of xanthan. Data from Charles[9].*

(9) Charles, M. *Adv. Biochem. Eng.*, **8**, 1 (1978).

5.2 Gas-Liquid Mass Transfer

In this section, we shall examine the mass transfer pathways involved in the transport of gaseous nutrients and products, primarily oxygen and carbon dioxide, between the gas phase and the cell. In aerobic bioreactors, the entire multiphase suspension is mixed, either mechanically or pneumatically, to suspend solids and enhance the transfer of these nutrients and products. The pathway for transport of oxygen is illustrated in Figure 5.5. Resistance to mass transfer can be encountered at eight possible locations (1) in the gas film; (2) at the gas/liquid interface; (3) in the liquid film surrounding the gas interface; (4) in the liquid phase; (5) in the liquid film surrounding the solid (microorganism); (6) at the liquid/solid interface; (7) in the solid phase; and (8) at the site of reaction (within the microorganism). These resistances occur in series and the largest of them will be rate controlling. Thus the entire mass transfer pathway can be modelled using a single mass transfer correlation.

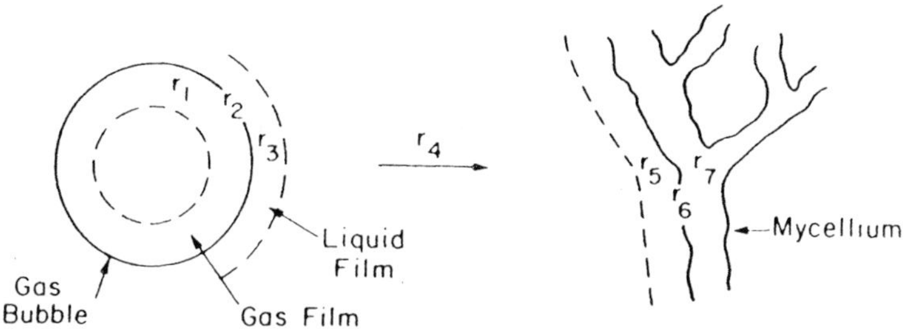

Figure 5.5. *The pathways of oxygen transfer from the gas phase to a microorganism in a bioreactor.*

We can eliminate some of these resistances from consideration as rate controlling. Since gas-phase mass diffusivities are typically much higher than liquid-phase diffusivities, the resistance of the gas film within the bubble (1) can be neglected relative to the liquid film surrounding the bubble. Similarly, the interfacial resistance to transport (2) is small and we may neglect resistance (6) as well. Provided the liquid is well-mixed, transport through the liquid phase is generally rapid and (4) may be neglected.

Three mass transfer resistances and the reaction rate (8) remain to be considered. The mass transfer resisitances are the two liquid film resistances and the intraparticle resistance. Depending on the size of the microbial particle (for example, either a single cell of 1-2 micron or a microbial pellet of several millimeters in diameter), any one of these resistances can be controlling. In the case of bacteria or yeast cells, their very small size (and hence large interfacial area) relative to that of a gas bubble will result in the liquid film surrounding the gas bubble being the rate determining step in oxygen transport. On the other hand, large microbial pellets or fungal hyphae may be of a size comparable to that of a bubble (approx. 4 to 5 mm) and resistance (5) in the liquid film surrounding the solid may dominate. The intraparticle resistance results from diffusion and reaction of oxygen within the pellet or fungal matte and depends on the effective diffusivity of oxygen and relative rates of reaction. We shall consider some examples of control by this type of resistance in Section 5.6.

The rheology of the liquid phase has a strong influence on the rate of mass transfer. With an inviscid liquid, very little agitation is required to maintain a well-mixed liquid phase and mass transfer through the liquid bulk is unhindered. With a more viscous broth, mixing is less efficient and the presence of a yield stress may result in regions of stagnant fluid near the walls and internals of the reactor. This can result in a substantial resistance to transport through the bulk liquid phase. The liquid phase rheology also influences the velocity of the gas bubble or microbial solid phase relative to that of the fluid. Thus the liquid film resistances will be increased. We shall consider two categories of mass transfer; first, when diffusion through the liquid film controls and second, when bulk mixing patterns influence the rate of transport.

5.2.1 Diffusion in Biological Media

Fick's law of diffusion defines the diffusion coefficient D_L. For unidirectional, steady-state diffusion in stagnant medium, the molar flux of a species A is given by

$$J_A = -D_L \frac{dC_A}{dx} \tag{5.24}$$

here dC_A/dx is the molar concentration gradient in the direction of diffusion. Values of the diffusivity D_L for low molecular weight solutes in water range from 0.5 to 2.0 x 10^{-5} cm^2sec^{-1}. In non-viscous liquids, the diffusivity may be estimated from the Stokes-Einstein equation:

$$D_L = \frac{kT}{6\pi r_o \mu} \tag{5.25}$$

where r_0 is the solute spherical radius, μ is the solvent viscosity, k is the Boltzmann constant and T the absolute temperature. For small solutes, the Wilke-Chang equation provides values of the diffusivity accurate to within 10 to 15%.

$$D_L \quad (cm^2 \sec^{-1}) \quad = 7.4x10^{-8} \frac{T(\chi M)^{1/2}}{\mu_L V_m^{0.6}} \tag{5.26}$$

where

M = solute molecular weight

V_m = molecular volume of solute at its boiling point (cm^3/gm mole)
μ_L = liquid viscosity (centipoise)
T = absolute temperature in K

Values of V_m may be estimated from Le Bas atomic volumes. Details may be obtained in any of the standard texts[10]. For water, the association parameter χ is 2.26, as a result of hydrogen bonding. For oxygen, the diffusion coefficient in water at 25°C, D_L has the value 2.25 x 10^{-5} cm^2/sec. As the viscosity of the solvent increases, the diffusivity decreases, however for polysaccharide broths or fungal broths the diffusivity may not decrease to the extent predicted by Equation (5.26) as the solution is "structured" and small solutes may diffuse through the interstices. Data for D_L in polymer solutions and values of D_L in albumin and globulin solutions may be found in Gainer et al.[11]; they are only slightly lower than the corresponding values for solutes diffusing in water.

Diffusion Coefficients for Proteins

In solution, proteins do not behave as rigid spheres or rods and attempts to predict their diffusion coefficients based on the Stokes-Einstein equation, with a rigid sphere diameter, are not always satisfactory. Proteins are solvated and swell; they may be quite flexible in solution and exhibit hydrodynamic behavior resembling that of a randomly coiled polymeric chain. Their spatial requirements in solution can be better described by their radius of gyration, r_g. The radius of gyration is very sensitive to molecular shape and size. Data for the radius of gyration can be obtained from light scattering techniques (e.g., small angle X-ray scattering, quasi-elastic laser-light scattering). Based on r_g, an equivalent hydrodynamic radius r_e can be defined on the assumption that the protein behaves as a randomly coiled chain, viz. $r_e = \xi\, r_g$. The Stokes-Einstein equation thus becomes

$$D = \frac{kT}{6\pi\mu\xi r_g} \tag{5.27}$$

This approach has been examined with a wide range of proteins by Tyn and Gusek[12], who found data for globular proteins could be correlated very well by

$$D = \frac{5.78x10^{-8}T}{\mu r_g} = \frac{1.69x10^{-5}}{r_g} \quad \text{at T = 298.3K} \quad \mu\text{= 1.002 cP} \tag{5.28}$$

where T is in K, the solution viscosity μ is in centipoise and r_g is in Angstrom. In the case of rod-like proteins, the above equation is not entirely satisfactory, underpredicting the experimental diffusivities. For such proteins, their length is a more appropriate correlator. When one-tenth the length L is used in place of r_g, considerably better agreement results.

---

(10) See Sherwood, T.K., Pigford, R. and C.R. Wilke, *Mass Transfer*, McGraw Hill, New York (1975).

(11) Gainer, J.L. and A. B. Metzner, *AIChE Inst. Chem. Eng. Joint Mtg. (London)* June 13-17, 1965; Gainer, J.L., *Ind. Eng. Chem. Fund.*, **9**, 381 (1970).

(12) M. T. Tyn and T.W. Gusek, *Biotech. Bioeng.*, **35** 327 (1990).

The length may be calculated from the radius of gyration from the relationship proposed by Tyn and Gusek; $L = \sqrt{12}\, r_g$. DNA behaves as a "quasi-coil" in solution, rather than as a rod-like particle at high molecular weights ($>10^6$) and is better treated as a sphere. Some representative values of protein diffusivities are shown in Table 5.1.

Table 5.1. *Experimental diffusion coefficients and radii of gyration for proteins (based on data summarized by Tyn and Gusek).*

Protein	Molecular Weight	D_{exp} ($\times 10^7$ cm^2/sec)	Radius of gyration r_g (Angstrom)
ribonuclease	12,640	13.1	14.8
lysozyme (chicken)	14,400	11.8	15.2
chymotrypsin B (bovine pancreas)	21,600	10.2	18.0
pepsin (swine)	34,160	8.7	20.5
ovalbumin	45,000	7.3	24.0
hemoglobulin (man)	64,500	6.3	24.8
bovine serum albumin	66,000	5.93	29.8
γ-globulin (human serum)	153,100	4.0	70.0
myosin (rabbit muscle)	570,000	1.0	468
DNA	4,000,000	0.13	1170
tobacco mosaic virus	39,000,000	0.44	924

5.2.2 Solubility of Gases in Biological Media

The solubility of gases in liquid media for slightly to moderately soluble gases is described by Henry's law;

$$C_A = \frac{p_A}{H} = \frac{P}{H} y_A \tag{5.29}$$

where C_A is the concentration in the liquid phase (gm/liter), p_A is the gas phase partial pressure of A (atm), H is the Henry's law coefficient (atm-liter/gm) and y_A is the mole fraction of A in the gas phase. Often solubilities are reported in terms of the *Bunsen coefficient*, which corrects the solubility data to a standard pressure assuming ideal behavior. The Bunsen coefficient α is defined as the volume of gas, measured at 0°C and 1 atmosphere, absorbed by unit volume of solvent at a partial gas pressure of 1 atm.

Oxygen solubility

For oxygen in water, the table below illustrates the variation in Henry's law constant with temperature.

Table 5.2. *The solubility of oxygen in water and values of the Henry's law coefficient.*

Temperature (°C)	Oxygen concentration (C_{O2} mg/l) in equilibrium with air (p_{O2} = 0.209 atm)	Henry's constant (atm-liter-mg^{-1})
25	8.10	0.0258
35	6.99	0.0299

Oxygen solubility in water can be correlated with temperature;

$$C_{O2} = 14.16 - 0.394T + 0.007714T^2 - 0.000646T^3 \quad (3.30)$$

where C_{O2} is in mg/l and T is in °C.

Table 5.3. *Solubilities of oxygen and carbon dioxide as given in the correlation of Wilhelm et al.*

Gas	Temperature range (K)	A (cal mol^{-1} K^{-1})	B (cal mol^{-1})	C (cal mol^{-1} K^{-1})	D (cal mol^{-1} K^{-2})
Oxygen	274-348	-286.942	15,450.6	36.5593	0.0187662
Carbon dioxide	273-353	-317.658	17,371.2	43.0607	-0.00219107

A general approach to describe the temperature dependence of gas solubilities is given by Wilhelm et al.[13], who propose the following equation for the equilibrium gas mole fraction X at 1 atmosphere pressure as a function of temperature T in K.

$$R \ln X = A + \frac{B}{T} + C \ln T + DT \quad (3.31)$$

Values of the constants for O_2 and CO_2 are given in Table 5.3.

Carbon Dioxide

Carbon dioxide is considerably more soluble than oxygen. The dissolution of carbon dioxide into water is more complex because of the liquid phase reactions of CO_2. The possible solute species are CO_2, HCO_3^-, H_2CO_3 and CO_3^{2-}. The equilibrium relationships between these species at 25°C are given by:

(13) Wilhelm, E., Battino, R., and Wilcock, R.J., *Chem. Rev.*, 77, 219 (1977).

$$\frac{[H^+][HCO_3]}{[CO_2]+[H_2CO_3]} = K_1 = 5.01\text{x}10^{-7}M$$

$$\frac{[H^+][CO_3^{-2}]}{[HCO_3^-]} = K_2 = 5.62\text{x}10^{-11}M \tag{5.32}$$

The dissolved carbon dioxide concentration $[CO_2]$ is thus pH sensitive:

$$[CO_2]_T = [CO_2]+[H_2CO_3]+[HCO_3^-]+[CO_3^{2-}] = [CO_2]\left(1+\frac{K_1}{[H^+]}+\frac{K_1K_2}{[H^+]^2}\right) \tag{5.33}$$

Below pH 5, nearly all the carbon dioxide is present as dissolved CO_2; bicarbonate dominates when 7<pH<9 and carbonate for pH>9. Depending on pH, the removal of CO_2 from the liquid to the gas phase may be controlled by chemical reaction (e.g., the bicarbonate to CO_2 reaction) or by the physical dissolution process. For example at 25°C the deprotonation of H_2CO_3 is quite rapid

$$H_2CO_3 \Leftrightarrow HCO_3^- + H^+$$

$$K_{eq} = \frac{[H^+][HCO_3^-]}{[H_2CO_3]} = 2.5\text{x}10^4\ M \quad \text{at} \quad 25^oC$$

and

$$H_2CO_3 \underset{k_{-1}}{\overset{k_1}{\Leftrightarrow}} CO_2 + H_2O$$

The second reaction is slower, with $k_1 = 20\ \text{sec}^{-1}$ and $k_{-1} = 0.03\ \text{sec}^{-1}$.

The Effect of Salts

In the presence of salts and organic solutes, the solubilities of sparingly soluble gases decrease (the salting-out effect). Only in a few cases, such as with short chain alcohols, does the solubility increase. As fermentation broths typically contain salts and sugars, corrections must be made to the pure solution solubilities. Schumpe et al.[14] review methods for measuring and predicting gas solubilities in fermentation media. For the case of strong electrolytes, the decrease in solubility of a solute may be correlated from a modification of the Sechenov equation (from Schumpe et al.)

$$\log\frac{\alpha_o}{\alpha} = \sum_{i=1}^{n} H_i I_i \tag{3.34}$$

where $I_i = \frac{1}{2}c_i z_i^2$

(14) Schumpe, A., Quicker, G., and W.D. Deckwer, *Adv. Biochem. Eng.*, **24**, 1 (1982).

I_i is the ionic strength attributable to a single ion, the parameter H_i is specific to the gas, the ion and the temperature, z_i is the charge on the ith salt species, c_i is the molar concentration of the ionic species (given by $x_i c_{elec}$, x_i being the number of ions of type i and c_{elec} being the molar concentration of the electrolyte), α is the Bunsen coefficient for the gas in the electrolyte solution and α_o refers to the Bunsen coefficient in pure water. For a single salt solution, the Sechenov coefficient can thus be related to H_i

$$K_s = \frac{1}{2}\sum_{i=1}^{n} H_i x_i z_i^2 \tag{3.35}$$

where $\log\frac{c_o}{c} = K_s c_{elec}$

Some typical values of H_i for oxygen at 25°C are given in Table 5.4.

***Table 5.4**. Values of H_i for various cations and anions.*

Cation	H_i (liter/mole)	Anion	H_i (liter/mole)
Na^+	-0.568	Cl^-	0.849
K^+	-0.587	NO_3^-	0.802
NH_4^+	-0.704	$H_2PO_4^-$	0.997
Ca^{2+}	-0.309	SO_4^{2-}	0.460
Mg^{2+}	-0.297	HPO_4^{2-}	0.477

The Effect of Organic Compounds

The reduction in gas solubility due to the presence of organic compounds can be represented in an analogous manner to that for salts.

$$\log\frac{\alpha_o}{\alpha} = \sum_{i=1}^{n} K_i c_{org,i} \tag{3.36}$$

K_i is an empirical constant that corresponds to the Sechenov constant. Some selected values of K_i for oxygen are given in the Table 5.5.

In media which contains both salts and organic compounds, the above solubility equations may be combined to include electrolyte and other solute effects:

$$\log\frac{\alpha_o}{\alpha} = \sum_{i=1}^{n} H_i I_i + \sum_{i=1}^{n} K_i c_{org,i} \tag{3.37}$$

Calculation of Oxygen solubility in a Fermentation Medium

Here we will calculate the Bunsen coefficient for oxygen in a simple glucose and salts medium. The medium has the composition indicated in Table 5.6 below, with the corresponding values of K for organics or salts determined from the expression $1/2H_i x_i z_i^2$. The medium is at 30°C.

***Table 5.5**. Values of the parameter K_i for oxygen for various organic compounds in water (from Schumpe et al.)*

Organic Compound	Concentration range (gm/l); temperature (°C)	K_i (x 10^{-4} liters/gram)
glucose	0-200 (37)	6.78
lactose	0-300 (25)	5.71
sucrose	0-200 (25)	4.36
citric acid	0-200 (25)	5.09
glycerol	0-300 (37)	4.07
glycine	0-200 (37)	12.46
albumin (bovine)	0-300 (37)	1.60
yeast extract	0-60 (30)	6.2

***Table 5.6**. Composition of a typical fermentation medium.*

Component	Concentration (gm/l)	moles/liter	K (liters/mole)
glucose	30	0.167	0.119
$(NH_4)_2HPO_4$	1.8	0.014	0.250
$(NH_4)_2SO_4$	6	0.045	0.216
$MgSO_4$	0.22	0.002	0.326
$CaCl_2$	0.32	0.003	0.231
KCl	0.87	0.012	0.131

The K values were calculated in the table above from the preceding data on H_i and K_i, modifying the values given at 25°C, as follows. For $(NH_4)_2HPO_4$ we find:

$$K = \frac{1}{2}\sum_{i=1}^{2} H_i x_i z_i^2 = \frac{1}{2}[-0.704x2x1^2 + 0.477x1x2^2] = 0.250 \quad \text{(liter/mole)} \tag{3.38}$$

The Bunsen coefficient for oxygen in the medium is then calculated as follows:

$$\log\frac{\alpha_o}{\alpha} = \sum_{i=1}^{6} K_i c_i = 0.167x0.119 + 0.014x0.250 + 0.045x0.216 + 0.002x0.326$$

$$+ 0.003x0.231 + 0.012x0.131 = 0.03601 \tag{3.39}$$

Using a standard Bunsen coefficient (α_o) at 30°C of 0.02635, we can determine the actual Bunsen coefficient:

$$\alpha = \alpha_o 10^{-0.03601} = 0.02635x10^{-0.03601} = 0.02425 \quad (3.40)$$

5.2.3 Mass Balances for Two-Phase Bioreactors

In this section we shall examine the transport of oxygen from the gas phase to the liquid phase in various bioreactor configurations. In most types of bioreactors, the flow behavior of both phases depends on the scale of the equipment employed. In small scale, laboratory equipment, the liquid phase can generally be considered to be well mixed, while the gas phase behavior may be well-mixed or plug flow, depending on whether mechanical agitation is employed or not. It is important to recognize these two types of behavior in writing the appropriate mass balances for oxygen transfer from one phase to the other. We shall develop balance equations for both types of systems. If the liquid phase is not well-mixed, we need to account for some degree of liquid dispersion. Above a critical impeller speed, most mechanically agitated reactors with low viscosity fluids approximate well-mixed behavior. In large bioreactors, the situation is less clear and care must be taken to delineate the liquid flow regime. In cases where the degree of gas absorption is high, the assumptions of well-mixed or plug flow behavior of the gas phase may predict absorption rates which differ by an order of magnitude. Thus it is apparent that care must be taken in developing models of these two-phase systems.

Well-Mixed Liquid and Well-Mixed Gas

When the gas phase and the liquid phase can both be considered to be well mixed, we can develop coupled equations which describe the transfer from the gas to the liquid phase. The situation is shown in Figure 5.6.

The rate of oxygen transfer from the gas to the liquid phase is given by

$$N_{O_2} \cdot a = K_g a\left(p_{O_2} - HC_{O_2}\right) = K_g aP\left(y - \frac{H}{P}C_{O_2}\right) \quad (5.41)$$

Each terms is defined as follows (with typical units shown):

N_{O2} molar oxygen flux {moles O_2 transferred-(m^2 interfacial area)$^{-1}$-(hour)$^{-1}$}

a interfacial area per unit reactor volume {m^2-(m^3 liquid volume)$^{-1}$}

K_g overall gas phase mass transfer coefficient {moles-(hour.m^2 interfacial area-atm pressure)$^{-1}$}

p_{O2} partial pressure of oxygen in the gas phase {atm pressure}

y mole fraction of oxygen in the gas phase {moles O_2-(total moles)$^{-1}$}

C_{O2} oxygen concentration in the liquid phase {moles O_2-(m^3 liquid volume)$^{-1}$}

H Henry's law constant {atm pressure-(moles/m^3)$^{-1}$}

P total pressure {atm}

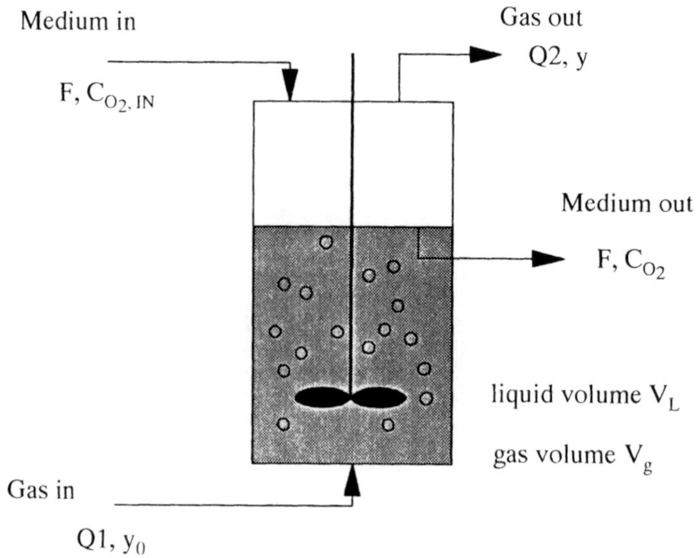

Figure 5.6. *A bioreactor with well-mixed gas and liquid phases. The mole fraction of oxygen in the gas is given by y and the concentration of oxygen in the liquid is C_{O2}. The volumetric flows of gas into and from the reactor are given by Q_1 and Q_2.*

The mass balance on the gas phase includes flow terms in and out of the reactor and a term describing transport from the gas phase.

$$\frac{d}{dt}\left(\frac{(NV_L V_b)P\,y}{RT}\right)=\left(\frac{P}{RT}\right)(Q_1 y_o - Q_2 y) - K_g a P V_L\left(y - C_{O_2}\cdot\frac{H}{P}\right) \tag{5.42}$$

where aV_L is the total area for mass transfer, NV_LV_b is the total gas volume, N is the number of bubbles per liquid volume, V_b is the bubble volume and the overall gas-phase mass transfer coefficient K_ga includes the gas and liquid film resistances, given by

$$\frac{1}{K_g a}=\frac{1}{k_g a}+\frac{H}{k_L a} \tag{5.43}$$

We have implicitly assumed that the bubble volume remains constant and in what follows, the pressure will be assumed constant. The assumption of constant bubble volume will be accurate for oxygen if the amount of oxygen transferred is small or for other gaseous solutes if their gas-phase concentration is small. If there is significant variation of pressure within the bubble column, its dependence on the height of the column must be included[15]. The liquid phase mass balance for oxygen yields:

(15) The general gas and liquid phase mass balance equations are developed in R. W. Schaftlein and T.W.F. Russell, *Ind. Eng. Chem.*, **60**, (5) 12 (1968).

$$\frac{d(V_L C_{O_2})}{dt} = F\left(C_{O_2,\text{in}} - C_{O_2}\right) + K_g a V_L P\left(y - C_{O_2}\frac{H}{P}\right) - r_{O_2} V_L \tag{5.44}$$

where r_{O2} is the volumetric rate of consumption of oxygen by reaction in the liquid phase. Well-mixed gas and liquid phases can be assumed only in small reactors with inviscid liquids. In larger reactors, the gas phase may be substantially depleted of oxygen as the gas bubbles rise through the liquid. We need then to consider the variation of oxygen concentration in the gas phase with vessel height.

Well-Mixed Liquid and Plug-Flow Gas

Here there is an axial dependence of the gas phase composition as well as time variation. The overall oxygen balance on the gas phase now becomes

$$\frac{\partial\left(\frac{PV_b}{RT} y\right)}{\partial t} = -v_b \frac{\partial\left(\frac{PV_b}{RT} y\right)}{\partial z} - K_g a' V_b P\left(y - C_{O_2}\frac{H}{P}\right) \tag{5.45}$$

The bubble rise velocity is v_b and a' is the bubble surface area to bubble volume ratio. (Note this is different to a, the interfacial area per liquid phase volume). We assume here that the pressure variation with height can be neglected[16]. The corresponding balance for a well-mixed liquid phase is identical to Equation (5.44), except that the gas phase composition y must be replaced with an average value $\hat{y}$, given by

$$\hat{y} = \frac{\int_0^{H_L} y(z,t)dz}{\int_0^{H_L} dz} = \frac{1}{H_L}\int_0^{H_L} y dz \tag{5.46}$$

If the gas phase oxygen concentration can be considered to be time invariant (as might be the case in a continuous stirred tank reactor at steady state), the gas phase mole fraction y depends only on position z (i.e., y = y(z)) and the liquid composition will only depend on time (i.e., $C_{02} = C_{02}(t)$), thus uncoupling the equations. We can now solve the gas phase balance by assuming the bubble volume V_b remains constant. In that case, the rise velocity v_b and a' will also be constant.

$$v_b \frac{d\left(\frac{PV_b}{RT} y\right)}{dz} = -K_g a' V_b P\left(y - C_{O_2}\frac{H}{P}\right) \tag{5.47}$$

Intergrating this over the vessel height

$$\int_{y_{\text{in}}}^{y} \frac{dy}{\left(y - \frac{H}{P} C_{O_2}\right)} = -\int_0^z \frac{K_g a' V_b RT}{v_b V_b} dz$$

(16) In very tall vessels this will be incorrect and the effect of hydrostatic pressure variation must be explicitly included.

$$\log\frac{\left(y-\frac{H}{P}C_{O_2}\right)}{\left(y_{\text{in}}-\frac{H}{P}C_{O_2}\right)}=-\frac{K_g a' RT}{v_b}z$$

$$y(z)=\frac{H}{P}C_{O_2}-\left(y_{\text{in}}-\frac{H}{P}C_{O_2}\right)\exp\left(-\frac{K_g a' RT}{v_b}z\right) \tag{5.48}$$

and the average gas phase oxygen mole fraction is thus

$$\hat{y}=\frac{H}{P}C_{O_2}\left(1-\frac{1-e^{-n}}{n}\right)+y_{\text{in}}\left(\frac{1-e^{-n}}{n}\right)$$

where $$n=\frac{K_g a' RTH_L}{v_b} \tag{5.49}$$

and thus the liquid phase oxygen balance becomes

$$\frac{d(V_L C_{O_2})}{dt}=F\left(C_{O_2,\text{in}}-C_{O_2}\right)+K_g a V_L P\left(\hat{y}-C_{O_2}\frac{H}{P}\right)-r_{O_2}V_L \tag{5.50}$$

At steady state the time derviative is zero and rearranging and dividing by F and noting that a can be written as $a'V_bN$ gives

$$C_{O_2}=C_{O_{2,\text{in}}}+\left(y_{\text{in}}-C_{O_2}\frac{H}{P}\right)(1-e^{-n})\left[\frac{PV_bNV_Lv_b}{RTFH_L}\right]-r_{O_2}\frac{V_L}{F} \tag{5.51}$$

which can then be solved if the volumetric oxygen uptake rate expression is known for the microbial or enzymatic system under consideration.

The extent to which these equations can be employed directly depends on the flow patterns in the vessel under consideration. In bubble columns which have large height to diameter ratios, the rise of the gas through the liquid induces a liquid circulation pattern which is not well-mixed and axial dispersion of the liquid must then be considered. We shall examine this in the following section. The second difficulty is in determining the value of K_ga. We shall thus develop expressions for the interfacial area and the mass transfer coefficient for bubble columns and stirred tanks in the following sections.

5.3 Bubble Columns

Bubble columns are widely used for production of antibiotics, bakers yeast and single cell protein. The height to diameter ratio of may vary, but values from 3 to 1 up to 6 to 1 are common. The gas phase rises in plug flow through the liquid, which may not be well-mixed in large tanks. In most cases of interest, the liquid film resistance controls the overall mass transfer rate and the coefficient K_ga can be replaced with the liquid film resistance $(1/H)k_La$. If the size of the bubbles in the vessel can be predicted, it is possible to calculate k_La directly.

Gas is introduced into the liquid through a sparger, which may be either a concentric ring or a tree-type device. The objective is to uniformly distribute the gas over the tank cross-sectional area. The size of the gas bubbles so formed and their subsequent coalescence and breakup determine the available interfacial area in the bubble column. We shall now examine the factors that influence the bubble size and gas phase holdup.

5.3.1 Bubble Generation at an Orifice

The introduction of gas into a liquid through a sparger is illustrated in Figure 5.7. The important dimensions are indicated on the figure. There are three distinct regimes of bubble formation, based on the gas flow rate through the orifice. At low gas flow rates, bubbles of constant volume are formed. The bubble size depends on the orifice diameter, surface tension and buoyancy. The bubble grows till the buoyancy force exceeds the surface tension force holding the bubble to the orifice. Inertial forces can be neglected.

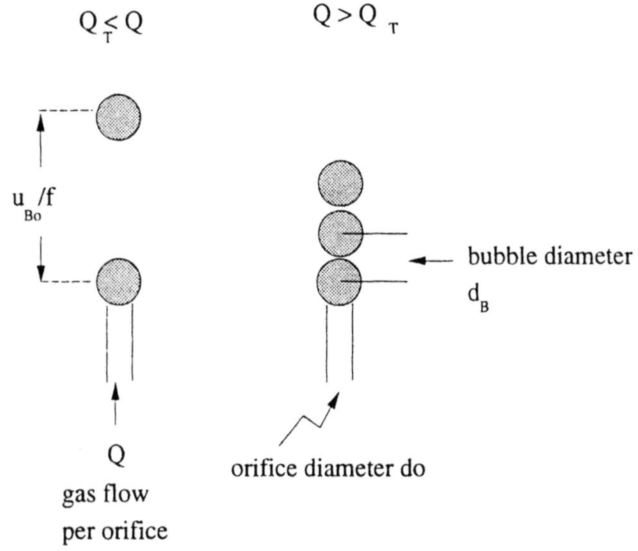

Figure 5.7. *Formation of gas bubbles at an orifice, illustrating the transition gas flow rate (Q_T) from the constant volume variable frequency regime to the constant frequency variable volume regime.*

A balance of buoyancy and surface tension forces yields:

$$g(\rho_L - \rho_G)\frac{\pi}{6}d_{Bo}^3 = \sigma(\pi d_o)$$

$$d_{Bo} = \left(\frac{6\sigma d_o}{g(\rho_L - \rho_G)} \right)^{1/3} \tag{5.52}$$

This equation will be valid as the orifice gas flow increases to a transition value Q_T. As bubbles of a constant volume rise at a constant rate, increasing gas flow will result in generation of bubbles at an increasing frequency. This low gas flow regime can thus also be considered as a constant volume, variable frequency regime. The distance between bubbles will be inversely related to the frequency of formation. At Q_T, bubbles will touch, as illustrated in Figure 5.7, and the gas can no longer be transported at in a variable frequency, constant volume manner. The transition gas flow rate is given by:

$$Q_T = \frac{\pi}{6} d_{Bo}^3 \cdot f_T \tag{5.53}$$

where

$$f_T = \frac{u_{Bo}}{d_{Bo}} \tag{5.54}$$

The rise velocity u_{Bo} of the bubble with a diameter d_{Bo} is given by Stokes equation

$$u_{Bo} = \left(\frac{g\rho_L}{18\mu_L} \right) d_{Bo}^2 \qquad \text{for} \qquad Re_B < 1 \tag{5.55}$$

or by the Mendleson relationship

$$u_{Bo} = \left(\frac{2\sigma}{\rho_L d_{Bo}} + \frac{g d_{Bo}}{2} \right)^{0.5} \qquad \text{for} \qquad Re_B \gg 1 \tag{5.56}$$

The bubble Reynolds number is defined with the liquid rather than the gas properties

$$Re_B = \frac{d_B u_B \rho_L}{\mu_L} \tag{5.57}$$

The transition orifice gas flow rates are thus

$$Q_T = \frac{\pi g(\rho_L - \rho_G)}{108\mu_L} \left(\frac{6\sigma d_o}{g(\rho_L - \rho_G)} \right)^{4/3} \qquad \text{for} \quad Re_B < 1 \tag{5.58a}$$

$$Q_T = 0.38 g^{1/2} \left(\frac{6\sigma d_o}{g(\rho_L - \rho_G)} \right)^{5/6} \qquad \text{for} \quad Re_B \gg 1 \tag{5.58b}$$

Above this orifice transition gas flow rate Q_T, the bubble size increases but bubbles are formed at a constant frequency. The bubble size in this regime has been correlated with the liquid properties and the gas flow rate[17]:

$$\frac{d_B}{d_o} = 3.23 Re_{oL}^{-0.1} Fr_o^{0.21} \tag{5.59}$$

(17) Bhavaraju, S.M., Russell, T.W.F. and H. W. Blanch, *AIChE Journal*, **24**, (3) 454 (1978).

$$\text{where} \qquad Re_{oL} = \frac{d_o\left(\frac{Q}{\pi d_o^2/4}\right)\rho_L}{\mu_L} = \frac{4\rho_L Q}{\pi d_o \mu_L} \qquad \text{and} \qquad Fr_o = \frac{Q^2}{d_o^5 g}$$

The modified orifice Reynolds number (Re_{oL}) describes the gas flow through the orifice with respect to the liquid properties and the Froude number (Fr_o) relates the inertial to gravity forces for flow through the orifice[18].

Equation (5.59) indicates that the bubble size depends on the gas flow rate to the 0.32 power, so that the frequency of bubble formation, given by Equation (5.60) below, is indeed constant in this regime.

$$f = \frac{Q}{\pi d_{Bo}^3/6} \sim \frac{6}{\pi}\frac{Q}{(Q^{0.32})^3} \sim \frac{6}{\pi} \tag{5.60}$$

In this regime we also see from the Reynolds number dependence in Equation (5.59) that the bubble size depends weakly on the liquid viscosity to the tenth power. However, as viscosity may rise by several orders of magnitude in polysaccharide fermentations, this dependence is significant.

Beyond these gas flow rates, an apparent gas jet forms at the orifice and this jet grows in size with entrainment of surrounding liquid. The bubble diameter that results is weakly dependent on the gas flow rate and is given by the following expression:

$$d_{Bo} = 0.71 Re_o^{-0.05} \quad \text{(in cm)} \qquad \text{where} \qquad Re_o = \frac{4Q\rho_g}{\pi d_o \mu_g} \tag{5.61}$$

The Reynolds number Re_o is based on gas flow *within* the orifice. The above equation applies for $Re_o > 10{,}000$.

As bubbles rise from the orifice, they are subject to breakup or coalescence events which may alter the bubble size. The extent to which this occurs depends on the size of the vessel. In a region close to the orifice, which can be conveniently defined as having a height equal to the vessel diameter, the bubble size is given by the size generated at the orifice. Above this region, coalescence and breakup, caused by bulk liquid motion, may determine the equilibrium bubble size. For small vessels, the entire vessel may be considered to be in region I. This is illustrated in Figure 5.8.

To calculate the overall mass transfer coefficient in a bubble column, the equilibrium bubble size needs to be determined. We shall examine the factors leading to bubble breakup and coalescence as these determine the resulting equilibrium bubble size and hence the interfacial area in the vessel.

---

(18) Care must be taken in working with dimensionless groups which appear in the bubble column and agitated tank literature. Often the gas properties are employed in the dimensionless groups, reflecting the gas flow *within* the orifice. More relevant is the bubble motion in the liquid above the orifice, characterized by the liquid properties. See the previous reference for details.

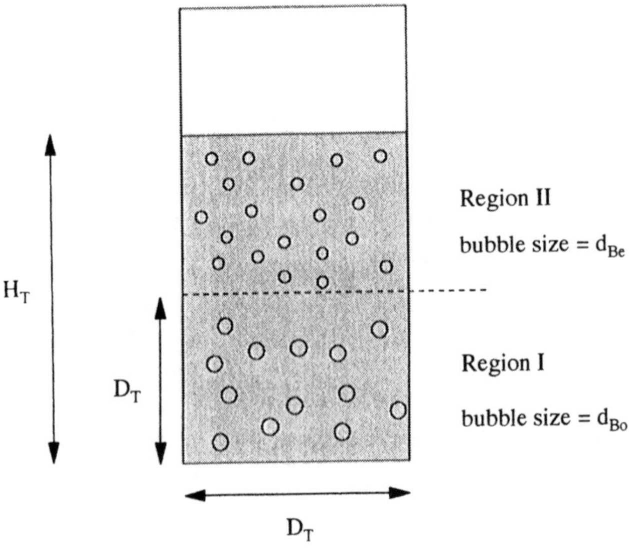

***Figure 5.8.** Regions I and II in a bubble column. In region I the bubble size is determined by the size generated at the orifice; in region II bulk liquid motions result in bubble coalescence and breakup, resulting in an equilibrium size d_{Be}.*

5.3.2 The Mass Transfer Coefficient k_La

In the region away from the orifice, the bubble size depends on the liquid properties and liquid motion generated by the rising gas stream. If the power input from the gas phase is insufficient to generate turbulence in the liquid phase, the bubble size in the tank will be that of the bubbles formed at the orifice and may increase due to bubble coalescence. Once the liquid is turbulent, however, bubble breakup may occur. This will result in an equilibrium bubble size in region II. We shall consider the factors that govern both these processes. With the bubble size determined, the mass transfer coefficient K_ga can be obtained, by considering K_ga to be composed of two terms, K_g and a, both of which depend on the bubble size, rise velocity and gas phase holdup. In most instances, the liquid film resistance is the most significant and the gas film resistance can be ignored. K_g can then be replaced by k_L/H.

Bubble Coalescence and Breakup

Bubble breakup is caused by the dynamic pressure forces exerted on the bubble by the turbulent liquid flow. These forces are opposed by the surface tension force and the resistance of the liquid phase to deformation, although the latter is usually small with respect to the first two factors. The shear stress at the bubble surface can then be considered to be opposed by surface tension, and at the point of bubble breakup these can be equated:

$$\tau = \text{constant} \cdot \frac{\sigma}{d_B} \tag{5.62}$$

The ratio of these forces is the Weber number

$$We = \frac{\tau d_B}{\sigma} \tag{5.63}$$

so that at equilibrium the Weber number is a constant. In isotropic turbulent flow, the dynamic pressure force can be expressed in terms of the fluctuating velocity v' and a length scale of a turbulent eddy, which can be assumed to be of order d_B, as smaller eddies will not cause breakup and larger eddies will transport bubbles rather than break them apart. Thus:

$$\tau \propto \rho(v')^2 \propto \rho_L\left(\frac{P\,d_B}{V\rho_L}\right)^{\frac{2}{3}} \qquad \text{where} \qquad v' \propto \left(\frac{P}{V}\right)^{\frac{1}{3}}\left(\frac{d_B}{\rho_L}\right)^{\frac{1}{3}} \tag{5.64}$$

The maximum bubble size can then be found by equating the forces

$$d_{B,\max} \propto \frac{\sigma^{0.6}}{\left(\frac{P}{V}\right)^{0.4}\rho_L^{0.2}} \tag{5.65}$$

P/V is the power input to the liquid phase per liquid phase volume. The equilibrium bubble size, as opposed to the maximum bubble size defined above, is usually quite close to $d_{B,max}$ (~ 95%) so that when a constant of proportionality is inserted, the above equation is commonly employed to represent the equilibrium size. In air-water systems this size is generally of order 0.45 cm.

The Effects of Surfactants and Viscous Liquids on the Bubble Size

When the liquid phase is viscous, the above equation needs to be modified to account for the viscous forces neglected in the analysis. We obtain

$$d_{Be} = 0.7\frac{\sigma^{0.6}}{(P/V)^{0.4}\rho_L^{0.2}}\left(\frac{\mu_{app}}{\mu_g}\right)^{0.1} \qquad \text{in meters} \tag{5.66}$$

The apparent viscosity μ_{app} will be the viscosity for purely viscous Newtonian fluids, but in the case of pseudoplastic fluids it will depend on the average fluid velocity, as this determines the shear rate. If the fluid velocity (v_L) can be determined (for example, from a liquid circulation model such as that described in Section 5.3.3), then the shear rate that the bubble experiences may be represented as v_L/d_{Be}, and this is then used to obtain the apparent viscosity.

Often fermentation broths contain considerable amounts of surfactant materials employed for foam control or produced by the organisms themselves. Surfactants tend to stabilize the bubble interface and alter the surface tension. A modification of the above equation to account for these effects has been proposed[19]:

---

(19) J.F. Walter and H.W. Blanch, Chem. Eng. Journal, **32** B7 (1986).

$$d_{Be} = \frac{(E+\sigma)^{0.6}}{(P/V)^{0.4}\rho_L^{0.2}}\left(\frac{\mu_{eL}}{3\mu_g}\right)^{0.1} \quad \text{in meters} \tag{5.67}$$

where E is the surface elasticity induced by surfactants present in the liquid and μ_{eL} is the elongational viscosity of the liquid phase. For a Newtonian fluid this is three times the Newtonian viscosity.

Breakup in Laminar Flows

In laminar flows, bubbles may also break up due to the shear field they experience as a result of their rise through the liquid phase. This situation will occur in bubble columns operated at gas flow rates insufficient to generate bulk liquid turbulence. As the bubble rises through the liquid (generally under potential flow), the surface of the bubble develops a disturbance in the form of a rippling wave. At a critical bubble size, this wave will grow on the upper surface of the bubble until it penetrates the cross-section of the bubble and the bubble breaks apart, forming two daughter bubbles. The maximum stable bubble size can be predicted by an analysis of the growth of instabilities on the bubble surface. By comparing the time required for instability growth to the instability residence time on the bubble surface, an equation for d_e, the equivalent maximum bubble diameter can be developed[20] and correlated with experimental data;

$$\frac{1}{Eo} = 3.2x10^{-3}[1 + 0.098Mo^{1/2} - 0.44Mo^{1/4}]^4 \tag{5.68}$$

where $Eo = \dfrac{g d_e^2 \rho_L}{\sigma}$ and $Mo = \dfrac{g\mu_L^4}{\rho_L \sigma^3}$

The Eotvos number (Eo) is equivalent to the Weber number for potential flow, and the Morton number (Mo) expresses the ratio of viscous to surface tension forces. At low Morton numbers, the Eotvos number has a constant value of 3.2 x 10^{-3}, predicting a stable bubble size in an air-water system of 4.8 cm.

$$Eo = \frac{g d_e^2 \rho_L}{\sigma} = \frac{1}{3.2x10^{-3}} \qquad d_e^2 = \frac{10^3\sigma}{3.2 g\rho_L} = \frac{73x10^3}{3.2x980x1} = 23.278 \tag{5.69}$$

thus $d_e = 4.825$ cm

As the viscosity of the liquid increases, Equation (5.68) predicts an increasing stable bubble size.

(20) See Prince, M.J., Walter, J.F. and H.W. Blanch, *Bubble Breakup in Air-Sparged Bioreactors*, in First Generation of Bioprocesses, ed. T.K.Ghose (Horwood Press) p160 (1990) for details of the stability analysis.

Power Input in Bubble Columns

In order to use the correlations for the bubble breakup (equations (5.65), (5.66) or (5.67)), the power input into the liquid phase must be estimated. A macroscopic energy balance on the gas phase can be written. The control volume is defined as including the gas just above the sparger up to the surface liquid. Hence the subscript 1 denotes conditions above the sparger and 2 denotes the gas leaving the liquid surface. The balance is written on a unit mass basis:

$$\Delta\left(\frac{1}{2}u^2\right)+\Delta\Phi+\int_{P_1}^{P_2}\frac{dP}{\rho_g}+\overline{W}+\overline{E}_v=0 \tag{5.70}$$

Neglecting the change of potential energy per unit mass ($\Delta\Phi$) and the frictional losses per unit mass ($\overline{E}_v$), we can express the work done by unit mass of the gas (assumed ideal, with a molecular weight of M) on the liquid as:

$$\overline{W}=-\int_{P_1}^{P_2}\frac{dP}{\rho_g}+\Delta\left(\frac{1}{2}u^2\right)=\frac{RT}{M}\ln\frac{P_1}{P_2}+\frac{1}{2}(u_1^2-u_2^2) \tag{5.71}$$

The velocity u_2 of the gas leaving the liquid is small and u_1 can be related to the orifice gas velocity u_o. We then obtain

$$\overline{W}=\frac{RT}{M}\ln\frac{P_1}{P_2}+\frac{\eta u_o^2}{2} \tag{5.72}$$

For air water systems η is of order 0.06 and may be neglected except at high gas velocities. The power input P can now be written

$$P=Q\rho_g\frac{RT}{M}\ln\frac{P_1}{P_2} \tag{5.73}$$

Since the pressure difference is due to the hydrostatic head, $P_1 = P_2 + \rho_L gH$, the power input per liquid volume can be more conveniently expressed as:

$$\frac{P}{V_L}=\frac{Q_M}{V_L}\rho_L gH$$

where $Q_M = Q\dfrac{P_2}{P_{LM}}$ and $P_{LM}=\dfrac{P_1-P_2}{\ln\frac{P_1}{P_2}}$ (5.74)

Here Q_M is the mean volumetric gas flow rate in the vessel and P_{LM} is the logarithmic mean pressure difference between the top and bottom of the vessel. This equation permits the power input to the liquid phase to be readily calculated from a knowledge of the liquid height and operating pressures.

5.3.3 Gas Holdup, Interfacial Area, k_L and Liquid Circulation Patterns

The gas phase holdup (i.e., volume of gas to volume of liquid) can now be estimated from a knowledge of the bubble size and gas flow rate into the bubble column. The average residence time of a bubble in the column is given by

$$t_g = \frac{H}{u_B} \tag{5.75}$$

where H is the aerated liquid height. Thus the total volume of gas in the vessel is the product of the gas flow rate and the residence time t_g:

$$V_g = Q_1 \frac{H}{u_B} \tag{5.76}$$

The total liquid volume is V_L and the holdup ϕ can be written as either the ratio V_g/V_L or in terms of the aerated liquid volume as $V_g/(V_g + V_L)$. Employing the former definition, we can express the gas flow rate on a superficial (i.e., per vessel cross-sectional area) basis:

$$\phi = \frac{V_g}{V_L} = \frac{(Q_1/A)\frac{H}{u_B}}{V_L/A} = \frac{u_s}{u_g} \tag{5.77}$$

noting that V_L/A is the liquid height H, where A is the vessel cross-sectional area. If the gas phase holdup is expressed on a total aerated volume basis, the inclusion of the $(V_g + V_L)$ term results in

$$\phi' = \frac{u_s}{u_s + u_B} \tag{5.78}$$

where u_s is the superficial gas velocity Q/A. The two definitions of holdup are related

$$\frac{1}{\phi} = \frac{1}{\phi'} - 1 \tag{5.79}$$

The interfacial area per unit liquid volume can now be obtained.

$$a = \frac{\text{area per bubble volume} \cdot \text{total bubble volume}}{\text{liquid volume}} = \left(\frac{6\pi d_B^2}{\pi d_B^3}\right)\phi = \frac{6\phi}{d_B} \tag{5.80}$$

Determination of the Mass Transfer Coefficient k_L.

The mass transfer coefficient k_L can be estimated by a number of methods. Penetration theory yields the simplest correlation. The exposure time of the liquid to a bubble is given by d_B/u_B and thus substitution into the penetration model yields:

$$k_L = \sqrt{\left(\frac{4D_{O_2}}{\pi t_{\exp}}\right)} = \sqrt{\left(\frac{4D_{O_2}u_B}{\pi d_B}\right)} \tag{5.81}$$

Equation (5.81) can be rewritten in terms of the bubble Reynolds number as

$$Sh = \frac{k_L d_B}{D_{O_2}} = \frac{2}{\sqrt{\pi}} Re_B^{1/2} Sc^{1/2} \tag{5.82}$$

where D_{O2} is the diffusivity of oxygen in the liquid phase. For bubbles of the equilibrium size in air-water systems, we find the following values:
$D_{O2} = 2 \times 10^{-5}$ cm^2/sec; $d_B = 0.45$ cm; $u_B = 18$ cm/sec (calculated from potential flow [Eq.(5.52)] and thus k_L is 0.10 cm.sec^{-1}.

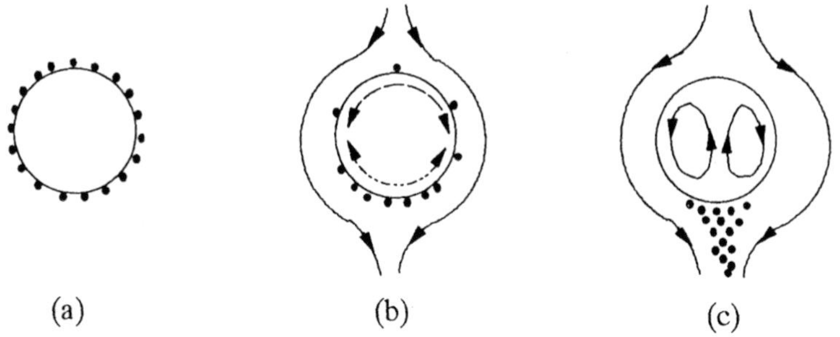

Figure 5.9. Possible extremes of interfacial mobility of bubbles rising through a liquid phase. In case (a) the interface is rendered immobile due to the presence of surfactants or salts. This will occur at low bubble velocities. (b) At intermediate velocities, restricted interfacial motion will occur as some of the surfactants are swept behind the bubble. The gas phase shows restricted internal circulation. (c) At high rise velocities, surfactants are completely swept to the rear of the bubble and internal circulation within the gas phase develops. The interface is then fully mobile.

Other approaches to determine k_L are based on the velocity profile for flow past a sphere. The form of these relationships depends on interfacial mobility. We shall consider both immobile and mobile gas-liquid interfaces, as both cases can occur in practice. The possible effects of surfactants or salts on interfacial mobility are illustrated in Figure 5.9.

Immobile Gas-Liquid Interfaces

If the gas-liquid interface is considered as immobile, for example when the liquid contains salts or surfactants, the mass transfer coefficient can be determined for creeping flow ($Re_B < 1$). For Péclet numbers ($Pe = Re\ Sc$) less than 10^4, a theoretical analysis by Brian and Hales[21] gives:

$$Sh = \frac{k_L d_B}{D_{O_2}} = (4.0 + 1.21 Pe^{2/3})^{1/2} \qquad \text{where} \quad Pe = Re \cdot Sc = \frac{d_B u_B}{D_{O_2}} \tag{5.83}$$

(21) P.L.T. Brian and H.B. Hales, *AIChE Journal*, **15**, 419 (1969).

This equation has limitations imposed by both the Re and Pe for liquid systems, where Sc is of order 10^2 to 10^3. Creeping flow will only occur with very small bubbles. In the limit of high Péclet numbers, it reduces to

$$Sh = 1.0Pe^{1/3} \tag{5.84}$$

which agrees with other theoretical treatments of creeping flow past a sphere. For higher flow velocities (2 < Re < 1300), Froessling correlated data in the following form[22]

$$Sh = 2 + 0.55Re^{1/2}Sc^{1/3} \tag{5.85}$$

At higher flows[23] (200 < Re < 4000)

$$Sh = 0.82Re^{1/2}Sc^{1/3} \tag{5.86}$$

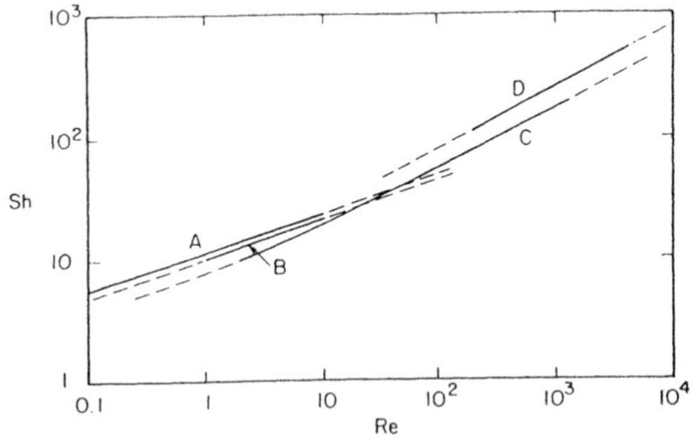

***Figure 5.10**. A comparison of various correlations of Sh as a function of the Reynolds number Re. The Schmidt number Sc is 10^3. Lines are A - Eq. (5.83), B - Eq. (5.84), C - Eq. (5.85) and D - Eq. (5.86). The length of the lines indicates the range of validity of the various equations.*

In turbulent flows, the characteristic velocity used to define the Reynolds number is that of the fluctuating eddies. Using this velocity, Calderbank and MooYoung[24] developed the following correlation for turbulent flow

$$Sh = 0.13Re_e^{3/4}Sc^{1/3} \qquad \text{where} \quad Re_e = \frac{d_B^{4/3}\rho_L^{2/3}(P/V)^{1/3}}{\mu_L} \tag{5.87}$$

(22) Froessling, N. *Gerlands Beitr. Geophys.*, **32**, 170 (1938).

(23) Rowe, P.N., Claxton, K.T. and J.B. Lewis, *Trans. Inst. Chem. Engnrs.*, **43**, 14 (1965).

(24) Calderbank, P. H. and M.M. MooYoung, *Chem. Eng. Sci.*, **16**, 39 (1961).

Thus k_L is dependent on the power input per unit volume to the 1/4 power. This dependence on power input is usually overshadowed by the influence of power input on interfacial area in determining k_La. A comparison of the various correlations for the Sherwood number is shown in Figure 5.10.

Mobile Gas-Liquid Interfaces

Gas bubbles in a liquid which is free of impurities have mobile interfaces and have internal gas circulation. Interfacial mobility can also occur at relatively high bubble rise velocities, where surface impurities are swept into the bubble wake, as illustrated earlier. The changes in the boundary layer behavior are reflected in a changed dependence of the Sherwood number on the Schmidt number, $Sh \propto Sc^{1/2}$, as compared with the one third power dependence for immobile interfaces. These are summarized below, and plotted as a function of the Reynolds number for $Sc = 10^3$ in Figure 5.11.

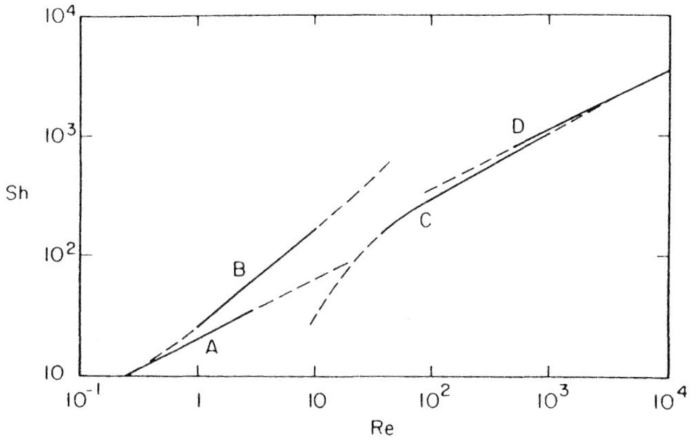

***Figure 5.11**. The mass transfer coefficient as given by the Sherwood number for bubbles with mobile interfaces as a function of the Reynolds number. Line A is Equation (5.88), line B Equation (5.89) and line C Equation (5.90). Calculations based on Sc=10³.*

$Re < 1$
$$Sh = 0.65\left(\frac{\mu_L}{\mu_L + \mu_g}\right)^{1/2} Pe^{1/2} \cong 0.65 Pe^{1/2} \tag{5.88}$$

$1 < Re < 10$
$$Sh = 0.65\left(1 + \frac{Re}{2}\right)^{1/2} Pe^{1/2} \tag{5.89}$$

$100 < Re < 1000$
$$Sh = 1.13\left(1 - \frac{2.9}{Re^{1/2}}\right)^{1/2} Pe^{1/2} \tag{5.90}$$

At high Reynolds numbers in turbulent flow

$$Sh = 1.13Pe^{1/2} \tag{5.91}$$

Small bubbles in dirty liquids tend to be perfectly spherical, while those in inviscid clean liquids tend to oscillate or wobble as they rise through the liquid phase. However, it has generally been found that the mass transfer coefficient does not deviate significantly from that observed with perfectly spherical bubbles. However, as the liquid phase viscosity increases, the stable bubble size increases significantly and above 70 cp very large spherical-cap bubbles appear. Due to the rather complex hydrodynamic patterns that appear behind such bubbles, the mass transfer characteristics differ from those found with spherical bubbles. For spherical cap bubbles

$$Sh = 1.79\frac{(3E^2+4)^{2/3}}{E^2+4}Pe^{1/2} \tag{5.92}$$

where E is the ratio of bubble width to bubble height (the eccentricity)[25]. Over a large range of flow conditions, spherical cap bubbles have a fairly constant value of E of 3.5, which simplifies the Equation (5.92) to

$$Sh = 1.31Pe^{1/2} \tag{5.93}$$

When the flow past the bubble is driven by buoyancy forces alone, Equation (5.88) can be employed to obtain the following expression for the Sherwood number for bubbles where the rise velocity is given by Stokes equation:

$$\begin{aligned} Sh &= 0.65Pe^{1/2} = 0.65\left(\frac{d_B u_B \rho_L}{\mu_L}\right)^{1/2} Sc^{1/2} \\ &= 0.65\left(\frac{d_B^3 \rho_L(\rho_L-\rho_g)g}{18\mu_L^2}\right)^{1/2}\left(\frac{\mu_L}{\rho_L D_{O_2}}\right)^{1/2} \\ &= 0.153\left(\frac{d_B^3 \rho_L(\rho_L-\rho_g)g}{\mu_L^2}\right)^{1/2}\left(\frac{\mu_L}{\rho_L D_{O_2}}\right)^{1/2} \\ &= 0.153Gr^{1/2}Sc^{1/2} \end{aligned} \tag{5.94}$$

Calderbank and Mooyoung[26] suggest the following expression for the mass transfer coefficient for "large" bubbles rising under gravitational forces:

$$Sh = 0.42Sc^{1/2}Gr^{1/3} \quad where \quad Gr = \frac{(\rho_L-\rho_g)\rho_L g d_B^3}{\mu_L^2} \tag{5.95}$$

(25) Calderbank, P.H. and A.C. Lochiel, *Chem. Eng. Sci.*, **19**, 485 (1964).

(26) P.H. Calderbank and M. Mooyoung, *Chem. Eng. Sci.*, **16**, 39 (1961).

Note that the dependence of the Sherwood number on the Grasshof number (Gr) is now 1/3, but the Schmidt number dependence is 1/2. "Large" bubbles are considered to have diameters greater than 2.5 mm. Smaller bubbles would only be produced, for example, from sintered glass frits with very small orifices.

Non-Newtonian Flow Effects

When the liquid phase has a complex rheological behavior, the boundary layer flows around the bubbles are altered and the mass transfer coefficient will be affected. If the liquid exhibits pseudoplastic behavior, the Sherwood number can be determined for the case of mobile interfacial behavior for creeping flows:

$$Sh = 0.65 f(n) Pe^{1/2}$$

where $$f(n) = \left\{1 - \frac{4n(n-1)}{2n+1}\right\}^{1/2} \quad \text{or} \quad f(n) = \{1 - 1.62(n-1)\}^{1/2} \qquad (5.96)$$

The first expression for f(n), where n is the power law index is due to Hirose and MooYoung; the second, to Bhavaraju et al[27]. Both were obtained by perturbation analyses. As n decreases (increasing pseudoplasticity), both theory and limited experimental data indicate an *increase* in the mass transfer coefficient. Theoretical predictions for Bingham fluids, viscoelastic fluids, bubble swarms and surface tension gradients are available[28]. For bubbles with immobile interfaces, momentum transfer analogies can be applied and drag coefficients have been derived by several investigators for various fluid rheologies.

Liquid Circulation Patterns

Liquid mixing in bubble columns is induced by the rise of bubbles through the liquid phase. In most bubble columns, the height to diameter ratio is quite large (3:1 up to 10:1), so that axial gradients are generally more important than radial concentration gradients. Axial liquid mixing can be characterized by an axial dispersion coefficient E_L, which necessitates the addition of a term into the mass balance for oxygen in the liquid phase. For example, a plug flow model of the liquid behavior is given in Equation (5.92) which includes an additional term for axial dispersion.

$$E_L \frac{d^2 C_{O_2}}{dz^2} - \bar{u}_L \frac{dC_{O_2}}{dz} - K_g a P\left(y - C_{O_2}\frac{H}{P}\right) + r_{O_2} = 0 \qquad (5.97)$$

When the flow can be approximated by homogeneous isotropic turbulence, Baird and Rice[29] provide a useful model for E_L:

---

(27) Hirose, T. and M.M. MooYoung, *Can. J. Chem. Eng.*, **47**, 265 (1969) and Bhavaraju, S.M., R.A. Mashelkar and H.W. Blanch, *AIChE Journal*, **24**, 1063, and **24**, 1070 (1978).

(28) Oolman, T.O. and H.W. Blanch, *CRC Critical Reviews in Biotechnology*, **4** (2), 133 (1986).

(29) Baird, M.H. and R.G. Rice, *Chem. Eng. J.*, **9**, 1701 (1975).

$$E_L = 0.35 D_T^{4/3} (g v_s)^{1/3} \tag{5.98}$$

where D_T is the tank diameter and v_s is the superficial gas velocity. A simple model is that of Ulbrecht and Baykava[30], who propose $E_L \sim D_T.u_L$, where u_L is the average liquid circulation velocity.

Models describing liquid flow in bubble columns can be divided into two groups; macroscopic balances which are used to obtain average circulation velocities; and microscopic balances which are used to obtain the shape of the velocity profiles in the liquid.

In the following treatment, we shall develop a simple macroscopic model which describes liquid velocities in the upflow and downflow regions of a bubble column. Consider the situation depicted in Figure 5.12. Following the analysis of power input from the gas phase to the liquid that was considered earlier [Eq.(5.74)], we write a macroscopic energy balance for the liquid phase.

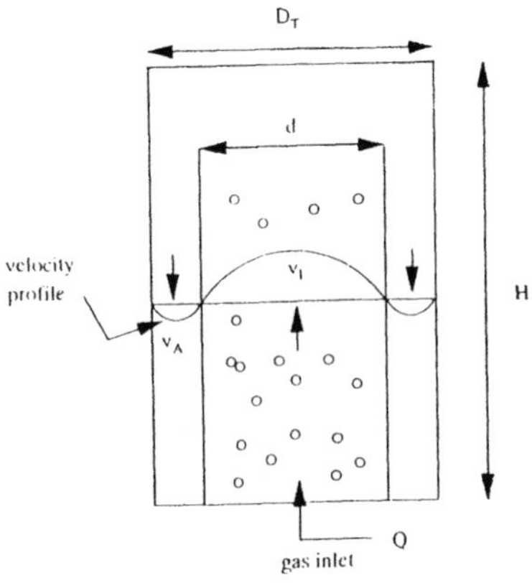

***Figure 5.12**. A simple liquid circulation model. The rise of gas bubbles induces an upflow of liquid in the center of the column (diameter d), which then flows down through an annular area given in terms of (D_T - d).*

At steady state, the energy balance per unit mass of liquid becomes

$$\Delta\left(\frac{1}{2}v^2\right) + \Delta\overline{\Phi} + \frac{\Delta p}{\rho} + \overline{W} + \overline{E}_v = 0 \tag{5.99}$$

Thus the rate at which energy is lost can be found if we neglect the change in potential energy and include only the kinetic energy and frictional loss terms from the above equation. Multiplying these by the mass flows gives

(30) Ulbrecht, J.J. and Baykava, Z.S., *Chem. Eng. Commun.*, **10** 165 (1981).

$$\Delta\left(\frac{1}{2}v^2\cdot\rho_L A v\right)+\Sigma\frac{1}{2}\rho_L v^2\cdot A f v \tag{5.100}$$

and equating this to the gas power input (Equation (5.74) yields

$$Q_M\rho_L g H=\left\{\frac{1}{2}\rho_L\pi d^2 v_I^3-\frac{1}{2}\rho_L\pi(D_T^2-d^2)v_A^3\right\}$$

$$+\left\{\frac{1}{2}\rho_L\pi d H v_I^3 f_I+\frac{1}{2}\rho_L(D_T+d)H\pi v_A^3 f_A\right\} \tag{5.101}$$

where v_I and v_A are the velocities in the inner and annular regions, and the friction factor f_I and f_A refer to the same regions. If the areas for upflow and downflow are equal ($D_T=d\sqrt{2}$, then the velocities v_I and v_A are equal, and d ~ $(1/\sqrt{2})\ D_T$ ~ 0.7 D_T. This agrees well with experimental measurements of the flow reversal point in bubble columns operating in the turbulent regime. We shall assume d/D_T ~ 0.7 in the examples that follow.

Equation (5.101) can be simplified to yield:

$$\frac{2Q_M g}{\pi d}=(2+\sqrt{2})f_I v_I^3 \tag{5.102}$$

When the flow areas are not equal, (5.101) can be rearranged and simplified to

$$\frac{2Q_M g}{\pi d}=v_I^3\left\{\left(1-\frac{1}{\left(\frac{D_T^2}{d^2}-1\right)^2}\right)\frac{d}{H}+\left(1+\frac{\left(\frac{D_T}{d}+1\right)f_A}{\left(\frac{D_T^2}{d^2}+1\right)^3 f_I}\right)f_I\right\} \tag{5.103}$$

If the flow is laminar and the fluid is Newtonian, we can insert a value for the friction factor.

$$f=\frac{16}{Re_L}\qquad\text{where}\quad Re_L=\frac{\rho_L v d}{\mu_L} \tag{5.104}$$

Thus (5.102) can be solved for the velocity v_I

$$v_I=\left[\frac{Q_M g\rho_L}{8\pi\mu_L(2+\sqrt{2})}\right]^{1/2} \tag{5.105}$$

For turbulent flows, f_I depends weakly on the Reynolds number, but we may assume f_I ~ 0.02 and thus Equation (5.102) reduces to

$$v_I=\left(\frac{2Q_M g}{\pi d(2+\sqrt{2})f_I}\right)^{1/3} \tag{5.106}$$

The macroscopic energy balance thus provides <u>average</u> liquid velocities in both the annular and inner regions of the bubble column.

The power input to the inner region where the bubbles rise is given by

$$P=Q_M\rho_L g H \tag{5.74}$$

and the volume of this inner region is

$$V_I = \frac{\pi}{4} d^2 H \tag{5.102}$$

thus the power per unit volume that applies to the breakup of bubbles in this region is

$$\frac{P}{V_I} = \frac{4 Q_M \rho_L g}{\pi d^2} \tag{5.107}$$

Thus the mean bubble diameter in the inner region can be found by inserting this power input into Equation (5.66) for the case of turbulent flow in the inner region.

$$d_{B_e} = 0.7 \frac{\sigma^{0.6}}{(P/V)^{0.4} \rho_L^{0.2}} \left(\frac{\mu_{app}}{\mu_g} \right)^{0.1} \quad \text{in meters} \tag{5.66}$$

In the case of non-Newtonian fluids, a characteristic shear must be employed to calculate the apparent viscosity in the above equation. This may be conveniently approximated by the ratio of the liquid velocity in the inner tube to the tube diameter. For a power law fluid, we have:

$$\mu_{app} = K \left(\frac{v_I}{d} \right)^{n-1} \frac{1}{F(n)} \tag{5.108}$$

where F(n) is given by

$$F(n) = 8 \left(\frac{n}{6n+2} \right)^n \tag{5.109}$$

Regimes of Bubble Size

In Figure 5.8 we have divided the bubble column into two regions, region I close to the orifice, where the bubble size is that generated at the orifice, and region II, where bubble breakup and coalescence may determine the equilibrium bubble size. We will consider these regions further in light of the above models of liquid circulation.

If the power input from gas sparging does not result in the generation of turbulence in the inner region (as predicted by the simple circulation model above), the liquid flow will be laminar and hence the bubble size in region II can be assumed to be equal to bubbles size in region I. Bubble breakup by turbulent eddies will be absent. With turbulent flow, the bubble size in region II will reflect a balance of bubble coalescence and breakup. The bubble size will be given by Equation (5.66). The liquid properties (surface tension and viscosity) enter this equation and significantly influence the bubble size. If coalescence is absent and bubbles are generated at the orifice at a size below the predicted equilibrium size, the bubble size in region II will be the same as in region I.

The height of region I is related to the sparger design and the vessel geometry. In the case of laminar circulation, the height of region I can be assumed to be the height of the tank. When bubble breakup occurs, the height of region I can be determined by examining the interaction of gas jets that issue from the individual orifices in the tank. As is illustrated in Figure 5.13, the apparent radius of a gas jet is defined as the distance from the axis of the jet where the jet velocity is half the velocity at the jet axis. The ratio of the apparent radius of the jet (r_j) to the axial distance from the orifice has been found experimentally to be given by

$$\frac{x_j}{r_j} = 2.25 \tag{5.110}$$

H_I can be defined as the distance x_j for which r_j is large enough to interact with the vessel walls or other jets, and hence depends on the sparger design. For a uniform distribution of sparger holes in the tank (as illustrated in Figure 5.13), the cross-sectional area per orifice is

$$A_o = \frac{A_T}{N} = \frac{\pi D_T^2}{4N} \tag{5.111}$$

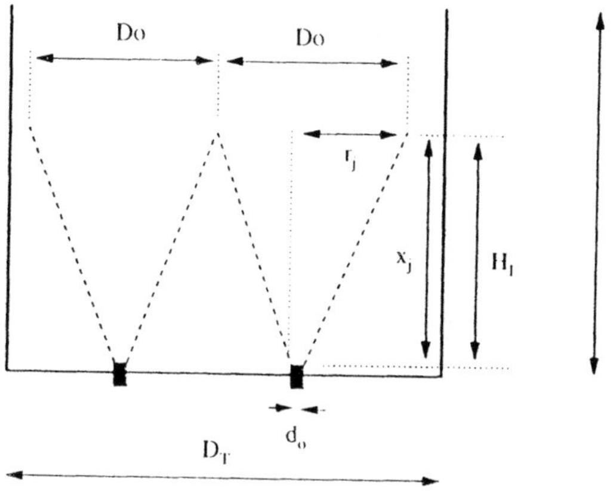

Figure 5.13*. The interactions of jets formed from spargers uniformly distributed at the bottom of a bubble column of diameter D_T.*

The diameter of the zone of influence of a single orifice can be estimated as

$$D_o = \sqrt{\frac{4A_o}{\pi}} = \frac{D_T}{\sqrt{N}} \tag{5.112}$$

and the height of region I will be the distance x_j at which r_j is equal to $D_o/2$.

$$H_I = 2.25\frac{D_T}{2\sqrt{N}} = 1.125\frac{D_T}{\sqrt{N}} \tag{5.113}$$

For a ring sparger, H_I would be the distance at which the jets interact at the center of the ring; i.e.,

$$H_I = 2.25\frac{D_S}{2} = 1.125 D_S \tag{5.114}$$

where D_S is the sparger diameter.

If the concentration driving forces for mass transfer in regions I and II can be considered equal, an overall k_La can be defined from a weighted mean of the k_La's in each of the regions.

$$k_La = \left(\frac{H_I}{H_T}\right)k_La_I + \left(1 - \frac{H_I}{H_T}\right)k_La_{II} \tag{5.115}$$

5.3.4 A Design Procedure for Bubble Columns

With the background equations developed in the preceding sections, we can now predict the mass transfer coefficient k_La in a bubble column, by determining the bubble diameter and gas-phase holdup. We then have all the necessary parameters to complete the design of a bubble column. In general, the tank volume is based on the desired reactor productivity and production kinetics, but the sparger type and tank geometry must be selected. The ratio of tank height to diameter is a variable to be specified. For bubble columns this may range from 1 to 10. The liquid and gas properties and tank geometry we need to consider are:

Gas flow rate	Q	m^3/sec (at atmospheric pressure)
Orifice diameter	d_o	m
Gas and liquid densities	ρ_L, ρ_g	Kg/m^3
Liquid viscosity	μ_L	$N.sec/m^2$
Surface tension	σ	N/m
Diffusivity	D_{O2}	m^2/sec
Tank diameter	D_T	m
Sparger diameter	D_S	m (for a ring sparger)
Number of holes	N	
Tank height	H_T	m

Step 1. Determination of bubble diameter generated at the orifice

The transition gas flow rate Q_T through the orifice is first calculated and compared with Q/N, the actual flow through the orifice. If $(Q/N) < Q_T$, bubble size can be calculated from the constant volume, variable frequency regime (Equation (5.52)), otherwise bubbles are formed in the variable volume, constant frequency regime. The steps involved are as follows.

(a). Calculate the bubble size generated at the orifice from Equation (5.52). Determine the terminal rise velocity corresponding to this size from Stokes' relationship (equation 5.55). Calculate the Reynolds number Re_B. If $Re_B<1$, the value of Q_T can be found from Equation (5.58a). If $Re_B>>1$, then the Mendelson Equation (5.56) must be used to calculate the rise velocity and the Reynolds number. Q_T can be found from Equation (5.58b).

(b). If $(Q/N) < Q_T$, the bubble size is obtained from Equation (5.52). If $(Q/N) > Q_T$, the bubble size is found from the correlation Equation (5.59)

$$\frac{d_B}{d_o} = 3.23 Re_{oL}^{-0.1} Fr_o^{0.21}$$

If the *orifice* Reynolds number $Re_o > 10{,}000$, jetting from the orifice is occurring and the bubble size is determined from Equation (5.61)

$$d_{Bo} = 0.71 Re_o^{-0.05} \qquad \text{in cm}$$

Step 2. Determination of bubble size in region II

(a). The liquid Reynolds number Re_L is now calculated from the liquid circulation model described in Section 5.3.3. The inner tube diameter d is assumed to be $0.7D_T$. This procedure is iterative, as it is first assumed the flow is laminar and the velocity v_I calculated and used to check the validity of $Re_L < 2000$. Otherwise turbulent flow is assumed and the Reynolds number calculated using the velocity given in Equation (5.105). Note that the gas flow Q_M used is based on the log mean total gas flow into the tank, not the flow per orifice.

(b). If the flow in region II is laminar, the bubble size in this region will be the same as that determined in region I. This may be altered by some breakup due to laminar shear or due to coalescence occurring if the liquid has a low ionic strength. If the flow is turbulent, the mean bubble size is determined from Equation ((5.66)

$$d_{Be} = 0.7 \frac{\sigma^{0.6}}{\left(\frac{P}{V}\right)^{0.4} \rho_L^{0.2}} \left(\frac{\mu_{app}}{\mu_g}\right)^{0.1} \qquad \text{in meters}$$

The power per volume is calculated following the procedures described in Section 5.3.3, based on the volume of the inner tube. If the bubble size predicted from the above equation is below the equilibrium size (0.45 cm in air-water systems), then the equilibrium bubble size should be employed unless bubble coalescence is strongly retarded in the system.

Step 3. Determination of $k_L a$ for regions I and II

(a). For each region, determine k_L from the liquid properties and the bubble diameter in each region, using the penetration theory result as a first approximation. If the liquid phase is viscous, or if information on the mobility of the gas-liquid interface is available, one of the Sh(Re,Sc) correlations should be employed. If the bubble size is large, it may not be spherical and its eccentricity could be approximated as 3.5; equation (5.93) can then be employed to calculate the Sherwood number.

(b). The gas holdup can now be calculated from the terminal rise velocity of the bubbles in each region, based on their respective diameters. Assume $Re_B<1$, use the Stokes flow result and check that the assumption of creeping flow was adequate. Otherwise the bubbles rise in the potential flow regime and the Mendleson equation is employed. The superficial gas velocity and holdup are then found from

$$v_s = \frac{4Q}{\pi D_T^2} \qquad \Phi = \frac{v_s}{v_B}$$

and the interfacial area per unit liquid volume for each region

$$a = \frac{6\Phi}{d_B}$$

Thus the mass transfer coefficient k_L and the interfacial area a are separately determine for each region in the tank.

Step 4. Determination of H_I and the overall $k_L a$

(a). Based on the sparger design, the height of region I is calculated following the procedures outlined in section 5.3.3. If the liquid Reynolds number Re_L<2000, the entire tank is in region I.

(b). An overall $k_L a$ can be calculated based on the assumption that the concentration driving force for oxygen transfer in both regions is the same. The weighted mean is

$$k_L a = \left[\frac{H_I}{H_T}\right] k_L a_I + \left[1 - \frac{H_I}{H_T}\right] k_L a_{II}$$

An Example of the Design Approach: The Prediction of $k_L a$ for Air Sparging into Water

Consider a 9.425 m^3 tank, 2m in diameter and 3m high, containing water. Air is sparged into the tank through a tree-type sparger at a rate of 0.071 m^3/sec. The tank is open to the atmosphere. The sparger has 45 holes of diameter 0.48 cm, distributed uniformly across the cross-sectional area of the tank. What is the overall mass transfer coefficient $k_L a$ for this tank?

The liquid and gas properties are:

Gas flow rate	Q	0.071 m^3/sec
Orifice diameter	d_o	0.48×10^{-2} m
Liquid density	ρ_L	1.0×10^3 Kg/m^3
Liquid viscosity	μ_L	1.0×10^{-3} N.sec/m^2
Gas viscosity	μ_g	1.84×10^{-5} N.sec/m^2
Surface tension	σ	72.7×10^{-3} N/m
Diffusivity of oxygen	D_{O2}	2.25×10^{-9} m^2/sec
Tank diameter	D_T	2 m
Number of holes	N	45
Tank height	H_T	3 m
Pressure P_2	P_2	1.01×10^5 N/m^2 (1 atm)

Step 1. Determination of bubble diameter generated at the orifice

(a). Calculation of Q_T. Assume bubbles are generated in the constant volume,- variable frequency regime and rise in potential flow.

$$d_{Bo} = \left[\frac{6\sigma d_o}{g(\rho_L - \rho_g)}\right]^{1/3} = 0.598 \quad \text{cm}$$

$$u_{Bo} = \left[\frac{2\sigma}{\rho_L d_{Bo}} + \frac{g d_{Bo}}{2}\right]^{0.5} = 0.231 \text{m/sec}$$

Check the bubble Reynolds number.

$$Re_B = \frac{d_{Bo} u_{Bo} \rho_L}{\mu_L} = 1382$$

since $Re_B >> 1$, the assumption of potential flow is justified. Q_T can now be obtained from Equation (5.58b).

$$Q_T = 0.38 g^{1/2} \left[\frac{6\sigma d_o}{g(\rho_L - \rho_g)} \right]^{5/6} = 0.323x10^{-5} \quad m^3/\sec$$

The gas flow per orifice Q/N is 0.157×10^{-2} m^3/sec and thus $Q>Q_T$. The bubble size in region I can now be calculated from Equation (5.59):

$$d_B = 3.23 d_o Re_{oL}^{-0.1} Fr_o^{0.21} = 3.23 d_o \left(\frac{4\{Q/N\}\rho_L}{\pi d_o \mu_L} \right)^{-0.1} \left(\frac{\{Q/N\}^2}{d_o^5 g} \right)^{0.21} = 4.88x10^{-2} \quad \text{m}$$

The bubble is assumed to rise in the potential flow regime and we again need to check the bubble Reynolds number to verify this assumption.

$$u_{Bo} = \left[\frac{2\sigma}{\rho_L d_{Bo}} + \frac{g d_{Bo}}{2} \right]^{0.5} = 0.489 \text{ m/sec} \qquad Re_B \gg 1$$

The bubble size generated at the orifice is thus an order of magnitude larger than the equilibrium bubble size in water (4.5 mm) and we expect the bubbles to rise and be broken to smaller bubbles in region II by the action of the liquid turbulence. The height of region I is first calculated and then we determine the effect of turbulence on bubble size in region II.

(b). The height of region I is given by Equation (5.113).

$$H_I = 2.25 \frac{D_T}{2\sqrt{N}} = 0.335 \text{ m}$$

and

$$k_L = \left[\frac{4 D_{O_2} u_B}{\pi d_B} \right]^{1/2} = 1.69x10^{-4} \text{ m/sec}$$

$$\Phi = \frac{u_s}{u_{Bo}} = \frac{4Q/(\pi D_T^2)}{u_{Bo}} = \frac{0.0226}{0.489} = 0.046$$

$$a = \frac{6\Phi}{d_B} = 5.68 \text{ m}^{-1} \qquad \text{and} \qquad k_L a_I = 9.6x10^{-4} \text{ sec}^{-1}$$

Step 2. Determination of bubble size in region II.

The liquid Reynolds number is first determined using the liquid circulation model. The pressure at the liquid surface is assumed to be atmospheric (i.e., $p_2 = 1.01 \times 10^5$ pascal). The mean gas flow rate Q_M is found from Qp_2/p_{LM}:

$$p_{LM} = \frac{p_1 - p_2}{\ln\frac{p_1}{p_2}} = \frac{\rho_L g H_T}{\ln\left(1 + \frac{\rho_L g H_T}{p_2}\right)} = 1.151x10^5 \text{ Pascal}$$

$$Q_M = \frac{Q p_2}{p_{LM}} = 0.062 \text{ m}^3/\text{sec}$$

The inner tube diameter d can be taken as approximately $0.7D_T$, i.e., 1.4 meters. We shall assume that the flow is turbulent and calculate Re_L. If Re_L is less than ~2,000 we would have to recalculate assuming laminar flow. We shall assume that f_I~0.02.

$$Re_L = \left[\frac{2Q_M g \rho_L^3 d^2}{\pi \mu_L^3 f_I (2 + \sqrt{2})}\right]^{1/3} = 2.2x10^5$$

Thus the assumption of turbulent flow is justified. The power input can now be calculated from:

$$\frac{P}{V_I} = \frac{4Q_M \rho_L g}{\pi d^2 H_T} = 132 \text{ watts/m}^3$$

The mean bubble size in region II is thus

$$d_B = 0.7\frac{\sigma^{0.6}}{\left(\frac{P}{V_I}\right)^{0.4}\rho_L^{0.2}}\left(\frac{\mu_L}{\mu_g}\right)^{0.1}(\text{meters}) = 7.71x10^{-3}(\text{meters})$$

The bubble size (7.71 mm) is thus slightly larger than that which would occur under coalescence/breakup equilibrium conditions (4.5 mm). We may now calculate a, k_L, and Φ using the same equations as for region I. Note u_B is 0.275 m/sec and k_L is 3.2 x 10^{-4} m/sec. The holdup is 8.22%, the interfacial area is 63.95 m^{-1} and k_La is 0.0205 sec^{-1} (73.7 hr^{-1}). The overall k_La for the tank is

$$k_La = \left(\frac{H_I}{H_T}\right)k_La_I + \left(1 - \frac{H_I}{H_T}\right)k_La_{II} = \frac{0.335}{3.0}x9.6x10^{-4} + \frac{(3.0 - 0.335)}{3.0}x0.0205 = 0.0182 \text{ sec}^{-1} \ \{65.5 \text{ hr}^{-1}\}$$

In this commercial scale tank, the height of region I is relatively small compared to the height of the tank. The overall k_La is determined primarily by the interfacial area created in region II by the turbulence induced by the rising gas stream. In a laboratory scale tank, however, the overall mass transfer coefficient may be determined solely from the bubble size generated at the sparger, as these tanks may not have a significant region II.

Prediction of k_La for air sparging when the liquid is viscous

We shall repeat the above calculations for the case of a liquid phase in the same tank having a viscosity of 1000 centipoise, as might be the case with a dense fungal broth or a viscous polysaccharide fermentation. In region I, the bubble size generated at the orifice is first calculated. As the liquid phase viscosity does not enter into the equation for d_{Bo}, we obtain the same value of 0.598 x 10^{-2} m, but the rise velocity will be considerably less. We shall assume Stokes flow to calculate u_{Bo}.

$$u_{Bo} = \left(\frac{g\rho_L}{18\mu_L}\right) d_{Bo}^2 = 1.95x10^{-2} \text{ m/sec}$$

The bubble Reynolds number Re_B is 0.116, and the assumption of Stokes flow is valid. The transition gas flow rate can be calculated and compared with the flow per orifice.

$$Q_T = \frac{\pi g(\rho_L - \rho_g)}{108\mu_L}\left[\frac{6\sigma d_o}{g(\rho_L - \rho_g)}\right]^{4/3} = 3.6x10^{-7} \text{ m}^3\text{/sec}$$

Thus (Q/N) > Q_T, which is generally the case even for highly viscous liquids. Assuming $d/D_T = 0.7$, the liquid Reynolds number must be calculated. In this case the flow is likely to be laminar. The diameter D_o in region I is calculated on the cross-sectional area per orifice [Eq. (5.112)]:

$$D_o = \sqrt{\frac{4A_o}{\pi}} = \frac{D_T}{\sqrt{N}} = 0.335 \text{ m} \tag{5.112}$$

The inner diameter d is 0.235 m. The Reynolds number is

$$Re_L = \left[\frac{Q_M g \rho_L^3 d^2}{8\pi\mu_L^3(2+\sqrt{2})}\right]^{1/2} = 2.66x10^3$$

The Reynolds number is very close to that for the transition from laminar to turbulent flow. If we assume the liquid circulation is laminar and bubble breakup is absent in the column, the bubble size in region I will be the same in the entire tank. This bubble size can now be calculated:

$$d_B = 0.7\frac{\sigma^{0.6}}{(P/V_I)^{0.4}\rho_L^{0.2}}\left(\frac{\mu_L}{\mu_g}\right)^{0.1} = 3.07x10^{-2} \text{ (meters)}$$

The bubble diameter is thus about five times that obtained in region I in the preceding case for water. the determination of the mass transfer coefficients then follows the approach illustrated earlier.

5.3.5 Correlations for k_La

There are a large number of correlations for predicting the overall mass transfer coefficient K_La in bubble columns and in columns which have internal devices, such as draft tubes, to enhance liquid circulation. The most useful of these are those obtained on relatively large scale equipment, as the bubble size variation from region I to II can be important in laboratory scale devices. Values of k_La are generally obtained using biological or chemical methods to deplete the oxygen content of the liquid (e.g., microbial growth or sulphite oxidation) and observation of the rate of increase in dissolved oxygen with aeration. These methods are reviewed later (Section 5.4.3) for agitated tanks, but apply equally as well to bubble columns. In bubble columns, the key parameters are the superficial gas velocity and the liquid properties. Often the vessel geometry is also considered. The considerable literature to 1965 is reviewed by Sideman *et al.*[31] Most studies have been restricted to systems of low liquid viscosity. Correlations have been developed for coalescing and non-coalescing systems. The dependency of k_La on superficial gas velocity in these systems is reported from 0.4 to 1.6. We shall examine some of these correlations.

Table 5.7. *Correlation of the overall mass transfer coefficient k_La as a function of the superficial gas velocity v_s, in the form $k_La = \alpha v_s^{\beta}$. k_La is given in 1/sec, v_s in cm/sec.*

Liquid phase	Sparger Type	D_T (cm)	H_T (cm)	v_s (cm/sec)	α	β
Bubble Columns						
sulphite solution (0.3N)	single orifice	7.7-60	90-350	3.0-22.0	0.42	0.9[32]
sulphite solution (0.3N)	single orifice	15.2	400	3.0-22.0	0.24	0.9[33]
water	multi-orifice	20	723	0.2-9.0	0.73	0.96[34]
0.7N Na_2SO_4	multi-orifice	20	723	0.2-9.0	0.75	0.89
Airlift Contactors						
yeast	multi-orifice	15	295	1.4-4.5	0.6	1.09[35]
yeast	multi-orifice	7.5	112	1.4-7.0	0.9	1.23
water	multi-orifice	5.5	165	2-10	1.09	1.06[36]

In a study employing tanks of diameters 7-, 14-, 30- and 60-cm diameters, Yoshida and Akita[37] proposed the following correlation;

$$\frac{k_LaD_T^2}{D_L} = 0.6Sc^{1/2}Bo^{0.62}Ga^{0.31}\Phi^{1.1} = 0.6\left(\frac{\mu}{\rho_LD_L}\right)^{1/2}\left(\frac{gD_T^2\rho_L}{\sigma}\right)^{0.62}\left(\frac{gD_T^3\rho_L^2}{\mu^2}\right)^{0.31}\Phi^{1.1} \tag{5.116}$$

The Bond ($Bo = gD_T^2\rho_L/\sigma$) and Galileo ($Ga = gD_T^3\rho_L^2/\mu^2$) numbers are referred to the tank diameter D_T. The gas holdup Φ is found from the correlation[38]

(31) S. Sideman, O. Hortacsu and J.W. Fulton, *Ind. Eng. Chem.*, **58** 32 (1966).
(32) F. Yoshita and K. Akita, *AIChE Journal*, **11** 9 (1965).
(33) F. Yoshita and K. Akita, *Ind. Eng. Chem. Proc. Des. Devpt.*, **12** 769 (1973).
(34) W. Deckwer, R. Burckhart and G. Zoll, *Chem. Eng. Sci.*, **29**, 2177 (1974).
(35) C. Lin, *Biotech. Bioeng.*, **18**, 1557 (1976).
(36) M. Mooyoung et al., *Proc. Int. Conf. Two Phase Flows*, p1049 Dubrovnik, Hemisphere Press (1979).
(37) F. Yoshita and K. Akita, *AIChE Journal*, **11** 9 (1965).
(38) K. Akita and F. Yoshida, *Ind. Eng. Chem. Proc. Des. Devpt.*, **13**, 583 (1971).

$$\frac{\phi}{(1-\Phi)^{1/4}} = 0.2\left(\frac{gD_T^2\rho_L}{\sigma}\right)^{1/8}\left(\frac{gD_T^3\rho_L^2}{\mu^2}\right)^{1/12}\left(\frac{v_s}{\sqrt{gD_T}}\right) \tag{5.117}$$

Correlations of the form $k_La = \alpha v_s^\beta$ are given in Table 5.7

5.4 Agitated Tanks

Agitated tanks are the most common types of industrial bioreactors. They are used in antibiotic production, yeast growth and in wastewater treatment. They are employed when high rates of mass or heat transfer are required or when the liquid phase is viscous. As was the case with bubble columns, we must consider a number of factors in agitated tanks, including the size distribution of gas bubbles, the gas phase hold-up, bubble breakup and coalescence and the liquid phase mixing pattern.

In an agitated tank without baffles, a vortex is formed in the center of the liquid which may, at sufficiently high agitator speeds, be sucked into the impeller blades and result in the dispersion of the headspace gas throughout the liquid. To avoid this, tanks are usually baffled.

When a gas is injected into an agitated tank, there is a minimum impeller speed required to disperse the gas throughout the liquid phase. This minimum speed, N_0, can be estimated from:

$$\frac{N_o D_i}{(\sigma g/\rho_L)^{1/4}} = A + B\left(\frac{D_T}{D_i}\right) \tag{5.118}$$

where

A = 1.22 and B = 1.25 for turbine agitators

A = 2.25 and B = 0.68 for paddles

Generally, agitated tanks are operated at impeller speeds well in excess of N_0.

Industrial agitated tanks often have standard geometries, where baffle and impeller sizes are related to the tank diameter. The number (n_B) and width (B_W) of baffles are related viz.

$$\left(\frac{B_w}{D_T}\right)n_B = 0.5 \tag{5.119}$$

Baffles may be attached to the tank wall or there may be a small clearance between the baffle and the wall. The impeller in aerated tanks is generally a vaned disk turbine, as this design suppresses the channeling of gas up the shaft and operates effectively in gas dispersions. The diameter of the impeller is usually one third of the tank diameter and the optimum width of the blades (b) can be found from

$$\left(\frac{b}{D_T}\right)n_P = 0.15\sim0.3 \tag{5.120}$$

where n_p is the number of blades. The impeller is often located one third of the total liquid height from the vessel bottom.

5.4.1 The Mass Transfer Coefficient and the Equilibrium Bubble Size

The rate of mass transfer from the gas to the liquid phase is generally governed by the liquid film mass transfer coefficient k_L, as we have seen earlier. Only in the case of a very fast reaction in the liquid phase will the gas film resistance be significant. We shall only examine the factors that influence both k_L and interfacial area in stirred tanks. The correlations presented earlier for the Sherwood number as a function of Re and Sc also apply to stirred tanks. Therefore only the bubble size needs to be determined.

In well-mixed tanks, the liquid phase is generally turbulent and the equilibrium bubble size depends on bubble breakup and coalescence events. As we saw earlier for bubble columns, this equilibrium bubble size can be found from the power input (equation 5.66):

$$d_{Be} = 0.7 \frac{\sigma^{0.6}}{\left(\frac{P}{V}\right)^{0.4} \rho_L^{0.2}} \left(\frac{\mu_{app}}{\mu_g} \right)^{0.1} \quad \text{in meters} \tag{5.66}$$

The power input per unit volume can be calculated from the tank geometry, impeller type and the agitation speed, as will be discussed in section 5.4.4. Determination of the gas phase hold-up is more difficult and generally makes the determination of k_La by independent estimation of k_L (from d_B) and a (from hold-up and d_B) impractical. We shall now turn our attention to various correlations.

5.4.2 Correlations for k_La

Values of k_La in stirred tanks have been obtained by a variety of experimental methods, the most common of which is the sulphite oxidation technique. This method involves the addition of a solution of sodium sulphite to the reactor, together with a catalyst such as Co^{2+}, so that dissolved oxygen reacts with the sulphite to form sulphate, depleting the liquid of dissolved oxygen. Provided the reaction is sufficiently fast, the bulk liquid oxygen concentration can be assumed to be zero and the volumetric rate of the chemical reaction will then equal $k_LaC_{O2}^*$. By measuring the concentration of sulphite with time, the chemical reaction rate can be determined, and hence k_La. This technique is subject to several restrictions (discussed in the following section), but it has been widely applied to determine k_La as a function of various operating parameters, such as agitation intensity and gas flow rate.

There are a number of correlations for the gas hold-up, interfacial area and bubble size that will be examined before we turn to general correlations for k_La. Using a light transmission technique to determine the bubble size, Calderbank[39] found the following dimensional correlation for gas hold-up (Φ) in agitated tanks:

$$\Phi = \left(\frac{v_s\Phi}{u_B}\right)^{1/2} + 0.0216\left(\frac{(P_g/V)^{0.4}\rho_L^{0.2}}{\sigma^{0.6}}\right)\left(\frac{v_s}{u_B}\right)^{1/2} \tag{5.121}$$

where

P_g/V is the gassed power per volume	(HP/ft^3)
v_s is the superficial gas velocity	(ft/sec)
u_B is the terminal rise velocity of the bubble	(ft/sec)
σ is the interfacial tension	(dyne/cm)
ρ_L is the liquid density	(gm/cm^3)

In the limit of low power input, the above equation reduces to $\Phi = v_s/u_B$, the same as that obtained in bubble columns. At high power inputs, the first term is small and assuming a constant bubble size, the hold-up has the form:

$$\Phi \propto (P/V)^{0.4} v_s^{0.5} \tag{5.122}$$

The interfacial area and bubble size may be obtained from the following correlations developed by Calderbank and Mooyoung[40], who considered two types of solution properties; "coalescing" and "non-coalescing". For clean ("coalescing") air-water dispersions:

$$a = 0.55\left(\frac{P}{V}\right)^{0.4} v_s^{0.5} \qquad d_B = 0.27\left(\frac{P}{V}\right)^{-0.17} v_s^{0.27} + 9\cdot 10^{-4} \tag{5.123}$$

For dirty ("non-coalescing") air-water dispersions, such as fermentation broths containing electrolytes:

$$a = 0.15\left(\frac{P}{V}\right)^{0.7} v_s^{0.3} \qquad d_B = 0.89\left(\frac{P}{V}\right)^{-0.17} v_s^{0.17} \tag{5.124}$$

In both these dimensional correlations, the units are (P/V) in watts/meter^3; v_s in meter/sec; a in meter^2/meter^3, and d_b in meters. It has also been found that the effect of surfactants, such as sodium lauryl sulphate, on a and d_B is similar to the effect of electrolytes in reducing bubble coalescence. This results in smaller bubbles.

Inviscid Systems

There are a number of correlations for k_La in stirred tanks. Many of these are reviewed

(39) P.H. Calderbank, *Trans. Inst. Chem. Engnrs. (London)*, **36**, 443 (1958).

(40) P.H. Calderbank and M. Mooyoung, *Chem. Eng. Sci.*, **16** 23 (1961) and P.H. Calderbank, *Trans. Inst. Chem. Engnrs.*, **36**, 443 (1958).

in van't Riet[41], who correlated extensive literature data and proposed the following dimensional correlations for inviscid systems:
(a) for coalescing (clean) dispersions

$$k_L a = 2.6 \cdot 10^{-2} \left(\frac{P}{V}\right)^{0.4} v_s^{0.5} \quad (\text{sec}^{-1}) \tag{5.125}$$

(b) for non-coalescing (dirty) dispersions

$$k_L a = 2.0 \cdot 10^{-3} \left(\frac{P}{V}\right)^{0.7} v_s^{0.2} \quad (\text{sec}^{-1}) \tag{5.126}$$

where (P/V) is in watts/meter^3 and v_s is in meter/sec. The range of power inputs over which these correlations were determined is 500 < (P/V) < 10,000 watts/meter^3. These are high power inputs; and generally apply to small vessels, up to several thousand liters. These equations agree with experimental data to ±20% and ±35% respectively. The mass transfer coefficients in coalescing systems are higher than those in non-coalescing systems. The opposite effect would be expected, due to the decreasing interfacial area which occurs with coalescence. Interfacial mobility is reduced with non-coalescing systems and k_L will thus decrease, but this effect is relatively small.

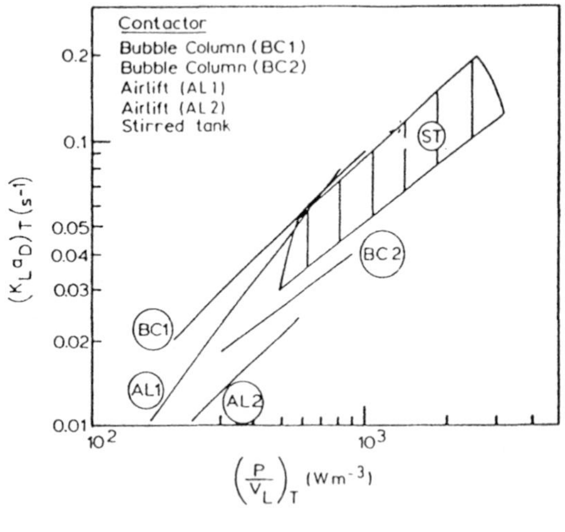

Figure 5.14. *Aeration efficiencies of various gas-liquid contactors in electrolyte solutions. From M. Mooyoung and H. W. Blanch, Adv. Biochem. Eng. **19** 1 (1981), which contains the references from which the data was obtained.*

(41) K. van't Riet, *Ind. Eng. Chem. Proc. Des. Devpt.*, **18**, 357 (1979).

A comparison of various types of gas-liquid contactors is illustrated in Figure 5.14. It can be seen that the effect of power input on the overall mass transfer coefficient k_La does not depend on whether the agitation is mechanical or pneumatic. The presence of internals, such as draft tubes, is also seen not to play a significant role. However, only mechanically agitated tanks are capable of attaining high values of k_La which may be required in highly aerobic fermentations (eg. antibiotic fermentations).

Viscous Systems

In viscous systems, two types of fermentation broth behavior can be distinguished:
(a) fungal cultivations where the broth viscosity is due to the structure of the mycelium, with essentially an aqueous medium as the continuous phase.
(b) systems such as polysaccharide fermentations where the viscosity is due to polymers excreted by the organism. Here the continuous phase is essentially homogeneous and viscous.

In both systems, dimensionless mass transfer coefficient correlations take the general form:

$$\frac{k_L a D_i^2}{D_{O_2}} = A\left(\frac{\mu_{app}}{\rho_L D_{O_2}}\right)^{\alpha}\left(\frac{\mu_{app} v_s}{\sigma}\right)^{\beta}\left(\frac{D_i^2 N \rho_L}{\mu_{app}}\right)^{\gamma}\left(\frac{\mu_{gas}}{\mu_{app}}\right)^{\delta} \tag{5.127}$$

The constant A accounts for geometric factors, such as impeller to tank diameters and height to diameter ratios. The apparent viscosity μ_{app} is obtained from the bulk shear rate for power law fluids:

$$\mu_{app} = K(11N)^{n-1}\left(\frac{3n+1}{4n}\right)^{n} \tag{5.128}$$

where K is the consistency index and n is the power law index, and the dispersed phase viscosity is μ_{gas}.

Yagi and Yoshida[42] examined viscous and viscoelastic solutions of carboxymethylcellulose and sodium polyacrylate and obtained the mass transfer coefficient by oxygen desorption. They found no effect of gas viscosity and proposed the following correlation, which incorporates an additional group (ND_i/v_s):

$$\frac{k_L a D_i^2}{D_{O_2}} = 0.06\left(\frac{\mu_{app}}{\rho_L D_{O_2}}\right)^{0.5}\left(\frac{\mu_{app} v_s}{\sigma}\right)^{0.6}\left(\frac{D_i^2 N \rho_L}{\mu_{app}}\right)^{1.5}\left(\frac{D_i N^2}{g}\right)^{0.19}\left(\frac{N D_i}{v_s}\right)^{0.32} \tag{5.129}$$

For viscoelastic fluids, the Deborah number (De) relates the characteristic material time to the process time. In a stirred tank De is given by λN, where λ is a characteristic material time. Yagi and Yoshida included this in their correlation for viscoelastic fluids to give:

(**42**) H. Yagi and F. Yoshida, *Ind. Eng. Chem. Proc. Des. Devpt.*, **14** 488 (1975).

$$\frac{k_L a D_i^2}{D_{O_2}} = 0.06\left(\frac{\mu_{app}}{\rho_L D_{O_2}}\right)^{0.5}\left(\frac{\mu_{app} v_s}{\sigma}\right)^{0.6}\left(\frac{D_i^2 N \rho_L}{\mu_{app}}\right)^{1.5}\left(\frac{D_i N^2}{g}\right)^{0.19}\left(\frac{N D_i}{v_s}\right)^{0.32}(1+2\mathrm{De}^{1/2})^{-0.67} \quad (5.130)$$

This equation reduces to Equation (5.129) when the Deborah number is zero.

Comparison of correlations for the mass transfer coefficients in stirred tanks.

A large stirred tank of 35,000 liter liquid capacity is used for the growth of a fungal species for antibiotic production. After the period of exponential growth, the mycelial concentration is constant and the broth shows power law behavior, with a power law index of 0.75 and a consistency index of 25 gm-cm^{-1}-sec$^{-1.25}$. Given the tank geometry, broth properties and air flow rate below, calculate the mass transfer coefficient k_La as a function of the impeller speed.

Data:

air flow rate	0.3	vvm
tank height to diameter	3:1	
D_i to D_{tank}	0.5	
liquid density ρ_L	1.05	gm/cm^3
surface tension σ	55	dyne/cm
D_{O2}	2.0x10^{-5}	cm^2/sec
g	980	cm/sec^2

Calculations:

liquid volume	= 35 meter^3
tank diameter	= $\{(4 \times 35)/(3\pi)\}^{1/3}$ meter
	= 245.8 cm
impeller diameter D_i	= 122.9 cm
apparent viscosity μ_{app}	=25 x (11 N)$^{0.75-1}$ $\{(0.75 \times 3 + 1)/(4 \times 0.75)\}^{0.75}$ (eqn 5.128)
	= 14.574 N$^{-0.25}$ (gm.cm^{-1}sec^{-1})
viscosity of water	= 0.01 (gm.cm^{-1}sec^{-1})
superficial gas velocity v_s	= $(0.3 \times 35)/(\pi \times 2.458^2/4)$
	= 2.213 meter/min
	= 3.688 cm/sec

The mass transfer coefficient can now be found by substitution into Equation (5.129)

$$\frac{k_L a D_i^2}{D_{O_2}} = 0.06\left(\frac{D_i^2 N \rho_L}{\mu_{app}}\right)^{1.5}\left(\frac{D_i N^2}{g}\right)^{0.19}\left(\frac{\mu_{app}}{\rho_L D_{O_2}}\right)^{0.5}\left(\frac{\mu_{app} v_s}{\sigma}\right)^{0.6}\left(\frac{N D_i}{v_s}\right)^{0.32} \quad (5.129)$$

$$Re = \frac{N \rho_L D_i^2}{\mu_{app}} = 1{,}088 N^{1.25} \qquad Fr = \frac{D_i N^2}{g} = 0.125 N^2 \qquad Sc = \frac{\mu_{app}}{\rho_L D_{O_2}} = 6.94 \cdot 10^5 N^{-0.25}$$

$$\frac{\mu_{app} v_s}{\sigma} = 0.977N^{-0.25} \qquad \frac{ND_i}{v_s} = 33.3N$$

thus

$$\frac{k_L a D_i^2}{D_{O_2}} = 0.06(1.088 \cdot 10^3 N^{1.25})^{1.5} (0.125N^2)^{0.19} (6.94 \cdot 10^5 N^{-0.25})^{0.5} (0.977N^{-0.25})^{0.6} (33.3N)^{0.32}$$

$$k_L a = 0.06 \cdot \left(\frac{2.0 \cdot 10^{-5}}{122.9^2} \right) \cdot 3.645 \cdot 10^6 N^{2.3} = 0.00483N^{2.3} \ (\text{sec}^{-1})$$

As we might expect, the mass transfer coefficient is quite sensitive to the impeller speed in this shear-thinning viscous broth. At an impeller speed of 90 rpm, the mass transfer coefficient would be 0.00735 sec^{-1} (26.5 hr^{-1}), and increases rapidly with rpm. Using the same correlation and operating conditions, we see that when the fermentation broth has a viscosity of that of water, the mass transfer coefficient at 90 rpm would be considerably higher, having a value of 0.186 sec^{-1} (671 hr^{-1}).

We can compare the above result with the correlations of van't Riet for coalescing inviscid sytems, by first calculating the power input P/V (the method is described later in Section 5.4.4). We assume that at the high impeller Reynolds number in water (2.3 x 10^6 for the above case) the power number is constant and has a value of 6 for a flat blade turbine impeller. Thus

Power number $= P/(\rho_L N^3 D_I^5) = 6$

Unaerated power $= 6\ (\rho_L N^3 D_I^5)$
$= 5.96 \times 10^{11}$ (gm-cm^2-sec^{-3})

Aerated power $= 0.4 \times 5.96 \times 10^{11}$ (gm-cm^2-sec^{-3})
$= 2.38 \times 10^{11}$ (gm-cm^2-sec^{-3})
$= 2.38 \times 10^4$ (Watts)

P/V $= (2.38 \times 10^4)/35$ (Watts/m^3)
$= 681$ watts/m^3

and hence

$k_L a$ $= 2.6 \times 10^{-2}\ (P/V)^{0.4}\ v_s^{0.5}\ (\text{sec}^{-1})$
$= 2.6 \times 10^{-2}\ (681)^{0.4}\ (3.688 \times 10^{-2})^{0.5}$
$= 0.0679\ \text{sec}^{-1}$
$= 244\ \text{hr}^{-1}$

which is lower than that predicted by the Yagi and Yoshida correlation, developed from data obtained in a 12 liter tank. Use of the van't Riet correlation for non-coalescing systems gives a value for $k_L a$ of 358 hr^{-1}, indicating the importance of coalescence in reducing the interfacial area and thus reducing the mass transfer coefficient. The actual value of $k_L a$ for water in this 35,000 liter tank is best taken from the non-coalescing systems correlation, as most practical systems have traces of contaminants which inhibit coalescence.

5.4.3 Experimental Determination of k_La

There are several approaches to determine k_La in aerated bioreactors. These are generally classified as static or dynamic methods. Both methods require a reaction in the liquid phase to reduce the dissolved oxygen concentration to a level below saturation. This reaction is generally the microbial consumption of oxygen. We do not need to know the kinetics of the reaction provided that the measurements can be taken over a sufficiently short period that the liquid phase oxygen concentration is essentially constant.

Static Method

For well-mixed gas and liquid phases, the starting point for experimentally determining k_La is based on the set of coupled gas and liquid phase oxygen mass balance equations:

Gas: $$\frac{d}{dt}\left(\frac{PV_g}{RT}y\right)=\left(\frac{P}{RT}\right)(Q_1y_o-Q_2y)-K_gaPV_L\left(y-C_{O_2}\cdot\frac{H}{P}\right) \tag{5.42}$$

Liquid: $$\frac{d(V_LC_{O_2})}{dt}=F\left(C_{O_2,\text{in}}-C_{O_2}\right)+K_gaV_LP\left(y-C_{O_2}\frac{H}{P}\right)-r_{O_2}V_L \tag{5.44}$$

In the static method, steady state is assumed and the gas phase oxygen balance reduces to:

$$\left(\frac{P}{RT}\right)(Q_1y_o-Q_2y)-K_gaPV_L\left(y-C_{O_2}\cdot\frac{H}{P}\right)=0 \tag{5.131}$$

Generally, the inlet mole fraction of oxygen in the inlet gas stream (y_o) is known and by measuring the exit mole fraction (y), the inlet and exit gas flow rates and the concentration of dissolved oxygen (C_{O2}) in the liquid phase, the overall mass transfer coefficient K_ga can be found. If we assume that all the transport resistance is in the liquid film around the bubble, then k_La is given by:

$$k_La=H\cdot K_ga=\frac{H(Q_1y_o-Q_2y)}{RTV_L\left(y-C_{O_2}\cdot\frac{H}{P}\right)} \tag{5.132}$$

Note that we have assumed that there is no significant variation of hydrostatic pressure within the vessel; thus Q_1 and Q_2 are evaluated at an average hydrostatic pressure. In laboratory scale equipment, the assumptions that the gas phase is well mixed and there are no pressure variations are quite reasonable. However, the hydrostatic pressure variation will be significant in vessels of 15 to 30 feet height. The average bubble volume will increase with height, the gas phase behavior will more closely resemble plug flow and pressure effects must be explicitly considered (see Problem 1).

When the gas phase behavior can be considered to be plug flow and we can neglect pressure variation with liquid height, Equation (5.45) can be integrated from the gas inlet to the exit:

$$v_b\frac{d\left(\frac{PV_g}{RT}y\right)}{dz}=-K_gaV_gP\left(y-C_{O_2}\frac{H}{P}\right) \tag{5.45}$$

integrating

$$\int_{y_{in}}^{y_{out}} \frac{dy}{\left(y - \frac{H}{P} C_{O_2}\right)} = -\int_0^{Z} \frac{K_g a RT}{v_b} dz$$

$$\log \frac{\left(y_{out} - \frac{H}{P} C_{O_2}\right)}{\left(y_{in} - \frac{H}{P} C_{O_2}\right)} = -\frac{K_g a RT}{v_b} Z$$

$$k_L a = \frac{H \cdot v_b}{ZRT} \log \frac{\left(y_{out} - \frac{H}{P} C_{O_2}\right)}{\left(y_{in} - \frac{H}{P} C_{O_2}\right)} \tag{5.133}$$

Thus k_La can be determined from the static (or steady state) method provided that the liquid and gas phase behavior is known. The inlet and exit gas pahse compositions are measured, together with the liquid phase dissolved oxygen concentration. This method has the advantage of simplicity, but in large tanks the liquid may not be well mixed and there may be significant spatial variations in the dissolved oxygen concentration.

Dynamic Method

The dynamic method for k_La determination is based on the response of the dissolved oxygen concentration to changes in the inlet gas phase oxygen concentration. In this case we employ the well-mixed liquid phase oxygen balance equation and represent the gas phase mole fraction of oxygen by y^{ave}, which we assume to be constant. This is tantamount to assuming that the dynamics of the gas phase response can be neglected; the change in y^{ave} with time is sufficiently small that it does not influence the liquid phase balance equation. As the saturation concentration of dissolved oxygen in fermentation broths is quite small, this is a reasonable assumption.

$$\frac{d(V_L C_{O_2})}{dt} = F\left(C_{O_2,in} - C_{O_2}\right) + K_g a V_L P\left(y^{ave} - C_{O_2}\frac{H}{P}\right) - r_{O_2} V_L \tag{5.44}$$

In batch reactors there is no liquid flow into the tank and F is zero. Even in continuous reactors the magnitude of the terms $F.C_{O2}$ for inlet and outlet are small with respect to the mass transfer and reaction terms in the above equation, as k_La (order 10^2 hr^{-1}) is usually much greater than the dilution rate F/V_L (order 10^{-1} hr^{-1}). Thus the above equation may be simplified for both batch and continuous operation to:

$$\frac{d(C_{O_2})}{dt} = K_g a P\left(y^{ave} - C_{O_2}\frac{H}{P}\right) - r_{O_2} = k_L a\left(C^*_{O_2} - C_{O_2}\right) - q_{O_2} X \tag{5.134}$$

If the dissolved oxygen concentration is monitored with a dissolved oxygen electrode, this equation can be employed to determine k_La by first halting aeration of the fermentation broth. If the gas phase disengages quickly from the liquid and there is no surface aeration (this can be ensured by sweeping the surface with an inert gas such as nitrogen), then the transport term disappears from the above equation and it reduces to:

$$\frac{d(C_{O_2})}{dt} = -q_{O_2}X \tag{5.135}$$

where $q_{O2}X$ is the microbial volumetric rate of oxygen consumption. Provided that this non-gassing period is short, the microbial suspension will continue to respire at the same rate as that obtained during gassing and the dissolved oxygen will fall linearly with time. We assume that C_{O2} is sufficiently high so that the specific oxygen consumption rate q_{O2} remains constant, independent of C_{O2}; i.e., it must remain above the critical concentration. If aeration is now resumed, Equation (5.134) can be rearranged to give:

$$C_{O_2} = C^{*}_{O_2} - \frac{1}{k_L a}\left(q_{O_2}X + \frac{dC_{O_2}}{dt}\right) \tag{5.136}$$

If C_{O2} is recorded as a function of time, a plot of C_{O2} against dC_{O2}/dt has a slope of $-1/k_La$. Hence k_La can be determined as shown in Figure (5.15).

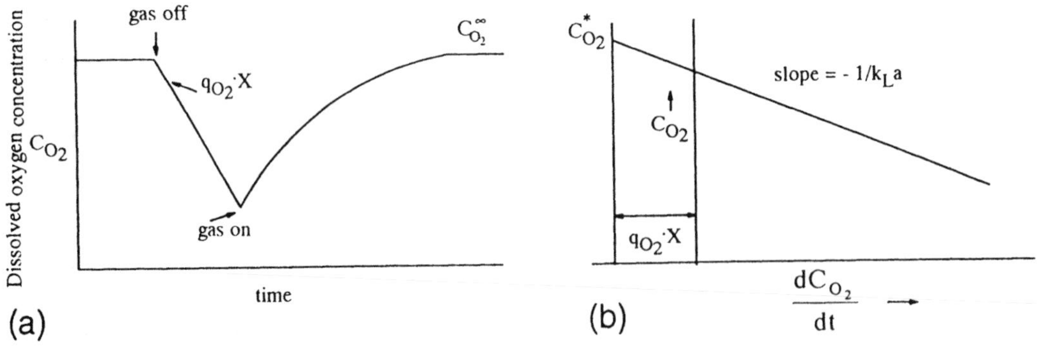

Figure (5.15). (a) Typical dissolved oxygen concentration profile obtained for continuous operation using the dynamic method for k_La. (b) A replot of the data shown on the left, with the slope (dC_{O2}/dt) estimated from the reaeration period.

Alternatively, we can avoid graphically differentiating the experimental C_{O2} versus time data by integrating a rearrangement of Equation (5.134).

$$\frac{d(C_{O_2})}{dt} = k_L a\left(C^{*}_{O_2} - C_{O_2}\right) - q_{O_2}X = k_L a\left\{\left[C^{*}_{O_2} - \frac{q_{O_2}X}{k_L a}\right] - C_{O_2}\right\} \tag{5.137}$$

and integrating:

$$\ln \frac{\left[C_{O_2}^{*} - \frac{q_{O_2}X}{k_L a}\right] - C_{O_2}^{t=0}}{\left[C_{O_2}^{*} - \frac{q_{O_2}X}{k_L a}\right] - C_{O_2}} = k_L a \cdot t \tag{5.138}$$

We note that upon reaeration, the value attained by the dissolved oxygen after the steady state is reestablished is given by

$$\left[C_{O_2}^{*} - \frac{q_{O_2}X}{k_L a}\right] = C_{O_2}^{t=\infty} \tag{5.139}$$

thus Equation (5.138) can be written in terms of measurable quantities:

$$\ln\left[\frac{C_{O_2}^{t=\infty} - C_{O_2}^{t=0}}{C_{O_2}^{t=\infty} - C_{O_2}^{t}}\right] = k_L a \cdot t \tag{5.140}$$

By plotting the log term against time, k_La can be directly obtained from the slope.

With either of these two graphical approaches for k_La determination, a value for C_{O2}^{*} can be obtained. This corresponds to an average mole fraction of oxygen in the gas phase ($C_{O2}^{*} = Py^{ave}/H$). If the value of y^{ave} obtained from this equation corresponds to the exit gas stream oxygen mole fraction, then the gas phase can be assumed to be well mixed. If it corresponds to a log mean of the inlet and exit gas phase mole fractions, then the gas phase behavior is plug flow. The dynamic method for obtaining k_La has the advantage of not requiring prior knowledge of the flow behavior of the gas phase. The oxygen balance equation for a well-mixed liquid simply requires an average value be used to describe the gas phase mole fraction.

The dynamic method is very commonly used in large and small scale equipment, as sterilizable oxygen electrodes permit k_La to be determined during a fermentation without causing significant disruption to the culture. This has the advantage that the k_La is found using the actual fermentation broth even when the broth is viscous. The method which was used prior to the development of sterilizable electrodes was the sulphite oxidation method, which is conducted using a solution of sodium sulphite. In the presence of a Cu^{2+} or Co^{2+} catalyst, sulphite is oxidized according to

$$Na_2SO_3 + 1/2\ O_2 \xrightarrow{Cu^{2+} \text{or} Co^{2+}} Na_2SO_4$$

This kinetics of this reaction are independent of sulphite concentration. The reaction consumes oxygen at a rate which is sufficiently fast so that transport of oxygen from the gas to the liquid through the liquid film around the gas bubbles is the limiting step. Thus k_La can be found by measuring the observed rate of reaction. Typically, a 1.0 N solution of sulphite is placed in the reactor with 10^{-3} M Cu^{2+} ion and the liquid is sparged with air. Samples of the solution are taken over a time period and the concentration of unreacted sulphite is determined by reacting the sulphite in each sample with excess iodine and then back-titration the iodine with thiosulphite. It is usually assumed that the dissolved oxygen is zero. The rate of oxygen transfer is then equated to the rate of decrease of sulphite.

$$k_L a = \frac{\text{rate of sulphite consumption}}{\left(C^*_{O_2} - C_{O_2}\right)} \sim \frac{\text{rate of sulphite consumption}}{C^*_{O_2}} \qquad (5.141)$$

Several assumptions are involved in determining k_La this way. First, the rate of reaction is assumed zeroth order in Na_2SO_3, but is not so fast that reaction occurs in the liquid film around the gas bubbles. This would decrease the apparent film thickness and give incorrectly high values of k_La. Second, the sulphite solution must be assumed to approximate the rheological characteristics of a fermentation broth. This latter assumption is clearly not always realistic. This method is fairly labor intensive and has fallen out of favor compared to static or dynamic methods.

There are, however, considerations that must be addressed with the dynamic method. Most dissolved oxygen electrodes have a substantial response time (of order 30 seconds to several minutes), so that changes in the actual value of C_{O2} in the broth are not recorded until some later time. The probe readings must therefore be reanalyzed to obtain the true values of k_La. The following section illustrates this.

The Dissolved Oxygen Electrode

There are two types of oxygen electrodes commonly employed in fermentations. These are steam sterilizable *polarographic* and *galvanic* types[43]. When an electrode of a noble metal such as platinum is made electronegative (0.6 to 0.8 V) with respect to a reference electrode (calomel or Ag/AgCl) in a neutral KCl solution, dissolved oxygen is reduced at the surface of the cathode. The polarographic probe supplies this 0.6 - 0.8 bias voltage and the current that results is proportional to the activity (or equivalent partial pressure) of dissolved oxygen. The cathode and anode are usually separated from the fermentation medium by means of a membrane permeable to dissolved gases but not ions. The reactions which occur are:

Cathiodic reaction	$O_2 + 2H_2O + 2e^- \text{--->} H_2O_2 + 2\ OH^-$
	$H_2O_2 + 2e^- \text{--->} 2\ OH^-$
Aniodic reaction	$Ag + Cl^- \text{--->} AgCl + e^-$
Overall	$4Ag + O_2 + 2H_2O + 4\ Cl^- \text{--->} 4AgCl + 4\ OH^-$

(43) For a detailed description of the oxygen electrode, see Y.H. Lee and G.T. Tsao, *Dissolved Oxygen Electrodes*, Adv. Biochem. Eng. **13**, 35 (1979).

The galvanic electrode differs from the polarographic electrode in that it does not require an external voltage source for the reduction of oxygen at the cathode. When a basic metal such as zinc, lead or cadmium is used as the anode and a noble metal such as silver or gold is used as the cathode, the voltage generated by the electrode pair is sufficient for oxygen reduction at the cathode. The reactions are:

Cathiodic reaction	$O_2 + 2H_2O + 4e^- \text{--->} 4\,OH^-$
Aniodic reaction	$Pb \text{--->} Pb^{2+} + 2\,e^-$
Overall	$2Pb + O_2 + 2H_2O \text{--->} 2Pb(OH)_2$

The electrolyte does not participate in the reaction, but the aniodic surface is gradually oxidized. Both types of electrodes contain a semi-permeable membrane over the cathode, through which dissolved oxygen must diffuse to react. The membrane provides the main resistance to oxygen transfer and this resistance is the principal determinant of the electrode response. The situation is illustrated in Figure (5.16).

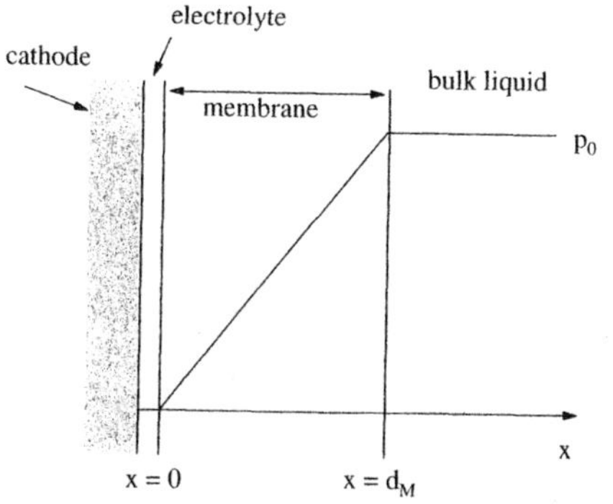

Figure 5.16. *Diagrammatic representation of oxygen transfer from the bulk liquid phase to the cathode of a dissolved oxygen electrode.*

Consider the partial pressure of oxygen in the bulk liquid initially at zero and increasing stepwise to a value p_0. The diffusion of oxygen through the membrane is governed by Fick's law

$$\frac{\partial p}{\partial t} = D_m \frac{\partial^2 p}{\partial x^2} \tag{5.142}$$

with the boundary conditions given by:

$p = 0$ at $t = 0$

$p = 0$ at $x = 0$

$p = p_0$ at $x = d_m$

The first boundary condition arises from assuming that the reaction at the cathode surface is very fast. The solution to this equation is

$$\frac{p}{p_o} = \frac{x}{d_m} + \sum_{n=1}^{\infty} \frac{2}{n\pi}(-1)^n \sin\left\{\frac{n\pi x}{d_m}\right\} e^{-n^2\pi^2 D_m t/d_m^2)} \tag{5.143}$$

The current output from the electrode is proportional to the rate of transport of oxygen to the cathode surface:

$$I = NFAD_m\left(\frac{\partial c}{\partial x}\right)_{x=0} = NFAP_m\left(\frac{\partial p}{\partial x}\right)_{x=0} \tag{5.144}$$

Here N, F, A and P_m are the number of electrons per mole of oxygen reduced, Farady's constant, the surface area of the cathode and the oxygen permeability of the membrane, repectively. The permeability of the membrane is related to the diffusivity, P_m is D_mS_m, where S_m is the solubilty of oxygen in the membrane and D_m is the diffusivity. The output of the electrode (I) can thus be found from equations (5.143) and (5.144)

$$I(t) = NFA\left(\frac{P_m}{d_m}\right)p_o\left[1 + 2\sum_{n=1}^{\infty}(-1)^n e^{-n^2\pi^2 D_m t/d_m^2}\right] \tag{5.145}$$

Thus at sufficiently long times the exponential term decays to zero and the probe reponse is directly proportional to the oxygen partial pressure. The time constant for the probe response is the most important result of this analysis. It is given by 1/k, where

$$k = \frac{\pi^2 D_m}{d_m^2} \tag{5.146}$$

i.e., the thinner the membrane the faster the time response of the electrode. This models could be extended to consider the liquid film resistance and diffusion through electrolyte at the cathode surface. The time constant thus will increase if these resistances are significant. Experimentally, k can be found from a plot of the fractional response of the electrode to a step change in the dissolved oxygen concentration:

$$\Gamma(t) = \frac{I(t) - I_0}{I_{ss} - I_0} \tag{5.147}$$

Substituting for the intital and steady state values of the current I, we obtain

$$\Gamma(t) = 1 - 2e^{-kt} + 2\sum_{n=2}^{\infty}(-1)^n e^{-n^2 kt} \tag{5.148}$$

When $\Gamma \geq 0.4$ and thus $kt \geq 1.2$, the series terms (n>2) can be neglected and k is obtained from a plot of ln(1-Γ) against time. A typical value of k for a 15 micron thick polypropylene membrane at 20°C is 0.4 sec^{-1}, with a range from 0.05 to 0.5 sec^{-1}, depending on the membrane thickness and composition. The truncation of the series terms is thus reasonable.

Determination of k_La when the oxygen probe response is significant

When the dynamic method for determination of k_La is employed and the time constant of the probe is significant (this is generally the case), the analysis presented above must be modified to account for the time varying boundary condition at the probe membrane:

$$p = p_o(t) \qquad at \qquad x = d_m \tag{5.149}$$

$p_o(t)$ can be found from solution of the liquid phase oxygen mass balance. Assuming that the gas phase dynamics can be neglected, we can rewrite this equation in terms of partial pressure for the case when no reaction occurs in the liquid:

$$\ln\frac{\left[C^{\bullet}_{O_2} - C^{t=0}_{O_2}\right]}{\left[C^{\bullet}_{O_2} - C_{O_2}\right]} = k_L a \cdot t$$

Rearranging
$$C_{O_2}(t) = C^{\bullet}_{O_2} - \left(C^{\bullet}_{O_2} - C^{t=0}_{O_2}\right)e^{-k_L a \cdot t} \tag{5.150}$$

For simplicity, the initial liquid phase dissolved oxygen concentration ($C_{O2}^{t=0}$) will be set to zero. Converting the above equation to oxygen partial pressures:

$$p_O(t) = p^{\bullet}_{O_2}\left(1 - e^{-k_L a \cdot t}\right) \tag{5.151}$$

The solution to the electrode response Equation (5.142) with this time varying boundary condition is more complex:

$$\Gamma = 1 - \frac{\pi B^{1/2}}{\sin(\pi B^{1/2})}\exp(-Bkt) - 2\sum_{n=1}^{\infty}(-1)^n\frac{\exp(-n^2kt)}{(n^2/B - 1)} \tag{5.152}$$

where B is k_La/k. Once again, the series terms can be neglected at large values of kt.

One simple approach, which overcomes the necessity of fitting parameter values to Equation (5.152) for determination of k_La, is to perform an aeration experiment in the vessel and a step change experiment in the laboratory. The data are used to obtain the normalized probe responses $\Gamma(t)$. If the step change was made by rapidly moving the electrode from liquid depleted of oxygen to liquid in equilibrium with the gas stream used in the aeration experiment, the difference in the integrals of the normalized responses yields k_La directly:

$$\int_0^{\infty}(1 - \Gamma)_{aeration}dt - \int_0^{\infty}(1 - \Gamma)_{step}dt = \frac{1}{k_L a} \tag{5.153}$$

This method has the advantage that any liquid film resistance around the probe is the same in both environments and cancels. This resistance can be quite significant with viscous broths. Typical data are shown in Figure 5.17.

5.4.4 Power Requirements and Mixing

The power requirements for dispersion of a gas in a liquid phase (e.g., aerobic fermentations) or for suspension of solids (e.g., immobilized enzyme systems) can be developed by analysis of the Navier-Stokes equations describing the liquid velocity profiles in the tank.

We shall develop dimensionless groups from the Navier-Stokes equation which can be generally used for correlation of power input with agitation speed, liquid properties and vessel geometry. Writing the Navier-Stokes equations:

$$\rho_L \frac{D\mathbf{v}}{Dt} = -g_c \nabla p + \mu_L \nabla^2 \mathbf{v} + \rho_L \mathbf{g} \tag{5.154}$$

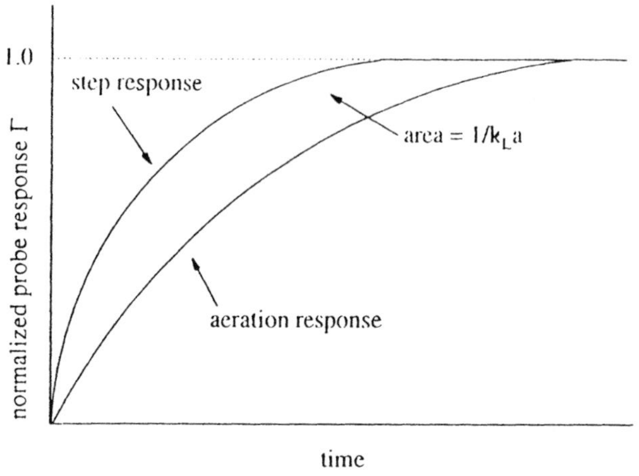

Figure 5.17. Graphical method for determination of k_La, based on Equation (5.153). The area between the normalized probe responses gives the reciprocal of the mass transfer coefficient.

Characteristic lengths and time scales can now be introduced based on the impeller diameter D_i and the agitator rotational speed 1/N. Thus, length and time become:

$$x^* = x/D_i \quad y^* = y/D_i \quad z^* = z/D_i \quad t^* = tN$$

Dimensionless velocity is $\mathbf{v}^* = \mathbf{v}/ND_i$, and a dimensionless pressure can now be defined as

$$p^* = \frac{(p - p_o)}{\rho_L N^2 D_i^2} \tag{5.155}$$

where a reference pressure p_o is selected to simplify boundary conditions. The Navier-Stokes equations now become, in dimensionless form:

$$\frac{D\mathbf{v}^*}{Dt^*} = -\nabla^* p^* + \left[\frac{\mu_L}{D_i^2 N \rho_L}\right] \nabla^{*2} \mathbf{v}^* + \left[\frac{g}{D_i N^2}\right] \frac{\mathbf{g}}{g}$$

$$= -\nabla^* p^* + \left[\frac{1}{Re}\right] \nabla^{*2} \mathbf{v}^* + \left[\frac{1}{Fr}\right] \frac{\mathbf{g}}{g} \qquad Re = \frac{D_i^2 N \rho_L}{\mu_L} \qquad Fr = \frac{D_i N^2}{g} \tag{5.156}$$

Here **g** is the acceleration due to gravity, g is the magnitude of **g**, g_c is the gravitational conversion factor. Thus the velocity and pressure distributions will depend on the Reynolds and the Froude numbers. When the tank is adequately baffled, the variation in liquid height, caused by vortex formation, will be minimized with respect to position in the tank, and thus the Froude number (which expresses the ratio of inertial to gravitational forces) dependence can be neglected.

The power imparted to the liquid phase results from the movement of the impeller blade through the liquid. Power is the product of rotational speed and applied torque. The torque is determined by integrating the pressure distribution over the surface of the impeller blade. Thus the pressure must be proportional to the transmitted power:

$$(p - p_o)_{blade} \propto \frac{P}{ND_i^3} \tag{5.157}$$

Substituting the expression for dimensionless pressure yields a relationship between dimensionless pressure and power.

$$p^{\bullet} \propto \left(\frac{P}{ND_i^3}\right) \frac{g_c}{\rho_L N^2 D_i^2} \propto \frac{P g_c}{\rho_L N^3 D_i^5} \tag{5.158}$$

The right hand expression in Equation (5.158) is the *power number*. It depends in general on the Reynolds and Froude numbers, but in baffled tanks gravitational effects are not significant and we can neglect the Fr dependence:

$$\frac{P g_c}{\rho_L N^3 D_i^5} = f(Re) \tag{5.159}$$

The power number is thus generally used for correlating agitator power. In turbulent flows the Reynolds number is large, inertial forces dominate viscous forces and thus the dependence of the power number on the Reynolds number will vanish. The terms involving viscous (order Re^{-1}) and gravitational forces in Equation (5.156) will vanish. The power number will thus be a constant. Under these conditions:

$$\frac{P g_c}{\rho_L N^3 D_i^5} = \text{constant} \qquad \text{and} \qquad P \propto \rho_L N^3 D_i^5 \tag{5.160}$$

This is typically the case in most applications involving flat blade agitators. Note that when the power is expressed in horsepower (HP), the gravitational constant must be included to make the power number dimensionless. In SI or cgs units, power can be expressed in Watts directly, without including g_c[44].

The second limiting case corresponds to small values of the Reynolds number. By neglecting the inertial and gravitational terms in Equation (5.156), the dimensionless Navier Stokes equation reduces to

---

(44) Note that 1 HP is 550 ft-lb_f-sec^{-1} and g_c has the value 32.17 lb_m-lb_f^{-1}-ft-sec^{-2}.

$$g_c \nabla p = \mu \nabla^2 \mathbf{v} \tag{5.161}$$

We must employ a different characteristic pressure in making the above equation dimensionless, as momentum has been neglected. Pressure is related to viscous force per unit area.

$$p^{**} = \frac{(p - p_o)g_c}{\mu N} \tag{5.162}$$

Substituting this dimensionless pressure into Equation (5.161) and employing the other dimensionless variables as defined earlier:

$$\nabla^* p^{**} = \nabla^{*2} \mathbf{v}^* \tag{5.163}$$

The dimensionless pressure and velocity distributions are thus constant, as the Reynolds number dependency has been eliminated. By substituting the expression for $(p - p_o)_{blade}$ into Equation (5.162), we obtain a "viscous" power number:

$$p^{**} \propto \left(\frac{P}{ND_i^3}\right)\frac{g_c}{\mu N} \propto \frac{P g_c}{\mu N^2 D_i^3} \tag{5.164}$$

As the dimensionless pressure is constant in this viscous regime, we obtain

$$P \propto \mu N^2 D_i^3 \tag{5.165}$$

In the viscous regime, the value of the turbulent power number will have the following form:

$$\frac{P g_c}{\rho_L N^3 D_i^5} \propto (\mu N^2 D_i^3)\frac{g_c}{\rho_L N^3 D_i^5} \propto \frac{\mu}{\rho_L N D_i^2} \propto Re^{-1} \tag{5.166}$$

Thus a plot of the logarithm of the turbulent power number against the logarithm of impeller Reynolds number will have a slope of -1 in the viscous flow regime.

Other important characteristics of stirred tanks can be derived from this dimensionless analysis. The fluid velocity in the viscous regime is independent of the Reynolds number. In turbulent flow the dimensionless velocity depends directly on the Reynolds number, so that a plot of dimensionless velocity (v/ND_i) against the Reynolds number will also be constant. Thus the dimensionless velocity will move through a transition regime between two constant values with increasing Reynolds number. This is illustrated in Figure 5.18(b).

A second characteristic of a particular impeller used for liquid agitation is the *pumping number*. It describes the volumetric flow of liquid that results from the motion of the impeller. An average fluid velocity in the tank can be defined as the volumetric flow per area:

$$v_{ave} \propto \frac{Q \text{ (volume per time)}}{A \text{ (area)}} \propto \frac{Q}{D_i^2} \tag{5.167}$$

where the characteristic area has been assumed to be proportional to the impeller diameter D_i. By making this average velocity dimensionless with respect to the impeller tip speed, we obtain the pumping number N_Q.

$$N_Q = \frac{(Q/D_i^2)}{ND} = \frac{Q}{ND_i^3} = f(Re) \tag{5.168}$$

The behavior of the pumping number is expected to follow that of the dimensionless velocity (Figure 5.18(c)).

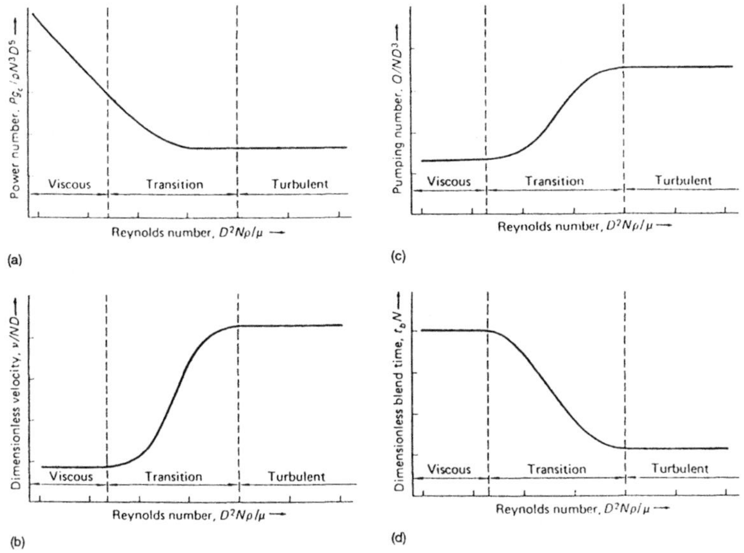

Figure 5.18. *Expected Reynolds number correlations of (a) the power number, (b) the dimensionless velocity, (c) the pumping number and (d) the dimensionless blending time for a turbine agitator in a stirred tank. [From Dickey, D. & J.G. Fenic, Chem. Eng. Jan 5, (1976)]*

Blending, the intermingling of fluid elements to produce a specified degree of uniformity, is also employed to characterize the behavior of an impeller. It is distinguished from *mixing*, which involves the production of uniformity between materials that may not be miscible. In the case of miscible fluids (e.g., a base solution added to control pH in a fermentation), the two concepts are synonymous. The factors that govern blending on a microscopic scale are not well understood, but a blend time is often employed to characterize the behavior of the agitation in the vessel. The blend time is the time required to introduce some degree of uniformity following the introduction of a tracer to the vessel. For example, a dye or solution of acid or base may be introduced a some point in the vessel and the time required to achieve a prespecified percentage (e.g., 99%) of the final tracer concentration is denoted as the blend time. Blend time is made dimensionless with respect to the impeller rpm.

$$t^* = t_{blend}N = f(Re) \tag{5.169}$$

Often the tank geometry is important and it has been found empirically that the ratio of the impeller to the vessel diameter can be incorporated into the dimensionless blend time:

$$t^*\left(\frac{D_i}{D_T}\right)^n = f(Re) \tag{5.170}$$

The value of n will depend on the type of agitator employed. For a pitched blade turbine, n has been found to be 2.3. The blend time dependence on the Reynolds number is illustrated in Figure 5.18(d).

The values of the power and pumping numbers as a function of Reynolds number depends on the type of impeller in the tank. Typically, the two main impeller types depend on the direction of the fluid discharged from the impeller. An axial flow impeller discharges fluid axially throughout the vessel. Marine-type propellers and pitched blade impellers have this property. Radial flow impellers discharge fluid toward the vessel walls. Flat-blade turbines show this characteristic. Examples of fluid patterns from these types are shown in Figure 5.19.

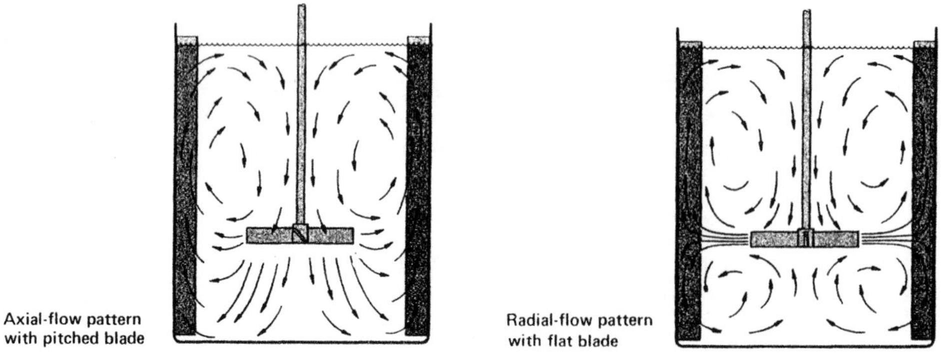

***Figure 5.19**. The flow patterns resulting from axial and radial flow impellers in a stirred tank.*

As expected from from the derivation of the power number in terms of the pressure distribution across the impeller blade, correlations of the power number as a function of the Reynolds number depend on the impeller type. Generally, impellers which pump the fluid axially have lower values of the power number (often less than 2), while flat blade turbines have values around 6. The actual values depend on the width to diameter ratio of the blades, the pitch of the propellers, etc. Figure (5.20) illustrates the power number correlations for various impellers.

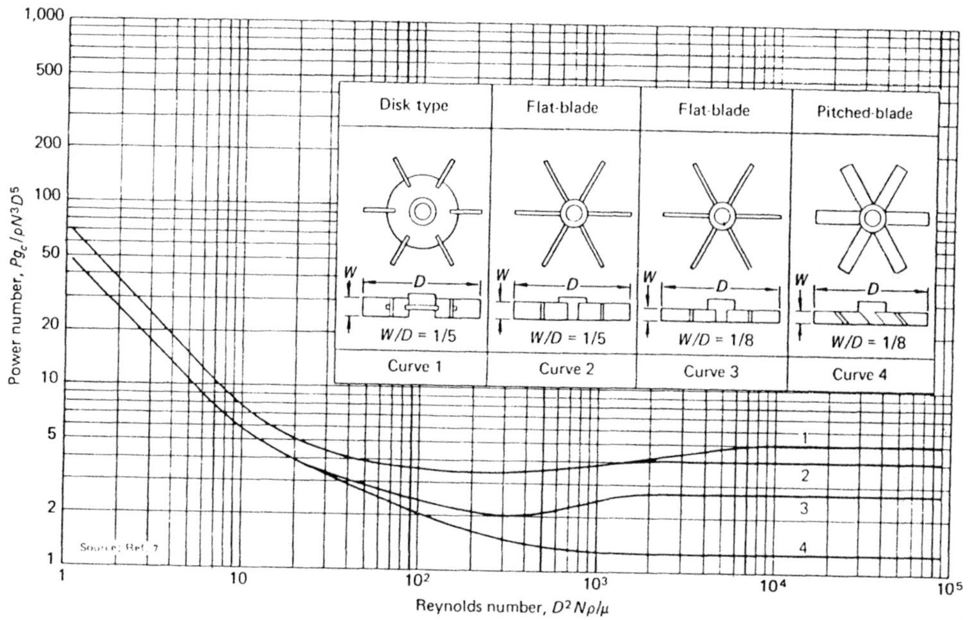

Figure 5.20. *The power number-Reynolds number relationship for various agitators. [From D.S. Dickey and J. Fenic et al., Chemical Engineering, Jan 5, Feb 2, May 24, July 19, Aug 2, Sept 27, Oct 25, and Nov 8 (1976); this series also presents correlations for blend times, power numbers, costs and scale-up methods].*

Determination of Power Requirements for Agitation of a Tank

We shall illustrate the use of Figure 5.20 to determine the power required to agitate a 10,000 liter tank filled with water. The diameter of the tank is 2.0 meters in diameter and is agitated at 100 rpm by a 6-blade turbine type agitator. The agitator is one half the tank diameter. We procede as follows:

Liquid properties	viscosity (μ_L)	= 0.001 newton-sec/meter^2
	density (ρ_L)	= 1000 Kg/meter^3
Reynolds number $(ND_i^2\rho_L)/\mu_L$		$= (100/60)(1)^2(1000)/(0.001) = 1.66 \times 10^6$
Power number $[P/(\rho_L.N^3.D^5_i)]$		= 4
	Power	$= 4\ (\rho_L.N^3.D_i^5) = 4\ (1000).(100/60)^3.(1)^5$
		$= 1.852 \times 10^4$ watts
		$= 1/746$ (HP/watts) 1.852×10^4 watts
		= 24.8 HP

Aerated Systems

When gas is introduced into an agitated tank, the bulk liquid density and viscosity in the tank decrease as a result of the formation of a gas dispersion. This alteration of the fluid properties is reflected in changes to the power required for agitation. The extent of the decrease in fluid density and viscosity depends on both the gas flow rate and the impeller speed. At low impeller speeds, the gas rises through the impeller region without being effectively dispersed. This condition is know as *flooding*. The power input required to overcome flooding depends on the superficial gas velocity and the impeller diameter to tank diameter ratio. Figure 5.21 illustrates this relationship for a radial flow impeller.

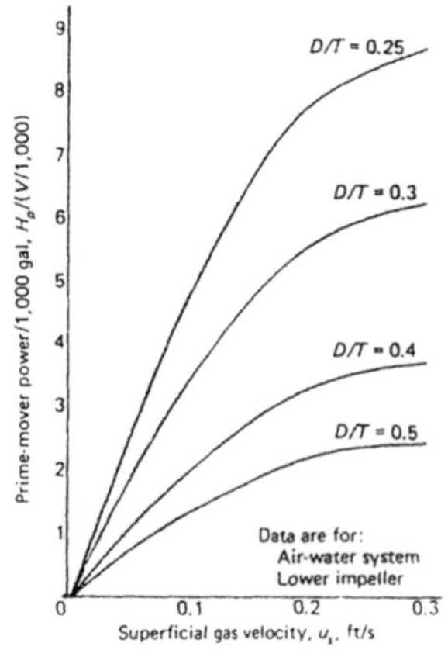

Figure 5.21*. The power required to overcome flooding for a six-bladed radially discharging impeller. The superficial velocity is calculated using the pressure at the bottom of the tank. [From D.S. Dickey and J. Fenic, Chem. Eng. Jan 5, (1976)].*

At higher impeller speeds beyond the point of flooding, there is a reduction in the power drawn by the impeller due to the change in the overall liquid density. This reduction in power has been correlated with the *aeration number*; the volumetric gas flow rate made dimensionless with respect to the impeller speed and diameter (Eq.5.171). The relationship between P/P_0 and the aeration number N_a (Q/ND_i^3) is given in Figure 5.22. Additional data for a wide variety of impellers are available[45].

$$\frac{P_{gassed}}{P_{ungassed}} = \frac{P}{p_o} = f\left(\frac{Q}{ND_i^3}\right) \tag{5.171}$$

(45) Y. Ohyamma and K. Endoh, *Chem. Eng. Japan*, **19**, 2 (1955) and Nagata *Mixing - Principles and Applications*, Halsted Press (1975).

A number of correlations are available to predict the gassed power consumption. Michel and Miller[46] propose

$$P_{gassed} = \mathrm{m} \cdot \left(\frac{P_{ungassed}^2 N D_i^3}{Q^{0.56}} \right)^{0.45} \tag{5.172}$$

where m is a constant (depending on the impeller type) and Q is the volumetric gas flow rate. Hughmark[47] has suggested the following equation based on data from several authors. It is the most reliable for estimation of gassed power in smaller tanks.

$$\frac{P_{gassed}}{P_{ungassed}} = 0.10 \left(\frac{Q}{NV_L} \right)^{-1/4} \left(\frac{N^2 D_i^4}{g B V_L^{2/3}} \right)^{-1/5} \tag{5.173}$$

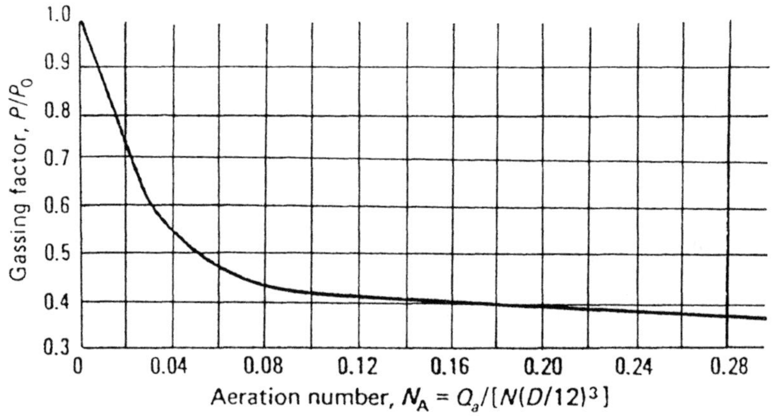

Figure 5.22. *The reduction in drawn power as a function of the aeration number, for a six-bladed turbine agitator and a ring sparger.*

The tank diameter ranged from 0.1 to 1.0 meters and the ratio of the impeller to tank diameter was 0.33 to 0.54 in the above correlation. B is the width of the impeller blades. An empirical fit of Figure 5.22 is provided by Oyama and Endo[48].

$$\left(\frac{P_{gassed} - P_\infty}{P_{ungassed} - P_\infty} \right) = e^{(-b/N_a)} \tag{5.174}$$

---

(46) B.J. Michel and S.A. Miller, *AIChE Journal*, **8**, 262 (1962).

(47) Hughmark, G., *Ind. Eng. Chem., Process Design Devpt.*, **19**,638 (1980).

(48) Y. Oyama and K. Endo, *Chem. Eng. Japan*, **19**, 2 (1955).

where b is a constant depending on the impeller type, P_∞ is the gassed power at very high gas flow rates.

In tall vessels, a second or third impeller is often employed to provide agitation in the upper part of the vessel. Gas is not introduced directly into these upper impellers and the reduction in their power requirements is often less than that in the lower impeller. In unaerated vessels, the power drawn by two impellers depends on the spacing of the impellers. This is illustrated in Figure 5.23. A distance between two impellers of 2 to 3 impeller diameters is necessary to draw twice the power of a single impeller.

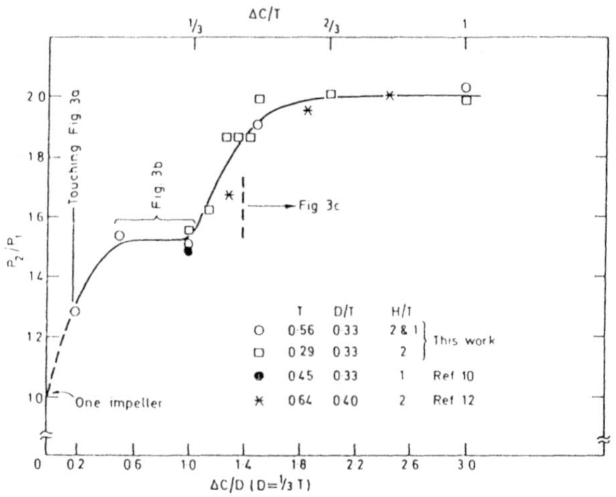

*Figure 5.23. Unaerated power drawn by two radial flow impellers in a stirred tank with a ratio of D_i/D_T of 1/3. P_2 is the power drawn by two impellers, P_1 is the power drawn by one impeller. [Data from V. Hudcova, V. Machon and A. Nienow, Biotech. Bioeng, **34**, 617 (1989)].*

When a two impeller system is aerated, the drawn power depends on impeller spacing and the aeration number. The upper impeller is much less prone to flooding than the lower impeller. The lower impeller behaves independently of the upper one and the design equations for a single impeller can be applied to it. The drawn power by the upper impeller depends on the impeller clearance ($\Delta C/D_i$), the impeller speed and the aeration rate. A useful approximation of the drawn power for two radial flow impellers ("Rushton" type), where the impeller spacing is greater than two impeller diameters is given by:

$$\left(\frac{P_{gassed}}{P}\right)_2 = 0.5\left\{\left(\frac{P_{gassed}}{P}\right)_1 + 1 - \varepsilon_g\right\} \tag{5.175}$$

where $(P_{gassed}/P)_1$ is the relative power drawn by a single impeller at the same aeration rate and agitation speed and ε_g is the gas hold-up. The justification for this equation is that the lower impeller behaves as an independent aerated impeller and that the upper one is in an unaerated liquid of density equal to the mean density of the gas-liquid mixture. For clearances less than two impeller diameters, Hudcova et al.[49] provide data for small tanks.

Determination of Power Requirements for Non-Newtonian Fluids

Unaerated Systems

In ungassed stirred tanks, the prediction of power consumption can be accomplished using the approach described earlier for Newtonian fluids. The difficulty arises in defining an apparent viscosity for calculation of the Reynolds number. Calderbank and Mooyoung[50] define an apparent viscosity in the vicinity of the impeller for pseudoplastic power-law fluids as

$$\mu_{app} = K(\gamma)^{n-1} = \frac{K}{(BN)^{1-n}}\left(\frac{3n+1}{4n}\right)^{n} \tag{5.176}$$

where B is a geometric parameter, with a value of ~ 10 for paddles and 6-blade turbine agitators and n is the power law index. By use of this apparent viscosity, the conventional power number - Reynolds number relationships in the laminar region can be employed. In cases where the fluid rheology does not follow a power law model, the approach of Metzner and coworkers[51] can be employed. By measuring shear rates directly and indirectly, the shear rate near the impeller was found to be proportional to the agitator speed:

$$\gamma = kN \tag{5.177}$$

where the constant of proportionality k has a value of 10 to 13 for 6-blade turbines and 10 for a propeller. With this value of the shear rate, the viscosity can be determined from rheological measurements on the fluid, where the shear stress corresponding to this shear rate in the tank is then divided by the shear rate to give the apparent viscosity, $\mu_{app} = \tau/\gamma$. This can then be used to calculate the Reynolds number and hence the power input. With many non-Newtonian fluids, agitation is in the laminar region, with Reynolds numbers ranging from 10 to 500. Typical power number-Reynolds number correlations for pseudoplastic fluids are shown in Figure 5.24.

---

(49) V. Hudcova, V. Machon and A. Nienow, *Biotech. Bioeng, 34*, 617 (1989).

(50) P. Calderbank and M.M. Mooyoung, *Trans. Inst. Chem. Engnrs.*, **37**, 26 (1959).

(51) A.B. Metzner and J.S. Taylor, *AIChE Journal*, **6** (1) 109 (1960) and A.B. Metzner and R.E. Otto, *AIChE Journal*, **3** (1) 3 (1957).

	Materials used	Baffles
⊙	Newtonians	no and yes
●	3.48%CMC sol.	no
○	3.26%CMC sol.	no
×	2.84%CMC sol.	no
+	2.51%CMC sol.	no
△	2.40%CMC sol.	no
▽	2.51%CMC sol.	no
◁	1.83%CMC sol.	no
□	1.67%CMC sol.	no
◇	1.50%CMC sol.	no
■	1.67%CMC sol.	4 Baffles
◆	1.50%CMC sol.	4 Baffles

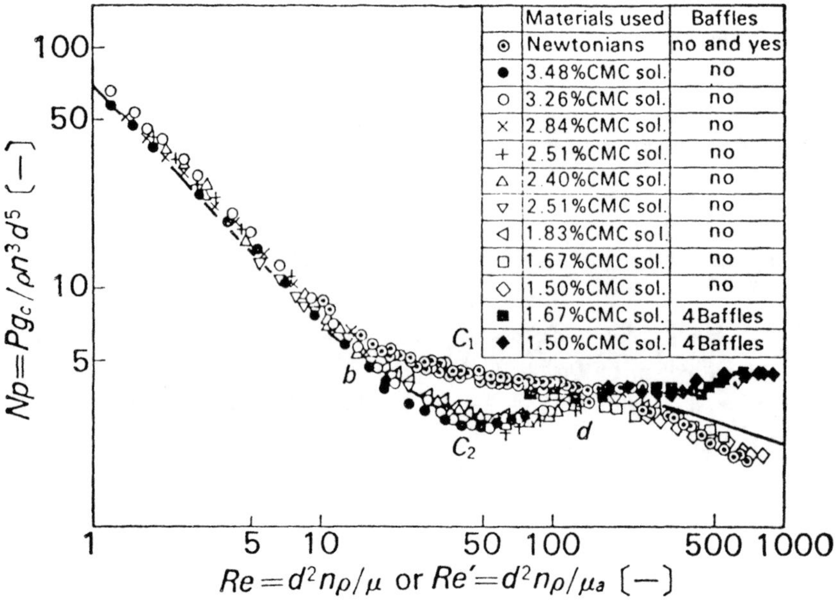

Figure 5.24. *Power number-Reynolds number correlation for pseudoplastic fluids; agitator is a 6-blade turbine with D_i/D_T of 0.5. [From S. Nagata, "Mixing, Principles and Application", Halsted Press (1975)].*

Aerated Systems

Bruijn et al.[52] observed the formation of gas cavities behind impeller blades with a rotating television camera and noted that the decrease in gassed power consumption with rising gas flow rates was due to an increasing number of large gas cavities formed behind the blades of the impeller. With viscous liquids (substantially Newtonian in nature) the authors reported a change in the shape of the gas cavities. This transition occurred over a range of viscosities of 5 to 300 cp. Once these cavities are formed in viscous liquids they are stable, even after the gas flow ceases. Thus, reduced power levels can be maintained even at very low gas flow rates; implying that the aeration number should not have a substantial effect on gassed power consumption. This indeed appears to be the case for viscous pseudoplastic and viscoelastic fluids.

Ranade and Ulbrecht[53] determined the power consumption of aerated polyacrylamide solutions (viscoelastic) and corn syrup (viscous Newtonian). Their results (shown in Figure

(52) W. Bruijn, K. Riet and J.M. Smith, *Trans. Inst. Chem. Engnrs.*, **52**, 88 (1974).

(53) V.R. Ranade and J. Ulbrecht, *AIChE Journal*, **24**, 796 (1978).

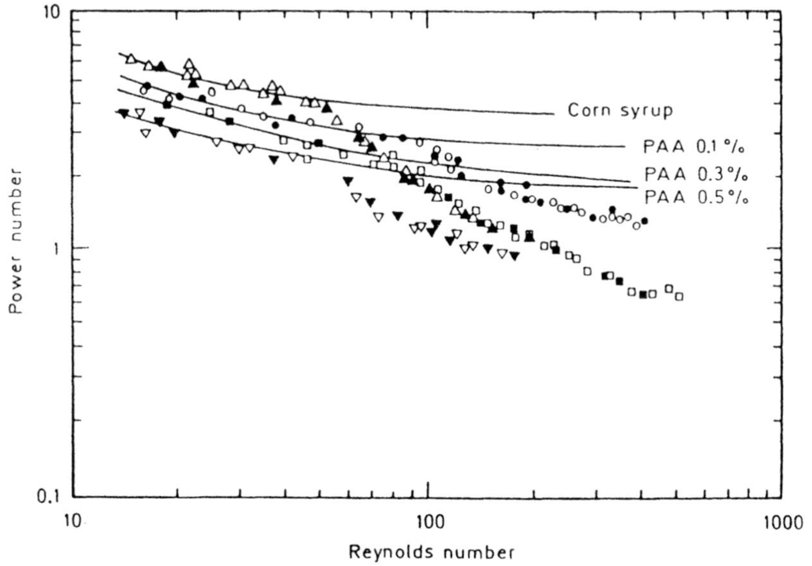

***Figure** 5.25. Data of Ranade and Ulbrecht of the power consumption in gassed non-Newtonian solutions. Corn syrup is viscous but Newtonian. The solid lines refer to unaerated solutions. The Reynolds number is calculated from an apparent viscosity defined by shear rate related to the impeller speed (11.5 N). Open symbols refer to 0.5 liters/min air flow, solid symbols are at 5.0 liters/min. The polyacrylamide concentrations employed are indicated on the figure.*

5.25) indicate that the aeration number has little effect on gassed power consumption. In addition to the viscosity dependence contained in the Reynolds number, there appears to be an additional effect of viscosity on the power number.

5.5 Heat Transfer

5.5.1 Sterilization of Media

Sterilization of microbial growth medium is required to prevent the growth of undesired microorganisms and viruses. It may be accomplished by a number of means, including radiation (such as UV and X-rays), filtration, chemical addition and heating. In large scale operations, filtration and heating are most commonly employed. For heat sensitive media, such as that required for the growth of mammalian cells, filtration is the sterilization technique of choice, while thermal treatment is more common for microorganisms grown in large stirred tanks. Filtration is generally more expensive than thermal treatment, and can be considered an "absolute" process, in the sense that the filter maximum pore size can be made smaller than the size of a microbe or virus particle. Heat sterilization is a probabalistic process, where the degree of contaminant removal depends on the time of thermal treatment.

This presents a potential problem in sterilizing canned foods, for example, where a single contaminating organism, such as *Clostridium botulinum* may grow in the can and be potentially lethal. Determination of the sterilization temperature-time cycle thus involves balancing the thermal destruction rate of microorganisms and spores with the thermal lability of the components in the medium. In the following discussions of heat transfer, we shall first consider the kinetics of thermal death of microorganisms.

Kinetics of thermal death of microorganisms

The thermal death of microorganisms typically follows a first order process, as described below:

$$\frac{dn_v}{dt} = -kn_v = -k_o e^{-[E_a/RT(t)]} n_v \qquad (5.178)$$

where n_v is the number of viable microorganisms per unit volume of medium. The rate constant k shows an Arrhenius dependence on the absolute temperature; when temperature is maintained constant, there is an exponential decrease in viable organisms with time:

$$n_v(t) = n_{vo} e^{-kt} \qquad (5.179)$$

Typical values of E_a (the activation energy for death of the organism) for bacterial spores range from 250 to 290 kJ-g mol^{-1} (60 to 122 Kcal/gmole). This energy of activation is higher than that of the heat sensitive nutrients found in the medium (70 to 100 kJ-g mol^{-1}). Enzymes have activation energies intermediate between microbial spores and vitamins (see Table 5.8). Spores are typically much more resistant to temperature than vegetative forms of bacteria or yeast. At 121 °C, for example, the value of k for vegetative cells may range up to 10^{10} min^{-1}, while k for spores is 0.5 to 5.0 min^{-1}. Mold spores are 2 to 10 times more resistant than vegetative forms of bacteria, viruses and bacteriophage are 1 to 5 times more resistant. Thus sterilization design is thus usually based on destruction of bacterial spores.

The large value of E_a for spores means that small increases in temperature have significantly greater effects on thermal death than is the case with vitamins, amino acids or enzymes. This is an important factor in the determination of the temperature-time profile for sterilization. The Maillard reaction is an important additional consideration; it results in the formation of a brown color as a result of the interaction of proteins and carbohydrates in the medium and can potentially remove important components for cell growth. Typical death rate data, shown in Figure 5.26, illustrate the sensitivity of temperature to the reduction of viable spores. At 121°C, only several minutes are required to reduce the viable number of spores to below 0.1% of the original number, at 108°C the time required is more than 30 minutes.

Table 5.8. Selected values of the activation energy for nutrients, enzymes and cells.

	Compound or reaction	Activation energy ($kJ. g\ mol^{-1}$)
spores	*B. stearothermophilus*	287.2
	B. subtilis	318.0
	Cl. botulinum	343.1
nutrients	vitamin B_{12}	96.6
	thiamine.HCl (B_6)	92.0
	riboflavin (B_2)	98.7
	folic acid	70.2
	d-pantothenyl alcohol	87.8
enzymes	trypsin	170.5
	peroxidase	98.7
	pancreatic lipase	192.3
reactions	Maillard (browning)	130.5

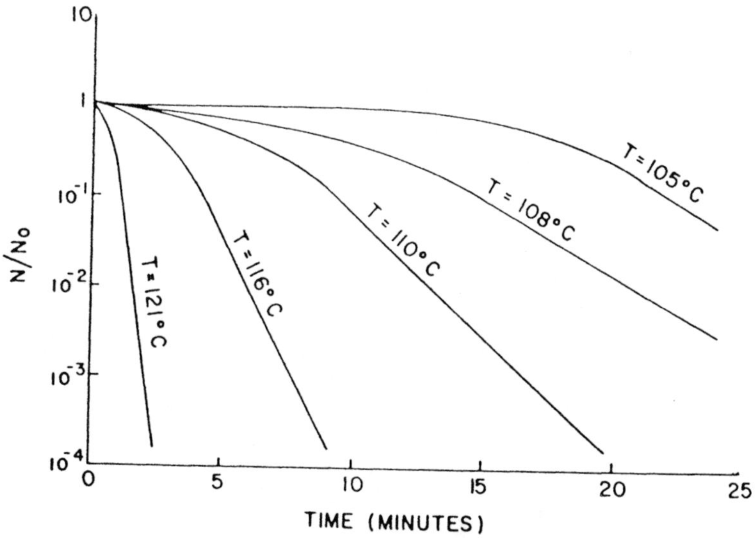

***Figure 5.26.** Typical death rate data for B. stearothermophilus spores.*

Table 5.9. *Typical values of k for B. stearothermophilus spores.*

Temperature (^oC)	*k (min^{-1})*
100	0.02
110	0.21
120	2.0
130	17.5
140	136
150	956

Another means of reporting thermal death rates is the use of the *decimal reduction time* (DRT), which is the time required to reduce the number of viable organisms by one $\log_{10}$ cycle (i.e., $\log_{10}(n_{vo}/n_v) = 1.0$). Use of this method is common in the food industry. Its use assumes that the DRT is temperature independent, in contrast to the Arrhenius relationship above. Its use over wide temperature ranges will thus result in errors which underpredict the required sterilization time.

The specification of a sterilization cycle is based on determining a desired degree of reduction in the number of bacterial spores. This may be taken as one spore being present in only, say, 100 or 1,000 fermentations. For a 10,000 liter vessel with 10^6 spores/ml initially present in the medium and a contamination rate of 1 in 100 fermentations:

$$\ln\left(\frac{n_{vo}}{n_v}\right) = \ln\left(\frac{10^6 \cdot 10^7}{10^{-2}}\right) = \ln(10^{15}) = 34.5 \tag{5.180}$$

We shall now examine the sterilization cycles for batch and continuous sterilizers required to achieve this desired reduction in microorganism or spore counts.

Batch Sterilization

Batch sterilization is usually accomplished in a well-mixed closed vessel. The fluid is heated to the desired temperature, maintained at that temperature for a specified period and then cooled. The number of viable organisms is reduced in each of these steps. The ratio of final to initial viable organisms (n_{vf}/n_{vo}) can be written as

$$\left(\frac{n_{vf}}{n_{vo}}\right)_{total} = \left(\frac{n_{v1}}{n_{vo}}\right)_{heating} \cdot \left(\frac{n_{v2}}{n_{v1}}\right)_{hold} \cdot \left(\frac{n_{vf}}{n_{v2}}\right)_{cooling} \tag{5.181}$$

This can be simplified:

$$\Delta_{total} = \ln\left(\frac{n_{vo}}{n_{vf}}\right) = \Delta_{heating} + \Delta_{hold} + \Delta_{cooling} \tag{5.182}$$

where $\Delta_{heating} = k_o \int_0^{t_1} e^{\frac{-E_a}{RT(t)}} dt \qquad \Delta_{hold} = k_o \int_{t_1}^{t_2} e^{\frac{-E_a}{RT(t)}} dt \qquad \Delta_{cooling} = k_o \int_{t_2}^{t_f} e^{\frac{-E_a}{RT(t)}} dt$

To solve this equation, we need to know the temperature-time profiles in each of the stages. The medium temperature is constant in the holding stage, but in the heating and cooling stages it depends on the type of heat transfer equipment employed, eg. jacketed tanks, tanks with internal coils or direct steam sparging. These can be calculated from heat balances. Table 5.10 illustrates the profiles for several common types of heat transfer.

***Table 5.10**. Temperature time profiles of media in batch sterilization. [From F.H. Deindoerfer and A.E. Humphrey, Appl. Microbiol., 7, 256 (1959)].*

Type of heating or cooling	Temperature-time profile	constants
direct steam sparging	$T = T_o\left(1 + \left\{\frac{\alpha t}{1+\delta t}\right\}\right)$	$\alpha = \frac{hS}{Mc_pT_o}$ $\delta = \frac{S}{M}$
steam (heat exchanger coils or tank jacket)	$T = T_{st}(1 + \beta e^{-\alpha t})$	$\alpha = \frac{UA}{Mc_p}$ $\beta = \frac{T_o - T_H}{T_H}$
electrical heating	$T = T_o(1 + \alpha t)$	$\alpha = \frac{q}{Mc_pT_o}$
cooling coils	$T = T_{co}(1 + \beta e^{-\alpha t})$	$\alpha = \left(\frac{Wc'_p}{Mc_p}\right)\left(1 - e^{\frac{UA}{Wc'_p}}\right)$ $\beta = \frac{T_o - T_{co}}{T_{co}}$

where:

T = temperature (°K)
T_o = initial temperature of medium (°K)
t = time (minutes)
h = enthalpy of steam relative to that of the medium (Kcal/Kg)
S = mass flow rate of steam (Kg/min)
M = mass of medium (Kg)
c_p = heat capacity of medium (Kcal/Kg.°C)
U = overall heat transfer coefficient (Kcal/m^2.min.°K)
A = heat transfer area (m^2)
T_H = temperature of the heat source, assumed constant (°K)
q = rate of heat transfer (Kcal/min)
c'_p = heat capacity of coolant (Kcal/Kg.°C)
W = mass flow rate of coolant (Kg/min)
T_{co} = temperature of coolant, assumed constant (°K)

With these profiles, the individual contributions to the overall thermal destruction of microorganisms can be determined. The heat transfer coefficients in Table 5.10 depend on the characteristics of the medium, such as viscosity and the presence of suspended solids. These are discussed later in section . Typical cycles are complete in 3 to 5 hours and the individual contributions of the portions of the cycle can be generalized below:

$$\Delta_{heating}/\Delta_{total} = 0.2 \qquad \Delta_{hold}/\Delta_{total} = 0.75 \qquad \Delta_{cooling}/\Delta_{total} = 0.05 \tag{5.183}$$

The major portion of thermal destruction is a result of the holding period of the cycle. Often the cooling period is shortened by sparging sterile air into the tank, removing heat by evaporation.

An Example of Sterilizer Design: Pasteurization in a Plate Heat Exchanger

Consider a liquid containing proteins which is to be used in the food industry. The material is to be pasteurized for shipment, but denaturation of the proteins places limits on the time and temperature for pasteurization. The thermal denaturation of proteins has an activation energy of 40 Kcal/gmole. The pasteurization is accomplished in a plate heat exchanger which can handle highly viscous solutions and provide turbulent conditions at very low Reynolds numbers (for example, Re~7). It has heating and cooling sections; the material is heated from T_1 to T_{hold}, the holding temperature, then cooled from T_{hold} to T_2. The flow pattern is countercurrent, as illustrated in Figure 5.27.

The flow rates and temperatures of the heating, cooling and product streams are given by

	product stream	*heating medium*	*cooling fluid*
inlet and exit temperatures (°K)	T_1, T_2	T_{h1}, T_{h2}	T_{c1}, T_{c2}
mass flow rates (Kg/hr)	m	w_h	w_c
heat capacities (kcal/Kg-°C)	c_p	c_{ph}	c_{pc}
heat exchange areas (m^2)		A_h	A_c

An overall energy balance on the heating section gives

$$w_h c_{ph}(T_{h1} - T_{h2}) = mc_p(T_{hold} - T_1)$$

The variation of heating medium temperature with distance (equivalently time) for plug flow in the exchanger is given by

$$T_h = T_{h1} - \left(\frac{mc_p}{w_h c_{ph}}\right)(T_{hold} - T) = T_{h1} - \lambda(T_{hold} - T)$$

Noting that for plug flow we can express the temperature-distance profile as a function of residence time t (distance/average product velocity, z/u), the product temperature is given by an energy balance, where U is the overall heat transfer coefficient.

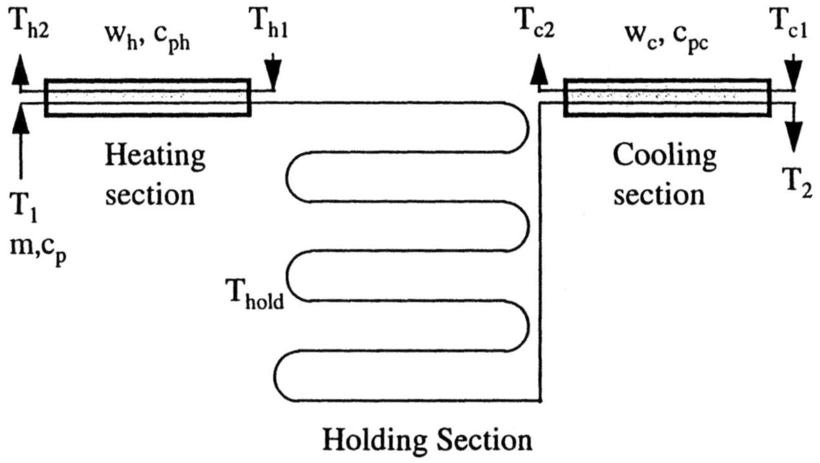

Figure 5.27. *A plate heat exchanger for thermal treatment of a protein solution. Inlet and exit temperatures and material properties are illustrated.*

$$mc_p \frac{dT}{dt} = UA(T_h - T) \qquad \frac{dT}{dt} = \frac{UA(\lambda - 1)}{mc_p}\left(T + \frac{(T_{h1} - \lambda T_{hold})}{\lambda - 1}\right)$$

This can be solved to give the following temperature profile for the product:

$$T = T_1 + \left(\frac{T_{h1} - \lambda T_{hold}}{\lambda - 1}\right)\left\{e^{(\lambda - 1)\frac{UA}{mc_p}t} - 1\right\} = a + be^{ct}$$

where $\quad a = T_1 + \dfrac{T_{h1} - \lambda T_{hold}}{\lambda - 1} \qquad b = \dfrac{T_{h1} - \lambda T_{hold}}{\lambda - 1} \qquad c = \dfrac{(\lambda - 1)UA}{mc_p}$

We see that the profile, while exponential, has a different form from those in Table 5.10, which assumes that the heating or cooling source has a constant temperature. The temperature profile is now inserted into the equation for thermal loss of microorganisms or for protein denaturation.

$$\frac{dn_v}{dt} = -k_d e^{-\frac{E_d}{RT}} \qquad \ln\frac{n_{v1}}{n_{vo}} = -k_o \int_0^{t_h} e^{-\frac{E_d}{RT(t)}} dt$$

The integral may be evaluated for this heating profile in terms of the exponential function Ei(x).

$$-\frac{1}{k_o}\ln\frac{n_{v1}}{n_{vo}} = \frac{1}{c}\left[Ei\left(\frac{Ea/R}{a + b \cdot e^{ct_h}}\right) - \exp(-Ea/a) \cdot Ei\left(\frac{-Ea/R \cdot be^{ct_h}}{(a + be^{ct_h}) \cdot a}\right)\right]$$

$$+\frac{1}{c}\left[Ei\left(\frac{Ea/R}{a + b}\right) - e^{\frac{-Ea/R}{a}} Ei\left(\frac{-Ea/R \cdot b}{a(a + b)}\right)\right]$$

The exponential function Ei(x) is defined as

$$Ei(x) = \int_{x}^{\infty} \frac{e^{-t}}{t} dt \cong \frac{e^{-x}}{x}\left\{1 - \frac{1}{x} + \frac{2!}{x^2} - \frac{3!}{x^3} + \ldots\right\}$$

From the above equation, $\Delta_{heating}$ can be found if the holding temperature T_{hold} and the inlet heating medium temperature T_1 are known. Δ_{hold} can be directly evaluated as T_{hold} is constant.

$$\Delta_{hold} = k_o \int_{t_h}^{t_{hold}} e^{\frac{-Ea}{RT_{hold}}} dt = k_o e^{\frac{-Ea}{RT_{hold}}} (t_{hold} - t_h)$$

The expression for $\Delta_{cooling}$ follows that for the heating section. The overall loss of micro-organisms can be found from Δ_{total} as a function of the sterilizer operating conditions. If denaturation of protein is of concern during the sterilization process, the extent can be obtained by substituting the corresponding rate parameters k_o and Ea for thermal protein denaturation into the above equations to obtain the fractional loss. The differences between the microbial thermal constants and those for the protein are such that a short-time, high-temperature process reduces the loss of protein while maintaining a desired level of microbial kill.

Continuous Sterilization

One of the disadvantages of batch sterilization is the high peak demand for steam to raise the contents of the tank to the holding temperature in a sufficiently short period. This difficulty is overcome by use of continuous sterilization. Continuous sterilization of medium, which has been "batched" in a holding tank and fed to either a tubular direct-steam-injection or plate-type heat exchanger, offers several advantages over batch sterilization:

(i) steam requirements are uniform and there is no peak steam load
(ii) process control is simplified
(iii) the sterilization cycle time is reduced and the shorter exposure time to intermediate temperatures during heating of the medium reduces the thermal degradation of medium components. This results in a more consistent medium composition.

The two main types of continuous heat exchangers used commercially are illustrated in Figure (5.28).

Continuous steam injection is very commonly employed for media sterilization in industrial fermentations. Steam is injected through a nozzle into liquid in turbulent pipe flow. This method results in rapid heating of medium without the use of a heat exchanger. Latent heat of vaporization is rapidly transferred to the medium by steam condensate. A fine dispersion of the steam into the liquid phase ensures rapid heating. The medium is then held in a tubular holding section for a predetermined period and then flash cooled through an expansion valve. To avoid two phase flow in the holding section, the exchanger is usually

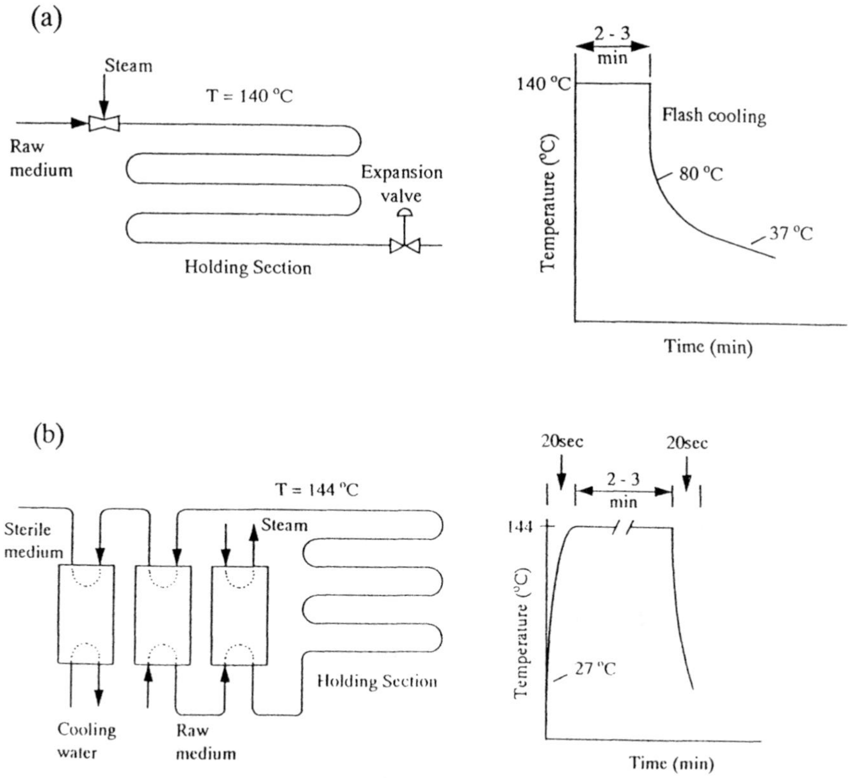

***Figure 5.28**. Two continuous heat exhanger designs used for medium sterilization (a) the direct-steam-injection exchanger and (b) the plate or spiral heat exchanger. The temperature-time profiles for both are shown.*

maintained at 5 to 10 psi above the bubble point. The disadvantages of this method are that the medium becomes somewhat dilute due to the condensation of steam and the steam must be clean to avoid particulate contamination of the medium.

The plate or spiral exchanger are high heat transfer coefficient exchangers which rapidly heat the medium to the holding temperature. Spiral exchangers are used with medium containing solids, as the high liquid velocities and wide gap between exchanger surfaces reduce fouling. Both these types of exchangers have higher captial costs than direct steam injection.

The reduced residence time in continuous sterilizers must be considered if the medium contains particles. The rate of temperature increase at the center of small (micron to centimeter) particles depends on the particle size, and large particles may not reach the desired sterilization temperature if the residence time is too short. As is illustrated in Table 5.11, particles larger than several millimeters may pass through the exchanger without being sterilized.

Table 5.11. *The effect of particle size on the time required to reach 99% of the final temperature. (Based on $\rho C_p/k = 250$)*

Diameter	*Time required (seconds)*
1 micron	10^{-6}
10 micron	10^{-4}
100 micron	10^{-2}
1 millimeter	1
1 centimeter	100

Design of Continuous Sterilizers

The liquid residence time is the main design variable for continuous sterilizers. It must be sufficiently long so that adequate sterilization occurs without thermal destruction of the medium components. The holding section of the sterilizer is tubular and in order to minimize radial temperature gradients, turbulent flow is desired. Generally this requires that the Reynolds number exceed 2 x 10^4 for fully developed turbulence. The actual distribution of residence times in the tubular holding section must be specified to determine the effective viable concentration of cells leaving the sterilizer.

We may characterize the residence time distribution in pipe flow by the axial dispersion coefficient E_z. Analogous to the molecular diffusion coefficient, it describes the motion of a specie resulting from the non-uniform flow. The axial Peclet number Pe_z expresses the axial dispersion in dimensionless terms as the ratio of mass transport by convection (uc) to that occurring by dispersion (E_zc/L). The Peclet number is defined as $Pe_z = uL/E_z$, where u is the average liquid velocity (0.5 u_{max} for laminar flow, 0.82 u_{max} for turbulent flow) and L is the length of the pipe.

Considering a section of medium in the holding coil of a sterilizer, we can write a mass balance on viable organisms as follows:

$$E_z \frac{d^2 n_v}{dz^2} - u \frac{dn_v}{dz} - kn_v = 0 \tag{5.184}$$

The first term represents axial dispersion, the second is convective transport and the third reflects the first order thermal death of viable cells. This equation can be made dimensionless as follows:

$$\bar{n} = \frac{n_v}{n_{vo}} \qquad Pe_z = \frac{uL}{E_z} \qquad \bar{z} = \frac{z}{L} \qquad Da = \frac{kL}{u} \quad \text{(Damkohler number)}$$

$$\frac{d^2\bar{n}}{d\bar{z}^2} = Pe_z \frac{d\bar{n}}{d\bar{z}} + Pe_z \cdot Da \cdot \bar{n} \tag{5.185}$$

Two boundary conditions are required for solution of Equation (5.185). These are typically

$\bar{z} = 1$ (exit) $$\frac{d\bar{n}}{d\bar{z}} = 0$$

$\bar{z} = 0$ (inlet) $$\bar{n} - \left(\frac{1}{Pe_z}\right)\frac{d\bar{n}}{d\bar{z}} = 1 \tag{5.186}$$

With these boundary conditions, the dispersion equation can be solved for the exit concentration of viable cells ($\bar{n}\,|_{\bar{z}=1}$):

$$\bar{n}_{\bar{z}=1} = \frac{4\delta}{(1+\delta)^2 e^{-Pe_z(1-\delta)/2} - (1-\delta)^2 e^{-Pe_z(1+\delta)/2}} \tag{5.187}$$

where $\delta = (1 + \frac{4Da}{Pe_z})^{1/2}$

For large values of the Peclet number (i.e., high liquid velocities), expanding the exponentials and dropping higher order terms reduces Eq. (5.187) to

$$\frac{n_v(L)}{n_{vo}} = e^{-Da + \frac{Da^2}{Pe_z}} \tag{5.188}$$

with the plug-flow limit occurring at large Pe_z values.

$$\frac{n_v(L)}{n_{vo}} = e^{-Da} \tag{5.189}$$

The length of the sterilizer holding section can now be determined. If the Reynolds number is selected such that the flow is turbulent, specification of a pipe diameter will then determine the liquid velocity provided that the medium viscosity at the sterilization temperature is known. The extent of axial dispersion can be obtained from correlations of E_z as a function of the Reynolds number. For laminar flow the correlation takes the form:

$$E_z = D_{mol} + \frac{u^2 d_t^2}{192 D_{mol}} \tag{5.190}$$

In the case of fully-developed turbulent flow, the value of the *modified* Peclet number[54] ($d_t \cdot u/E_z$) is approximately 3.33. It then remains to specify the degree of tolerance for contamination (n_v/n_{vo}) and calculate the Damkohler number from Equation (5.185), which reflects the thermal death constant and the length of the sterilzer holding section. By specifying the holding temperature, k can be found and hence L is determined from the value of Da. This approach is illustrated below.

---

(54) Correlations for the axial Peclet number Pe_z are generally based on the pipe diameter d_t as the characteristic length [e.g., O. Levenspiel, Ind. Eng. Chem. **50**, 343 (1958)]. The mass balance is made dimensionless with respect to the pipe length; thus the value of ($E_z/d_t.u$) must be multiplied by the ratio (d_t/L) to obtain the axial Peclet number we have employed above.

Determination of the length of a continuous sterilizer

Sterile medium for a 20,000 liter fermenter is to be prepared using a continuous direct steam injection sterilizer. The fermenter is to be filled in a two hour period. A level of contamination of one batch in 1000 will be considered acceptable for the sterilizer design. What should the length of the holding section of the sterilizer be?

The acceptable level of contamination provides for 10^{-3} spores per fermenter. We shall assume an initial level of 10^6 spores per ml in the incoming medium, thus the ratio n_{vo}/n_v is

$$n_{vo}/n_v = (10^6 \text{ x } 10^3 \text{ x } 2 \text{ x } 10^5)/(10^{-3}) = 2 \text{ x } 10^{17}$$

The flow rate through the sterilizer is

$$\left(\frac{\pi D^2}{4}\right)u = \frac{20,000}{2 \cdot 60} = 166.67 \quad (\text{liters.min}^{-1}) \qquad D^2 u = 3536.8 \quad (\text{cm}^3.\sec^{-1})$$

The pipe diameter will be set at 3 inches (7.62 cm), to insure a reasonable pressure drop. Thus the average medium velocity will be 60.9 cm/sec. The Reynolds number for the holding section can now be calculated, assuming the medium has the same properties as water.

$$\left(\frac{D\bar{u}\rho}{\mu}\right) = \frac{1.0 \quad (gm/cm^3) \cdot 7.62(cm) \quad 60.9 \quad (cm/\sec)}{10^{-2}(gm.cm^{-1}.\sec^{-1})} = 4.6410^4$$

The flow will thus be turbulent. The axial Peclet (Pe_z) number can be assumed to be the limiting value of $(1/0.3)(L/d_t)$, where we have used the value of 3.3 for the modified Peclet number. The simplified expression for n_v/n_{vo} (equation 5.188) can now be employed to calculate the relationship between the forst order rate constant for thermal death k and L.

$$\ln\left(\frac{n_v(L)}{n_{vo}}\right) = -Da + \frac{Da^2}{Pe_z}$$

$$\ln\left(\frac{1}{2 \cdot 10^{17}}\right) = -Da + \frac{Da^2}{3.3\left(\frac{L}{d_t}\right)} = -30.6$$

$$\frac{k^2 L}{(60.9)^2} - \frac{3.3kL}{7.62 \cdot 60.9} + \frac{30.6 \cdot 3.3}{7.62} = 0$$

Thus we can determine the length as a function of the holding temperature, based on values of k(T) obtained from Table 5.9. For a holding temperature of 140°C, k is 136 min^{-1} (2.27 $\sec^{-1}$). Thus the holding length L can be found from:

$$\frac{(2.27)^2 L}{(60.9)^2} - \frac{3.3 \cdot 2.27 \cdot L}{7.62 \cdot 60.9} + \frac{30.6 \cdot 3.3}{7.62} = 0 \qquad \text{thus} \qquad L = 892.9 \ (\text{cm})$$

In the absence of axial dispersion, we find that

$$\ln\left(\frac{n_v(L)}{n_{vo}}\right) = -\frac{kL}{u} = -30.6 \qquad L = \frac{30.6 \cdot 60.9}{2.27} = 820.9 \ (\text{cm})$$

indicating that axial dispersion increases the length of the holding section by only ~8% under these conditions.

5.5.2 Heat Transfer in Agitated Tanks and Columns

Heat may be either added or removed from biological reactors to maintain temperature in a range suitable for microbial or enzymatic processing. In aerobic systems, the biological combustion of the carbon-containing substrate to carbon dioxide generates heat, as discussed in Chapter 3. Heat removal from rapidly-growing cells can present significant problems in large scale equipment. On the other hand, in anaerobic sewage sludge treatment, the optimum temperature for digestion of the sludge is 55-60°C, requiring the addition of heat to the system. In agitated tanks, heat may be added or removed from the liquid phase by contact with the tank walls or by use of immersed tubes which provide additional surface area required for heat transfer. The types of heat transfer surfaces are shown in Figure 5.29.

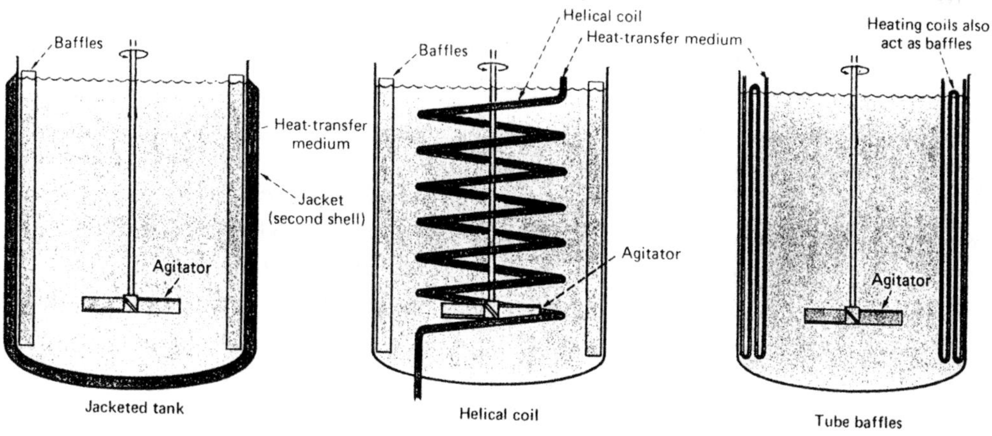

Figure 5.29. *Heat transfer surfaces in agitated tanks may be the vessel walls or helical or vertical coils.*

Jacketed tanks may have heat transfer surfaces which cover just the walls or may include the tank bottom. The jacket may be a second shell placed on the tank or coils attached to the outer surface of the tank. Internal helical coils provide a larger surface area for heat transfer and are placed coaxially with the agitator. Tube baffles serve the purpose of heat exchange and as well as acting as tank baffles. In extremely large tanks with high rates of heat generation, such as single cell protein production from methanol, sufficient transfer area cannot be attained by jackets or coils and an external heat exchanger must be added. The fermentation broth is circulated at high flow rates through this external exchanger, which must have a large surface area and a high heat transfer coefficient.

The overall heat transfer coefficient (U_i) is determined by the geometry of the heat transfer surface and the operating conditions. In general, the overall coefficient, based on the inside (with respect to the tank) surface area, or on the process-liquid side of the surface is given by:

$$Q = U_i A_i (T_o - T_i) \tag{5.191}$$

$$\frac{1}{U_i} = \frac{1}{h_i} + \frac{\Delta r}{k_w}\left(\frac{A_i}{A_w}\right) + \left(\frac{1}{h_o}\right)\left(\frac{A_i}{A_o}\right) \tag{5.192}$$

where the first term in the above equation is the inside film coefficient, the second term involves conduction through the tube or jacket wall (of thickness Δr) and h_o refers to the outside heat transfer coefficient. A_i and A_o are the inner and outer surface areas for heat transfer. The thermal conductivity of the tube or jacket walls (k_w) depends on the material of construction. From dimensional analysis, the film heat transfer coefficients h_i and h_o depend on the fluid properties and the appropriate length scales. The general form of the correlation for transfer from a jacketed tank is:

$$\left(\frac{hD_T}{k}\right) = a\left(\frac{D_i^2 N \rho_L}{\mu_L}\right)^{\alpha}\left(\frac{C_p \mu_L}{k}\right)^{\beta}\left(\frac{\mu_L}{\mu_{wall}}\right)^{\gamma} \quad \text{or} \quad Nu = aRe^{\alpha}Pr^{\beta}\left(\frac{\mu_L}{\mu_{wall}}\right)^{\gamma} \tag{5.193}$$

The Nusselt number (hD_T/k) is defined with respect to the tank diameter, whereas the Reynolds number, used to characterize the liquid flow in the vessel employs the impeller diameter D_i. The last term, the viscosity ratio, is required when there is a significant difference in the jacket wall temperature and the bulk fluid temperature which results in a difference in viscosity. The exponent usually has a value between 0 and 0.4, with 0.14 being generally employed. For a turbine agitator in a baffled tank, the inner heat transfer coefficient from the vessel walls to the liquid can be found from:

$$\left(\frac{h_i D_T}{k}\right) = 0.74\left(\frac{D_i^2 N \rho_L}{\mu_L}\right)^{0.66}\left(\frac{C_p \mu_L}{k}\right)^{0.33}\left(\frac{\mu_L}{\mu_{wall}}\right)^{0.14} \tag{5.194}$$

This correlation applies to tanks of standard geometry, where $D_i/D_T = 1/3$ and the liquid height to vessel diameter is unity. The above equation can be extended for tanks of different geometry by incorporating geometric factors:

$$\left(\frac{h_i D_T}{k}\right) = 0.85\left(\frac{D_i^2 N \rho_L}{\mu_L}\right)^{0.66}\left(\frac{C_p \mu_L}{k}\right)^{0.33}\left(\frac{\mu_L}{\mu_{wall}}\right)^{0.14}\left(\frac{H}{D_T}\right)^{-0.56}\left(\frac{D_i}{D_T}\right)^{0.13} \tag{5.195}$$

where H is the height of the liquid in the tank. For helical coils, correlations are difficult to generalize, as it is difficult to maintain geometric similarity as the scale of the vessel increases. In addition to length dimensions associated with the tank and the impeller, the diameter and length of the coil and the spacing between the coils must also be considered. A correlation developed by Oldshue and Gretton[55] replaces the tank diameter with the coil diameter d_t in the Nusselt number:

$$\left(\frac{h_i d_t}{k}\right) = 0.17\left(\frac{D_i^2 N \rho_L}{\mu}\right)^{0.67}\left(\frac{c_p \mu}{k}\right)^{0.37}\left(\frac{D_i}{D_T}\right)^{0.1}\left(\frac{d_t}{D_T}\right)^{0.5} \tag{5.196}$$

The effect of viscosity is not included in the above correlation and extrapolation outside the geometric ratios for which the correlation was developed should be done with care. When the coils take the form of vertical tubes, when they often act as tank baffles, the following correlation[56] includes the effect of impeller to tank diameter ratio and the number of baffles (n_b), but excludes effects due to changes in the tube to tank diameter ratio.

$$\left(\frac{h_i d_t}{k}\right) = 0.09\left(\frac{D_i^2 N \rho_L}{\mu}\right)^{0.65}\left(\frac{c_p \mu}{k}\right)^{0.3}\left(\frac{D_i}{D_T}\right)^{0.33}\left(\frac{2}{n_b}\right)^{0.2}\left(\frac{\mu_L}{\mu_w}\right)^{0.14} \tag{5.197}$$

The inner tube heat transfer coefficient for the coils (either vertical or helical) can be found from the Seider-Tate equation:

$$\left(\frac{h_i d_t}{k}\right) = 0.027(Re_{tube})^{0.8}\left(\frac{c_p \mu}{k}\right)^{0.33}\left(\frac{\mu_L}{\mu_w}\right)^{0.14} \tag{5.198}$$

The Reynolds number refers to flow within the tube. For impellers other than turbine (e.g., helical ribbon, propellers) and for placement of the impeller above or below helical coils, the text of Nagata[57] provides details of heat transfer correlations.

Heat Transfer to Dense Suspensions

When the rheological behavior of the fermentation broth is complex, as is the case with filamentous fungal growth or broths containing a high concentration of suspended solids, heat transfer rates are strongly influenced by the apparent broth viscosity near the vessel walls or coils, rather than the bulk viscosity. In solid suspensions, or filamentous growth of *Penicillium chrysogenum* or *Streptomyces kanamyceticus* for example, the solution often exhibits a yield stress, behaving as a Bingham plastic fluid. This rheological behavior can be represented by a shear-dependent viscosity:

$$\tau - \tau_o = \mu_B \gamma \qquad \text{for} \qquad \tau > \tau_o$$

$$\gamma = 0 \qquad \text{for} \qquad \tau \leq \tau_o \tag{5.199}$$

(55) J.Y. Oldshue and A.T. Gretton, *Chem. Eng. Progr.*, p615, Dec, 1954.

(56) I.R. Dunlap and J.H. Rushton, *Chem. Eng. Prog. Symp.*, **49**, (5) 137 (1953).

(57) S. Nagata, *Mixing - Principles and Applications*, Halsted Press (1975).

Based on Levich's three zone model for the velocity distribution in turbulent flow close to a surface, Kawase and Mooyoung[58] proposed a velocity distribution for turbulent flow of a Bingham plastic which consisted of a laminar sublayer and a turbulent core. A generalized relationship for the vessel wall-fluid heat transfer coefficient (h) was derived from this velocity profile (in SI units):

$$h = 0.075\left(1-\frac{\tau_o}{\tau}\right)^{5/6}\left\{11.6\left(1-\frac{\tau_o}{\tau}\right)^{-1.27}\right\}^{1/4}\cdot\left(\frac{k}{\rho C_p}\right)^{2/3}\left(\frac{\mu_B}{\rho}\right)^{-2/3}(\mu_B\rho)^{1/4}\varepsilon^{1/4} \qquad (5.200)$$

where ε is the energy dissipation rate per unit mass in the liquid phase, k is the thermal conductivity and C_p is the specific heat capacity. This equation indicates that the heat transfer coefficient decreases in turbulent flow as the yield stress increases. In stirred tanks, the rate of energy dissipation is given by

$$\varepsilon = C\cdot\frac{N^3D_i^5}{D_T^2H} \qquad (5.201)$$

In fully developed flow, the constant C is independent of the rheological properties of the fluid, and depends only on the type of agitator. For a six-blade turbine in a baffled tank, C is 5.5 and for an anchor impeller C is 0.29. Inserting this value into the expression for ε, for a six-blade turbine the correlation becomes

$$Nu = \frac{hD_T}{k} = 0.276\left(1-\frac{\tau_o}{\tau}\right)^{0.516} Pr_B^{1/3}Re_B^{3/4} \quad \text{where} \quad Re_B = \frac{\rho D_i^2 N}{\mu_B} \quad Pr_B = \frac{C_p\mu_B}{k} \qquad (5.202)$$

The shear stress ratio a may be approximated viz:

$$\frac{\tau_o}{\tau} = \frac{\tau_o}{\mu_B\gamma_{ave}} \quad \text{where} \quad \gamma_{ave} = 1.15BN$$

$$B = 11 \quad (turbines) \qquad \text{and} \qquad B = 9.5+\frac{9S^2}{S^2-1} \quad S = \frac{D_T}{D_i} \quad (anchors) \qquad (5.203)$$

In gassed systems, the aerated power input is required. If this is approximated by 40% of the unaerated power, the correlation can predict the heat transfer coefficient in aerated tanks quite well. In bubble columns, the energy dissipation per unit mass is given by

$$\varepsilon = u_{sg}g \qquad (5.204)$$

where u_{sg} is the superficial gas velocity. The heat transfer correlation becomes

$$Nu = \frac{hD_T}{k} = 0.138\left(1-\frac{\tau_o}{\tau}\right)^{0.516} Pr_B^{1/3}Re_{BC}^{3/4}Fr^{-1/4} \qquad (5.205)$$

$$\text{where} \quad Re_{BC} = \frac{\rho D_T u_{sg}}{\mu_B} \quad Pr_B = \frac{C_p\mu_B}{k} \quad Fr = \frac{u_{sg}^2}{D_T g}$$

---

(58) Y. Kawase and M.Mooyoung, *Chem. Eng. Journal*, **41**, B17 (1989).

In order to evaluate the yield stress to shear stress ratio, a correlation for the average shear rate in a bubble column must be employed. For superficial gas velocities between 0.04 to 0.1 meter/sec, Nishikawa *et al.*[59] propose

$$\gamma_{ave} = 5000u_{sg} \tag{5.206}$$

This can then be inserted into Equation (5.203) to determine the shear stress ratio.

5.6 Mass Transfer Coupled with Biological Reaction

In this section, we shall consider the special case of the rate of mass transfer occurring at a rate comparable to the rate of biological reaction. Generally, the transport of substrate to cells or enzymes occurs at a rate which is considerably faster than the rate of the biological reaction. The overall rate of substrate conversion is thus governed soley by the kinetics of the reaction. However, if mass transfer is slow relative to reaction, transport may influence the observed kinetic rates. In comparing the rates of transport and reaction, the time constants associated with reaction and transport provide a useful means of distinguishing the various ways in which these factors can be coupled.

The diffusional time constant governs mass transfer. For substrate transport through a liquid film, for example, it is defined as

$$t_D = \frac{D_s}{k_L^2} \tag{5.207}$$

where D_s is the diffusivity of the substrate in the liquid and k_L is the mass transfer coefficient. The reaction time constant for an irreversible reaction is defined:

$$t_R = \frac{C_o'}{r(C_o')} \tag{5.208}$$

where C_0' is concentration of the reacting species (eg. substrate) that would be in equilibrium with the concentration of that specie in the phase where the reaction occurs. For example, in the case of oxygen transfer from the gas to the liquid phase, C_0' would be the concentration of dissolved oxygen in the liquid that is in equilibrium with the concentration of oxygen in the gas phase (i.e., from Henry's law $C_o' = Py_o/H$). If the reaction occurring in the liquid phase is first order, $r(C_o') = k_1C_o'$. The time constant is thus the reciprocal of the first order rate constant k_1.

In the case of a slow reaction, the diffusion time is much less that the time required for reaction to proceed to an appreciable extent at a concentration of C_o'. Thus $t_D << t_R$. Conversely the reaction will be fast if $t_D >> t_R$. We shall examine these limiting cases of

(59) N. Nishikawa, H. Kato and K. Heshimoto, *Ind. Eng. Chem. Process Des. Devpt.*, **16** 133 (1977).

slow and fast reactions, using gas-liquid mass transfer as an example.[60] We consider oxygen diffusing from the gas phase into a liquid phase where it is consumed by a microbial or enzymatic reaction. The possible reaction/diffusion regimes are defined as follows.

Slow reaction (Regime I)

In the slow reaction regime, $t_R > t_D$, and we consider that no reaction occurs while the substrate diffuses from the gas through the liquid hydrodynamic boundary layer. All the reaction takes place in the bulk liquid phase. We shall write the liquid film mass transfer coefficient k_L as k_L^o, to indicate that it has the same value as that which obtained in the absence of reaction in the liquid phase. A liquid phase mass balance on the reacting species yields:

$$\frac{dC_o}{dt} + r(C_o) = k_L^o a(C_o' - C_o) \tag{5.209}$$

where a is the interfacial area per volume of the liquid phase. If the substrate concentration C_o is approximately constant, we can employ the quasi-steady state assumption to set to the time derivative zero. This is reasonable as the rate of reaction in the liquid is slower than the rate of mass transfer from the gas phase and a constant liquid concentration will result. The overall concentration driving force can be divided into two parts; $(C_o' - C_o)$ for diffusion and $(C_o - 0)$ for reaction. Two subregimes can now be distinguished on the basis of the relative rates of substrate diffusion per unit interfacial area $(k^o_L C_o')$ and the reaction rate per unit area $(r(C_o)/a)$. The overall rate is governed by these two processes, which occur in series.

Diffusional Subregime (Ia)

If the reaction rate per unit interfacial area is greater than the absorption rate, under conditions of equal (maximum) driving forces we have

$$\left(\frac{1}{a}\right) r(C_o') \gg k_L^o C'_o \tag{5.210}$$

and the reaction will be limited by diffusion through the aqueous film. In this case, the reaction rate is sufficiently high to reduce the aqueous substrate concentration to zero or its equilibrium value. The observed rate of uptake of substrate from the aqueous phase is given by

$$V = k_L^o a C_o' \tag{5.211}$$

In this regime the enzymatic or microbial reaction rate would be sufficient to maintain the aqueous substrate concentration at its equilibrium value (or zero) and reaction does not

(60) The general area of mass transfer with reaction is very well described in the text by G. Astarita, *Mass Transfer and Chemical Reaction*, Elsevier, Amsterdam, (1967).

influence the value of the physical mass transfer coefficient k_L^o. The observed reaction rate is thus proportional to the interfacial area and substrate concentration (C_o') and independent of the reaction kinetics.

Kinetic Subregime (Ib)

If the enzymatic or microbial reaction is slow however, such that

$$\left(\frac{1}{a}\right) r(C_o') \ll k_L^o C'_o \tag{5.212}$$

mass transport is rapid enought so that the aqueous phase will be almost saturated with substrate (i.e., $C_o = C_o'$). The observed rate of reaction will be given by

$$V = r(C_o') = kC_o' \tag{5.213}$$

and the observed rate is thus independent of the interfacial area and the mass transfer coefficient, but depends on the substrate concentration C_o'.

Intermediate Subregime (Ic)

Consider the simplest case of a first order reaction (kC_o). The steady state substrate concentration can be found from the solution of Equation (5.209), with the time derivative set to zero. Substituting this into the rate expression yields:

$$V = \frac{k_L^o a \cdot C'_o}{1 + \frac{k_L^o a}{k}} \tag{5.214}$$

The magnitude of the term ($k^o_L a/k$) determines the subregime in which the process falls. If this term is small (<0.1), the reaction is in the diffusional subregime; If it is large (>10), the reaction is in the kinetic subregime. For values between these extremes, both mass transfer and kinetics influence the observed rate.

Fast Reaction Regime (II)

In the fast reaction regime, most of the substrate reacts before it diffuse across the entire thickness of the hydrodynamic boundary layer. Thus most of the concentration gradient is be confined to a small region, well within the boundary layer. The rate of substrate transfer is thus independent of the film thickness and the hydrodynamic conditions in the aqueous phase. The mass transfer rate in the fast reaction regime is directly proportional to the interfacial area, but depends on the reaction rate, which sets the concentration gradient.

$$V \propto a\sqrt{r} \tag{5.215}$$

An increase in the reaction rate causes the concentraton gradient to increase, which in turn increases the rate of mass transfer. Thus the fast reaction regime can be differentiated from the slow reaction/diffusion regime by the effects of the hydrodynamics and intrinsic kinetics on the observed rate of substrate depletion. Problem 15 provides an example of the enhanced rate of transport of gaseous CO_2 into a solution containing carbonic anhydrase, illustrating the dependence of the observed reaction rate on the intrinsic kinetics to the 0.5 power.

Table 5.12 illustrates the influence of kinetics (r(E)), interfacial area (a), mass transfer and hydrodynamics (k_L^o) and susbtrate concentration (C_o') on the observed rate of reaction in each of the regimes discussed. There is a third regime, the instantaneous reaction regime, where substrate and enzyme react so rapidly that the substrate concentration is reduced to zero at some distance within the hydrodynamic boundary layer. Both substrate and enzyme (or cells) must diffuse from opposite sides boundary layer to react. The observed reaction rate depends only on the rates of diffusion of substrate and of enzyme into the boundary layer.

***Table 5.12**. The dependence of the observed reaction rate on system parameters when mass transport is significant.* $V = constant\,(k_L^o)^u\,(a)^v\,(r(C_o,E))^x\,(C_o')^y$

Regime	*u*	*v*	*x*	*y*
I. Slow Reaction				
(a) diffusion	1	1	0	1
(b) kinetic	0-1	0-1	0-1	1
(c) intermediate	0	0	1	1
II. Fast Reaction	0	1	0.5	1
III. Instantaneous Reaction	1	1	0	0

5.7 Microbial Growth in Films and Flocs

There are a number of important situations where microbial growth occurs in films or in flocs. Microorganisms in water become attached to surfaces, held by extracellular polymeric materials produced by the microorganism. Biofilms are found in groundwater aquifers, streambeds, ships hulls and in water pipes. Microbial corrosion of plastics, metals and concrete pipes and surfaces result from such attachment. Microbially-induced corrosion of metal pipes has enormous economic consequences. In contrast to this, the treatment of waste-water in trickle beds relies on the attachment of microorganisms to the support surfaces for removal of organic waste materials. The thickness of the microbial film depends on the flux of substrate into the film and on the rates of bacterial growth and decay within the film. In waste water trickle bed reactors, the steady state film thickness is of order 0.01 cm.

In microbial films, the transport of nutrients through the film is hindered by the presence of the polymeric material in the film. For example, in the slime of trickle filters, dissolved oxygen has a diffusivity less than the corresponding value in water, typically around 88% (1.76×10^{-7} cm^2/sec) of its aqueous value. For organics, the value is generally around 80% of the aqueous diffusivity.

Due to the presence of microoganisms in the film, we encounter a situation similar to that illustrated in Chapter 2, namely substrate diffusion into an immobilized enzyme system. Diffusion occurs simultaneously with reaction and may mask the true kinetics of substrate uptake. We shall briefly examine the problem of diffusion in one dimension into a microbial film. Assuming the microbial consumption of substrate S can be represented by the Monod equation, a mass balance on the substrate in the film can be written.

$$\frac{\partial S}{\partial t} = D_f \frac{\partial^2 S}{\partial x^2} - \frac{v_{max} X S}{K_S + S} \tag{5.216}$$

Here D_f is the diffusivity of the substrate in the biofilm, which contains bacteria at a concentration X. The substrate concentration depends on both position and time in the film S(x,t). If steady state is achieved within the film, we can remove the time dependence and the above equation becomes

$$D_f \frac{d^2 S}{dx^2} - \frac{v_{max} X S}{K_S + S} = 0 \tag{5.217}$$

This may be integrated once; the solution depends on the boundary conditions employed. If the film is thick, then substrate may not penetrate to the solid where the film is attached. Only microorganisms close to the surface of the film will be supplied with substrate. The boundary condition in this case is

$$\frac{dS}{dx} = 0 \quad at \quad S = 0 \tag{5.218}$$

and the result of the integration is

$$\frac{dS}{dx} = \sqrt{\frac{2 v_{max} X}{D_f} \left\{ S - K_S \ln\left(\frac{K_S + S}{K_S} \right) \right\}} \tag{5.219}$$

When the film is thin, substrate is able to be transported to the solid surface, where it is assumed to have a concentration S_o. The boundary condition is

$$\frac{dS}{dx} = 0 \quad at \quad S = S_o \tag{5.220}$$

and integration yields

$$\frac{dS}{dx} = \sqrt{\frac{2 v_{max} X}{D_S} \left(S - S_o - K_S \ln\left\{ \frac{K_S + S}{K_S + S_o} \right\} \right)} \tag{5.221}$$

The observed rate of substrate uptake by the microbial film may be calculated if the surface area and the substrate concentration at the film-water interface (S_i) are known:

$$rate = A D_f \left(\frac{dS}{dx} \right)_{S = S_i} \tag{5.222}$$

The substrate concentration profiles within the film can be analytically determined for some limiting cases of the Monod equation. If the substrate concentration at the interface S_i is low, the Monod expression can be approximated as first order (an equivalent situation results when K_S is large). The steady state mass balance (5.217) can be made dimensionless if a film thickness L is specified.

$$\frac{d^2\hat{S}}{d\xi^2} = \frac{v_{max}XL^2}{D_f K_S}\hat{S} = \alpha^2\hat{S}$$

where $\hat{S} = \frac{S}{S_i}$ $\quad \xi = \frac{x}{L}$ $\quad \alpha = \left(\frac{v_{max}XL^2}{D_f K_S}\right)^{1/2}$ (5.223)

This has a general solution

$$\hat{S} = C_1 \cosh(\alpha\xi) + C_2 \sinh(\alpha\xi) \tag{5.224}$$

where the constants C_1 and C_2 can be evaluated from the boundary conditions:

$$\frac{d\hat{S}}{d\xi} = 0 \quad at \quad \xi = 1 \quad \text{and} \quad \hat{S} = 1 \quad at \quad \xi = 0 \tag{5.225}$$

The substrate concentration profile within the film is thus

$$\hat{S} = \alpha\cosh(\alpha\xi) - \tanh\alpha \cdot \sinh(\alpha\xi) \tag{5.226}$$

The observed rate of substrate removal per unit area by the film can be directly determined.

$$rate = -D_f\left(\frac{dS}{dx}\right)_{x=0} = D_f\left\{\frac{\alpha S_o}{L}\cdot\tanh\alpha\right\} = \left(\frac{v_{max}XL}{K_S}\right)S_o \cdot \frac{\tanh\alpha}{\alpha} \tag{2.227}$$

Typical parameter values for a wastewater trickle filter are

$L = 0.01$ cm $\qquad K_S = 10$ mg/liter

$S_i = 3\text{~}5$ mg/liter $\qquad D_f = 0.64$ cm^2/day

Typical microbial growth parameters are

$X_{ss} = 40$ mg/liter $\qquad Y_{X/S} = 0.5$ gm cells/gm substrate

$v_{max} = 8$ day^{-1} $\qquad m = 0.1$ day^{-1} (maintenance coefficient)

These parametersresult in α^2 being 5.0. The effect of the diffusional limitation within the film is to reduce the rate of substrate removal to 44% ($\alpha^{-1}\tanh\alpha$) of its maximum value in the absence of diffusion.

5.8 Cell Motility and Chemotaxis

Many situations in nature are far removed from the spatially homogeneous case of a stirred tank reactor. In soils or in aquatic systems, significant gradients in nutrients may exist, resulting in spatially-varying cell concentrations. Approximately half of all bacterial orders contain at least one species which is motile. Cell movement results from the rotation of the flagellum (clockwise when viewed from the back of the bacterium) at speeds of up to 50 micron per second. The bacterial flagellum is composed of a filament, a hook and a basal structure. The filament is helical, composed of 11 parallel strands of the protein flagellin. The basal strucure is held in the cell envelope by a series of rings. The flagellum arises from the hook region. Bacteria may have a single flagellum at one of the poles of the cell, or may have many flagella distributed around the cell surface (peritrichate flagella). These peritrichously-flagellated bacteria typically move in a straight line for a short period of time (usually around 1 second), then stop and randomly change direction by a tumbling action (duration of 0.1 second).

Many motile bacteria exhibit *chemotaxis*, the movement of cells towards or away from chemicals. Specific chemo-receptors (at least 20 different proteins in *E. coli*) detect attractant molecules and activate the chemotaxis pathway. In chemotatic bacteria, tumbling remains random in direction but the frequency of tumbling decreases when the cell moves toward the chemical species to which it is attracted, or toward a lower concentration of chemical to which it is repelled. This leads to longer run times in the direction of the attractant, and results in a net migration in the desired direction. In an optimal nutrient gradient, the net migration velocity which may be up to one half the linear swimming velocity of the bacterium. The mechanism by which the bacteria sense the chemical gradient is thought to result from the reversible binding of stimulus molecules to chemo-receptors on the cell membrane. Some attractants and repellents for common chemotactic bacteria are given in Table 5.13.

It is thought that chemotaxis and motility may provide some survival advantages over non-motile bacteria. Some studies have shown that chemotatic bacteria grown in quiescent media or on semi-solid agar indeed show higher rates of growth than non-motile strains.

Table 5.13. *Classification of some bacterial chemotactic repsonses [from D. Lauffenburger, in "Foundations of Biochemical Engineering", ACS Symposium Series **207**, 265 (1983)]*

Genus	*Attractants*	*Repellents*
Escherichia	sugars, amino acids	pH extremes, aliphatic alcohols
Pseudomonas	sugars, amino acids, nucleotides, O_2, vitamins	inorganic ions, pH extremes, amino acids
Bacillus	sugars, amino acids, O_2	inorganic ions, pH extremes
Salmonella	sugars, amino acids, O_2	aliphatic alcohols
Vibrio	amino acids, O_2	
Spirillum	sugars, amino acids, O_2	inorganic ions, pH extremes
Erwinia	sugars	inorganic ions, pH extremes

We shall analyze the behavior of a chemotatic bacterium confined to one-dimensional movement in a region which is non-mixed, with substrate diffusing from one of the boundaries[61]. The situation is illustrated in Figure 5.30. Although the assumption of one-dimensional movement is overly restrictive, it does illustrate some of the features of chemotaxis.

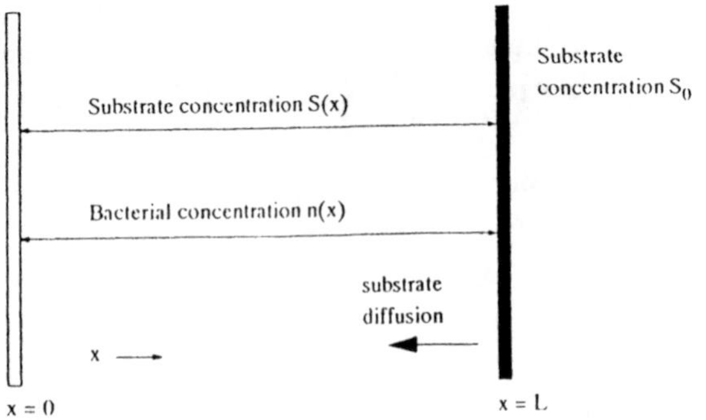

Figure 5.30*. Growth of a chemotatic bacterium in a confined one-dimensional system with substrate diffusion.*

The cell concentration (n(x)) and substrate (S(x)) mass balances can be written using the Monod expression for growth, modified to include a first order endogenous metabolism (or death) term given by $k_e n$:

$$\frac{\partial n}{\partial t} = -\frac{\partial J_n}{\partial x} + \frac{\mu_{max} S n}{K_S + S} - k_e n \qquad \frac{\partial S}{\partial t} = -\frac{\partial J_S}{\partial x} - \frac{1}{Y_{n/x}} \frac{\mu_{max} S n}{K_S + S} \tag{5.227}$$

J_n and J_S are the viable cell and substrate fluxes, respectively. The boundary conditions are

x=L: $J_n = 0$ and $S = S_0$ (the substrate concentration is maintained constant)

x=0: $J_n = 0$ and $J_S = 0$

To permit an analytical solution, we shall replace the Monod expression with a first order growth rate viz.

$$\left[\frac{\mu_{max} S}{K_S + S} - k_e\right] \cdot n \Rightarrow (\mu_{max} - k_e) n \qquad S > S_{crit}$$

$$\Rightarrow -k_e \qquad S \le S_{crit}$$

(61) From D. Lauffenberger, *Effects of Cell Motility Properties on Cell Populations in Ecosystems*, ACS Symp. Ser. **207**, 265 (1983)

$$\frac{1}{Y_{n/S}}\left[\frac{\mu_{max}S}{K_S+S}\right]\cdot n \Rightarrow \frac{1}{Y_{n/S}}\mu_{max}n \qquad S > S_{crit}$$

$$\Rightarrow 0 \qquad S \le S_{crit} \tag{2.228}$$

The substrate flux is a result of simple diffusion and is given by Fick's law

$$J_S = -D_S\frac{\partial S}{\partial x} \tag{5.229}$$

where D_S is the substrate diffusion coefficient. An expression for the cell flux J_n can now be develped. We assume it contains a random motility coefficient (α) and a term intended to describe chemotaxis (a "chemotaxis coefficient" χ). This latter term depends on the substrate gradient.

$$J_n = -\alpha\frac{\partial n}{\partial x} + \chi n\frac{\partial S}{\partial x} \tag{5.230}$$

The mass balances can be made dimensionless as follows:

$$u = \frac{S}{S_o} \qquad v = \frac{n}{n_o} \qquad \xi = \frac{x}{L} \qquad \tau = \frac{D_S t}{L^2} \qquad n_o = \frac{Y_{n/S}S_oD_S}{\mu_{max}L^2} \tag{5.231}$$

$$\lambda = \frac{\alpha}{D_S} \qquad \kappa = \frac{\mu_{max}L^2}{D_S} \qquad \theta = \frac{k_eL^2}{D_S} \qquad \delta = \frac{\chi S_o}{\alpha}\cdot F(u)$$

where $F(u) = 1 \quad u > u_{crit}$ and $F(u) = 0 \quad u \le u_{crit}$

and thus in the region $0 < \xi < 1$

$$\frac{\partial v}{\partial \tau} = \lambda\frac{\partial^2 v}{\partial \xi^2} - \delta\lambda\frac{\partial}{\partial \xi}\left(v\frac{\partial u}{\partial \xi}\right) + [\kappa F(u) - \theta]v$$

$$\frac{\partial u}{\partial \tau} = \frac{\partial^2 u}{\partial \xi^2} - F(u)v \tag{5.232}$$

and the boundary conditions now become

at $\xi = 1 \qquad \frac{\partial v}{\partial \xi} - \delta v\frac{\partial u}{\partial \xi} = 0$ and $u = 1$

at $\xi = 0 \qquad \frac{\partial v}{\partial \xi} = 0 \qquad \frac{\partial u}{\partial \xi} = 0$ (5.233)

From the mass balances, the steady state solution can be directly derived.

$$n_{ss} = \frac{Y_{n/S}S_oD_S}{\mu_{max}L^2}\cdot\int_0^1 v\,d\xi \tag{5.234}$$

A steady state results when cell growth in the substrate-rich zone near the boundary and loss of viability in the substrate-depleted zone are balanced by the cell movement from the substrate-rich zone to the substrate-depleted zone. The steady-state profiles are illustrated in Figure 5.31. The division of the growth zone from the depleted zone occurs when the substrate concentration falls to the critical value S_{crit}. At this position, $\xi = \omega$, where ω is obtained from the largest root, less than unity of

$$\sqrt{\left(\frac{\theta}{\lambda}\right)}\tanh\left(\sqrt{\left\{\frac{\theta}{\lambda}\right\}}\,\omega\right) = \sqrt{([\kappa-\theta]/\lambda)}\,\tan\{\sqrt{([\kappa-\theta]/\lambda)}\cdot(1-\omega)\} \qquad (5.235)$$

Hence ω increases as κ increases and λ decreases.

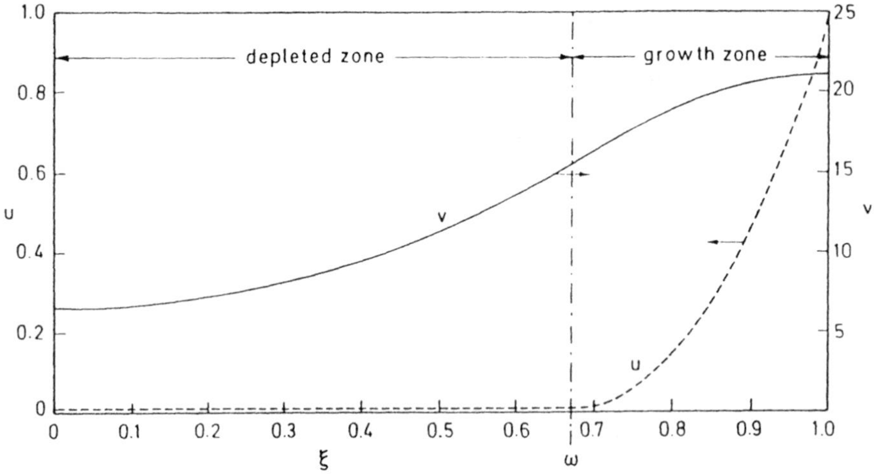

Figure 5.31. *Typical steady state profiles of dimensionless bacterial concentration (v) and dimensionless substrate concentration (u). In the growth zone $\omega < \xi < 1$, $u > u_c$ and bacterial growth can be supported. In the depleted zone, $0 < \xi < \omega$, $u = u_c$ and bacterial growth cannot be supported. [From D. Lauffenberger, Effects of Cell Motility Properties on Cell Populations in Ecosystems, ACS Symp. Ser.* ***207****, 265 (1983)].*

5.9 Nomenclature

Symbol	Meaning	Typical units
a	interfacial area per reactor volume	cm^{-1}
a'	surface area to volume of a bubble	cm^{-1}
Bo	Bond number (Eq. 5.116)	
C_A	concentration of A	gm $liter^{-1}$
c_p'	heat capacity	kcal kg^{-1} oC^{-1}
Da	Damkohler number (Eq. 5.185)	
d_B	bubble diameter	cm
d_{Be}	equilibrium bubble size	cm
d_{Bo}	bubble diameter at orifice	cm
De	Deborah number (Eq. 5.129)	
D_i	impeller diameter	cm
D_L	diffusivity	cm^2 sec^{-1}
d_o	orifice diameter	cm
D_T	tank diameter	cm
E	surface elasticity	dyne cm^{-1}
E	bubble eccentricity (bubble width to height)	(-)
E_a	activation energy	kcal $mole^{-1}$
E_L	axial dispersion coefficient	cm^2 sec^{-1}
Eo	Eotvos number (Eq. 5.68)	
$\overline{E}_v$	frictional loss per unit mass of liquid	erg gm^{-1} (ft $lb_f lb^{-1}$)
f	friction factor $(\Delta P D/(2\rho v^2 L))$	
F	Faraday's constant (23,062 cal $volt^{-1}$ $equiv^{-1}$)	
f, f_T	frequency of bubble formation	sec^{-1}
F	vollumetric flow of medium	liters hr^{-1}
Fr	Froude number (Eq. 5.156)	
Fr_o	orifice Froude number (Eq. 5.59)	
g	acceleration due to gravity	cm sec^{-2}
Ga	Galileo number (Eq. 5.116)	
Gr	Grasshof number (Eq. 5.95)	
h	heat transfer coefficient	BTU $hr^{-1}ft^{-2}oF^{-1}$
H	Henry's law coefficient	atm liter gm^{-1}
H	height of aerated liquid	cm
H_i	constant in the Sechenov equation (Eq. 3.34)	liter $mole^{-1}$
H_I	height of region I in tank	cm
I	ionic strength	molar
I	current	amperes
J	geometric ratio of particles	

J_A	flux of A	moles cm^{-2} sec^{-1}
k	Boltzmann constant (1.38×10^{-16})	erg $molecule^{-1}K^{-1}$
k	first order rate constant	sec^{-1}
k	thermal conductivity	BTU $hr^{-1}ft^{-1}hr^{-1}$
K	consistency index (rigidity)	gm cm^{-1} sec^{n-2}
K_{eq}	equilibrium constant	
k_g	gas phase mass transfer coefficient	mole $m^{-2}hr^{-1}atm^{-1}$
K_g	overall gas phase mass transfer coefficent	mole $m^{-2}hr^{-1}atm^{-1}$
k_L	liquid phase mass transfer coefficient	cm $hour^{-1}$
K_S	Monod constant	moles $liter^{-1}$
L	length of tube	cm
L	length of rod-like protein	Angstrom
m	maintenance coefficient	hr^{-1}
M	measured torque	gm cm^2 sec^{-2}
M	solute molecular weight	
M	molecular weight	
Mo	Morton number (Eq. 5.68)	
n	power law index	
N	impeller tip speed	revolutions sec^{-1}
N	number of holes in orifice	
n_B	number of baffles	
N_{O2}	molar flux of oxygen	moles O_2 $m^{-2}hr^{-1}$
n_v	number concentration of viable cells or spores	cells ml^{-1}
P	pressure	atm (bar)
P	power	HP (watts)
p_A	partial pressure of A in the gas phase	atm (bar)
Pe	Peclet number (Eq. 5.83)	
P_m	membrane permeability	gm $liter^{-1}$ cm^2 sec^{-1}
Pr	Prandtl number (Eq. 5.193)	
Q	volumetric flow rate	liters sec^{-1}
q_{O2}	specific rate of microbial oxygen consumption	moles gm dry $weight^{-1}$
R	spherical radius	cm
R	gas constant (1.987 cal g-$mole^{-1}$ K^{-1})	
Re_B	bubble Reynolds number (Eq. 5.57)	
r_g	radius of gyration	Angstrom
r_j	radius of gas jet	cm
r_{O2}	volumetric rate of oxygen consumption	moles $liter^{-1}$ hr^{-1}
r_{O2}	volumetric rate of oxygen consumption	moles $liter^{-1}$ hr^{-1}
S	substrate concentration	moles $liter^{-1}$
Sh	Sherwood number (Eq. 5.83)	

T	temperature	K (°C)
t_{blend}	blending time	sec
U	overall heat transfer coefficient	BTU $hr^{-1}ft^{-2}hr^{-1}$
u_{Bo}	bubble rise velocity	cm sec^{-1}
u_i	velocity at position i	cm sec^{-1}
$\overline{u}_L$	average liquid velocity	cm sec^{-1}
V	volumetric rate of reaction	mole $liter^{-1}$ hr^{-1}
v'	fluctuating velocity	cm sec^{-1}
v_b	bubble rise velocity	cm sec^{-1}
V_g	gas volume	liters
V_L	liquid volume	liters
V_m	molecular volume of solute at boiling point	cm^3 gm^{-1} $mole^{-1}$
v_s	superficial gas velocity	cm sec^{-1}
v_t	settling velocity	cm sec^{-1}
v_x	velocity	cm sec^{-1}
$\overline{W}$	work done by liquid on gas, per unit mass liquid	erg gm^{-1} (ft $lb_f lb^{-1}$)
w_B	width of a baffle	cm
We	Weber number (Eq. 5.63)	
X	biomass concentration	gm $liter^{-1}$
x_j	distance from orifice of jet	cm
y_A	mole fraction of A in the gas phase	
$\hat{y}$	average gas phase mole fraction of oxygen	
z	height or length	cm
z_i	charge on ith species	
α	Bunsen coefficient	cm^3
Δ	shear stress tensor	
$\Delta\Phi$	change in potential energy, per unit mass liquid	erg gm^{-1} (ft $lb_f lb^{-1}$)
η	apparent viscosity	gm cm^{-1} sec^{-1}
μ_s	viscosity of suspension	gm cm^{-1} sec^{-1}
μ_{el}	elongational viscosity	centipoise
μ	viscosity	gm $cm^{-1}sec^{-1}$ (or centipoise)
Ω	angular velocity	radians sec^{-1}
Π_{osm}	osmotic pressure	bar
Φ	volume fraction of particles	
Φ	gas phase hold-up, volume gas per volume liquid	(-)

Φ'	gas phase hold-up, volume gas per aerated volume	(-)
σ	surface tension	dyne cm^{-1}
ρ_L	density of liquid	gm cm^{-3}
ρ_g	density of gas	gm cm^{-3}
ρ	density	gm cm^{3}
τ_{xy}	shear stress	gm cm^{-1} sec^{-2}
τ_o	yield stress	gm cm^{-1} sec^{-2}
τ_w	shear stress at the wall	gm cm^{-1} sec^{-2}

5.10 Problems

1. *Oxygen Transfer in Tall Vessels*

This problem illustrates the development of gas and liquid phase mass balances for oxygen transfer when the vessel is tall and the variation of hydrostatic pressure results in a change in bubble volume. We will develop the gas and liquid phase mass balances for transfer of oxygen from air sparged into a fermentation broth in a tall agitated tank.

(i) Assuming that the gas phase can be modelled as plug flow and that the liquid is well mixed, show that the gas phase oxygen balance can be written as:

$$-K_G a' P V_b\left(y - C\frac{H}{P}\right) - v_b\frac{d\left(\frac{PV_b}{RT}y\right)}{dz} = 0$$

where V_b is the bubble volume, v_b is its rise velocity, and a' is the ratio of surface area to volume for a single bubble, y is the mole fraction of oxygen in the gas phase. Show that the liquid phase mass balance for oxygen is

$$F\left(C_{O_2,\text{in}} - C_{O_2}\right) + K_G\overline{a}' P V_{\overline{b}} N v_L\left(\overline{y} - C_{O_2}\frac{H}{P}\right) - r_{O_2}V_L = 0$$

(ii) Noting that the bubble volume and pressure depend on z, rewrite the gas phase mass balance in terms of oxygen mole fraction, y. V_b can be related to y via a balance on the inert gas component (N_2), from the sparger to a height z in the tank. The interfacial area a' can also be related to the bubble volume and hence the oxygen mole fraction, y. Propose an appropriate expresion for the relationship between bubble volume and its rise velocity, v_b.

(iii) Write the equations that define the average bubble volume and interfacial area that appear in the liquid phase oxygen balance.

(iv) Describe the approach you would take to solving the above set of equations in order to determine the exit gas phase composition when the kinetic of microbial oxygen uptake are known.

2. *Dynamic Response Method for k_La Determination*

The following $C_{O2}(t)$ data were obtained in a 10 liter, air-sparged laboratory fermenter during the continuous cultivation at 0.5 hr^{-1} of *K. aerogenes* on glucose-based medium, at 37°C. The air was turned off at zero time and the surface of the liquid was swept with a nitrogen gas stream. After 2 minutes, aeration was recommenced. The concentration of *K. aerogenes* remained constant at 5.0 gm DCW/liter during this period. Determine the specific oxygen uptake rate of *K. aerogenes* and the mass transfer coefficient under the agitation conditions employed. List the assumptions made in analyzing the data.

It was later found that the dissolved oxygen probe used to obtain the above data had a response time constant k of 0.2 sec^{-1}. Recalculate the mass transfer coefficient k_La taking the dynamic probe response into consideration.

Time (seconds)	*Probe reading (% sat)*	*Time (seconds)*	*Probe reading (% sat)*
0	80	130	23.7
10	73.6	150	49.2
20	67.2	170	63.0
30	60.8	190	70.7
40	54.4	210	74.9
60	41.2	230	77.2
80	29.2	250	78.5
100	15.5	290	79.5
110	10.2	350	80
120	4.1	400	80

3. *Determination of Mass Transfer Coefficients*

Consider two stirred tank reactors with a liquid height to diameter ratio of unity. both are equipped with flat-blade turbine agitators ($D_i = 0.4\ D_T$). One tank has a liquid volume of 10 liters, the other 10 meters^3. Each tank is equipped with an agitator motor providing 3 kW/m^3 to the liquid phase and each tank is sparged at the rate of 1.0 vvm (volume gas/liquid volume.minute). Assuming the properties of the fermentation liquid to be similar to those of "dirty" water, determine the mass transfer coefficients k_La for each tank by using the correlations provided in this chapter, noting the range of applicability of these correlations. Repeat this calculation for the larger tank, assuming that the fermentation broth has pseudoplastic properties given by the equation

$$\eta = k\gamma^{1-n}$$

Here η is the apparent viscosity, and n = 0.5 and k is 0.1 Pa-sec$^{1.5}$. Comment on the results you have obtained.

4. *Estimating Protein Diffusivities*

Tabulated below are physical properties of the proteins myoglobin (sperm whale) and phosphoglycerate kinase (yeast).

Protein	Dimensions (Å)	R_e (Å)	R_g (Å)	D_{exp} (x10^7 cm^2 s^{-1})
Myoglobin	(44 x 44 x 25)	13.1	15.3	11.3
Phosphoglycerate kinase	(70 x 45 x 35)	16.9	23.3	6.38

a) Estimate the diffusivities at 25°C of each protein using both the Stokes-Einstein equation and the Tyn-Gusak equation.
b) Explain why better agreement between the two estimates is obtained for myoglobin. Why is the Stokes-Einstein equation less accurate for phosphoglycerate kinase?

c) Estimate the effective diffusivity of each protein in a single, straight membrane pore 100 Å in diameter.

5. *Methane Production in Continuous Culture*

Consider the thermophilic methanogen, *Methanococcus jannaschii*, grown in continuous culture at steady state. *M. jannaschii* produces methane according to the overall stoichiometric equation:

$$4H_2 + CO_2 \rightarrow CH_4 + 2H_2O$$

The gaseous substrates, H_2 and CO_2, are supplied in the entering gas stream, and the product methane is removed in the exiting gas stream.

(a) Write unsteady state mass balances for hydrogen and methane in both the gas and liquid phases. Write an unsteady state mass balance for biomass.

(b) In practice, a readily measurable parameter is p_{CH4}, the partial pressure of CH_4 in the exiting gas stream. Assuming steady state and neglecting the loss of methane in the exiting liquid phase, derive an expression for the specific methane production rate (q_{CH4}, in moles CH_4 (cell-hr)$^{-1}$) in terms of p_{CH4}, Q (the volumetric gas flow rate), V_L (the liquid volume), and X (the steady state biomass concentration).

6. *Dynamic Measurement of k_La in Animal Cell Culture Medium*

As discussed in Chapter 3, animal cell culture medium generally consists of a complex mixture of amino acids, vitamins, and mineral salts. In addition, it is often supplemented with serum. Lavery and Nienow have measured k_La values for oxygen transfer in animal cell culture medium [M. Lavery and A.W. Nienow, *Biotechnology and Bioengineering*, **30**, 368 (1987)] using dynamic measurements with an oxygen electrode. Deoxygenated culture medium was either headspace- or sparge-aerated with a 5% CO_2 in air mixture and the increase in dissolved oxygen level was followed with time. The response to a step change in aeration is summarized by the data below.

Time (min)	Percent response	Time (min)	Percent response	Time (min)	Percent response	Time (min)	Percent response
0	0	20	68.3	40	93.7	70	99.2
5	15.9	25	77.8	45	95.2	80	99.2
10	34.9	30	84.1	50	96.8	90	100
15	50.8	35	90.5	60	97.6	100	100

a) Based on the data shown, what was the k_La value for these conditions?

b) Neglect of the oxygen electrode response time is a common source of error in k_La measurements. The response time of the oxygen electrode used by Lavery and Nienow ranged from 18 to 21 seconds. Would you expect time constants in this range to influence the k_La measurements?

7. *Temperature Dependence of k_La*

Aerobic cultivation of thermophilic microorganisms at elevated temperatures (e.g., from ca. 40-90°C) is becoming more common as interest in thermophiles continues to grow. In this context, it is important to consider the possible dependence of k_La on temperature. S. Aiba and co-workers have addressed this issue (S. Aiba *et al.*, *Biotechnology and Bioengineering*, **26**, 1136 (1984)) by measuring the effect of temperature on k_La and viscosity values, μ, of LG medium. Their data are given below.

Temperature (°C)	k_La (h^{-1})	μ (mPa-secs)
45	212	0.648
50	222	0.591
55	241	0.531
60	246	0.497
65	255	0.463

Show that the above data are consistent with the temperature dependence of k_La predicted by a simple model based on penetration theory (section 5.3.3) and the Stokes-Einstein Equation for the diffusivity of a sphere (D_s) of radius r_p (section 6.4.3):

$$D_s = \frac{kT}{6\pi\mu r_p}$$

8. *Measuring k_La in Large-Scale Fermentors*

Imai and co-workers recently developed a simple Na_2SO_3 feeding method for measuring k_La in large-scale fermentors. The method is based on the measurement of the dissolved oxygen concentration while Na_2SO_3 is contiuously supplied to an aerated, mechanically-agitated fermentor. Under these conditions, oxygen absorption is accompanied by sulfite oxidation in the vessel. In this problem we shall develop the equations necessary to calculate k_La by this method.

a) Begin by writing unsteady state mass balances for both oxygen and Na_2SO_3 in the fermentor. Assume the kinetics of sulfite oxidation are mth-order in the dissolved oxygen concentration, C_O, and nth-order in sulfite concentration, C_S. The reactor volume is V_L, the volumetric feed rate of Na_2SO_3 solution is Q, the equilibrium concentration of dissolved oxygen is C_{Oi}, and the feed concentrations of oxygen and sulfite are C_{OF} and C_{SF}, respectively.

b) Assume steady state and derive and expression for k_La from the mass balances. Simplify the expression by assuming, as was the case in the experiments, that both the dissolved oxygen concentration in the feed and the concentration of Na_2SO_3 in the vessel are very small, and that the dissolved oxygen concentration in the vessel is much smaller than the Na_2SO_3 concentration in the feed.

c) Calculate k_La for the case of $V_L = 4000$ liters, $Q = 0.649 \times 10^{-7}\ m^3/s$, $C_{Oi} = 0.266\ mol/m^3$, $C_{SF} = 92.8\ mol/m^3$, and $C_O/C_{Oi} = 0.733$.

9. Oxygen Transfer in Hydrocarbon Fermentation

Consider the hydrocarbon fermentation with yeast growing aerobically on hexadecane. (a) Estimate the liquid phase oxygen transfer coefficient from gas bubbles to the aqueous liquid phase in an airlift fermentor in which the aqueous liquid phase density is 1 g/ml, the aqueous liquid phase viscosity is 0.009 g/cm-sec, the diffusivity of oxygen is 2.5×10^{-5} cm^2/sec, the mean diameter of the bubbles is 1.3 mm, the gas flow rate is 1 volume of gas per volume of liquid per minute, and the pressure is 17 psia entering and 14.7 psia in the effluent.
(b) Estimate the oxygen transfer coefficient for oxygen transfer from dissolved oxygen in the aqueous liquid phase to yeast cells suspended in the aqueous liquid phase. Assume the yeast cells are spherical with a diameter of 0.006 mm.
(c) What is the rate controlling resistance in the overall transport of oxygen from the gas bubbles to the cells? (Based on a problem provided by Professor L.E. Erickson, Kansas State University.)

10. Batch Sterilization With and Without Heating/Cooling Periods

Batch sterilization is being considered as a means to prevent contamination of fermentations by spores of *Bacillus stearothermophilus*. (a) Assuming a holding temperature of 120°C, calculate the time required to achieve a probability of *contamination*, 1 - P(t), of 10^{-3} in a 1-liter fermentor (where P(t), the probability of *extinction*, is equal to $[1 - n_v(t)/n_{vo}]^{N_o}$, and N_o is the initial *number* of viable spores). Assume $n_{vo} = 10^4$ spores/liter, $T_o = 20°C$, and neglect the heating and cooling periods. Table 5.9 provides data on the thermal death constants for the spores. (b) What is the probability of contamination if a 100-liter fermentor is heated for the same time at 120°C? (c) Redo the calculation for part (a), assuming constant-temperature steam heating in the heating stage (i.e., $T_H = T_{st}$) and constant-temperature cooling in the cooling stage. Heat exchanger coils are used for both operations. For the steam heating coils, the values of α and β in Table 5.10 are 0.05 sec^{-1} and -0.25. For the cooling coils, α and β have the values 0.065 sec^{-1} and 0.20. The heating and cooling periods are 3 minutes and 10 minutes respectively.

11. Constant Temperature Continous Sterilization

Consider sterilization of a fermentation medium in the holding section of a continuous sterilizer. Assuming constant temperature, the specific death rate of the contaminant is 10 sec^{-1}. If the the average residence time in the holding section is 5 seconds, calculate the following:

a) the reduction of contaminant level, n_v/n_{vo}, for $Pe_z = 200$

b) the reduction of contaminant level for $Pe_z = 200$, assuming plug flow

c) the reduction of contaminant level for $Pe_z = 50$

d) the reduction of contaminant level for $Pe_z = 50$, assuming plug flow

Explain the results.

12. Determination of Power Input

A 100 liter, fully baffled bioreactor has a liquid height-to-diameter ratio of unity and is agitated by a six-bladed flat-blade turbine impeller ($W/D = 1.5$), with a D_i/D_T ratio of 0.33. The tank contains a fermentation broth with rheological properties described by the power law equation

$$\tau = K\gamma^n$$

where n = 0.6 and K = 0.5 Pa-$sec^{0.6}$.

For an impeller speed of 150 rpm and an air sparging rate of 0.8 vvm, estimate

(a) the power input to the impeller

(b) the k_La value for the vessel.

13. Kinetics of Hydrogen Consumption by Lake Sediments

As discussed by Robinson and Tiedje (Appl. Environ. Microbiol. 1982. 44:1374-1384), hydrogen is a key intermediate in the degradation of organic matter, affecting both the rate of the process and the nature of the end products. An important mechanism of biological hydrogen consumption is methanogenesis, which occurs in methanogenic habitats such as rumen fluid, digestor sludge, and eutrophic lake sediments. Reported below are laboratory data for the consumption of hydrogen by lake sediment samples taken from Wintergreen Lake in Hickory Corners, Michigan.

(a) Use these data to determine a V_{max} and K_m for H_2 consumption by the Wintergreen sediment.

(b) When the sediment was studied in a more concentrated form, Michaelis-Menten kinetics were not observed. Instead, the kinetics of H_2 consumption were first order. Provide an explanation for this result, and discuss what may be governing the kinetics under such conditions.

Time (hours)	Dissolved H_2 Concentration (μmol l^{-1})
0.0	20.5
2.3	16.2
4.3	12.7
6.1	9.42
8.1	7.00
10.0	4.31
12.4	2.70
14.4	1.62
17.3	0.538

14. Mass Transfer and Enzyme Reactions[62]

Many enzymes adsorb preferentially at gas-liquid interfaces. Consider the adsorption of carbonic anhydrase to the gas interface of a stirred reactor. The gas contains carbon dioxide, such that its interfacial concentration is maintained constant at C^*. In the region of thickness δ near the surface, the enzyme is present at a concentration E_0, higher than that present in the bulk, E. As the CO_2 diffuses through the surface region, the rate of CO_2 hydration is given by

$$r(E_o, C) = \frac{kE_oC}{K_M + C}$$

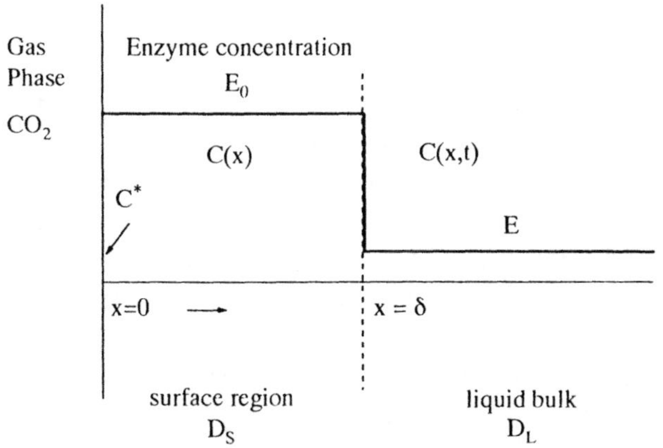

(62) From G.T. Tsao, *The effect of carbonic anhydrase on carbon dioxide absorption*, Chem. Eng. Sci. **27**, 1593 (1972)

We shall assume that CO_2 is transported by molecular diffusion (with a diffusivity D_s) in the surface region ($x < \delta$), and that the CO_2 concentration is at steady state. In the liquid bulk ($x > \delta$), little reaction occurs and CO_2 transport can be represented by an effective bulk diffusivity D_L. The situation is depicted in the figure.

(i) Develop the mass balance equation for the concentration of CO_2 in the surface region; write appropriate boundary conditions for C(x) to solve this equation, with attention being given to the continuity condition at $x = \delta$.

(ii) Write the unsteady state mass balance and boundary conditions for C(x,t) in the liquid bulk. Show that the rate of transport of CO_2 at $x = \delta$ can be written using penetration theory:

$$-D_L\left(\frac{\partial C}{\partial x}\right)_{x=\delta} = C_\delta\sqrt{D_L s}$$

where s is the surface renewal rate.

(iii) Integrate the mass balance in the surface region and show that the rate of absorption is given by:

$$R = -D_s\left(\frac{\partial C}{\partial x}\right)_{x=0}$$

$$= \sqrt{D_L s C_\delta^2 + 2kD_sE_o(C^* - C_\delta) - 2kD_sE_oK_M \ln\left(\frac{K_M + C^*}{K_M + C_\delta}\right)}$$

Express this result as an enhancement factor E, where E represents the ratio of the enhanced rate of CO_2 absorption with reaction to that which would occur in the absence of enzyme.

15. Bacterial Density at a Nutrient Surface

Consider a system in which bacteria can multiply *only* at a surface which acts as a nutrient source (a situation relevant to biofilms and microbially-induced corrosion). Let the bacteria be motile and chemotactically attracted to the nutrient. Also, assume the bacteria are present in a suspension at a constant density ρ_o at a distance δ away from the surface, while the suspension flows very slowly past the surface at a velocity v. Finally, assume also that the chemotaxis coefficient, χ, is constant, and that the nutrient concentration gradient is linear from $x=0$ to $x=\delta$.

(a) Formulate the appropriate model equation for the bacterial density near the surface.

(b) Assuming steady state, solve for the bacterial density at the surface.

(Based on a problem provided by D.A. Lauffenburger, University of Illinois.)

Chapter 6. Product Recovery

6.1 Introduction

Separating and purifying a desired product from the fermentation broth or cell culture supernatant is a crucial and challenging component of commercial biotechnology. Product recovery usually accounts for a large portion of the product cost and in some cases is *the* major manufacturing cost. The ratio of fermentation to product recovery costs is about 60:40 for many older antibiotics. For recombinant DNA fermentation products, the cost of downstream purification is very important and can amount to 80-90% of the total manufacturing costs.

The high cost of product recovery stems from many factors; for example, in many cases the product is present in small amounts in an aqueous fermentation broth containing intact cells, soluble extracellular products, and residual substrates. The desired product must often be isolated from such a complex mixture and purified to homogeneity. Further, bioproduct recovery can require several discrete steps. (Isolating the product through a series of separation and recovery steps is often referred to as *downstream processing*.) Multiple separations demand more equipment and labor, and generally lower the yield of the final product. The exact number of steps involved will depend on the original materials used, the concentration and physicochemical properties of the product, and the final purity required. Thus, the ultimate objective is to devise an overall recovery strategy that satisfies the purity requirements and minimizes the product cost.

The first task in formulating a purification strategy is to define or acknowledge the required purity of the product. The allowable ranges of impurity concentrations, and the specific impurities which may be tolerated, will be dictated by the end-use of the product. For example, very stringent purity requirements apply to recombinant DNA-derived proteins produced for therapeutic applications. As of 1990, U.S. standards required that the final product, which is usually administered intravenously or subcutaneously, must contain less than 0.1 % protein impurities and less than 100 pg nucleic acid per dose. Figure 6.1 illustrates the required purity for different types of products. On the other hand, enzymes can be used as industrial catalysts in relatively crude form. In addition to defining the end-use criteria, it is also important to characterize the starting material (e.g., what contaminants are known to be present?) in as much detail as possible.

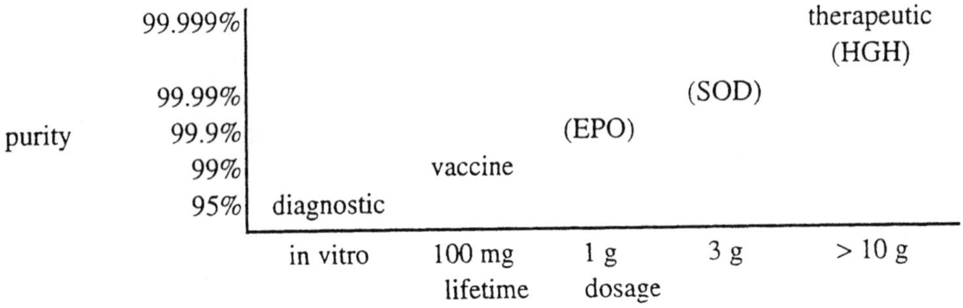

Figure 6.1. Required purity as a function of lifetime dosage for various products of biotechnology. EPO = erythropoietin, SOD = superoxide dismutase, HGH = human growth hormone. [From S.M. Wheelwright, The Design of Downstream Processes for Large-Scale Protein Purification, J. Biotechnol., 11, 89 (1989).]

Once the purity criteria have been established, the specific purification procedures can be selected. For products derived from living cells, such as recombinant proteins, recovery procedures will be influenced to some extent by the producing organism. The location of the protein upon completion of the fermentation is particularly important. In the case of bacteria, the protein is often found as an inclusion body (an insoluble protein aggregate) in the cell, whereas in the case of cell cultures, it is secreted in soluble form. Some advantages and disadvantages of different host organisms for the production of foreign proteins, including their impact on product recovery, are summarized in Table 6.1.

Individual recovery operations can be grouped into different categories, depending on their general purpose. These groupings are certainly neither definitive nor mutually exclusive in that some separation techniques fit into more than one category. Nonetheless, these generalizations provide a useful organizational framework for this chapter and define the sequence of recovery operations typically employed in practice:

1. *Separation of Insolubles.* Insoluble materials include whole cells, cell debris, pellets of aggregated protein, and undissolved nutrients. Common operations for this purpose are sedimentation, centrifugation, and filtration.

2. *Isolation and Concentration.* Generally refers to the isolation of the desired product from unrelated impurities. Significant concentration is achieved in the early stages, but concentration accompanies purification as well. This category includes extraction, ultrafiltration, precipitation, and ion exchange.

Table 6.1. Different hosts as producers of foreign proteins (characteristics that can strongly impact product recovery are highlighted in italic).

	Escherichia coli	*Bacillus subtilis*	*Saccharomyces cerevisiae*	Mammalian Cells
Advantages:	Fast growing (t_d ca. 20-45 min); easy to grow on an industrial scale.	Fast growing.	Easy to grow on a large scale.	Natural protein produced: correct post-translational modification carried out (e.g., proteolytic cleavage of precursor protein, macromolecular assembly, glycosylation).
	Extensive understanding of genetics and physiology.	Genetics are well understood.	Genetics are well understood.	*Product secreted into medium; no inclusion bodies.*
	More genetic engineering tools and procedures available than for any other organism.	Several promoters are available.	Strong promoters are available and convenient expression vectors have been developed.	
		Secretes proteins.	*Can glycosylate and secrete proteins.*	
Disadvantages:	Cannot glycosylate proteins or form disulfide bonds.	Instability and lack of versatility of plasmids.	Grows more slowly than *E. coli*: 1-2 hr doubling time.	Extremely slow growing (12-24 hr generation times); expensive media required.
	Usually doesn't secrete proteins, although a number of strategies to achieve secretion have been developed	*Secretes large amounts of proteases, which can cause degradation of the secreted foreign protein.*	*Heterologous genes are expressed relatively poorly (but it is still possible to produce tens of mg per L of a desired protein).*	Molecular biology still being worked out; long term stability of some cell lines not clear.
	Eukaryotic proteins overproduced in E. coli typically form inclusion bodies.			*Product recovery complicated by extraneous proteins and other process contaminants secreted by cells or included in medium.*
	Produces endotoxins, which must be removed from food and health care products (levels below 1 ng/ml can cause fever or even death).			*May contain or express hazardous agents, e.g., viruses, oncogenes.*

3. *Primary Purification.* More selective than isolation; some purification steps can distinguish between species having very similar chemical and physical properties. Primary purification techniques include chromatography, electrophoresis, and fractional precipitation.

4. *Final Purification.* Necessitated by the extremely high purity required of many bioproducts, particularly pharmaceuticals and therapeutics. After primary purification the product is nearly pure but may not be in the proper form. Partially pure solids may still contain discolored material or solvent. Crystallization and/or drying are typically employed to achieve final purity.

Table 6.2 summarizes the results of some of these steps for antibiotics and a recombinant protein. Note the dramatic difference in final product concentrations. Moreover, in the case of recombinant human leukocyte α A interferon (rIFN-α A), a column of immobilized monoclonal antibodies was used to isolate and purify oligomers of rIFN-alphaA in a single step.

Table 6.2. *Purification of recombinant interferon (rIFN-α A) and antibiotics.*

	Concentration (g/l)		Quality (%)	
Step	rIFN-αA	Antibiotics[a]	rIFN-αA[c]	Antibiotics
Harvest Broth	16[b]	0.1-5	2.8	0.1-1
Removal of Insolubles (Clarification)	3.7	1.0-5	6.4	0.2-2
Primary Purification	5.0	50-200	~100[b]	50-80
Final Purification	NR	50-200	--	90-100

NR = not reported

[a]Data for antibiotics taken from P.A. Belter, E.L. Cussler, and W.-S. Hu, *Bioseparations--Downstream Processing for Biotechnology* (Johy Wiley & Sons, N.Y., 1988), p. 5. In this case, "quality" may denote chemical purity, activity, or efficacy.

[b]Refers to a crude extract prepared by adding acid-treated cell paste to 24 liters of extraction buffer

[c]For rIFN-αA, the quality refers to purity estimated from sodium dodecyl sulfate-polyacrylamide gel electrophoresis (SDS-PAGE). The gel was stained with Coomassie brilliant blue. From S.H. Tarnowski et al., "Large Scale Purification of Recombinant Human Leukocyte Iterferons," *Meth. Enzymol.*, **119**, 153 (1986).]

We begin this chapter by considering some general aspects of bioproduct recovery, particularly with regard to protein production by recombinant DNA technology. Next, the four stages of a recovery sequence will be discussed in detail, emphasizing both theory and practice. The logic used to integrate individual steps into an overall strategy for downstream processing is also described. Finally, several examples of bioproduct separation and purification, presented as process flow diagrams, are included at the end of the chapter.

6.1.1 General Aspects of Bioproduct Recovery

The products of biotechnology are considerably diverse. The desired product of the bioprocess may be either the cells themselves (as in the production of single-cell protein) or a specific metabolite of the cell. Metabolic products are either *intracellular* or *extracellular*. Intracellular products of interest include vitamins, certain enzymes, and a few antibiotics (for example, sisomicin and griseofulvin). Examples of important extracellular products are citric acid, alcohols, hydrolytic enzymes (for example, amylases and proteases), and most antibiotics (for example, penicillin and streptomycin). We shall see later that the location of the product (extracellular or intracellular) has a major impact on the purification scheme.

Typically, biological products are present in fermentation broths and cell culture supernatants in low concentrations. Low product concentrations coupled with large amounts of interfering species can seriously complicate the task of purification. A striking example of how the broth or medium composition can affect purification is shown in Table 6.3, which compares recombinant tissue plasminogen activator (tPA) purified from supernatants of animal cells grown in medium with and without 10% calf serum. In the latter case, the initial purity of tPA in the supernatant was substantially higher (due to the absence of serum), resulting in a much purer product (95% versus 60%) after the same three-step purification process. In general, there is a strong correlation between the concentration of a desired substance in the crude biomass or synthesis mixture and its selling price, as shown in Figure 6.2.

Table 6.3. *Influence of protein concentration in the medium on the recovery and purity of recombinant tissue plasminogen activator (tPA) purified from tissue culture supernatant*[a].

	Medium containing 10% calf serum	Serum-free medium
Starting specific activity (units/mg)	711	5,600
Final specific activity (units/mg)	330,000	560,000
Purification factor	460	74
Purity (estimated from HPLC) (%)	60	95
Yield (%)	54	68

[a]The same three-step purification process was used for the supernatants of cells grown in serum-containing and serum-free media. The starting concentration of tPA in units per ml was effectively the same in each case. [From T. Cartwright, TIBTECH, 5, 25 (1987).]

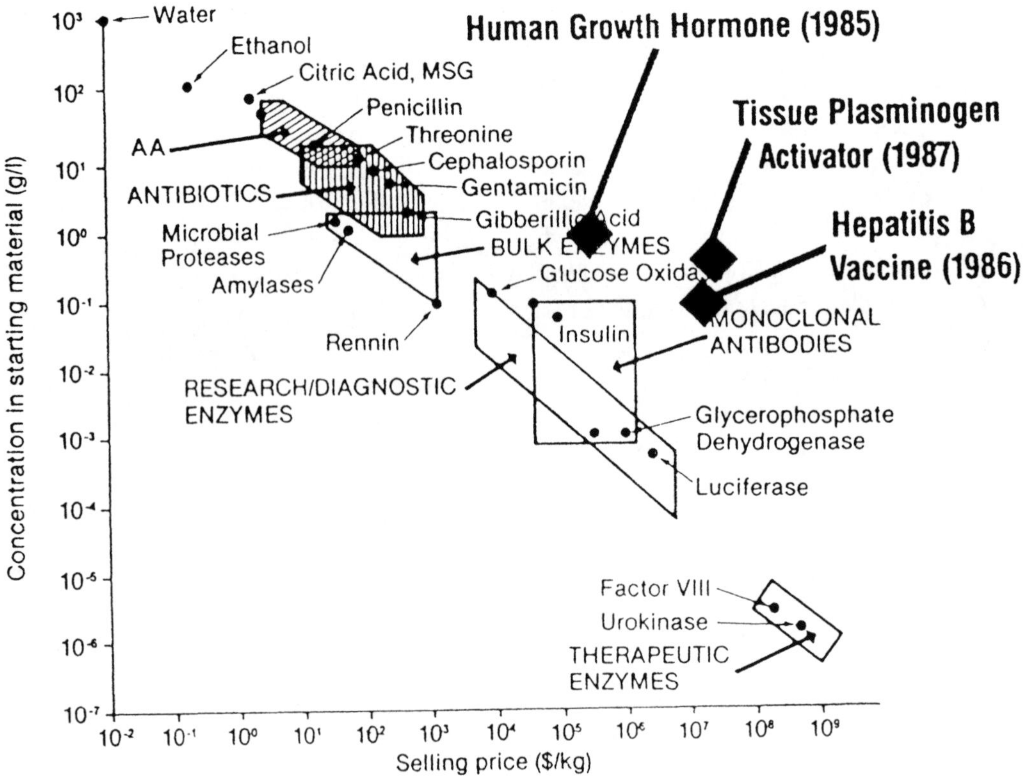

Figure 6.2. Correlation between concentration in starting material (pre-purification) and approximate or estimated selling price of various biological products. [From J.L. Dwyer, Bio/Technology, 2, 957 (1984)]. The original graph dates from 1984; products introduced later (solid squares) lie above the line and command prices even higher than predicted.]

6.1.2 Impact of Recombinant DNA Technology on Protein Recovery

The advent of recombinant DNA technology enabled the production of many important proteins that were previously extremely difficult or impossible to produce in commercial quantities. For example, in 1989, 400 grams of human growth hormone (sufficient to treat 550 growth-hormone deficient children at a maximum dose of 15 mg per week for a year) could be purified from 4,000 liters of an *E. coli* fermentation. Previously, it would have required 100,000 cadaver pituitary glands to obtain an equivalent amount of growth hormone.

The world-wide annual manufacturing demands for recombinant-DNA derived proteins can be grouped into three categories:

1. **Gram to kilogram scale.** This group includes tPA and the leukokines such as the interferons, the colony stimulating factors, and the interleukins.
2. **Kilograms scale**. Included in this category are such therapeutic proteins as insulin, human growth hormone, erythropoeitin, monoclonal antibodies, and factor VIII.
3. **Multi-ton scale**. By far the largest of the three scales, this category includes products for agricultural use such as porcine, bovine, and ovine growth hormones (somatotropins).

The large difference between the production scales of human and animal growth hormones stems from the size and nature of their potential markets. Young animals given somatotropin grow faster and leaner, and their mothers produce more milk. Therefore, unlike the relatively uncommon child suffering from stifled growth, every market animal is a potential recipient of somatotropin.

Products in categories one and two can be produced profitably because small doses can be marketed at high prices. In 1988, the cost of medication to treat a 50 kg growth-hormone deficient child with 15 mg of growth hormone per week amounted to about $35,000 per year. Clearly, the animal somatotropins are priced very differently to accommodate their larger markets.

Recombinant DNA technology can also have a significant impact on the purification process. Recombinant DNA methods lead to higher production levels of the desired protein compared to the natural source and create opportunities for genetic manipulations to facilitate recovery. Moreover, as mentioned previously, the details of a recovery sequence depend partly on the characteristics of the host cell type, as different organisms (e.g., the bacterium *Escherichia coli*, the yeast *Saccharomyces cerevisiae*, or a mammalian cell line) present specific advantages and disadvantages for the recovery of recombinant proteins.

Inclusion Bodies

Recombinant DNA techniques can be used to achieve very high expression levels of prokaryotic and eukaryotic proteins in *E. coli*. Accumulation of recombinant proteins to levels ranging up to 25% of the total cell protein are typical, and levels up to 50% have been reported. However, in the majority of cases, overproduced proteins aggregate in an insoluble form inside the cell. These aggregated proteins are called *inclusion bodies or refractile bodies* because they appear as dark, refractile areas when the cells are examined in a phase contrast microscope (Figure 6.3). Examples of recombinant eukaryotic proteins that form inclusion bodies in *E. coli* include bovine growth hormone, urokinase, IFN-β, IFN-γ, and interleukin-2 (IL-2). However, normal *E. coli* proteins can also aggregate if they are synthesized to high levels using recombinant DNA techniques. The sequestering of proteins into inclusion bodies is therefore not simply a response of *E. coli* to 'foreign' proteins. Nor does it appear that such aggregation is just a precipitation phenomenon resulting from the high concentration of recombinant protein, since stringent chemical conditions are required to solubilize inclusion bodies. What is evident, however, is that inclusion body formation is promoted by hydrophobic, ionic, and in some cases covalent interactions (i.e., disulfide bonds) between protein molecules.

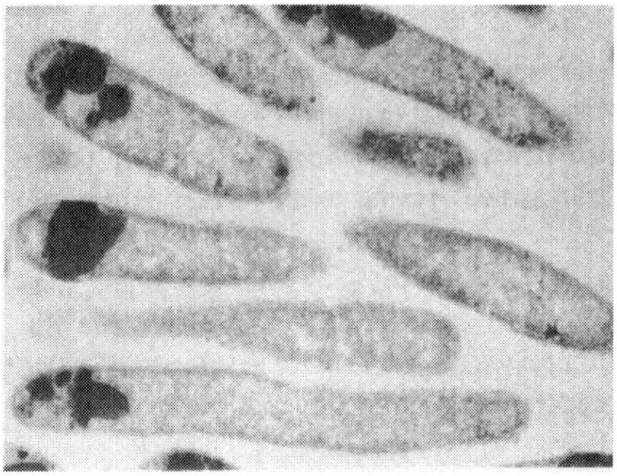

Figure 6.3. Transmission electron micrograph of genetically engineered E. coli containing a runaway-type plasmid with the gene for human α-interferon analogue. Inclusion bodies containing large concentrations of interferon appear as dense dark spots. [From J. Fieschko et al., Biotechnol. Prog., 1, 205 (1985)].

Inclusion bodies can facilitate protein purification. Because of their size and density, inclusion bodies can be separated from other cellular components by low-speed centrifugation (speeds as low as 500 g are sometimes sufficient but values of 5000-12,000 g are more generally used). Aggregation also protects the protein product from proteolytic degradation. (Indeed, inclusion bodies have been likened to proteins encased in stainless steel shells!) On the other hand, inclusion bodies present the problem of solubilizing the recombinant protein. The situation is somewhat similar to developing a method for isolating egg albumin from hard-boiled eggs. Common solubilization agents include 5-8 M guanidinium chloride, 6-8 M urea, detergents, alkaline pH (>9.0), and organic solvents. Note that these are conditions which typically denature native proteins. If covalent disulfide bonds exist in the aggregates, thiol reagents can be used in conjuction with denaturants. Following solubilization and unfolding of the protein, the denaturant must be diluted or removed by dialysis so that the protein can refold to its native, active form. The complexity of the solubilization-reactivation process varies with different proteins; key variables in published protocols include time, pH, ionic strength, the choice of denaturant, and the ratio of denaturant to protein. In most cases the recovery of biologically active proteins is incomplete.

Fusion Proteins

In some cases, accumlation of a eukaryotic protein expressed directly in *E. coli* is limited because the protein is recognized as foreign and is degraded. However, expression levels can sometimes be improved by linking a eukaryotic gene with a bacterial gene to produce a so-called *fusion protein*. If the end use of the final product requires that the eukaryotic polypeptide be isolated from the fusion protein, a cleavable peptide sequence, often called a *linker peptide*, can be inserted between the C-terminus of the prokaryotic sequence and the N-terminus of the eukaryotic sequence. Cleavage can then be effected by chemical methods or by proteolytic enzymes. Another fusion method is one in which multiple copies of the gene are linked in tandem. For example, enhanced expression of proinsulin in *E. coli* can be achieved by linking two or more proinsulin genes and expressing them directly or in conjuction with a small segment of the N-terminus of β-galactosidase.

Fusion proteins can also be constructed to simplify purification. One approach is to select a bacterial sequence that codes for a polypeptide that can be isolated by affinity chromatography (Section 6.4.2). An alternative is to fuse a synthetic gene sequence coding for polyarginine to the eukaryotic gene of interest. Cation-exchange chromatography can then be used to purify the positively charged recombinant fusion protein. In any event, the following points should be kept in mind when designing a genetic construction for purification purposes:

(a) The fusion should allow a simple, rapid, and inexpensive purification by ion exchange or affinity chromatography.

(b) If a linker peptide is used, the efficiency of the cleavage must be evaluated.

(c) The fusion peptide must have a negligible effect on the protein's folding and no permanent effect on its biological activity.

(d) The fusion peptide must be readily and specifically removed after purification.

Having discussed some of the special opportunities presented by recombinant DNA techniques, we now move on to consider in detail the different recovery operations introduced earlier.

6.2 Separation of Insolubles

6.2.1 Sedimentation and Centrifugation

Many bioreactors contain microorganisms suspended in liquid culture media as single cells, flocks, or particles formed by mycelia. In such cases the first separation step is the removal of insoluble biomass by a solid-liquid separation process such as sedimentation, centrifugation, flotation, or filtration. Sedimentation and centrifugation are both based on density differences between insoluble particles and the surrounding fluid. Sedimentation relies on gravity and settling to achieve solid-liquid separation, and is generally performed in rectangular or circular flow tanks. Centrifugation, on the other hand, involves mechanical application of a centrifugal force to obtain a solid concentrate and clarified supernatant.

The earliest uses of centrifugation were in sugar manufacture and in separating cream from milk. In fact, cream separation is thought to have been the primary application of the world's first continuous centrifugal separator, purportedly invented in 1877 by a Swedish engineer, Dr. Carl Gustaf Patrick DeLaval. Since then, and particularly during and after World War II, many additional applications have emerged, including the preparation of blood plasma, the separation of penicillin solvents, centrifugal extraction of edible proteins from animal and vegetable products, and the clarification of fermentation beers by the removal of yeast and other solid matter. Needless to say, centrifugation has become a widespread technique for the removal of solids, particularly under circumstances where filtration is ineffective or undesirable. One such example is the removal of precipitated protein, which is often gelatinous or colloidal in nature. In some cases, the proteinaceous material causes severe blockage of cloths or meshs and therefore precludes the use of filtration equipment.

Centrifuges may be classified in several ways, for example, on the basis of bowl design (Figure 6.4). Each configuration includes machines that discharge solids intermittently or continuously. Also important is whether separation depends only on density differences between phases (as it does in *centrifugal settling machines*), or whether filtration also takes place (as it does in *centrifugal filters* or *filtering centrifuges*, in which the centrifugal field acts as a pressure force). The principle of centrifugal filtration is illustrated by the basket centrifuge in Figure 6.4(f). Some characteristic advantages and disadvantages of the different centrifuge designs are summarized in Table 6.4.

We shall now discuss the basic theory of sedimentation, under both gravity and centrifugal forces. Consider an isolated spherical particle in a gravitational or centrifugal force field. The rate at which the particle settles will depend on the net force acting on the particle. This force will be the sum of the external force F_E, a buoyancy force F_B, the drag force F_D exerted by the fluid on the particle, and the particle-particle or particle-wall interaction forces F_P. If we ignore the latter, which is equivalent to assuming the particle is in an infinite fluid, a force balance on the falling particle is

$$(F_E - F_D - F_B) = m\frac{dv}{dt} \tag{6.1}$$

where m is the mass of the particle, and v is the free stream velocity relative to the particle (for settling, this velocity is the terminal velocity since the fluid is stationary). The external force F_E may be expressed by Newton's law:

$$F_E = ma_E \tag{6.2}$$

were a_E is the acceleration of the particle due to the external force. For settling, the acceleration is due to gravity and $a_E = g$; for centrifugation, $a_E = \omega^2 r$, where ω is the angular rotation in rad/sec, and r is the radial distance from the center of the centrifuge to the particle. The drag force acting on the particle can be expressed in terms of a drag coefficient, C_D, defined as

$$F_D = C_D A \frac{\rho_m v^2}{2} \tag{6.3}$$

where A is the cross-sectional area of the particle normal to the flow (= $(\pi d_p^2)/4$). The buoyancy force can be related to the mass of fluid displaced by the solid particle, $(m/\rho_p)\rho_m$, where ρ_p and ρ_m are the densities of the solid particle and fluid, respectively. Therefore,

$$F_B = \left(\frac{m}{\rho_p}\right)\rho_m a_E \tag{6.4}$$

Combining Eqs. (6.1), (6.2), (6.3), and (6.4) yields an equation for the total force acting on a particle in a centrifugal force field:

$$\frac{dv_c}{dt} = \left(1 - \frac{\rho_m}{\rho_p}\right)\omega^2 r - \frac{C_D v_c^2 \rho_m A}{2m} \tag{6.5}$$

where v_c is the settling velocity of the particle.

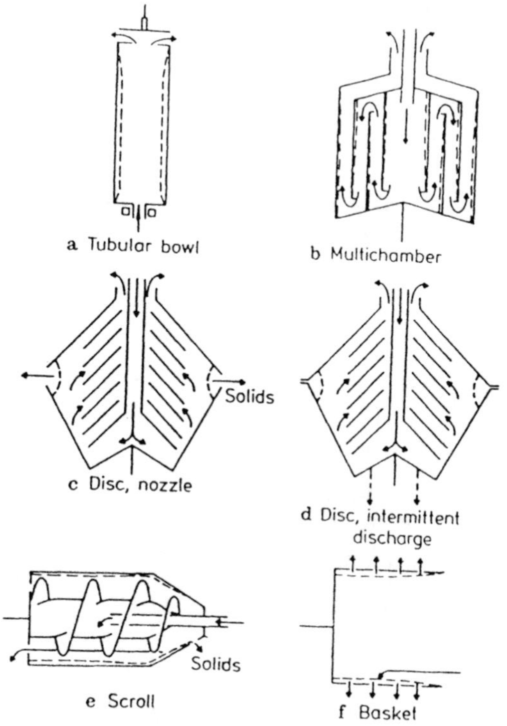

Figure 6.4. Common centrifuge configurations. Arrows show the flow pattern of the liquid phase. Dashed lines indicate regions of solid build-up. [From D.I.C. Wang et al., Fermentation & Enzyme Technology, John Wiley & Sons, NY, (1979). p. 263.]

Table 6.4 *Characteristics of different centrifuge designs*[a].

System	Advantages	Disadvantages
Tubular Bowl	a) High centrifugal force b) Good dewatering c) Easy to clean d) Simple dismantling of bowl	a) Limited solids capacity b) Foaming unless special skimming or centripetal pump used c) Recovery of solids difficult
Chamber Bowl	a) Clarification efficiency remains constant until sludge space full b) Large solids holding capacity c) Good dewatering d) Bowl cooling possible	a) No solids discharge b) Cleaning more difficult than tubular bowl c) Solids recovery difficult
Disc-centrifuges	a) Solids discharge possible b) Liquid discharge under pressure eliminates foaming c) Bowl cooling possible	a) Poor dewatering b) Difficult to clean
Scroll discharge	a) Continuous solids discharge b) High feed solids concentration	a) Low centrifugal force b) Turbulence created by scroll

[a]From D.J. Bell et al., *Adv. Biochem. Eng. Biotechnol.*, **26**, 1 (1983).

Drag coefficients for spherical particles in various flow regimes are given by

$$C_D = \frac{24}{Re} \qquad \text{Re} < 0.4 \qquad \text{(Stokes regime)} \tag{6.6}$$

$$C_D = \frac{10}{\sqrt{Re}} \qquad 0.4 < \text{Re} < 500 \qquad \text{(Allen regime)} \tag{6.7}$$

$$C_D = 0.44 \qquad 500 < \text{Re} < 2 \times 10^5 \qquad \text{(Newton regime)} \tag{6.8}$$

where Re [$= (d_p \rho_m v_c)/\mu$] is the Reynolds number. Substituting the above expressions for C_D into eqn. (6.5) and letting $dv_c/dt = 0$ (i.e., assuming steady state) leads to expressions for the terminal velocity of a particle in the different flow regimes:

$$v_c = \frac{(\rho_p - \rho_m) d_p^2 \omega^2 r}{18\mu} \qquad \text{Re} < 0.4 \tag{6.9}$$

$$v_c = \left[\frac{4}{225} \frac{(\rho_p - \rho_m)^2 \omega^4 r^2}{\rho_p \mu} \right]^{1/3} d_p \qquad 0.4 < \text{Re} < 500 \tag{6.10}$$

$$v_c = \left[3\omega^2 r(\rho_p - \rho_m)\frac{d_p}{\rho_m} \right]^{1/2} \qquad 500 < \mathrm{Re} < 2 \times 10^5 \qquad (6.11)$$

In general, centrifugation of proteins and whole cells will fall within the Stokes regime, whereas larger crystals may have higher Reynolds numbers that exceed the Stokes limit.

For sedimentation under gravity, we can replace $\omega^2 r$ by g in Eq. (6.9) (the drag force on a biological particle settling under gravity can be described by Stokes law, although you may wish to convince yourself of this). Figure 6.5 demonstrates the dependence of v_g (the velocity of a particle settling under the influence of gravity) of a spherical particle in water at 10°C on the particle diameter, d_p, and the density ratio (ρ_p/ρ_m). Although this plot applies to gravity driven settling, plots of similar shape can be generated for centrifugal sedimentation.

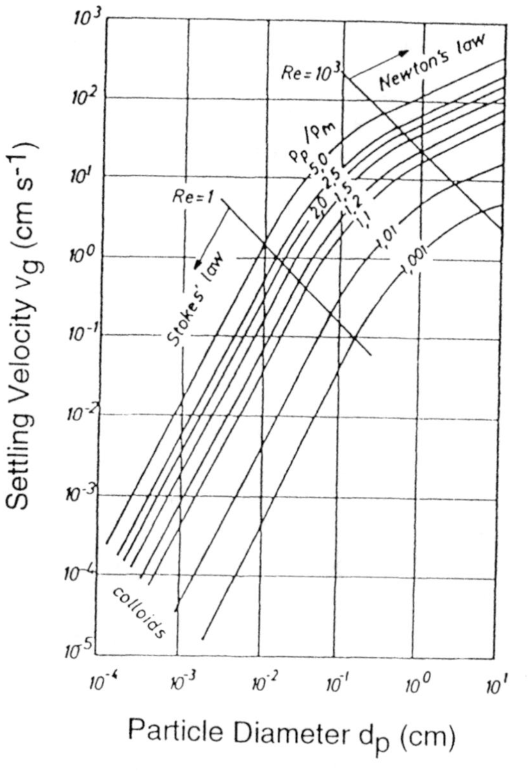

Figure 6.5. Settling velocities of spherical particles in water at 10 °C. [From G.M. Fair et al., Water and Wastewater Engineering, Vol. 2: Water Purification and Wastewater Treatment and Disposal, John Wiley & Sons, NY, (1968).]

The above analysis is applicable for dilute suspensions; however, in concentrated slurries, particle-particle interactions lead to so-called hindered settling. In this case, the sedimentation velocity can be modified by an empirical expression of the form:

$$v_h = v_c(1-\phi_v)^\alpha \tag{6.12}$$

where v_h is the hindered sedimentation velocity, ϕ_v is the volume fraction of particles in suspension, and α is an empirical geometric factor. The geometric factor, α, has a value of about 4.6 for monosized, rigid, spherical particles, and varies from about 10 to 100 for non-rigid, non-spherical particles.

One way to relate sedimentation velocities to the operation and scale up of centrifuges is through the use of Σ (sigma) factors. These factors relate the liquid flow rate through the centrifuge, Q, to the sedimentation velocity of the particle. The general equation for flow in the Stokes region is

$$Q = v_c\left(\frac{g}{\omega^2 r}\right)[\Sigma] \tag{6.13}$$

where Σ depends on the type of centrifuge (note that Σ has units of $[\text{length}]^2$). An expression for Σ can be obtained directly from the equation describing the trajectory of captured particles traveling the greatest distance in the centrifuge.[1] For a tubular bowl centrifuge, such particles enter at $r = R_1$ and traverse the full length of the unit before reaching $r = R_0$ at $z = l$ (Figure 6.6). The volumetric flow rate, Q, can thus be expressed in terms of the particle velocity, v_c, and centrifuge characteristics (i.e., l, R_0, R_1, and ω). Beginning with the definition of Q,

$$Q = \pi(R_0^2 - R_1^2)\frac{dz}{dt}$$

and the equation for particle movement in the r direction

$$\frac{dr}{dt} = v_c = v_g\left(\frac{\omega^2 r}{g}\right)$$

we can relate Q to the particle trajectory dr/dz. Integrating this expression for the most-traveled particles and assuming $R_1 \cong R_0 \cong R$, we obtain

$$Q = v_c\left(\frac{g}{\omega^2 r}\right)\left[\frac{2\pi l R^2\omega^2}{g}\right]$$

which gives

$$\Sigma = \left[\frac{2\pi l R^2\omega^2}{g}\right] \tag{6.14}$$

(1) P.A. Belter, E.L. Cussler, and W.-S. Hu, *Bioseparation--Downstream Processing for Biotechnology* (John Wiley & Sons, N.Y., 1988), Chapter 3.

The longest distance particles can travel in a disc bowl centrifuge before reaching the disc surface is from $x = y = 0$ to $x = (R_0 - R_1)/\sin\Theta$ and $y = l$. For these hard-to-trap particles, Σ is found to be

$$\Sigma = \left[\frac{2\pi n \omega^2}{3g} (R_0^3 - R_1^3) \cot\Theta \right] \tag{6.15}$$

where n is the number of discs and Θ is the angle at which the discs are tilted from the vertical. The aim is to determine the required Σ value that satisfies the process requirements of v_c and Q. Alternatively, the centrifugation can be analyzed in terms of v_g by replacing $v_c(g/\omega^2 r)$ with v_g in Eqn. (6.13).

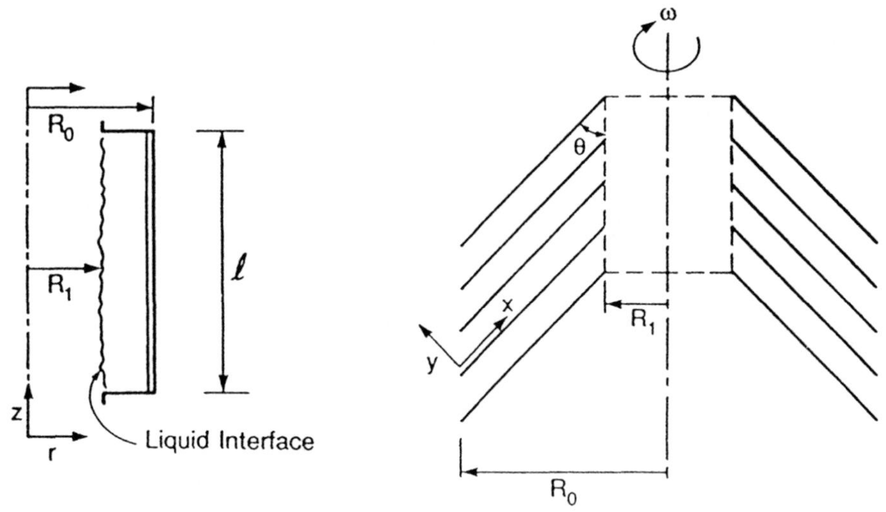

Figure 6.6. Simplified schematics of a tubular bowl centrifuge (left) and a disc centrifuge (right). [Adapted from P.A. Belter, E.L. Cussler, and W.-S. Hu, Bioseparations--Downstream Processing for Biotechnology, John Wiley & Sons, N.Y., (1988).]

6.2.2 Conventional Filtration

Filtration, like fermentation, has evolved from an ancient art into a powerful tool. A workhorse of many industries, filtration is probably the most common means for separating solids. Filtration offers many advantages for separating solids from liquids, as it separates mainly on the basis of a single physical parameter--the size of the solute. Moreover, filtration can be very cost effective; compared to centrifugation it consumes less energy and requires considerably less capital investment. Versatility is another advantage: some filtration membranes exclude solutes as small as 10 Å or so in diameter, whereas other membranes allow passage of particles in the hundred-micron (million-Å) size range. In addition, many different biocompatible filter materials are available.

Filtration processes are generally categorized by the pore size and retention characteristics of the filter (or membrane). However, differentiation among various filtration techniques is somewhat arbitrary, and different definitions can be found in the literature. One common classification scheme is summarized in Table 6.5. Another important characteristic of different filtration techniques is the flow pattern of the feed stream. For example, in *conventional* or *ordinary filtration*, the flow is usually perpendicular to the filter surface and the filtered solids accumulate as a cake on the filter (filtrations in which the feed stream flows perpendicularly to the filter are generally known as *dead-end filtrations*). The deposited cake can then be removed by a variety of methods. As evidenced by the size range shown in Table 6.5, biomass, amorphous precipitates, and crystalline products are among the solid materials effectively removed by conventional filtration.

Table 6.5. *General classes of membrane filters and their properties.*

	Pore size (microns)	Retains	Materials of Construction	Applications
Microporous	0.1-10	Bacteria, cell debris	Cellulose esters, polyvinylidene fluoride, polycarbonate	Sterile filtration, cell processing
Ultra-filtration	0.001-0.1	Macrosolutes (e.g., viruses, pyrogens, proteins, peptides)	Polysulfone, cellulosics	Cell and macrosolute processing, pyrogen removal
Reverse Osmosis	0.0005-0.001	Ions	Cellulose acetate, polyamides, polysulfone	Water purification, concentration of small organics

In *microfiltration* and *ultrafiltration*, build-up of solid on the membrane is minimized by vigorous mechanical stirring of the *upstream* liquid, or by employing *cross flow* or *tangential flow*. The term upstream refers to the side of the retentate, whereas *downstream* refers to the side of the filtrate. In cross flow filtration the feed stream runs parallel to the membrane surface and only a small fraction of the liquid permeates the membrane during a single pass. Further discussion of tangential flow filtration is deferred to Section 6.3.2.

Filtration Equipment

The most common apparatus for large-scale dead-end filtration is the rotary vacuum filter (Figure 6.7). Commonly employed in the fermentation industry, the rotary vacuum filter consists of a large, perforated, horizontal drum, the inside of which is under partial vacuum. The drum is partially immersed in a bath that contains a filter aid, an inert, microporous material derived from diatomaceous earth or from perlite, a processed volcanic rock. As the drum rotates through the bath, the vacuum draws liquid through the filter and solids are retained on the cylinder surface (the filter is thus *precoated*). The bath is then filled with product-containing broth to which filter aid has been added. Rotation of the precoated filter and maintenance of the vacuum results in the build up of biomass on the precoat, just as a cake would accumulate on a laboratory filter. As the cake rotates out of the broth, it is washed, dewatered, and removed by continuous scraping.

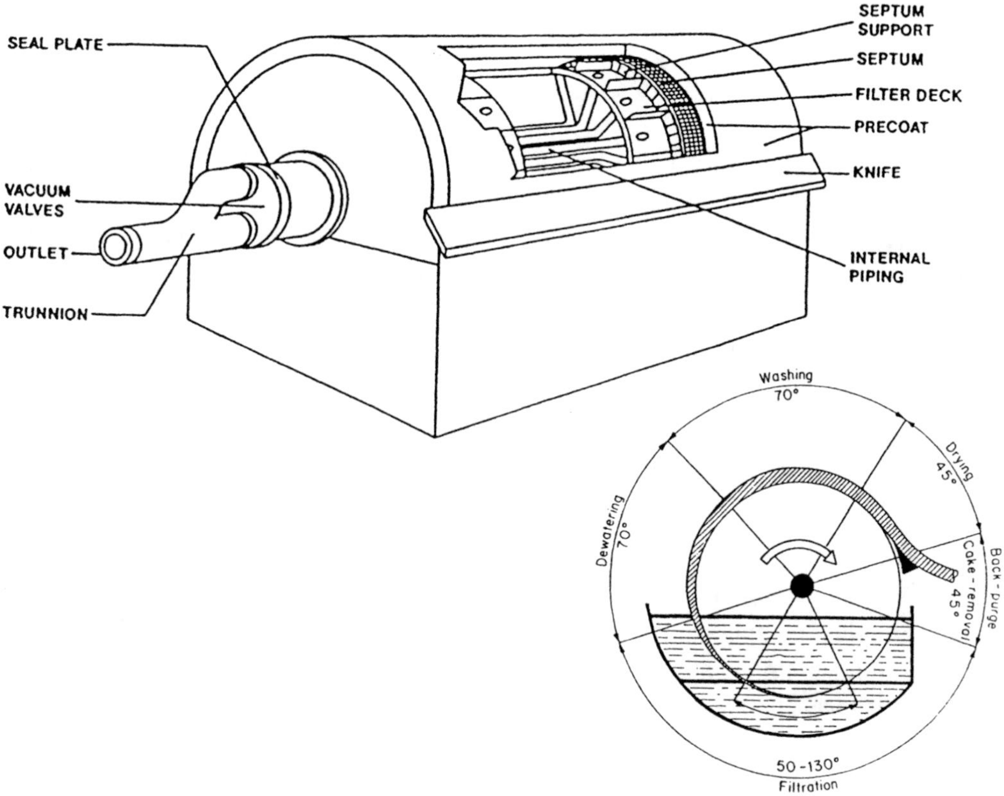

Figure 6.7. *Rotary vacuum filter, including an end-view showing precoat and removal of filtrate (inset). [Adapted from T.H. Meltzer, Filtration in the Pharmaceutical Industry Marcel Dekker, Inc., NY, (1987), p. 22, and (inset) G. Schmidt-Kastner and C.F. Gölker, in Basic Biotechnology, J. Bu'lock and B. Kristiansen, eds. Academic Press, NY, (1987).]*

The Behavior of Filtration Devices

The basic equation for filtration can be derived readily from Darcy's law, which relates the flow rate through a packed bed of solids to the pressure drop driving the flow

$$u = \frac{k\Delta P}{\mu l} \tag{6.16}$$

where u is the velocity of the liquid, k is a proportionality constant (often called the Darcy's law permeability), ΔP is the pressure drop across the bed of thickness l, and μ is the viscosity of the liquid. If the filtration is performed in batch mode, the velocity can also be described in terms of the changing filtrate volume V (*filtrate* or *permeate* refers to the liquid that passes through the filter):

$$u = \frac{1}{A}\frac{dV}{dt} \tag{6.17}$$

Combining Eqs. (6.16) and (6.17) gives

$$\frac{1}{A}\frac{dV}{dt} = \frac{k\Delta P}{\mu l} \tag{6.18}$$

Not surprisingly, the right-hand side of Eq. (6.18) consists of a driving force, ΔP , divided by a resistance to flow, l/k. This resistance term can be rewritten to include explicit contributions from both the filter and the accumulated cake, giving

$$\frac{1}{A}\frac{dV}{dt} = \frac{\Delta P}{\mu(R_M + R_C)} \tag{6.19}$$

where R_M and R_C are the resistances of the membrane (filter) and cake, respectively. R_M is a constant that depends on the type of filter used, whereas R_C will vary with the volume of filtrate V. The resistance of a compressible cake can be expressed as

$$R_c = \alpha\rho_0\left(\frac{V}{A}\right) \tag{6.20}$$

where α is the specific cake resistance (in units of length/mass) and ρ_0 is the mass of cake solids per volume of filtrate. The specific cake resistance α for a particular system can be determined by inserting Eq. (6.20) into Eqn (6.19) and integrating from t = V = 0 to t = t and V = V to obtain

$$\left(\frac{At}{V}\right) = \left(\frac{\mu\alpha\rho_0}{2\Delta P}\right)\left(\frac{V}{A}\right) + \left(\frac{\mu R_M}{\Delta P}\right) \tag{6.21}$$

A plot of (At/V) versus (V/A) for a given pressure drop Δp will thus provide α and R_M.

On the other hand, if the cake is compressible, as cakes comprised of biological materials typically are, the cake resistance varies with the pressure raised to the s power:

$$\alpha = \alpha'(\Delta P)^s \tag{6.22}$$

Values of s encountered in practice range from 0.1 to 0.8.

6.2.3 Cell Disruption

If the desired product is an intracellular substance, such as an intracellular enzyme, cell disruption is an important early step in product recovery. Methods used for disrupting microorganisms can be classified as physical (mechanical), chemical, or biological (Figure 6.8). Physical methods can employ shear (e.g., passage through a homogenizer under high pressure), agitation with abrasives (grinding in a ball mill with glass or metal beads), and repeated freezing and thawing (freeze-thaw). Non-mechanical methods include treatment with alkali, detergents, or organic solvents, osmotic shock, and enzymatic digestion. The effectiveness of a particular disruption technique is usually assessed in terms of the degree of cell breakage and/or the level of enzyme activity recovered in the disrupted suspension. Of the wide range of techniques available, not all are appropriate for large-scale application. For example, although the use of ultrasound is popular in the laboratory, the large amount of energy required and problems with heat removal generally prevent this procedure from being used on an industrial scale.

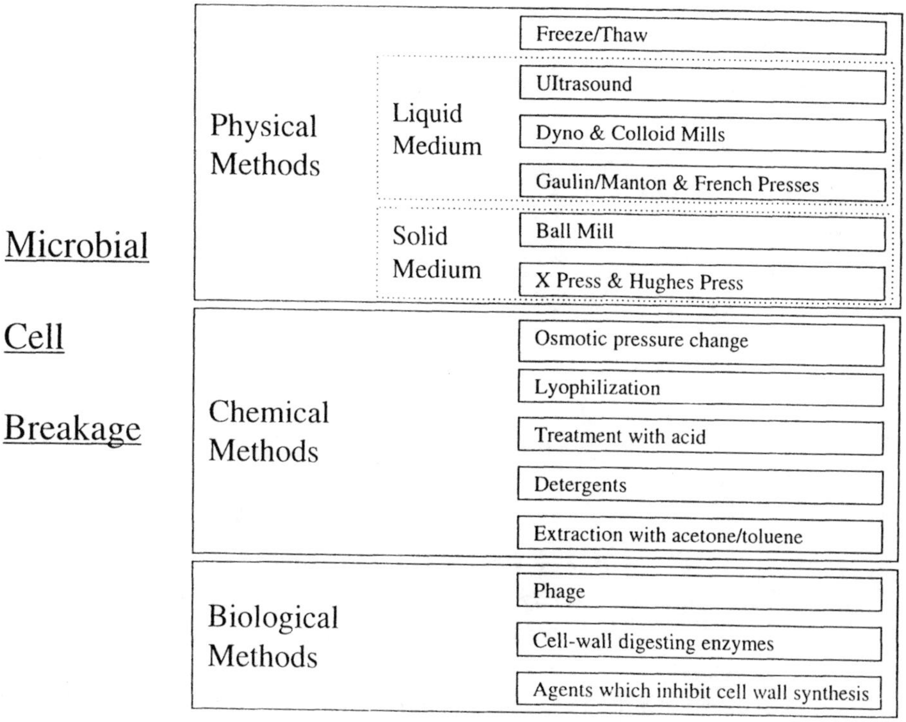

Figure 6.8. *Methods for disrupting cells. [From W. Crueger and A. Crueger, Biotechnology: A Textbook of Industrial Microbiology (Sinauer, New York, 1982).]*

Physical Methods

Exposure of cells to high liquid shear rates by passing cells through a restricted orifice under high pressure (Figure 6.9) are generally favored for the large scale disruption of microbial cells. In addition to the orifice valve, the equipment used (usually referred to as a homogenizer) consists of an air-driven or motor-driven pump capable of attaining pressures up to 200 MPa (about 2000 atm). Disintegration of cells is due primarily to hydrodynamic shear, with the magnitude of the pressure drop across the orifice being the most important factor. Cell disruption in an industrial homogenizer is considered further in Problem 6.1.

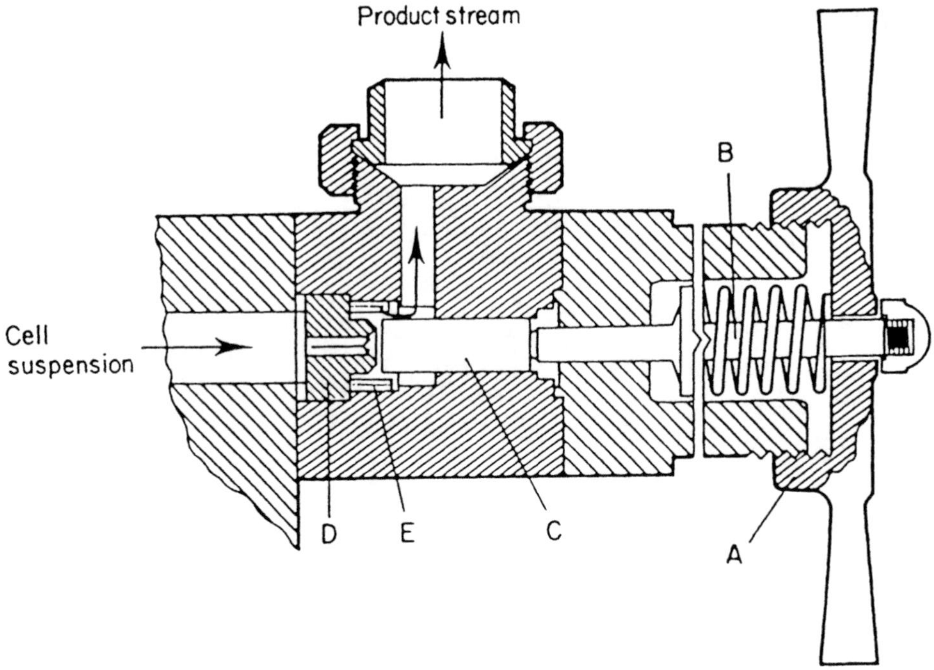

Figure 6.9. *Details of homogenizer valve assembly. A, handwheel; B, rod for valve adjustment; C, valve; D, valve seat; E, impact ring. [From G. Schmidt-Kastner and C.F. Gölker, in Basic Biotechnology, J. Bu'lock and B. Kristiansen, eds. (Academic Press, NY, 1987).]*

Another powerful cell disruption technique involves rapid agitation of a microbial cell supension with glass beads or similar abrasives. Continous processing can be achieved with a flow-through agitator mill (such as the Dyno-Mill shown in Figure 6.10). The cell suspension is pumped though a chamber containing the beads and several disc impellers, and the product exits through an orifice small enough to retain the beads without impeding the cell mash. Factors governing the performance of an agitator mill include impeller speed, flow rate, cell density, and bead size. Laboratory studies of agitator mills have shown that enzyme release can be expressed as an approximately first-order process.

Chemical and Biochemical Methods

Cells can also be broken by chemical or enzymatic means. For example, under the appropriate conditions of pH, ionic strength, and temperature, ionic detergents (sodium dodecyl sulphate, cetyltrimethyl ammonium bromide) or non-ionic detergents (Tween, Triton X-100) can effectively lyse cells. Alternatively, lytic enzymes, such as hen eggwhite lysozyme, microbial glucanases, and various proteases, can attack the cell walls of certain cells and thus effect gentle disruption. Organic solvents (acetone, ethanol, propanol, toluene) have also proven useful in some cases, particularly in the isolation of enzymes from yeast cells. However, when using an additive, consideration must be given to how its presence will affect subsequent purification steps. For example, salt precipitation and hydrophobic chromatography cannot be performed in the presence of detergents. Moreover, detergents, organic solvents, and proteases may denature or degrade the product of interest. Thus, when choosing a method of cell disruption (or any individual recovery step), the potential impact of the method on the overall purification scheme must be kept in mind.

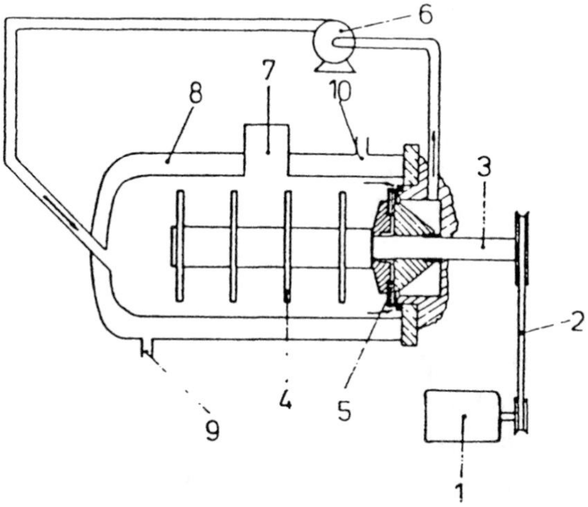

Figure 6.10. *Dyno-Mill agitator mill. 1, drive motor; 2, variable V-belt drive; 3, shaft; 4, agitator discs; 5, microseparator; 6, recirculating pump; 7, temperature measuring pocket; 8, temperature jacket; 9, temperature jacket inlet; 10, temperature jacket outlet. [From F. Marffy and M.-R. Kula, Biotechnol. Bioeng., 16, 623 (1974).]*

6.3 Initial Isolation and Concentration

6.3.1 Extraction

Solvent Extraction

Extraction of lipophilic substances with water-immiscible organic solvents is a well-established separation process, or unit operation, of the chemicals industry. Extraction also plays a major role in the isolation of some biologicals, particularly antibiotics. A typical antibiotic extraction involves transfer of the solute from a clarified fermentation broth into an organic phase, followed by reextraction of the concentrated product into aqueous buffer. The two-step process thus combines product concentration with purification. Final recovery is often achieved by precipitation, crystallization, or evaporation. A major advantage of solvent extraction is that it can be performed quickly on a large scale; for example, it takes about 90 seconds to cycle penicillin through a two-step, aqueous-to-organic-to-aqueous extraction process.

An alternative to conventional extraction is *reactive extraction*. In this case, extraction of a charged species is facilitated by a carrier species present in the extractant phase. For example, a penicillin acid anion, P^-, can be removed from the aqueous phase by displacing the anion X^- of an ammonium salt dissolved in the organic phase (typical of an ion exchange reaction):

$$NR_4^+X^- \text{ (org)} + P^- \text{ (aq)} \leftrightarrow NR_4^+P^- \text{ (org)} + X^- \text{ (aq)}$$

Reactive extraction can also be effected by a free amine, A, in which case the acid anion is extracted into the organic phase along with a proton (ion pair extraction):

$$A \text{ (org)} + P^- \text{ (aq)} + H^+ \text{ (aq)} \leftrightarrow AHP \text{ (org)}$$

The general details of solvent extraction and the design of extraction processes are covered extensively in standard texts[2] ; therefore, our coverage of these topics is brief. We consider, however, a variation of conventional aqueous-organic extraction that is emerging as an important method for protein recovery.

Aqueous Two-Phase Systems

A relatively new method for separating mixtures of biomolecules under mild conditions is aqueous two-phase extraction. This technique is based on the partitioning of biomolecules between the two liquid phases that arise when two incompatible polymers [e.g., polyethylene glycol (PEG) and dextran] are dissolved in water, with one polymer predominating in each phase. Similarly, two aqueous phases can result when a polymer is added to an aqueous salt solution above a certain concentration. Examples of aqueous two-phase systems are listed

---

(2) C.J. King, *Separation Processes* (McGraw-Hill, NY, 1980); P.A. Belter, E.L. Cussler, and W.-S. Hu, *Bioseparations--Downstream Processing for Biotechnology* (John Wiley & Sons, N.Y., 1988), Chapter 5.

in Table 6.6. When a mixture of biomolecules, e.g., a fermentation broth, cell-culture supernatant, or solution of lysed cells, is added to an aqueous two-phase system, each type of biomolecule partitions uniquely between the two phases. A wide range of biomolecules (proteins, lipids, nucleic acids, viruses, and whole cells) have been separated using this approach. A partial list of enzymes separated from cell homogenates by aqueous two-phase extraction is given in Table 6.7.

Table 6.6 *Two-phase systems formed by incompatibility of neutral polymers in water.*

Polymer 1	Polymer 2
Polypropylene glycol	- dextran - hydroxypropyldextran - methoxypolyethylene glycol - polyethylene glycol - polyvinyl alcohol - polyvinylpyrrolidone
Polyethylene glycol	- polyvinyl alcohol - polyvinylpyrrolidone - dextran - Ficoll
Polyvinyl alcohol	- methylcellulose - hydroxypropyldextran - dextran
Polyvinylpyrrolidone	- methylcellulose - hydroxypropyldextran - dextran
Methylcellulose	- hydroxypropyldextran - dextran
Ethylhydroxyethylcellulose	- dextran
Hydroxypropyldextran	- dextran
Ficoll	- dextran

Advantages of aqueous two-phase systems include: 1) high biocompatibility and low interfacial tension (since each phase is usually 75-95 wt% water), thereby minimizing product degradation often observed in nonaqueous systems; 2) good resolution and yields, which can be dramatically improved, if needed, by covalently binding affinity ligands to the polymers (resulting in *affinity partitioning*); 3) high capacity; 4) linear scale-up of up to 2 x 10^4 times that of lab scale; and 5) direct use of available chemical engineering technology (liquid-liquid extraction equipment) for industrial-scale separations.

Table 6.7 Enzymes extracted from cell homogenates of various microorganisms[3].

Enzyme	Organism	Two-phase system	Purification factor	Yield (%)
Catalase	*Candida boidinii*	PEG dextran	-	81
Formaldehyde dehydrogenase		PEG dextran	-	94
Formate dehydrogenase		PEG salt	1.5	94
Isopropanol dehydrogenase		PEG salt	2.6	98
α - Glucosidase	*Saccharomyces cerevisiae*	PEG salt	3.2	95
Glucose-6-phosphate dehydrogenase		PEG salt	1.8	91
Hexokinase		PEG salt	1.6	92
Glucose isomerase	*Streptomyces* species	PEG salt	2.5	86
Leucine dehydrogenase	*Bacillus* species	PEG salt	1.3	98
Alanine dehydrogenase		PEG salt	2.6	98
Glucose dehydrogenase		PEG salt	2.3	95
β -Glucosidase	*Lactobacillus* species	PEG salt	2.4	98
D-Lactate dehydrogenase		PEG salt	1.5	95
L-Hydroxy isocaproate dehydrogenase		PEG salt	1.6	98
Glucose-6-phosphate dehydrogenase	*Leuconostoc mesenteroides*	PEG salt	7.3	94
Fumarase	*Brevibacterium* species	PEG salt	7.5	83
Phenylalanine dehydrogenase		PEG salt	1.5	99
Aspartate-β -decarboxylase	*Pseudomonas dacunhae*	PEG salt	1.1	100
Fumarase	*Escherichia coli*	PEG salt	3.4	93
Aspartase		PEG salt	6.6	96
Penicillin acylase		PEG salt	8.2	90
Pullulanase	*Klebsiella pneumoniae*	PEG dextran	2.0	91
1,4-α -Glucanphosphorylase		PEG dextran	1.2	85

(3) All data from the Department of Enzyme Technology, GBF (H. Hustedt, K.H. Kroner, U. Menge, H. Schutte, published and unpublished results).

These practical advantages, along with the increasing desire to optimize extractive separations of biomolecules, have resulted in the development of both constitutive and predictive thermodynamic models of aqueous two-phase systems. The objective of these models is to describe both the phase separation and the partition coefficients of different biomolecules. The partition coefficient K_p of a macromolecule or particle may be represented as the sum of several contributions:

$$lnK_p = lnK_{el} + \ln K_{hphob} + \ln K_{hphil} + \ln K_{conf} + \ln K_{lig} \quad (6.23)$$

where *el*, *hphob*, *hphil*, *conf*, and *lig* denote electric, hydrophobic, hydrophilic, conformational, and ligand interactions. While these individual contributions cannot be determined theoretically or experimentally, they provide an indication that partitioning of biomolecules in aqueous two-phase systems is governed by many factors, including the physicochemical properties of the biomolecule, the nature of the phase polymers, the nature and concentration of the added salts, and the pH and temperature of the system.

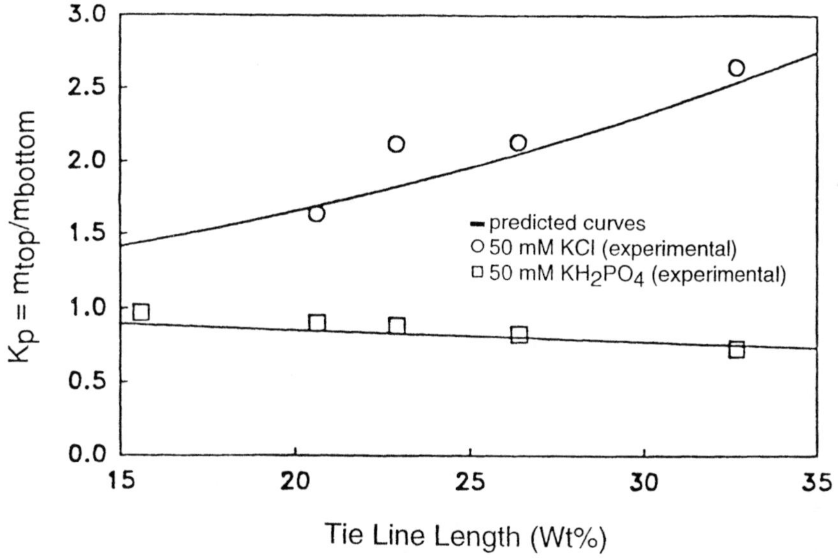

Figure 6.11. *Experimental and predicted protein partition coefficient vs. tie line length for lysozyme in PEG 3350/dextran T-70/salt-water system. [From R.S. King et al., AIChE J., 34, 1585 (1988).]*

Phase behavior and protein partition coefficients in two-phase systems can be predicted by a molecular-thermodynamic model, based on the osmotic virial equation. The virial coefficients contain contributions from electrostatic, hydrophobic and excluded volume forces between the species in solution. The model is developed in Example 6.1, with the result that the protein partition coefficient is given by

$$\ln K_p = \ln \frac{m_p'}{m_p''} = a_{2p}(m_2'' - m_2') + a_{3p}(m_3'' - m_3') + \frac{z_p F(\Phi'' - \Phi')}{RT} \quad (6.24)$$

where a_{ij} = second virial coefficient (kg/mol) for interaction between i and j in a solvent, m_i = molal concentration (mol/kg solvent), z_p = net surface charge (valence) of the protein, Φ = electrical potential relative to some reference (mV), F = Faraday constant, R = gas constant, T = temperature (K), the superscripts ′ and ″ denote lighter and heavier phase, respectively, and the subscripts 1, 2, 3, and p denote solvent, polymer, polymer, and protein, respectively. Encouraging agreement between experimental and predicted protein partition coefficients is shown in Figure 6.11. Partition coefficients are plotted against the tie-line length, shown in Figure 6.12. The tie-line length is a useful parameter for correlating properties of aqueous two-phase systems; it is representative of the composition difference between the two liquid phases at equilibrium.

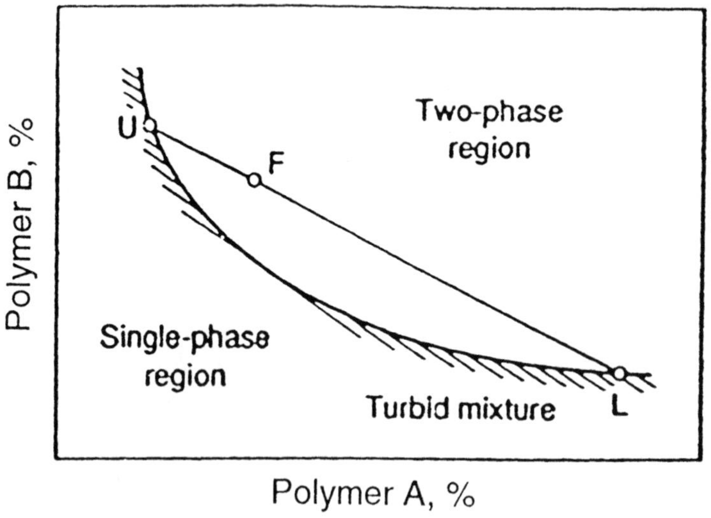

Figure 6.12. Phase diagram of a mixture of two polymers and water. A system with total composition F separates into upper phase U and lower phase L, which are connected on the diagram by the tie line UL.

Example: Determination of protein partition coefficients for aqueous two-phase systems.

Phase diagrams and protein partition coefficients in aqueous two-phase systems can be obtained from application of expression for the chemical potentials of all species in solution. We begin with general expressions for the chemical potentials of each component:

$$\mu_i(T,P,x_i) = \mu_i^0(T,P=0) + \bar{v}_{m_i}P + RT\ln\gamma_i X_i$$

where $\bar{v}_{m_i}$ is the average molar volume of the pure component (from pressure 0 to P bar), γ_i is the activity coefficient, X_i is the mole fraction, and the standard state is defined as that of pure component at P = 0 bar. At constant temperature and pressure, the equilibrium condition for the chemical potentials is expressed by the Gibbs-Duhem equation:

$$SdT - VdP + \sum_i n_i d\mu_i = 0$$

which reduces to

$$n_1 d\mu_1 + n_2 d\mu_2 + n_3 d\mu_3 = 0$$

where n_i is the number of moles of component i. Substituting for the chemical potentials and differentiating with respect to the mole fraction of solvent (X_1) gives

$$n_1\frac{\partial}{\partial X_1}(\ln\gamma_1 X_1) + n_2\frac{\partial}{\partial X_1}(\ln\gamma_2 X_2) + n_3\frac{\partial}{\partial X_1}(\ln\gamma_3 X_3) = 0$$

$$n_1\frac{\partial\ln\gamma_1}{\partial n_1} + n_2\frac{\partial\ln\gamma_2}{\partial n_1} + n_3\frac{\partial\ln\gamma_3}{\partial n_1} = 0$$

the multicomponent Gibbs-Duhem equation. The subscripts denote the following:

1 = solvent (water), molecular weight MW_1
2 = polymer A (e.g., PEG) molecular weight MW_2
3 = polymer B (e.g., dextran) molecular weight MW_3

In terms of molalities, m_i (moles/kg solvent), the above equation becomes

$$\frac{1000}{MW_1}\partial\ln\gamma_1 + m_2\partial\ln\gamma_2 + m_3\partial\ln\gamma_3 = 0$$

where MW_1 is the molecular weight of the solvent. Taking the derivatives of the above equation with respect to each molality m_j (j = 2,3,) yields expressions of the form

$$\frac{1000}{MW_1}\frac{\partial\ln\gamma_1}{\partial m_j} + m_2\frac{\partial\ln\gamma_2}{\partial m_j} + m_3\frac{\partial\ln\gamma_3}{\partial m_j} = 0$$

Edmond and Ogston[4] has shown that the activity coefficients γ_2, γ_3 may be expressed as a series of terms of the form $bm_2^p m_3^r$. Noting that

(4) Edmond, E. and Ogston, A.G., "An Approach to the Study of Phase Separation in Ternary Aqueous Systems", Biochem. J. **109** 569 (1968).

$$\frac{\partial \ln \gamma_3}{\partial m_2} = \frac{\partial \ln \gamma_2}{\partial m_3}$$

and that the activity coefficients must tend to unity at low molalities, the following expressions can be obtained

$$\ln \gamma_2 = a_{22}m_2 + a_{23}m_3$$

$$\ln \gamma_3 = a_{23}m_2 + a_{33}m_3$$

Noting that

$$\Delta\mu_i = \mu_i - \mu_i^o = RT \ln m_i \gamma_i$$

we substitute expressions for the activity coefficients and obtain

$$\Delta\mu_2 = RT(\ln m_2 + a_{22}m_2 + a_{23}m_3)$$

$$\Delta\mu_3 = RT(\ln m_3 + a_{23}m_2 + a_{33}m_3)$$

Rewriting the Gibbs-Duhem equation in terms of molalities and taking partial derivatives with respect to m_2 and m_3

$$\frac{1000}{MW_1}\frac{\partial \mu_1}{\partial m_2} + m_2\frac{\partial \mu_2}{\partial m_2} + m_3\frac{\partial \mu_3}{\partial m_2} = 0$$

$$\frac{1000}{MW_1}\frac{\partial \mu_1}{\partial m_3} + m_2\frac{\partial \mu_2}{\partial m_3} + m_3\frac{\partial \mu_3}{\partial m_3} = 0$$

The expressions for the chemical potentials can be substituted in the above equations and by integration of the two resulting equations and evalutating the constants of integration, we obtain the following expression for the chemical potential of the solvent

$$\Delta\mu_1 = -\frac{RTMW_1}{1000}\left(m_2 + m_3 + \frac{a_{22}}{2}m_2^2 + \frac{a_{33}}{2}m_3^2 + a_{23}m_2m_3\right)$$

$$\Delta\mu_2 = RT(\ln m_2 + a_{22}m_2 + a_{23}m_3)$$

$$\Delta\mu_3 = RT(\ln m_3 + a_{23}m_2 + a_{33}m_3)$$

The interaction coefficients a_{ij} (liters/mol) are related to the traditional virial coefficients A_{ij} (ml·mol/gm^2) via:

$$2A_{22} = 1000\frac{a_{22}}{MW_2^2}$$

$$2A_{23} = 1000\frac{a_{23}}{MW_2MW_3}$$

$$2A_{33} = 1000\frac{a_{33}}{MW_3^2}$$

Phase diagrams can now be obtained by equating the chemical potentials of each species in the top and bottom phases. This gives the three equations

$$\mu_1' = \mu_1'', \quad \mu_2' = \mu_2'', \quad \mu_3' = \mu_3''$$

There are four unknowns, $\mu_2', \mu_2'', \mu_3', \mu_3''$ and three equations. By specifying the molality of species 2 or 3 in one of the phases, the other molalities can be determined. Repeating this procedure yields the phase diagram. If the solution to the above three equations gives identical values in each phase, then only one phase exists for that selected value of the specified component.

When a protein is placed in the two-phase system, the expressions for the chemical potentials can be extended viz:

$$\Delta\mu_1 = -\frac{RTMW_1}{1000}\left(m_2 + m_3 + m_p + \frac{a_{22}}{2}m_2^2 + \frac{a_{33}}{2}m_3^2 + \frac{a_{pp}}{2}m_p^2 + a_{23}m_2m_3 + a_{2p}m_2m_p + a_{3p}m_3m_p\right)$$

$$\Delta\mu_2 = RT(\ln m_2 + a_{22}m_2 + a_{23}m_3 + a_{2p}m_p)$$

$$\Delta\mu_3 = RT(\ln m_3 + a_{23}m_2 + a_{33}m_3 + a_{3p}m_p)$$

$$\Delta\mu_p = RT(\ln m_p + a_{2p}m_2 + a_{3p}m_3 + a_{pp}m_p)$$

If there is uneven partitioning of salts between the two phases, an electric potential develops between the phases, which affects the partitioning of the protein between the phases (the standard state for the chemical potentials will also be different in each phase). This can be accounted for most simply by adding the term $z_pF\Phi$ to the expression for the chemical potential of the protein, and assuming the standard states are the same.[5] In solutions dilute in the protein, $m_p<m_2$ and $m_p<m_3$, and the protein partition coefficient is given by

$$\ln K_p = \ln\frac{m_p'}{m_p''} = a_{2p}(m_2'' - m_2') + a_{3p}(m_3'' - m_3') + \frac{z_pF(\Phi'' - \Phi')}{RT}$$

The phase equilibrium calculations and the protein partition coefficent can be determined from these equations as previously described, by equating the chemical potentials of the species in each phase. The virial coefficients may be readily determined from membrane osmometry (see Appendix A) or from low-angle laser light scattering.

In practice, aqueous two-phase extraction consists of three basic operations: 1) mixing of the phase components; 2) phase separation; and 3) enzyme recovery and polymer removal. Adequate mixing can be achieved in conventional agitated vessels, and equilibrium between the two phases is normally reached within a matter of minutes. Such rapid equilibration is due to the low interfacial tension of aqueous two-phase systems. Low interfacial tension

---

(5) A complete treatment of the electrostatic potential difference is given in Haynes, C.A., Benitez, F.S., Blanch, H.W., and J.M. Prausnitz, AIChE Journal, **39**(9), 1539 (1993).

also minimizes the energy input required to achieve rapid phase dispersal during mixing. Following the mixing step, phase separation can be carried out by centrifugation (as is often done for the removal of cell debris) or by settling under gravity.

Separation of protein and polymer, which usually entails separating the protein from PEG, can be accomplished in many ways. Some proteins can be removed directly from PEG-rich phases by ultrafiltration. Alternatively, adding salt to a protein-containing, PEG-rich phase will lead to a new phase, and reextraction or concentration of the protein into the salt phase is often possible. Salt can be then be removed by common unit operations such as ultrafiltration, dialysis, or gel-filtration chromatography. Adsorption of the protein to ion exchangers or other suitable adsorbents, followed by a wash step to remove phase-forming polymers, is also feasible.

The largest added expense associated with aqueous extractions is the cost of the salts and polymers, particularly dextran. Dextran recovery is therefore an important process objective. Nonetheless, compared to many other techniques, liquid-liquid separation processes can lead to major cost savings due to reductions in process time, labor requirements, and energy demand. Furthermore, the invariance of the protein partition coefficient with protein concentration (up to about 30 wt %) and process scale facilitates scale-up from laboratory data. Given these significant advantages, aqueous two-phase extraction has the potential to become a major unit operation in the biochemical industry.

6.3.2 Microfiltration and Ultrafiltration

Microfiltration retains particles 0.1 - 10 microns in diameter, for example, whole cells or cell debris. Common materials used for the manufacture of microfilters include paper, polymers [for example, nylon, cellulosics, polysulfones, polyethylenes, polytetrafluoreethylene (i.e., PTFE or Teflon)], and ceramics. In addition, most microporous membranes are *symmetric* or *isotropic*, that is, the membrane pores are the same size throughout the depth of the filter. On the other hand, *ultrafiltration membranes*, which exclude particles and macromolecules of 1,000 - 500,000 Daltons, are usually *asymmetric* or *anisotropic*. Anisotropic membranes consist of an extremely thin "skin" of homogeneous polymer supported upon a much thicker, spongy substructure. The pores of the skin layer are markedly smaller than the pores through the rest of the membrane. Consequently, the thin surface layer constitutes the major transport barrier and governs the filtration characteristics of the entire membrane.

Tangential Flow Equipment

To alleviate the build-up of material that can cause filter blocking and/or clogging during conventional filtration, liquid on the upstream side can be circulated tangential to or across the filter surface. This technique serves to sweep the membrane clean of deposited particles and is known as tangential flow filtration. A variety of membrane configurations are available for tangential flow filtration. Plate and frame devices, hollow fiber cartridges, and spiral-wound cartridges are common examples (Figure 6.15). Each design aims to combine high cross flow and a large filter area with other important features, such as ease of cleaning, repair, or scale-up. Which configuration is best depends on the particular situation, but in general, plate and frame systems are better suited for dirty feed streams but give lower fluxes per volume than spiral wound and hollow fiber modules.

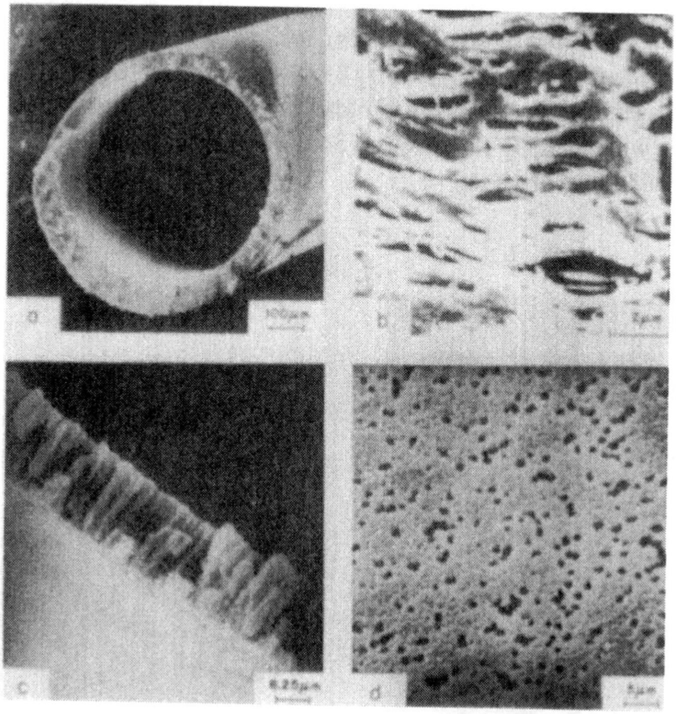

Figure 6.13. *Scanning electron micrographs of symmetric membranes. The membranes are (a) porous glass, (b) stretched polypropylene sheet, and the side (c) and plan (d) views of a polycarbonate sheet membrane. Reproduced from G. Belfort, "Membrane Separation Technology: An Overview," in Advanced Biochemical Engineering, H.R. Bungay and G. Belfort, eds., John Wiley & Sons, N.Y., (1987).*

An Analysis of Ultrafiltration

For dilute solutions under pressure on one side of an essentially homogeneous, isotropic membrane, the transmembrane fluxes at steady state can be approximated by the relations:

$$J_w \cong L_p(\Delta P - \sigma\Delta\pi) \tag{6.25}$$

$$J_s \cong P_s \Delta c_s + J_w(1-\sigma)\bar{c}_s \tag{6.26}$$

where J_w is the solvent (water) flux [cm^3 ($cm^2 \cdot s)^{-1}$], J_s is the solute flux [gm ($cm^2 \cdot s)^{-1}$], L_p is the *membrane permeability* for the solvent [cm^3 ($cm^2 \cdot s \cdot atm)^{-1}$], P_s is the membrane permeability for the solute (cm/s), ΔP is the hydrostatic pressure difference across the membrane (atm), $\Delta\pi$ is the osmotic pressure, σ is the *reflection coefficient* (not to be confused with the *rejection coefficient*, to be defined later), Δc_s is the solute-concentration difference across the membrane (gm/cm^3) (sometimes referred to as the solute-concentration difference between the *upstream* and *downstream* solutions), and $\bar{c}_s$ is the average concentration of solute in the upstream solution. The quantity $(1-\sigma)$ represents the fraction of the solvent flux carried by pores large enough to pass the solute. If the reflection coefficient is small ($\sigma \cong 0$), the membrane will be highly permeable to both solute and solvent; if σ is large ($\sigma \cong 1$), the membrane will reject all solute.

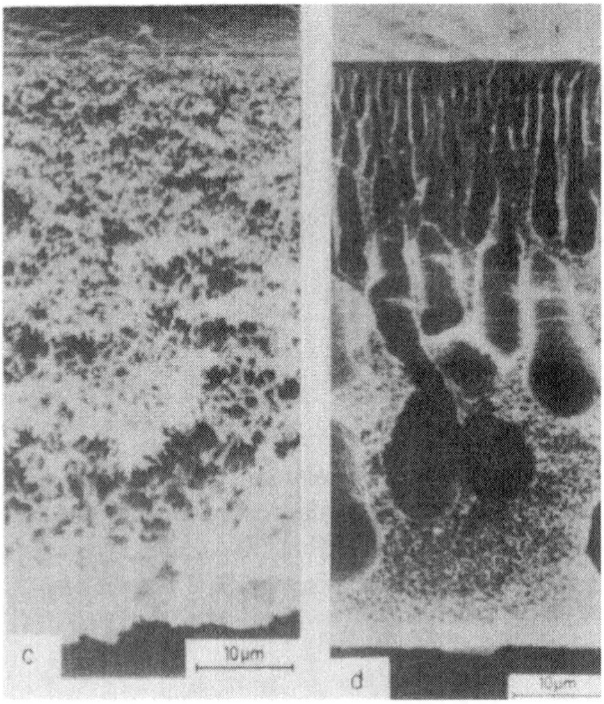

Figure 6.14. Scanning electron micrographs of asymmetric polysulfone (left) and RC100 composite (right) membranes. Reproduced from G. Belfort, "Membrane Separation Technology: An Overview," in Advanced Biochemical Engineering, H.R. Bungay and G. Belfort, eds., John Wiley & Sons, N.Y., (1987).]

The first term on the right hand side of Eq. 6.26 corresponds to the diffusive flux of solute through the membrane. The second term represents the "convective flux" or "coupled transport" of solute driven by the net flux of solvent (it is sometimes interpreted as a consequence of frictional drag between moving solvent molecules and solute molecules within the membrane). In addition, a high diffusive flux of solvent will result in momentum exchange between solvent and solute molecules, due to collisions, which could serve to increase the net flux of solute. Such momentum interchange would not be expected to cause a significant reduction in the diffusive flux of the more abundant solvent molecules, however, and no term accounting for coupled transport is included in Equation 6.25.

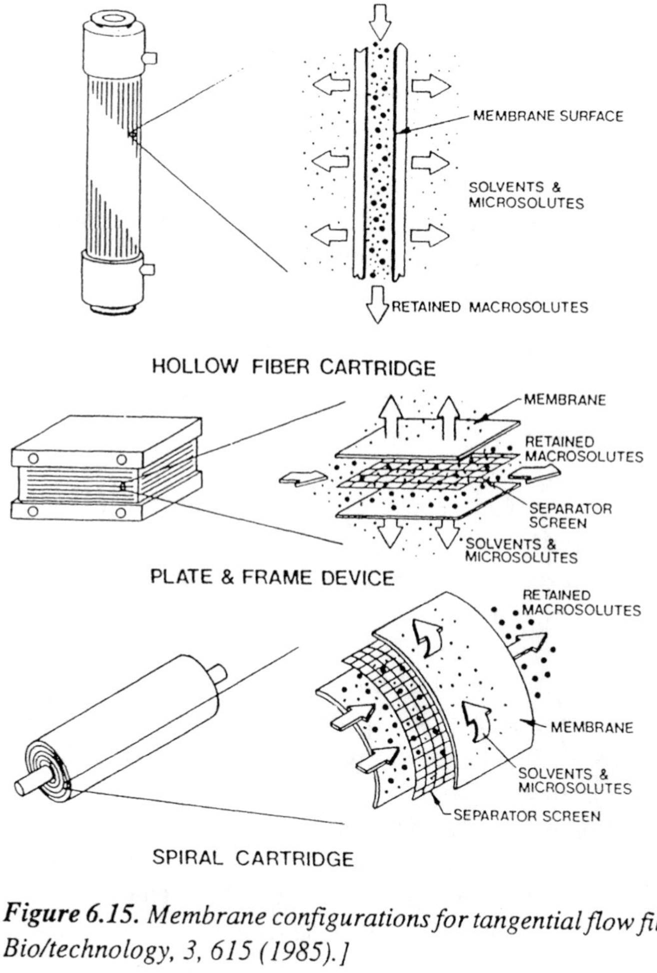

Figure 6.15. *Membrane configurations for tangential flow filtration. [From R.S. Tutunjian, Bio/technology, 3, 615 (1985).]*

From mass conservation, the solute flux, J_s, can also be written in the form

$$J_s = c_{2s} J_v \tag{6.27}$$

where c_{2s} is the concentration of permeate or ultrafiltrate (the concentration in the downstream solution), and J_v is the total flux of permeate (usually assumed to equal J_w for dilute solutions). Alternatively, J_s can be expressed in terms of c_{1s}, the concentration of solute in the upstream solution:

$$J_s = (1 - R) J_v c_{1s} \tag{6.28}$$

where R is the *rejection coefficient*, which is equal to the fraction of solute present in the upstream solution that is rejected by the membrane. In terms of concentrations, the rejection coefficient is defined as $R = (c_{1s} - c_{2s})/c_{1s}$.

Finally, we should note that for a purely diffusive-type membrane (for which J_w is negligible), the solute mass flux is

$$J_s = \left(\frac{K_s D_e}{l} \right) (c_{1s} - c_{2s}) \tag{6.29}$$

where K_s is the distribution coefficient of solute between the membrane and solution (assumed to be constant), D_e is the effective diffusivity of solute through the membrane, and l is the membrane thickness. Equation (6.29) models the membrane as a continuous "solvent" in which the solubility of solute is described by the distribution coefficient K_s. An alternative approach is to treat the membrane as a sieve with distinct pores and a specific porosity.

Concentration Polarization

During concentration, pressure exerted on the (upstream) solution in contact with the membrane causes solute to flow toward the membrane surface. Initially, the convective flux of solute to the membrane will exceed the rate at which solute passes through the membrane, resulting in accumulation of solute at the membrane, with the maximum solute level at the membrane surface. This phenomenon is known as *concentration polarization* (Figure 6.16). If the membrane is not completely impermeable to solute, a consequence of such polarization is solute leakage, or an anomalously low rejection efficiency. In addition, the high concentration of solute can increase the osmotic pressure difference and thus reduce the effective driving force for solvent transport through the membrane [Eq. (6.25)]. In many cases (particularly with macromolecular solutes), concentration polarization is the limiting factor governing flux rates; hence, cross-flow filtration or vigorous mechanical stirring of the upstream liquid is employed to alleviate its effects.

If the concentration of solute at the membrane surface is high enough, the polarization layer displays gel-like properties and *gel polarization* is said to occur. The layer of concentrate assumes the form of a "slime" or "cake" at the membrane surface, and this adherent layer presents a hydraulic barrier in series with the membrane. In fact, the gel usually contributes the dominant resistance to mass flow. Concentration levels at which gel formation takes place vary with the size, shape, and degree of solvation of the solute particles.

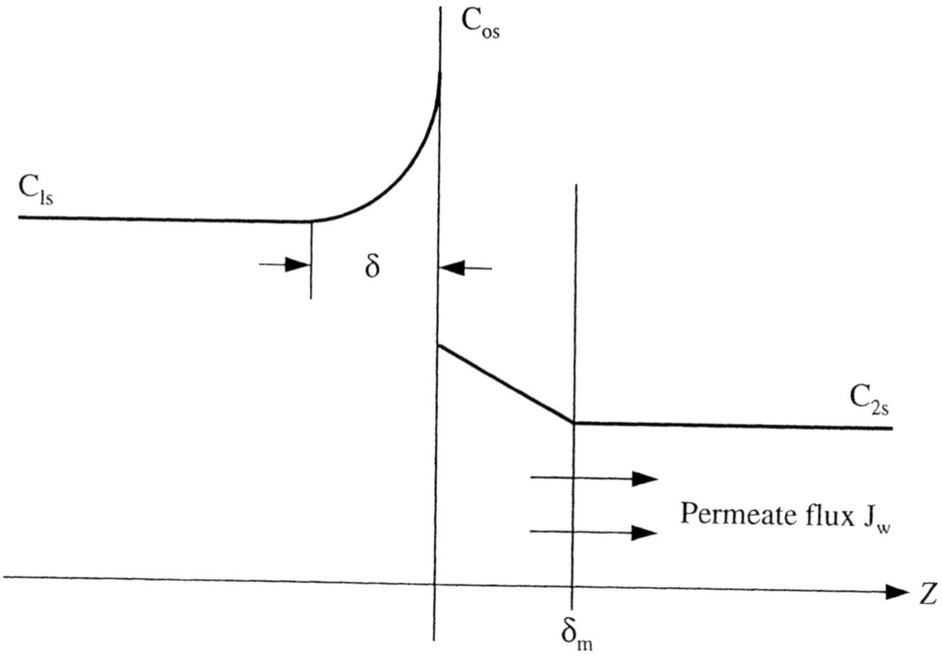

Figure 6.16. *Concentration gradient during concentration polarization.*

For proteins and nucleic acids, concentrations as high as 10-30 percent by weight are often required before gelation is observed. In contrast, for rigid-chain, solvated macromolecules like polysaccharides, concentrations below 1 percent by weight may be sufficient.

To analyze the effects of polarization on solvent flux, we recall that the volumetric flux of solvent, J_w, can be written as

$$J_w = L_p(\Delta P - \sigma\Delta\pi) \quad \Delta\pi = \mathrm{RT}c_{os} \tag{6.30}$$

where the osmotic pressure, $\Delta\pi$, has been expressed in terms of c_{os}, the concentration of solute at the surface of the membrane. Integration of a steady-state mass balance on solute in the polarization layer (Problem 7) leads to

$$J_w = \frac{D_s}{\delta}\ln\frac{c_{os} - c_{2s}}{c_{1s} - c_{2s}} \tag{6.31}$$

where D_s is the diffusivity of solute (assumed here to be independent of solute concentration). Furthermore, if R = 1, that is, if $c_{2s} = 0$, then

$$J_w = \frac{D_s}{\delta}\ln\frac{c_{os}}{c_{1s}} \tag{6.32}$$

This equation provides a relationship between the boundary layer thickness, δ, and the concentration of solute at the membrane surface, c_{os}. Note that if gel polarization occurs, $c_{os} = c_g$, the gel concentration of the macrosolute. The gel concentration, of course, places an upper limit on the solute concentration at the membrane surface.

At this point we should consider another subtle yet important distinction between concentration polarization and gel polarization. In the former, c_{os} is less than c_g, and c_{os}/c_{1s} will adjust itself to an imposed J_w. Nonetheless, the polarized region presents a resistance to mass transfer and can be modeled as a membrane in series with the original membrane. The resulting flux is given by

$$J_w = \frac{\Delta P}{(1/L_p) + R_P} \tag{6.33}$$

where $(1/L_p)$ is again the membrane resistance and R_P is the resistance of the polarized boundary layer. Increasing the transmembrane pressure drop will increase c_{os} and, consequently, R_p. Thus, concentration polarization causes the flux, J_w, to increase *less than proportionately* with increasing ΔP. This effect is shown in Figure 6.17 with ultrafiltration data obtained for bovine serum albumin in a thin-channel, laminar-flow cell. On the other hand, gel polarization results in a constant value of $c_{os} = c_g$. Equation (6.32) then predicts that the flux will be independent of the applied pressure. Such behavior has indeed been observed for many systems, particularly at high pressures, and is shown by the high-pressure, low-shear rate data of Figure 6.18.

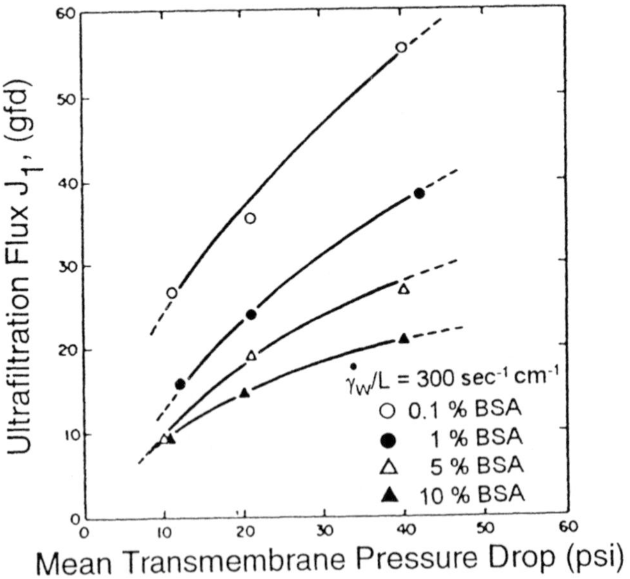

Figure 6.17. *Effect of pressure on ultrafiltration rates of solutions of bovine-serum albumin in a thin-channel recirculating system. [From W.F. Blatt et al., in Membrane Science and Technology. Industrial, Biological, and Waste Treatment Processes, Plenum Press, NY, (1970) p. 67.]*

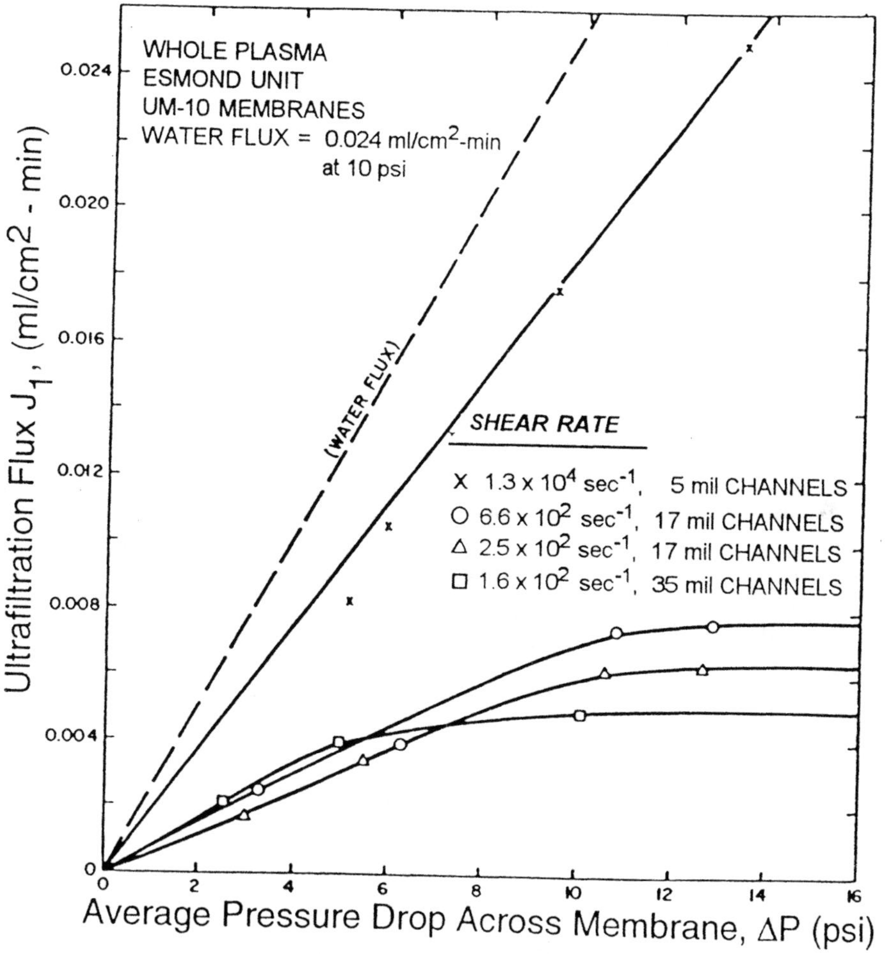

Figure 6.18. *Dependence of ultrafiltration rates on pressure for different shear rates. The data were obtained by recirculating human blood plasma through thin channels of triangular cross-section. [From W.F. Blatt et al., in Membrane Science and Technology. Industrial, Biological, and Waste Treatment Processes, Plenum Press, NY, (1970) p. 68.]*

Returning to Eq. (6.32), the leading term on the right hand side, D_s/δ, can be viewed as a mass transfer coefficient, k_s, which allows us to write

$$J_w = k_s \ln \frac{c_{os}}{c_{1s}} \tag{6.34}$$

Mass transfer coefficients for fluid flow in narrow channels conventionally are described by expressions of the form

$$\frac{k_s d_h}{D_s} = ARe^a Sc^{1/3} \tag{6.35}$$

where

d_h = equivalent diameter of the channel

$= 4\left(\dfrac{\text{cross-sectional area}}{\text{wetted perimeter}}\right)$

Re $= d_h u\rho / \mu$ = Reynolds number

u = average velocity of the fluid along the channel

ρ = density of the fluid

μ = viscosity of the fluid

Sc $= \mu / \rho D_s$ = Schmidt number

and A and a are constants. For laminar flow parallel to a flat plate, A is 0.33 and a is 0.5; for turbulent flow in a smooth pipe, A and a have been determined empirically to be 0.023 and 0.80, respectively. However, several investigators have found that during the ultrafiltration of solutes, a is about 1.0 in turbulent flow, and the value of 0.5 still holds for laminar flow. On the other hand, during the cross-flow microfiltration of cells or particles, a has been shown to vary from about 0.8 in laminar flow to about 1.3 in turbulent flow.

Different values of both A and a have been noted for mass transfer to a stationary supporting surface in an agitated cylindrical vessel. For this situation, the following correlations have been proposed[6]:

(1) Laminar boundary layer over the membrane surface:

$$\frac{k_s r}{D_s} = 0.285\left(\frac{\omega r^2}{\nu}\right)^{0.55}\left(\frac{\nu}{D_s}\right)^{0.33} \tag{6.36}$$

when $8{,}000 < (\omega r^2/\nu) < 32{,}000$

(2) Turbulent boundary layer over the membrane surface:

$$\frac{k_s r}{D_s} = 0.0443\left(\frac{\omega r^2}{\nu}\right)^{0.75}\left(\frac{\nu}{D_s}\right)^{0.33} \tag{6.37}$$

where

r = cell radius, cm

ω = stirrer speed, radians/sec

ν = kinematic viscosity, cm^2/sec

D_s = solute diffusivity, cm^2/sec.

(6) C.K. Colton, Ph.D. Thesis, MIT, Dept. of Chem. Eng. (1969).

6.4 Primary Purification

6.4.1 Precipitation

Precipitation is a very valuable and relatively simple technique for recoverying many biological products. Precipitation is most commonly used in the purification of antibiotics, biopolymers, and proteins; our discussion will focus primarily on proteins. Proteins currently isolated by precipitation include analytical and industrial enzymes, blood-plasma proteins, and food proteins. These products include proteins produced by genetic engineering as well as those produced naturally from microbial, plant, and mammalian sources.

Protein precipitation is usually an intermediate step toward final purification, since precipitates are typically impure. They may be aggregates of several proteins and/or contain large amounts of adsorbed salts. Impurities may be particularly troublesome when the original mixture contains many proteins that have similar properties with respect to precipitation. In such a case, precise fractionation of one protein from another may be possible, albeit difficult. Indeed, recovering a target protein from a mixture is one of the most demanding challenges in designing a precipitation scheme.

A protein's tendency to precipitate is governed by many factors: solvent environment (e.g, salt concentration, dielectric constant, pH), temperature, and the size, shape, charge, and hydrophobicity of the protein. Therefore, protein precipitation can depend on a complex combination of environmental parameters. In comparison, existing theories of precipitation are relatively simple and usually apply only to an isolated protein precipitating by a single mechanism. Some of these theories are discussed in the following subsections.

One of the most common strategies to induce precipitation is to alter the solvent's properties and reduce the protein's solubility. For example, protein solubility is often greatly reduced by adding an electrolyte such as ammonium sulfate (so called *salting-out*). Precipitation is also encouraged by adding miscible organic solvents such as ethanol or acetone. The addition of such organics decreases the dielectric constant of the medium and enhances the electrostatic dipole-dipole attraction between protein molecules at their isoelectric point.

An alternative to changing the solvent's properties is to add agents that interact directly with the protein and thereby promote precipitation. Non-ionic polymers, which reduce the amount of water available for protein solvation, are useful for this purpose. Charged polyelectrolytes are also capable of precipitating proteins by acting as flocculating agents. Furthermore, protein solubility is a strong function of solution pH and is usually lowest at the isoelectric pH. Varying the pH changes the number of ionized groups on the protein. At its isoelectric point, a protein has no net charge and the charge-charge repulsive electrostatic force is absent. These ideas will now be described in more detail.

Methods of Precipitation

<u>Salting-out</u>

Salting-out refers to the precipitation of proteins at high salt concentrations. Although the exact nature of precipitation is not completely understood, salting-out results from a number of effects, the primary one being the decrease in available free volume of the solvent resulting from the excluded-volume effect of the added salt. Other mechanisms include screening by salts of charge-charge repulsion, osmotic attraction due to depletion of salt in the region between proteins as they approach one another, and specific ion effects related to the structure of the water surrounding the protein. The relative effectiveness of salts in salting-out follows the lyotropic (Hofmeister) series, which, for some common anions, is phosphate > sulfate > acetate ~ chloride > bromide > nitrate > thiocyanate. The Hofmeister effect appears to be due to the ability of the ions to change the structure of water.

Typically, a precipitate is a solid phase and the solubility of a material in a given solution is the concentration of that material in equilibrium with the solid. In the case of proteins, however, the precipitate is an amorphous phase enriched in protein, containing 70-80% water, 20-25% protein, and salt. For some proteins, the "solubility" depends on the initial concentration of protein, and is better described as a phase equilibrium with a corresponding partition coefficient for the protein between the phases.[7]

The solubility of a protein, S (g/l), at ionic strength I (mol/l) has been correlated by the empirical equation of Cohn and Edsall[8]:

$$\log S = \beta - K_s I \tag{6.38}$$

where β is a constant (roughly equivalent to the logarithm of the protein's solubility at zero ionic strength), and K_s is the salting-out constant. β varies markedly with pH and is generally a minimum at or near the isoelectric point of the protein. On the other hand, the salting-out constant is independent of pH and temperature but varies with the salt and protein involved. The relative magnitude of K_s for different salts follows the Hofmeister series presented earlier. For a particular electrolyte, K_s varies over a two-fold range for different proteins and is greatest for large, asymmetric molecules. Thus, proteins of similar chemical composition but increasing molecular weight require less salt for precipitation.

Melander and Horvath[9] have presented a model, based on solvophobic theory, which has the form of the Cohn equation and correctly accounts for the Hofmeister series effects of the salts employed in precipitation. The theory proposes that a hole is made in the solvent to accomodate a protein molecule, and the energy required to add a protein to the solvent can be described by several terms. The Gibbs energy (erg/mol) associated with transferring a protein from a hyppothetical gas phase to the solution is given by:

$$\Delta\overline{G}_p^o = \Delta\overline{G}_{cav} + \Delta\overline{G}_{es} + \Delta\overline{G}_{VDW} + RT\ln(RT/PV)$$

(7) Shih, E., Prausnitz, J.M., and H.W. Blanch, *Biotechnol. Bioeng.*, **40**, 1155 (1992).

(8) E. Cohn and J.T. Edsall, *Proteins, Amino Acids, and Peptides* (Reinhold Pub. Corp., New York, 1943)

(9) W. Melander and C. Horvath, "Salt Effects on Hydrophobic Interactions in Precipitation and Chromatography of Proteins: An Interpretation of the Lyotropic Series," *Arch. Biochem. Biophys.*, **183**, 200 (1977).

V is the molar volume of the solvent (cm^3/mol), $\Delta\overline{G}_{cav}$ is the contribution of formation of the cavity in the solvent to accomodate the protein, $\Delta\overline{G}_{es}$ accounts for the electrostatic interactions between the protein and the electrolyte, $\Delta\overline{G}_{VDW}$ expresses the contribution of van der Waals interactions between the protein and electrolyte and the term RT ln(RT/PV) accounts for the entropy change arising from the free volume of the solvent. It is assumed that the energetics of protein salting-out are only affected by $\Delta\overline{G}_{cav}$ and $\Delta\overline{G}_{es}$, as the remaining terms are essentially constant. We shall examine each of these terms.

$\Delta\overline{G}_{cav}$

The energy required to form a cavity is assumed to be due to the surface tension of the solvent:

$$\Delta\overline{G}_{cav} = RT\Omega\gamma$$

where
$$\Omega = [N_A\Phi + 4.84N_A^{1/3}(k^c - 1)V^{2/3}]/RT$$

where N_A is Avagadro's number, Φ is the hydrophobic surface area of the protein (cm^2/molecule), and V and γ are the molar volume (cm^3/mol) and surface tension (dyne/cm) of the solvent, respectively (note that the constant, 4.84, has units of $(\text{molecules})^{-1/3}$). k^c is a dimensionless constant related to the ratio of the energy required for the formation of a cavity in the solvent and the energy required for the formation of a planar surface of the same area in the solvent. For a protein-salt solution, we may assume k^c is close to unity and the second term is negligible. The surface tension of the solvent is determined by the nature of the added salt. This can be approximated by a linear expansion:

$$\gamma = \gamma^o + \sigma m$$

where γ^o is the surface tension of pure water (72.0 dyne/cm at 25°C) and m is the molality of the salt. The molal surface tension increment σ (dyne kg cm^{-1} mol^{-1}) depends on the nature of the salt, and is positive for most salts.

$\Delta\overline{G}_{es}$

The electrostatic interactions between the protein and the salt solution depend on the salt concentration. At low ionic strengths, the protein can be considered as a charged spherical ion and Debye-Hückel theory applied. At high salt concentrations, there is considerable shielding of the charges on the protein molecule and the protein behaves a a neutral dipole. Thus the electrostatic contributions to the Gibbs energy can be considered to be comprised of two parts:

$$\Delta\overline{G}_{es} = \Theta - RT\Lambda I \tag{6.39}$$

$$\Theta = \frac{z_p^2 e^2 N_A}{2\varepsilon_s}\left(\frac{1}{b} - \frac{\kappa}{1+\kappa a}\right) \qquad \Lambda = \frac{2\pi F^3 g(\lambda_o)\mu}{(\varepsilon_s RT)^2}$$

where ε_s is the dielectric constant of the solvent, z_p is the net charge number (valence) of the protein (generally obtained from titration data); e is the elementary charge (esu); b is the radius of the protein (cm); a is the distance of closest approach between the center of the protein and the salt ions; κ is the reciprocal of the double layer thickness, which depends on the square root of the ionic strength, I; F is the Faraday constant (esu/mol); $g(\lambda_o)$ is a function of the eccentricity of the protein (dimensionless); and μ is the dipole moment of the protein (esu-cm). The first term in Equation (6.39) represents the electrostatic contribution to the free energy based on Debye-Hückel theory and is valid only at low ionic strength. At high ionic strength, $\kappa a \gg 1$ and the Debye-Hückel term approaches a constant value.

$$\Theta = \frac{Z^2 e^2 N_A}{2Da} = -\beta_o \tag{6.40}$$

The second term in Equation (6.39) represents the Kirkwood contribution, which considers the protein as a neutral dipole. This term includes the ionic strength, I, and is dependent on the eccentricity of the protein[10] through the function $g(\lambda_o)$. The dipole moment of the protein, μ, can be determined experimentally or by calculation if the crystal structure is known. We see from this expression that the shielded charge-charge contribution vanishes at the isoelectric point as the net charge on the protein becomes zero (z_p=0). Hence the electrostatic contribution is directly proportional to the ionic strength. At pH's distant from the isoelectric point, the relationship is non-linear at low ionic strengths and becomes linear as the ionic strength increases.

We now turn to using the expression for the Gibbs energy to predicting the solubility (S) of a protein in a salt solution. For an ideal solution we write

$$\Delta\overline{G}_p = \Delta\overline{G}_p^o + RT\ln S$$

which can be rearranged

$$\ln S = \frac{1}{RT}\left(\Delta\overline{G}_p - \Delta\overline{G}_p^o\right)$$

Inserting only the Gibbs energy terms associated with cavity formation and electrostatics, and ignoring the entropic contribution, we see

$$\ln S = x_o + \beta_o - (\Omega\sigma - \Lambda)I$$

---

(10) The eccentricity is the ratio of the major to the minor axes, where the protein is assumed to be an ellipsoid.

where x_0 incorporates all the salt-independent terms. This equation has the form of the Cohn equation, with

$$K_s = \Omega\sigma - \Lambda \qquad \beta = x_o + \beta_o$$

A salt with a high effectiveness to precipitate proteins will have a high salting-out constant K_s. The approach of Melander and Horvath provides a quantitative basis for the hofmeister series. A salt with a large value of σ will have a high K_s and hence be an effective salting-out agent. This can be seen in the table below, which lists some values of the surface tension increments for a number of salts.

Table 6.8. *Molal surface tension increments for several salts in water at 25°C. [From Melander and Horvath, Arch. Biochem. Biophys., **183**, 200 (1977).]*

Salt	$\sigma \times 10^3$ (dyne·g/cm·mol)
KSCN	0.45
NH_4NO_3	0.85
NaCl	1.64
Na_2HPO_4	2.02
$(NH_4)SO_4$	2.16
Na_2SO_4	2.73
$CaCl_2$	3.66

Reduction of Solvent Dielectric Constant

Adding a miscible organic cosolvent to an aqueous solution will reduce the dielectric constant of the medium. A lower dielectric constant enhances the charge-dipole and dipole-dipole attraction between proteins and promotes precipitation. Like salting-out, this approach is effective and simple, and solvent-induced precipitation has become very popular. Furthermore, when combined with parameters such as pH, ionic strength, temperature, and protein concentration, the dielectric constant can be used to fine tune a precipitation procedure and thus achieve a higher degree of protein fractionation.

The solubility of some proteins at their isoelectric points have been related to the dielectric constant of the solvent by the following expression[11]:

$$\log\frac{x}{x_o} = \frac{K''}{\varepsilon_s^2} \tag{6.41}$$

(11) N.A. Frigerio and T.P. Hettinger, "Protein Solubility in Solvent Mixtures of Low Dielectric Constant," *Biochim. Biophys. Acta*, **59**, 228 (1962).

where ε_s is the dielectric constant of the solvent-water mixture, and x_0 is the mole-fraction solubility extrapolated to infinite dielectric constant. K″ is a function of the protein's shape and charge and is constant at fixed ionic strength and temperature.

A potential disadvantage of organic solvents is their tendency to denature proteins; therefore, acetone and ethanol are usually preferred over longer-chain, more hydrophobic organics. Flammability of organic solvents is also a concern, particularly for large-scale operation. For acetone-initiated precipitation, the following equation has proven useful:

$$\begin{bmatrix} \text{(v/v)\%} \\ \text{producing} \\ \text{precipitation} \end{bmatrix} = 1.8 - 0.12 \ln \begin{bmatrix} \text{solute} \\ \text{molecular} \\ \text{weight} \end{bmatrix}$$

Addition of Non-ionic Polymers

This method is rapidly gaining favor for protein recovery. Non-ionic polymers frequently employed for protein precipitation include dextrans and polyethylene glycols, i.e., the same reagents used in aqueous two-phase extraction. In this case, protein-protein interactions due to the polymer arise primarily from excluded volume, although there may also be electrostatic effects at low ionic concentrations.[12] The polymer occupies volume and causes the proteins to approach each other and ultimately phase separate.

Studies with purified proteins indicate that the dependence of protein solubility on polymer concentration is often described by a simple equation analogous to Eq. 6.38 for salting out:

$$\log S = \beta'' - K_s''C \tag{6.42}$$

where S is the protein solubility (g/l), C is the polymer (e.g., PEG) concentration (w/v), and β'' and K_s'' are constants. The linear dependence of logS on PEG concentation predicted by Eq. 6.42 is shown for various proteins in Figure 6.19.

In some cases, particularly at high protein concentrations and at pH values far from the isoelectric point, protein-protein interactions will influence protein solubility in a solution of polymer. These effects are included in a more complex for of Eq. 6.42, obtained by utilizing expressions for the chemical potentials of the polymer (component 2) and protein (component 3). In terms of molarities, rather than molalities, these expressions are

$$\mu_2 = \mu_2^0 + RT(\ln C_2 + a_{22}C_2 + a_{23}C_3) \tag{6.43}$$

$$\mu_3 = \mu_3^0 + RT(\ln C_3 + a_{33}C_3 + a_{23}C_2) \tag{6.44}$$

(12) Mahadevan, H. and C.K. Hall, AIChE J., **36**(10) 1517 (1990) and AIChE J., **38**(4) 573 (1992) present a statistical mechanical theory of precipitation based on the mean spherical approximation for a mixture of charged hard spheres.

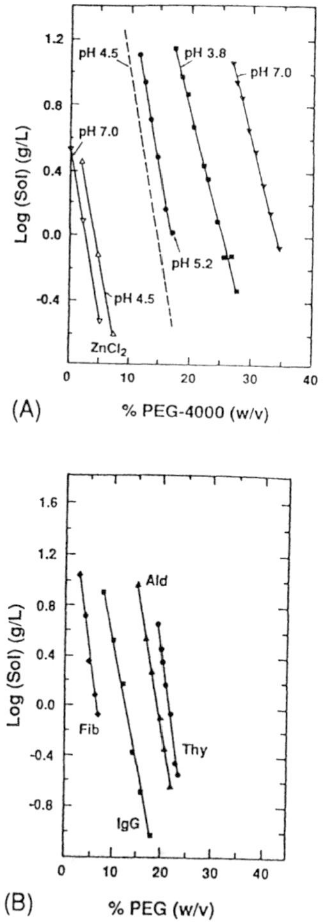

Figure 6.19. *(A) Effect of pH on the solubility of human serum albumin in PEG-4000. (B) Solubility of various proteins in PEG-4000. Fib: human fibrinogen; IgG: human γ globulin; Ald: aldolase; Thy: thyroglobulin. [From D.H. Atha and K.C. Ingham, "Mechanisms of Precipitation of Proteins by Polyethylene Glycols. Analysis In Terms of Excluded Volume," J. Biol. Chem., 256, 12108 (1981).]*

where a_{ij} refers to apparent second virial coefficient for interaction between i and j (liters/mol), and C_2 and C_3 have units of moles/liter. From Eq. 6.44, we can obtain expressions for the chemical potentials of the protein in the absence (μ_3') and presence (μ_3) of polymer:

$$\mu_3' = \mu_3^0 + RT(\ln S_3' + a_{33}S_3') \quad (6.45)$$

$$\mu_3 = \mu_3^0 + RT(\ln S_3 + a_{33}S_3 + a_{23}C_2) \quad (6.46)$$

where S_3 and S_3' represent the corresponding solubilities under condition of saturation. The chemical potentials of the protein in solution and the precipitate phase must be equal at equilibrium; thus, if we assume the chemical potential of the amorphous solid phase is constant over the range of polymer concentrations, we can equate Eqs. 6.45 and 6.46 to obtain:[13]

$$\ln S_3 = \ln S_3' + a_{33}S_3' - a_{33}S_3 - a_{23}C_2 \tag{6.47}$$

$$\ln S_3 + a_{33}S_3 = \ln S_3' + a_{33}S_3' - a_{23}C_2 \tag{6.48}$$

or

$$\ln S_3 + a_{33}S_3 = \text{constant} - a_{23}C_2 \tag{6.49}$$

Eq. 6.49 is similar in form to Eq. 6.42 but differs by the inclusion of the protein-protein interaction term, $a_{33}S_3$. The importance of this term is illustrated by the three solubility plots for alcohol dehydrogenase shown in Figure 6.20

Large Scale Precipitation

Large scale precipitation, generally defined as precipitation on an industrial rather than laboratory scale, can be idealized as a six-step process:

1. *Initial mixing*. Mixing of the precipitant (e.g., PEG or ammonium sulfate) with the solute-containing feed.
2. *Nucleation*. Small particles rapidly appear and begin to grow.
3. *Diffusion Limited (Perikinetic) Growth*. Initially, growth occurs as solute (protein) diffuses to the nucleated particles. When the particles are small, their growth is limited by diffusion. In a mono-sized dispersion the initial rate of change of the molar concentration of aggregating particles, c_s, can be written as

$$\left(\frac{dc_s}{dt}\right) = -\kappa c_s^2$$

which, upon integration, gives

$$\frac{1}{c_s} = \frac{1}{c_{so}} + \kappa t \tag{6.50}$$

where c_{so} is the initial concentration of aggregating particles, and κ is the second-order rate constant of the process. Interestingly, at relatively long times the particle concentration decreases inversely with time and is independent of the initial concentration.

(13) For further details, see D.H. Atha and K.C. Ingham, "Mechanisms of Precipitation of Proteins by Polyethylene Glycols. Analysis In Terms of Excluded Volume," *J. Biol. Chem.*, **256**, 12108 (1981).

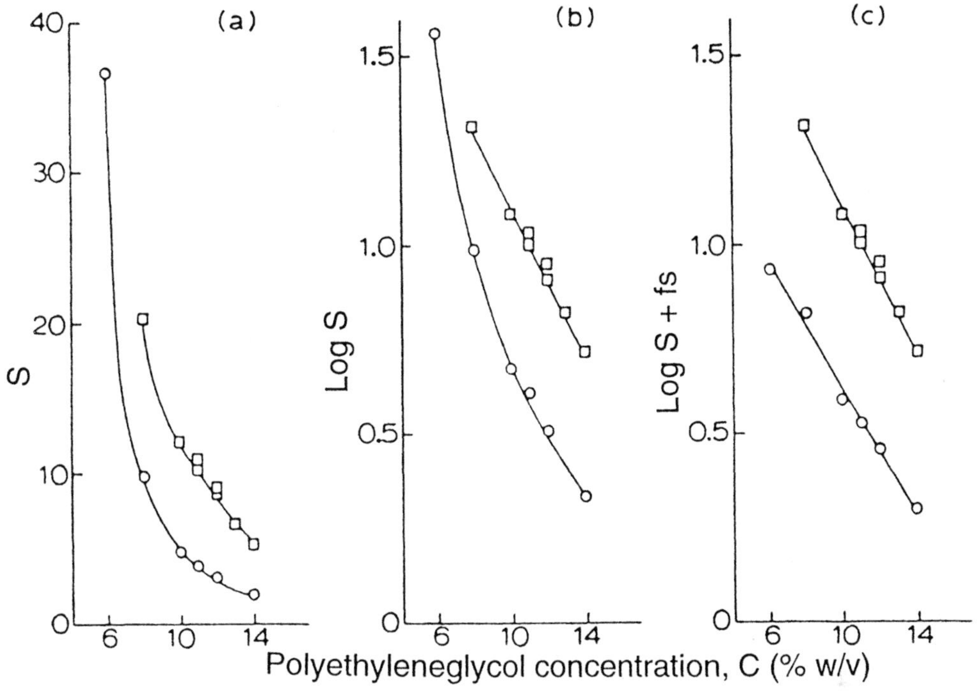

Figure 6.20. *Alcohol dehydrogenase solubility (S_3) as a function of PEG concentration (C_2). (a) S_3 versus C_2; (b) $\log S_3$ versus C_2; (c) $\log S_3 + a_{33}S_3$ versus C_2. [From P.R. Foster, P. Dunnill, and M.D. Lilly, "The Precipitation of Enzymes from Cell Extracts of Saccharomyces cerevisiae by Polyethyleneglycol," Biochim. Biophys. Acta, 317, 505 (1973).]*

The rate constant can be determined from the diffusion equation for the dispersed particles.[14] Solving for the flux of particles across the surface of a sphere of diameter d_p leads to an expression for the total number of particle contacts per unit volume of dispersion per unit time. Writing this expression in terms of molar concentration yields the rate constant:

$$\kappa = 8\pi D_s d_p N_{av} \tag{6.51}$$

where D_s is the diffusivity of the particles and N_{av} is Avagadro's number. Equation (6.50) can be rewritten in terms of the weight-average molecular weight of the aggregate, $\overline{M}_w$, by noting, from a mass balance, that

(14) This analysis is based on Smoluchowski's theory and is described in V.G. Levich, *Physicochemical Hydrodynamics* (Prentice-Hall, Inc., N.J., 1962), pp. 207-211.

$$c_s\overline{M}_w = c_{so}\overline{M}_{wo}$$

where $\overline{M}_{wo}$ is the initial weight-average molecular weight of the precipitate. Therefore,

$$\overline{M}_w = \overline{M}_{wo}(1 + c_{so}\kappa t) \tag{6.52}$$

Eq. (6.52) is more practical than Eq. (6.50) because $\overline{M}_w$ can be readily determined by quasi-elastic light scattering, whereas the molar concentration of aggregates is difficult to measure.

The growth of casein particles in a salting-out process is summarized in Figure 6.21. The experimental results followed perikinetic theory reasonably well at short times where minimal settling occurred; however, at larger "ageing times" significant deviation from theory was observed (note the logarithmic dependence between particle size and time). The poor agreement at larger times may have been due to the increase in viscosity of the thickened sludge.

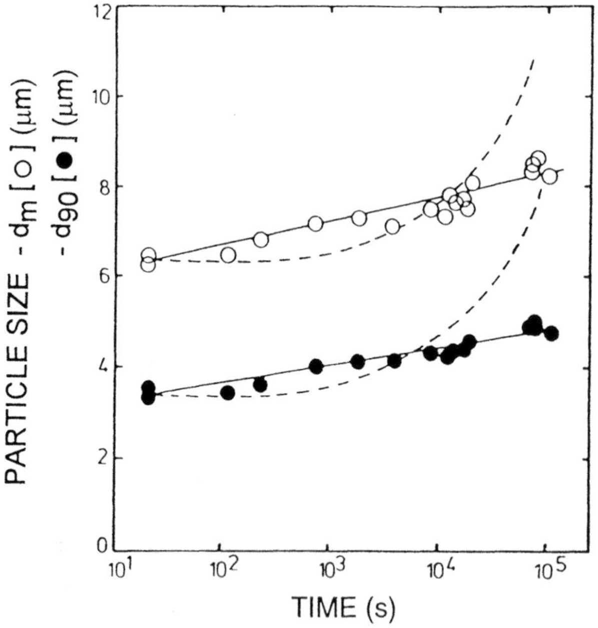

Figure 6.21. Change with time of particle size for casein precipitate prepared by continuous salting-out with 1.8 M ammonium sulfate. The dashed lines represent theoretical predictions obtained by assuming perikinetic growth. [From D.J. Bell et al., Advances in Biochemical Engineering/Biotechnolgy, 26, 1 (1983).]

4. *Shear Induced (Orthokinetic) Growth.* Following perikinetic growth, further aggregation involves collisions between larger particles. The frequency of these collisions increases with the shear rate; therefore, shear-induced, or *orthokinetic* growth is promoted by fluid motion. This can be pictured as particles aggregating in a velocity gradient, where particles moving at higher velocities collide with particles moving at lower velocities. Orthokinetic aggregation becomes important once the particles reach a limiting size, about 0.1 and 10 μm for high and low shear fields, respectively. Growth under these conditions also obeys second-order kinetics:

$$\left(\frac{dc_s}{dt}\right) = -\kappa c_s^2 \qquad (6.53)$$

Assuming that there is a continuous distribution of particle sizes, the rate constant depends on the particle volume and the average shear rate, and is given by

$$\kappa = \frac{2}{3}\alpha N_{av} d_p^3 \bar{\gamma} \qquad (6.54)$$

where d_p is the average aggregate size, $\bar{\gamma}$ is the mean shear rate, and α is the collision effectiveness factor, or the fraction of collisions resulting in permanent aggregates. Because the total volume of aggregating particles is constant, their volume fraction, ϕ, is independent of time. Therefore, ϕ can be written as

$$\phi = \left(\frac{1}{6}\pi d_p^3\right) c_s N_{av} \qquad (6.55)$$

Inserting Eq. (6.55) into Eq. (6.53) yields a first-order rate expression, which we integrate to obtain the desired expression for c_s:

$$\frac{c_s}{c_{si}} = \exp\left(-\left(\frac{4\alpha\phi\bar{\gamma}}{\pi}\right)t\right) \qquad (6.56)$$

Ideally, Eqs. (6.55) and (6.56) should also include a term for perikinetic aggregation, but this mechanism becomes less important as the particle size increases.

In an isotropic turbulent flow field, the mean shear rate, $\bar{\gamma}$, is related to the power input per unit volume by

$$\bar{\gamma} = \left[\frac{P/V}{\rho\nu}\right]^{1/2} \qquad (6.57)$$

where ρ is the solution's density, ν is its kinematic viscosity, and (P/V) is the power input per unit volume[15]. The term $\bar{\gamma}t$ is a dimensionless quantity known as the Camp number, Ca; it is also referred to as the *ageing parameter*.

It is worth re-emphasizing that the above derivation is based on average properties of

(15) In aqueous solutions stirred at moderate rpm, the power input is of order 1 hp/10^3 gal, i.e., P/V is 1.97 x 10^3 gm cm^{-1} s^{-3} and $\rho\nu$ is 10^{-2} gm cm^{-1} s^{-1}. Hence, $\bar{\gamma}$ is ca. 4.4 x 10^2 s^{-1}.

the particle-size distribution. A rigorous description of orthokinetic growth requires a population balance to describe the aggregation kinetics. The rate of change of the particle concentration is found by integrating the volume distributions of colliding particles. A similar final result is obtained, however, in that the growth kinetics turn out to be first-order in the particle concentration c_s. This analysis is presented in Appendix B.

5. *Flocculation*. Ultimately, existing precipitates agglomerate into larger flocs. The mechanisms of flocculation are uncertain, and adequate theories for describing this step have not yet been developed. It has been observed, however, that particle-particle collisions induced by shear can serve to compress large flocs and enhance the formation of compact particles stable to centrifugation. Therefore, flocculation is sometimes carried out in a moderate shear field. On the other hand, excessive shear may be too destructive, and an unstirred tank may be favored for the ageing process. Experiments are needed to determine the optimal reactor system.

6. *Centrifugation*. The flocs are removed by the methods described in Section 6.2.1.

6.4.2 Chromatography and Fixed Bed Adsorption

Basic Concepts

In its broadest sense the term chromatography is used to describe any separation process that depends on partitioning between a flowing fluid and a solid adsorbent. In analytical applications and some preparative uses, the mixture to be separated is applied as a pulse to a packed column of adsorbent particles (known as the *stationary phase*). The pulse is introduced via a flowing solvent, or carrier stream (the *mobile phase*), and is resolved into its different components as it travels through the column. Different solutes exit the column at different times, as shown in Figure 6.23. Chromatography in which a pulse is injected at one end of the column and eluted from the other is referred to as *elution chromatography*.

In contrast to elution chromatography, *fixed-bed adsorption* (also known as *frontal chromatography*) is a multi-step process. In the first step, feed is continuously contacted with a solid phase that has a high affinity for the solute of interest. The feed is maintained until the desired product appears in the effluent. At this point, the column is nearly saturated with bound solute and *breakthrough* is said to occur (Figure 6.24). An intermediate wash step to remove non-specifically adsorbed material may follow, after which the desired solute is recoverd by elution with buffer at a different pH or with a different solvent. Unlike elution chromatography, fixed-bed adsorption typically does not resolve a multicomponent mixture into its individual constituents.

To avoid confusion, we will refer to elution chromatography as chromatography and to frontal chromatography as fixed-bed adsorption. Although these more precise designations are generally favored by engineers, they are not universally applied. In fact, in the biochemistry and biochemical engineering literature, the term "chromatogaphy" is associated with just about all packed bed techniques, regardless of the operating mode. We will refer to specific techniques by their commonly accepted names. Examples include ion-exchange chromatography and affinity chromatography, which are actually selective, fixed-bed adsorption processes.

Figure 6.22. Production scale HPLC column 7.5 feet in length. Shown beneath the column is a piston for packing and unpacking the column with chromatographic medium. [Courtesy ProChrom, Champigneulles, France. From Biotechnology, 7, 248 (1989).]

Many different adsorbents, or stationary-phase packings, are used in chromatography and fixed bed adsorption. Among the most common are charged resins (e.g., polymers of sulfonated styrene, or polymer gels containing diethylaminoethyl (DEAE) groups), dextran gel, cross-linked agarose gel (untreated or derivatized), beaded cellulose, hydroxyapatite, alumina, and silica gel. The different adsorbents give rise to different chromatographic techniques, which are generally classified according to the mechanism of separation. For example, ion-exchange chromatography employs ion-exchange resins to separate species on the basis of charge, whereas gel filtration chromatography (or gel permeation or size exclusion) chromatography utilizes different gel types (Figure 6.25) to separate molecules of different sizes. Another popular method is hydrophobic interaction chromatography. The salient features of these methods are summarized in Table 6.9.

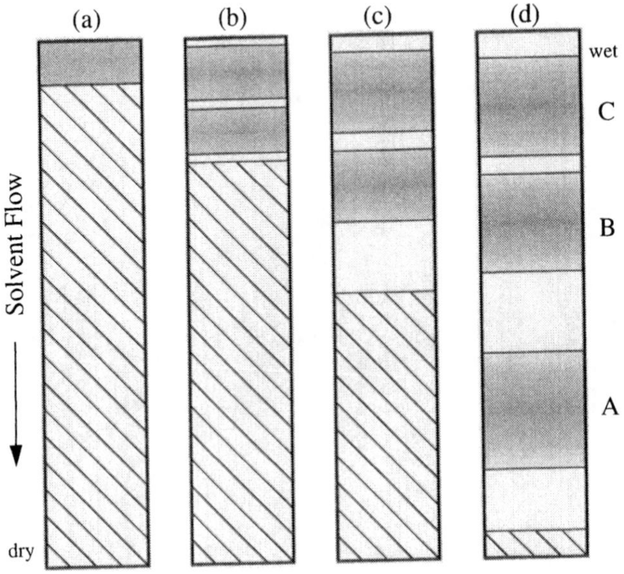

Figure 6.23. *Hypothetical chromatographic separation of 3-component sample consisting of compound A, compound B, and compound C.*

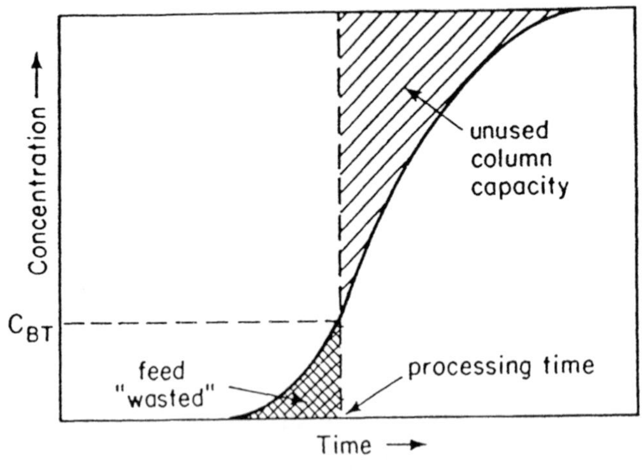

Figure 6.24. *Breakthrough curve for fixed-bed adsorption or frontal chromatography. If the feed is stopped when the effluent concentration is C_{BT} (corresponding to the processing time), the column will not have been used to its full capacity. The effluent will also contain a small amount of solute, indicated by the hatched area.*

The usual goal of chromatography is to achieve adequate separation, or resolution, of a sample mixture. The resolution, R_s, of two adjacent peaks is defined quantitatively as the distance between the two peak centers divided by the average band width:

$$R_s = \frac{t_{r,2} - t_{r,1}}{(1/2)(w_1 + w_2)} \tag{6.58}$$

where $t_{r,1}$ and $t_{r,2}$ are the retention times, that is, the times of emergence of the maximum concentrations of eluates from the column, and w_1 and w_2 are the peak widths measured by the baseline intercept, as shown in Fig. 6.26.

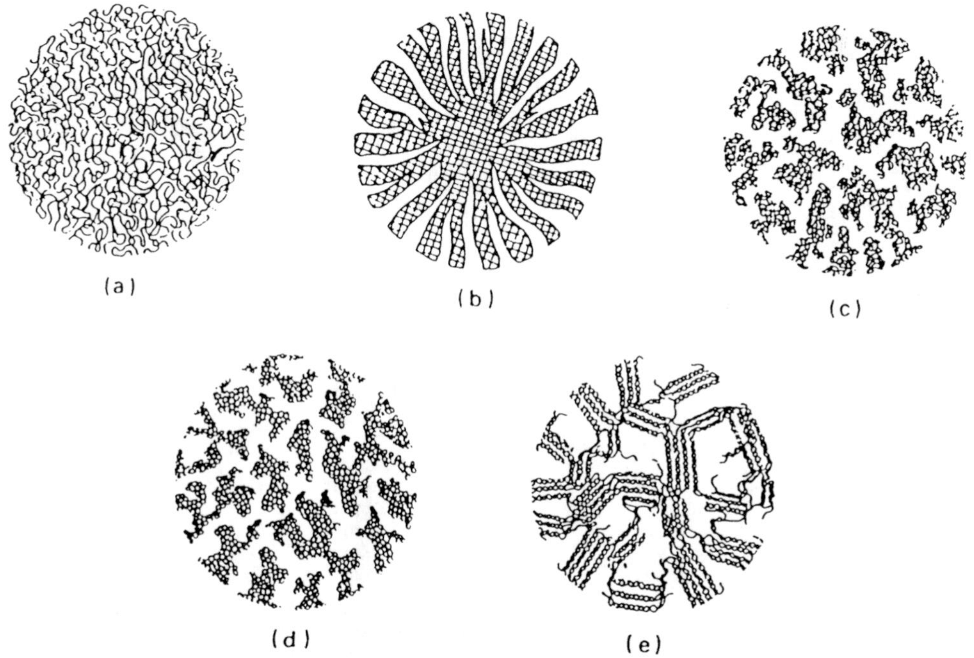

Figure 6.25. *Schematic representation of different gel types used in chromatography: (a) xerogel; (b) aerogel (porous glass); (c) aerogel (porous silica); (d) organic macro-reticular polymer (hydroxyethyl methacrylate); (e) agarose. [From Chromatography of Synthetic and Biological Polymers, Vol. 1, Column Packings, G.P.C. and Gradient Elution (Ellis Horwood, Chichester, 1978).]*

Table 6.9. Chromatographic methods for the purification of biologicals.

Basis of separation	Chromatography type	Characteristics	Application
Charge	Ion-exchange	*Resolution* can be very high. *Capacity* is high. *Speed* can be high, depending on the matrix.	Most widely used method. Is most effective at early stages where large volumes are handled. Can also be used in batch mode.
van der Waals interactions	Hydrophobic interaction	*Resolution* is good. *Capacity* is good. *Speed* is high.	Can be applied at any stage, but is most usefully applied when the ionic strength is high (after ion exchange or salt precipitation). Possibility of some denaturation. *May involve organics during elution.*
Size and shape	Gel filtration	*Resolution* moderate for fractionation. Good for buffer exchange and desalting. Relatively *low capacity* for loaded protein, but *high flow rates* possible with some packings.	Fractionation is best left to later stages of purification where loads are lower. Buffer can be exchanged at any time.
Biological affinity	Affinity	*Selectivity* can be very high. *Capacity* can be high, depending on the ligand. *Speed* is high.	Can be used at any stage, but the materials are *expensive*; best to use when protein loads and fouling substances have been reduced by less-costly methods.

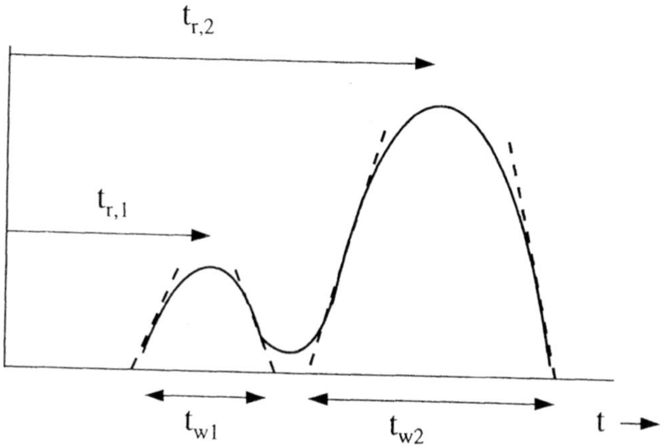

Figure 6.26. *Resolution in chromatography.*

Differential migration in chromatography results from the compounds to be separated having different equilibrium distributions between the stationary phase and the mobile phase. Such equilibria are described by *adsorption isotherms*, which relate the solute concentration in solution (usually expressed in units of mass of solute per volume of solution) to the solute concentration on the adsorbent's surface (the *sorbate concentration*, commonly in mass of solute per mass or volume of adsorbent). For example, many adsorption systems are characterized by the familiar hyperbolic Langmuir isotherm:

$$q^* = \frac{q_{\max} K_L c}{1 + K_L c} \tag{6.59}$$

where q^* is the sorbate concentration at equilibrium with c, the concentration of solute in the bulk solution, q_{max} is the maximum sorbate concentration, and K_L is the Langmuir constant. Alternatively, the adsorption of many antibiotics, steroids, and hormones is described by the empirical Freundlich isotherm:

$$q^* = Kc^n \tag{6.60}$$

where the constants n and K must be determined experimentally. With regard to units, in this chapter we will assume that q has units of mass per mass of dry adsorbent.

A more general classification scheme for equilibrium isotherms is based on values of the so-called *separation factor*, R_{eq}, defined as $R_{eq} = 1/(1 + K_L c_0)$, where c_0 is the feed concentration of solute:

$R_{eq} = 0$	$0 < R_{eq} < 1$	$R_{eq} = 1$	$R_{eq} > 1$
Irreversible	Favorable	Linear	Unfavorable

The different types of isotherms, *irreversible*, *favorable*, *linear*, and *unfavorable*, can be understood by reference to Figure 6.27, which distinguishes the three systems on an equilibrium diagram. An isotherm that is concave toward the abscissa is termed *favorable* because significant adsorption occurs even at low solute concentrations. Important implications of the different types of isotherms will be discussed later.

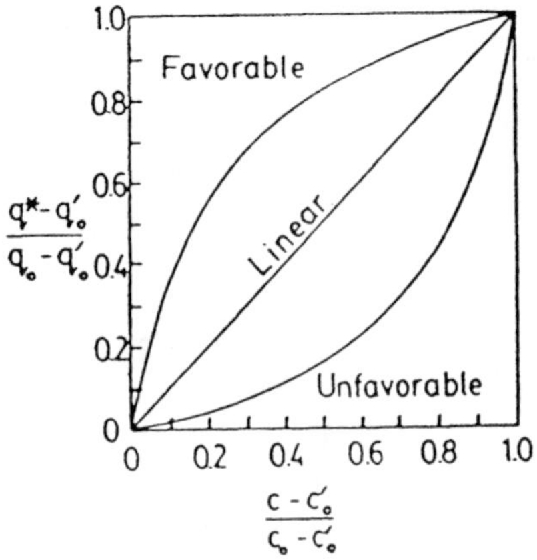

Figure 6.27. *Equilibrium diagram showing distinction between favorable, unfavorable, and linear isotherms. The X-Y diagram is a nondimensional representation of the equilibrium relationship, expressed in terms of the reduced variables $X = (c - c'_o)/(c_o - c'_o)$ and $Y = (q^* - q'_o)/(q_o - q'_o)$. The sorbate concentrations q_o and q'_o are in equilibrium with the limiting fluid-phase concentrations, c_o and c'_o, where $(c_o - c'_o)$ represents the change in fluid-phase concentration over a limited portion of the isotherm.*

Example: **Adsorption Isotherms for Ion Exchangers**

Adsorption isotherms of various proteins on hydrophilic ion exchangers are shown in Figure 6.32. Note that the isotherms depend on the ionic strength, the pH, and the types of ion exchangers and proteins. However, in all cases the isotherms are linear at sufficiently low protein concentrations. A rigorous thermodynamic description of the binding equilibrium between an ion exchanger and a protein in an electrolyte solution involves many variables and is fairly complex. Here we will adopt a simpler approach that illustrates the salient features of ion-exchange equilibria. Assuming that uniform bonds are formed between the ion exchanger with fixed ionic groups and an oppositely charged protein, the binding of protein to the resin and accompanying displacement of counterions initially present on the resin can be represented by:

$$P + \left(\frac{z_P}{z_B}\right)\overline{B} \leftrightarrow \overline{P} + \left(\frac{z_P}{z_B}\right)B$$

where P and B signify the protein and the counterion displaced from the ion exchange resin, the overbar indicates the ion exchanger-bound state, and z_P and z_B are the valences of the protein and oppositely charged counterion, respectively. The equilibrium constant can be expressed in terms of the activity of each species a_i by

$$K_e = \frac{(\overline{a}_p)^{|z_B|}(a_B)^{|z_P|}}{(a_P)^{|z_B|}(\overline{a}_B)^{|z_P|}} \qquad (E6.1)$$

In terms of molarities, Eq. (6E2.1) can be rewritten as

$$K_e = \left(\frac{\overline{\gamma}_P\overline{m}_P}{\gamma_P m_P}\right)^{|z_B|}\left(\frac{\gamma_B m_B}{\overline{\gamma}_B\overline{m}_B}\right)^{|z_P|} \qquad (E6.2)$$

where m_i and γ_i = molarity and activity coefficients of the species i, respectively.

Electroneutrality requires that $m_R = (|z_P|/|z_B|)\overline{m}_P + \overline{m}_B$, where m_R is the molarity of the fixed ionic groups on the ion exchanger. Substitution of this expression into Eq. (E6.1) and rearrangement yields an expression for the ratio of the resin-bound protein to that in the solution. For the case of a monovalent counterion B (i.e., $|z_B| = 1$) and a relatively low concentration of protein net charge (i.e., $z_P m_P << m_R$), this expression is

$$K_{i-e} = \frac{\overline{m}_P}{m_P} = \frac{(K_e')(m_R/m_B)^{|z_P|}}{1 + (K_e')(m_R/m_B)^{|z_P|}z_P^2 m_P/m_R} \qquad (E6.3)$$

where

$$K_e' = \frac{\overline{m}_P}{m_P}\left(\frac{m_B}{\overline{m}_B}\right)^{|z_P|} \qquad (E6.4)$$

which can be assumed to be constant at a fixed ionic strength.

From Eq. (E6.3) it is clear that for a given pH and ionic strength, K_{i-e} will depend on the valence of the protein and the molarities of the counterion, the protein, and the fixed ionic groups on the ion exchanger. Eq. (E6.3) also predicts that a plot of $\overline{m}_P$ versus m_P for fixed values of m_B, m_R, and z_P will follow a saturation curve of the kind shown in Figure 6.28. Furthermore, if m_P is very much smaller that m_R, Eq. (E6.3) simplifies to

$$K_{i-e} = (K_e')\left(\frac{m_R}{m_B}\right)^{|z_P|} \qquad (E6.5)$$

Note that the K_{i-e} value expressed by Eq. (E6.5) is independent of m_P but strongly depends on the concentration and/or ionic strength of the outer solution (m_B). This behavior is consistent with the linear isotherms of Fig. 6.28.

One special case that warrants further attention is that of irreversible adsorption, or a rectangular isotherm (Figure 6.29). Irreversible adsorption occurs in many types of affinity chromatography, especially when antibody-antigen interactions are involved, and in some applications of ion-exchange and hydrophobic chromatography. As explained previously, highly specific irreversible adsorption (of the kind achieved with a column of immobilized monoclonal antibodies, for example) favors an operating mode similar to that of conventional fixed bed adsorption. Feed containing a solute concentration of c_0 is continuously passed through the column until the desired product appears in the effluent. Following a wash step to remove non-specifically adsorbed contaminants, the solute is eluted under conditions designed to disrupt the specific solute-support interactions.

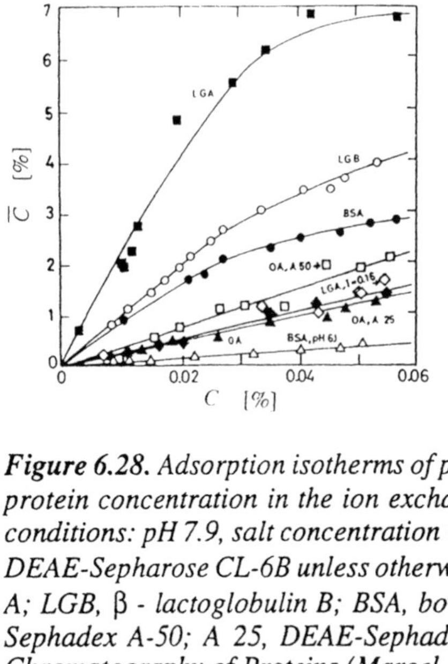

Figure 6.28. *Adsorption isotherms of proteins on ion exchangers. Here $\overline{C}$ and C denote the protein concentration in the ion exchanger and bulk solution, respectively. Experimental conditions: pH 7.9, salt concentration = 0.11 M, temperature 20° C, and the ion exchanger DEAE-Sepharose CL-6B unless otherwise indicated. Abbreviations: LGA, β - lactoglobulin A; LGB, β - lactoglobulin B; BSA, bovine serum albumin; OA, ovalbumin; A 50, DEAE-Sephadex A-50; A 25, DEAE-Sephadex A 25. [From S. Yamamoto et al., Ion-Exchange Chromatography of Proteins (Marcel Dekker, Inc., New York, 1988), p. 129.]*

Different species migrate through a chromatography column at different rates because of differences in their equilibrium distributions between the two phases; however, different molecules of the *same* species also travel at different velocities. Such differences in molecular migration rate are caused by physical and/or rate processes and are responsible for *band broadening* through the column. These processes, shown schematically in Figure 6.30, can include longitudinal molecular and/or eddy diffusion of eluate molecules during their journey through the column (Figure 6.30B), mobile phase mass transfer arising from nonuniform flow profiles within the column (Figure 6.30C), and mass transfer (diffusion) within pores of the stationary phase (Figure 6.30D) and within the stationary phase itself (Figure 6.30E).

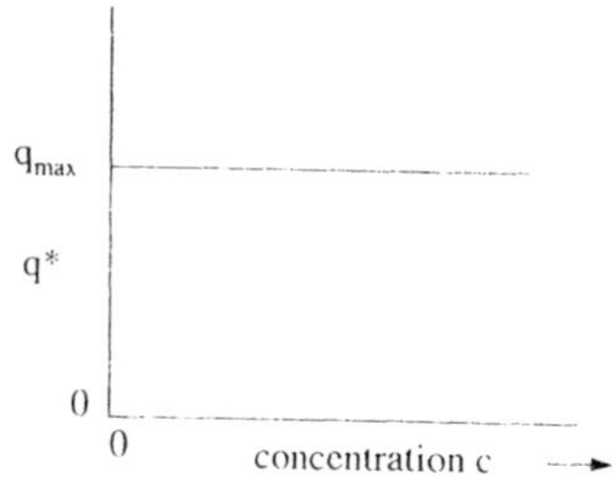

Figure 6.29. *The irreversible or rectangular isotherm.*

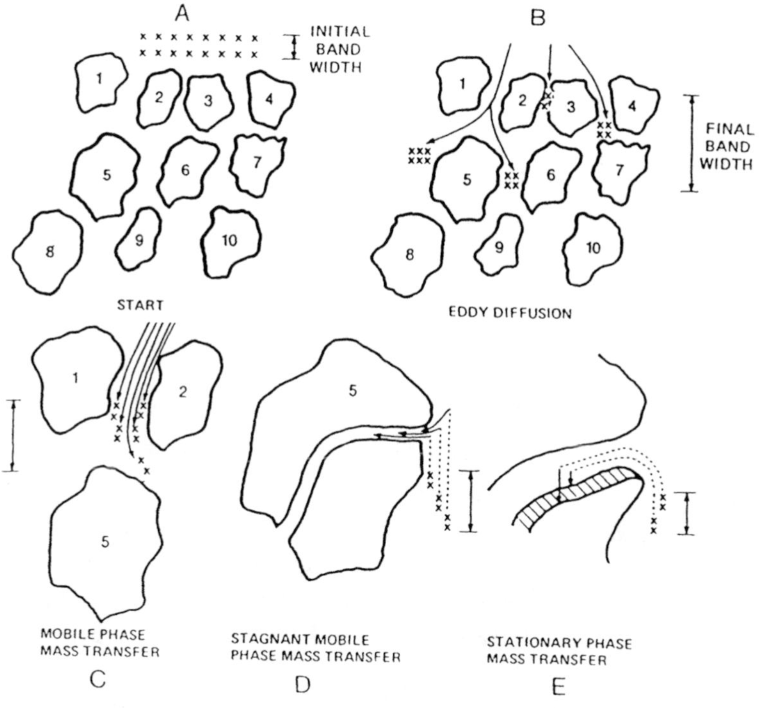

Figure 6.30. *Contributions to band broadening in chromatography. [From L.R. Snyder and J.J. Kirkland, Introduction to Modern Liquid Chromatography (John Wiley & Sons, N.Y., 1974).]*

Mathematical Analysis of Chromatography

Now let's consider chromatography from a more mathematical viewpoint. If the column is sufficiently long, in the absence of mass transfer effects each exiting peak will be Gaussian in shape. For a pulse input that yields a Gaussian peak on the chromatogram, the eluate concentration, C, is given by the relationship

$$C = \frac{1}{F}\frac{M}{\sigma_t\sqrt{2\pi}}e^{-0.5[(t_r - t)/\sigma_t]^2} = C_{max}e^{-0.5[(t_r - t)/\sigma_t]^2} \tag{6.61}$$

where F is the volumetric flowrate of solvent, M is the amount of solute injected into the column, σ_t is the standard deviation of the peak measured in time units, t_r is the retention time of the eluate in the column, and C_{max} is the maximum concentration, i.e., the height of the Gaussian peak. Properties of a Guassian peak are shown in Figure 6.31.

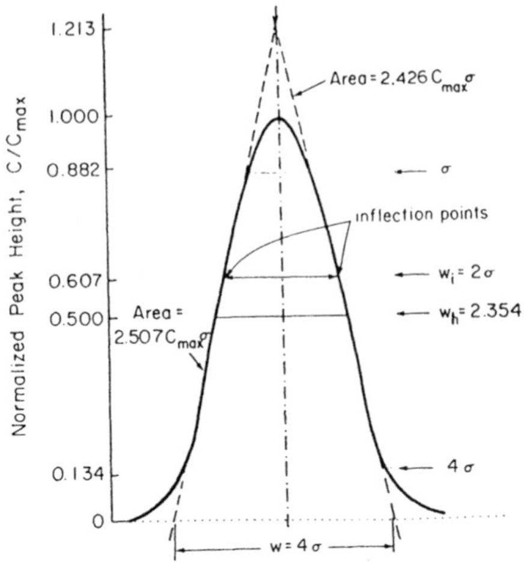

Figure 6.31. *Properties of a Gaussian peak.* C_{max}, *maximum peak height;* σ_t, *standard deviation;* w_{in}, *peak width at inflection points;* w_h, *peak width at half-height; w, peak width at base (base intercept). The area under the triangle formed by the tangents to the inflection points and the baseline as well as the peak area (zeroth moment) are also shown. [From C. Horvath and W.R. Melander, in Chromatography. Fundamentals and Applications of Chromatographic and Electrophoretic Methods. Part A., E. Heftmann, ed. (Elsevier Scientific, The Netherlands, 1983) p. A41.]*

Chromatographic peaks can also be characterized by their *statistical moments*, which are defined as follows:

$$\text{Zeroth moment (peak area)} = m_0 = A = \int_0^\infty C(t, z = L)dt \tag{6.62}$$

$$\text{First absolute moment} = \mu_1 = t_r \quad \text{(average retention time)}$$

$$= \frac{1}{A}\int_0^\infty C(t, z = L)t\,dt \tag{6.63}$$

$$\text{Second moment} = \mu_2 = \frac{1}{A}\int_0^\infty C(t, z = L)t^2 dt$$

$$\text{Second central moment} = \mu_2' = \sigma_t^2 \quad \text{(measure of peak spreading)}$$

$$= \frac{1}{A}\int_0^\infty C(t, z = L)\,(t - t_r)^2 dt$$

$$= \mu_2 - \mu_1^2 \tag{6.64}$$

Here the concentrations of eluate c(t,z) are measured at the end of the column, z = L. For a Gaussian peak, the second central moment is equal to the variance (σ_t^2), and moments higher than the second have value zero. The third and fourth moments of non-Gaussian peaks are measures of the deviation from a Gaussian, or symmetrical, shape.

Having a relationship for the concentration profile of solute, we can readily calculate the *yield* of solute obtained between two times, t_1 and t_2. In general, the yield is given by

$$\text{yield}\,\Big|_{t_1}^{t_2} = \frac{\text{(amount solute eluted)}}{\text{(total solute)}} = \frac{\int_{t_1}^{t_2} C \cdot F dt}{\int_0^\infty C \cdot F dt} \tag{6.65}$$

By substituting Eq. (6.61) into Eq. (6.65), we obtain

$$\text{yield}\,\Big|_{t_1}^{t_2} = \frac{1}{2}\left\{\text{erf}\left[\frac{(t_2 - t_r)}{\sqrt{2}\sigma_t}\right] - \text{erf}\left[\frac{(t_1 - t_r)}{\sqrt{2}\sigma_t}\right]\right\} \tag{6.66}$$

in which erf is the error function, defined as

$$\text{erf}(x) = \frac{2}{\sqrt{\pi}}\int_0^x e^{-u^2} du \tag{6.67}$$

Similarly, for a multicomponent mixture, the purity of solute i can be defined as

$$\text{purity of solute i} \Big|_{t_1}^{t_2} = \frac{C_{\max}(i)\text{yield}(i)\Big|_{t_1}^{t_2}}{\sum_j C_{\max}(j)\text{yield}(j)\Big|_{t_1}^{t_2}} \tag{6.68}$$

where the summation is over all j solutes in the system.

Now that we are familiar with the important features of Gaussian peaks, we can proceed to analyze the factors that control chromatographic separations. Two basic approches have been employed in the literature: first, the column can be described in terms of differential mass balances for the fluid and adsorbent phases; second, the column can be treated as a set of theoretical "plates," analogous to a distillation column. We will develop both approaches and show the relationship between them.

Dynamics of Adsorption Columns

We begin by considering a thin section of an adsorption column of cross-sectional area A, as shown in Figure 6.32, through which a fluid stream containing concentration c(t,z) of an adsorbable species is flowing. The concentration of adsorbed solute is denoted by q(t,z,r). Writing a differential mass balance on the solute in the liquid and solid phases, with volumes V_L and V_s, respectively, gives

$$\frac{\partial(V_L c)}{\partial t} + u_s \frac{\partial(Vc)}{\partial z} + \frac{\partial(V_s \bar{s})}{\partial t} = D_L \frac{\partial^2(Vc)}{\partial z^2} \tag{6.69}$$

(accumulation in liquid) (convective flow) (accumulation in solid) (axial dispersion)

where V is the column volume [= ($V_L + V_s$)], u_s is the superficial liquid velocity [= (F/A), the total volumetric flow rate divided by the column cross section], and $\bar{s}(t,z)$ is the total concentration of solute within the adsorbent, equal to

$$\bar{s} = \bar{c}_i \beta + \rho_p \bar{q} \tag{6.70}$$

where $\bar{q}(t,z)$ is the total concentration in the solid and $\bar{c}_i(t,z)$ is the total concentration in the pore. For spherical particles of radius R, these parameters can be written as

$$\bar{c}_i(t,z) = \frac{\int_0^R c_i(t,r,z) 4\pi r^2 dr}{\frac{4}{3}\pi R^3} = \frac{3}{R^3}\int_0^R r^2 c_i dr$$

$$\bar{q}(t,z) = \frac{3}{R^3}\int_0^R r^2 q dr \tag{6.71}$$

and β is the particle porosity. The void fraction of the adsorbent bed is ε, which is equal to V_L/V (where V_L is the interstitial liquid volume, which does not include the liquid in the pores of the solid phase). The effect of non-ideal axial liquid flow is represented by the axial dispersion coefficient D_L (which is based on the column cross-sectional area).

At this point it is clear that our analysis of chromatography will include many variables with precise definitions. To help clarify the nomenclature, some of the more important definitions are summarized below:

$c(z,t)$	solute concentration in the interstitial liquid (mass/volume)
$c_i(z,r,t)$	solute concentration within the adsorbent pores (mass/pore volume)
$\bar{c}_i(z,t)$	average solute concentration within the adsorbent (pores) (mass/pore volume)
$q(z,r,t)$	local solid phase (sorbate) concentration (mass/mass dry solid)
$q_i^*(z,r,t)$	solid phase concentration in equilibrium with c_i (mass/mass dry solid)
$q^*(z,t)$	solid phase concentration in equilibrium with c (mass/mass dry solid)
$\bar{q}(z,t)$	average solid phase concentration (mass/mass dry solid)
$\bar{s}(z,t)$	$\bar{c}_i\beta + \rho_p\bar{q}$, average solute concentration in the particle, including pore liquid (mass/volume wet adsorbent)
ρ_p	particle density (mass dry solid/volume of wet particle)
β	particle porosity (pore volume/volume of wet particle)
V_L	interstitial liquid phase volume of bed
V_S	solid phase volume (solid + pore volume) in bed
ε	column void fraction $[V_L/(V_L + V_S)]$

The above mass balance can now be simplified as

$$\varepsilon\frac{\partial c}{\partial t} + u_s\frac{\partial c}{\partial z} + (1-\varepsilon)\frac{\partial \bar{s}}{\partial t} = D_L\frac{\partial^2 c}{\partial z^2} \tag{6.72}$$

$$\frac{\partial c}{\partial t} + \left(\frac{u_s}{\varepsilon}\right)\frac{\partial c}{\partial z} + \left(\frac{1-\varepsilon}{\varepsilon}\right)\frac{\partial \bar{s}}{\partial t} = \frac{D_L}{\varepsilon}\frac{\partial^2 c}{\partial z^2} \tag{6.73}$$

or, equivalently,

$$\frac{\partial c}{\partial t} + u_i\frac{\partial c}{\partial z} + \left(\frac{1-\varepsilon}{\varepsilon}\right)\frac{\partial \bar{s}}{\partial t} = \frac{D_L}{\varepsilon}\frac{\partial^2 c}{\partial z^2} \tag{6.74}$$

where u_i is the *interstitial* fluid velocity of solvent (mobile phase) in the column.

In practice, the sorbate concentration is usually much greater than the liquid phase concentration, and the accumulation of solute in the pore liquid can be neglected. Thus, we can assume $\partial\bar{c}_i/\partial t \cong 0$ and $\partial\bar{s}/\partial t \cong \rho_p\partial\bar{q}/\partial t$. Furthermore, if we neglect axial dispersion and assume plug flow, Eq. (6.74) reduces to

$$\frac{\partial c}{\partial t} + u_i\frac{\partial c}{\partial z} + (\rho_p)\left(\frac{1-\varepsilon}{\varepsilon}\right)\frac{\partial \bar{q}}{\partial t} = 0 \tag{6.75}$$

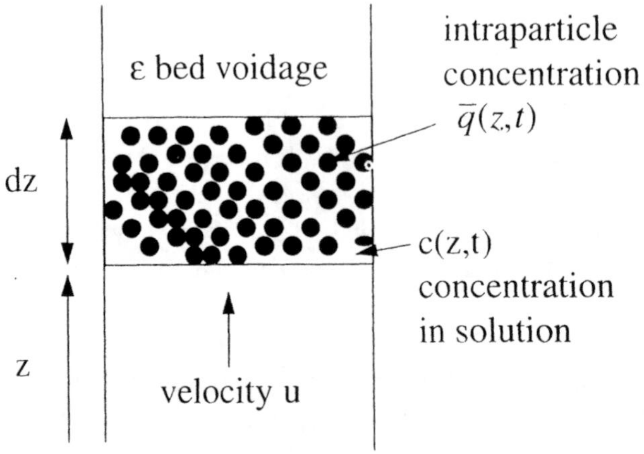

Figure 6.32. *Differential volume element of packed bed, indicating the liquid film and intraparticle mass transfer resistances.*

The simplest model of adsorption assumes an instantaneous equilibrium between solute in the pore liquid and sorbate in the solid phase at all points in the column. Representing the equilibrium isotherm in general form as

$$q^* = f(c) \tag{6.76}$$

and assuming negligible resistance to mass transfer (i.e., $\bar{q}$ = q and the absence of a liquid film resistance) and rapid equilibrium throughout the column (i.e., q^* = q) we can write

$$\left(\frac{\partial \bar{q}}{\partial t}\right)_z = \left(\frac{\partial q^*}{\partial t}\right)_z = \frac{dq^*}{dc}\left(\frac{\partial c}{\partial t}\right)_z \tag{6.77}$$

or, combining Eq. (6.77) with Eq. (6.75),

$$\left(\frac{\partial c}{\partial t}\right)_z + \frac{u_i}{\left[1+\left(\frac{1-\varepsilon}{\varepsilon}\right)\left(\frac{dq^*}{dc}\right)\rho_p\right]}\left(\frac{\partial c}{\partial z}\right)_t = 0 \tag{6.78}$$

which has the form of the kinematic wave equation:

$$\left(\frac{\partial c}{\partial t}\right)_z + w(c)\left(\frac{\partial c}{\partial z}\right)_t = 0 \tag{6.79}$$

A solution to Eq. (6.79) that yields c(t,z) can be obtained by the *method of characteristics*[16].

The velocity at which the species propagates through the column (w) is thus given by

$$w(c) = \left(\frac{\partial z}{\partial t}\right) = -\frac{(\partial c/\partial t)_z}{(\partial c/\partial z)_t} = \frac{u_i}{\left[1+\left(\frac{1-\varepsilon}{\varepsilon}\right)\left(\frac{dq^*}{dc}\right)\rho_p\right]} \tag{6.80}$$

and the mean retention time is equal to

(16) See R. Aris and N. Amundsen, *Mathematical Methods in Chemical Engineering, Vol. II, First Order Partial Differential Equations and Applications* (Prentice Hall, NJ, 1973).

$$\bar{t} = \frac{L}{u_i}\left[1 + \left(\frac{1-\varepsilon}{\varepsilon}\right)\left(\frac{dq^*}{dc}\right)\rho_p\right] \tag{6.81}$$

If the equilibrium isotherm is linear, $(dq^*/dc) = q^*/c = K$, where K is constant and is called the *partition ratio* or *distribution coefficient.*

The complexity of mathematical models for adsorption systems depends markedly on the type of isotherm involved and on the degree of mass-transfer resistance encountered. We are generally only interested in the exit concentration profile c(t, z = L) for either a pulse or a step-change input of solute at the column inlet. So far we have focused on the case of negligible mass transfer combined with rapid equilibrium. However, these conditions are rarely met in practice. Moreover, the adsorption of biological species seldom obeys a linear isotherm. Thus, more general models of chromatography encompass some combination of mass transfer resistances (as many as three different resistances, in some cases), axial dispersion effects, and nonlinear adsorption. Such models consist of a material balance for the fluid phase (either Eq. 6.74 or Eq. 6.75) together with a rate expression for solute adsorption.

To solve Eq. (6.74) or (6.75), we thus require an expression for $\partial\bar{q}/\partial t$ that describes the movement of solute into the solid phase. The rate of change of the average concentration of solute in the particle, $\bar{q}$, is

$$(1-\varepsilon)\rho_p\frac{\partial\bar{q}}{\partial t} \;=\; \frac{3}{R}(1-\varepsilon)D_{eff}\frac{\partial c_i}{\partial r}\Big|_{r=R} \tag{6.82}$$

where D_{eff} is the effective diffusivity of solute in the particle and $c_i(r,z,t)$ is the intraparticle liquid-phase concentration of solute. The term $[3(1 - \varepsilon)/R]$ is the total particle surface area per unit bed volume.

Diffusion of solute within a particle is described by the following unsteady-state mass balance:

$$D_{eff}\left(\frac{\partial^2 c_i}{\partial r^2} + \frac{2}{r}\frac{\partial c_i}{\partial r}\right) - \beta\frac{\partial c_i}{\partial t} - \rho_p\frac{\partial q}{\partial t} = 0 \tag{6.83}$$

where β is the void fraction *within* the particle and ρ_p is the density of the particle. Recall that q is the *adsorbed* solute concentration (gm/gm dry solid).

To proceed further we must relate the intraparticle liquid-phase concentration (c_i) to the external solute concentration (c). If there is a liquid film surrounding the particle through which the solute must diffuse, we can write

$$k_s(c - c_i|_{r=R}) \;=\; D_{eff}\frac{\partial c_i}{\partial r}\Big|_{r=R} \tag{6.84}$$

where k_s is the mass transfer coefficient for diffusion through the film. If the film mass transfer resistance is small [i.e., if $(k_sR)/D_{eff}$ is large], then $c_i(R,z,t) = c(z,t)$ and we can use this expression as a boundary condition in Eq. (6.83).

We now need to relate the intraparticle solute concentration c_i to that adsorbed on the solid (q). The Langmuir isotherm (Eq. 6.59) provides us with one such relationship:

$$q_i^* = \frac{q_{\max} K_L c_i}{1 + K_L c_i} \tag{6.59}$$

Here we assume that the solute in the pores is always in equilibrium with that on the solid surface. If, instead, solute binding is not in equilibrium, we can employ the binding kinetics that correspond to the Langmuir isotherm:

$$\frac{\partial q}{\partial t} = k_f c_i (q_{\max} - q) - k_r q \tag{6.85}$$

The forward (k_f, second-order) and reverse (k_r, first-order) rate constants for binding are related by the Langmuir constant:

$$K_L = \frac{k_f}{k_r} \tag{6.86}$$

Thus, the general adsorption problem must consider the following:

a) a solute mass balance on the liquid phase and the effects of axial dispersion [Eq. (6.74) or (6.75)],

b) diffusion of solute through a liquid film around the particles [Eq. (6.84)],

c) diffusion of solute within the particle to the adsorption sites [Eq. (6.83)],

d) the kinetics of solute binding [Eq. (6.85)].

Clearly, addressing all of these effects makes the mathematics quite complex, and simplifications are usually possible. For example, if the adsorption isotherm is linear, i.e., q = Kc, analytical solutions for the exiting concentration c(L,t) can be obtained, at least in principle. Solutions for a step change of solute feed, plug flow of fluid, and a linear isotherm are given in Table 6.10.

In lieu of a complete solution to the above problem, considerable information on axial dispersion, liquid-film and intraparticle mass transfer resistances, and the binding kinetics can be obtained from the analysis of the first and second moments of the exiting peaks from a pulse of solute fed to a column. We shall consider such an analysis next.

Table 6.10. *Analytical solutions for breakthrough curves for linear, isothermal dilute systems. [Adapted from D.M. Ruthven, Principles of Adsorption & Adsorption Processes (John Wiley & Sons, New York, 1984), pp. 236-239.]*

1. No intraparticle gradient (i.e., $\bar{q} = q$)	Linearized Rate Expression: $\frac{\partial \bar{q}}{\partial t} = k_f q_{\max} c - k_r \bar{q} = k_r(q^* - q) = k_f q_{\max}(c - c^*) = kK(c - c^*)$ where $q^* = \frac{k_f q_{\max}}{k_r} c$ and $q = \frac{k_f q_{\max}}{k_r} c^*$

Flow Model	Solution for Breakthrough Curve
Plug Flow [Eq. (6.75)]	$\frac{c}{c_0} = e^{-\zeta}\int_0^{\tau} e^{-\theta} I_0(2\sqrt{\zeta\theta})\,d\theta + e^{-(\tau+\zeta)} I_0(2\sqrt{\zeta\theta})$ $\tau = k(t - z/u_i), \quad \zeta = \frac{kKz}{u_i}\left(\frac{1-\varepsilon}{\varepsilon}\right)$ I_o = modified Bessel function (zero order) of the first kind $\frac{c}{c_0} = \frac{1}{2} erfc\left(\sqrt{\zeta} - \sqrt{\tau} - \frac{1}{8}\sqrt{\zeta} - \frac{1}{8}\sqrt{\tau}\right)$ (approximate solution; error < 0.6% for $\zeta > 2.0$) $\frac{c}{c_0} = \frac{1}{2} erfc(\sqrt{\zeta} - \sqrt{\tau})$ (asymptotic form for large ζ)
Dispersed Plug Flow [Eq. (6.69)]	$\frac{c}{c_0} = \exp\left(\frac{u_i z}{2D_L}\right)\left(F(t) + k\int_0^{t} F(t)\,dt\right)$ $F(t) = e^{-kt}\int_0^{t} I_0\left\{2k\left[K\left(\frac{1-\varepsilon}{\varepsilon}\right)\theta(t-\theta)\right]^{1/2}\right\}\frac{z}{2\sqrt{\pi D_L \theta^3}}\exp\left[\frac{-u_i^2}{4D_L\theta} - \frac{u_i^2\theta}{4D_L} - kK\theta\left(\frac{1-\varepsilon}{\varepsilon}\right) - k\theta\right]d\theta$ $\frac{c}{c_0} = \frac{1}{2} erfc\left\{\left(\frac{\bar{t}-t}{2\bar{t}}\right)\left(\frac{D_L}{u_i z}\left(\frac{t}{\bar{t}}\right)\right)^{-1/2}\right\}, \quad \bar{t} = \frac{z}{u_i}\left\{1 + K\left(\frac{1-\varepsilon}{\varepsilon}\right)\right\}$ (asymptotic form for k → ∞, long column)

2. Intraparticle Diffusion Control (no external mass transfer resistance):

$$\rho_p \frac{\partial q}{\partial t} = \frac{1}{r^2}\frac{\partial}{\partial r}\left(D_{eff} r^2 \frac{\partial q}{\partial r}\right), \quad q\left(R, t - \frac{z}{u_i}\right) = q^* = Kc$$

$$\bar{q} = \frac{3}{R^3}\int_0^R q r^2\,dr, \quad \frac{\partial q}{\partial r}\left(0, t - \frac{z}{u_i}\right) = 0$$

Flow Model	Solution for Breakthrough Curve
Plug Flow [Eq. (6.75)]	$\frac{c}{c_0} = \frac{1}{2} + \frac{2}{\pi}\int_0^{\infty} \exp[-\zeta H_1(\lambda)/5]\sin[2\lambda^2\tau/15 - \zeta H_2(\lambda)/5]\frac{d\lambda}{\lambda}$ $H_1(\lambda) = \lambda[\sinh(2\lambda) + \sin(2\lambda)]/[\cosh(2\lambda) - \cos(2\lambda)] - 1$ $H_2(\lambda) = \lambda[\sinh(2\lambda) - \sin(2\lambda)]/[\cosh(2\lambda) - \cos(2\lambda)]$ $\zeta = \frac{15D_{eff}Kz}{R^2\rho_p u_i}\left(\frac{1-\varepsilon}{\varepsilon}\right), \quad \tau = \left(\frac{15D_{eff}}{R^2}\right)(t - z/u_i)$ $\frac{c}{c_0} = \frac{1}{2}\mathrm{erfc}\left(\frac{\zeta - \tau}{2\sqrt{\zeta}}\right)$ (asymptotic form for large ζ)

Table 6.10 Continued.

3. Intraparticle Diffusion Control with External Film Resistance:	$\rho_p \frac{\partial q}{\partial t} = \frac{1}{r^2}\frac{\partial}{\partial r}\left(D_{eff} r^2 \frac{\partial q}{\partial r}\right),\quad D_{eff}\frac{\partial q}{\partial r}(R,t) = \frac{3k_s}{R_p}[c(z,t) - q(R_p,t)/K]$ $\bar{q} = \frac{3}{R_p^3}\int_0^{R_p} qR^2 dR,\quad \frac{\partial q}{\partial R}(0,t) = 0$
Flow Model	Solution for Breakthrough Curve
Plug Flow [Eq. (6.75)]	As in 2 (above) with $H_1(\lambda)$ and $H_2(\lambda)$ replaced by $H_1'(\lambda, \nu)$ and $H_2'(\lambda, \nu)$ $H_1'(\lambda,\nu) = [H_1 + \nu(H_1\lambda^2 + H_2^2)]/[(1+\nu H_1)^2 + (\nu H_2)^2]$ $H_2'(\lambda,\nu) = H_2/[(1+\nu H_1)^2 + (\nu H_2)^2] \qquad \nu = \frac{D_{eff}K}{R_p k_s}$

Moment Analysis of a Solute Pulse

We shall consider a chromatography column where axial dispersion and external and internal mass transfer resistances are important. For simplicity, let us assume that the number of binding sites on the particle is large with respect to the solute concentration, i.e., $q_{max} >> q$, and the solute binding kinetics can be considered as reversible first order. The mass balances are

$$\frac{\partial c}{\partial t} + u_i \frac{\partial c}{\partial z} + \rho_p\left(\frac{1-\varepsilon}{\varepsilon}\right)\frac{\partial \bar{q}}{\partial t} = \frac{D_L}{\varepsilon}\frac{\partial^2 c}{\partial z^2} \tag{6.74}$$

$$\frac{\partial \bar{q}}{\partial t} = \left(\frac{3}{R}\right)\frac{k_s}{\rho_p}(c - c_i(r=R)) = \left(\frac{3}{R}\right)\frac{D_{eff}}{\rho_p}\frac{\partial c_i}{\partial r}\Big|_{r=R} \tag{6.87}$$

$$D_{eff}\left(\frac{\partial^2 c_i}{\partial r^2} + \frac{2}{r}\frac{\partial c_i}{\partial r}\right) - \beta\frac{\partial c_i}{\partial t} - \rho_p\frac{\partial q}{\partial t} = 0 \tag{6.83}$$

$$\frac{\partial q}{\partial t} = k_f q_{max}\left(c_i - \left(\frac{1}{K_L q_{max}}\right)q\right) \tag{6.88}$$

The boundary conditions for the pulse input of duration t_0 are

$$\frac{\partial c_i}{\partial r} = 0 \quad \text{at} \quad r = 0 \quad \text{for} \quad t > 0$$

$c = 0$ at $z > 0$ for $t = 0$

$c_i = 0$ at $r \geq 0$ for $t = 0$

and the pulse is described by

$c = c_0$ at $z = 0$ for $0 \leq t \leq t_0$

Rather than attempt to solve this set of equations, we will calculate the first and second moments of the solution, which give the mean residence time and the extent of band broadening. This is most conveniently done by taking the Laplace transforms of the mass balance equations. We need not invert the transform since the nth moment can be obtained from

$$m_n = (-1)^{n-1} \lim_{p \to 0} \frac{\partial^n \bar{c}(p,L)}{\partial p^n} \tag{6.89}$$

where p is the Laplace variable. Alternatively, the Laplacian $\bar{c}(p,L)$ can be related to the moments by the expansion

$$\bar{c}(p,L) = m_0 - m_1 p + m_2 \frac{p^2}{2!} - m_3 \frac{p^3}{3!} \tag{6.90}$$

Thus, if $\bar{c}(p,L)$ can be expressed as a polynomial, m_n can be obtained from Eq. (6.90) by equating coefficients of p^n. Proceeding in this way, the first absolute moment and the second central moment are found to be

$$\mu_1 = \frac{m_1}{m_0} = \frac{L}{u_i}\{1 + \delta_0\} + \frac{t_0}{2} \tag{6.91}$$

where

$$\delta_0 = \frac{1-\varepsilon}{\varepsilon}\beta\left\{1 + \frac{\rho_p}{\beta}(K_L q_{max})\right\} \tag{6.92}$$

and

$$\mu_2' = \mu_2 - \mu_1^2 = \frac{2L}{u_i}\left\{\delta_1 + \frac{D_L}{\varepsilon}(1+\delta_0)^2 \frac{1}{u_i^2}\right\} + \frac{t_0^2}{12} \tag{6.93}$$

where

$$\delta_1 = \frac{1-\varepsilon}{\varepsilon}\beta\left[\frac{\rho_p}{\beta}\frac{(K_L q_{max})^2}{k_f} + \frac{R^2\beta}{15}\left(1 + \frac{\rho_p}{\beta}K_L q_{max}\right)^2\left(\frac{1}{D_{eff}} + \frac{5}{k_s R}\right)\right] \tag{6.94}$$

The first absolute moment shows the influence of the equilibrium constant and the bed and particle porosities. The second central moment contains terms accounting for axial dispersion, external and internal mass transfer, and the kinetics of adsorption. The bed characteristics and mass transfer effects can thus be determined by using a tracer pulse of a non-adsorbing substance (i.e., a tracer for which $K_L = 0$). In this case the first absolute moment becomes (assuming $t_0 \equiv 0$)

$$(\mu_1)_{inert} = \frac{L}{u_i}\left\{1 + \frac{1-\varepsilon}{\varepsilon}\beta\right\} \tag{6.95}$$

Therefore, a plot of (μ_1 - $t_0/2$) for an adsorbing material versus L/u_s (i.e. $\varepsilon L/u_s$) can be used to determine the binding constant, provided an inert pulse is used to obtain $(1 + \beta(1 - \varepsilon)/\varepsilon)$. For example, the case of bovine serum albumin (BSA) adsoption to a non-affinity monoclonal antibody-controlled pore glass column is illustrated in Figure 6.33.

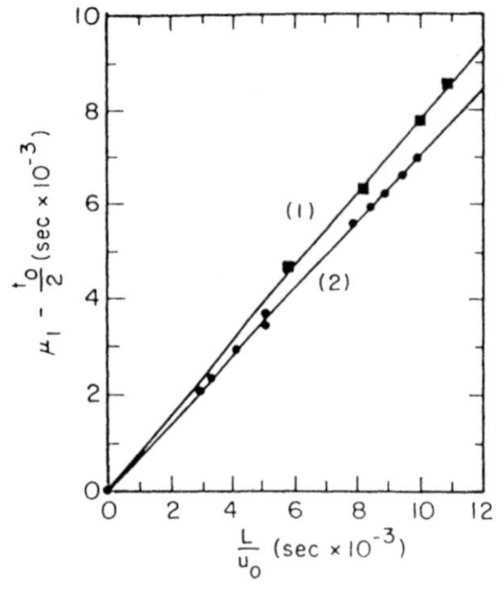

Figure 6.33. *First moments for 1 ml pulses of bovine serum albumin on a column packed with an immobilized antibody immunosorbent. Monoclonal antibody was immobilized to controlled pore glass (550 Å pore size; 100/200 mesh). (1), buffer with 0.1 M NaCl, slope = 0.78. (2), buffer with no NaCl, slope = 0.70. The presence of NaCl leads to increased adsorption due to decreased electrostatic repulsion. [From F.H. Arnold, H.W. Blanch, and C.R. Wilke, Chem. Eng. J., 30, B25 (1985).]*

The second central moment can be rewritten as

$$\frac{\mu_2' - t_0^2/12}{2(L/u_i)} = \delta_{Ad} + \delta_D + \delta_f + \frac{D_L}{\varepsilon}(1+\delta_0)^2\frac{1}{u_i^2} \tag{6.96}$$

where

$$\delta_{Ad} = \frac{1-\varepsilon}{\varepsilon}\beta\frac{\rho_p}{\beta}\frac{(K_L q_{\max})^2}{k_f} \quad \text{(adsorption term)}$$

$$\delta_D = \delta_0\frac{R^2\beta}{15}\left(1+\frac{\rho_p}{\beta}K_L q_{\max}\right)\frac{1}{D_{eff}}$$

(intraparticle diffusion term)

and

$$\delta_f \;=\; \delta_0 \frac{R^2\beta}{15}\left(1+\frac{\rho_p}{\beta}K_L q_{\max}\right)\frac{5}{k_s R}$$

(external film mass transfer term)

The axial dispersion term of the second central moment is sensitive to the bed packing and geometry. The axial dispersion coefficient has been correlated with the superficial velocity in the following manner:

$$D_L = \eta D_{mol} + l u_s \tag{6.97}$$

where D_{mol} is the molecular diffusion coefficient of the solute, and the "diffusibility" η and scale of dispersion l are characteristics of the column and the particle diameter and shape. The second term of Eq. (6.97) contains the velocity-dependent dispersion contribution, whereas the first term--the molecular diffusion contribution--is usually negligible in liquid chromatography.

Substituting Eq. (6.97) for D_L into Eq. (6.96) and setting $\eta D_{mol} = 0$ gives

$$\frac{\mu_2' - t_0^2/12}{2(L/u_i)} \;=\; \delta_{Ad} + \delta_D + \delta_f + \frac{l}{u_i}(1+\delta_0)^2 \tag{6.98}$$

Since the axial dispersion term contains a dependence on $1/u_i$, a plot of the left-hand side of Eq. (6.98) against $1/u_i$ yields the axial dispersion contribution as the slope. For chromatography of proteins, this is usually quite small. The intercept at $1/u_i = 0$ gives the sum of the particle contributions $(\delta_{Ad} + \delta_D + \delta_f)$ to the second central moment.

Miller and King[17] have shown that the product of the voidage and the axial Peclet number, εPe_{axial}, has a value of about 0.2 for $0.01 < Re' < 100$, where Re' is a modified Reynolds number:

$$\varepsilon Pe_{axial} \;=\; \varepsilon\frac{2Ru_s}{D_L} \;=\; 0.2 \tag{6.99}$$

$$Re' \;=\; \frac{2Ru_s}{\nu(1-\varepsilon)} \tag{6.100}$$

Thus, Eqn (6.98) can also be written as

$$\frac{\mu_2' - t_0^2/12}{2(L/u_i)} \;=\; \delta_{Ad} + \delta_D + \delta_f + \frac{2R\varepsilon^2}{0.2u_i}(1+\delta_0)^2 \tag{6.101}$$

Analysis of Breakthrough Behavior

When a step change in solute feed concentration is applied to an initially unsaturated column, the breakthrough behavior can be determined for the case of favorable equilibria (i.e., $R_{eq} < 1$). As the solute passes through the bed a "constant pattern" of liquid phase

(17) S.F. Miller and C.J. King, *A.I.Ch.E. J.*, **12**, 767 (1966).

concentration develops and the average solid and liquid phase concentrations no longer need to be considered separately. In the absence of axial dispersion, the shape of the constant pattern breakthrough curve depends only on the rate-determining mass transfer step. The slower this step, the more disperse the breakthrough curve.

The breakthrough curve can be described in terms of a dimensionless effluent concentration, $X = c/c_0$, and a dimensionless mean solid phase concentration, $Y = q/q_0$, where c_0 is the feed solute concentration and q_0 is the sorbate concentration in equilibrium with c_0. The assumption of a constant pattern in the bed leads to X=Y everywhere. In addition, a dimensionless throughput parameter T is defined as

$$T = \frac{(V - \varepsilon V_B)}{\Gamma V_B} \tag{6.102}$$

where V is the throughput volume, V_B is the bed volume, and Γ is a partition coefficient defined as

$$\Gamma = \frac{q_o^* \rho_B}{c_0} \tag{6.103}$$

With the above definitions, the area behind the breakthrough curve is equal to unity. Mass transfer resistances can now be introduced as numbers of transfer units, N_t. For the different possible mechanisms they are defined as[18]

pore diffusion $$N_{t,pore} = \frac{15 D_{eff}(1-\varepsilon) L}{R^2 u_s}$$

solid homogeneous particle diffusion $$N_{t,p} = \frac{15 \psi_p D_p \rho_B q_0^* L}{R^2 u_s c_0}$$

film mass transfer $$N_{t,f} = k_s \frac{3}{R} \frac{L}{u_s}$$

axial dispersion $$N_{t,d} = \frac{Pe_{axial} L}{2R} = \frac{u_s L}{D_L}$$

The values of these dimensionless numbers can be estimated from known mass transfer correlations and a knowledge of the particle size and properties. The rate-controlling mechanism will be the one with the smallest N_t value. If several mechanisms have

(18) For further discussion of the these equations, see F.H. Arnold et al., *ACS Symp. Ser.*, **271**, 113 (1985); and T. Vermeulen, G. Klein, and N.K. Heisler, "Adsorption and Ion Exchange," in *Chemical Engineers' Handbook*, R.H. Perry, C.H. Chilton, eds. (McGraw-Hill, New York, 1973).

comparable values, they must all be considered. Generally, $N_{t,d}$ is large (i.e., axial dispersion is not important) for proteins and other large molecules. Pore diffusion is likely to be rate controlling in most porous packings.

The dimensionless breakthrough curve has the general form

$$X = f(N_t, T) \tag{6.104}$$

As we have seen, the definition of N_t depends on the rate-determining mechanism. The exact form of Eq. 6.104 also varies, as shown by the solutions given in Table 6.11.

Table 6.11. *Dimensionless breakthrough curves for irreversible adsorption*

Rate Determining Mechanism(s)	Dimensionless Breakthrough Curve[a]
Pore Diffusion	$N_{t,pore}(T-1) = 2.44 - 3.66(1-X)^{1/2}$ $N_{t,pore} = \dfrac{15D_{eff}(1-\varepsilon)L}{R^2 u_s}$
Solid Homogeneous Particle Diffusion	$N_{t,p}(T-1) = -1.69[\ln(1-X^2)+0.61]$ $N_{t,p} = \dfrac{15\psi_p D_p \rho_B q_0 L}{R^2 u_s c_0}$ $\psi_p = 0.590$ for irreversible adsorption
External Fluid Film Mass Transfer and Pore Diffusion	$T-1 = \left(\dfrac{1}{N_{t,pore}} + \dfrac{1}{N_{t,f}}\right)\left\{\phi(X) + \dfrac{N_{t,pore}}{N_{t,f}}(\ln X + 1)\right\}\left(\dfrac{N_{t,pore}}{N_{t,f}} + 1\right)^{-1}$ $N_{t,f} = k_s \dfrac{3}{R}\dfrac{L}{u_s}$ $\phi(X) \equiv 2.44 - 3.66(1-X)^{1/2}$
External Fluid Film Mass Transfer and Solid Homogeneous Particle Diffusion	$N_{t,f}(T-1) = -mN_{t,p}(T-1) = \ln\dfrac{X}{\beta} + 1 + m \quad 0 \le X \le \beta$ $= m\left[\ln\dfrac{(1-X)}{(1-\beta)}\right] + 1 + m \quad \beta \le X \le 1$ $m = -\dfrac{N_{t,f}}{N_{t,p}};\ \beta = 1/(1 + N_{t,f}/N_{t,p})$

[a]Assumes $c(0,z) = \overline{q}(0,z) = 0$ for $0 < z < L$, and $c(t,0) = c_0$ for $t \ge 0$.

The assumption of a single rate-limiting step has been substantiated in practice. For example, in the case of anti-arsanilic acid monoclonal antibodies immobilized to controlled-pore glass, good agreement between experimental and theoretical breakthrough curves was obtained using the pore diffusion equations with $N_{t,pore} = 8$ (Problem 6.14).

Plate Theory

Although the differential mass balances of Section 6.4.2.2 may represent the most rigorous and accurate description of chromatography, alternative models do exist. In fact, the first mathematical model of chromatography, published by Martin and Synge in the *Biochemistry Journal* in 1941, takes a very different approach. This description, known as plate theory, models the column as a series of discrete, well-mixed stages. Plate theory is analogous to the "tanks-in-series" model for a nonideal flow reactor. The number of stages is a direct measure of axial dispersion and mass-transfer resistance in the system. Although plate theory is not a rigorous description of chromatography, it can be very useful for elucidating key factors that govern the efficiency and performance of chromatographic columns. Thus, the plate theory approach has been widely used, and many variations of Martin and Synge's original treatment have appeared over the years. The following presentation of plate theory combines the original concept with some of these more recent developments.

Assuming the column is comprised of individual stages, or plates, we can write

$$H = \frac{L}{N_p} \tag{6.105}$$

where H is the *height equivalent of a theoretical plate* (HETP), L is the column length, and N_p is called the *theoretical plate number*, the *number of theoretical plates*, or simply the *plate count of the column*.

A general expression for the plate count in terms of moments can be obtained by considering flow through N equal-size ideal stirred tanks (the tanks-in-series model common to reaction engineering). If a pulse of inert tracer is injected into the first vessel, the output of tracer is described by the residence time distribution of the fluid, $\xi(t)$[19]. Thus for an impulse input, $\bar{c}(p) = \bar{\xi}(p)$, where $\bar{c}(p)$ is the Laplacian of the exiting tracer concentration. The Laplace transform of a material balance on the jth tank in the series gives

$$\frac{\bar{c}_j(p)}{\bar{c}_{j-1}(p)} = \frac{1}{1 + p(\bar{t}/N)} \tag{6.106}$$

where $\bar{t}$ is the total mean residence time in the N-tank series. Eq. (6.106) indicates that

$$\bar{c}_N(p) = \bar{\xi}(p) = (1 + p\bar{t}/N)^{-N} \tag{6.107}$$

If N is sufficiently large, Eqn (6.107) can be rewritten as

(19) O. Levenspiel, *Chemical Reaction Engineering, Second Edition* (John Wiley & Sons, New York, 1972), pp. 253-325.

$$\bar{c}_N(p) = 1 - p\bar{t} + \frac{(-N)(-N-1)}{2}\left(\frac{p\bar{t}}{N}\right)^2 \tag{6.108}$$

where we have neglected the higher order terms of the expansion. The second moment of the output curve can now be calculated from Eq. (6.108) and Eq. (6.89):

$$m_2 = \frac{N(N+1)\bar{t}^2}{N^2} = \frac{(N+1)}{N}\bar{t}^2 \tag{6.109}$$

Since the second central moment is equal to

$$\mu_2' = \sigma_t^2 = \mu_2 - \mu_1^2 \tag{6.110}$$

the variance of the tracer output, σ_t^2, is given by

$$\sigma_t^2 = \left(\bar{t}^2 + \frac{\bar{t}^2}{N}\right) - \bar{t}^2 = \frac{\bar{t}^2}{N} \tag{6.111}$$

This result also applies to plate theory, provided the plate count is large enough for the exiting peak (i.e., the chromatogram in response to a pulse input) to be Gaussian. For such a case the plate count can be expressed as the squared average retention time (the squared first moment) divided by the variance (the second central moment):

$$N_p = \frac{t_r^2}{\sigma_t^2} \tag{6.112}$$

and H can be written as

$$H = \frac{L\sigma_t^2}{t_r^2} \tag{6.113}$$

Strictly speaking, plate theory is applicable only when the exiting peaks are Guassian. However, plate theory can also provide useful guidelines for the optimal choice of operating conditions under other, less ideal circumstances, such as those often encountered in preparative chromatography. Moreover, the concept of plate height can be very useful in the interpretation of chromatographic results and in the diagnosis of chromatographic columns. The height equivalent of a theoretical plate is also a useful measure of column efficiency.

For plate theory to be useful in predicting how a particular operating variable will affect the chromatographic process, we must be able to relate N_p or H to that operating variable. Understanding the dependence of plate height on flow velocity is particularly important. The relationship between H and u is expressed by various *plate height equations*. We will now describe a plate height equation derived from the moment method.

Combining Eqs. (6.91), (6.93), and (6.110) we can write, for a short pulse (i.e., $t_0 \cong 0$):

$$\frac{L\sigma_t^2}{t_r^2} = \frac{2L^2}{u_i}\frac{\left\{\delta_1 + \frac{D_L}{\varepsilon}(1+\delta_0)^2\frac{1}{u_i^2}\right\}}{(L/u_i)^2\{1+\delta_0\}} \tag{6.114}$$

Therefore,

$$H = \frac{\left(\frac{u_i}{L}\right)^2 \left\{ \frac{2L^2}{u_i^3}\frac{D_L}{\varepsilon}(1+\delta_0)^2 + \frac{2L^2\delta_1}{u_i} \right\}}{(1+\delta_0)^2} \tag{6.115}$$

$$H = \frac{2D_L}{\varepsilon u_i} + \frac{2u_i\delta_1}{(1+\delta_0)^2} \tag{6.116}$$

$$H = \frac{2D_L}{\varepsilon u_i} + \frac{\frac{2(1-\varepsilon)}{\varepsilon}\left[\frac{\rho_p(K_L q_{max})^2}{k_f} + \frac{R^2}{15}(\beta + \rho_p K_L q_{max})^2\left(\frac{1}{D_{eff}} + \frac{5}{k_s R}\right)\right]u_i}{\left[1 + \left(\frac{1-\varepsilon}{\varepsilon}\right)\beta + \left(\frac{1-\varepsilon}{\varepsilon}\right)\rho_p(K_L q_{max})\right]^2} \tag{6.117}$$

In terms of the superficial velocity, u_s, H becomes

$$H = \frac{2D_L}{u_s} + \frac{2(1-\varepsilon)\left[\frac{\rho_p(K_L q_{max})^2}{k_f} + \frac{R^2}{15}(\beta + \rho_p K_L q_{max})^2\left(\frac{1}{D_{eff}} + \frac{5}{k_s R}\right)\right]u_s}{[\varepsilon + (1-\varepsilon)\beta + (1-\varepsilon)\rho_p(K_L q_{max})]^2} \tag{6.118}$$

Because of its complexity, Eq. (6.118) is not very convenient to use for the interpretation of experimental results of the analysis of column performance. A simpler form of Eq. (6.118) is

$$H = \frac{2D_L}{u_s} + \frac{2[K'/k_d + (\alpha - \varepsilon + K')^2/K_{OL}a_p]u_s}{(\alpha + K')^2} \tag{6.119}$$

where k_d is a modified desorption rate constant [$k_d = k_f/q_{max} = k_f/(K_L q_{max})$] and

$K' = (1-\varepsilon)\rho_p(K_L q_{max})$ = (maximum bound solute)/(column volume) $\cdot K_L$

$\alpha = \varepsilon + (1-\varepsilon)\beta$ = fraction of column volume available to solute

$K_{OL}a_p = \left[\frac{R^2}{15(1-\varepsilon)}\left(\frac{1}{D_{eff}} + \frac{5}{k_s R}\right)\right]^{-1}$ = (column mass transfer coefficient) $\cdot$ (particle surface area).

If we again relate the dispersion coefficient to the molecular diffusivity and the superficial velocity [Eq. (6.97)], Eq. (6.119) becomes

$$H = 2l + \frac{2\eta D_{mol}}{u_s} + \frac{2[K'/k_d + (\alpha - \varepsilon + K')^2/K_{OL}a_p]u_s}{(\alpha + K')^2} \tag{6.120}$$

If the slight velocity dependence of k_s is neglected, Eq. (6.120) assumes the same form as the so-called *Van Deemter equation*[20], written in simplified form as

$$H = A + \frac{B}{u} + Cu \tag{6.121}$$

where

(20) J.J. Van Deemter, F.J. Zuiderweg, and A. Klinkenberg, *Chem. Eng. Sci.*, 5, 271 (1956).

$$A = 2l, \quad B = 2\eta D_{mol}, \quad C = 2\frac{[K'/k_d + (\alpha - \varepsilon + K')^2/K_{OL}a_p]}{(\alpha + K')^2} \tag{6.122}$$

The influence of molecular diffusion in liquid chromatography is usually quite small; hence, a suitable plate height equation for liquid chromatography is

$$H = 2l + \frac{2[K'/k_d + (\alpha - \varepsilon + K')^2/K_{OL}a_p]u_s}{(\alpha + K')^2} \tag{6.123}$$

Alternatively, in terms of the axial Peclet number (Eq. 6.99),

$$H = \frac{2d_p}{Pe_{\text{axial}}} + \frac{2[K'/k_d + (\alpha - \varepsilon + K')^2/K_{OL}a_p]u_s}{(\alpha + K')^2} \tag{6.124}$$

Eqs. (6.123) and (6.124) predict a linear dependence between H and u_s, which, as shown in Figure 6.34, is typical for chromatographic separations of proteins.

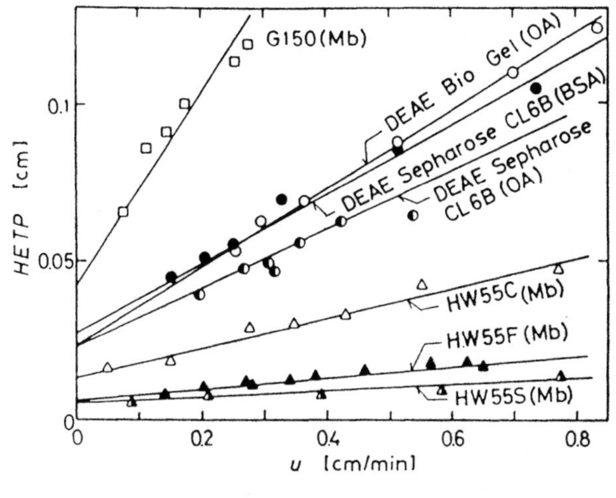

Figure 6.34. *HETP versus linear mobile-phase velocity us for various gel filtration columns (G150, HWFFC, and HW55F) and for ion-exchange gel columns at high ionic strength (DEAE Bio Gel and DEAE Sepharose). Mb, OA, and BSA: myoglobin, ovalbumin, and bovine serum albumin, respectively. [From S. Yamamoto et al., Ion Exchange Chromatography of Proteins, Marcel Dekker, Inc., New York, 1988.]*

Plate theory is also useful for evaluating the resolution of a chromatographic separation. Specifically, for two closely spaced peaks the resolution (Eq. 6.58) can be approximated as the product of three terms that reflect the *selectivity*, *efficiency*, and *capacity* of the chromatographic system. To derive this expression, we begin with Eq. (6.58) for the resolution of two adjacent peaks:

$$R_s = \frac{t_{r,2} - t_{r,1}}{(1/2)(w_1 + w_2)} \tag{6.58}$$

For Gaussian peaks, the baseline peak widths can be related to the standard deviation, σ_t, by $w = 4\sigma_t$ (Fig. 6.35). Assuming that the peaks have the same width, Eq. (6.58) becomes

$$R_s = \frac{t_{r,2} - t_{r,1}}{4\sigma_t} = \frac{1}{4}\frac{t_{r,1}}{\sigma_t}\left(\frac{t_{r,2} - t_{r,1}}{t_{r,1}}\right) \tag{6.125}$$

Combining Eq. (6.125) with the general definition of the plate count [Eq. (6.112)] yields

$$R_s = \frac{\sqrt{N_p}}{4}\left(\frac{t_{r,2} - t_{r,1}}{t_{r,1}}\right) \tag{6.126}$$

We now introduce the *retention factor* (also referred to in the literature as the *capacity factor*), k_i, defined for species *i* as

$$t_{r,i} = t_0(1 + k_i) \tag{6.127}$$

Therefore, Eq. (6.126) can be expanded as

$$R_s = \frac{\sqrt{N_p}}{4}\left[\frac{t_0(1 + k_2) - t_0(1 + k_1)}{t_0(1 + k_1)}\right] = \frac{\sqrt{N_p}}{4}\left[\frac{k_2 - k_1}{1 + k_1}\right] \tag{6.128}$$

$$R_s = \frac{\sqrt{N_p}}{4}\left(\frac{k_1}{1 + k_1}\right)\left(\frac{k_2 - k_1}{k_1}\right) = \frac{\sqrt{N_p}}{4}\left(\frac{k_1}{1 + k_1}\right)\left(\frac{k_2}{k_1} - 1\right) \tag{6.129}$$

Finally, defining the *separation factor* $\alpha = k_2/k_1$ for solutes 1 and 2, and recognizing that $k_1 \cong k_2 \cong k'$ (the average retention factor), leads to

$$R_s = \underbrace{\left(\frac{\sqrt{N_p}}{4}\right)}_{\text{(Efficiency)}} \cdot \underbrace{\left(\frac{k'}{1 + k'}\right)}_{\text{(Retention)}} \cdot \underbrace{(\alpha - 1)}_{\text{(Selectivity)}} \tag{6.130}$$

The *selectivity* is related to the ratio of retention factors and is a measure of the chromatographic system's discriminatory power. This parameter can be varied by changing the composition of the mobile and/or stationary phases. The *retention* measures the fraction of eluate present in the stationary phase and therefore expresses the retentive power of the system. The average retention factor k′, and hence the retention, is related to the *solvent strength*, i.e., the ability of the solvent to provide a small or large k′ value (strong solvents give small k′ values; weak solvents, large k′ values)[21]. The *efficiency* measures the relative narrowness of the peaks, and is governed by the column length L and the various factors affecting H (Eqs. (6.105) and (6.123)]. The different ways in which each of the parameters of Eq. (6.130) affects resolution are shown graphically in Figure 6.35.

---

(21) For listings of solvent strengths and further discussion of resolution in liquid chromatography, see L.R. Snyder and J.J. Kirkland, *Introduction to Modern Liquid Chromatography* (John Wiley & Sons, New York, 1974).

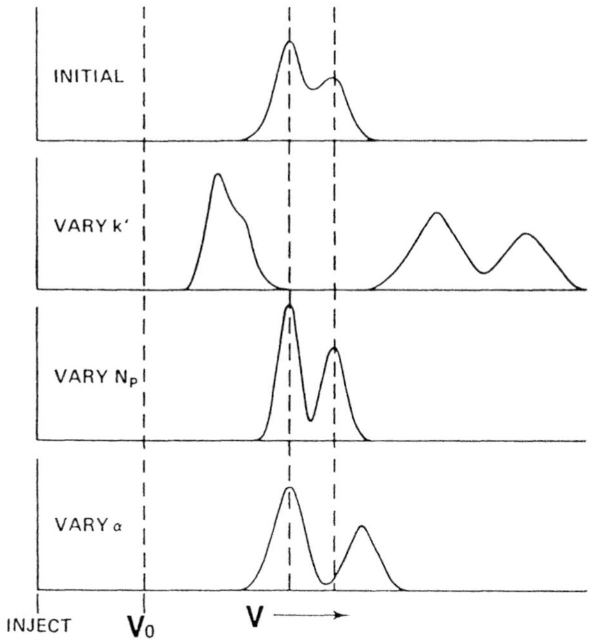

Figure 6.35. Effects of different parameters on chromatographic resolution. If k' is increased, the resolution improves but the peaks become broader, as shown. Increasing N_p improves resolution due to the narrowing of peak widths. Increasing α will also increase resolution, and α is the most powerful parameter for improving resolution. [Adapted from L.R. Snyder and J.J. Kirkland, Introduction to Modern Liquid Chromatography (John Wiley & Sons, New York, 1974).]

Summary of Key Equations

Our analysis of chromatography has included many equations. Below we restate some of the more important equations for the design and interpretation of chromatographic separations. The parameters are defined in the text where the equations originally appeared and in the *Nomenclature* section.

The resolution, R_s, of two adjacent peaks in a chromatogram is defined quantitatively as

$$R_s = \frac{t_{r,2} - t_{r,1}}{(1/2)(w_1 + w_2)} \tag{6.58}$$

Alternatively, R_s can be written as

$$R_s = \underset{\text{(Efficiency)}}{\left(\frac{\sqrt{N_p}}{4}\right)} \cdot \underset{\text{(Retention)}}{\left(\frac{k'}{1+k'}\right)} \cdot \underset{\text{(Selectivity)}}{(\alpha - 1)} \tag{6.130}$$

Adsorption in many systems follows the hyperbolic Langmuir isotherm:

$$q^* = \frac{q_{\max} K_L c}{1 + K_L c} \tag{6.59}$$

For a pulse input that yields a Gaussian peak on the chromatogram, the eluate concentration, C, is given by

$$C = \frac{1}{F}\frac{M}{\sigma_t\sqrt{2\pi}} e^{-0.5[(t_r - t)/\sigma_t]^2} = C_{\max} e^{-0.5[(t_r - t)/\sigma_t]^2} \tag{6.61}$$

For a thin section of an adsorption column, we can write a differential mass balance on the solute in the liquid and solid phases as

$$\frac{\partial c}{\partial t} + u_i \frac{\partial c}{\partial z} + \left(\frac{1-\varepsilon}{\varepsilon}\right)\frac{\partial \bar{s}}{\partial t} = \frac{D_L}{\varepsilon}\frac{\partial^2 c}{\partial z^2} \tag{6.74}$$

Neglecting axial dispersion and assuming plug flow, Eq. (6.74) becomes

$$\frac{\partial c}{\partial t} + u_i \frac{\partial c}{\partial z} + (\rho_p)\left(\frac{1-\varepsilon}{\varepsilon}\right)\frac{\partial \bar{q}}{\partial t} = 0 \tag{6.75}$$

Assuming the number of binding sites on the particle is large with respect to the solute concentration, i.e., $q_{\max} >> q$, and the solute binding kinetics can be considered as reversible first order, the mass balances are

$$\frac{\partial c}{\partial t} + u_i \frac{\partial c}{\partial z} + \left(\frac{1-\varepsilon}{\varepsilon}\right)\frac{\partial \bar{q}}{\partial t} = \frac{D_L}{\varepsilon}\frac{\partial^2 c}{\partial z^2} \tag{6.74}$$

$$\frac{\partial \bar{q}}{\partial t} = \left(\frac{3}{R}\right)\frac{k_s}{\rho_p}(c - c_i(r = R)) = \left(\frac{3}{R}\right)\frac{D_{\mathit{eff}}}{\rho_p}\frac{\partial c_i}{\partial r}\Big|_{r=R} \tag{6.87}$$

$$D_{\mathit{eff}}\left(\frac{\partial^2 c_i}{\partial r^2} + \frac{2}{r}\frac{\partial c_i}{\partial r}\right) - \beta\frac{\partial c_i}{\partial t} - \rho_p\frac{\partial q}{\partial t} = 0 \tag{6.83}$$

$$\frac{\partial q}{\partial t} = k_f q_{\max}\left(c_i - \left(\frac{1}{K_L q_{\max}}\right)q\right) \tag{6.88}$$

The boundary conditions for the pulse input of duration t_0 are

$$\frac{\partial c_i}{\partial r} = 0 \quad \text{at} \quad r = 0 \quad \text{for} \quad t > 0$$

$c = 0$ at $z > 0$ for $t = 0$

$c_i = 0$ at $r \geq 0$ for $t = 0$

and the pulse is described by

$c = c_0$ at $z = 0$ for $0 \leq t \leq t_0$

Assuming negligible resistance to mass transfer (i.e., $\bar{q} = q$ and the absence of a liquid film resistance) and rapid equilibrium throughout the column (i.e., $q^* = q$), the velocity at which a species propagates through a packed column (w) is given by

$$w(c) = -\left(\frac{\partial z}{\partial t}\right) = -\frac{(\partial c/\partial t)_z}{(\partial c/\partial z)_t} = \frac{u_i}{\left[1+\left(\frac{1-\varepsilon}{\varepsilon}\right)\left(\frac{dq^*}{dc}\right)\rho_p\right]} \tag{6.80}$$

and the mean retention time is equal to

$$\bar{t} = \frac{L}{u_i}\left[1+\left(\frac{1-\varepsilon}{\varepsilon}\right)\left(\frac{dq^*}{dc}\right)\rho_p\right] \tag{6.81}$$

The dimensionless breakthrough curve for a step change in solute feed has the general form

$$X = f(N_t, T) \tag{6.104}$$

Specific forms of Eq. 6.104 are given in Table 6.11.

Assuming the column is comprised of individual stages, or plates, and the exiting peaks are Gaussian, the height equivalent of a theoretical plate, H, can be expressed as

$$H = \frac{L\sigma_t^2}{t_r^2} \tag{6.113}$$

In expanded form, the plate height equation is

$$H = \frac{2d_p}{Pe_{\text{axial}}} + \frac{2[K'/k_d + (\alpha - \varepsilon + K')^2/K_{OL}a_p]u_s}{(\alpha + K')^2} \tag{6.124}$$

where k_d is a modified desorption rate constant [$k_d = k_f/q_{max} = k_f/(K_L q_{max})$] and

$K' = (1-\varepsilon)\rho_p(K_L q_{max}) = $ (maximum bound solute)/(column volume) $\cdot K_L$

$\alpha = \varepsilon + (1-\varepsilon)\beta = $ fraction of column volume available to solute

$$K_{OL}a_p = \left[\frac{R^2}{15(1-\varepsilon)}\left(\frac{1}{D_{eff}} + \frac{5}{k_s R}\right)\right]^{-1} = \text{(column mass transfer coefficient)} \cdot \text{(particle surface area)}.$$

6.4.3 Electrophoresis

Electrophoresis is a leading method for resolving mixtures of charged macromolecules, either proteins or nucleic acids, on a laboratory scale. Electrophoretic separations are influenced by many factors, including the size (or molecular weight), shape, secondary structure, and charge of the macromolecule. These features can influence electrophoretic properties either separately or jointly. In general, electrophoresis is carried out by placing a solution of biopolymers (e.g., proteins) in a channel or gel, and applying a voltage at the ends of the channel. An electric field is thus formed and direct current passes through the solution. Proteins then migrate toward the cathode or the anode depending on their net charge. Upon completion of gel electrophoresis, the gel can be cut into pieces and individual proteins eluted from the separate slices. Alternatively, continuous operation is made possible by flowing an elution stream perpendicular to the applied field.

The electrophoretic mobility of a biopolymer (e.g., a protein or nucleic acid) in a gel is determined by the net charge to mass ratio of the migrating species. In the case of proteins, the ionization state of each amino acid varies with the pH (Figure 6.36). Thus, the charge to mass ratio of the protein, and hence the velocity and direction of protein motion, are sensitive to the pH of the electrophoretic medium. The optimum pH is that which maximizes the *difference* in charge between component proteins rather than the charge itself. Optimal pH values for acidic proteins fall in the neutral or slightly alkaline region (in which the pH is greater than the pI of the protein), and such proteins migrate toward the anode. Slightly acidic buffers (pH 4-5) tend to work best for basic proteins (pH < pI), inducing migration toward the cathode.

$$H_2N----\overset{\overset{\textstyle H}{|}}{\underset{\underset{\textstyle R}{|}}{C}}----COO^- \rightleftharpoons \quad ^+H_3N----\overset{\overset{\textstyle H}{|}}{\underset{\underset{\textstyle R}{|}}{C}}----COO^- \rightleftharpoons \quad ^+H_3N----\overset{\overset{\textstyle H}{|}}{\underset{\underset{\textstyle R}{|}}{C}}----COOH$$

High pH, z = -1   Neutral pH, z = 0   Low pH, z = +1

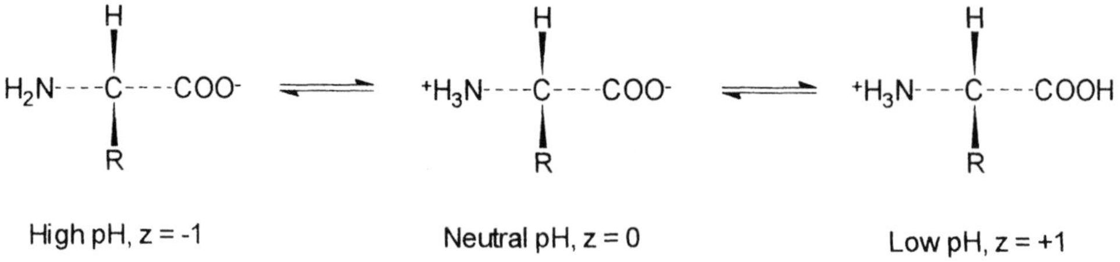

Figure 6.36. *Ionized forms of an amino acid as a function of pH.*

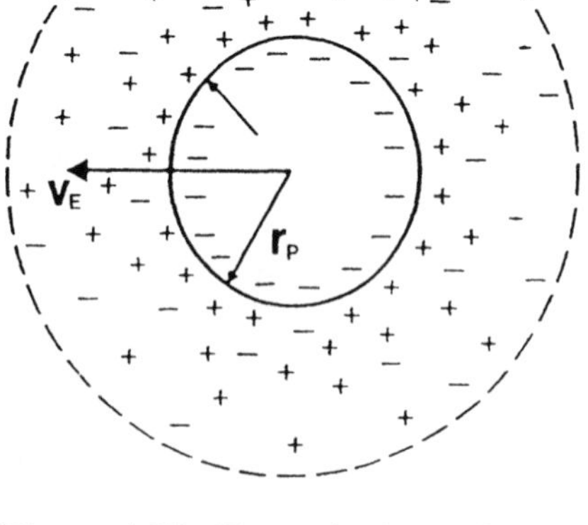

Figure 6.37. *Charged spherical particle moving at an electrophoretic velocity v_E.*

Friction against the gel network is also an important factor in electrophoretic mobility. Frictional resistance is determined by the size, shape, and rigidity of the protein, and by the pore size within the gel. Polyacrylamide is the most commonly employed gel, and polyacrylamide concentrations of 5 - 20% are typically used to achieve the proper gel porosity.

Electrophoresis can also be used to characterize proteins. In particular, molecular weights can be determined if the shape, rigidity, and charge to mass ratio are essentially uniform among different proteins. In practice these criteria are met by treating proteins with sodium dodecyl sulphate (SDS). The ionized detergent binds to the denatured protein and eliminates the influence of all factors other than molecular weight.

Elementary transport equations for electrophoresis can be developed by treating the migrating species as a charged particle of net charge q (Figure 6.37). At steady state, the Stokes drag on a charged sphere will be balanced by the electrical force; hence, in an electric field E,

$$qE = 6\pi\mu v_E r_p \tag{6.131}$$

where r_p is the radius of the particle and μ is the viscosity of the medium (assumed to be Newtonian). Solving for the velocity, v_E, we obtain

$$v_E = \frac{qE}{6\pi\mu r_p} \tag{6.132}$$

Eq. 6.132 can also be written in terms of the diffusion coefficient, D_s, which, for a sphere of radius r_p, is equal to $kT/6\pi\mu r_p$ (the Stokes-Einstein equation). Therefore,

$$v_E = \frac{kT}{6\pi\mu r_p}\left(\frac{zN_A e}{RT}\right)E = D_s\left(\frac{zF}{RT}\right)E \tag{6.133}$$

where z_p is the charge number (valence) of the particle, e is the elementary charge (note that q = ze), N_A is Avagadro's number, and *F* is Faraday's constant (= $N_A e$, the charge of one mole of singly ionized molecules). Moreover, since the electric field is the negative gradient of the electrostatic potential:

$$E = -\nabla\phi \tag{6.134}$$

we can write

$$v_E = -D_s\left(\frac{zF}{RT}\right)\nabla\phi \tag{6.135}$$

In the electrophoresis literature, Eq. 6.135 is often written in terms of the *electrophoretic mobility*, or the velocity per unit field:

$$m = \frac{v_E}{E} = D_s\left(\frac{zF}{RT}\right) \tag{6.136}$$

The mobility, m (= v_E/E), is a measure of how mobile the particle is in an electric field. Electrophoretic mobilities of normal human plasma components are given in Table 6.12.

Table 6.12. Mobilities of normal human plasma components[a].

Component	Concentration (g/100 ml)	Percent Total Solutes	Mobility m (10^5 cm^2/V sec)
Albumin	4.04	60	-5.9
α_1 -globulin	0.31	5	-5.1
α_2 -globulin	0.48	12	-4.1
β -globulin	0.81	12	-2.8
Fibrinogen	0.34	5	-2.1
γ-globulin	0.74	11	-1.0

[a]All data are in 0.1 M ionic strength diethylbarbiturate buffer of pH 8.6. [From P.A. Belter, E.L. Cussler, and W.-S. Hu, *Bioseparations: Downstream Processing for Biotechnology* (John Wiley & Sons, New York, 1988) p. 247.]

Eq. (6.132) is valid for particles only at very low ionic strengths, I, i.e., when $\kappa r_p \ll 1$, where κ is the *inverse Debye length.* The Debye length (λ_D in Figure 6.38) is a measure of the thickness of the double layer surrounding the particle, and is defined as[22]

$$\lambda_D = \kappa^{-1} = \left(\frac{\varepsilon RT}{2F^2 z^2 c} \right)^{1/2} \tag{6.137}$$

where c is the average molar concentration of counterion, and ε is the permittivity of the medium, which is equal to the permittivity of a vacuum, $\varepsilon_0 = 8.854 \times 10^{-12}$ C^2 N^{-1} m^2 (C V^{-1} m^{-1}), multiplied by the relative permittivity ε_r (also known as the dielectric constant).

For an aqueous solution of a symmetrical electrolyte at 25°C,

$$\lambda_D = \frac{9.61 x 10^{-9}}{(z^2 c)^{1/2}} \tag{6.138}$$

with λ_D in meters and c in mol m^{-3}. For example, the enzyme trypsin has a molecular weight of 23,200 and a radius of ca. 20Å. In NaCl near 25°C, κ^{-1} is equal to about 10Å for a 0.1 M solution, hence $\kappa r_p \cong 2$ and Eq. (6.132) is not applicable. (How does this situation change if the salt concentration is reduced to 1 mM?)

A more general relation, applicable for double layer thicknesses in the range defined by $0.2 < \kappa r_p < 50$, is

$$m = \frac{ze X_1(\kappa r_p)}{6\pi\mu r_p(1+\kappa r_p)} + \text{correction terms} \tag{6.139}$$

(22) R.F. Probstein, *Physicochemical Hydrodynamics, An Introduction* (Butterworths, Boston, 1989), p. 187.

where the function, $X_1(\kappa r_p)$, varies between unity in the low salt limit and 3/2 in the asymptotic limit of $\kappa r_p >> 1$. Correction terms for Eq. (6.139) have been determined by Booth[23].

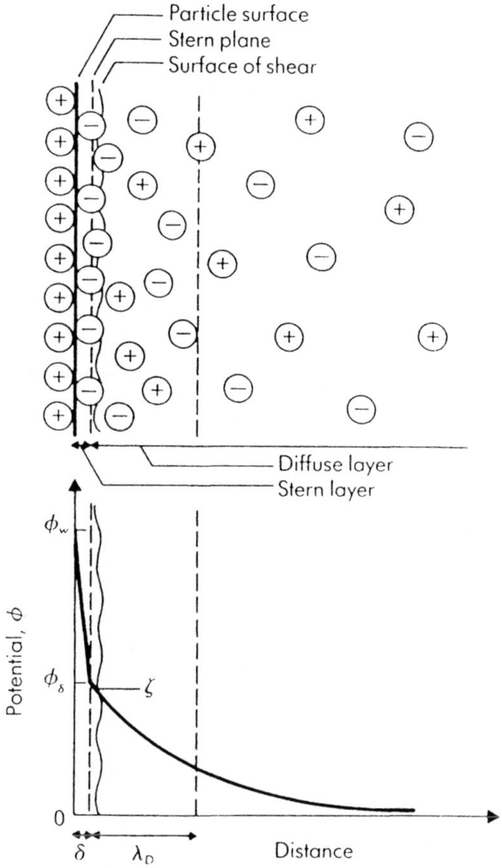

Figure 6.38. *Structure of electric double layer with ζ potential. [From R.F. Probstein, Physicochemical Hydrodynamics, An Introduction (Butterworths, Boston, 1989), p. 189.]*

The electrophoretic mobility can also be expressed in terms of the zeta (ζ), defined as the potential at the "shear surface" between the charge surface and the electrolyte solution (Figure 6.38):

(23) F. Booth, "The Catophoresis of Spherical Particles in Strong Fields," *J. Chem. Phys.*, **18** (10) 1361 (1950).

$$m = \frac{2}{3}\frac{\zeta\varepsilon(1 + r_p/\lambda_D)}{\mu} \tag{6.140}$$

However, for practical purposes, the ζ potential remains an empirical parameter since it cannot be related to the titratable charge of a protein *a priori*.

Scale-up of gel electrophoresis is hindered by ohmic heating of the gel. The heat generated is equal to the product of the current and voltage, and this heating can cause free convection and mixing within the gel, thereby destroying the separation. Too much heat can also denature the protein. Although the mixing problem can be alleviated on a small scale by using gels of high viscosity, electrophoresis equipment designed and built prior to the mid-1970's did not translate well to an industrial scale. However, recent designs have addressed the poor scaling characteristics of electrophoresis and the future of industrial-scale electrophoretic processing appears promising.

A popular variant of conventional electrophoresis is isoelectric focusing (IEF). This technique employs a medium (e.g., a gel) of changing pH (i.e., a pH gradient) to fractionate proteins of different isoelectric points. Application of an electric field through the medium causes each protein to migrate until it reaches a region in the gradient where the isoelectric point of the protein is equal to the pH at that point. At this point, the net charge of the protein is zero and the protein is no longer influenced by the electric field. IEF can be used by itself or in combination with conventional electrophoresis. The latter approach, in which IEF and electrophoresis are performed perpendicular to each other, is an example of two-dimensional fractionation. This technique is extremely powerful for analyzing complex mixtures of proteins (Figure 6.39).

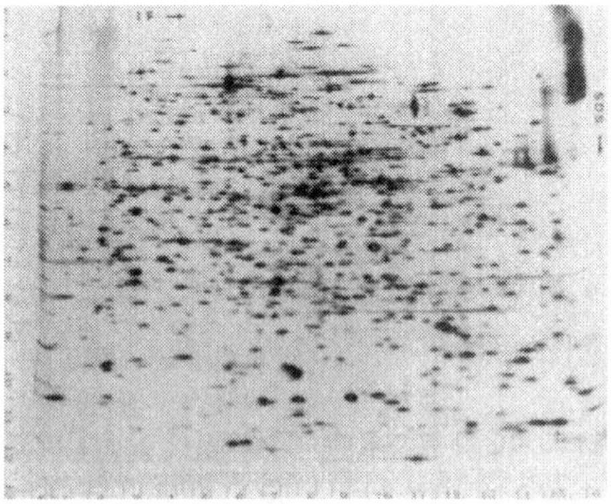

Figure 6.39. *Two dimensional electrophoresis of E. coli proteins labeled with* ^{14}*C-amino acids. It is possible to count 1,000 spots on the original autoradiogram [P.H. O'Farrell, J. Biol. Chem., 250, 4007 (1975)].*

6.5 Final Purification

6.5.1 Crystallization

Crystallization of a substance from solution is the oldest known method for obtaining a highly pure solid material. Under the appropriate conditions, crystallization will occur when the concentration of solute exceeds its solubility limit, that is, when the solution is *supersaturated.* Cooling a warm, concentrated solution below its saturation temperature is a familiar laboratory procedure for generating a supersaturated state and initiating crystallization. Exceptional purity, ease of handling, and a pleasant appearance combine to make crystals a highly desirable end product, and many pharmaceuticals and fine chemicals are marketed in crystalline form. Large scale crystallization is performed in either batch or continuous crystallizers. However, like snowflakes, crystals can differ widely in size and shape. Therefore the product from a crystallizer must be characterized by a distribution of sizes, or a *crystal-size distribution* (the shape of crystals can often be described by a characteristic shape factor). In this section, we will briefly discuss a few of the topics which are central to the modeling of crystallization processes, namely, one-dimensional distribution functions, population balances, and crystallization kinetics[24].

If we assume that the function n(L) represents the population density of particles with a characteristic length L (with dimensions of number $(\text{length})^{-4}$), then the total concentration of particles between size zero and L is given by[25]

$$N_L = \int_0^L n(L)dL \tag{6.141}$$

Likewise, the total concentration of particles is

$$N_T = \int_0^\infty n(L)dL \tag{6.142}$$

Therefore, the fraction of particles in size range L_1 to L_2 is

$$F(L_1,L_2) = \int_{L_1}^{L_2} n(L)dL/N_T \tag{6.143}$$

The mean particle size in the suspension, weighted on a population basis, is given by

$$\overline{L}_L = m_1/m_0 \tag{6.144}$$

---

(24) Our analysis will follow that of A.D. Randolph and M.A. Larson, as presented in their text entitled *Theory of Particulate Processes: Analysis and Techniques of Continuous Crystallization* (Academic Press, New York, 1988).

(25) Note that, unlike many distribution functions, n(L) is not normalized over the entire size range (0, ∞); thus, the integral in Eq. (6.141) does not have the value of unity. We could, however, define a normalized distribution function $f(L) = n(L)/N_T$ having this property.

where m_j is the jth moment of the distribution, defined as

$$m_j = \int_0^\infty L^j n(L) dL \tag{6.145}$$

Similarly, the *total* particle surface area per unit volume of suspension (in units, for example, of m^2/m^3) is

$$A_T = k_a m_2 \tag{6.146}$$

and the total mass concentration (kg/m^3) is given by

$$M_T = \rho k_v m_3 \tag{6.147}$$

where k_a and k_v terms are shape factors characteristic of the cystal geometry. For example, for a cube $k_a = 6$ and $k_v = 1$. For particles of length L, the crystal area A_c and the crystal mass M_c are

$$A_c = k_a L^2 \tag{6.148}$$

$$M_c = \rho k_v L^3 \tag{6.149}$$

We shall now analyze a continuous, constant-volume, isothermal, well-mixed crystallizer (Figure 6.40). We assume the crystallizer is fed continuously with a solution free of suspended solids and crystals, and that breakage of crystals is negligible. In general, a mass balance for crystals within a given size range Δ L can be stated as "accumulation = input - output + net generation." In particular, a population balance on crystals of size L to L + ΔL gives

$$\frac{\partial}{\partial t}[n(L)V\Delta L] = \frac{\partial L}{\partial t}[Vn(L)] \quad - \frac{\partial L}{\partial t}[Vn(L+\Delta L)] \quad -Qn(L)\Delta L$$

accumulation	influx of crystals by growth into range	loss as crystals grow out of range	outflow of crystals in given range

As Δ L approaches zero the mass balance can be written as

$$\frac{\partial n}{\partial t} = G\frac{\partial n}{\partial L} + \frac{Qn}{V} = 0 \tag{6.150}$$

where G (= dL/dt) is the growth rate of the crystals, *assumed to be independent of the crystal size*, V is the crystallizer volume, and Q is the volumetric flow rate of slurry outflow. The first term of Eq. (6.150) represents the accumulation of crystals, the second term describes the net efflux of crystals from the size range due to growth, and the third term represents the removal of crystals by bulk flow. Assuming steady state and introducing τ , the space time[26] (= V/Q), gives

---

(26) Here we have employed the terminology of reaction engineering; however, τ is often referred to as the *drawdown time* in the crystallization literature.

$$G\frac{dn}{dL} + \frac{n}{\tau} = 0 \tag{6.151}$$

When L is very small, the population density will consist primarily of nucleating crystals. Defining the population density of these vanishingly small crystals as n_o allows us to solve Eq. (6.151) as L -> 0, giving

$$n = n_0 \exp(-L/G\tau) \tag{6.152}$$

Eq. (6.152) expresses the relationship between population density and crystal size for a continuous crystallizer satisfying the assumptions discussed above.

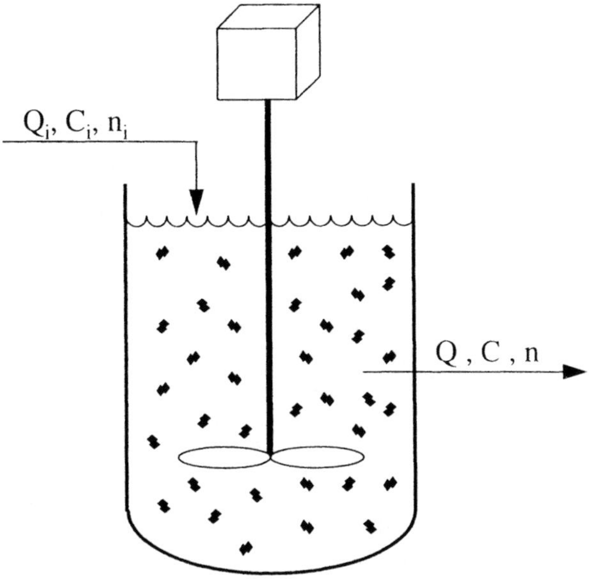

Figure 6.40. *Constant-volume continuous crystallizer.*

We now turn our attention to the mechanism of crystal growth to determine an expression for dL/dt. Crystal growth from solution proceeds through two successive steps: transport of solute to the crystal surface followed by integration into the crystal lattice. In many nonagitated systems, the first step is rate limiting and the growth rate can be expressed as a simple mass transfer equation:

$$\frac{dM_c}{dt} = k_s A_c (c - c_s) \tag{6.153}$$

where k_s is a mass transfer coefficient, A_c is the surface area of the crystal, and $(c - c_s)$ is the difference between the bulk-phase concentration and the concentration of crystallizing solute at saturation. In terms of the characteristic length L, Eq. (6.153) can be rewritten as

$$k_v \rho \frac{d(L)^3}{dt} = k_s k_a L^2 (c - c_s)$$

$$3k_v \rho L^2 \frac{dL}{dt} = k_s k_a L^2 (c - c_s)$$

or

$$\frac{dL}{dt} = \frac{k_s k_a}{3k_v \rho}(c - c_s) \tag{6.154}$$

6.5.2 Drying

Removal of residual water or organic solvents is often required to stabilize otherwise pure bioproducts for storage and handling. The diversity of available drying methods and equipment is considerable; however, the heat sensitivity of most bioproducts necessitates that water be removed with a minimal increase in temperature. Heat transfer can be effected by conduction (i.e., through contact with a heated surface), convection (e.g., by spraying into a hot dry gas), radiation, or a combination of these. In this chapter we will not analyze the different methods in detail, but instead will briefly describe only two of the more important drying techniques.

Lyophilization (Freeze Drying)

Lyophilization, also called freeze drying, is the most gentle form of drying, but it is also the most complex and expensive. In lyophilization, drying is achieved by freezing the wet substance and subliming the ice directly to water vapor under very low pressure (often about 0.5 mbar). Heat is transferred to the frozen solid primarily by conduction from a heated plate; about 700 kcal is needed to sublime 1 kg water from a frozen product at -40°C. The water vapor is removed by low-temperature condensation and the solid temperature is regulated by controlling the pressure in the drying chamber. Under optimum temperature and pressure conditions, the typical primary drying time for a 1-cm thick cake is between 10 and 20 hours. Primary drying removes the free water. Bound moisture (for example, water of crystallization or water dispersed in a glassy material) is typically removed by heating the product to between 15°C and 30°C for a period of about one-third the primary drying time.

Lyophilization is a favored method for drying many sensitive biological materials, for example, vaccines, pharmaceuticals, blood fractions, enzyme preparations, as well as labile and costly ingredients of diagnostics. In addition, some food products (e.g., instant coffee) are lyophilized on a very large scale to retain volatile flavoring agents that are lost in traditional drying processes. As a point of interest we might note that a few less common applications of lyophilization include restoration of water-damaged books and manuscripts, and preservation of museum specimens (plants and animals), archeological artifacts, and tissues for transplantation surgery. The method does have disadvantages, however, such as high capital costs, high energy costs, and long process times.

Spray Drying

Spray drying is perhaps the most important example of a convective drying method for biological materials. The feed solution or slurry to be dried is atomized by a nozzle or rotating disc, then sprayed into a hot dry gas (150 - 250°C). Evaporation proceeds rapidly enough that the temperature of the particles remains relatively low. Nonetheless, the time-temperature profile of an aqueous product is potentially more damaging in spray drying than in freeze drying (Figure 6.41).

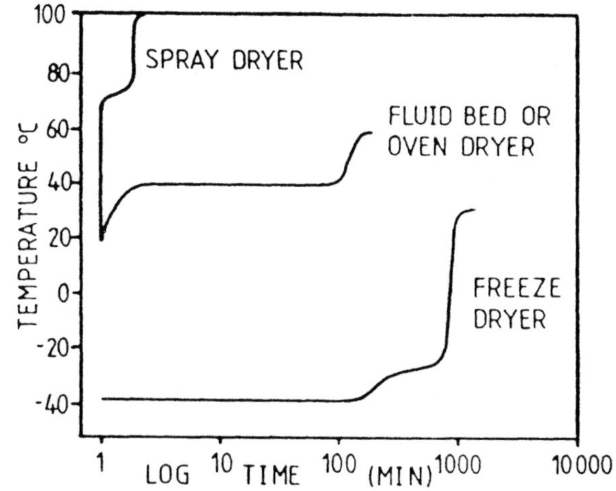

Figure 6.41. *Time-temperature profiles for aqueous products dried by different methods [From J. W. Snowman, Lyophilization Techniques, Equipment, and Practice, in Downstream Processes: Equipment and Techniques, A. Mizrahi, Ed. (Alan R. Liss, Inc., New York), p. 319.]*

Enzymes and antibiotics are among the products dried by spray drying. Detergent-grade enzymes and feed-grade antibiotics can even be obtained by drying a whole fermenter broth, provided the other substances present pose no harm. Spray drying is also widely employed in the food and detergent industries. For example, in 1990 the U.S. production of spray-dried detergent totaled about 1 million tons.

6.6 Putting It All Together

Now that we have discussed the various operations involved in the recovery of biological products, it is time to see how individual steps are combined in practice to achieve the required purity of a given product. Specific examples of product recovery trains are presented below, and these will illustrate just a few of the many ways one can effect the general recovery sequence described in Section 6.1. These different purification schemes share common goals, however, and some relatively simple rules-of-thumb (also known as *heuristics*) can be very helpful in selecting which steps to use. Likewise, choosing the appropriate sequence of steps is largely a matter of common sense.

Guidelines for devising a recovery scheme have been proposed by S.M. Wheelright[27]. Although these six principles do not apply without exception, they represent a useful starting point for the design and scale-up of a downstream process.

1. *Choose separation processes based on different physical properties.* Remember that successive steps based on different physical properties, e.g., charge (ion exchange) and size (gel filtration), will generally yield a better separation than repeated steps based on the same property.

2. *Choose processes that exploit the greatest differences in the physical properties of the product and the impurities.* This point underscores the importance of knowing as much as possible about the physical characteristics of the product and the impurities.

3. *Do the biggest step first.* The "biggest" steps are those which result in the largest reductions of mass or volume. Mass or volume reduction should be achieved early, because the more material one has, the more it costs to process through multiple steps. Therefore, the first step should reduce the quantity of material to be processed by ca. 30% or more.

4. *Do the most expensive step last.* The cost of a step increases with the amount of material treated; therefore, steps requiring the most expensive media, such as affinity chromatography, should be performed at a point where the volume or mass of processed material is relatively small.

5. *Just because it works in the lab doesn't mean it's right for the factory.* In a manufacturing plant, labor costs often exceed the cost of supplies and amortized equipment; generally the opposite is true in the laboratory. This is one reason why the best process in the laboratory (where the time required is less important) may not be the most economical process in the plant.

6. *Whenever possible, keep it simple.* The fewer the steps, the higher the yield; the simpler the process, the easier for manufacturing.

---

(27) S.M. Wheelwright, "Designing Downstream Processes for Large-Scale Protein Purification," *Biotechnology*, **5**, 789 (1987).

Other authors have proposed somewhat different heuristics for the order of recovery steps, and some of these are compared in Table 6.13. The different guidelines are all fairly similar, however, and follow the basic pattern of low-resolution steps followed by high-resolution steps. An apparent exception to this trend is the use of gel filtration as the last step, since gel filtration is certainly not the highest resolution step available. Nonetheless, the high cost of gel filtration per unit of protein processed, together with the technique's greater suitabilility for a sample containing only a few species, dictate that gel filtration be used near the end of the recovery scheme.

Table 6.13. *Design heuristics for the recommended order of separation steps. [From S.M. Wheelwright,* The design of downstream processes for large-scale protein purification, *J. Biotechnol., 11, 89 (1989).]*

Reference[28]	1. Choose first step based on:	2. Choose next step based on:	3. Choose last step based on:
Anonymous (Pharmacia)	Differential solubility (crude adsorption)		Selectivity (high resolution chromatography)
Belter et al. (1988)	Nonspecific separation	High selectivity	Polishing (i.e., crystallization)
Bonnerjea et al. (1986)	Precipitation	Ion-exchange, affinity chromatography	Gel filtration
Östlund (1986)	Intermediate purification (ion-exchange, adsorption, affinity)	Further purification (i.e., chromatography)	Gel filtration
Werner and Berthold (1988)	Precipitation		Gel filtration
Wheelwright (1987)	Largest volume or mass removed		Most expensive (per unit-of-product basis)

(28) P.A. Belter, E.L. Cussler, and W.-S. Hu, *Bioseparations: Downstream Processing for Biotechnology* (John Wiley & Sons, New York, 1988); J. Bonnerjea, S. Oh, M. Hoare, and P. Dunnill, "Protein purification: the right step at the right time," *Bio/Technology* **4**:954-958 (1986); C. Östlund, "Large-scale purification of monoclonal antibodies," *Trends in Biotechnology* **4**:288-293 (1986); R.G. Werner, and W. Berthold, "Purification of proteins produced by biotechnological process," *Arzneimittel Forschung* **38**:422-428 (1988); S.M. Wheelwright, "Designing downstream processes for largescale protein purification," *Bio/Technology* **5**:789, 791-793 (1987).

Figure 6.42 illustrates a generalized recovery scheme for purifying recombinant proteins. The order of the steps is consistent with the sequence outlined earlier (Section 6.1) and with the sampling of research papers summarized in Figure 6.43. Recovery trains for specific proteins are shown in Figures 6.44 to 6.47.

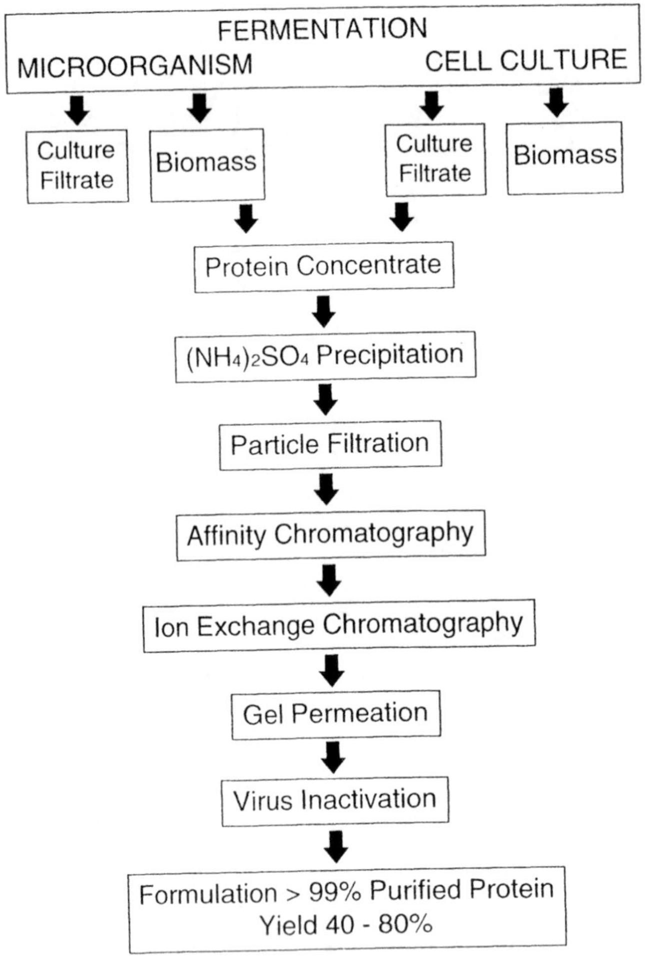

Figure 6.42. *Generalized flow chart for the purification of recombinant proteins. [Adapted from R.G. Werner and W. Berthold, Purification of Proteins Produced by Biotechnological Process, Arzneim.-Forsch./Drug Res., 38 (I), 422 (1988).]*

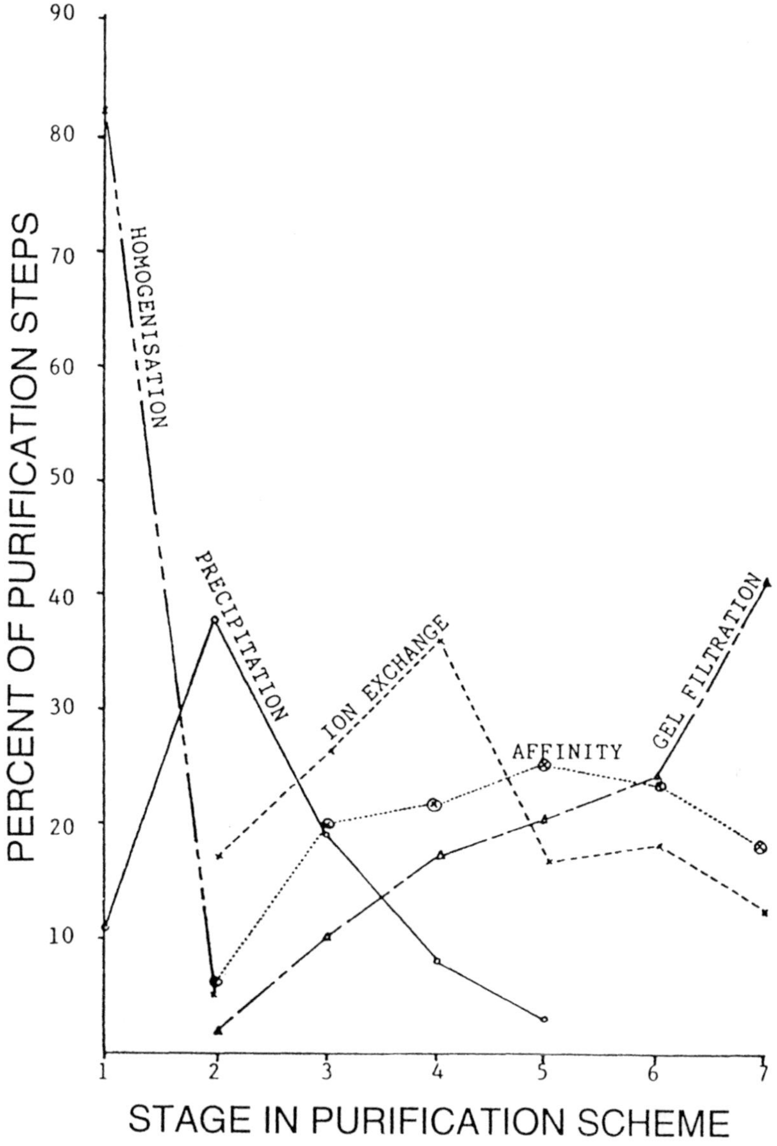

Figure 6.43. *Purification methods used at successive steps of 100 reported purification schemes. Of the 100 papers analyzed, 46 described protein purification from microbial sources, 49 from animal tissue, and 5 from plant material. Just under 10 percent of the papers reported protein isolation from genetically engineered organisms. The average number of steps required to obtain homogeneous protein was 4, with an average overall yield of 28 percent and a purification factor of 6,380. [Reproduced from J. Bonnerjea, S. Oh, M. Hoare, and P. Dunnill, Protein Purification: The Right Step at the Right Time, Bio/Technology, 4, 954 (1986).]*

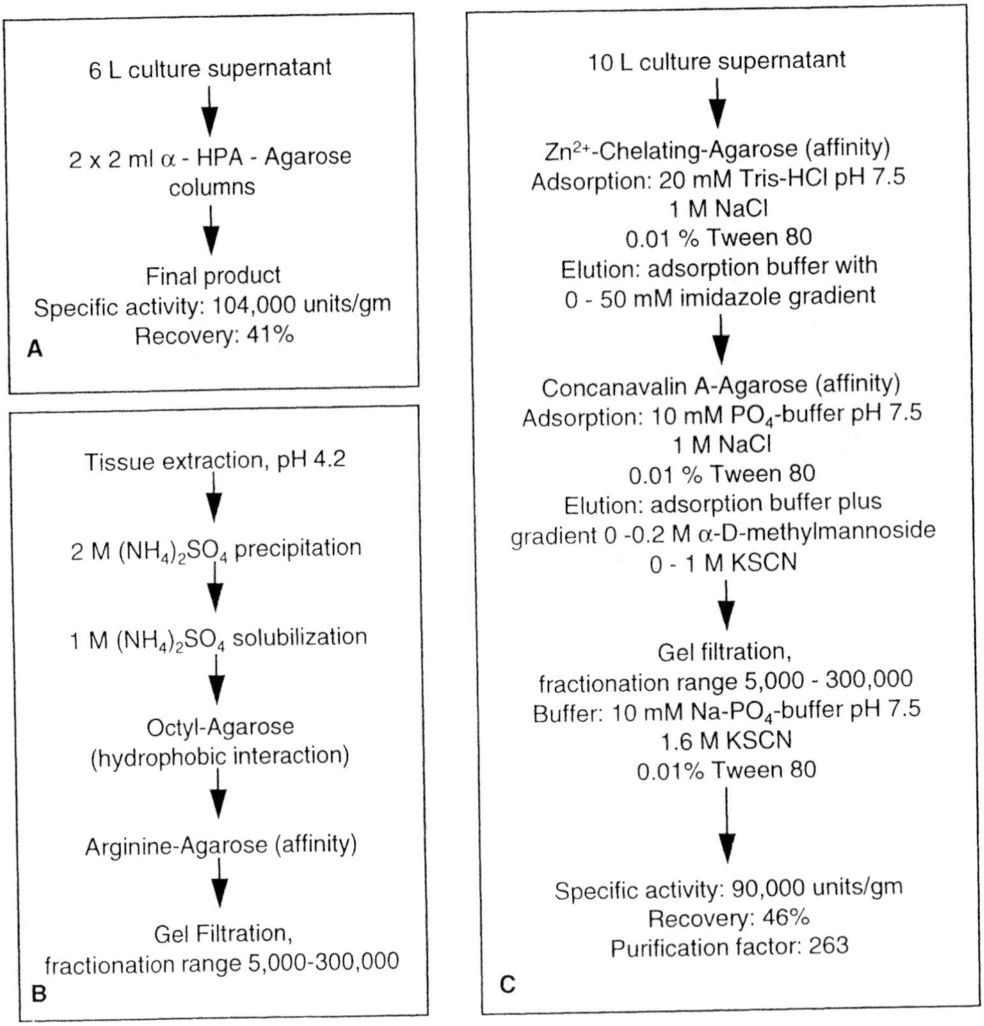

Figure 6.44. *Purification of plasminogen activator from three different sources. A: from human glioblastoma cells; B: from human heart; C: from human melanoma cells. Further examples of chromatographic separation schemes can be found in G. K. Sofer, Current Applications of Chromatography in Biotechnology, Bio/Technology, 4, 712 (1986).*

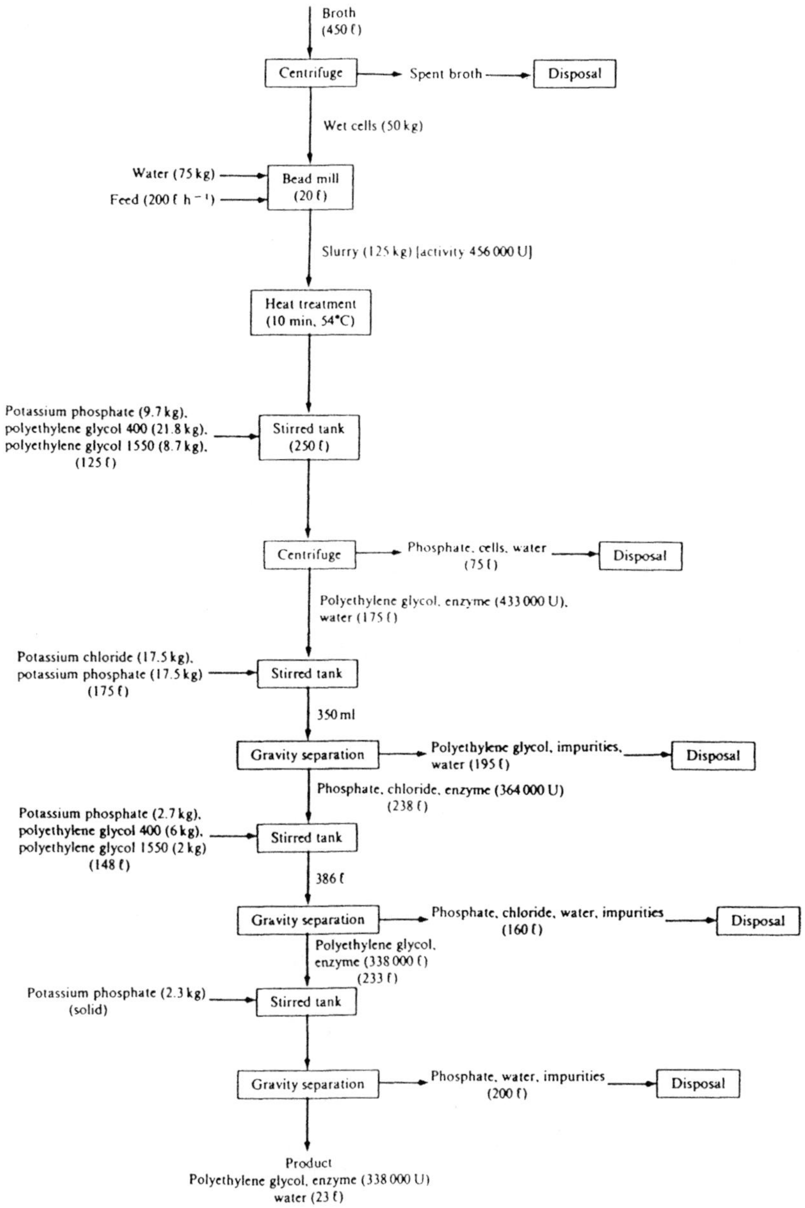

Figure 6.45. *Aqueous two-phase extraction process for an intracellular enzyme, based on a 0.5 m*^{3} *fermenter producing 111 g of enzyme. Final separation of the enzyme from polyethylene glycol is performed by ultrafiltration. [From B. Atkinson and F. Mavituna, Biochemical Engineering and Biotechnology Handbook (Nature Press, New York, 1983), pp. 914-915.]*

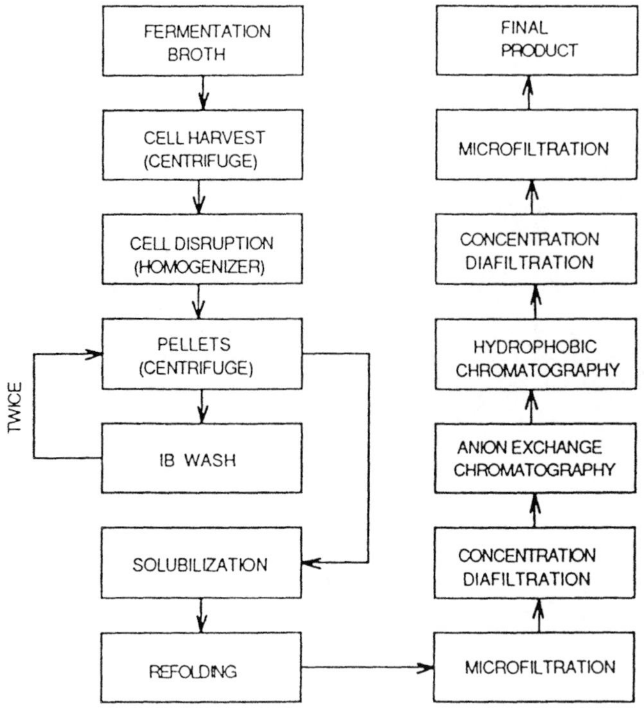

STEP NO.	PURIFICATION STEP	VOLUME (L)	CONC. (mg/ml)	PROTEIN (kg)	PERCENT RECOVERY	PERCENT PURITY
1	REFOLDED SOMATOTROPHIN	230,000	0.9	207.0	80	27
2	ANION EXCHANGE FEED	18,500	9.0	166.5	80	30
3	ANION EXCHANGE ELUATE	34,000	3.0	102.0	61	90
4	HYDROPHOBIC ELUATE	15,000	5.0	75.0	74	98
5	FINAL PRODUCT	838	80.0	67.0	90	98

Figure 6.46. *Top: Process block diagram for the purification of bovine growth hormone (somatotropin) produced by E. coli. Bottom: Purification summary for processing 260 kg of inclusion bodies (IB). (Courtesy of A. Shahidi.)*

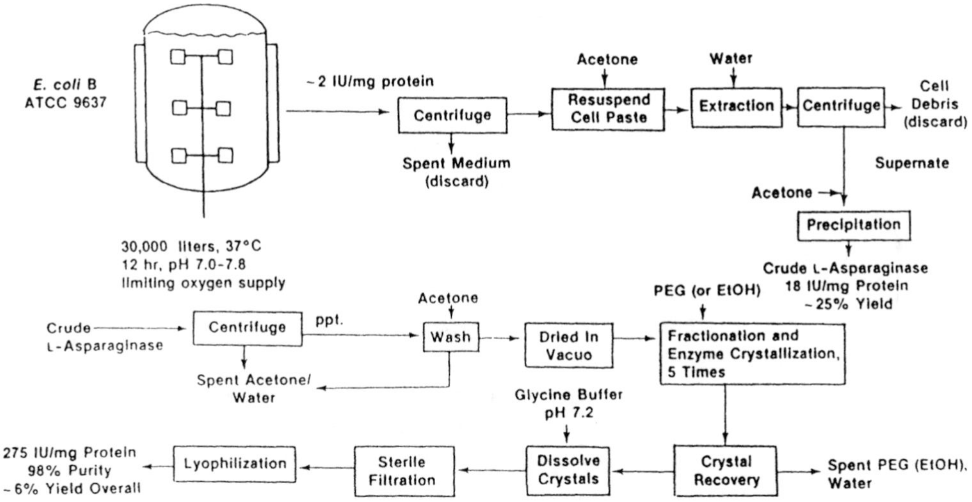

Figure 6.47. *Recovery process for E. coli L-asparaginase EC 2, an enzyme which has been used in the treatment of acute lymphocytic leukemia in children. From M.C. Flickinger, Anticancer Agents, in Comprehensive Biotechnology, Volume 3, H.W. Blanch, S. Drew, and D.I.C. Wang, eds. (Pergamon Press, New York, 1985), p. 245.]*

6.7 Appendix A. The Osmotic Pressure of Macroions in Electrolyte Solutions

In the procedures for separation of proteins in this chapter, we have frequently encountered aqueous solutions of these proteins which also contain salts. Examples include precipitation, aqueous two-phase extraction, ultrafiltration, dialysis and reverse osmosis. Analysis of the separation process generally requires expressions for the chemical potential of the proteins and solvent in the solution. Such expressions can be developed from the osmotic virial equation, and we shall briefly derive it in this section.

The addition of macroions (proteins) and salts alters the activity of the solvent. For example, if a semi-permeable membrane (i.e., one allowing passage of solvent and small ions but not the protein) is used to separate pure solvent from the protein-containing solution, the pure solvent will flow to the protein-containing solution, attempting to equilibrate the solvent activities on both sides of the membrane. The salts will also partition on both sides of the membrane. This flow causes an increase in the hydrostatic pressure which would be reflected in an increase in the height of liquid in a capillary on the protein-containing side. When equilibrium is attained, the increased pressure just balances the solvent flow through the membrane. This increased pressure is the *osmotic pressure*, Π. The situation is illustrated below.

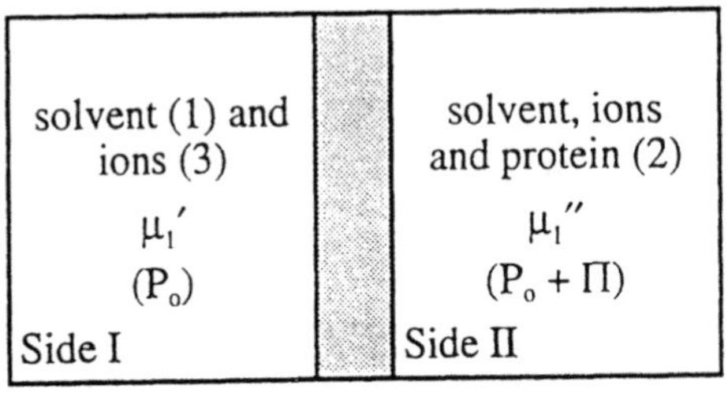

Component 1 is the solvent, 2 is the protein, and 3 is the salt, which are designated by the subscripts below.

We designate side I properties with a single prime and side II with a double prime, referring to properties relative to a pressure $(P_o + \Pi)$. Unprimed quantities refer to conditions at pressure P_o. At equilibrium we have

$$\mu_1' = \mu_1'' \tag{A.1}$$

The solvent chemical potential on side II is related to that at pressure P_o by

$$\mu_1'' = \mu_1 + \int_{P_o}^{P_o+\Pi} \left(\frac{\partial \mu_1}{\partial P} \right)_T dP$$

$$= \mu_1 + \overline{V}_1 \Pi \tag{A.2}$$

where $\overline{V}_1$ is the partial molar volume of the solvent, assumed independent of pressure. If we now assume that the salt is very dilute, we may assume that side I is essentially pure solvent, such that $\mu_1' = \mu_1^o$. Hence

$$\Pi = -\frac{\mu_1 - \mu_1^o}{\overline{V}_1} \qquad (A.3)$$

In the limit of dilute protein concentrations, we can express the chemical potential of the solvent as

$$\mu_1 - \mu_1^o = RT\ln X_1 = RT\ln(1 - X_2)$$

$$= -RT(X_2 + X_2^2 + \ldots)$$

$$= -RTV_1^o\left(\frac{c_2}{M_2} + \left(\frac{V_1^o}{2M_2^2}\right)c_2^2 + \ldots\right) \qquad (A.4)$$

where V_1^o is the solvent molar volume (ml), c_2 is the protein concentration in gm/ml, and M_2 is its molecular weight. As c_2 tends to zero, the partial and molar solvent volumes become equal. We can substitute for the chemical potential and rearrange the above equations to obtain

$$\Pi = -\frac{\mu_1 - \mu_1^o}{\overline{V}_1} = RT\left(\frac{c_2}{M_2} + \left(\frac{\overline{V}_1}{2M_2^2}\right)c_2^2 + \ldots\right) \qquad (A.5)$$

The second order term can be neglected in the limit of very dilute solutions, and we recover van't Hoffs relationship

$$\frac{\Pi}{c_2} = \frac{RT}{M_2} \qquad (A.6)$$

Equation A.5 can be generalized in a manner analogous to the virial expansion for a non-ideal gas

$$\Pi = -\frac{\mu_1 - \mu_1^o}{\overline{V}_1} = RTc_2\left(\frac{1}{M_2} + Bc_2 + Cc_2^2 + \ldots\right) \qquad (A.7)$$

Thus a plot of Π/c_2 against c_2 can be used to determine the molecular weight of the protein, by extrapolation of the osmotic pressure data to a zero intercept for c_2. Provided the third and higher terms can be neglected, i.e. the solution is dilute in protein, the slope of this plot can be used to determine the second virial coefficient B. Statistical mechanics provides a relationship between B and the potential of mean force between the proteins in the solution. The interaction potential contains terms accounting for excluded volume, charge repulsion, dipole and hydrophobic forces. If only excluded volume is important, i.e. proteins may approach only within a distance of twice their radius, the second virial coefficent is given by

$$B = \frac{4v_2}{M_2} \tag{A.8}$$

v_2 is the specific volume of the protein, i.e. the reciprocal of its density. The density of most proteins is ~ 1.33 gm/ml, and thus $BM_2 \sim 3.0$. We might expect this to hold at moderate to high salt concentrations, where the electrostatic effects are shielded. This can be seen in the table below, where the agreement with the prediction of Equation A.8 indicates that the proteins considered are indeed approximately spherical.

Table A.1. Molecular weights and virial coefficents obtained from osmotic pressure measurements at high salt concentrations. [From Tanford, C. "The Physical Chemistry of Macromolecules in Solution," John Wiley & Sons, Inc. (1961).]

Protein	Molecular weight	BM_2 (cm^3/gm)
β- lactoglobulin	39,000	3.2
Ovalbumin	45,000	1.4
Hemoglobin	67,000	3.7
Bovine serum albumin	69,000	1.5

6.8 Appendix B. Population Balance Model of Protein Aggregation

A more rigorous analysis of the kinetics of aggregating particles requires that we start by defining the population density of particles, $n(v,t)$, such that $n(v,t)dv$ equals the number of particles per unit volume of suspension at time t with particle volumes between v and dv. Therefore, the total number of particles per unit volume of suspension, N(t), is given by

$$N(t) = \int_0^\infty n(v,t)dv$$

As in our previous discussion of orthokinetic growth, the volume fraction of particles, ϕ, remains constant, and is equal to

$$\phi(t) = \int_0^\infty vn(v,t)dv = \text{constant}$$

The average particle volume, $\langle v \rangle$, is thus

$$\langle v \rangle = \frac{\phi}{N}$$

A rate equation for the kinetics of particle aggregation can now be written as

$$\frac{dN}{dt} = -\frac{1}{2}\int_0^\infty \int_0^\infty \beta(v,\tilde{v})n(v,t)n(\tilde{v},t)d\tilde{v}dv$$

where β is the rate constant for aggregation. This expression describes the coagulation a particle of volume v with another particle of volume $\tilde{v}$ to form a larger particle, integrated over all possible volumes (note the factor of 1/2, which is included to ensure that colliding particles are not counted twice). This rate equation can be rewritten in a more conventional form by introducing a normalized volume, η, and a normalized volume distrubution function, $\Psi(\eta)$:

$$\eta = \frac{vN}{\phi} \qquad \Psi(\eta) = \frac{n\phi}{N^2}$$

Assuming that β has the properties of a homogeneous function, that is,

$$\beta(cv_1, cv_2) = c^P \beta(v_1, v_2)$$

the time rate of change of N can now be written in terms of an average rate constant, $\langle\beta\rangle$:

$$\frac{dN}{dt} = -\frac{1}{2}\langle\beta\rangle\left(\frac{\phi}{N}\right)^P N^2$$

where

$$\langle\beta\rangle = \int_0^\infty \int_0^\infty \beta(\eta,\tilde{\eta})\Psi(\eta)\Psi(\tilde{\eta})d\eta d\tilde{\eta}$$

The value of P depends on the driving force for aggregation. For shear induced aggregation, P=1; thus, the kinetics reduce to first order in N, as we showed before by starting with Eq. 6.53.

A population balance can also be written for the population density of particles, $n(v,t)$, which can then be solved using Laplace transforms. This analysis leads to the interesting result that at long times, $n(v,t)$ is independent of the initial particle density, $n_o(v)$, and its history at relatively short times.

6.9 Nomenclature

Symbol	*Meaning (Typical Units)*
A	cross-sectional area of particle [Eq. (6.2)], filter area [Eq. (6.17)], peak area [Eq. (6.62)]
A, B, C	coefficients of the van Deemter equation [Eq. (6.121)]
A_c	area of a single crystal (cm^2)
A_T	total particle surface area per unit volume of suspension (m^2/m^3)
a_E	particle acceleration due to external force (cm/sec^2)
a_{ij}	second virial coefficient for interaction between i and j (kg/mol or liters/mol)
A_{ij}	traditional virial coefficient ($ml{\cdot}mol/gm^2$)

a_{pp}, a_{1p}	protein-protein, protein-polymer interaction coefficients (kg/mol)
C	eluate concentration (gm/liter)
C_{max}	maximum eluate concentration (gm/liter)
c_{1s}	solute concentration upstream (gm/liter)
c_{2s}	solute concentration in permeate (ultrafiltrate) (gm/liter)
C_D	drag coefficient
c_g	gel concentration (gm/liter)
c_i, c_s	concentration of solute i in fluid phase, solute concentration (gm/liter)
c_o	feed concentration of solute (gm/liter)
c_{os}	solute concentration at membrane surface (gm/liter)
$\overline{c}_s$	average solute concentration upstream (gm/liter)
c_s	molar concentration of aggregating protein [Eq. (6.50)]
D_{eff}, D_s	effective diffusivity, bulk-phase diffusivity (cm^2/sec)
D_p	effective solid-phase diffusivity (cm^2/sec)
D_L	axial dispersion coefficient (cm^2/sec)
d_p	particle diameter (μm)
d_p	particle diameter (μm)
e	elementary charge (1.6021×10^{-19} coulomb)
E	electric field (volt/cm)
F	volumetric flowrate, Faraday constant [Eq. (6.133) (esu/mole)]
$F(L_1, L_2)$	fraction of particles in size range L_1 to L_2
F_B	buoyancy force on the particle [dyne or newton (N)]
F_D	drag force exerted by the fluid on the particle (dyne or newton)
F_E	external force acting on a particle in a gravitational or centrifugal force field (dyne or newton)
F_P	particle-particle or particle-wall interaction forces (dyne or newton)
g	acceleration due to gravity (9.81 m/sec^2)
G	linear growth rate of crystals (= dL/dt), assumed to be independent of cystal size
H	height equivalent of a theoretical plate
I	ionic strength (M)
J_s	solute flux (gm/cm^2-sec)
J_w	solvent flux (cm^3/cm^2-sec or gm/cm^2-sec)
k	Darcy's law permeability [Eq. (6.16) (cm^2-cp/s-atm)], Boltzman constant [Eq. (6.133)]
K	solubility constant $(M)^{-1}$
K″	solubility constant for a protein at its isoelectric point [Eq. (6.41)] (dimensionless)

K_{conf}, K_{el}, K_{lig}, K_{hphil}, K_{hphob}	contributions to aqueous two-phase partition coefficient, K_p, of conformational, electric, ligand, hydrophilic, and hydrophobic interactions
k_a	shape factor relating area to size squared
K_i	distribution coefficient of sorbate i
k_i, k'	retention factor of species i, average retention factor
K_{i-e}	distribution coefficient of sorbate for an ion exchange resin
K_L	Langmuir constant (cm^3/gm)
K_p	partition coefficient or a macromolecule between two aqueous phases
k_s	mass transfer coefficient (cm/sec)
K_s	distribution coefficient of solute between a membrane and solution [Eq. (6.29)], salting-out constant [Eq. (6.38) $(M)^{-1}$]
k_v	volumetric shape factor relating volume to size cubed
l	centrifuge length [Eq. (6.14)], bed thickness [Eq. (6.16)], membrane thickness [Eq. (6.29)]
L	particle size, i.e., independent variable of the distribution function (μm
$\overline{L}_L$	mean particle size in a crystal suspension
L_p	membrane permeability for the solvent (cm^3 $(cm^2 \cdot sec \cdot atm)^{-1}$)
m	mass of particle [Eq. (6.1)], molality [Eqs. (6.24) and (6.39)], molarity [Eq. (6E2.3)], mobility [Eq. (6.136) (cm^2/volt-sec)]
$\overline{m}$	molarity of resin-bound protein (moles $(\text{liter resin})^{-1}$)
M	total amount of solute injected onto a chromatography column (gm)
m(L)	mass density distribution function
M(L)	mass of particles per solids-free volume between size zero and L
m_1	polymer molality
M_c	mass of a single crystal
m_j	j^{th} moment of the distribution function [Eq. (6.145)]
m_n	n^{th} moment of an exiting peak [Eq. (6.89)]
m_p	protein molality
$\overline{M}_w$	average molecular weight of precipitate
$\overline{M}_{wo}$	initial average molecular weight of precipitate
n(L)	population density distribution function
N(L)	number of particles per solids-free volume between size zero and L
N_{av}	Avagadro's number
n_o	population density of nuclei
N_p	theoretical plate number
N_t	number of transfer units
N_T	total concentration of crystal particles (per liquid volume)

N_T	total mass concentration per unit volume of suspension
P	pressure (atm or N/m^2)
P/V	power input per unit volume (hp/gal)
P_s	membrane permeability for the solute (cm/sec)
q	sorbate concentration (mass or moles per unit volume of adsorbent)
$\bar{q}$	value of q averaged over an individual pellet, particle charge (= ze)
Q	volumetric flow rate (liters/min)
q_i^*, q_o^*	equilibrium sorbate concentration (per unit volume of adsorbent), sorbate concentration in equilibrium with c_o
q_{max}	maximum sorbate concentration (per volume of adsorbent)
r	radial distance, cell radius [Eq. (6.36)]
R	gas constant, rejection coefficient [Eq. (6.28)], particle radius [Eq. (6.71) and others, Eq. (6.100)]
R, R_o and R_1	centrifuge radii [Eq. (6.14)], [Eq. (6.15)]
R_C	cake resistance (m^{-1})
Re	Reynolds number
R_{eq}	separation factor
R_M	membrane resistance (m^{-1})
R_p	resistance of polarized boundary layer $cm^2 \cdot s \cdot atm$ (cm^{-3})
R_s	chromatographic resolution
S, S_o	protein solubility, protein solubility at zero ionic strength
Sc	Schmidt number
T	temperature, dimensionless throughput parameter [Eq. (6.102)]
t_o	duration of pulse input
t_r	chromatographic retention time
u	liquid velocity of permeate through filter [Eq. (6.16)], average velocity of fluid in narrow channel [Eq. (6.35)], interstitial velocity of mobile phase in column (u_i) [Eq. (6.75)]
V	filtrate volume [Eq. (6.17)], fluid volume fed to column [Eq. (6.102)], crystallizer volume [Eq. (6.150)]
V_T	total column volume
v, v_c, v_g, v_h, v_E	particle velocity, settling velocity in centrifugal field, settling velocity under gravity, hindered sedimentation velocity, velocity of charged particle in electric field
w	peak width measured a baseline intercept
w_h	width of Gaussian peak at half-height
w_{in}	width of Gaussian peak at inflection point

X, X	dimensionless local fluid-phase concentration (Figure 6.27); dimensionless breakthrough concentration [Eq. (6.104)]
Y, Y	dimensionless local solid-phase concentration (Figure 6.27); dimensionless mean solid-phase concentration
z	valence of a particle
z_p	valence of a protein

Greek Symbols

α	geometric factor for hindered settling [Eq. (6.12)], specific cake resistance [Eq. (6.20) (m/g)], collision effectiveness factor for precipitation [Eq. (6.54)], chromatographic separation factor [Eq. (6.130)]
β	Cohn constant [(Eq. 6.38)], void fraction of particle [Eq. (6.83)].
γ	activity coefficient, solvent surface tension (dyne/cm)
γ°	surface tension of water (72.0 dyne/cm at 25°C)
$\bar{\gamma}$	mean shear rate (sec^{-1})
δ	boundary layer thickness adjacent to membrane
$\delta_{Ad}, \delta_D, \delta_f$	components of second central moment [Eq. (6.96)]
κ	reciprocal of the double layer thickness (cm^{-1}), second order rate constant for particle aggregation (M^{-1} sec^{-1})
ε	column void fraction
ε_s	solvent dielectric constant
λ_o	protein eccentricity
μ	solution viscosity (g/cm-sec), protein dipole moment (esu-cm)
μ_1, μ_2	first absolute moment, second moment
μ_2'	second central moment
μ_p	chemical potential of protein (erg/mol)
μ_p^0	standard chemical potential of protein (erg/mol)
ν	kinematic viscosity (cm^2/sec)
ω	angular rotation rate in a centrifuge [Eq. (6.5)], stirrer speed in an ultrafiltration cell [Eq. (6.36), (6.37)] (rad/sec)
Ω	relative surface hydrophobicity of protein [Eq. (6.40)] (cm/dyne)
$\Delta\pi$	solute osmotic pressure difference across a membrane
ρ_B	particle density per bed volume [Eq. (6.103)] (gm/cm^3)
ρ_m, ρ_p	density of liquid and solid particle, respectively (gm/cm^3)

σ	reflection coefficient [Eq. (6.25)], molal surface tension increment of salt (dyne-kg/cm-mol)
σ_t	standard deviation of a Gaussian peak
Σ	sigma factor for centrifugation [Eq. (6.14)] ($length^2$)
τ	space time for a continuous crystallizer (= V/Q)
ϕ	volume fraction of solute [Eq. (6.55)], electrostatic potential [Eq. (6.134)]
Φ	electrical potential of aqueous phase in two-phase system (mV)
ψ	electrostatic potential (mV)
ψ_P	correction factor [$=0.894/(1-0.106R_{eq}^{1/4})$ for $R_{eq}^{1/4} < 1$; for $R_{eq}^{1/4} > 1$, $\psi_P \sim 1$]
ξ, Ξ	residence time distribution functions

Subscripts

1, 2, 3	solvent, polymer, and polymer in aqueous two-phase system
B, R	counter ion displaced from ion exchanger, ionic groups fixed on ion exchanger
P	protein, particle

Superscripts

′, ″	lighter and heavier phase, respectively, of aqueous two-phase system

6.10 Problems

1. Release of Protein from Baker's Yeast by Disruption in an Industrial Homogenizer

Hetherington *et al.* (Trans. Instn Chem. Engrs, 49, 142 (1971)) have investigated the release of protein from suspensions of *Saccharomyces cerevisiae* by disruption in an industrial homogenizer. The homogenizer was operated on a recycle basis, and the dependence of protein release on the number of passes, N, and the pressure, P, was studied. If the disruption process is first-order, it can be represented by an equation of the form:

$$\log\left(\frac{R_m}{R_m - R}\right) = KN[fn(P)]$$

where R is amount of released protein (mg/g yeast), R_m is the maxium amount of protein available for release, and K is a temperature-dependent dimensional constant. The relationship between log $[R_m/(R_m - R)]$ and N at 5°C is plotted below. Pressures are given on the graph.

Assuming that the overall performance of the homogenizer can be represented by the equation

$$\log\left(\frac{R_m}{R_m - R}\right) = KNP^C$$

where C is a constant, determine the values of K and C.

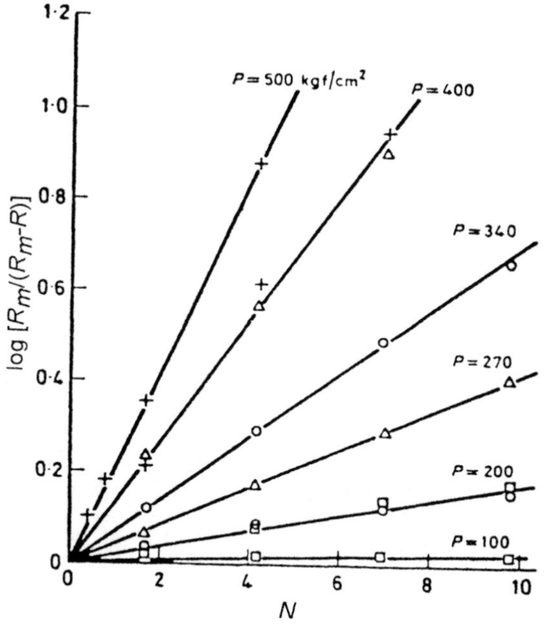

2. *Solvent Extraction of Biological Products*

a) An aqueous mixture containg 6.8 mg/liter of a steroid is extracted with initially pure methylene dichloride. The equilibrium constant for the steroid is 170 and the ratio of water to solvent is 82. What is the steroid concentration in the organic phase after a single batch extraction? What fraction of the steroid has been removed?

b) For the extraction of the amino acid glycine, the equilibrium relationship between toluene (phase′) and water (phase″) is

$$(C')^2 = (0.001)C''$$

where C′ and C″ are the molar concentrations of glycine in the two phases. We plan to contact 4.7 liters of toluene containing 0.006 M glycine with 1 liter of water. What fraction of glycine would be extracted? What additional amount would be removed in a second extraction with 1 liter of water?

3. *Protein Partitioning in Aqueous Two-phase Systems*

A two-phase system is to be employed to recover the serine protease α-chymotrypsin (molecular weight 28,500). The phase forming polymers are polyethylene glycol (PEG 3350) and dextran (T-70), in a 50 mM KCl solution. Values of the osmotic virial coefficients for the polymer-polymer and polymer-protein interactions are available from the low-angle light-light scattering (King, R.S., Blanch, H.W. and J.M. Prausnitz, AIChE Journal, **34** 1585 (1988)). These are

PEG 3350 (M_n 3860)	A_{22}	36.36×10^{-4} (ml·mol/gm^2)
Dextran T70 (M_n 74,540)	A_{33}	4.04×10^{-4} (ml·mol/gm^2)
α-chymotrypsin - PEG 3350	A_{2p}	5.10×10^{-4} (ml·mol/gm^2)
α-chymotrypsin - Dextran T70	A_{3p}	2.40×10^{-4} (ml·mol/gm^2)
PEG 3350 - Dextran T70	A_{23}	17.0×10^{-4} (ml·mol/gm^2)

(a) Using the equations given in Section 6.3.1 for the chemical potentials of the polymers and the solvent (50mM KCl, assumed to be a pseudo-solvent where the KCl partitions evenly between the phases) determine the phase diagram for the polymer-polymer-solvent system in the absence of protein. Indicate the tie-lines on the diagram.

(b) By assuming that the presence of the protein does not disturb the phase system, calculate the partition coefficent of α-chymotrypsin between the phases, using the equations in Section 6.3.1.

4. *Partitioning of Proteins in Aqueous Two-Phase Systems*

Partitioning of proteins in aqueous two-phase systems is often dominated by the electrical states of the two equilibrium phases. An electrostatic-potential difference, $\Delta\Phi$, between the two equilibrium phases can develop as a result of the partitioning of a strong electrolyte. This potential difference can be calculated by arbitrarily splitting the electrochemical potential, μ_i, of each ionic species i into a chemical term and an electrical term:

$$\mu_i = \mu_i^o + RT\ln(m_i\gamma_i) + z_iF\Phi$$

where Φ is the electrostatic potential (volts), z_i is the charge number of ion i, F is Faraday's constant, γ_i is the chemical activity coefficient for ion i (which is independent of the electrical state of the phase), m_i is the molality of species i, and μ_i^o is the standard state chemical potential of species i.

Described below is an aqueous two-phase system containing a small amount of KCl. The salt has completely dissociated into ν_+ cations and ν_- anions.

Phase′	
$m_2' = 4.5 \times 10^{-2}$	$T = 25°C$
$m_3' = 1.34 \times 10^{-5}$	$P = 1$ atm
	Polymer 2 = PEG 3350
	Polymer 3 = Dextran T-70
Phase″	$\gamma_+' = 1.29$
$m_2'' = 6.0 \times 10^{-3}$	$\gamma_+'' = 1.05$
$m_3'' = 3.36 \times 10^{-3}$	$\gamma_-' = 1.00$
	$\gamma_-'' = 1.30$

a) Calculate the electrostatic potential difference $\Phi'' - \Phi'$ that develops across the liquid-liquid interface due to the partitioning of ions.
b) A protein P is now added to the above two-phase system at a total concentration of 1.0 mMolal. Calculate the partition coefficient of the protein, if the net charge on the protein is +5 at the pH of the system. The second virial coefficients are a_{2p} = 37.7 liters/mol and a_{3p} = 37.7 liters/mol.

5. *Gel Concentration of Casein in an Ultrafiltration Cell*

The table below reports declining ultrafiltration rates with increasing bulk concentrations of casein in a thin-channel recirculating flow cell. Assuming a gel layer accumulates at the membrane surface, what is the gel concentration, C_g, of the protein?

Concentration (weight percent)	Ultrafiltration flux ($ml\ cm^{-2}\ min^{-1}$)
1.0	0.057
2.0	0.042
5.0	0.023
10	0.00043

6. *Time Required for Ultrafiltration*

Suppose we wish to concentrate a protein solution by ultrafiltration in a batch system. The permeability of the membrane, L_p, is 3.5 liter $m^{-2}\ hr^{-1}\ atm^{-1}$, and the total membrane area is 100 cm^2. The initial volume of solution is 1.0 liter and the amount of protein present is 0.01 mmoles. The pressure drop is 145 psi and the temperature of operation is 25°C. Assume that the membrane completely rejects the solute and that concentration polarization is negligible.

a) Derive an expression for the rate of change of the batch retentate volume with time (dV/dt).

b) Calculate the time required to reduce the volume of solution from 1.0 liters to 10 ml.

7. *Solvent Flux with Concentration Polarization*

Concentration polarization is a problem sometimes encountered in ultrafiltration. It arises when a concentration gradient of solute develops near the membrane surface, as shown in Figure 6.16 (C_{1s} is the solute concentration on the upstream side at a distance greater than $-\delta$ from the membrane, C_{os} is the concentration at the membrane surface, and C_{2s} is the concentration of the permeate). The build-up of solute near the membrane reduces the flux of permeate (which is mainly solvent) through the membrane. In this problem we will derive expressions for the solute concentration and the permeate velocity (solvent flux) through the membrane.

(a) Starting with a steady-state mass balance for solute through the polarization layer, derive an expression for the concentration profile of solute on the upstream side near the membrane. (Hint: remember that the mass balance includes a velocity term, which corresponds to the permeate velocity.)

(b) Now derive an expression relating the permeate velocity to the solute diffusivity, the thickness of the polarization layer, and the various concentrations (i.e., Eq. (6.31)). Hint: it is useful to recognize that the flux at the membrane surface can be written as

$$D_s[dc/dz]_{z=0} = v_m(C_{0s} - C_{2s})$$

8. *Retention of Enzyme in a Continuous Membrane Reactor*

A continuously operated membrane reactor is shown by the schematic below. Substrate solution is fed at constant flux to a recycle loop that contains, for example, a tubular ultrafiltration module. A biocatalyst (enzyme) is physically immobilized in the loop, since it is mostly retained by the membrane. Products of low molecular weight leave the system with the filtrate. In practice, some enzyme is also lost due to incomplete retention. As the senior biochemical engineer at a pharmaceutical company planning to develop a large-scale membrane reactor, your job is to calculate the rate at which enzyme will be lost from the system.

a) Start by formulating a time-dependent mass balance for the concentration of enzyme, E. Use the following notation:

J	$m^3\,m^{-2}\,s^{-1}$	solvent flux
A_m	m^2	membrane area
V_R	m^3	reactor volume
E	$kg\,m^{-3}$	active enzyme concentration (assuming homogeneous distribution)
E_f	$kg\,m^{-3}$	enzyme concentration in the filtrate

b) Assuming that E_f can be related to an apparent retention coefficient, R, by

$$E_f = (1 - R)E$$

rewrite your mass balance from part (a) in terms of R and the mean residence time, t_r, of the reactor. Solve this equation for E as a function of t and the enzyme concentration at time zero, E_0. Prepare a plot of E/E_0 versus t/t_r for high and low values of the retention factor, R.

c) Derive an expression for an apparent first-order deactivation constant of enzyme, k_d.

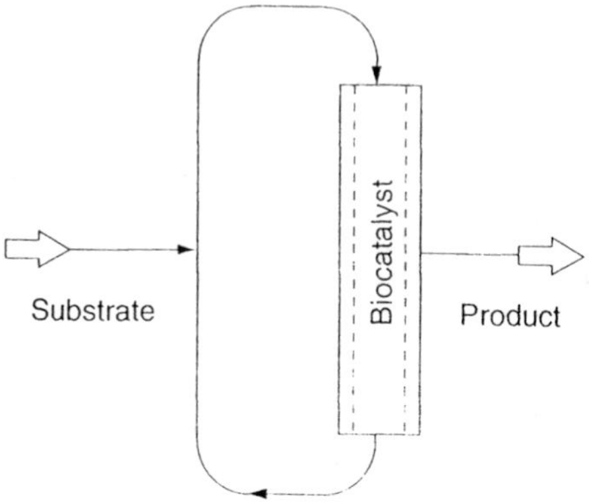

9. *Cell Washing by Cross-Flow Filtration*

During product recovery, cell washing by cross-flow filtration is one way to change the liquid, or solvent, suspending the cells. The pH or ionic strength can be changed or a particular component can be separated from the bulk cell suspension. Cell washing is thus an alternative to dialysis or centrifugation. A schematic representation for a cell washing setup is shown below. In the cell washing mode, the cells are retained by a membrane filter, and fresh wash solution is added to the cell suspension at the same rate filtrate is removed. Therefore the suspension volume remains constant. With time, the old cell-suspension medium is replaced by fresh solvent.

a) Let C_o be the concentration of new solvent in the fresh wash solution and C(t) be the concentration of new solvent in the filtrate (t is time). Derive an expression for F (= $C(t)/C_o$), the fraction of wash completed, in terms of V_o, the initial volume of cell suspension, and V_F, the volume of the filtrate.

b) What is the percent completion of cell washing when $V_F = V_o$? $V_F = 3V_o$? $V_F = 5V_o$?

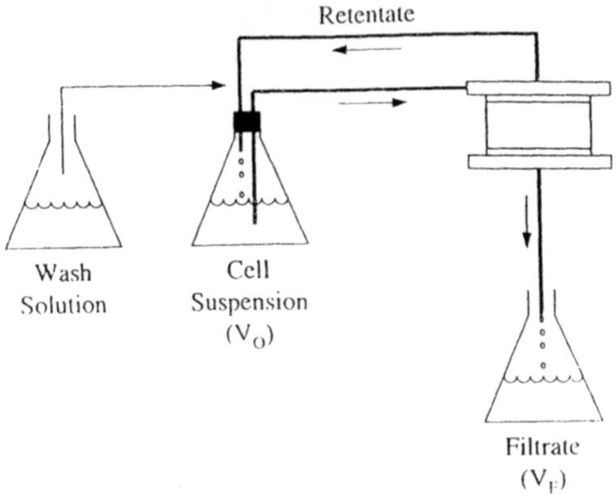

10. *Selective Precipitation of the Enzyme Catalase*

We wish to precipitate catalase from an enzyme mixture that contains cholesterol oxidase, our desired product. Catalase has a molecular weight of 250,000, a precipitate density of 1.3 g/cm^3, a diameter in solution of 10.4 nm, and a diffusion coefficient of 4.1 x 10^{-7} cm^2/sec. It is to be precipitated by a sudden pH change in a 100 liter tank stirred with a 0.1 hp motor. The feed concentration is 0.2 g/liter and the sticking coefficient is about 0.05 sec. Estimate the following:

i) the time over which diffusion will limit growth,

ii) the concentration of particles at the end of this time,

iii) the time required to grow 100 micron particles for centrifugation.

11. *Protein Solubility in Mixtures of Low Dielectric Constant*

Solubility data for the enzyme bovine serum albumin (BSA) at its isoelectic point in mixtures of ethanol and distilled, deionized water are given in the table on the following page.

Dielectric constant (ε_s)	Mole-fraction solubility (X)
81.6	1.55×10^{-5}
79.8	1.0×10^{-5}
75.6	3.16×10^{-6}
70.7	6.31×10^{-7}
68.9	3.16×10^{-7}
66.1	1.0×10^{-7}

a) Based on these data, estimate the mole-fraction solubility of BSA in a water-ethanol mixture of $\varepsilon_s = 63.0$.
b) If the water-ethanol mixtures are adjusted to a mole-fraction ionic strength of 0.00867 by addition of the neutral salt LiCl, the solubility parameters of Eq. (6.41) become $K'' = -9.33 \times 10^3$ and $\log X_o = -2.77$. What, then, is the mole-fraction solubility of BSA in such a mixture of water-ethanol-LiCl, D = 63.0?
c) In low concentration, salts increase the solubility of many proteins. This phenomenon is known as *salting-in*. We have also seen that the solubility of a protein will decrease at sufficiently high ionic strength (giving rise to so-called *salting-out*, as described in Section 6.4). What do the results of parts (a) and (b) above indicate about the relative effects of salting-out and salting-in as the dielectric constant of the solvent is reduced? Can you explain these results in terms of possible interactions between ions and the protein molecules? In general, the action of small amounts of salts can be used to refine fractional precipitation by ethanol and other solvents.

12. Elution of Amino Acids from an Ion Exchange Column

The amino acids below have the following pK_a values:

	α-COOH	ω-COOH	α-NH_3^+	ε-NH_3^+
Val	2.32		9.62	
Asp	1.88	3.65	9.60	
Lys	2.18		8.95	10.53
Glu	2.19	4.25	9.67	
Gly	2.34		9.60	

a) At pH 3.25, what would be the order of emergence of these amino acids from a column filled with the cation exchange resin Dowex 50? Would glycine and valine be well separated? Why or why not?
b) Predict the order of elution at pH 9 from the anion exchange resin Dowex 1.

13. Equilibrium Binding of Proteins to Ion Exchange Resins

Protein binding to an ion exchange resin is accompanied by the displacement of counterions, as shown by the reaction

$$P + \left(\frac{z_P}{z_B}\right)\overline{B} \leftrightarrow \overline{P} + \left(\frac{z_P}{z_B}\right)B$$

where P, B = protein and counterion in solution, the overbar designates the exchanger-bound state, z = the valence of the species, and K_e = equilibrium constant for the reaction. The equilibrium constant can be expressed in terms of the activity of each species by

$$K_e = \frac{(\overline{a}_p)^{|z_B|}(a_B)^{|z_P|}}{(a_P)^{|z_B|}(\overline{a}_B)^{|z_P|}} \qquad (E6.1)$$

i) Invoking the electroneutrality condition in the ion exchanger, show that the partition coefficient for a monovalent counterion B, K_{i-e}, can be written as

$$K_{i-e} = \frac{K_e}{\Gamma_1}\left(\frac{m_R - |z_P|\overline{m}_P}{m_B}\right)^{|z_P|} \qquad (P6.1)$$

where

$$\Gamma_1 = \frac{\overline{\gamma}_P}{\gamma_P}\left(\frac{\gamma_B}{\overline{\gamma}_B}\right)^{|z_P|}$$

and

$$\gamma_i = \frac{a_i}{m_i}\, i = B, P$$

where m_i and γ_i = molarity and activity coefficients of species i, respectively, and m_R is the molarity of the charged species on the resin.

ii) Show that, when $z_p m_p < m_R$, Eq. (6P.1) reduces to Eq. (E6.3):

$$K_{i-e} = \frac{(K_e\Gamma_1)(m_R/m_B)^{|z_P|}}{1 + (K_e\Gamma_1)(m_R/m_B)^{|z_P|} z_P^2 m_P/m_R} \qquad (E6.3)$$

As we saw before, as m_p/m_R becomes even smaller, Eq. (E6.3) reduces to

$$K_{i-e} = (K_e\Gamma_1)\left(\frac{m_R}{m_B}\right)^{|z_P|} \qquad (E6.5)$$

and the isotherm is no longer sensitive to the protein concentration (Figure 6.28).

14. Chromatography Under "Overload" Conditions

When using chromatography to separate a minor contaminant from a protein product, it is often practical to overload the column. This can result in plateau formation and peak asymmetry, as shown in the chromatographic separation profile on the next page.

A minor component with elution volume V_{e2} is separated from a major component with elution volume V_{e1}. Peak asymmetry is represented by a, the asymmetry factor, and σ, the standard deviation of a symmetrical Gaussian peak profile.

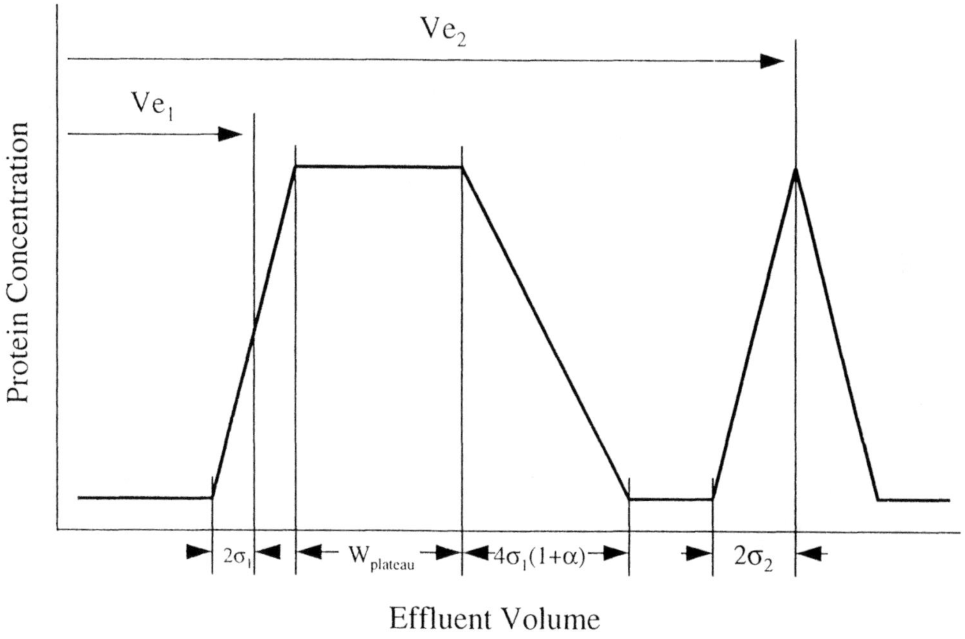

If we assume that the plateau width is proportional to the sample feed volume V_s:

$$W_{plateau} = \psi \cdot V_s$$

and if $\sigma_1 = \sigma_2$, then the resolution R_s can be defined as

$$R_s = \frac{V_{e2} - V_{el}}{2\sigma + \psi V_s + 4\sigma(1+a) + 2\sigma}$$

Furthermore, the chromatographic cycle time, t_c, is defined as

$$t_c = \frac{V_{equil} + V_s + V_{wash} + V_{grad}}{A_{cross} F}$$

where V_{equil} is the column equilibration volume, V_{grad} is the chromatographic gradient volume, and V_{wash} is the combined volume of column washings.

Now consider the *chromatographic production rate*, P. This practical parameter is the quantity of protein product that can be separated per unit time and is defined as the ratio of sample feed volume per cycle time:

$$P = \frac{V_s}{t_c}$$

(i) Using the following assumptions:

$$V_{e1} \cong V_T(1 + k'_1), \quad V_{e2} \cong V_T(1 + k'_2), \quad \text{and} \quad \sigma = \frac{V_{e1}}{\sqrt{N}}$$

derive an expression for P in terms of A_{cross}, F, θ, ψ, α, k'_1, R_s, a, and N (the plate number). The column effluent proportionality factor, θ, is defined by

$$V_T\theta = V_{equil} + V_{wash} + V_{grad}$$

A complete list of notation is given below.

- A_{cross} column cross section area
- F linear flow-rate
- θ column effluent proportionality factor
- ψ peak width proportionality factor
- α selectivity factor $(V_{e2} - V_T)/(V_{e1} - V_T)$
- k′ capacity factor $(V_e - V_o)/V_o$
- R_s resolution
- a peak asymmetry factor
- N plate number
- L column bed height
- t time
- t_c chromatographic cycle time
- W peak width between the base line intercepts
- V_{equil} column equilibration volume
- V_{grad} chromatographic gradient volume
- V_{wash} combined volume of column washings
- α standard deviation of the Guassian concentration profile
- V_e peak elution volume
- V_T column bed volume

(ii) Using the following values,

$A_{cross} = 1\ ft^2$, F = 1 ft/hr, $\theta = 6$, $\psi = 5/3$, $k'_1 = 10$, and a = 1,

plot P as a function of N (e.g., for N ranging from 0 to 3000) for selectivity factor values of 1.5 and 2.5 and for resolution values of 1 and 1.5 (i.e., 4 plots in total).

(iii) Based on the plots of P versus N, which parameter would you vary in practice to increase the production rate? How might you alter the purification system to effect such a change. Answer this question for the following three cases: ion exchange chromatography, affinity chromatography, and gel filtration.

15. Affinity Chromatography With an Immobilized Dye

Cibracon blue F3G-A is a competitive inhibitor of many enzymes, and you wish to determine whether Cibracon blue F3G-A covalently attached to agarose would be useful for the chromatographic separation of the enzyme aldehyde dehydrogenase. The inhibition constant (K_{In}) for inibition of aldehyde dehydrogenase (from *S. cerevisiae*) by Cibracon blue F3G-A is 4.2 μM (the inhibition constant is the dissociation constant for the enzyme-inhibitor complex). Assuming this same value applies for the reaction between the enzyme and immobilized dye, estimate the velocity of the enzyme through a column packed with Cibracon blue F3G-A immobilized to agarose ($\varepsilon = 0.35$). Assume that the effective dye concentration is 5 μmols Cibracon blue F3G-A per milliliter bed volume, that the enzyme is applied at a concentration of 5 mM and travels through the column in plug flow, and that the solvent velocity through the column is 0.01 cm/sec. State any other assumptions you must make to complete your analysis.

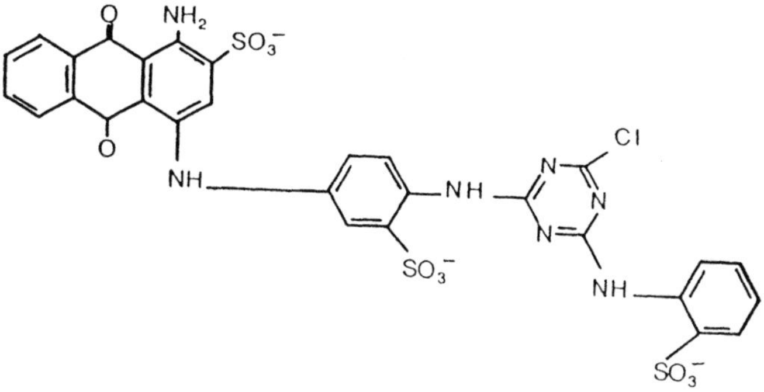

Cibracon blue F3G-A

16. Binding Capacity of an Immunosorbent Column

The total amount of protein that can be isolated in a given number of adsorption/elution cycles of immunoaffinity chromatography can be calculated once the immunosorbent capacity is known. Typically, the capacity per cycle of an immunosorbent, C(n), decreases as the number of cycles, n, increases. It has been reported in some cases that the decrease of C(n) with n is first order in C(n), that is,

$$\frac{dC(n)}{dn} = -\alpha C(n)$$

a) Assuming this to be the case, derive an expression for C(n) in terms of $n_{1/2}$, the number of cycles at which the capacity is half its initial value. Define the intial column capacity by an expression that includes the total amount of immobilized antibody (g L^{-1}) and the yield, or fraction, of active immobilized antibody.

b) Show that, if $n_{1/2} >> \ln 2$, the total amount of protein that can be isolated in n cycles, C(T), is

$$C(T) = \frac{C(O)}{\exp\left(\frac{-0.693}{n_{1/2}}\right)}\left[1 - \frac{\exp\left(\frac{-0.693n}{n_{1/2}}\right)}{\exp\left(\frac{-0.693}{n_{1/2}}\right)}\right]$$

17. *Effective Diffusivity of Derivatized Protein in Controlled Pore Glass*

Shown below are experimental and calculated dimensionless breakthrough curves for the isolation of derivatized bovine serum albumin by immunoaffinity chromatography. Arsanilic acid was conjugated to bovine serum albumin (BSA) to prepare an antigen for the anti-benzenearsonate antibodies used in the column. The column contained anti-benzenearsonate monoclonal antibody immobilized to controlled pore glass (CPG). The breakthrough curve was modeled by the pore diffusion equation assuming $N_{pore} = 8$. Based on this result, what was the effective diffusivity for conjugated BSA through the CPG? How does your answer compare to the bulk diffusivity of unmodified BSA, 6.7 x 10^{-7} cm^2/sec? *Additional data:* $d_p = 0.01$ cm, $u_s = 0.011$ cm/sec, L (column length) = 16.5 cm, $\varepsilon = 0.3$.

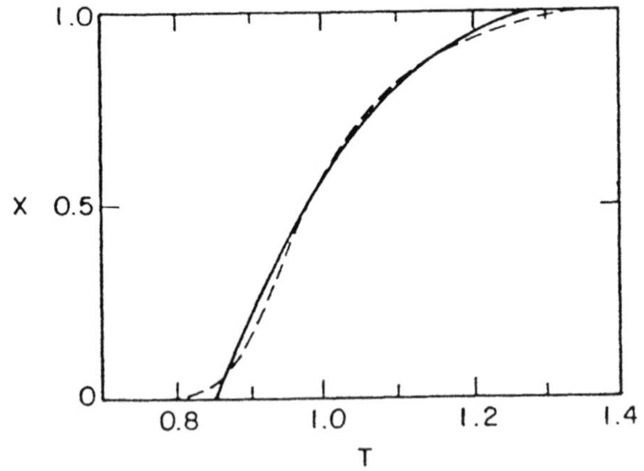

18. *Plate Height Equation for Non-adsorbing Systems*

The height equivalent of a theoretical plate, H, can be calculated from the first and second moments of a Guassian chromatographic peak according to Eq. 6.113:

$$H = \frac{L\sigma_t^2}{t_r^2} \tag{6.113}$$

On the next page is a plot of HETP versus the superficial velocity determined for 1 ml pulses of bovine serum albumin on Sepharose CL-4B.

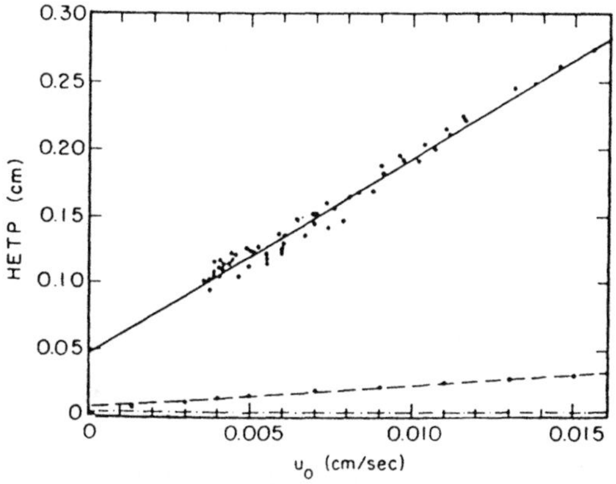

HETPs for 1 ml pulses of bovine serum albumin on Sepharose CL-4B (column diameter, 1.6 cm, L = 26 - 31 cm): ---, HETP contribution from fluid phase mass transfer; - · -, HETP contribution from end effects. [From Arnold et al., 1985.]

Bovine serum albumin does not adsorb to Sepharose CL-4B. Thus, we must derive an expression for H that applies to non-adsorbing systems. We begin by defining an effective equilibrium constant, K, as

$$K = \frac{\bar{s}}{c}$$

where $\bar{s}$ equals the average solute concentration inside the particle (a combination of adsorbed solute plus solute in the pores). Therefore, we can write

$$\bar{s} = \beta c + (1 - \beta)q'$$

where β is the porosity of the adsorbent particles, and q' is the sorbate concentration in units of mass per volume of solid adsorbent. Defining an intraparticle equilibrium adsorption coefficient, K_p, as q'/c, K becomes

$$K = \beta + (1 - \beta)K_p$$

For a non-adsorbing system, $K_p = 0$.

a) Show that, for a non-adsorbing system, the coefficient C of the van Deemter equation, Eq. 6.121, becomes

$$C = \left(\frac{2}{K_{OL}a_p}\right)\frac{\{(1-\varepsilon)\beta\}^2}{\{\varepsilon+(1-\varepsilon)\beta\}^2}$$

b) The overall mass transfer coefficient, k, often contains contributions due to fluid film and intraparticle mass tansfer, for example,

$$\frac{1}{K_{OL}a_p} = \frac{R}{3k_f} + \frac{R^2}{15\varepsilon_p D_{eff}}$$

The contribution of fluid mass transfer to H is shown by the dashed line in the plot below. After subtracting the contribution of fluid mass transfer from the overall H values, use the data below to determine the effective diffusivity of BSA in Sepharose CL-4B. For comparison, the bulk-phase diffusivity of BSA is about 6.7 x 10^{-7} cm^2/s.

Other pertinent data:

R(particle radius) = 0.005 cm, ε = 0.4, β = 0.53

19. Value of H for Maximum Separation Efficiency

Assume that H for a chromatographic column can be approximated by Eqs. 6.121 and 6.122, and that K′ >> 1.

a) Derive an expression for the minimum value of H (= H_{min}) that corresponds to the most efficient separation process.

b) Based on your answer in part a, explain why the column efficiency decreases as the column diameter increases (as illustrated in the figure below). The decrease in efficiency can pose a problem in the scale-up of chromatography.

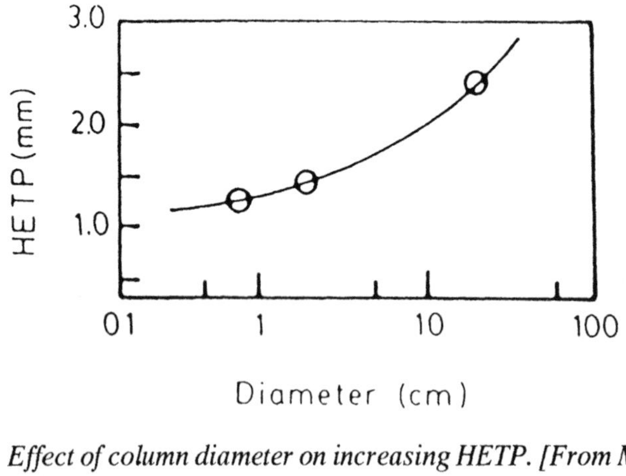

Effect of column diameter on increasing HETP. [From M. Seko, H. Takenchi, and T. Inada, Ind. Eng. Chem. Prod. Res. Develop., 21, 656 (1982).]

20. Optimal Operating Conditions for Preparative Chromatography

In general, achieving maximum resolution is the primary goal when selecting operating conditions for analytical chromatography. For preparative chromatography, however, the major objective is generally to maximize the production rate with a given column or to minimize the column volume required for a given production rate, subject to the allowable limits of product purity. Economic factors are also important, such as capital costs (determined mainly by the column volume) and energy costs (dictated largely by the carrier flow rate and the pressure drop through the column). Because the production rate increases with both the quantity of feed injected and the injection frequency, preparative columns are often run under "overload" conditions (i.e., outside the Henry's law region), and the sample is typically injected as a square pulse of finite duration rather than as an ideal delta function. Operating conditions are then selected to obtain the minimun acceptable resolution with the maximum possible injection frequency. In this problem we shall develop criteria for determining such conditions.

Let us assume we are injecting successive rectangular pulses of width $4\sigma_i$ which consist of compounds A and B, as shown in the figure below. Compound B travels through the column first, and we wish to achieve a situation where A and B are just resolved both within a given sample and between adjacent pulses. For a rectangular injection pulse, the width of response peak B will be $4(\sigma_i + \sigma_B)$ and the width of response peak A will be $4(\sigma_i + \sigma_A)$.

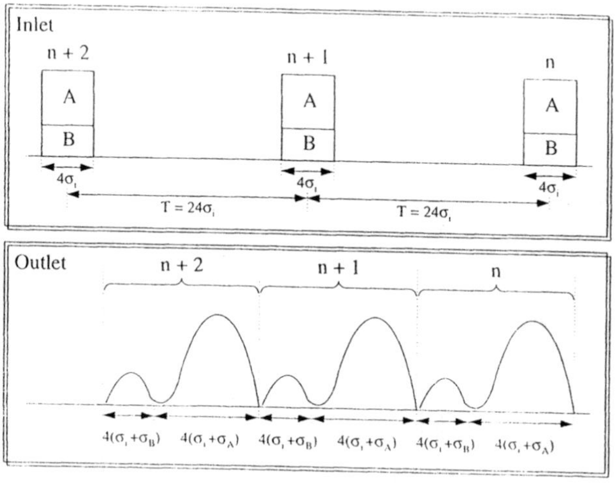

Optimal pulse sequence for production chromatography. Successive pulses of feed are introduced at intervals T. [Adapted from D.M. Ruthven, Principles of Adsorption & Adsorption Processes (John Wiley & Sons, New York, 1984).]

a) If $t_{r,A}$ and $t_{r,B}$ are the average retention times of compounds A and B, respectively, derive expresssions for the following quantities:

- time required for B to first emerge from the column
- time required for all of B to travel through the column
- time required for A to first emerge from the column
- time required for all of A to travel through the column

Now show that, if the separation between peaks introduced in successive injections is the same as the minimum separation between adjacent peaks of the same injection (the situation shown above), the time interval between successive injections, T, is given by

$$T = 2(t_{r,A} - t_{r,B})$$

b) Show that T can also be expressed as

$$T = 8R_s(\sigma_i + \sigma_{AB})$$

Thus, the relative duration of the injection pulse is given by

$$\theta = \frac{4\sigma_i}{T} = \frac{1}{2R_s(1 + \sigma_{AB}/\sigma_i)}$$

where

$$\sigma_{AB} = \frac{\sigma_A + \sigma_B}{2}$$

c) In the case of a rectangular injection pulse, the number of theoretical stages required to achieve a given resolution, R_s, is given by

$$N = 4R_s^2\left(\frac{\alpha + 1}{\alpha - 1}\right)^2\left(1 + \frac{\sigma_i}{\sigma_{AB}R_s}\right)^2$$

At $R_s = 1.0$ successive peaks are almost resolved while resolution is essentially complete for $R_s \geq 1.5$.

The required number of theoretical stages can now be expressed as a function of θ with α and R_s as parameters:

$$N(\theta) = 4R_s^2\left(\frac{\alpha + 1}{\alpha - 1}\right)^2(1 - 2\theta)^{-2}$$

(i) Derive an expression for the production rate per unit column volume, which is equal to the production rate per unit cross-sectional area (= $\theta u \varepsilon$) divided by the column length (= $H(u)N(\theta)$).

(ii) Show that the maximum volumetric production rate is obtained when θ is 1/6.

(iii) From your expression for $\theta u \varepsilon /NH$, what conclusion can be drawn about the relationship between the fluid velocity and volumetric productivity? What about the particle size?

21. *Electrophoretic Separation of Histidine from a Peptide*

It is necessary to separate histidine by paper electrophoresis from a peptide. The ionization characteristics are as follows:

	pK COOH	pK Imidazole	pK NH_3^+	pI
Histidine	1.8	6.0	9.2	7.6
Peptide	3.0	6.5	9.5	

In answering the following questions, assume
i) independent ionization of the groups,
ii) charge alone determines electrophoretic motion,
iii) the difference in charge for separation must be ± 0.1,
iv) the limit of the pH range is 3 to 10.

a) What is the pH or range of pH over which separation will occur?
b) What is the approximate pI of the peptide?
c) Toward which electrode will the peptide move at pH 6.5?
d) Is there a pH at which the compounds will move in opposite directions? If so, what is the pH or range of pH?

22. *Growth Rate of Crystals in a Continuous Crystallizer*

Consider a continuous crystrallizer satisfying the assumptions and constraints discussed in Section 6.5.1. The crystal product has been sized and the population density determined. The data are summarized in the table below.

Population density, n	L (μM)
80	428
400	360
1000	325
4000	274
80,000	150
200,000	116
1×10^6	56

If the space time is equal to 15 minutes, what is the growth rate of the crystals?

Chapter 7. Microbial Interactions

In natural environments, there are several microbial species which coexist and interact. These microbial interactions are important in the natural cycles of carbon, nitrogen and other elements, and play a key role biodegradation and bioremediation, in industrial processes such as wastewater treatment, biodegradation, and in commercial processes such as cheese manufacturing. The behavior of mixed culture systems is not merely a composite of the pure culture behavior of the individual species present, but reflects the spatial and physiological relationships among the component species and strains.

In ecosystems, the microbial population is dominated by a relatively small number of species. This results from selective environmental factors, such as growth components, pH or temperature, which impose limits on the the possible taxonomic heterogeneity of the microbial population. In addition, nutrients are transferred from lower levels to higher levels in the food chain. One of the interests of population ecologists has been to examine the stability of these multiple-level systems as a function of the number of species in the system. In this chapter we shall introduce and analyze some of the more important interactions found in microbial systems and their effect on the stability of the system.

7.1 Microbial Interactions

In examining the range of possible microbial interactions, it will be useful to consider a system of just two or three species, as these illustrate the important features without imposing great complexity. Interactions can be classified by whether an organism is benefited, harmed, or not affected by the presence of another organism. Table 7.1 lists the common types of interaction.

Neutralism connotes a lack of interaction between species, so that the growth characteristics of one species are unaffected by the presence of another. This implies that the species must have different growth limiting substrates and that their by-products do not influence other species. As one might expect, this is not frequently observed and few instances of pure neutralism have been reported in experimental systems.

Table 7.1. Microbial interactions and their definitions.

		Effect of Interaction	
Interaction	*Definition*	*Population A*	*Population B*
Neutralism	Lack of interaction	0	0
Commensalism	One member benefits while the other is unaffected	0	+
Mutualism	Each member benefits from the other	+	+
Competition	A "race" for nutrients or space	-	-
Amensalism	One species adversely changes the environment for the other	0 or +	-
Parasitism	One organism steals from the other	+	-
Predation	One organism ingests the other	+	-

0: unaffected; +: benefited; -: impaired

In *commensalism*, one organism requires a product from the other to grow, while not affecting the other organism to an appreciable extent. Another type of commensal interaction occurs when one organism removes a toxic product from the environment, enabling the second species to grow. The second species provides no benefit to the first, distinguishing this interaction from *mutualism*. An example of the first type of commensal interaction is that of *Proteus vulgaris* and *Saccharomyces cerevisiae*. The yeast produces niacin, which is required for the growth of the *Proteus* species. If niacin is supplied to a mixed culture of both species, the dependence vanishes. Table 7.2 lists some common types of commensal interactions that have been observed.

Mutualistic interactions result in systems where growth and survival of both organisms are possible only by their association; neither can survive without the other. Mutualism may involve several mechanisms. One is the exchange of growth factors between the species, so that each benefits from the presence of the other. For example, when a phenylalanine-requiring strain of *Lactobacillus* is grown with a folic-acid requiring strain of *Streptococcus*, the mixed cultures grow well because each species produces the nutrient required by the other. However, when grown individually, neither species is able to grow to an appreciable extent[1]. Likewise, in the carbon cycle, algae produce carbohydrates from CO_2 and oxygen; bacteria consume carbohydrates and oxygen to produce CO_2.

When mutualism is an essential requirement for growth it is termed *symbiosis*. For example, termites ingest cellulose, which they are unable to hydrolyze themselves. However,

(1) See V. Nurmikko, *Biochemical Factors affecting Symbiosis among Bacteria*, Experientia **12**, 245 (1956).

Table 7.2. Examples of Commensal Interactions. [From J.L. Meers "Growth of Bacteria in Mixed Cultures", p156 in "Microbial Ecology", eds. A. Laskin and H. Lechevalier].

A. Interactions where a compound is supplied by one organism and required by another.

Compound	Species producing compound	Species requiring the compound
nicotinic acid	*Saccharomyces cerevisiae*	*Proteus vulgaris*
Hydrogen sulfide	*Desulfovibrio*	Sulphur bacteria
methane	anaerobic methane bacteria	methane oxidizing bacteria
nitrate	*Nitrobacter*	denitrifying bacteria
fructose	*Acetobacter suboxydans*	*Saccharomyces carlsbergensis*

B. Interactions where a compound is removed by one organism.

Compound	Interrelationship
oxygen	aerobic organisms may reduce the oxygen tension thus allowing anaerobes to grow
hydrogen sulfide	toxic H_2S is oxidized by photosynthetic sulfur bacteria and the growth of other species is then possible
food preservatives	the growth inhibitors benzoate and sulfur dioxide are destroyed biologically
mercuric germicides	Desulfovibrium form H_2S from sulfate and the sulfide combines with Hg-containing germicides to permit bacterial growth

they contain flagellated protozoa which host bacteria which, in turn, produce cellulases that provide their hosts with glucose by hydrolysis. Similar symbiotic relationships exist in the stomachs of the cow, where rumen microorganisms aid in digesting plant material. Other types of mutualism involve one species destroying a toxin for the other species, while the second species provides a nutrient for the first.

Competition results from the struggle between organisms for a common essential resource, such as nutrients, water, light, or space, that is present in the environment in a limited amount. In microbial systems, competition is intense due to their high population densities and their short generation times. In the presence of a sufficient amount of nutrients, competition ceases. Competing species do not directly harm each other by producing toxic compounds or by feeding on rivals. Mutants of a microbial species may arise which are better able to consume nutrients or which are antibiotic resistant; these mutants will then be able to outgrow the original organism. In chemostats, natural mutation and the resulting competition can be employed to develop enriched microbial populations with desirable traits.

In *parasitic* relationships, a small organism (the parasite) lives at the expense of the larger one (the host). The host organism is not necessarily killed by the parasite, and this distinguishes this relationship from predation. The dynamic behavior of parasitic relationships is, however, similar to predation. Bacteriophage systems are typical of parasitic interactions. In *predation*, the predator feeds on the prey and kills it. The predator-prey interaction has been examined in animal systems (e.g., the lynx-hare and owl-lemming systems), as well as microbial systems. The predator reduces the population of prey to such an extent that it is unable to support the large predator population. The predator population then declines and permits the prey to then increase in number so that the predator population once again increases. Such a situation thus may exhibit sustained oscillations in both populations.

Amensalism is a form of interaction in which organic or inorganic metabolites, generated as a result of metabolism of an organism, are inhibitory to other organisms and act to reduce the growth of these other populations. Amensalism is sometimes termed inhibition, antibiosis or antagonism. Microbially synthesized toxins such as antibiotics are thus classified as causing amensal interactions.

We shall examine the mathematical descriptions of the two most common types of interaction, competition and predation. Although we shall focus on microbial systems, the conclusions also apply to other types of systems.

7.2 Competition

In strict competition, the competing species do not directly harm each other; rather their interactions arise indirectly through the struggle for common food or other limiting resources. In a closed system, only one of two competing species will survive, this being an extreme case of Darwin's concept of natural selection. One of the earliest mathematical treatments of competition is that of the Italian mathematician Vito Volterra, who considered two species n_1 and n_2 in a closed system, competing for a common food resource. It was assumed that both species could grow at maximum rates μ_1 and μ_2 when nutrients are present in excess. When nutrients are limited, the growth rate is reduced by an amount which depends on the concentration of both species and the rates at which each species consumes nutrients, these rates being given by $\lambda_1 n_1$ and $\lambda_2 n_2$ respectively. The batch mass balances for each species thus become:

$$\frac{dn_1}{dt} = (\mu_1 - \gamma_1\{\lambda_1 n_1 + \lambda_2 n_2\})n_1 \qquad (7.1a)$$

$$\frac{dn_2}{dt} = (\mu_2 - \gamma_2\{\lambda_1 n_1 + \lambda_2 n_2\})n_2 \qquad (7.1b)$$

The constants γ_1 and γ_2 can be thought of as effective yield coefficients for the growth of both species on the limiting nutrient. By multiplying Equation (7.1a) by γ_2/n_1 and (7.1b) by γ_1/n_2 and adding the resulting equations, we obtain:

$$\gamma_2 \frac{d \ln n_1}{dt} - \gamma_1 \frac{d \ln n_2}{dt} = \gamma_2\mu_1 - \gamma_1\mu_2 \tag{7.2}$$

which can be integrated to yield

$$\frac{n_1^{\gamma_2}}{n_2^{\gamma_1}} = Ce^{(\gamma_2\mu_1 - \gamma_1\mu_2)t} \tag{7.3}$$

where C is a constant of integration. The ratio of n_1 to n_2 thus increases or decreases with time, depending on the sign of $(\gamma_2\mu_1 - \gamma_1\mu_2)$. If this is negative, then at large times either n_1 tends to zero or n_2 grows without bound. Examination of Equation (7.1b) shows that the quadratic term in n_2 $(-\gamma_2\lambda_2 n_2^2)$ will become large as n_2 increases and thus make dn_2/dt negative. Thus n_2 cannot grow without bound and n_1 must thus approach zero with increasing time. A similar argument holds when $\gamma_2\mu_1 - \gamma_1\mu_2$ is positive; n_2 must tend to zero with time. Thus one species dies out and the other survives. This is an example of the competitive exclusion principle (survival of the fittest). Only if $\gamma_2\mu_1 = \gamma_1\mu_2$ can both populations coexist. By further applying this approach to multiple species in a closed environment, all but one species will disappear, provided that there is sufficient nutrient available to ensure that the surviving species (say n_1) can grow at a rate given by $(\mu_1 - \gamma_1\lambda_1 n_1)$. By neglecting the dependence of the growth rate on nutrient concentration, some of the details in the analysis are thus lost. We shall now turn to microbial systems which exhibit a Monod-type substrate dependence, and show interesting dynamics as a result.

7.2.1 Microbial Competition in a Chemostat

Consider two microorganisms exhibiting a Monod type growth rate dependence on a single limiting nutrient S. The three possible shapes of the $\mu(S)$ relationship are shown in Figure 7.1. In case (a), organism 1 exhibits a higher specific growth rate at all values of S; case (b) is the converse, organism 2 exhibits the higher growth rate. In both cases, we can write the steady-state chemostat equations for each species viz:

$$\frac{dX_1}{dt} = -DX_1 + \mu_1 X_1$$

$$\frac{dX_2}{dt} = -DX_2 + \mu_2 X_2 \tag{7.4}$$

If steady-state populations of both X_1 and X_2 are to coexist, then coexistence requires $D = \mu_1 = \mu_2$. Because $\mu_1 > \mu_2$ in case (a) for all values of S, coexistence cannot occur and organism 2 must be washed out of the reactor. A similar result is also true in case (b). In case (c), we see that there is a single value of the substrate concentration where $D = \mu_1 = \mu_2$. The value of the dilution rate at which coexistence occurs is denoted D_c.

At dilution rates greater than D_C, organism 1 will have the higher growth rate and organism 2 will be washed out. At $D < D_C$ however, organism 2 will dominate. If we consider the set of mass balances for both organisms and substrate:

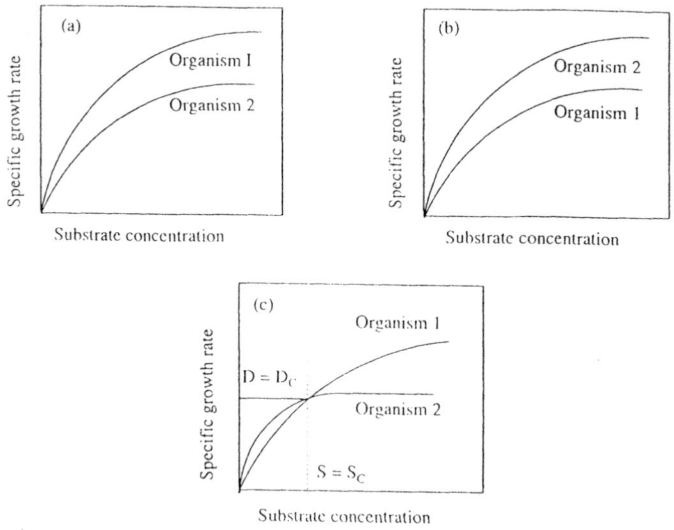

Figure 7.1. *Possible forms of the specific growth rate dependence of organisms 1 and 2 on the single substrate S. Case (a) has* $\mu_1 > \mu_2$ *for all S; case (b) has* $\mu_1 < \mu_2$ *and case (c) indicates that at low S, organism 1 has a lower growth rate, but at higher values of S organism 2 has a larger specific growth rate.*

$$\frac{dX_1}{dt} = -DX_1 + \mu_1(S)X_1 \tag{7.5a}$$

$$\frac{dX_2}{dt} = -DX_2 + \mu_2(S)X_2 \tag{7.5b}$$

$$\frac{dS}{dt} = D(S_o - S) - \frac{1}{Y_1}\mu_1 X_1 - \frac{1}{Y_2}\mu_2 X_2 \tag{7.5c}$$

Coexistence can occur at $D = D_C$ when

$$D_C = \mu_1(S) = \mu_2(S) \tag{7.6a}$$

$$D_C(S_o - S) - \frac{\mu_1 X_1}{Y_1} - \frac{\mu_2 X_2}{Y_2} = 0 \tag{7.6b}$$

Equation (7.6a) will specify the value of S at which coexistence occurs when the forms of the $\mu(S)$ relationships for each species are given. Equation (7.6b) provides a relationship between X_1 and X_2 at steady state, but cannot provide the actual steady state values. Possible steady states lie on the line given in Figure 7.2.

$$S_o - S = \frac{X_1}{Y_1} + \frac{X_2}{Y_2} \tag{7.7}$$

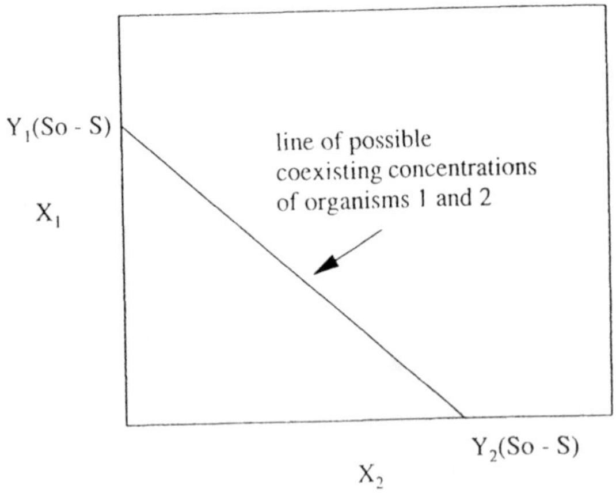

***Figure** 7.2. Steady state populations of the competing organisms X_1 and X_2 lie on the line.*

From Equation (7.6a) we see that the inlet substrate concentration does not affect the steady state substrate concentration in the reactor, but does affect the total population concentrations of the organisms. If the substrate concentration in the feed is less than the substrate concentration obtained from solution of Equation (7.6a), then both organisms will wash out. If we insert the Monod expression for the specific growth rates of both organisms, we can solve for the steady state value of S:

$$\mu_1 = \frac{\mu_{m1} S}{K_1 + S} \qquad \mu_2 = \frac{\mu_{m2} S}{K_2 + S}$$

$$S_{ss} = \frac{K_1\mu_{m2} - K_2\mu_{m1}}{\mu_{m1} - \mu_{m2}}$$

$$D_C = \frac{K_1\mu_{m2} - K_2\mu_{m1}}{K_1 - K_2} \tag{7.8}$$

Denoting organism 1 as having the higher maximum specific growth rate, the criterion for positive values of D_C and S_{ss} is

$$K_1 > \frac{\mu_{m1}}{\mu_{m2}} K_2 \tag{7.9}$$

If this requirement is not satisfied, then at all dilution rates only one organism can exist, as was shown in Figure 7.1 (a) and (b). The behavior of this system can be conveniently represented on a plot of S_o as a function of D, the operating diagram shown in Figure 7.3. An analysis of Region IV shows that the coexistence steady state is stable to small perturbations.

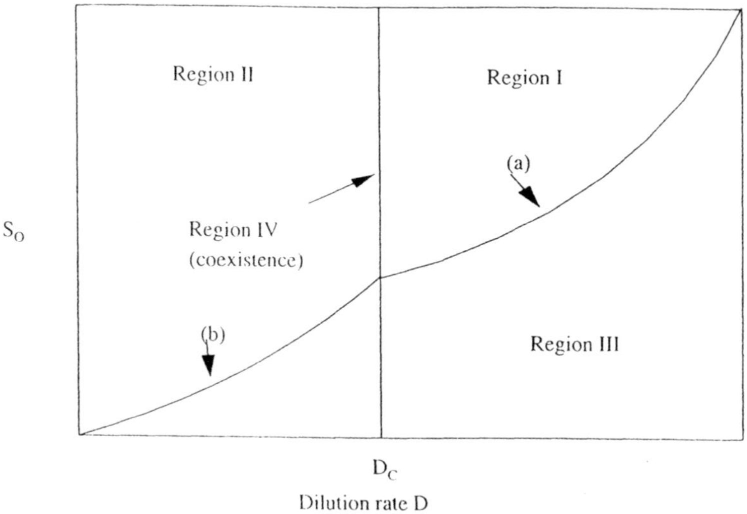

Figure 7.3. *The operating line of a competing two organism-one substrate system. In Region I, organism 1 is dominant and organism 2 is washed out. In Region II, organism 2 dominates and organism 1 is washed out. In Region III both organisms are washed out as there is insufficient substrate in the feed stream (S_o) to support steady state populations. Coexistence of both organisms occurs along the line at dilution rate D_c (denoted as Region IV), above the intersection of lines (a) and (b). The equation of lines (a) and (b) are given by:*

Line (a)
$$S_o = -K_1 + \frac{\mu_{m1} K_1}{\mu_{m1} - D}$$

Line (b)
$$S_o = -K_2 + \frac{\mu_{m2} K_2}{\mu_{m2} - D}$$

7.2.2 Selection and Competition in the Chemostat

There are a number of situations in which two species may be present in the chemostat. Some of these produce desirable outcomes; others do not. When a contaminating microorganism is present in the culture, it may outgrow the desired species and the resulting fermentation may need to be abandoned. A similar situation may arise in the case of revertant strains which loose ability to produce a desired product. As we saw in Chapter 3, loss of plasmid from a recombinant strain results in a plasmid-free organism that may have a higher maximum growth rate and outgrow the desired plasmid-containing organism. On the other hand, competition in the chemostat for the limiting nutrient can be advantageous if selection of a fast-growing strain is desired. If an inhibitory substrate is fed to a mixed microbial population, the chemostat will select for the fastest-growing species, generally the one which is most resistant to the inhibitory material. This is often used to advantage in developing bacterial strains for treatment of toxic waste streams.

The dynamics of competition are governed by the differences in the specific growth rates of each organism. The governing set of mass balance equations is given by Equation (7.5). If the "contaminant" is present at a low initial concentration, it can take considerable time before it outgrows the desired organism in the reactor. This is illustrated in Figure 7.4.

This prediction is compared with the experimental data on the competition between *E. coli* and *Azotobacter vinelandii* in a glucose limited chemostat in Figure 7.5. Washout of the slower growing *A. vinelandii* occurs after a period of three days, as Equation (7.9) for coexistence is not satisfied.

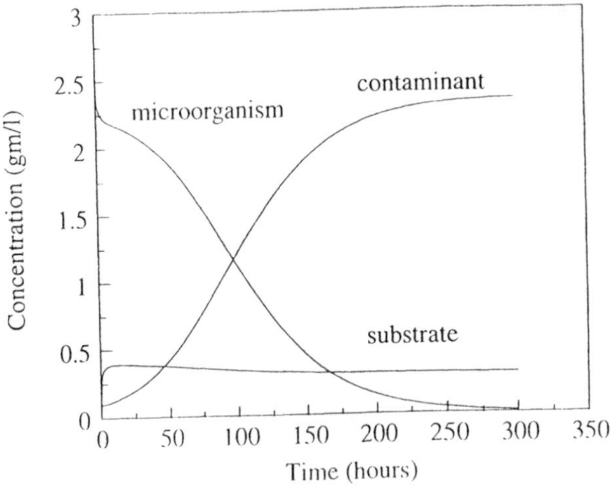

Figure 7.4. *The transient response of cell and substrate concentrations when a contaminating organism is present at low concentration in a culture previously operating at steady state. The parameter values for the original organism are* $\mu_{max\,1} = 0.5\ hr^{-1}$, $K_{S1} = 100$ *mg/liter,* $Y_1 = 0.5$ *gm/gm; those for the contaminant are* $\mu_{max\,2} = 0.6\ hr^{-1}$, $K_{S2} = 150$ *mg/liter,* $Y_2 = 0.5$ *gm/gm. The initial contaminant concentration is 0.1 gm/liter, the inlet substrate concentration is 5 gm/liter and the dilution rate is* $0.4\ hr^{-1}$.

7.2.3 Competition for Two Growth-Limiting Substrates

In natural ecosystems, there may be more than one substrate available to competing organisms. In continuous culture, such availability affects the possible range of dilution rates that may result in coexistence of both species. We shall examine the situation where two organisms may grow on either of two substrates, S_1 and S_2. The growth rate of each organism depends on both substrates and we shall employ a model developed in Chapter 3 to describe this dependence. Mass balances around the chemostat can be written as follows:

$$\frac{dX_1}{dt} = -DX_1 + \mu_1 X_1 \quad ; \quad \mu_1 = \mu_1(S_1, S_2) \tag{7.10a}$$

$$\frac{dX_2}{dt} = -DX_2 + \mu_2 X_2 \quad ; \quad \mu_2 = \mu_2(S_1, S_2) \tag{7.10b}$$

$$\frac{dS_1}{dt} = D(S_{1o} - S) - \frac{\mu_1 X_1}{Y_{11}} - \frac{\mu_2 X_2}{Y_{12}} \tag{7.10c}$$

$$\frac{dS_2}{dt} = D(S_{2o} - S) - \frac{\mu_1 X_1}{Y_{12}} - \frac{\mu_2 X_2}{Y_{22}} \tag{7.10d}$$

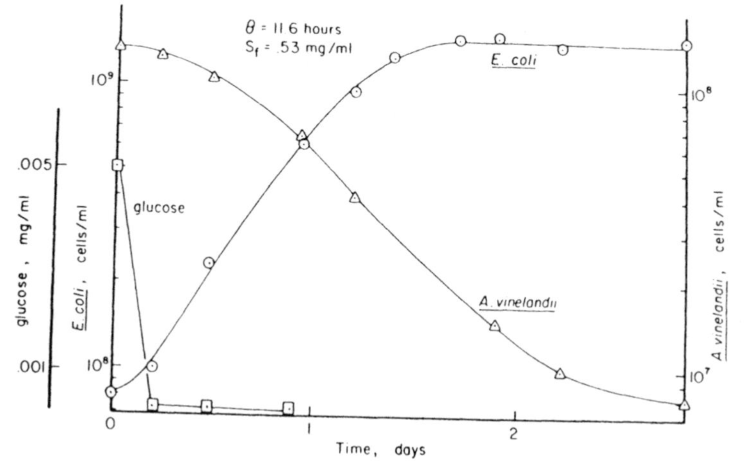

***Figure** 7.5. Illustration of the competition between the bacteria E. coli and A. vinelandii for the growth-limiting nutrient glucose in a chemostat operated at a dilution rate of 0.086 hr^{-1}. The maximum specific growth rates for E. coli and A. vinelandii were 0.32 and 0.23 hr^{-1}, resp. The K_s values were $1x10^{-7}$ mg/ml and $1.2x10^{-2}$ mg/ml, respectively. The inlet glucose concentration was 0.53 gm/liter. [Data from Jost, J.L., Drake, J.F., Tsuchiya, H.M. and A.G Fredrickson, J. Bacteriol, **113** 834 (1973)]*

At steady state, the first two equations do not suffice to determine the values of S_1, S_2 or D. These equations show $\mu_1(S_1, S_2) = \mu_2(S_1, S_2) = D$. This illustrates the difference between single and double substrate systems. There may be a set of substrate concentrations and dilution rates which will give a steady state rather than the single value of D that supports the coexistence steady state with one substrate. If μ_1 and μ_2 are known functions of S_1 and S_2, the steady state values of both substrate concentrations can be determined. There are four possible types of steady states in this system; one in which both organisms coexist, two situations where only one organism is present at steady state, and a steady state in which both organisms are washed-out.

If the functional forms of $\mu_1(S_1, S_2)$ and $\mu_2(S_1, S_2)$ are known, contour plots describing constant values of μ_1 and μ_2 can be drawn on a graph of S_1 against S_2. Possible steady states are at the intersection of these contour plots of μ_1 and μ_2[2]. We shall examine this in the following section.

Coexistence of Competing Species with Two Substrates

Consider the specific growth rates of each microorganism described by the extended Monod model for two substrates developed in Chapter 3.

$$\mu_1 = \frac{\mu_{m11} S_1}{K_{11} + S_1 + a_{11} S_2} + \frac{\mu_{m12} S_2}{K_{12} + S_2 + a_{12} S_1} \tag{7.11a}$$

$$\mu_2 = \frac{\mu_{m21} S_1}{K_{21} + S_1 + a_{21} S_2} + \frac{\mu_{m22} S_2}{K_{22} + S_2 + a_{22} S_1} \tag{7.11b}$$

	Organism 1	*Organism 2*
Maximum specific growth rates μ_{max} (hr^{-1})	$\mu_{m11} = 0.5$	$\mu_{m21} = 0.4$
	$\mu_{m12} = 0.5$	$\mu_{m22} = 0.4$
Saturation constants (mg/liter)	$K_{11} = 6$	$K_{21} = 12$
	$K_{12} = 6$	$K_{22} = 12$
Yield coefficents (gm/gm)	$Y_{11} = 0.15$	$Y_{21} = 0.15$
	$Y_{12} = 0.15$	$Y_{22} = 0.15$
Mutual inhibition constants (mg/liter)	$a_{11} = 75$	$a_{12} = 0.75$
	$a_{21} = 1/a_{11}$	$a_{22} = 1/a_{12}$

We shall examine the behavior of this model numerically, employing values of the constants typical for bacteria. Consider the the growth of both microorganisms in a chemostat at steady state. Equations (7.10) and (7.11) can be solved provided the inlet substrate concentrations are known. The feed concentrations of both substrates are set at 500 mg/liter to simplify the calculations.

(2) An example of this method is given in Bader, F.G., Meyer, J.S., Fredrickson, A.G., and H. Tsuchiya, *Biotech. Bioeng.* **17**, 279 (1975)

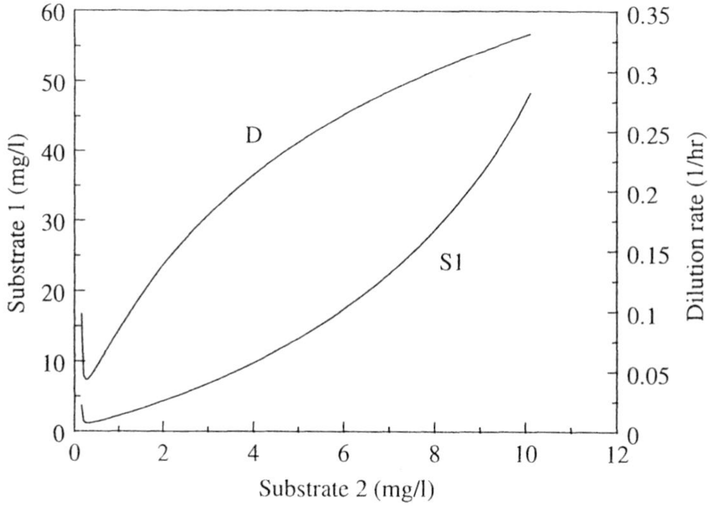

Figure 7.6. *The range of substrate concentrations and dilution rates which permit coexistence of both microbial species, X_1 and X_2. At low values of S_1, there are two possible values of S_2 that satisfy the coexistence condition. Similarly, at low dilution rates there are two substrate concentrations of S_2 that permit coexistence.*

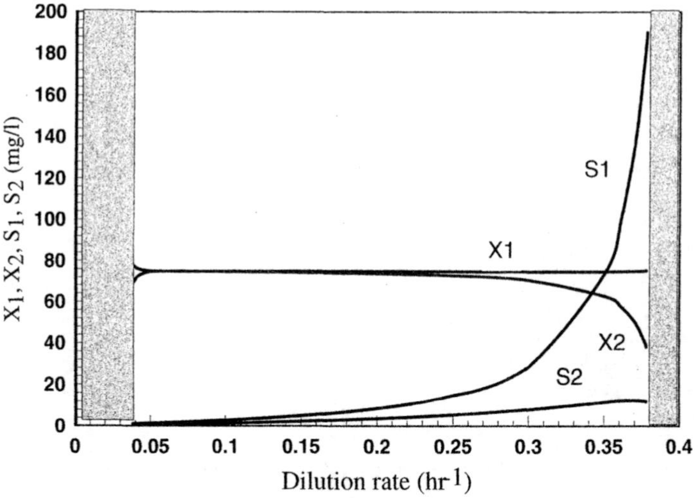

Figure 7.7. *The steady state behavior of the system of two microorganisms competing for two substrates. At dilution rates below 0.039 hr^{-1}, organism II predominates; at dilution rates above 0.378 hr^{-1}, organism I predominates. At intermediate dilution rates, both organisms are present in the reactor.*

The range of substrate concentrations that will support the coexistence of both species is given by the solution of $\mu_1 = \mu_2$. These are shown in Figure 7.6. The steady state substrate and cell concentrations are shown in Figure 7.7. At low and high dilution rates, only one species is present. The region of coexistence is much larger with two limiting substrates than the single dilution rate capable of supporting both populations and only one substrate.

7.3 The Effect of Chemotaxis on Competition

Many ecosystems are not well-mixed and cannot be analyzed using the assumptions employed in the preceding section. For example, nitrifying bacteria in soils may compete with other species for limited carbon resources. In sewage treatment, there is competition in microbial films for essential nutrients. In Chapter 5, we saw that directed cell movement provides a means by which cells can obtain nutrients in environments that are not well-mixed. Thus slow-growing bacteria may be able to survive if they are sufficiently chemotactic. The equations describing chemotaxis effectively provide a convective-like term in the resulting mass balance equations. When several chemotactic species compete for a common resource, the resulting population concentrations depend on the rates of diffusion of the nutrient in the system and on the growth and chemotactic properties of both species involved. We shall illustrate this with a simple model of competition by two microbial species for a nutrient which diffuses from a source[3].

The system considered is one-dimensional and the limiting nutrient is present at a fixed concentration at one of the boundaries (x = L in Figure 7.8).

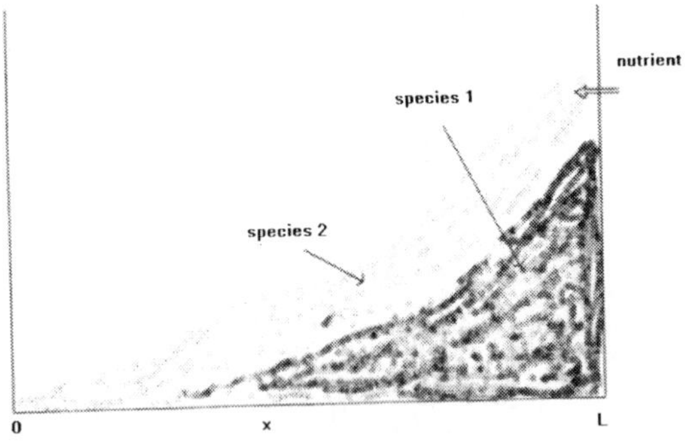

Figure 7.8. A one-dimensional systems in which species n_1 and n_2 compete for a limiting nutrient S. The region considered lies between x = 0 and x = L.

(3) A complete description is provided in Kelly, F.X., Dapsis, K.J. and D.A Lauffenburger, *Microbial Ecology*, **16** 115-131 (1988).

The conservation equations for both bacterial species and for the nutrient S can be written as follows:

$$\frac{\partial n_i}{\partial t} = \frac{\partial}{\partial x}\left\{-\alpha_i \frac{\partial n_i}{\partial x} + \chi n_i \frac{\partial S}{\partial x}\right\} + \mu_i n_i \qquad (7.12a)$$

$$\frac{\partial S}{\partial t} = \frac{\partial}{\partial x}\left\{-D_s \frac{\partial S}{\partial x}\right\} - \frac{1}{Y_1}\mu_1 n_1 - \frac{1}{Y_2}\mu_2 n_2 \qquad (7.12b)$$

The expression for the cell flux contains two terms, the first represents random motility (α_i) and the second represents chemotaxis (χ_i), which depends on the substrate gradient. Models of chemotaxis have been developed which assume χ_i has an an inverse quadratic dependence on subtrate concentration. This assumption arises from considerations of substrate binding to receptors on the cell surface[4]. The simplest model would assume that the growth rate of both bacterial populations could be described by zeroth order kinetics (μ_i are constant) and that the random motility and chemotaxis coefficients are also constant. Typical values for *E.. coli* are $\alpha = 1 \times 10^{-5}$ cm^2/sec and $\chi = 8 \times 10^{-5}$ cm^2/sec. The diffusivity of the substrate (D_s) will be of similar order, typically 1×10^{-5} cm^2/sec. The bacteria are confined to the system; this is reflected in the boundary conditions below.

$$-\alpha_i \frac{\partial n_i}{\partial x} + \chi_i \frac{\partial S}{\partial x} n_i = 0 \quad \text{at} \quad x = 0, L$$

$$-D_s \frac{\partial S}{\partial x} = 0 \quad \text{at} \quad x = 0$$

$$S = S_o \quad \text{at} \quad x = L \qquad (7.13)$$

Various outcomes of this type of competition are possible. Even if one of the bacterial species has a growth rate advantage, it may not be able to outcompete the second species if that species has a higher chemotactic coefficient. There are ranges of specific growth rate values for coexistence when both species are chemotactic; when one species is immotile and the other is chemotactic; or when both species is are motile and only one is chemotactic. The result also depends on the length of the system, L. The situation is thus more complex than the well-mixed populations considered earlier. Solution of the mass balance equations is quite difficult due to the presence of the chemotactic term; finite element methods must generally be employed. Figure 7.9 provides an indication of the types of outcomes that are possible.

(4) The constitutive equations used to describe both random motility and chemotaxis are discussed in Ford, R.M. and D.A. Lauffenburger, Biotech. Bioeng. **37** 661 (1991); we have used the simplest descriptions, although other models may give a better representation of the experimental data.

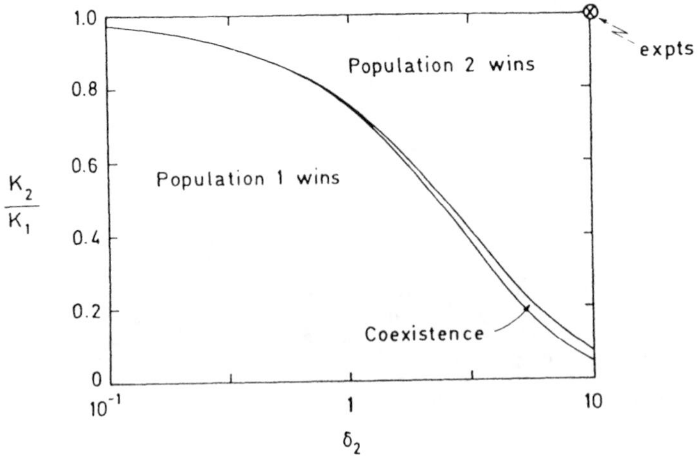

Figure 7.9. The effect of chemotaxis on the possible steady states in a system with two populations with equal random motility and with one species being chemotactic. The abscissa is a normalized chemotactic constant ($\delta_2 = \chi_{o2}(K/S_o)/\mu_2$), where the chemotaxis coefficient has been assumed to be of the form $\chi_{o2}K/(K+S)^2$. The ordinate is the ratio of the cell specific growth rate constants for population 2 to population 1. There is a range of parameter values which support coexistence of both species. The details of the simulations are found in Kelly, F.X., Dapsis, K.J and D.A. Lauffenburger, Microb. Ecol. 16, 115 (1988).

7.4 Predation

Predation is an important factor in food chains, where it serves as a mechanism of transferring energy and carbon along the chain. Activated sludge wastewater treatment plants provide a practical example of the importance of predation. The microbial population in the sludge is comprised of bacteria, protozoa and small metozoa, with bacteria serving as prey for the protozoa. The protozoa serve to minimize the amount of excess sludge produced by injesting bacteria. The low cell yields (gm protozoa/gm bacteria) are an important factor in this reduction. In natural ecosystems, predation serves to stabilize the population of fast-growing species lower in the food chain, and may prevent depletion of resources that support other species. If two species compete for one resource, a predator may stabilize the populations by consuming the superior bacterial species at a higher rate than it consume the inferior species. We shall examine this stabilization later.

In many ecosytems, steady states are not attained, and the populations of all species may cycle with time. A simple example of predation serves to illustrate this possibility.

7.4.1 The Lotka-Volterra Model of Predator-Prey Interactions

A simple model which describes the cyclic behavior of many ecosystems was developed by Lotka, based on the the ideas proposed by the Italian mathematician Volterra[5]. We consider the predator and prey number concentrations to be given by n_2 and n_1 respectively. The prey are assumed to grow at a first order rate and the predators grow at a rate which depends on the frequency of predator-prey encounters; thus a second order mechanism can represent these two-body collisions. We can write the prey mass balance in a batch system as follows:

$$\frac{dn_1}{dt} = an_1 - \gamma n_1 n_2 \tag{7.14}$$

Prey thus grow exponentially with a growth rate given by the constant a. The coefficient γ represents the efficiency of injestion of prey by the predators. The mass balance for predators has a growth term which modifies this expression by the constant ε, which is an effective yield of predators from prey.

$$\frac{dn_2}{dt} = -bn_2 + \varepsilon n_1 n_2 \tag{7.15}$$

predators are asssumed to die at a first-order rate, described by the constant b. The steady state solution where both species are present is

$$\hat{n}_1 = \frac{b}{\varepsilon\gamma} \qquad \hat{n}_2 = \frac{a}{\gamma} \tag{7.16}$$

The dynamic behavior of the system can be described by examining the phase-plane i.e. $n_1(t)$ versus $n_2(t)$. This can be obtained by dividing the mass balances.

$$\frac{dn_2/dt}{dn_1/dt} = \frac{-bn_2 + \varepsilon\gamma n_1 n_2}{an_1 - \gamma n_1 n_2} = \frac{(-b + \varepsilon\gamma n_1)\, n_2}{(a - \gamma n_2)\ \ n_1} \tag{7.17}$$

Multiplying both sides of this equation by $(a - \gamma n_2)(dn_1/dt)/n_2$ and rearranging gives

$$\frac{a}{n_2}\frac{dn_2}{dt} - \gamma\frac{dn_2}{dt} = -\frac{b}{n_1}\frac{dn_1}{dt} + \varepsilon\gamma\frac{dn_1}{dt} \tag{7.18}$$

This rearrangement permits the equation to be integrated:

$$a \ln n_2 - \gamma n_2 + b \ln n_1 - \varepsilon\gamma n_1 = C$$

$$\left(\frac{n_2^a}{(e^{n_2})^\gamma}\right)\left(\frac{n_1^b}{(e^{n_1})^{\varepsilon\gamma}}\right) = e^C \tag{7.19}$$

---

(5) Volterra, V., "Leçons sur la Theorie Mathematique de la Lutte pour la Vie", Gauthiers-Villars, Paris (1931); Volterra, V., "Variazioni e fluttuazioni del numero d'individui in specie animali conviventi", Memorial Acad. Lincei, Ser. IV **2**, 31-113 (1926); Lotka, A.J. "Undamped Oscillations Derived from the Law of Mass Action", J. Am. Chem. Soc. **45** 1595 (1920).

The constant of integration is determined from the inital values of n_1 and n_2. This equation is transcendental, and must be solved numerically. There are zero, one or two solutions, depending on the ranges of n_1 and n_2. As can be seen in Figure 7.10, the solutions describe a closed trajectory, the size of which is dependent on the intital values of n_1 and n_2, and thus C. If we translate the phase plane behavior to temporal values of n_1 and n_2, the closed trajectory implies that both variables describe cycles, i.e., sustained oscillations in both species are predicted.

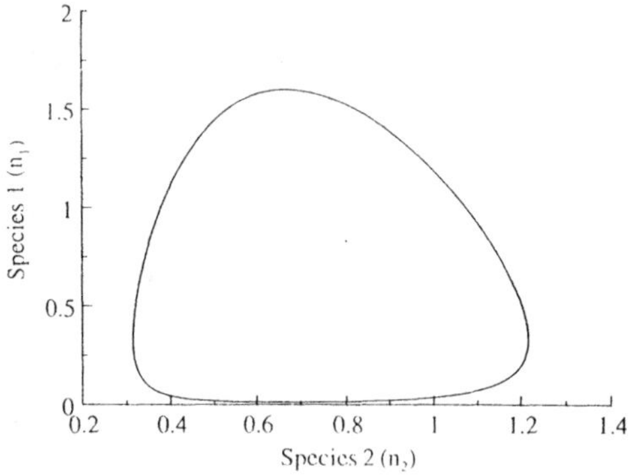

***Figure 7.10**. The phase plane behavior of n_1 and n_2. The stable steady state is at the focus of the closed trajectories ($\hat{n}_1 = 0.444$, $\hat{n}_2 = 0.667$). Other trajectories can be formed for different values of the constant C. Parameter values are $a = 0.5$, $b = 0.05$, $\gamma = 0.75$, $\varepsilon = 0.15$.*

We can obtain additional information about the behavior of this system by integrating Equation (7.14) over the period of one cycle, from t = 0 to t = T:

$$\int_0^T \frac{1}{n_1}\frac{dn_1}{dt}dt = \int_0^T (a - \gamma n_2)dt$$

$$\ln\left(\frac{n_1(T)}{n_1(0)}\right) = aT - \int_0^T \gamma n_2 dt \qquad (7.20)$$

Over the period of a cycle, $n_1(T)$ equals $n_1(0)$, so that Equation (7.20) can be rewritten as

$$\frac{1}{T}\int_0^T \gamma n_2 dt = a \qquad (7.21)$$

Noting the steady state value of n_2 is a/γ we see that

$$\frac{1}{T}\int_{0}^{T} n_2 dt = \hat{n}_2 \tag{7.22}$$

This indicates that although the value of n_2 may vary over the cycle, the value of n_2 averaged over this cycle is just the steady state concentration. By integration of Equation (7.15) for the predator we obtain a similar result for n_1.

7.4.2 Predator-Prey Interactions in Microbial Populations

The Lotka-Volterra description of predator-prey interactions can be extended to microbial systems by incorporating appropriate equations describing growth of both species and that of substrate. The Monod expression provides a simple constitutive model when both predators and prey are can be considered to have only one limiting substrate. In a well-mixed continuous flow system, we can write mass balance equations to indicate that the limiting substrate for the predator is the prey, and that for the prey is the organic substrate S.

$$\frac{dn_2}{dt} = \frac{\mu_{max2} n_1 n_2}{K_2 + n_1} - Dn_2 \tag{7.23a}$$

$$\frac{dn_1}{dt} = \frac{\mu_{max1} S n_1}{K_S + S} - \frac{1}{Y_1}\frac{\mu_{max2} n_1 n_2}{K_2 + n_1} - Dn_1 \tag{7.23b}$$

$$\frac{dS}{dt} = D(S_o - S) - \frac{1}{Y_S}\frac{\mu_{max1} S n_1}{K_S + S} \tag{7.23c}$$

This model is more realistic than the Lotka-Volterra equations from a biological standpoint, in that there are maximum rates of growth of both prey and predator and there is a finite nutrient supply rate. A difficulty with the above set of equations is that they predict very low concentrations of prey during the oscillations. Such low concentrations in practice might result in the elimination of prey. In this case, the prey behavior might be better described by a stochastic model. Furthermore, the numerical solution of Equations (7.23) is subject to some computational difficulties if the step size in the numerical procedure employed is not sufficiently small. Typical concentration profiles are shown in Figure 7.11, where the sustained nature of the oscillations is evident from the limit cycle behavior illustrated in Figure 7.12.

Such oscillations have been experimentally observed in nature and in the laboratory. The predation of the ciliated protozoan *Colpoda steinii* on the bacterium *E. coli B/r* in a chomostat is shown in Figure 7.13. Glucose was the limiting nutrient for the growth of the bacteria. Higher concentrations of both predators and prey result from a higher feed concentration of glucose. When the residence time was increased to 16 hours, it was found that the oscillations in species concentrations were damped. At a residence time of 32 hours, the damping was very rapid. This damping cannot be explained by the simple Monod relationships employed in Equations (7.23). More complex models must be considered.

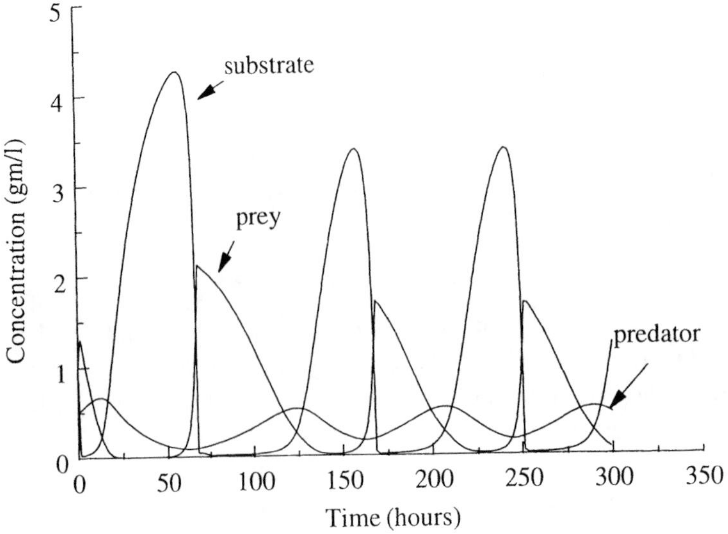

Figure 7.11*. The solution of the Equations (7.23) describing predator-prey interactions in the chemostat. The values of the constants describing the growth are* $\mu_{max2} = 0.1\ hr^{-1}$, $\mu_{max1} = 0.5\ hr^{-1}$, $K_2 = 0.25$ *gm/l*, $K_s = 0.10$ *gm/l*, $Y_1 = 0.25$ *gm/gm*, $Y_2 = 0.5$ *gm/gm. The dilution rate in the simulation is* $0.05\ hr^{-1}$*, and the inlet substrate concentration is 5.0 gm/l.*

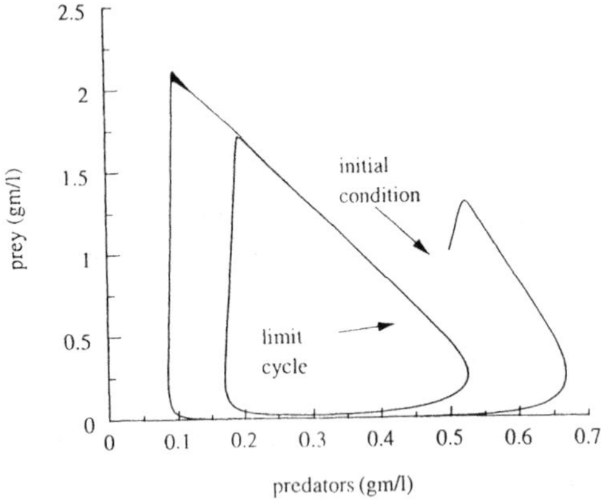

Figure 7.12*. Phase plane plot of the behavior of the predator-prey system shown in Figure 7.11, illustrating the limit cycle behavior.*

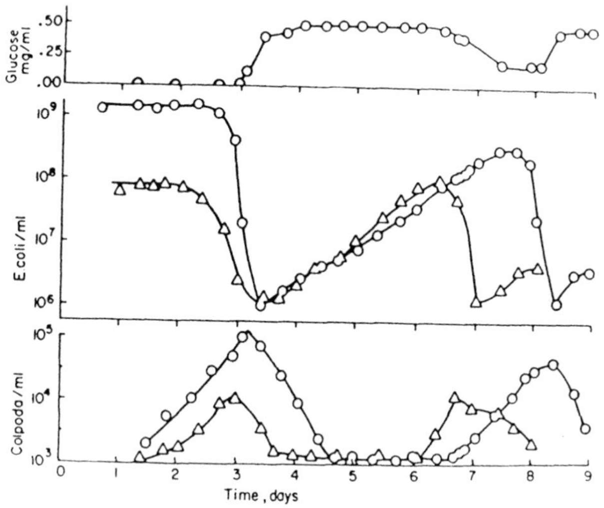

*Figure 7.13. Predator-prey interactions between Colpoda steinii and E. coli B/r in a glucose-limited chemostat. The residence time was 8 hours, and the glucose feed concentration was 0.5 mg/ml (O) or 0.025 mg/ml (Δ). [From Drake, J.F. and H.M Tsuchiya, Appl. Environ. Microbiol. **31** (6) 870-874 (1976)].*

7.5 The Stability of Large Food Webs

Considerable insight into the behavior of large, interactive food webs can be obtained without the need to specify the actual kinetics of each interaction in the system. An interesting approach to this problem has been developed by Ashby and by May[6]. The question posed is that of the stabilty of the system as the size and nature of the interaction between species increase. We will apply the linearized stability analysis developed in Chapter 4 to examine this question. The population dynamics in a system of m species can be described by a set of equations of the form

$$\frac{dn_i}{dt} = f_i(n_1, n_2, \ldots n_m) \qquad i = 1, .., m \tag{7.24}$$

where n_i is the population density of the ith species, and the function f_i describes the possible interactions between the ith species and all other species. At steady state, the population numbers can be determined from

$$f_i(\hat{n}_1, \hat{n}_2, \ldots, \hat{n}_m) = 0 \qquad i = 0, .., m \tag{7.25}$$

(6) Gardner, M.R. and W.R. Ashby, Nature **228** 784 (1970); May, R.M. "Stability and Complexity in Model Ecosystems", Princeton University Press (1973).

Writing the population numbers of each species in terms of deviations from these steady-state values $(\hat{n}_i + x_i)$ and substituting into the balance equations, we can express the deviations from steady state in a Taylor series.

$$\frac{dx_i}{dt} = \sum_{j=1}^{m} a_{ij} x_i \qquad (7.26)$$

The values of a_{ij} can be found from

$$a_{ij} = \left(\frac{df_i}{dn_j} \right)_{\hat{n}_1, \hat{n}_2, \ldots, \hat{n}_m} \qquad (7.27)$$

Thus the set of balance equations can be rewritten in matrix form

$$\frac{d\underline{x}}{dt} = \underline{\underline{A}}\underline{x} \qquad (7.28)$$

where $\underline{\underline{A}}$ is the Jacobian or "community" matrix; the sign and magnitude of the elements a_{ij} describe the interactions between the species. The magnitude of the deviations from the steady state in response to small disturbances in any of the species numbers is given by the solution of Equation (7.28):

$$x_i(t) = \sum_{j=1}^{m} c_{ij} \exp(\lambda_j t) \qquad i = 1, .., m \qquad (7.29)$$

where the eigenvalues of $\underline{\underline{A}}$ are λ_j . The size of the community matrix $\underline{\underline{A}}$ is thus directly related to the size of the food web, and the "connectance" C describes the extent of interactions between species in the sytem.

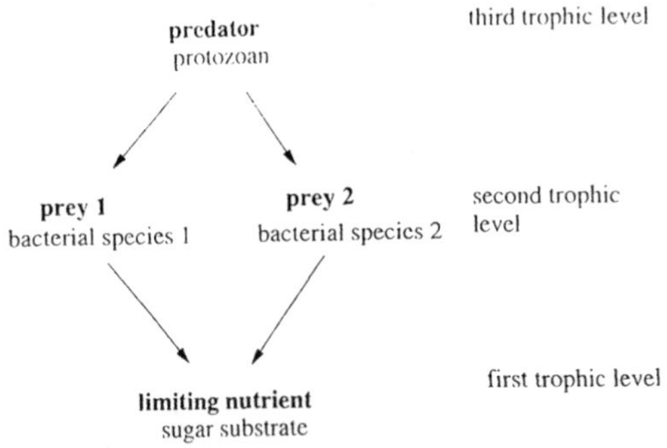

Figure 7.14. *A simple food web containing two prey which compete for a common nutrient; both bacteria can be consumed by a pedator.*

C is the probability that any pair of species will interact. It is measured as the fraction of non-zero elements in the community matrix. A web with little connectance will have a large number of zero terms in the community matrix. It is thus more accurately defined as the ratio of the actual links to the topologically-possible links in the trophic web. The position of the non-zero elements in the community matrix is determined by the system "hierachy", i.e., the type of interactions (competition, predation etc.). This is reflected in the number of trophic levels. As we move up the food chain, we move from a lower to a higher trophic level. This is illustrated in Figure 7.14.

Let us consider the situation described by equations (7.5), for two bacteria competing for a common substrate in a chemostat. The elements of the community matrix are given by

$$\begin{pmatrix} a_{11} = -D + \mu_1 & a_{12} = 0 & a_{13} = \hat{X}_1\left(\dfrac{d\mu_1}{dS}\right)_{ss} \\ a_{21} = 0 & a_{22} = -D + \mu_2 & a_{23} = \hat{X}_2\left(\dfrac{d\mu_2}{dS}\right)_{ss} \\ a_{31} = -\dfrac{\mu_1 \hat{X}_1}{Y_1} & a_{32} = -\dfrac{\mu_2 \hat{X}_2}{Y_2} & a_{33} = -D - \dfrac{\hat{X}_1}{Y_1}\left(\dfrac{d\mu_1}{dS}\right)_{ss} - \dfrac{\hat{X}_2}{Y_2}\left(\dfrac{d\mu_2}{dS}\right)_{22} \end{pmatrix} \tag{7.30}$$

Here the competing bacteria are species 1 and 2, and the substrate is 3. Inserting the steady state values and simply indicating the signs, we have

$$\begin{pmatrix} 0 & 0 & + \\ 0 & 0 & + \\ - & - & - \end{pmatrix}$$

The stability of the system is determined by the eigenvalues of $\underline{\underline{A}}$. If all the eigenvalues have negative real parts, the system is stable. The criteria for negative real parts of the eigenvalues can be determined by the signs of the a_{ij}. The necessary and sufficient conditions for stability thus are

(i) $a_{ii} \leq 0$ for all i
(ii) $a_{ii} < 0$ for at least one i
(iii) $a_{ij}a_{ji} \leq 0$ for all $i \neq j$
(iv) $a_{ij}a_{jk}a_{kl} \ldots a_{st} = 0$ for all sequences of three or more unequal indices
(v) $\det \underline{\underline{A}} \neq 0$

The case of simple competition satisfies all the conditions and the coexistence steady state is thus stable to small perturbations. If, however, we consider competition in a closed or batch system, the community matrix would not meet condition (ii) unless it is assumed that the species are self-regulating. The Monod model, for example, would predict that a_{11} and a_{22} would be zero, satisfying (i) but not (ii). To obtain a steady state in a closed ecosystem, there must be at least one self-regulating species. An example of this is provided by the Pearl-Verlhulst (the logistic) equation for bacterial growth, described in Chapter 3.

$$\frac{dn}{dt} = an(b - n) \tag{7.31}$$

The constant b is the steady state population density and a is a constant. The community matrix thus contains the term

$$a_{11} = \frac{\partial}{\partial n}\left(\frac{dn}{dt}\right)_{n=b} = -ab \tag{7.32}$$

In non-microbial systems, it may not be realistic to assume that all species would reach population densities sufficiently high to be self-regulating. In these cases, a system with simple competition would not be stable.

Let us now consider a large food web where each species has a self-regulating specific growth rate. The diagonal elements of the community matrix would thus all have negative values. By scaling the mass balance equations, these elements can be made to have values of -1. We now follow the development of May[7] to examine the effects of connectance and size on stability.

If there are n species in the food web, the community matrix (n x n) can be constructed so that the diagonal elements are negative by writing it as

$$\underline{\underline{A}} = \underline{\underline{B}} - \underline{\underline{I}} \tag{7.33}$$

where $\underline{\underline{I}}$ is the identity matrix with zeros everywhere except for values of 1 along the diagonal. The connectivity of the web is reflected in the fraction of non-zero elements in $\underline{\underline{B}}$. If the connectivity is C, then a fraction (1-C) of b_{ij} will be zero. The non-zero elements of $\underline{\underline{B}}$ may be either positive or negative, and for a random web it is assumed that these two possibilities are equally likely. The mean of b_{ij} is thus zero, and the mean square value of all non-zero b_{ij}s is σ^2. By randomly selecting values of b_{ij} for a specified connectance, the community matrix can be constructed as a function of the size of the food web. The eigenvalues of the matrix can then be computed. Proceding in this manner, May has shown that the probability that a system of n >> 1 components will be stable and tend to unity when

$$\sigma\sqrt{nC} < 1 \tag{7.34}$$

Conversely, the system will be unstable if

$$\sigma\sqrt{nC} > 1 \tag{7.35}$$

We see from these equations that increasing the connectivity (C) increases the likelihood of instability, as does an increasing number of species (n). This is contrary to the views held by many ecologists, who believe that increasing species diversity is important in increasing the stability of the ecosystem. In nature, the stability of an ecosystem may be damaged if one key species is removed. This is not necessarily inconsistent with the mathematical analysis we have developed, as we have not considered the hierachy of the system, i.e. the nature of the interactions. Species interactions place constraints on the location of the positive and negative values of the elements of $\underline{\underline{A}}$. This effect can be seen if

(7) May, R.M., "Stability and Complexity in Model Ecosystems", Princeton Univ. Press (1973).

we consider a multispecies predator-prey system. The community matrix will have positive and negative regions. Such a system is illustrated in Figure 7.15, if Monod relationships are assumed for the rates of growth of predators and prey.

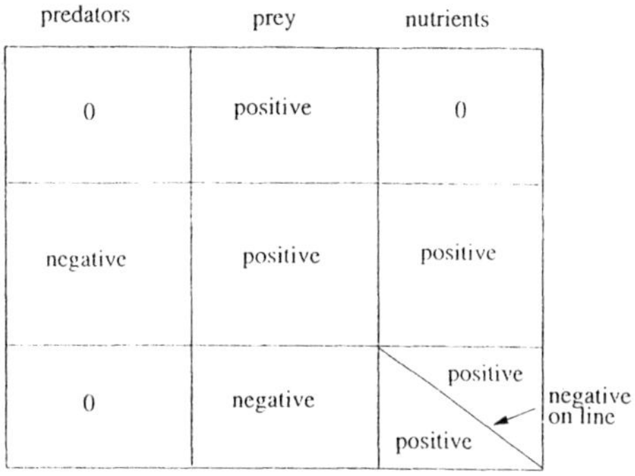

***Figure 7.15**. Location of positive and negative elements of a community matrix for a well-mixed continuous flow system with multiple predators, prey and limiting nutrients. The system has well defined interactions and in this case, there is no direct interaction between predator and the nutrient for the prey. Thus the predator-nutrient regions contain zero values.*

In contrast to the results of May, the multispecies predator-prey system does not show increasing instability with increasing number of components[8]. Rather, the stability is unaffected by the number of either predators, prey or nutrients. The nature of the species interactions has the strongest influence on the stability of the system. McMurtrie[9] has shown that hierachy strongly influences the stability of randomly constructed webs. Even in a complex ecosystem, the assumption that the food flow direction is random may not be appropriate and the nature of the species interactions must be examined more closely. The structure of the food web is a prime determinant of stability. Saunders and Bazin[10] provide a very readable overview on the stability of food webs from a microbiological perspective.

7.6 Other Interactions

Although predator-prey interactions have been extensively examined from both theoretical and experimental viewpoints, other interactions may play significant roles in natural ecosystems. Mutualism, where each partner in the interaction produces a substance

---

(8) Yoon, H.Y. and H.W. Blanch, "The Stability of Predator-Prey Interactions in Microbial Ecosystems", Math. Biosciences 35, 85 (1977).

(9) McMurtrie, R.E. J. Theoret. Biol. **50** 1 (1975).

(10) Saunders, P.T. and M.J. Bazin, "On the Stability of Food Chains", J. Theoret. Biol. **52** 121-142 (1975).

that is essential for the growth of the other, and commensalism, where only one of the species benefits, are significant forms of natural interactions occurring in the carbon and nitrogen cycles. For example, in aerated waste-treatment processes, organic material containing nitrogen is converted to ammonia, which must be oxidized to nitrite and then nitrate prior to discharge. The conversion of NH_3 to NO_2^- is accomplished by *Nitrosomonas* species, while *Nitrobacter* convert this nitrite to nitrate. This is thus a commensal relationship. A schematic of a mutalistic interaction is given in Figure 7.16.

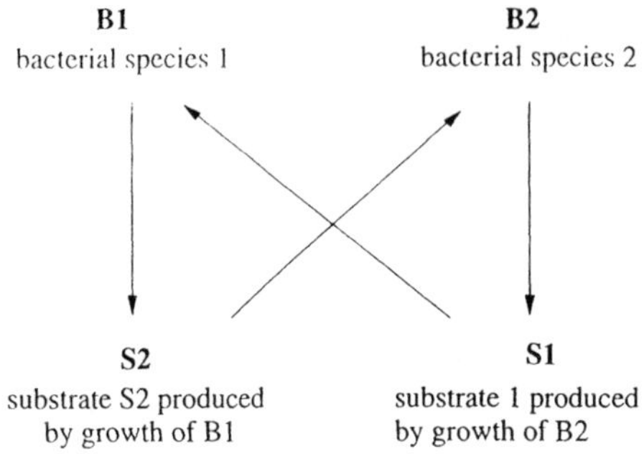

Figure 7.16. *Complementary interaction of two microbial populations, where each population produces a product essential for the growth of the other population.*

Commensal interactions are important in soil ecology and in the cycles of carbon, nitrogen and sulphur. In milk spoilage, lactose is fermented by *Streptococci*, *Bacilli* and other microorganisms with a resultant decrease in pH. This slows the activities of these microorganisms sufficiently, so that *Lactobacilli*, which are acid tolerant, can now grow. At lower pHs, the caseins are denatured and precipitate, forming a curd. The resultant lactic acid produced can serve as a substrate for yeast and molds, raising the pH. This subsequently permits bacteria and fungi to grow on the casein and fats in the milk. If this growth is active, the solution may become anaerobic and permit the growth of anaerobic bacteria, causing putrefaction. This complex set of microbial interactions is typical of those occurring in other situations; e.g., sausage manufacture and pickling processes.

The production of growth factors by one species to the benefit of another is another common commensal interaction in microbial populations. In soils, many bacteria depend on the production of vitamins by other microorganisms. This permits the growth of fastidious microorganisms that might not otherwise survive. Insoluble compounds may be microbially converted to soluble forms, making them available to other species. For example, methanogenesis may result from a commensal relationship in some situations. *Desulfovibrio* can supply *Methanobacterium* with acetate and hydrogen, arising from anaerobic fermentation on sulphate and lactate. The *Methanobacterium* can then use these products from the *Desulfovibrio* to reduce carbon dioxide to methane.

Closely realted to mutualism is synergism, in which the relationship between the species is not obligatory; either species can survive in the absence of the other. The relationship permits microbial populations to perform activites which neither population could accomplish alone. The degradation of cyclohexane by *Norcadia* and *Pseudomonas* shown in Figure 7.17 provides an example of this interaction.

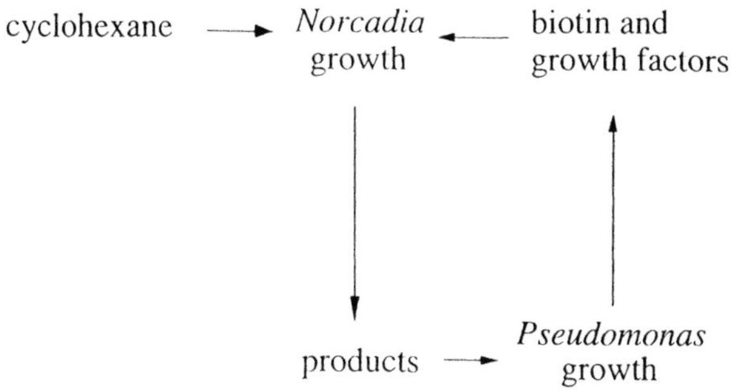

Figure 7.17. *The degradation of cyclohexane by Norcadia and Pseudomonas species. [From Slater, J.H. "Microbial Communities in the Natural Environment", in K.W. Chater and H.S. Somerville, "The Oil Industry and Microbial Ecosystems", Heyden & Sons, London. pp. 137-154, (1978)].*

7.7 Nomenclature

a	first order rate constant	hr^{-1}
a_{ij}	inhibition constants in the competition model	(-)
b	rate of death of predators	hr^{-1}
C	connectance of a food web	
D	dilution rate	1/hr
D_S	substrate diffusion coefficient	cm^2/sec
K_i	Monod constant for ith substrate	gm/liter
n_1	number concentration of species 1	number/ml
n_2	number concentration of species 2	number/ml
S_i	ith substrate concentration	gm/liter
S_o	inlet substrate concentration	gm/liter
T	duration of predator-prey cycle	hrs
X_i	concentration of species i	gm/liter

Y_i	substrate yield coefficient	gm/gm
α_i	random motility coefficient	cm^2/sec
χ_i	chemotaxis diffusion coefficient	cm^2/sec
ε	effective yield of predators from prey	
γ_i	effective yield coefficents for species i	number/number
γ	second order rate constant in Lotka-Volterra model	$(gm/liter)^{-1}hr^{-1}$
λ_i	constants in Lotka Volterra equations	1/number
λ_i	eigenvalues of the community matrix	
μ_i	specific growth rate of species i	1/hr
$\mu_{max,i}$	maximum specific growth rate of species i	1/hr

7.8 Problems

1. Stability of predator-prey interactions in the chemostat.

This problem examines the effect of a predator on the stability of bacterial species competing for a common resource in a well-mixed continuous flow system, as illustrated in Figure 7.14. A similar set of equations for a batch system would result if we include death terms for all species and a source term for the substrate. For the case of two prey (bacteria) and one predator (protozoan), show that the mass balance equations are

$$\frac{dp}{dt} = -Dp + \mu_p(n_1, n_2, p)p$$

$$\frac{dn_1}{dt} = -Dn_1 + \mu_1(S)n_1 - \frac{1}{Y_1}\mu_p(n_1, n_2, p)p$$

$$\frac{dn_2}{dt} = -Dn_2 + \mu_2(S)n_2 - \frac{1}{Y_2}\mu_p(n_1, n_2, p)p$$

$$\frac{dS}{dt} = D(S_o - S) - \frac{1}{Y_{S1}}\mu_1(S)n_1 - \frac{1}{Y_{S2}}\mu_2(S)n_2 \tag{7.36}$$

The specific growth rate of the predator (μ_p) includes a dependence on p, representing a possible self-regulation.

(a) Consider first the case where this dependency is not present, and we have only one prey species. Show that the steady state concentrations of the species are those given below, where a Monod relationship has been assumed for substrate consumption by prey and prey consumption by predator.

$$\hat{p} = Y_1 Y_{S1}(S_o - \hat{S}) - Y_1 \hat{n}_1$$

$$\hat{n}_1 = \frac{DK_n}{\mu_{mp} - D}$$

$$\hat{S} = \frac{1}{2}\left(S_o - K_S - \frac{\mu_{mn}\hat{n}_1}{Y_{S1}D}\right) + \frac{1}{2}\left(\left(S_o - K_S - \frac{\mu_{mn}\hat{n}_1}{Y_{S1}D}\right)^2 + 4K_S S_o\right)^{\frac{1}{2}} \tag{7.37}$$

(b) Determine the elements of the community matrix, and show these have the following signs.

$$\begin{pmatrix} 0 & + & 0 \\ - & + & + \\ 0 & - & - \end{pmatrix}$$

What can be said about the stability of this system? By assuming typical values of the constants, calculate the eigenvalues of the community matrix for a range of values of the dilution rate and the inlet substrate concentration. Comment on the range of stable solutions.
(c) If the second prey species is now introduced, how will the community matrix be altered? What effect does this have on the range of operating parameters of the system? Determine the eigenvalues of the community matrix for this system using values of the constants given in Figure 7.4, and select reasonable values for the predator-related constants. If the predator species growth rate had a self-regulating term, such as that in the Pearl-Verhulst equation, how would the stability of the sytem be affected? Quantify you answer by calculating the range of (D, S_o) values that give a stable steady state.

2. Commensal Interactions

The growth of *Acetobacter suboxydans* on the monosaccharide mannitol results in the production of fructose. When a yeast, such as *Saccharomyces carlsbergensis* is grown in continuous fermentation with the *Acetobacter*, it is able to consume the fructose for growth. It is unable to consume mannitol and thus has a commensal relationship with *Acetobacter*. Chao and Reilly[11] showed that this system demonstrated damped oscillations in both sugars and cell concentrations after increases or decreases in the dilution rate.
(a) Develop the set of unsteady state mass balances (for bacteria, yeast and both substrates) which describe this commensal interaction in a CSTR. Use the Monod kinetic model to describe the growth of the *Acetobacter* on mannitol, and the growth of the yeast on fructose. Denote the *Acetobacter* kinetic constants by 1 and those for the yeast by 2. Assume that the fructose is produced at a rate which depends on the product of the specific growth rate of the *Acetobacter* and its concentration, and a stoichiometric coefficient. This coefficient is the same as the yield coefficient for *Acetobacter* production from mannitol (Y_{X1}), based on the reaction below.

$$2\ \textit{mannitol} \rightarrow \textit{fructose (yeast growth)} + \textit{mannitol (bacteria growth)}$$

(b) Show that there are three possible steady states, with one of these corresponding to the case of coexisting populations. What are the concentrations of substrates and cells at this steady state? What is the range of dilution rates over which this steady state is possible?
(c) Show the community matrix has the following form at the coexistence steady state.

$$\begin{pmatrix} 0 & 0 & Y_{X1}B & 0 \\ 0 & 0 & 0 & Y_{X2}C \\ -D/Y_{X1} & 0 & -D-B & 0 \\ D/Y_{X1} & -D/Y_{X2} & B & -D-C \end{pmatrix}$$

(11) Chao C.C. and P.J. Reilly, Biotech. Bioeng. **14** 75 (1972) and Biotech. Bioeng. **16** 1373 (1974).

where

$$B = \frac{(\mu_{max1} - D)(\mu_{max1}S_{10} - DS_{10} - DK_{S1})}{\mu_{max1}K_{S1}}$$

$$C = \frac{(\mu_{max2} - D)^2}{\mu_{max2}K_{S2}}\left(S_{10} - \frac{DK_{S1}}{\mu_{max1} - D} - \frac{DK_{S2}}{\mu_{max2} - D}\right)$$

(d) Is this system stable to small perturbations? If the microbial constants have the values below, numerically solve the set of equations and plot X_2 against S_2 to illustrate the phase plane behavior of the sytem. Comment on the differences in the simulations for the two sets of initial conditions.

$D = 0.1\ hr^{-1}$

$S_{10} = 1.0$ gm/l

$\mu_{max1} = 0.2\ hr^{-1}$ $\qquad \mu_{max2} = 0.2\ hr^{-1}$

$K_{S1} = 0.01$ gm/l $\qquad K_{S2} = 0.01$ gm/l

$Y_{S1} = 0.01$ gm cells/gm substrate $\qquad Y_{S2} = 0.01$ gm cells/gm substrate

Initial conditions

(i) $X_1 = 2 \times 10^{-2}$ gm/l $\quad X_2 = 0.98 \times 10^{-2}$ gm/l
$S_1 = 2 \times 10^{-2}$ gm/l $\quad S_2 = 2 \times 10^{-2}$ gm/l

(ii) $X_1 = 0.5 \times 10^{-2}$ gm/l $\quad X_2 = 0.98 \times 10^{-2}$ gm/l
$S_1 = 2 \times 10^{-2}$ gm/l $\quad S_2 = 2 \times 10^{-2}$ gm/l

3. *Analysis of the Stability of a Predator-Prey System*

The behavior of a model food chain glucose-*Escherichia coli-Dictyostelium discoideum* (a myxameba) has been experimentally examined by Tsuchiya et al.[12] The bacterium consumes the glucose and is itself consumed by the myxameba. Using Monod expressions for the growth rates of the bacteria and the myxameba on their respective substrates, show that the resulting set of mass balances in the chemostat can be written in dimensionless form as follows:

$$\frac{dy}{d\tau} = \gamma(y_o - y) - \frac{xy}{1+y}$$

$$\frac{dx}{d\tau} = -\gamma x + \frac{xy}{1+y} - \frac{\mu xz}{K+x}$$

$$\frac{dz}{d\tau} = -\gamma z + \frac{\mu xz}{K+x}$$

(12) Tsuchiya, H., Drake, J.F., Jost, J.L. and A.G. Fredrickson, *J. Bacteriol.* **110** 1147 (1972).

where y is the dimensionless substrate concentration (S/K_1), x is the dimensionless prey concentration (n_1/Y_1K_1) and z is the dimensionless predator concentration ($n_2Y_1Y_2K_1$), γ is the dimensionless dilution rate (D/μ_{max1}) and y_o is the dimensionless inlet substrate concentration (S_o/K_1).

(i) Find the expressions for the scaling constants μ, τ and K.

(ii) Find the solutions for the three possible steady states of this set of equations.

(iii) We shall examine the stability of the solutions by letting

$$a = \frac{\hat{y}}{1+\hat{y}} \qquad b = \frac{\mu\hat{x}}{K+\hat{x}} \qquad c = \frac{K\mu\hat{z}}{(K+\hat{x})^2} \qquad d = \frac{\hat{x}}{(1+\hat{y})^2}$$

Write the Jacobian for this system. Show that the eigenvalues of this equation can be found from the solution of the characteristic equation

$$\lambda^2 + (\gamma + d + c - a)\lambda + (\gamma + d)c = 0$$

What are the conditions for the eigenvalues to have imaginary roots?

(iv) The experimental data of Tsuchiya et al. give the following values of the microbial constants

Organism	$\mu_{max\,i}$ (hr^{-1})	K_i	Y_i
E. coli	0.25	5×10^{-4} (mg/ml)	3×10^{9}
D. discoideum	0.24	9×10^{8} (bacteria/ml)	7.1×10^{-4}

Assuming that the inlet substrate concentration is 360 times the K_s value, what is the value of the dimensionless dilution rate that makes each of the steady states stable. For the case of the coexistence steady state, what is the dimensionless dilution rate that results in oscillations (imaginary parts of the eigenvalues are present) and in unstable oscillations (real parts are positive and imaginary parts exist)?

Chapter 8. Bioproducts and Economics

The variety of compounds that are or can be manufactured by biological processes is impressive. Products include simple gaseous products, such as methane, to complex therapeutic proteins, such as hormones. The scale of manufacture ranges from fuel ethanol, produced at the level of billions of gallons at costs of one to two dollars per gallon, to high-value proteins, such as interferons and antibodies, that have manufacturing costs of hundreds to thousands of dollars per gram. In many cases, the regulations and constraints placed on the manufacture of health care and food products make the economic assessment of these processes quite different from that undertaken in the more traditional chemical industries.

In this chapter, we shall examine some of the broad classes of biological products currently manufactured. A comprehensive examination is beyond the scope of this text, but those described have been selected to represent typical products. The processing steps for various classes of products are similar. We shall then turn our attention to estimation of the cost of manufacture of these products. The scale of the process has a significant influence on the manufacturing costs, particularly if the product is for therapeutic use. Labor and quality control also play a major role, unlike traditional large-scale chemical products. Cost estimation is the subject of a number of engineering texts[1]; it is not the intention of this chapter to repeat this material, but rather to discuss the unique aspects of biological processes. The design and economic analysis of biological processes has been discussed in several reviews, and further details can be found there[2].

---

(1) Happel, J. and D.G. Jordan, "Chemical Processes Economics", Marcel Dekker (1975); Peters, M.S. and K.D. Timmerhaus, "Plant Design and Economics for Chemical Engineers", McGraw Hill 3rd Edition (1980).

(2) Reisman, H.B. "Economic Analysis of Fermentation Processes", CRC Press Baton Rouge, Florida (1988); Bartholomew, W.H. and H.B. Reisman, "Economics of Fermentation Processes", in *Microbial Technology* Vol. 2 (2nd Ed) H.J Peppler and D. Perlman, eds., Academic Press, NY (1978); Kalk, J.P. and A.F. Langlykke, "Cost Estimation for Biotechnology Projects", American Society for Microbiology, Manual of Industrial Microbiology and Biotechnology (1985); Comprehensive Biotechnology, Volume 3, H. Blanch, S. Drew, D. Wang eds. Pergamon Press (1985).

8.1 Manufacture of Biological Products

Biological processes and products are familiar components of everyday life, from foods and food additives, medicines and specialty chemicals to fuels and waste treatment. We shall consider these products based on their chemical nature and review methods of manufacture. In many cases there are several manufacturing routes available; e.g. direct fermentation of a sugar to final product, use of a precursor fermentation, or use of an enzymatic conversion process. The actual process employed commercially depends on a variety of factors, including economics, process and product patents, raw material availability, and manufacturing equipment. In some cases, traditional processes are employed even in the face of technological advances, for example the manufacture of cheese and fermented milk products, and beer and wine. With products regulated by Food and Drug agencies, changing the manufacturing process may involve costly clinical trials, and such processes are often not optimized from a production standpoint.

We shall now examine the major classes of biological products, highlighting important examples of each.

8.1.1 Organic Acids

8.1.1.1 Citric Acid

A variety of organic acids are produced by either fermentation or by enzymatic conversion of precursors. Of these, citric acid is produced in the largest amount and has been produced commercially since 1826, being initially obtained from lemon juice. Citric acid is used in foods, confectionery and beverages (75% of total) as an acidulent and flavor enhancer, and in pharmaceutical preparations (10%), antacids and aspirins. It also has various industrial uses (15%), including use as a stabilizer of oils and fats, due to its ability to complex with heavy metals. The world market is in excess of 300,000 tonnes[3] per year. The fermentation process for citric acid has an interesting history. In 1917, the Englishman Currie found a strain of *Aspergillus niger* that produced better yields of citric acid than did an earlier strain of *Penicillium*, which had been found to produce citric acid when grown on sugar solution. Currie joined Chas. Pfizer Inc. and citric acid was produced with this strain in the U.S. in 1923. Plants in England, Belgium and Germany entered production using surface cultivation with beet molasses as a substrate. Submerged fermentation processes were introduced in the late 1940's and 1950's, with glucose syrups, beet or cane molasses serving as carbon sources.

Citric acid is formed by the action of citrate synthetase in the mitochondria. The normal metabolism of citric acid is altered by a reduction in the levels of Fe^{3+}. Iron is important in the function of aconitate hydratase, an enzyme that interconverts isocitric acid and aconitate.

(3) A tonne is 1000 kg.

The medium is thus treated by cation exchange to reduce iron, manganese and zinc to low levels. Productive strains of *A. niger* are thought to have active glycolytic enzymes, with pyruvate carboxylase and citrate synthetase providing a high rate of citrate formation, which is then transported across the cell membrane.

The fermentation typically uses molasses or glucose-based medium, at pH 5-7 and at 30°C. Both air-lift and agitated tank fermenters are employed, with growth of the *Aspergillus* preceding the phase of citric acid production. Typical yields of citric acid are 0.70 to 0.90 gm citrate/gm glucose; the maximum yield from anhydrous glucose is 1.17 gm/gm. Citric acid is recovered from the fermentation broth by heating the broth and adding lime. Calcium citrate (tetrahydrate) is precipitated, and the precipitate is treated with H_2SO_4 to yield soluble citric acid and an insoluble precipitate of gypsum ($CaSO_4$). The solution of citric acid is decolorized with activated carbon and crystallized and dried. The process is illustrated in Figure 8.1.

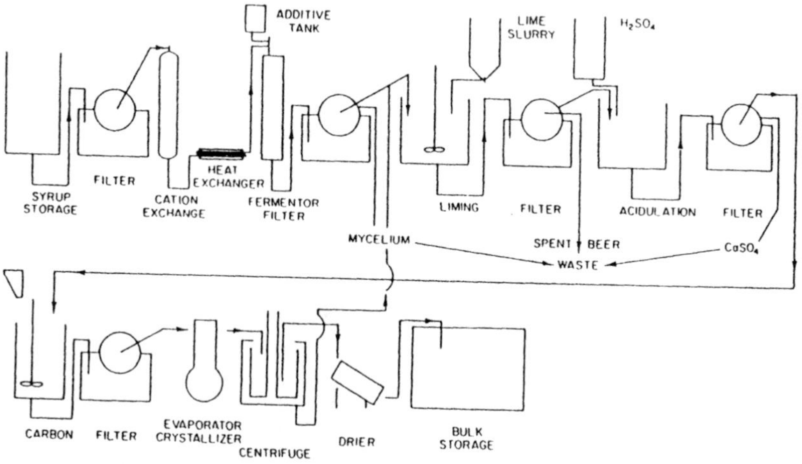

Figure 8.1. *Flow sheet for the manufacture of citric acid, illustrating the fermentation and recovery steps [from Lockwood, L.B. and L. Schweiger, In "Microbial Technology", H.J. Peppler ed., pp 183-199, Van Nostrand-Reinhold (1967)].*

Other approaches to recover citric acid have been examined. Solvent extraction has been proposed, as it avoids the use of lime and sulfuric acid in the process, and hence the disposal costs of the gypsum. Butan-2-ol has been used as an extractant, as has tri-butyl phosphate. Ion pair extraction, using secondary or tertiary amines dissolved in a water-immiscible solvent (e.g. octyl alcohol), provides an alternative route. With recent developments in electrodialysis membranes, the use of this technique to recover citric acid directly from the fermentation broth becomes a feasible alternative. Here the ammonium citrate in the broth is electrodialyzed, and ammonium hydroxide and citric acid are formed on either side of the membrane. The reduction in chemical consumption is balanced by the increased electrical costs. Other developments in the citric acid process are the use of yeast in place of the *Aspergillus*, and improvements aimed at shortening the fermentation time.

8.1.1.2 Itaconic Acid

Itaconic acid is a dicarboxylic acid (structure below). The methylene group can participate in polymerization reactions, yielding products with many carboxyl groups. Acrylonitrile/itaconic acid copolymers have improved dyeing characteristics. Styrene-butadiene lattices which incorporate itaconic acid are used in carpet backings and paper coatings, due to their superior adhesive properties.

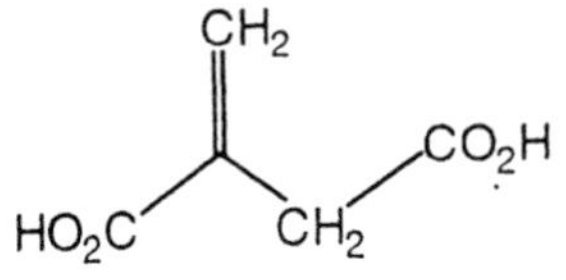

Structure of Itaconic Acid

Itaconic acid also reacts with N-substituted amines to form N-pyrrolidones which have uses in detergents, shampoos, herbicides and pharmaceuticals. Itaconic acid is formed by the action of aconitate decarboxylase on cis-aconitic acid. The sugar source enters the TCA cycle via acetyl CoA. Citric acid appears to be an intermediate in the fermentation pathway.

Itaconic acid can be made by either surface or submerged batch fermentation of *Aspergillus terreus*, although the submerged process is preferred. The carbon source may be glucose, sucrose or beet or cane molasses. The fermentation is similar to that for the production of citric acid, with ammonium sulphate or nitrate being the preferred nitrogen sources. The carbon sources are treated by cation exchange to remove iron from the medium and control the level of copper, because Cu^{2+} restricts growth and itaconic acid production. The sugar source is batched at 30 wt% (i.e. initial concentration) and the resultant itaconic acid concentration is about 85 gm/liter (the yield being 0.55 to 0.65 gm itaconic acid/gm sugar source). The maximum theoretical yield from anhydrous glucose is 0.72 gm/gm. Control of pH in the process is important and the pH is permitted to fall in the early stages of the fermentation to around pH 3. Itaconic acid production commences at this low pH. The temperature for growth of *A. terreus* is quite high, ranging from 35 - 42°C. The fermentation is aerated at 0.25 to 0.5 vvm (volume of air per reactor volume per minute), and supply of oxygen is critical to maintenance of high itaconic acid levels. The duration of the fermentation is 72 hours, with the initial period of growth being ~ 24 hours.

When all the sugar has been consumed from the fermentation broth, itaconic acid can be directly crystallized from the fermentation broth, once it has been clarified and decolorized. If a base such as NaOH, NH_4OH or $Ca(OH)_2$ has been employed to maintain the pH, sulfuric acid is added prior to filtration and clarification. The acidified broth is concentrated and itaconic acid crystallized by temperature reduction. The process flow sheet for itaconic acid is shown in Problem 8.1.

8.1.1.3 Gluconic Acid and Glucono-δ-lactone

Gluconic acid and glucono-δ-lactone are produced by a variety of bacteria (*Pseudomonas, Vibrio, Gluconobacter*) and fungi (*Aspergillus, Penicillium* and *Endomycopsis*). The sodium salt of gluconic acid is used in washing glass bottles, because it is able to sequester calcium and iron; the formation of calcium and magnesium hydroxides otherwise results in the formation of a scum, particularly in hard water. Its sequestering abilities with iron are exploited in alkaline derusting of metals. The calcium salt of gluconic acid is widely used to treat calcium deficiencies and in intravenous therapies, and iron gluconate is used for treating anemia. Iron and other trace metal salts of gluconate are used in horticultural applications. When added to cement, gluconates influence the time of setting and the water resistance of the cement. Gluconolactones are in equilibrium with the acid in aqueous solutions of gluconic acid, as the δ-lactone and the γ-lactone. Glucono-δ-lactone is used as an ingredient in baking powders, bread mixes and foodstuffs where its acidogenic properties are useful. The world market is approximately 50,000,000 kg/yr.

While chemical routes for the conversion of glucose to glucono-δ-lactone are available, including electrochemical oxidation, the fermentation method is the commercial method of choice. Most commercial processes employ *Aspergillus niger* as the production organism, with glucose or dextrose solutions as the carbon source. Nitrogen is supplied in the form of ammonium salts, corn steep liquor or urea. A vegetative form of *A. niger* is used as the inoculum. The biochemical steps in gluconic acid production are illustrated in Figure 8.2.

Typical yields of gluconic acid from glucose are generally over 0.9 gm/gm. The maximum theoretical yield, assuming no production of cell mass or CO_2, would be 1.09 gm/gm (based on glucose monohydrate). Thus, in practice, excellent yields are obtained in the fermentation. A vegetative inoculum of the production organism is used in the main fermentation vessel, at about a 10% volume basis. The batched medium contains glucose at a concentration around 200-220 gm/liter, and is controlled at pH 6-7 by addition of NaOH. Typical batch fermentation conditions are given below.

temperature	30-33°C
pH	pH 6-7
aeration	1.5 vvm
batch fermentation time	20 hours
yield	0.95 gm gluconic acid/gm glucose
Na gluconate solubility	39.6 wt% (30°C); 34.6 wt% (15°C)

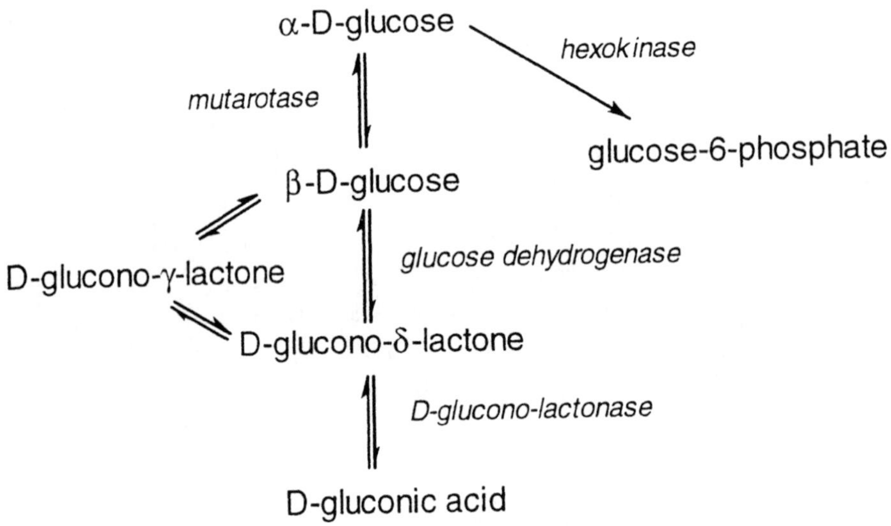

Figure 8.2. *The metabolism of A. niger resulting in the the production of gluconic acid and glucono-δ-lactone.*

The recovery of sodium gluconate employs evaporation of a previously filtered and decolorized broth. This may be dried to produce the technical grade, or crystallized to produce pure grade product. If calcium salts have been used to maintain the pH in the fermentation, calcium gluconate is recovered by cooling to 10°C after decolorization and collecting the crystals.

8.1.1.4 Lactic Acid

Lactic acid occurs widely in nature and is found in sour milk and distilleries. It is produced from sucrose, whey (lactose), and maltose and dextrose from hydrolyzed starch. An anaerobic batch fermentation using *Lactobacillus delbrueckii* or *Lactobacillus bulgaricus* is generally employed. The batch process is not subject to significant contamination, apart from butyric acid bacteria, and often the fermenters are simply steamed or washed with boiling water. Lactic acid finds use in technical, food and pharmaceutical grades. It is used medically as an intermediate in pharmaceutical manufacture, for adjusting the pH of preparations, and in topical wart preparations. Polylactic acid is used in biodegradable (dissolving) sutures and for implants. The main food application is for production of stearoyl-2-lactylates. This ester is used as a dough conditioner in baked goods, where it combines with the gluten, making the mixture more tolerant of mixing conditions. Both calcium and sodium lactylates are sold for this purpose, as well as for starch conditioning in dehydrated potatoes. Worldwide lactic acid production is approximately 30 million kg/yr, of which ~20% is used in stearoyl-2-lactylates. Lactic acid is used directly as a food ingredient, as it is a naturally-occurring acidulent and has no strong flavor of its own. It is also used as a preservative and can be found in brines used for pickling olives, pickles and sauerkraut.

Fermentation conditions with *L. delbrueckii* are given below.

temperature	45-60°C
pH	pH 5.0-6.5
inoculum	5-10 vol%
batched sugar concentration	5% with whey
	15% with dextrose or glucose solutions
duration of fermentation	1-2 days for 5% sugar
	2-6 days for 15% sugar
reactor productivity	1-3 kg m^{-3} hr^{-1}
yield	0.9-0.95 gm lactic acid/gm sugar

Recovery of lactic acid from the fermentation broth requires consideration of the corrosive nature of lactic acid. Usually 316 stainless steel vessels are employed. When lime or slaked lime is used to control pH, the fermentation broth is first heated to 80-100°C to kill the bacteria and solubilize calcium lactate. The fermentation broth is then filtered. Crude product can be made at this stage by acidification with sulfuric acid; products of high purity require additional treatment. Such products are usually based on fermentation of relatively pure sugars with minimal complex nitrogen sources, as this reduces the formation of colored products during the fermentation. The broth is filtered and treated with activated carbon to remove colored impurities. The calcium lactate is evaporated to a 37% concentration at 70°C and acidified with sulfuric acid. The resultant calcium sulfate precipitate is removed by filtration and the lactic acid solution is treated with activated carbon and evaporated to a concentration (50 - 80%) depending on final use. Other proposed recovery methods include liquid-liquid extraction using alkylamines and back-extraction into aqueous base or electrodialysis when ammonia solutions are employed to control pH. Lactic acid may also be crystallized when the carbon source is crude, e.g. molasses. It may also be distilled from sugars and other impurities by first esterifying it with ethanol and subsequently hydrolyzing the ester. A general problem is the corrosion of equipment by concentrated lactic acid solutions. Problem 8.2 compares the economics of two proposed recovery methods.

8.1.1.5 Acetic Acid

Acetic acid, or vinegar, has been produced since the early practice of winemaking. Vinegar is referred to in the Old and New Testaments, and was an early popular nostrum.

It was probably first produced as spoiled wine[4]. Its production from alcoholic mashes predates the discovery of the acetic acid bacteria. Acetic acid can also be produced chemically from ethylene, ethanol and acetaldehyde.

The bacteria responsible for acetic acid production from alcohol are *Acetobacter*; there are a number of groups distinguished on the basis of catalase activity. The oxidation of ethanol by Acetobacter is a two-step process; the first step is a partial oxidation of ethanol to acetaldehyde and acetic acid. Under aerobic conditions the second step converts one mole of acetaldehyde to 0.5 mole of ethanol and 0.5 mole of acetic acid, viz.

$$2CH_3CHO + H_2O \rightarrow C_2H_5OH + CH_3COOH$$

The ethanol so formed is further oxidized to acetaldehyde and reenters the reaction scheme. Thus under adequate aeration conditions there are simultaneous oxidation and dismutation reactions. Supply of sufficient oxygen is critical to maintain high rates of acetic acid formation. In the original or "Orleans" process wooden barrels were partially filled with vinegar, providing an inoculum, and wine was added over a one month period. Air was supplied through holes just above the liquid level, and the acetic acid bacteria grew as a gelatinous mat.

Modern vinegar processes trickle wine through a bed packed with wood shavings to ensure adequate aeration. Air is sparged through perforations in the bottom of the bed. The Frings aerator was introduced in 1929, when temperature control and forced aeration became part of the standard operation. Submerged culture is used to a lesser extent commercially, and at the laboratory scale immobilized cell reactors have been examined. Typical commercial Frings processes convert 12% alcohol to acetic acid in about 5 days, at conversion efficiencies of ~ 98%. High concentrations of acetic acid are inhibitory to the fermentation process.

The recovery of acetic acid from aqueous solution has been accomplished by a number of routes, including fractional distillation, azeotropic dehydration distillation, solvent extraction and extractive distillation. Vinegar is sold as a dilute acetic acid solution, containing 5-6 wt% acid. For processes where acetic acid is to be used as a feedstock, the recovery method must be evaluated. For vinegar, the aerobic fermentation is the preferred route, but for chemical production, an anaerobic process has the advantage of lower energy requirements.

8.1.1.6 Other Organic Acids

A number of other organic acids can be produced by fermentation, although they are not produced by this method commercially. These include propionic and butyric acids,

(4) The word vinegar is derived from the French vin (wine) and aigre (sour). The Latin word *acetum* means sour wine.

fumaric, and gallic acids. Because the fermentation results in concentrations of 20-30 gm/l of these acids, the cost of recovery of these acids has precluded them from competing with their petrochemically-derived counterparts. A number of *Propionibacteria* produce propionic acid, and butyric acid is formed by *Clostridia, Butyvibrio,* and some eubacteria. The simplified stoichiometry for the production of propionic acid from glucose via lactic acid and the dicarboxylic acid pathway is

$$\text{glucose} + 2H_2O \rightarrow 2\text{ acetic acid} + 2CO_2 + 8H$$
$$2\text{ glucose} + 8H \rightarrow 4\text{ propionic acid} + 4\ H_2O$$

$$3\text{ glucose} \rightarrow 4\text{ propionic acid} + 2\text{ acetic acid} + 2CO_2 + 2H_2O\ (+\ 12\text{ ATP})$$

The maximum yield of propionic acid from glucose is thus 54.8% (wt/wt). Butyric acid is formed from the same pathway in all bacteria. It is generally formed together with acetate, as this is needed to balance the reducing equivalents in the fermentation. The maximum yield is 36.7 wt% butyric acid from glucose. The typical reactions are

$$2\text{ glucose} + 4H_2O \rightarrow 4\text{ acetic acid} + 4CO_2 + 16H$$
$$2\text{ glucose} + 16H \rightarrow 3\text{ butyric acid} + 6H_2O + 2H_2$$

$$4\text{ glucose} \rightarrow 3\text{ butyric acid} + 4\text{ acetic acid} + 4CO_2 + 2H_2O + 2H_2$$

Fumaric acid was formerly manufactured by fermentation, but is currently made chemically from aromatic hydrocarbons. A simple inorganic salts/sugar medium with *Rhizopus* was employed. Calcium fumarate precipitated from the fermentation broth, simplifying the recovery process.

8.1.2 Alcohols and Ketones

Industrial solvents, such as ethanol, acetone and butanol, were commercially produced by fermentation until the favorable economics of producing ethylene and propylene as by-products of natural gas recovery and gasoline production resulted in the demise of this route in the 1960s. Acetone and butanol are no longer manufactured in commercial quantities by fermentation, but with the increase in crude oil prices in 1973, fuel ethanol has enjoyed somewhat of a resurgence. Other solvents can be produced by fermentation, but are not commercially attractive. These include 2,3 butanediol, methyl ethyl ketone, glycerol, propylene glycol, and ethlyene glycol. We shall examine the three main products in this group: ethanol, acetone, and butanol.

8.1.2.1 Ethanol

Distilled alcohol has been used in non-beverage applications as an incendiary, a solvent, and for medicinal purposes for many centuries. Prior to the 1940's, fermentation was the prime route for industrial alcohol manufacture. The subsequent advent of inexpensive ethylene led to a synthetic industrial alcohol industry and the dismantling of many alcohol fermentation facilities. Increasing crude oil prices caused a resurgence of interest in the production of ethanol from renewable resources, primarily sugar, corn and cellulosic materials, from 1975 to the present.

The raw materials for ethanol production include cane and beet sugar (sucrose), molasses (high test and blackstrap[5]), whey, starches, and cereal grains (corn, wheat, barley and sorghum), in which glucose is the carbohydrate source. Cellulosic materials include corn stover, rice straw, newsprint and wood residues. Cellulosics must first be pretreated to remove the lignin from the cellulose and hemicellulose components of the biomass. The cost of the raw material has a significant impact on the economics of the process.

Yeast are the most widely employed organisms for ethanol production, although some thermophilic bacteria such as *Clostridium thermosaccharolyticum* have been examined. Yeast have high growth rates and high ethanol tolerance, and produce ethanol in high yields. The reaction involved in the conversion of glucose to ethanol under anaerobic conditions can be summarized as

$$C_6H_{12}O_6 \rightarrow 2\ C_2H_5OH + 2\ CO_2$$

The maximum yield of ethanol from glucose is thus 0.511 gm/gm and in practice yields are typically 90-95% of this value. Yeast are generally inhibited completely by ethanol concentrations above 110 gm/l, although sake yeast strains (*Saccharomyces sake*) can tolerate ethanol concentrations up to 160 gm/l. Temperature is an important factor in the ethanol fermentation. Most distillery yeast have an optimum temperature of 30-35°C. Brewing and wine yeast generally operate at lower temperatures (12-18°C).

In the conventional batch fermentation using sugars from corn or sugar cane, the inoculum size is typically 2-5% by volume. A more concentrated inoculum can by employed in the semi-aerobic process, where 20-25% of a previous batch is retained in the fermenter. Nutrients are added at high concentration and the fermentation time decreases. For blackstrap molasses, the fermentation time is 36 hours, while grain derived sugars require 40-50 hours. Final ethanol concentrations are 70 gm/liter and the overall productivity is ~1.9 gm EtOH/(liter-hour). Variants on this process, e.g., the Mellé-Boinot process, recycle around 80% of the final yeast at harvest and the reduced fermentation time increases productivity

(5) Blackstrap molasses is concentrated liquor from the crystallization of of sugar cane juice. It contains 50-55% sucrose. High test molasses is concentrated sugar cane juice, where 67% of the sucrose has been converted to glucose and fructose. This prevents crystallization.

to as high as 6 gm EtOH/(liter-hour). The use of continuous processes improves the overall ethanol productivity considerably, and some beer and grain fermentations can achieve ~11 gm EtOH/(liter-hour). Other processes are described in the review by Maiorella[6], which should be consulted for detailed descriptions of processes and economics.

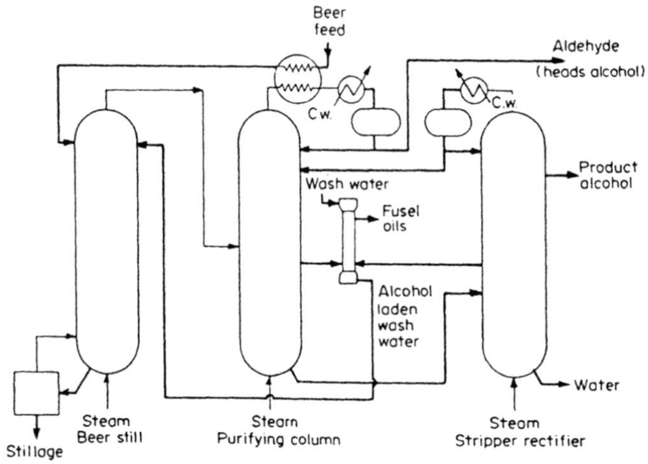

Figure 8.3. *The three column Barbet distillation system for the production of 190 proof ethanol. (From Maiorella[6]).*

There are many methods to recover ethanol from fermentation broths. Distillation may consume up to 50% of the energy in the ethanol production process. Industrial ethanol for use as a solvent is produced at 190 proof (95 volume %), since ethanol and water form an azeotrope at 25°C of 95.7 wt% (89 mol%). Distillation is thus composed of two steps: stripping of ethanol from the dilute fermentation broth, and rectification of the stripped broth. In rectification, the azeotrope sets the energy requirements. In one or two-column distillation, which is used to produce lower quality industrial ethanol, a broth of 6-8 wt% ethanol is fed to the top of the stripping section and alcohol and fusel oils[7] are stripped from the broth as it descends. The ethanol is then rectified and concentrated. Higher quality ethanol can be obtained by use of a three column process (see Figure 8.3) or the Othmer system, where vapor reuse reduces steam consumption (2.4 - 3.0 kg/liter of 94 wt% ethanol). Using

---

(6) Maiorella, B. "Ethanol", in Comprehensive Biotechnology, Chapter 43, Volume 3, eds. Blanch H.W., Drew S. and D.I.C. Wang, Pergamon Press (1985).

(7) Fusel oils are produced as a fermentation by-product. They are a mixture of amyl and propyl alcohol isomers, found at concentrations of 1-5 liters/1000 liters ethanol, depending on the type of substrate. Pure fusel oils are high boilers (128-137°C), but in aqueous solutions their volatility is increased and they are found in the stripping section and cannot be recovered in the bottoms product.

vacuum rectification can reduce the energy requirement to 1.8 - 2.2 kg/liter of 94 wt% product. Other recovery processes that have been considered include vapor-phase water adsorption onto solids, molecular sieves, and membrane based separations.

By-products of the ethanol fermentation are extremely important in the overall process economics. The most important by-product stream arises from the bottoms of the distillation process, referred to as stillage. The volume of stillage is quite high as the fermentation broth is relatively dilute in ethanol, and if no yeast are recycled (such recycle is known as *back-setting*) around 10 to 15 liters of stillage are produced per liter of ethanol product. This stillage has a high biological oxygen demand, around 50,000 ppm BOD for blackstrap molasses and 20,000 ppm BOD for corn-dervied stillage. The stillage contains proteins and vitamins (2.3% in the case of corn-derived stillage) and can be dried for use as a cattle feed.

8.1.2.2 Acetone and Butanol

The acetone-butanol fermentation has a fascinating history. Interest in developing the fermentation as a commercial process arose in 1909 in England (Strange and Graham, Ltd.) as a route to the manufacture of butadiene for use as a raw material in the production of synthetic rubber. Among those working on the process was Chaim Weizmann, who later found a strain, *Clostridium acetobutylicum*, which produced acetone, butanol and ethanol from starchy grains. The onset of the first World War caused renewed interest in the fermentation in the U.K., this time for the production of acetone, an important material in the production of the naval explosive cordite. Strange and Graham, Ltd. undertook production of acetone, but this was not a success until the process developed and patented by Weizmann was installed. Due to the shortage of corn in the U.K., production moved to Canada in 1916, and a supplemental plant was constructed in Terre Haute, Indiana, in 1918. Interest in butanol waned subsequent to the war, and it was not until DuPont found that butyl acetate was the preferred solvent for nitrocellulose lacquers for the automobile industry that butanol fermentation enjoyed a resurgence. The Terre Haute plant was operated by Commercial Solvents from 1920; however by the 1960's few commercial fermentation manufacturers could compete with the synthetic butanol manufacturers. At the height of its commercial use, the scale of the fermentation facilities was impressive. A Peoria, Illinois plant was capable of grinding 25,000 bushels of corn per day to supply raw materials for ninety-six 50,000 gallon fermenters. In 1940, 45 million kg of butanol and 90 million kg of acetone were produced.

Clostridium acetobutylicum is able to ferment a variety of sugars, including black strap molasses, high test molasses, whey, and various lignocellulosic hydrolysates, to yield a mixed solvent. The composition of acetone, butanol and ethanol depends on the nature of the raw material, the strain and the fermentation temperature. Table 8.1 illustrates typical ratios.

Table 8.1. *Typical solvent compositions and yields on carbohydrate sources. The yield is expressed as lb of solvents produced per lb of dry substrate sugar. Fermentation times are generally 50-56 hours.*

Organism	*Substrate*	*Overall yield (%)*	*Acetone (%)*	*Butanol (%)*	*Ethanol (%)*
C. acetobutylicum	corn	26.5	30	60	10
C. saccharo-butyl-acetonium-liquefaciens	corn	25	19	78	3
C. saccharo-butyl-acetonium-liquefaciens	sucrose	30-33	20-35	76-61	4
Butyl culture	pentose (13.5%)	25.4	3	67	0

The typical fermentation based on corn involved removal of the corn oil by steeping and grinding, with the oil being removed from the germ. The corn meal is added to a concentration of 8.5% (dry corn basis) to the fermentation liquid, which consists of 60% water and 40% of stillage. This fermentable starch concentration insures that the final butanol concentration does not exceed 13 gm/liter, so that the sugars are completely converted. This concentration results in ~50% inhibition of the growth of *C. acetobutylicum*. Final solvent concentrations are thus ~20 gm/liter. Inoculum is typically prepared in 1,000 gallon seed fermenters and grown for 26-28 hours. Fermentation was conducted in 50,000 gallon vessels, with 90-95% working volume. A slop-back (i.e. recycle of yeast from an earlier fermentation) of ~25-50% provided additional nutrients. The anaerobic batch fermentation required 50-60 hours. The fermentation broth was sent to a holding tank at the conclusion of the fermentation and fed continuously to a distillation column to recover the solvents. The energy costs associated with the recovery of the acetone and butanol fractions are substantial and make the process uneconomical today. The bottoms from the beer still have a high biological oxygen demand (20,000 to 40,000 BOD) and high solids content. This can be evaporated and dried in either a drum or spray drier to yield a material suitable for swine or poultry feed.

8.1.3 Amino Acids

The manufacture of amino acids owes much to the initial finding in 1908 that the flavor-enhancing component of the kelp-like seaweed *konbu* was due to L-glutamic acid. The Ajinomoto company entered into mono-sodium glutamate (MSG) production in 1909, based on hydrolysis of wheat gluten or soybean protein and recovery of the glutamic acid. Subsequently, microbial routes were discovered, and fermentations for almost all the amino acids were developed. The market for amino acids has expanded considerably and alternative routes, based on enzymatic conversion of precursors, have been developed. The largest markets for amino acids are MSG as a flavor enhancer, lysine and methionine for animal feed supplements, and aspartic acid and phenylalanine as components of the sweetener *Aspartame*. The worldwide markets for amino acids are shown in Table 8.2.

Commercial fermentations for amino acids rely on strain selection to achieve high concentrations and good conversions of carbohydrate substrates to product. We will review the glutamic acid and lysine fermentations, noting that the other fermentations are quite similar in nature.

8.1.3.1 Glutamic Acid

There are several bacterial strains that overproduce glutamic acid, including *Corynebacterium, Brevibacterium, Microbacterium*, and *Arthrobacter. Corynebacterium glutamicum* is typically used in commercial production. Worldwide production exceeds 340,000 tons. Most of the production strains can use a variety of carbon sources, including glucose, fructose, sucrose, maltose and pentose sugars. Molasses and hydrolyzed starch are typical industrial substrates. The key metabolite that controls the fermentation is *biotin*, which is a cofactor of acetylCoA carboxylase, the first enzyme in the synthesis of oleic acid, which is incorporated into the cell membrane phopholipids. Cells with deficient phospholid membranes excrete glutamic acid (this effect has also been found with aspartic acid). Accumulation of L-glutamic acid is a maximum when biotin is suboptimal for growth. Beyond a concentration of ~0.5 mg/gm DCW, biotin reduces L-glutamic acid production. Strain improvement beyond alteration of membrane permeability was accomplished by selection of temperature-sensitive leaky mutants.

The usual fermentation process for glutamic acid is batch, with continuous medium sterilization and temperature, pH and antifoam control. Nitrogen is provided as ammonium chloride or sulfate; the broth may become acidic as NH_4^+ ions are assimilated and glutamic acid is excreted. Optimum pH is 7.0 - 8.0, and the temperature is ~30°C. Supply of oxygen is critical in maintaining high glutamate productivity. The duration of the fermentation is around 24 hours, with 100-200 gm/liter sugar being converted to glutamic acid at 50% yield.

Glutamic acid is recovered by first separating the cells from the broth by continuous centrifugation and a polishing filtration. Then glutamic acid, in the form of monosodium glutamate (MSG), is recovered by multi-effect crystallization, often with multi-effect evaporation to concentrate the solution of MSG and reduce the amount of water to be handled in the process. The production of MSG is cost-sensitive, and strain improvements made manufacture competitive. MSG sells at less than $1/lb. In 1984, the only U.S. manufacturer of MSG, Stauffer Chemical Co., closed its fermentation facility, and the market is dominated by Japanese companies.

Table 8.2. Worldwide and Japanese production of amino acids. [From the Japan Essential Amino Acids Association, Inc., Amino Acids (1987).]

Amino acid	*Production (ton/yr)*		*Production Methods*				
	Japan	World	F	E	C	C&R	Ex
glycerine	3,500	6,000			•		
L-alanine	150			•		•	
DL-alanine	1,500				•		
L-aspartic acid	2,000	4,000		•			
L-asparagine	30			•			•
L-arginine	700	1,000	•				•
L-cysteine	300	1,000		•			•
L-glutamic acid	80,000	340,000	•				
L-glutamine	850		•				
L-histidine	250		•				
L-isoleucine	200		•				
L-leucine	200						•
L-lysine	30,000	70,000	•	•			
L-methionine	150					•	
DL-methionine	30,000	250,000			•		
L-ornithine	70		•				
L-phenylalanine	1,500	3,000	•			•	
L-proline	150		•				•
L-serine	60		•				•
L-, DL- threonine	200		•		•		
L-, DL-tryptophan	250		•	•	•	•	•
L-tyrosine	60						•
L-valine	200		•			•	

KEY: (F) fermentation; (E) enzymatic methods; (C) chemical routes; (R) optical resolution of DL-amino acids; (Ex) extraction.

8.1.3.2 Lysine

Lysine is an essential amino acid that is not present in adequate quantities in cereal grains used for animal feed, thus the market for lysine is primarily for supplementation of grains for cattle, poultry and swine. Lysine is produced both by chemical and fermentation routes, although fermentation accounts for over 80% of the production. Annual production is ~70,000 tons. An enzymatic process, based on conversion of DL-α-amino-ε-caprolactam

(DL-ACL), has been recently developed. The metabolic pathway to lysine in bacteria is via aspartate and diaminopimelic acid. Lysine is one of a family of amino acids synthesized from aspartate (these include arginine, methionine, threonine, lysine, and isoleucine). The lysine precursor diaminopimelic acid is a cell wall precursor in bacteria. High-yielding auxotrophs of *Corynebacterium* and *Brevibacterium* are used commercially. The use of homoserine auxotrophs in commercial lysine production was in fact the first instance of the use of an auxotrophic strain in amino acid production.

The carbon source for lysine fermentation is usually cane molasses, though other materials can be employed. Urea or ammonia are used to maintain the pH near neutrality and complex nitrogen sources can be added. Maintenance of adequate aeration is essential for lysine production. Yields of lysine from the sugar source are ~ 0.3-0.4 gm lysine/gm sugar, with a fermentation time of 60-70 hours. High concentrations of lysine, greater than 75 gm/liter, can be achieved. A typical production medium contains 130 gm/l glucose equivalent, and supply of biotin is important. Lysine is recovered from the fermentation broth by adsorption onto a cation exchange resin and elution with alkali. The basic lysine solution is then neutralized with hydrochloric acid and lysine is recovered by concentrating the solution, forming a precipitate of lysine hydrochloride, the form in which it is typically sold.

The enzymatic route is based on the action of L-aminocaprolactam hydrolase, produced by *Cryptococcus laurentii*. DL-ACL is the starting material, and a racemase produced by *Achromobacter obae* converts the D to the L form. The reaction sequence is

D-α-aminocaprolactam ⇔ L-α-aminocaprolactam ⇔ L-lysine

The first racemization step and the subsequent hydrolysis reaction can be conducted at the same pH, as both enzymes have comparable pH optima. Lysine is decolorized with activated carbon and crystallized, and is sold at $1.50 to $2.00 per lb.

8.1.3.3 Other Amino Acids

Aspartic acid is a component of the sweetener Aspartame and is produced by the action of aspartase on fumaric acid and ammonia. The enzyme is present in immobilized cells of *E. coli* and the reaction is conducted in a continuous operation with a high conversion of fumaric acid to aspartic acid (90-95%). The reaction is

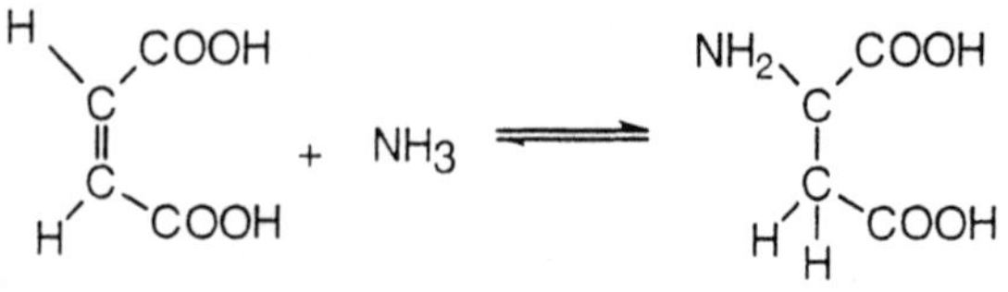

A solution of ~ 1M ammonium fumarate typically passes through a packed column of immobilized *E. coli* (in polyacrylamide or carrageenan) at pH 8.5 with a residence time of ~ 0.8 hr at 37°C. The effluent is acidified and adjusted to pH 2.8 with sulphuric acid. At 15°C the aspartic acid is crystallized from the effluent, centrifuged, and washed.

Phenylalanine, the other component of Aspartame, is produced by either fermentation of glucose by *Cornyebacterium* or *E. coli* mutants, or by enzymatic conversion of a precursor. Routes considered include (a) transamination of phenylpyruvate, the coupled action of phenylalanine and formate dehydrogenases on phenylpyruvate and formate, (b) phenylalanine ammonia lyase with phenylsuccinate and ammonia, and (c) phenylalanine hydantoin with hydantoinase.

With some amino acids, strain development for a direct fermentation route based on glucose as carbon source has been difficult, due to tight metabolic regulatory control. In these cases, the addition of a precursor provides a means of increasing productivity. For example, production of L-tryptophan from the precursor anthranilic acid gives high yields and concentrations of tryptophan in *Corynebacterium* and *Brevibacterium* strains. Final concentrations of tryptophan can reach 30-40 gm/l. Other precursor fermentations include the production of L-serine using glycine as a precursor, isoleucine using 2-hydroxybutyrate as a precursor, and methionine using 2-hydroxy-4-methylthiobutyrate as a precursor.

8.1.4 Antibiotics

Since the discovery by Alexander Fleming in 1928 of the inhibition of microbial growth by a penicillin-producing mold, the production of antibiotics has been a major component of the healthcare industry. Antibiotics are the largest selling group of drugs in human healthcare, followed by cardiovascular agents and anti-inflammatory compounds. There is a tremendous variety of known antibiotics (over 4,000), with uses in both human and animal healthcare. Many chemically-derivatized antibiotics show greater efficacy than the original natural product. Many antibiotics can be grouped by their chemical structure (see Table 8.3).

The main production route for antibiotics is by fermentation, often followed by chemical modification of the resulting antibiotic structure. Antibiotic fermentations are typically batch or fed-batch with withdrawal processes. Fermentation vessels are 30,000 to 100,000 gallon size and because antibiotic fermentations are highly aerobic, the aeration and agitation levels are high. Most antibiotics are produced by filamentous microorganisms, and the viscosity of the broth may increase substantially during the course of the fermentation, placing further demands on agitation.

Strain development has been key to increasing antibiotic productivity. The improvements in penicillin titers, from the level of 200 units/ml in the original culture isolated by Fleming, to current concentrations of over 50,000 units/ml, are illustrative (1 Oxford unit is 0.6 mg of Na penicillin G). Initial strain selection was based on chemical mutagenesis. When the metabolic pathways of the major antibiotics were subsequently elucidated, more

Table 8.3. *Classification of antibiotics by their structure.*

Structure	*Antibiotic or Class*	*Examples of antibiotics*
Peptides & amino acids	β-lactams	penicillins fermentative (G, V, F) & semisynthetic (ampicillin, amoxicillin etc.)
		cephalosporin (cephalothin, cefaclor etc)
		cephmycins (A, B, C, cefoxitin) thienamycin clavulanic acid norcardicins monobactams
	peptides	bacitracin blasticidin S gramicidins bleomycins
	cyclosporins	
	polymixins	
	depsipeptide	valinomycin
aminoglycosides	streptomycin dihydro-streptomycin neomycin	
	kanamycin	(kanamycins A, B, tobramycin, habeka-cin)
	gentamycin	(gentamycin C_1, C_2)
	sisomycin streptothricin (aminoglycoside-like)	
macrolides	erythromycin	
	tylosin	(tylacotones)
	oleandomycin spiramycin kitasatamycin	
polyenes	candicidin fungimycin nystatin	
quinones	tetracycline (7-chlortetracycline, 5-oxytetracycline)	
alicyclics	cycloheximide	
aromatics	chloramphenicol griseofulvin novobiocin	

logical mutation schemes that amplified these important pathways were developed. Many of the commercially important antibiotics are derived from fungi. Cloning is more difficult in fungi than in bacterial and this has hindered the development of high-yielding strains.

8.1.5 Vitamins

Vitamin B_{12} (cyanocobalamin) is a hematopoietic factor in mammals. It is synthesized by many microbes and is required by all animals. It is supplied by meat in the animal diet or is synthesized by the intestinal flora of the animal. It was originally supplied from beef liver, but is currently obtained from *Propionibacteria*. Various other species produce B_{12}, including *Pseudomonas*. The complexity of the B_{12} molecule means that its chemical synthesis is not an economically viable alternative to fermentation. Fermentation with *Propionibacteria* employs carbohydrates such as glucose or molasses at ~100 gm/l, and salts, including cobalt salts. Nitrogen is supplied as either yeast extracts, casein hydrolysates or corn steep liquor. Growth of the bacteria on the carbohydrate occurs first, with little production of the B_{12} precursor, 5,6-dimethylbenzimidazole, as the culture is conducted under nearly anaerobic conditions. In the second stage of the fermentation, aeration induces precursor formation and the synthesis of 5'-deoxyadenosylcobalamin (coenzyme B_{12}) from etiocobalamin. Concentrations range from 50 to hundreds of mg/liter B_{12}. The product is isolated by adsorption onto supports such as amberlite or alumina, or by extraction of the aqueous solutions by phenol or cresol. B_{12} is then purified by chromatography and is precipitated from aqueous alcohol solution with ether. The world market is not large, about 10,000 kg/yr.

8.1.6 Enzymes

As we saw in Chapter 2, enzymes are used in a wide variety of applications in the food, pharmaceutical and chemical industries, and are extensively used for analytical purposes. While some of the processes involving enzyme catalysts are large scale, the amounts of enzymes required are generally modest. The commercial production of enzymes originated with the introduction of glucoamylase in the early 1960's as a replacement for acid hydrolysis of starch. Enzymatic hydrolysis increased the yield of glucose by reducing by-product formation. Following this, proteolytic enzymes were introduced into detergents around 1965. In the late 1960's it was found that workers involved in handling enzyme concentrates experienced allergic reactions. By the end of the decade, enzymes were subsequently removed from the detergents and it was not till dust-free encapsulated enzymes were developed that enzyme use in powdered detergents was resumed. The next major application of enzymes arose in the early 1970's with the introduction of the process for isomerization of starch-derived glucose to fructose. The product, high fructose corn sweetener (HFCS), is used as a sugar substitute. Today, high fructose corn sweeteners are an extremely large volume product.

Alkaline proteases for detergent use account for 25% of enzyme sales, and proteases are the largest group of enzymes in commercial use. Table 8.4 presents the industrial production of enzymes.

Enzymes are produced from microbial, plant and animal sources. In the case of animal enzymes, the organ is obtained from a slaughterhouse and the enzyme is recovered by extraction. Calf rennet, used in cheese manufacture, is recovered from the fourth stomach of the calf and is comprised of chymosin (88-94%) and pepsin (6-12%). It acts on the κ-casein fraction of casein micelles in milk, releasing a soluble peptide and leaving an insoluble fraction, p-κ-casein, which destabilizes the micelle and causes milk to clot. Papain is a plant protease prepared by water extraction of crude *Carica papaya*. Ficin is similarly obtained from *Ficus carica*. These enzymes are then filtered, solvent precipitated and dried.

Table 8.4. *Production of industrial enzymes in 1981. (From O.P. Ward, Comprehensive Biotechnology, Volume 3, Chap. 40 (1985), Pergamon Press). World market sales expressed in millions of dollars (U.S.).*

Enzyme	*Type*	*World market ($)*	*Market share (%)*
Proteases	alkaline (detergents)	100	25.0
	other alkaline	24	6.0
	neutral	48	12.0
	animal rennet	26	6.5
	microbial rennet	14	3.5
	trypsins	12	3.0
	other acid proteases	12	3.0
α-amylases		20	5.0
β-amylases		52	13.0
Glucose isomerases		24	6.0
Pectinase		12	3.0
Lipase		12	3.0
All others		44	11.0

Microbial enzymes may be either intracellular or extracellular and the recovery method depends on the type of enzyme (see Chapter 6 for details). The economics of microbial enzyme production are strongly influenced by the productivity of the organism employed, and strain selection is an important first step. The production process may employ submerged or surface culture techniques; several fungal proteases are produced in semi-solid cultures on media such as wheat bran. In submerged culture, carbohydrate sources include starch, ground barley, and lactose. Complex nitrogen sources are required for protease production, and include corn steep liquor, soybean meal and distillers' solubles. Depending on the enzyme, the carbohydrate may be added continuously throughout the fermentation since high initial carbohydrate concentrations are often inhibitory (e.g., amylase production is repressed by glucose). The pH is generally controlled around neutral pH.

Enzyme recovery usually starts with cooling the fermentation broth to ca. 5°C. Cells are removed by filtration or centrifugation, after the pH has been adjusted to assist in removal of colloidal particles. Inorganic salts (e.g., Ca^{2+} or Mg^{2+}) can also be added to assist in colloid precipitation. These solids are then removed by filtration (vacuum drum or filter press) or by centrifugation. Depending on the degree of purity of the final product, the enzyme solution may be dried under vacuum or further purified. The enzyme may be precipitated with alcohol, acetone or inorganic salts. Fractional precipitation provides a higher degree of purity. If the enzyme is located intracellularly, the cells are disrupted with a homogenizer or bead mill and the resulting broth is subjected to the same purification steps as extracellular enzymes. The lower final concentrations and additional processing step result in higher costs. A general processing scheme is illustrated in Figure 8.4.

Proteolytic Enzymes

The market for detergent enzymes is growing rapidly and alkaline proteases are common constituents of pre-wash, liquid and heavy-duty detergents. The washing conditions are such that the enzyme must be stable at high temperatures and pHs. Ideally, a detergent enzyme should have broad specificity. The alkaline serine protease from *Bacillus licheniformis* (known as subtilisin Carlsberg) is the preeminent commercial detergent protease. Other proteases from *Bacilli* are also used in detergent preparation as they are stable up to pH 12. These enzymes are similar to subtilisin.

Proteases are also used in protein hydrolysis. Defatted soybean has an excellent amino acid composition and is widely used as a foodstuff. The soy protein is hydrolysed to modify the protein into a form suitable for applications, including soft drinks, dietetic foods, non-dairy ice creams and whips. Subtilisin is widely used for these purposes. Soybeans are also fermented with *Aspergillus oryzae* in the production of soy sauce, where protease and lipase activities are involved. After about three days, brine is added with yeast and a lactic acid fermentation completes the processes. The amount of soy sauce produced places it behind only alcohol and milk products in terms of volume. Meat and fish proteins are also recovered by use of proteases. Meat hydrolyzates are incorporated into canned meats and soups, while fish hydrolyzates are used as fish solids and solubles. Other applications include the dehairing of hides, and the improvement of loaf volume in baking.

Calf and microbial rennets are used in cheesemaking, and the market is split between the two sources, with a majority comprised of calf rennet. Typically about 200 ml of rennet solution is added to milk at 31°C, and coagulation occurs in about 15 minutes. After 45

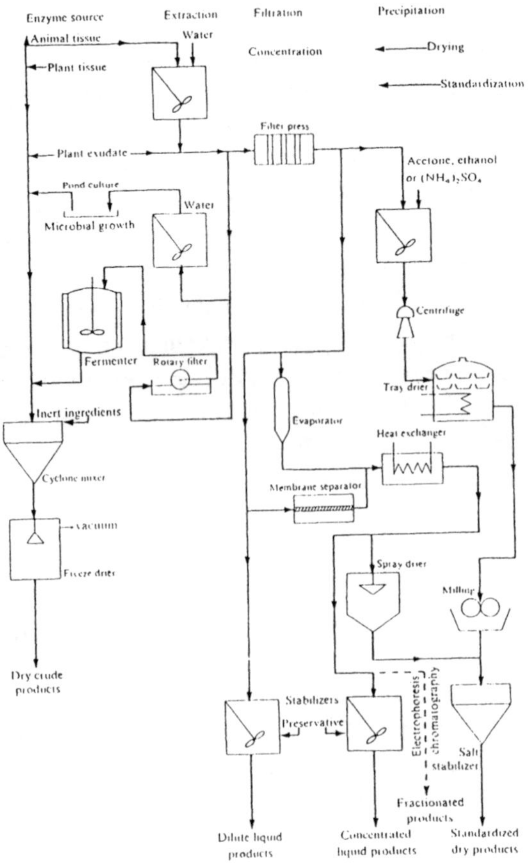

Figure 8.4. *Summary of processing operations used in the manufacture of commercial enzymes.*

minutes the curd can be cut. At lower temperatures a soft curd forms, while higher temperatures favor a firmer curd. Control of pH is important as curd firmness increases below pH 5.8; typically the pH is at 6.4 to 6.6.

Hydrolytic Enzymes

The conversion of starch to dextrose (glucose)[8] is an integral part of the production of high fructose corn sweeteners. It involves a thermostable α-amylase and a subsequent glucoamylase step. A starch slurry is initially gelatinized by cooking at high temperature, then it is liquified and dextrinized by treatment with α-amylase. The continuous process

(8) Starch hydrolysis products are described by their reducing sugar content, as depolymerization results in a mixture of glucose monomers and oligomers. The reducing sugar assay of the mixture is expressed as dextrose equivalents (DE). This is related to the reducing power of glucose. The larger the DE the higher the degree of hydrolysis.

produces a solution with a DE of 0.5-1.5 which is then further liquified in the second stage at 95°C. About 0.05 to 0.10% (dry basis starch) of α-amylase is added in the first stage, where the pH is 6.0-6.5 and the temperature is 104-107°C. The resulting solution has a DE of 10-15. It is then saccharified by glucoamylase, typically in a continuous process. A reaction time of 65-75 hours is required for this stage, with conditions at 60-62°C and pH 4.0-4.4. Approximately 1 liter of glucoamylase solution is required per ton of dry starch treated. About 96% (dry basis) of the final solution is dextrose.

Bacterial α-amylases are produced by *Bacillus amyloliquefaciens* and *B. lichenformis*. The fermentation is conducted at 30-40°C and neutral pH. The fermentation time is 100-150 hours. The amylase preparations are marketed as liquids which contain 2% active enzyme and 20% NaCl as a preservative. Glucoamylase is produced by *Rhizopus, Aspergillus* and *Endomyces* species. The *Aspergillus* fermentation requires 5 days and liquified and hydrolyzed starch serves as substrate. It is marketed as a liquid preparation containing 5% active enzyme.

Pectinases are used to modify the pectin (acidic polysaccharides) that appears when fruit tissue is crushed in the preparation of fruit juices. Pectin increases the viscosity of the juice and confers a haze or cloudiness on the final product. Pectinase hydrolyzes the pectin, reduces the solution viscosity and permits cloudy particles to settle out. Pectinases are also used in winemaking to enhance the yield of juice and to clarify the juice or wine. Commercial pectinases are produced by *Aspergillus niger*; the product contains pectin esterase, polygalacturonase and endopolymethylgalacturonate lyase. Cellulases are also used in conjunction with pectinases in fruit extraction and winemaking.

8.1.7 Therapeutic proteins

The first therapeutic protein products that were widely available were blood products, vaccines and a limited number of enzymes. The advent of recombinant DNA technology opened new routes for the production of protein-based pharmaceuticals that are present in very small amounts in human or animal tissue. The production of monoclonal antibodies has provided a new class of products for diagnosis and disease treatment. While many of these products are still in clinical trails, the potential markets are very large and provide continuing incentive for product development. Materials derived from whole cells are often termed "biologics" and require extensive approval processes for use.

The first step in this process, once the biologic has been tested *in vitro* and in laboratory animals, is the filing of an investigational new drug application (IND) with the FDA. All preclinical information is supplied at this stage. If further studies are appropriate, IND status is granted for the material and human clinical trials commence. These are divided into three categories. In Phase I trials, volunteers are employed to assess pharmacological responses and dose levels, to test for side effects. If results are satisfactory, Phase II trials commence. These involve volunteer patients to test for efficacy and safety. Such trials may be of one to two years in duration. Phase III trials involve a large number of patients, perhaps several thousand, and safety and therapeutic response are further determined. Phase III trials may last up to three years. If successful, a new drug application (NDA) can be submitted, and it is then reviewed by the FDA. Once the NDA is approved, the drug may be marketed. The total time for the whole process may be 10 years or more, at a cost of up to $200 million. The potential market for the drug must therefore be sufficiently large to warrant this expense. The current markets for therapeutic proteins are illustrated in Table 8.5.

***Table 8.5**. Major biopharmaceutical products. Sales in 1993 & 1994 in $ million. [Source: Genetic Engineering News, p6, Jan 1, (1994) and p6, March 15, (1995)].*

Segment	*Sales (1993)*	*Est. Sales (1994)*
Cardiovascular		
Erythropoietin (EPO)	825	1,430
Tissue plaminogen activator (tPA)	240	260
Blood factors	20	35
Subtotal	1,085	1,725
Cancer		
Granulocyte colony stimulating factor	735	870
Interferons (α, β, γ)	625	835
Interleukins	40	40
Subtotal	1,400	1,745
Hormones/Growth Factors		
Human growth hormone (hGH)	290	235
Human insulin	350	600
Subtotal	640	835
Vaccines		
Hepatitis B	520	650
Monoclonal Antibody Therapeutics	35	125
Total ($ million)	***3,680***	***5,080***

Blood Products

Blood proteins have long been commercially available, and have typically been obtained by fractionation of donated whole blood. Recombinant DNA techniques have opened opportunities to produce materials that are virus-free and thus avoid possible contamination from infectious agents. Whole blood is a source of red blood cells, platelets, clotting factors, immunoglobulins and other plasma components. Plasma protein fractions, produced by fractionation of blood serum or plasma, are used to expand blood volume in cases of internal bleeding, dehydration or shock. Some of the proteins in plasma are part of the blood clotting process and can be separated further into blood clotting factors, designated as Factors I through XIII. Except for Factor IV (Ca^{2+} ions), all are proteinaceous, and many exhibit proteolytic activity. There are two coagulation pathways (extrinsic and intrinsic) that involve these factors. Poor coagulation may be a result of a genetic deficiency in these factors. Over 90% of such defects are a result of a deficiency in Factor VIII. Production of this factor in Chinese hamster ovary (CHO) or baby hamster kidney (BHK) cells may provide alternative routes to correctly-glycosylated Factor VIII product, avoiding manufacture from whole blood.

The removal of blood clots (thrombi) is important in the treatment of several diseases. When a thrombus forms, blood flow in vessels is restricted and may lead to death of part of the heart muscle in the case of thrombi in the coronary artery (myocardial infarction), or strokes in the case of blood flow to the brain. Removal of clots is part of the wound healing process and is mediated by the serine protease plasmin, which is derived from its zymogen plasminogen. The activation of plasminogen is primarily a result of tissue plasminogen activator (tPA). Human tPA is a serine protease that is manufactured in murine cells, and it has found wide use in treatment of heart attacks.

Other important blood products include human serum albumin and red blood cell concentrates. Serum albumin is used to treat kidney and liver diseases, and as a plasma expander. It is prepared in concentrations of 5 to 25% from serum, plasma or placentas by precipitation and chromatography. It has been produced as a heterologous product in yeast and bacteria. Red blood cell concentrates are used to treat certain haemolytic diseases and anemia. Hemoglobin preparations that have been chemically modified to increase stability are potential alternatives to this product, and have been produced by recombinant methods in yeast and *E. coli.*

Insulin

The peptide hormone insulin plays a central role in the regulation of carbohydrate metabolism. Insulin increases the level of glucose uptake by liver, muscle and other tissues, and secretion of insulin from beta cells of the islets of Langerhans is regulated by the blood glucose level. When insulin is not produced, blood glucose levels are elevated, with subsequent increases in fatty acid oxidation, ketone body and urea formation. This disease state is diabetes mellitus, and results in reduced fatty acid and protein production; it is controlled by parenteral administration of insulin.

Insulin has been typically obtained from the pancreas of pigs or cattle. It is crystallized from crude pancreatic tissue extracts with the addition of zinc. It is then subject to gel filtration to separate higher molecular weight proteolytic enzymes, insulin dimers and proinsulin. Gel filtration may be followed by ion exchange to further reduce contaminant levels. Human insulin differs from porcine insulin by one amino acid; threonine at position 30 (the carboxyl-terminus) in the human B chain is replaced by alanine in porcine insulin. Porcine insulin is commercially converted to human insulin by trypsin digestion; cleavage occurs between residues 22 and 23, and 29 and 30 of the B chain. A synthetic octapeptide with the correct human sequence is then coupled to the porcine digest. An alternative to modified porcine insulin, recombinant insulin, was the first product of genetic engineering approved for human use.

Proinsulin is produced in recombinant *E. coli* in the cytoplasm, and purified by gel filtration and ion exchange. The purified proinsulin is subject to proteolytic cleavage and the resultant insulin purified by ion exchange chromatography. It is then crystallized and subject to reverse phase HPLC and gel filtration. This approach is preferred to separate production of the A and B chains and joining of these chains by a disulphide bridge. Insulin is usually injected subcutaneously, as its half-life in the bloodstream is only of order minutes. Zinc may be added to promote insulin crystallization.

Erythropoietin

Erythropoietin (EPO) is a glycoprotein hormone that enhances the production of red blood cells from stem cells. It is produced in the kidneys and is highly glycosylated, containing 60% carbohydrate, much of which is sialic acid, with hexosamines and hexoses. EPO has a molecular mass of approximately 34,000 Da, and is one of a family of haemopoietic growth factors, that influence proliferation and differentiation of haematopoietic stem cells. Stem cells produce all blood cell types.

EPO is regulated by the dissolved oxygen concentration in tissues. In anaemic individuals, EPO production sharply decreases due to the decreased supply of oxygen, which leads to secondary anaemia. EPO administration can be used to treat anaemia associated with chronic renal failure, whether patients are on dialysis or not, and for anaemia resulting from AZT-treatment of HIV-infected patients. Recombinant human EPO is produced by CHO cells.

Cytokines

Cytokines comprise a large class of proteins which influence cell-modulated immunity. They are produced by a number of cell types; cytokines produced by lymphocytes are termed lymphokines, those produced by monocytes are termed monokines. Other important cytokines include the interferons (IFN), interleukins (IL-1, IL-2), colony stimulating factor (CSF), and tissue necrosis factor (TNF). Cytokines have been shown to to be effective in treating various types of cancer, arthritis, multiple sclerosis, asthma and allergies.

The first cytokine approved for clinical use was interferon-α in 1986. There are three types of interferons, IFN-α, IFN-β and IFN-γ. When a virus attacks a cell, it may produce and release interferons. The interferon then induces other cells to become resistant to the virus. Treatment with IFN may be effective with established viral infections, such as hepatitis B and genital warts. Recombinant IFN-α or lymphoblastoid interferon is produced in *E. coli*, as the unglycosylated form of IFN-α retains its biological activity. IFN-β or fibroblast interferon was produced in fibroblasts exposed to viruses. It can now be produced in *E. coli*, yeast and CHO cells, as can IFN-γ, although most emphasis on commercial production centers on recombinant *E. coli*. The interferons are purified from cell homogenates by affinity chromatography (using lectins, anti-IFN MAbs, or concanavalin A as ligands), ion exchange and HPLC.

The interleukins are a large subfamily of cytokines, comprising at least 10 types. They function as regulators of the immune response and bind to receptors at various cell surfaces, where they modulate immunological activity. IL-2 plays a key role in immune functioning, and is produced by T lymphocytes when activated by an antigen or mitogen. It stimulates growth and differentiation of T and B lymphocytes, and controls the activity of natural killer cells. Its function is thus key to both cell-mediated and humoral immunity. IL-2 has a molecular weight of 15,500 Da, and contains one intrachain disulphide linkage. The protein is arranged in four antiparallel α-helical regions. It is produced in *E. coli* in an unglycosylated form, but retains its biological activity. It has been used to treat several forms of cancer and infectious diseases, including AIDS. IL-1 is naturally produced by phagocytic cells and it elicits a variety of responses, including proliferation of thrombocytes, fibroblasts and lymphocytes. It also promotes wound healing. It is produced in recombinant *E. coli*.

Monoclonal Antibodies

The advent of hybridoma cell technology opened the possibility of large-scale production of monoclonal antibodies (MAbs). They have become a major diagnostic product and are also finding therapeutic use. MAbs are used in diagnosis of cancer types, including breast, colorectal, lung and ovarian cancers. They are also used as affinity ligands in some affinity chromatography separations. The antibody OKT3 has been approved for clinical use in the reversal of acute kidney rejection.

Most monoclonal antibodies are produced in submerged cultures of hybridoma cells. Serum-free medium formulations are available, simplifying recovery and purification. Fed-batch cultures can be highly productive, with final concentrations of antibody over 1 gm/liter. In perfusion cultures, perfusion rates of 2 reactor volumes per day can result in high volumetric productivities. Monoclonal antibody sales for diagnostics reached $575 million in 1992, and this area is projected to grow substantially.

8.1.8 Polysaccharides

The exopolysaccharides produced by microorganisms are widely used as emulsifiers, stabilizers, coagulants, gelling agents, lubricants, and thickening and suspending agents in the food, pharmaceutical and chemical industries. Marine algae and plant gums are being displaced by some of these microbially-produced materials. Microbial polysaccharides are present intracellularly as cell wall components and for energy storage. Exopolysaccharides can sometimes form capsules around the cell or can be present as slime. Table 8.6 summarizes the production and uses of polysaccharides in the USA.

Table 8.6. *Estimated consumption and production of industrial polysaccharides in the USA in 1975. [From A. Margaritis and G. Pace, Comprehensive Biotechnology, Vol 3, Chapter 49, Pergamon Press (1985)].*

Polysaccharide	*Food use (tonnes)*	*Industrial use (tonnes)*	*Price ($/kg) (1975)*
cornstarch	203,000	1,013,000	0.20
carboxymethylcellulose	6,100	40,000	2.00
methylcellulose	820	21,500	3.30
alginate	3,700	3,600	5.50
pectin	4,900	0	4.85
xanthan	950	2,500	6.90
gum arabic	9,340	2,850	1.65
guar gum	6,070	14,180	0.96
carrageenan	3,700	90	4.40
tragacanth	526	81	26.50
locust bean gum	3,650	1,620	2.00
karaya	410	2,900	2.10
ghatti	4,050	410	1.15
agar	125	165	16.60
Total	*247,341*	*1,102,896*	*$596,500,000*

The most important microbial polysaccharide is xanthan gum, whose use has grown considerably since 1975. Of particular interest is the use of xanthan in secondary and tertiary oil recovery, due to its excellent rheological properties. It is produced by *Xanthamonas campesteris*, and is a branched anionic heteropolysaccharide of five repeating sugars. Its molecular weight may reach 10^6, depending on the fermentation conditions. Some of its uses are summarized in Table 8.7.

Other important microbial polysaccharides include dextran, alginate and pullulan. Dextran is produced by *Leuconostoc mesenteroides* by the action of dextran sucrase. It can also be produced directly from sucrose by the action of dextran sucrase. Dextran is a branched neutral polysaccharide consisting primarily of 1,6 linkages of glucose monomers. Branching occurs at the 3 position. Dextran is used in molecular sieves for separating molecules on the basis of size. Alginate is primarily obtained from seaweed, but microbial routes are available from *Azotobacter vinlandii*. Alginates are linear polysaccharides containing mannopyranosyluronic and guluronic acid residues. Pullulan and scleroglucan are also produced by fermentation.

***Table 8.7**. Major food and industrial uses of xanthan gum [From A. Margaritis and G. Pace, Comprehensive Biotechnology, Vol 3, Chapter 49, Pergamon Press (1985)].*

Food Applications
Dressings
Relishes and sauces; syrups and toppings
Starch base products (canned desserts, sauces, fillings)
Dry mix products (desserts, gravies, beverages, sauces, dressings)
Farinaceous foods (cakes)
Beverages; dairy products (ice cream, cakes, cheese spreads)

Industrial Chemical Applications
Flowable pesticides; liquid feed supplements
Cleaners, abrasives and polishes
Metal working; ceramics; foundry coatings
Slurry explosives; dye and pigment suspensions

Oil Field Applications
Drilling muds; workover and completion fluids; enhanced oil recovery (polymer flooding)

The production of exopolysaccharides is hampered by the high viscosities that are reached during the fermentation. Xanthan is typically formed under carbon-limited conditions, and the composition of the medium employed has a considerable influence on the rates of product formation and on its final composition. The effects of broth viscosity on mass and heat transfer in the fermentation have been extensively studied. The fermentation is batch and the product is recovered from the dilute broth (15-30 gm/l). If the product is to be used in applications which require it to be clear, cell removal is accomplished by filtration or centrifugation. The polymer may then be isolated by precipitation with methanol, ethanol, isopropanol or acetone, or by addition of salt ($Ca(OH)_2$) or acid (HCl). Isopropyl alcohol precipitation is most common. The wet precipitate may then be dewatered prior to drying. It is then milled and packaged. The yields of polymers from glucose are ~0.7 gm/gm, with product concentrations at the conclusion of a 40 hour fermentation being ~25 gm/l.

8.1.9 Single Cell Protein (SCP)

The term "single cell protein" was coined to refer to microbial cells grown for food or feed applications. Consumption of certain microorganisms has occurred since ancient times; top-fermenting yeast were employed as leavening agents in bread; fermented milk and cheese products contain *Lactobacillus* and *Streptococcus* species; and mushrooms were prized by the Romans of the first century B.C. Blue-green algae of the genus *Spirulina*, which grow in alkaline lakes, were eaten by the Aztecs when the Spanish explorers of the 16th century arrived, and the peoples of the Lake Chad region have eaten *Spirulina* for many generations. Bakers' yeast was grown on molasses for food in Germany during World War I. Since then, there have been expanded efforts to employ microbial protein as a food or feed supplement. Algae, bacteria, yeast and filamentous fungi have all been cultivated for these purposes. The raw materials for fermentative production of SCP have ranged from hydrocarbons, methane, cellulosic materials, and saccharides to carbon-containing wastes. The cost of these materials has fluctuated and production today depends on these costs relative to other sources of protein, such as soybeans.

Bacterial Biomass

Bacteria have attractive attributes as sources of protein; for example, they can employ a wide range of carbon sources, and they have rapid growth rates. Carbon sources include simple sugars and disaccharides (glucose, sucrose, galactose, lactose), and lignocellulosic residues including cellulose and hemicellulose (generally after some pretreatment). Hydrocarbons have also served as substrates when their cost was sufficiently low. Methane, ethane and propane have all been considered. Shell Research (U.K.) extensively examined bacterial SCP production from methane in the 1970's. Methanol has advantages in that it is not potentially explosive and ICI England developed a large scale SCP process (70,000 tonnes per year) based on this raw material. N-alkane fractions (mainly C_{16}) have also been considered, although yeast were generally employed as SCP source. Table 8.8 lists some of the bacteria, their substrates and growth characteristics for SCP production.

Bacterial SCP production may be conducted as a batch or continuous process. The large-scale facilities that were constructed in the 1970's and 1980's were generally operated batchwise. Heat evolution at high growth rates in large fermenters requires special considerations, such as an external heat exchanger. Recovery of bacterial cells presents difficulties due to their small size; they are first agglomerated or flocculated and then centrifuged to yield a concentrated product. The resulting protein may be used for animal feed or human food, or as a functional protein (a whippable or thickening agent), or for gel formation. As the bacteria have high nucleic acid contents (up to 16% dry weight), their use in foods requires removal of nucleic acids, which otherwise lead to gout and kidney stone formation.

Yeast and Filamentous Fungi

Bakers' yeast (*Saccharomyces cerevisiae*) was used for food and as animal feed during WWI in Germany, and in WWII *Candida utilis* was grown on wood hydrolyzates containing pentose sugars. *Candida* has the advantage that it does not require amino acids or B-group vitamins for growth, making it an attractive candidate for SCP production. *Candida* species can use a variety of carbon sources, including sugars (starch hydrolyzates, molasses sulphite

Table 8.8. Growth characteristics of selected bacteria and Actimomycetes for SCP production[9].

Organism	*Carbon source*	*Growth rate (hr^{-1})*	*Dry weight (gm/l)*	*Cell yield (gm/gm substrate)*
Achromobacter	diesel oil	-	10-15	-
Acinetobacter	gas oil	0.4-1.0	8-10	0.1-0.12
Acinetobacter	n-hexadecane	1.1-2.0	8-10	0.8-0.9
Cellulomonas	bagasse	0.2-0.29	16	0.44-0.5
Corynebacterium	propane	0.046	0.9	0.3
Methylococcus	methane	0.14	0.4	1.00
Norcadia	n-alkanes	1.25	14.7	0.98
Pseudomonas	methane	0.06	0.8	0.99
Thermomonospora	cellulose	0.48	2.3	0.44

waste liquor and whey), ethanol, methanol, n-alkanes, gas oil, and LPG. In batch cultures, the carbohydrate concentration is typically around 1-5% and the C:N ratio is 7:1 to 10:1. Heat removal in batch and continuous processes make heat-tolerant yeast advantageous, as most strains have optimum temperatures in the range 30-34°C, and cooling water is generally not available at a sufficiently low temperature to make growth at the optimum temperature feasible. Yeast have high aeration requirements to provide optimal rates of protein production. In the 1970's British Petroleum developed a gas oil based *Candida* process using an air-lift fermenter at Lavera, France, with a capacity of 16,000 metric tons/yr[10]. In a typical yeast process, anhydrous ammonia is employed for pH control and the system may be operated sterilely or not. Specific growth rates are 0.15-0.25 hr^{-1} and 10-20 gm DCW/l are typical. The cell yield depends on the substrate, ranging from 0.5 gm/gm for carbohydrate-based materials to ~1.0 gm/gm for hydrocarbons. Due to their larger size, yeast can be readily recovered from the fermentation broth by centrifugation. When gas oil is used, the cells must be washed to remove residual hydrocarbons. The cells can be drum or spray dried and used directly, or functionalized into a variety of products.

The use of fungi in human foods is widespread. For example, fungi serve as flavor enhancers in cheese and in soybean products such as miso and tempeh. Fungi can grow on a wide variety of carbon sources and this spurred interest in their use as SCP. The structure resulting from hyphal growth can be used as a basis for food fabrication, and the cells are

(9) From J.H. Litchfield, "Bacterial Biomass", in Comprehensive Biotechnology, Vol 3, (1985), Pergamon Press.

(10) The status of yeast-based SCP processes is discussed in J.H. Litchfield, Chapter 4, *Microbial Technology*, Vol 1, ed. Pepler, H.J. and D. Perlman, Academic Press (1979).

easy to recover from the fermentation broth. While fungal growth rates are slower than those of yeast or bacteria, they can be adequate for SCP production purposes. *Fusarium, Penicillium, Aspergillus*, and *Trichoderma* have been examined on such carbon sources as cassava, carob extracts, whey and coffee wastes. The quality of fungal biomass has been found to be adequate for SCP, and a Finnish Pekilo process was approved for animal feed production. The Rank Hovis process was cleared by the British government in 1980 for production of myco-protein. This process is shown in Figure 8.5. In the 1980s, interest in single cell protein decreased considerably, due to increasing costs of large-scale fermentation equipment and raw materials, coupled with low prices for other protein sources such as soybeans.

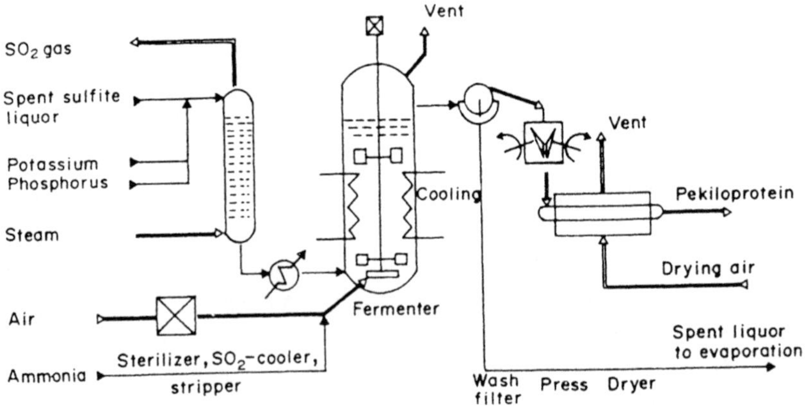

Figure 8.5. *The Finnish Pekilo process for the production of Paecilomyces variotii on stripped suphite waste liquor. Two 360 m^3 fermenters are employed, operating at pH4-5, and 38°C. The fermenters are operated at dilution rates of 0.14 to 0.20 hr^{-1} and contain 13 gm/l biomass. The biomass productivity is 2.7 gm/l/hr. [From J. Laine and R. Kuoppamaki, Ind. Eng. Chem. Proc. Des. Dev.* ***18****, 501 (1979)].*

8.1.10 Steroids

Steroids have been employed for the treatment of inflamatory diseases since the late 1940's. The sources of steroids are plant and animal tissues, and most steroids are derived by either chemical or microbial conversion of starting materials, typically phytosterols, such as diosgenin and stigmasterol. Diosgenin is obtained from the root of the barbasco plant, grown in Mexico and Central America; stigmasterol is obtained from soybean seed oil. Diosgenin is a starting material in the formation of pregnenolone, and stigmasterol is the precursor for progesterone. The 3-one-4-ene A-ring of corticosteriods can be obtained from pregnenolone which is converted to progesterone. The most therapeutically important steroids (cortisone, hydrocortisone, prednisone, prednisolone etc) have an 11-hydroxyl or 11-keto function, in addition to a 17a-hydroxyl function. The ring numbering system of steroids and the carbon numbering scheme are illustrated in Figure 8.6.

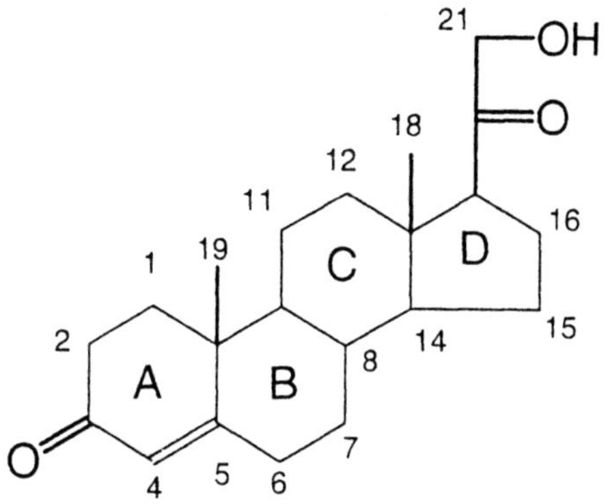

Figure 8.6. *The structure of deoxycorticosterone, illustrating the ring and carbon atom numbering schemes.*

The development of microbial transformations to introduce an 11α-hydroxyl group was an important step in steroid conversions. Chemical methods were subsequently developed to convert this hydroxyl group to the desired 11-keto or 11β-hydroxyl groups. The microbial transformation of progesterone to 11α-hydroxyprogesterone occurs with *Rhizopus nigricans* and *Aspergillus ochraceus*. Small amounts of side products, such as 6β-11α-dihydroxyprogesterone are produced. The ususal yields are 70 - 90% of the starting material, which is usually present at ~20 gm/liter concentration.

The second most important bioconversion is 1-dehydrogenation, catalyzed by *Arthrobacter simplex, Bacillus sphaericus*, and other organisms. A typical example of this reaction is the conversion of hydrocortisone to prednisolone. In this reaction, substrate is present at 0.5 to 2.0 gm/liter and the conversion yield is 80-90%. Increased concentrations of substrates can be achieved using organic solvents or semi-solid reactions.

16α-hydroxylations are the third most important class of steroid bioconversions. The introduction of this hydroxyl permits the use of 9α-fluoro steroids, which otherwise have the undesirable effect of increasing salt retention in humans. These 9α-fluoro derivatives show greatly enhanced anti-inflamatory activity. Conversion levels of ~50% are reported, with fairly low substrate concentrations.

The above steroid bioconversions are generally conducted after the growth period of the microorganism having the desired bioconversion activity. This is typically 20-30 hours. In some cases the enzyme is induced and the substrate is added to the fermenter as a slurried powdered solid or dissolved in a suitable solvent (DMSO, N,N-DMF, ethanol etc). The duration of the conversion process may be 30 to 100 hours. 11-hydroxylations have been conducted with spores rather than with vegetative cells. The low solubility of the substrates can be overcome by conducting the conversion in an organic solvent, using either whole cells or enzymes. Two-phase organic-aqueous systems have also been investigated for steroid bioconversions. Products are recovered from the fermentation by extraction into organic solvents when aqueous systems are employed. In some bioconversions, the substrate and product are present as solids and the solids (product and cells) are contacted with solvent.

The market for steroids is large. They are used in contraceptives, sedatives, anti-inflamatories, antitumor agents, and therapeutically as hormones. As of the mid-1980s, the steroids prednisone, prednisolone and methyl prednisolone accounted for 10% of the total drugs prescribed in hospitals.

8.1.11 Insecticides

Control of insect pests by chemical means has been a very successful commercial undertaking. There are significant disadvantages associated with the use of such chemical insecticides, however, including development of resistance, the residue of the pesticide on the food product, and residues that may enter the ground water supply. For these reasons there is an increasing emphasis on biological control agents. It has been long recognized that insect pathogens may be used to control insect pests. Among the pathogens are bacteria, fungi, viruses, rickettsiae, protozoa and nematodes. Table 8.9 illustrates the major microbial insecticides currently produced.

Table 8.9. *Microbial Insecticides in Production [From H. Stockdale, Comprehensive Biotechnology, Vol 3, chapter 46 (1985)].*

Organism	*Target*	*Characteristics*
Bacillus thuringiengsis	Lepidopteran larvae	fermentation production
Bacillus thuringiengsis israelensis	Dipteran larvae	fermentation
Bacillus popilliae	Japanese beetle	in collected larvae
Hursutella thompsonii	citrus mites	fermentation
Verticillium lecanii	glasshouse aphids	fermentation
cotton bollworm nuclear polyhedrosis virus (NPV)	*Heliothis sp.*	in mass-reared larvae

The major microbial insecticide is *Bacillus thuringiengsis*, which forms a crystalline proteinaceous spore. This spore is a lepidopteran pathogen, which causes paralysis of the larval gut. *B. thuriengensis* also produces an exotoxin which is capable of killing flies and other insects. The bacterium is produced by fermentation, using complex carbon and nitrogen sources, such as starch, tryptone, fish meal etc. The duration of the batch fermentation depends on the nature of the medium, ranging from 14 to 72 hours, at 30°C, and pH 7.2-7.6. The spore content of the broth can reach 4 x 10^9 spores/ml in optimized fermentations. Spores of *Bacillus polilliae* are produced similarly. When these spores are ingested, they become infectious.

Nematodes have insecticidal properties and are of interest due to the large variety of insects which are attacked. Some nematodes can be grown on solid nutrient media, while non-entomogenous nematodes have been cultivated in liquid medium and may be capable of being cultivated by fermentation.

While fungi provide a broad class of pathogens, their commercial uses have been restricted. The infectious elements of some fungi are the conidia, which are difficult to produce in submerged fermentation. Thus these are grown on solid media. For other fungi, growth by fermentation is possible, but the physiology of growth must be controlled by appropriate carbon and nitrogen sources.

Insect viruses can only be produced by infecting insects or by use of established cell lines. These can be grown in submerged cultures; however, the insect cells are shear sensitive and aeration in large scale equipment is difficult. Microbial insecticides represent a small (~ 1%) fraction of the worldwide insecticide market sales. It is likely that they will become increasingly important as the problems associated with chemical insecticides become more apparent.

8.2 Economic Analysis of Bioprocesses

The analysis of the economic feasibility of a fermentation process is a complex undertaking that involves a variety of factors, including the nature of the company involved in manufacture of the product, the economic climate, and regulatory aspects. The typical questions that arise are

- what is the amount of product required and will the demand grow?
- will the plant be new, or can an existing facility be used?
- will the plant be multipurpose or dedicated?
- what is the availability of raw materials?
- what is the nature of the regulatory requirements for the product?
- how quickly should the plant be constructed to meet anticipated market demand?
- what is the patent position of the company for the product?

One of the important components in answering many of these questions is the cost estimate - the amount of capital required to construct the plant and the operating costs associated with manufacturing of the product. The accuracy of this estimate depends on the extent to which the process is defined; many processes in the research and development phase will undergo changes before they are implemented in the plant. In this section we will focus only on the development of material and energy balances for biological processes and on approximate means for estimating the final capital and manufacturing costs. In this phase of cost estimation, changes in the process can be evaluated, alternative technologies can be compared, and the effect of manufacturing scale on operating costs can be determined.

8.2.1 Components of the Cost Estimate

The total cost of production is comprised of various elements, including both the direct operating costs, the fixed costs, and plant overhead. We shall follow the outline employed by Kalk and Langlykke, as it provides a convenient framework for discussing these elements. Table 8.10 indicates the elements of the total production cost and provides a rough estimate of the percentage contribution of each to the production cost.

The total product cost has been divided into the manufacturing cost and general expenses. The manufacturing cost includes those elements that contribute directly to the cost of production (operating costs, fixed costs, and plant overhead costs). The direct operating costs include the raw materials, utilities and supplies; these costs will generally scale with the plant size, although labor costs may not. Fixed costs relate to the physical plant and do not change with productivity levels. The category denoted as plant overhead includes charges for services that are not directly attributable to the cost of the product, such as janitorial services, accounting, personnel etc. These can be related to the number of personnel involved in the production operation and thus can be related to the salaries and wages component of labor costs.

The category denoted as general expenses includes those charges for marketing, research and development, and general administration charges. These charges are often difficult to estimate, and depend on the usual practice in the company. When therapeutic products are manufactured, these costs can be significant if the expenditures for clinical trials, for example, are factored into this category.

The category for capital investment includes the costs for the purchased equipment, its installation, and the costs of land, building construction, and engineering design. Also included here are start-up costs, reflecting modifications that may be required when the plant is commissioned. Also included here is the working capital, those funds used to provide an inventory of raw materials, and supplies and cash to pay salaries and vendors. The sum of the fixed and working capital is the total capital investment. The elements of the total capital investment are summarized in Table 8.11.

There are a number of methods which use the information from Table 8.11 to determine whether the proposed process will be profitable. One common approach is to calculate the return on investment or ROI, determined before or after taxes. This measure is the ratio of the annual profit made by the plant to the total capital investment. Other methods include the cash flow analysis, which relates the net cash in (profits plus depreciation) to the cash outflow (initial investment), the payback time, present worth, and internal rate of return. Details of these methods are found in the texts described earlier. We now turn to examine the methods for evaluating the elements of the capital and operating costs.

***Table 8.10**. Elements of total production cost. [From J. Kalk and A. Langlykke, ASM Manual of Industrial Microbiology and Biotechnology (1985)].*

I. DIRECT OPERATING COSTS
- A. Raw materials and supplies
  - 1. Raw materials (30-80% of manufacturing cost)
    - a. Primary
    - b. Secondary
    - c. Freight
  - 2. Supplies
    - a. Operating (3-5% of direct labor and supervision)
    - b. Maintenance (100% of maintenance labor and supervision; 2-5% of fixed capital investment)
    - c. Laboratory (20-40% of laboratory labor)
    - d. Other (10-20% of plant overhead)
  - 3. By-product credits/debits
- B. Labor and Supervision (10-40% of manufacturing cost)
  - 1. Base salaries and wages from manpower estimate
  - 2. Overtime (6% of base hourly wages)
  - 3. Fringe benefits (30-40% of salaries and wages)
- C. Utilities (5-20% of manufacturing cost)
  - 1. Steam
  - 2. Electricity
  - 3. Water
  - 4. Waste treatment

II. FIXED COSTS
- A. Depreciation and interest (8-12% of fixed capital for depreciation, 10-15% for interest)
- B. Taxes (1-4% of fixed capital)
- C. Insurance (1-3% of fixed capital)
- D. Rent (variable)

III. PLANT OVERHEAD

10-70% of labor and supervision, depending on level of detail of manpower estimate

IV. ADMINISTRATION

V. MARKETING

VI. RESEARCH and DEVELOPMENT

MANUFACTURING COST = I + II + III

GENERAL EXPENSES = IV + V + VI

TOTAL PRODUCT COST = MANUFACTURING COST + GENERAL EXPENSES

Table 8.11. Outline of the Total Capital Investment. [From J. Kalk and A. Langlykke, ASM Manual of Industrial Microbiology and Biotechnology (1985)].

Total Capital Investment	% of Fixed Capital
Fixed Capital	
A. Direct Costs	
(i) Land - property, surveys, recording, fees, commissions, taxes, etc.	2-3%
(ii) Site development - utility hook-ups, site clearing, grading, excavating, roads, walkways, landscaping, railway	4-6%
(iii) Buildings - foundations, offices, shops, warehouses, processing and utility areas, labs, locker rooms, cafeteria, services to buildings (plumbing, HVAC, painting, etc.)	10-15 %
(iv) Processing - equipment, installation, piping, electrical, instrumentation, special foundations, insulation, paint, insurance, taxes	40-70 %
(v) Services - as above, plus distribution systems. Non-processing equipment also included, e.g., lab and office equipment	20-30 %
B. Indirect Costs	
(i) Engineering - design, PI & D, procurement, administration, cost control, etc.	5-15%
(ii) Construction - field supervision, temporary facilities, tools, equipment	5-15%
(iii) Contingency - depends on the level of detail of the estimate	3-50%
(iv) Fees - engineering, construction, contractors	4-6%
C. Start-up Costs	5-20%
Equipment and construction modifications, personnel training, technical support, operating expenses	
Working Capital	15%
A. Inventory	
Raw materials, and supplies in warehouse, raw materials and supplies in finished product, finished product	
B. Accounts receivable	
Shipped finished product waiting payment	
C. Accounts payable	
Salaries and wages due, raw material and supply payments due, utilities, etc.	

8.2.2 Process Flow Sheets

The starting point for estimating the operating and capital costs of a product is the process flow sheet. At the early stages of process development, the elements of the process may not be well defined, and estimates will be only approximate. Even later in the process development, an operation may not be defined, e.g., whether a protein should be recovered by ultrafiltration, precipitation, or aqueous two-phase extraction. At the early stages, on order-of-magnitude estimate can be made if there is some knowledge of the capital and operating costs of an earlier process. This approach is described later.

8.2.3 Material and Energy Balances

Once the conceptual design of a process is developed, a flow chart indicating the various plant operations can be derived. This indicates all liquid, gas and solid flows in the process. In fermentation processes and in processes with enzymatic conversions, most of the flow streams are liquid, generally dilute in solutes. The size of the equipment is determined by the targeted annual production. Information on the rates and yields of the biological processes set the size of the reactors and recovery equipment. With these flows set, the raw materials and utilities flows can be specified.

Consider a batch process for the manufacture of citric acid. The following biological information is available.

batch fermentation time	120 hours (including turnaround)
initial glucose concentration	150 gm/l
citric acid yield	0.8 gm citric acid/gm glucose
temperature	30°C
pH	5-7
aeration	1.5 vvm

We can now determine the material and energy requirements for the process, based on a desired annual production capacity of 10 million pounds, and an onstream plant operating time of 8,000 hours/yr.

broth volume per year	=	(10×10^6 lb /yr) x (454 gm/lb)/ (0.8 gm/gm x 150 gm/l)
	=	37,833,333 liters/ yr
total number of batches	=	(8000 hrs/yr)/ (120 hrs/batch)
	=	66.7 batches/yr
total volume of batches	=	567,472 liters
number of fermenters	=	5 (arbitrary)
liquid volume each	=	113,494 liters
broth flow to recovery	=	37,833,333/8 ,000 liters/hour
	=	113,500 liters/day
sterilization steam	=	(121-20 °C) x (113,494 kg/day) x (1.0 Kcal/kg-°C)/(544 Kcal/kg)
	=	21,072 kg/batch
sterilization time	=	3 hours

steam requirement	=	7,024 kg/hr
air flow rate (fermenter)	=	(1.5 vvm) x (113,494 liters)/(1000 liters/m^3)
	=	170 m^3/min

The cooling requirements for the fermentation can be determined from knowledge of the heat transfer area in the fermenter, as described in Chapter 5. The requirements for recovery and purification are determined by the process employed. For example, calcium citrate precipitation with $Ca(OH)_2$ as a first step in recovery requires an excess of $Ca(OH)_2$ over the stoichiometric requirement. Similarly, an excess of sulphuric acid is then required to precipitate $CaSO_4$ and form citric acid. Other media requirements, for example the nitrogen source, can be determined from a knowledge of the medium composition.

8.2.4 Equipment Sizing

Once the material and energy requirements of the process are specified, the equipment can be sized. In most cases, the sizing is based directly on the flow rates obtained from the material and energy balances, recognizing that some equipment sizes may be discrete; for example, centrifuges, sterilizers, ultrafilters etc, may only be sold in sizes that increase stepwise.

In developing the process flow sheet, holding tanks need to be considered, and for fermenters and tanks a working volume of 70% of the total tank capacity is usual. For the citric acid example above, the complete equipment list is summarized in Table 8.12, based on the flow sheet in Figure 8.1.

8.3 Capital Cost Estimates

The total cost of construction of a manufacturing facility can be broken into the components listed in Table 8.11. The fixed capital is comprised of direct costs, which are those covering the equipment, and the labor required to construct the facility, and indirect costs, which include engineering design costs, construction and fee. The third component of direct costs is the cost associated with start-up of the facility. Working capital is generally included here. It accounts for the capital involved in inventory and accounts receivable. It can usually be estimated from consideration of supplies, etc., required for three months plant operation.

There are various levels of detail in developing plant capital costs. The simplest is an order-of-magnitude estimate, based on a knowledge of the costs of similar plants of different size. This estimate provides only a rough guideline and is generally applicable only to large plants. For an antibiotic production facility we might develop a very rough estimate based on the total fermentation volume required. The fixed capital could be determined on the

Table 8.12. *Process equipment summary for citric acid production. [From H.B. Reisman, Economic Analysis of Fermentation Processes, CRC Press (1988)].*

Equipment Item		*Specifications*
Fermenters	5	agitated with internal coils
Seed fermenters	3	agitated with internal coils
Continuous sterilizer	1	with heat recovery
Harvest tank	1	clean in place
Rotary vacuum filter	2	one standby
Precipitation tank		agitated internal coils
Rotary vacuum filter		
Acidulation tanks		
Rotary vacuum filter		for $CaSO_4$ removal
Evaporator feed tank		agitated
Evaporator		water removal
Concentrate tank		agitated
Ion exchange columns		carbon steel, rubber lined
Carbon decolorization		
Filters		automatic, leaf type
Crystallizer		agitated, evaporative
Centrifuge		automatic discharge
Mother liquor tank		agitated
Dryer		indirect heat
Storage bins		
Classifier		continuous screening
Packaging line		
Glucose storage tanks		
Media makeup tank		
Hydrated lime feed		
Lime storage silo		
Sulphuric acid storage		
Air compressors		

basis of $200 per gallon of fermenter capacity required[11], using 1992 capital cost indices. If a previous estimate is available at a different scale or time, this can be used to determine the capital cost of the new plant as follows.

Assume that a company constructed an antibiotic plant (e.g., penicillin) ten years earlier and the total capital cost was $35,000,000 for a 1×10^6 kg/year capacity. The cost to build another facility of 2×10^6 kg/yr capacity at present can be estimated by scaling the plant cost to present day costs by use of an appropriate cost index (e.g. the Marshall & Swift Index or the Chemical Engineering plant cost index). The increase in size can be taken into

(11) Based on Swartz, R., Chapter 2, Comprehensive Biotechnology Volume 3, Pergamon Press (1986).

account by use of a power index. For chemical production facilities, this power index generally lies between 0.5 and 1.0, with 0.6 being often employed. For fermentation facilities, however, a value of 0.75 is recommended. The new capital costs can thus be estimated viz.

$$\text{new plant cost} = \frac{MSI_{present}}{MSI_{10yrs\,ago}}\left(\frac{\text{new capacity}}{\text{old capacity}}\right)^{0.75}(\text{old plant cost}) \qquad (8.1)$$

The MS index[12] shows an increase over a ten year per iod of around 1.8, thus the new plant cost will be $1.8 \times 2^{0.75}$, about 3 fold higher than the original cost. To obtain better estimates, we must develop a flow diagram for material and energy flows and calculate amounts and concentrations in each stream.

The next level of detail requires a list of all capital equipment, based on material and energy balances. The price of each piece of equipment is determined as the f.o.b. (free on board) price plus the cost of freight. From these costs the overall plant cost can be estimated. The sum of all the equipment costs can be multiplied by a factor which is characteristic of the materials (solids, liquids or gases) in the plant. One factor commonly used is the Lang factor, which was derived by considering the elements involved in installing the equipment, including labor, supervision and materials. For plants which involve liquids processing, the Lang factor is 5.69. This Lang factor method may overestimate the cost of capital in fermentation facilities, where the equipment is often of stainless steel and has a high capital cost relative to carbon steel. The installation costs of equipment of either material may be comparable, however. This approach can be improved by use of a "module" method developed by Guthrie[13]. Each piece of equipment is multiplied by a factor which is characteristic of that type of equipment. Directs costs (materials and labor) and indirect costs (engineering and construction) are obtained from the f.o.b. equipment cost by appropriate multipliers. The sum of the the indirect and direct costs is then the "bare module" cost. The sum of all the bare module costs for all pieces of equipment in the plant then serves as the basis for deriving the other elements of the overall plant fixed capital cost. A slightly simpler approach is to use a factor to directly obtain the bare module cost for equipment, rather than break it into the direct and indirect components. This factor is the bare module factor, or "bmf". We shall use this approach in subsequent examples and Appendix B provides values of bmfs for fermentation equipment. Table 8.13 illustrates how these bare module costs can be used to obtain the total capital investment.

(12) See the appendix for values of the Marshall and Swift and Chemical Engineering plant indices.

(13) Guthrie, K.M. Capital Cost Estimating, Chem. Eng. **76** 116 (1969).

Table 8.13. Use of bare module costs to determine the total capital investment.

Capital equipment costs	
Detailed list of all major equipment pieces	A (bare module costs)
Ancillary equipment (pumps, boilers, air compressors, cooling towers etc.)	B (bare module costs)
Subtotal	(A + B)
Contingency and fees	40% of (A + B)
Total Equipment	1.4 (A + B)
Buildings (15% of total equipment)	0.15 x {1.4(A + B)}
Land and site development (7% of total equipment)	0.07 x {1.4(A + B)}
Startup costs (10% of total equipment)	0.10 x {1.4(A + B)}
Fixed Capital Costs	C (sum of above elements)
Working capital (3 months raw materials and supplies, 1 month labor)	W
Total Capital Investment	C + W

8.3.1 Large-scale equipment and utilities

Utilities costs include equipment for steam generation, electricity supply, process and cooling water, chillers (e.g., brine), and waste treatment. If the plant is to be a new facility, these costs are included as capital equipment. If the facility is added to an existing plant, then these costs are usually treated as direct operating costs, and charged on a per unit cost basis (e.g., $15 per 1000 kg steam), which includes the depreciated capital costs of the equipment used to generate the utility.

The most accurate method to determine the costs of large-scale equipment is a vendor quotation. However, there are published sources of common equipment costs[14] in addition to charts and tables in texts. In extrapolating to larger equipment scales, the exponent method

(14) Hall, R.S., Matley, J., and K.J. McNaughton, Chem.Eng. **89** 80-116 (1982).

may be used, provided there is sufficient confidence in the value of the exponent for that piece of equipment, and the capacity range for scaling should generally not exceed a factor of 2 to 3.

8.3.2 Skid-mounted equipment

Some components of fermentation facilities are purchased as skid-mounted units, such as fermenters (up to 40,000 liters), filters, centrifuges and ion exchange and gel filtration units. These units have all the piping, instrumentation and controls in place, and their installation only requires transportation to the site, location on a pad or foundation, and attachment to utilities. Typically the skid-mounted cost thus includes charges otherwise made for direct material and labor costs. Installation costs of skid-mounted equipment may be estimated as 10 to 20% of the f.o.b modular unit cost.

8.4 Manufacturing Cost Estimates

8.4.1 Operating Costs

Raw materials

The importance of raw material costs depends on the scale of the fermentation process. For high-value therapeutic products, the raw materials may not comprise a significant part of the production costs, but selection of raw materials may nevertheless be important in maintaining a consistent product quality. The important raw materials for fermentation are the carbon source, the nitrogen source and to a lesser extent the sources of phosphorus and sulphur. Microorganisms can use a variety of carbon sources, from complex ones such as starch and wood hydrolyzates to refined sources such as glucose syrup and sucrose. Trace metal requirements need to be met with salt solutions. Complex nitrogen and carbon sources are required in some fermentations where the slow release of nutrients may be important in regulating the metabolism leading to the product. This may be especially important in antibiotic fermentations, where the metabolic pathway may be poorly understood and a simple, defined medium may not produce high levels of product.

Costs of raw materials can often be found in the *Chemical Marketing Reporter*, although for some specialty items vendor quotes are required. The cost of transportation must be considered in obtaining raw materials prices. Some typical sources of raw materials are summarized below[15].

The main source of carbohydrate in fermentations is glucose, which in the United States is mainly derived from corn. Approximately 67 lb of starch and 7.1 lb of cornsteep liquor are produced per 100 lb of shelled corn. When hydrolyzed, the glucose syrups so produced are characterized by their DE, or dextrose equivalent, the percentage of the solids that is present as glucose. Molasses is a by-product of sugar refining and its availability and price

(15) From Reisman, H.B., Economic Analysis of Fermentation Processes, CRC Press, (1988).

depend on local circumstances. It is produced at about 30% of the total sucrose production, but its price fluctuates. Brazil, Russia, and India produce about one third of the world molasses output. The fermentable sugar content of molasses may be taken as about 50% of the molasses by weight, i.e., one short ton (2,000 lb) of molasses contains 1000 lb of fermentable sugar. At a price of $100/short ton, this would correspond to $0.10/lb. Corn-derived glucose costs range from $0.12 to $0.25 per lb.

Raw material	*Source*
Carbon sources	carbohydrates, hydrocarbons, alcohols, proteins, fats, oils, fatty acids, organic acids
Nitrogen sources	ammonia, nitrate, nitrite, amino acids, proteins, urea
Sulphur	sulphate, sulphide, amino acids (cysteine, methionine), proteins
Phosphorous	phosphate
Carbohydrate sources	glucose syrup, dextrose, sucrose, high DE corn syrup, enzose, starch, lactose, cane molasses, beet molasses, whey, corn meal, sulfite waste liquor, wood hydrolyzates, n-parafins
Oils and Alcohols	soybean oil, corn oil, methanol, ethanol
Nitrogen sources	cornsteep liquor, distillers dried solubles, cottonseed flour, soybean meal, brewers' yeast, casein, meat or milk peptone, yeast autolysate, ammonium hydroxide, amines, yeast extract, fish meal, peanut meal, urea, nitrate or nitrite

The largest source of protein is soy (~80% of the world demand for protein cake), and soy oil comprises 25% of all vegetable oils. Soybean meal, fish meal, cornsteep liquor and distillers' dried solubles are all used as complex nitrogen sources. Anhydrous ammonia may be directly sparged into a fermentation or supplied as ammonium hydroxide or an ammonium salt.

In addition to the medium components, raw materials for the recovery process must be included. These may be acids and bases, ion exchange resins, membranes etc. Other supplies include those related to personnel and record keeping. These can be estimated as 3-5% of salaries and wages. Laboratory supplies can be estimated at 20-40% of associated personnel salaries. In some high-value therapeutic products, the cost of analytical tests required to meet FDA requirements may be extremely high, up to 40 to 50% of the final product cost. Such costs should be estimated based on the actual test required.

Labor

The estimation of labor requirements is difficult in fermentation plants, particularly small capacity operations where the correlations developed for large petrochemical plants are no longer valid. The simplest approach is to prepare a table indicating the number of workers per shift for each major piece of equipment and base the labor requirements on this table. Most fermentation facilities are operated on a 21 shifts, 7 days per week basis. Four operators can cover 20 shifts, with a rotating overtime assignment. Labor costs can be a very large fraction of the operating costs of a small capacity plant.

In downstream operations, one operator may cover several operations, particularly if the equipment is well automated. Supervision of the plant depends on its complexity; one supervisor might be able to cover a plant on a single shift if the operation is relatively straightforward.

Beyond the direct labor, plant personnel associated with payroll, maintenance, accounting, etc., are lumped as plant overhead. This is estimated as 50-70% of the direct labor costs. A more accurate approach would be to detail the fringe benefits, overtime and accounting requirements for plant labor, if data are available.

Table 8.14. *Operator requirements for fermentation and product recovery. [From H. B. Reisman, Economic Analysis of Fermentation Processes, CRC Press, (1988)].*

Plant Operation	*Operators per shift*	
Fermenter	0.2-0.3	
Centrifuge	0.1-0.2	
Rotary vacuum filter	0.1-0.2	
Evaporator	0.1	
Reactor, batch	0.5	(also for media make up)
Reactor, continuous	0.2	
Crystallizer	0.1-0.2	
Membrane system	0.2	
Adsorption operations	0.2	
Dryer		
spray	1.0	
drum	0.5	
tray	0.4	
rotary	0.4	
Boiler room	1	(includes other utilities)

Utilities

In all plants there is a base utility loading required for interior heating and lighting, sanitation, communications, etc. If the plant throughput is low, this base loading can be an important component of the utility costs. In larger capacity plants, steam, water, electricity and waste treatment costs are significantly higher. Utility requirements are estimated from the plant material and energy balances; the actual costs depend on the geographical location of the plant as utility rates vary widely.

Steam is not generally purchased directly, but is generated on site from some other energy source. The cost of the steam generation equipment may be included in the overall capital cost estimate for the process. Steam generation costs then include water treatment costs, fuel, labor and steam distribution costs. These lumped costs are assigned as an overall steam cost to the process. These costs are typically \$13-\$20 per 1000 kg steam. Fermentation facilities require both low and high pressure steam. Low pressure steam (30-50 psig) is used for steam sterilization of seals, vessels, and piping. Higher pressure steam is used in evaporators and sterilizers. This steam is generated at high pressures, ranging from 200 to 500 psig. Steam requirements for sterilization in batch processes is around 0.2 to 0.4 kg/liter broth, and in continuous sterilizers this is reduced by up to 75%. Sterilization costs are not usually a large component of utility costs in fermentation, as the electrical requirements for aeration and agitation are usually much larger than this.

Electrical power consumption costs are usually the largest component of the overall utility costs for fermentation processes. Aeration, agitation and pumping electrical requirements are the most significant elements of this cost. Agitation of highly aerobic fermentations requires power inputs from 1 to 2 HP/100 gal (1 to 3 kW/1000 liters). For a 100-hour fermentation, at \$0.05 per kW-hr, the electrical cost for agitation is thus ~\$10/1000 liters. Aeration requirements range from 0.5 to 2.0 vvm. Air compressor costs at 1.0 vvm thus are ~5 kW/1000 liters of broth, for 100 psia delivery air compressors. For a 100-hour fermentation, the aeration costs are thus ~\$25/1000 liters. The total electrical costs are thus quite high, about \$35/1000 liters. Electrical costs for pumping are more difficult to estimate, as they depend strongly on the process considered. Most fermentations involve large water volumes and pumping can be an important consideration. Generally, 10 to 20 kW-hr /1000 liters can be assumed for pumping costs.

Although water is used in large quantities in fermentations, primarily for cooling of the fermenter, it is not a large cost relative to the other utilities. Process water and cooling water are generally obtained from the same source, the municipal water supply. Unless there is an on-site source of cooling water, for example a lake or well, cooling water may need to be recycled using cooling water towers or a closed cooling system. The cost of water may be as high as \$1/1000 liters. Costs of \$0.005 to \$0.05/1000 liters for well or river water are significantly lower. High purity water obtained by reverse osmosis is \$1 to \$5/1000 liters, and that obtained by distillation is \$15 to \$25 /1000 liters.

Waste treatment costs can be an important component of the overall process costs. It may be considered as a utility cost or as a raw material cost. The biomass from fermentation processes can present disposal problems, or may be a valuable by-product that can be sold (e.g., distillers' dried grains). Fermentations produce waste streams with high BOD and COD levels. If a batch is contaminated, it is generally sent to the effluent treatment system. Cleaning of fermenters also produces considerable liquid wastes. In large plants, on-site waste treatment facilities may be constructed, particularly if the municipal facilities are unable to handle the load imposed. The costs of such on-site facilities may be from 10 to 20% of the total plant cost. Typical waste treatment costs are

landfill	2-5 cents/kg
wet oxidation	\$10 to \$20/m^3 (50,000 to 100,000 ppm COD)
biological oxidation	10 to 20 cents/m^3 (500 to 1,000 ppm BOD)

The composition of fermentation wastes is illustrated in Table 8.15. The BOD depends on the nature of the fermentation plant. For fermentation plants producing antibiotics, Table 8.16 presents some typical characteristics.

Table 8.15. *Wastes from fermentation plants*

Fermentation Waste	*Solids (%)*	*BOD (ppm)*
Brewery press liquor	3	10-25,000
Yeast plant	1-3	7-14,000
Industrial alcohol	5	22,000
Distillery slops	4.5-6	15-20,000

Table 8.16. *Characteristics of some antibiotic wastes. [From Industrial Water Pollution, N.L. Nemereow, Addison-Wesley, (1978)].*

Characteristic	*General antibiotic waste*	*Terramycin*	*Penicillin*
BOD (ppm)	1,500-1,900	20,000	8,000-13,000
Suspended solids (ppm)	500-1,000	10	
pH	1-11	9.3	2-4

8.4.2 Fixed Costs and Overhead

The elements of fixed costs and overhead are the depreciation of the plant capital equipment and the associated buildings, local and federal taxes and other expenses such as insurance and rent. Generally the largest component of these is the depreciation on the capital. The depreciation is based on the life of the plant. The useful life of a plant is generally determined by the applicable tax codes, as is the method of depreciation (straight line, declining balance, etc.). A guideline is a life of 5 to 15 years for process equipment and 30 years for buildings. Land costs are not depreciated.

If the capital required for the construction of the plant is borrowed, the interest costs are considered as part of the fixed costs. Thus the charge would include depreciation and interest. An alternative approach is to use a technique for capital recovery costs in place of the depreciation and interest costs. An annual charge is made which would recover the cost of equipment and buildings at a specified interest rate. Thus the plant is made responsible for repayment of all the capital charges associated with its construction. This type of "break-even" analysis and other discounted cash flow analyses are described in various texts on engineering economics. In the examples we shall encounter, the simple depreciation method will be employed.

Taxes may be paid to local and state authorities; these are based on the value of the plant and the buildings. Property taxes may vary from 1 to 5% of the fixed capital investment. Insurance costs are variable, but a figure of 1 to 3% is representative. If the site and buildings are not purchased, but are rented, these costs are considered as part of the fixed costs. If rental values can be reasonably estimated, they should be used. Otherwise the depreciation of the capital costs provides a more realistic approach.

Plant overhead is a large part of the salaries and wages of direct labor and supervision (50 to 70% of these costs). This overhead charge accounts for personnel, medical, janitorial, cafeteria and related costs that cannot be directly assigned to the product manufacturing cost.

8.5 Case Studies

In the following sections we shall illustrate the procedures for developing a cost estimate of several types of fermentation processes. In the first case, we will determine the manufacturing costs for production of penicillin in an existing facility. The production of penicillin is well-established and the techniques for penicillin recovery are characteristic of many antibiotic fermentations. The example provides some insight into the major cost elements. A more complex process is the fermentation and recovery of an intracellular protein, such as those produced by recombinant organisms. Here the product must be recovered by lysis of the cells. Finally we shall examine various approaches for the production of the amino acid phenylalanine, and provide rough estimates as a guide for selection of the final process.

8.5.1 Penicillin

We consider the costs for production of non-sterile bulk potassium penicillin G from *P. chrysogenum*, using an existing fermentation plant. The design assumptions, fermentation yield and fungal growth parameters for this example are taken from Swartz[16]. The design presented here is not as detailed as that in the following section; our objectives are to illustrate the major components of the cost of a well-established process.

Commercial production of penicillin makes use of a fed-batch process, in which the starting fermentation volume is supplemented by additional medium later in the growth cycle of the *Penicillium*. The fermenter volume increases and broth withdrawals are made.

---

(16) R. Swartz, *The use of economic analysis of penicillin manufacturing costs in establishing priorities for fermentation process improvement*, Ann. Rept. Ferment. Processes, **3**, 75-110 (1979) and Queener, S. and R. Swartz, *Penicillins; biosynthetic and semisynthetic*, Economic Microbiology, **3**, pp 35-123 Academic Press (1979).

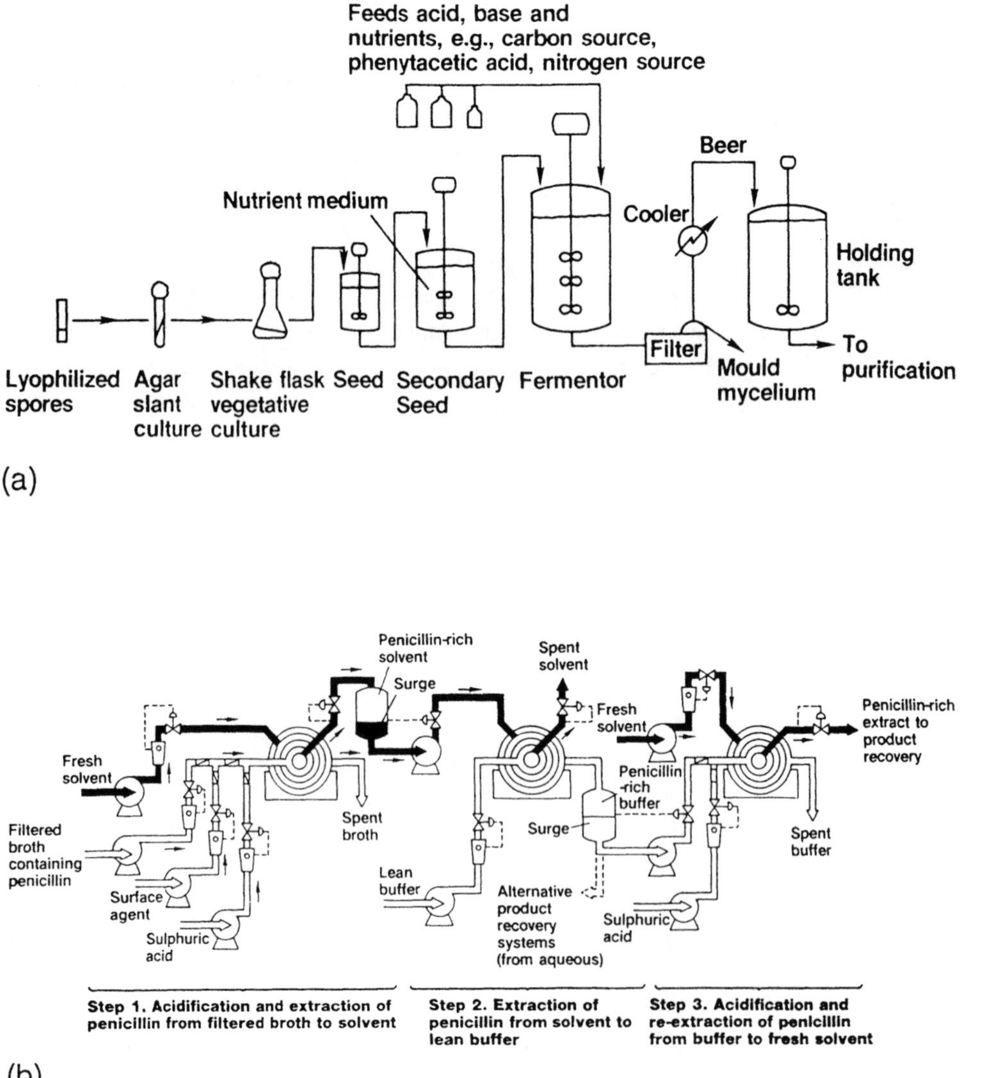

Figure 8.7. *(a) Penicillin fermentation flow sheet, illustrating the inoculum stages. (b) The recovery train for penicillin.*

This design assumes that 50,000 gallon vessels are available and the initial volume of medium is 40,000 gallons in each vessel. At harvest, the total volume sent to the recovery train is 55,000 gallons per vessel. The process flow sheet is given in Figure 8.7. The steps involved in developing the cost estimate and the assumptions are described in the following sections.

(1) Penicillin yield and medium formulation

The yield of potassium penicillin G from the carbon source is an important factor in the overall process economics. Various carbon substrates have been employed for penicillin production, including glucose, lactose, sucrose, ethanol, and vegetable oils. Typical yields for biomass production and for synthesis of penicillin from glucose are:

$Y_{P/S}$ = 1.13 gm/gm[17] (based on the potassium salt of penicillin G; MWt 372.5)

$Y_{X/S}$ = 0.45 gm/gm

We can now determine the glucose requirement for a typical fermentation. Assume that a typical final cell concentration is 35 gm/l and the penicillin concentration at harvest is 30 gm/l. The maintenance coefficient for *Penicillium* is 0.027 gm glucose/(gm cells-hr) and the fermentation is assumed to have a 25 hour growth phase and a 180 hour synthetic phase, during which penicillin is formed. The total glucose required per liter of final broth is

glucose for cells	35 (gm/l) x (1/0.45) = 77.8 gm
glucose for maintenance	0.027 (gm/gm.hr) x (35 gm/l) x 180 hr = 170 gm
glucose for penicillin	30 (gm/l) / 1.13 (gm/gm)= 26.5 gm
total requirement per liter of broth	274.3 gm
effective overall yield (gm pen G per gm glucose)	*30/274.3 = 0.109 gm/gm*

Based on this yield estimate, the raw material components can be specified, and the overall raw materials cost for penicillin production estimated based on the total volume of broth processed. The costs of raw materials are based on 1979 prices.

(2) Fermentation

We assume that the plant has five 50,000 gallon vessels, with a working volume of 40,000 to 45,000 gallons. The agitators are 700 HP each (1.5 HP/100 gallon) and the vessels are 304 stainless steel. Aeration is provided by three compressors at 10,500 ft^3/min, based on an aeration requirement of 0.82 vvm. One of the compressors is on standby. The fermentation vessels are supplied by three inoculum tanks, and there are tanks for supply of phenylacetic acid[18], a precursor for penicillin G, and tanks for supply of acid and base for pH control.

The cycle time is 205 hours (25 hours growth and 180 hours of production) and a 10 hour turnaround time. If the plant is on-line for 8,000 hours per year, the annual broth produced is thus 10.23 x 10^6 gallons.

---

(17) From Cooney, C. and F. Acevedo, Biotech. Bioeng. **19**, 1449 (1977), assuming that amino adipate is recycled.

(18) See Chapter 3, Section 3.2.7 for the general structure of penicillins, and that of penicillin G.

Raw Material Requirements

Component	*Amount (gm/liter)*	*Unit cost ($/Kg)*	*Cost ($/liter)*
glucose	274.3	0.22	0.0603
corn steep liquor	44	0.11	0.0048
$(NH_4)_2SO_4$	10.7	0.25	0.0027
Na phenylacetate (80% incorporated)	15.9	2.95	0.0469
KH_2PO_4	2.74	0.985	0.0027
other			0.0119
Total			*0.1294 ($/liter)* *496.76 ($/1000 gallon)*

(3) Product recovery

The overall recovery of penicillin from the fermentation broth is typically around 85%. The recovery train is illustrated in Figure 8.7 (b), and consists of belt filtration, extraction in Podbielniak extractors, crystallization and vacuum drying. The costs of the purification train are fairly independent of the concentration of penicillin in the broth; they depend on the volume of broth handled. The recovery train is able to handle 1.6 times the fermentation broth volume. Raw material use includes surfactant, sulphuric acid, buffer and solvent make-up (e.g. some methyl iso-butyl ketone used as the extractant solvent may be lost in the aqueous phase).

(4) Utility Costs

The major utility costs are those for steam for sterilization of the broth and the electrical costs for agitation. Waste treatment costs are based on ~1,000 tons dry material produced per year, treated by landfill, and primary treatment of liquid wastes. The direct production costs are summarized in Table 8.17, with the utility rates as indicated.

It can be seen from Table 8.17 that the fermentation raw materials comprise almost half of the direct production costs, and that electrical costs account for the majority of the utilities costs. We can estimate the overall production costs by adding fixed charges for capital depreciation and for plant overhead (costs of sales etc.). The cost of the penicillin produced, assuming 85% recovery is thus $15.96 per kg. These cost details are provided in Table 8.18.

Table 8.17. *Direct production costs for penicillin, based on 1,000 gallon of fermentation broth.*

Item	*Cost per 1,000 gallons ($)* Fermentation	*Recovery*	Total
Raw materials	496.76	32.99	529.75
Operating labor at $10/hr	27.45	52.45	79.90
Direct supervision at $15/hr	15.00	15.00	30.00
Maintenance	116.26	35.91	152.17
Laboratory	1.27	2.02	3.29
Utilities			
Steam at $3 /1,000 lb	17.75	28.83	46.58
Electricity at 3.75 cents/Kw-hr	127.77	18.18	145.95
Water at $0.47/1,000 gallon	2.43	0.92	3.35
Waste treatment at $250,000/yr	11.00	11.00	22.00
Supplies	9.66	9.66	19.32
Total	*825.35*	*206.96*	*$1,032.31*

Table 8.18. *Manufacturing costs for penicillin, based on 1,000 gallon of fermentation broth.*

Item	*Fermentation*	*Cost per 1,000 gallons ($)* Recovery	*Total*
Direct production costs	825.35	206.96	1,032.31
Fixed charges	351.80	90.16	378.00
Plant overhead	103.89	46.00	131.00
Total	*1,281.04*	*343.12*	*$1,541.31*

8.5.2 Recombinant Proteins

We consider here a batch process to manufacture an intracellular protein of molecular weight ca 40,000 Dalton, produced in a recombinant yeast. This example is based on the process analyzed by Kalk and Langlykke[19]. The target production capacity is 5,000 kg per year. The fermentation parameters typical for yeast growth are as follows: a final dry cell weight of 40 gm/l is assumed, and a fermentation time of 24 hours with a turn-around time of 6 hours. The target protein concentration is 10% of the total intracellular protein content. Based on a protein content of 40% of the dry cell weight, this corresponds to a target protein

(19) J. Kalk and A. Langlykke, *Cost Estimation for Biotechnology Projects*, ASM Manual of Industrial Microbiology and Biotechnology (1986).

concentration of 1.6 gm/l. It is assumed that the recovery efficiency is 40%, as efficiencies higher than 50% for intracellular proteins are uncommon. The cell yield coefficient based on glucose as the carbohydrate source is 0.4 gm DCW/gm glucose. An inoculum volume ratio of 5% is assumed for all stages.

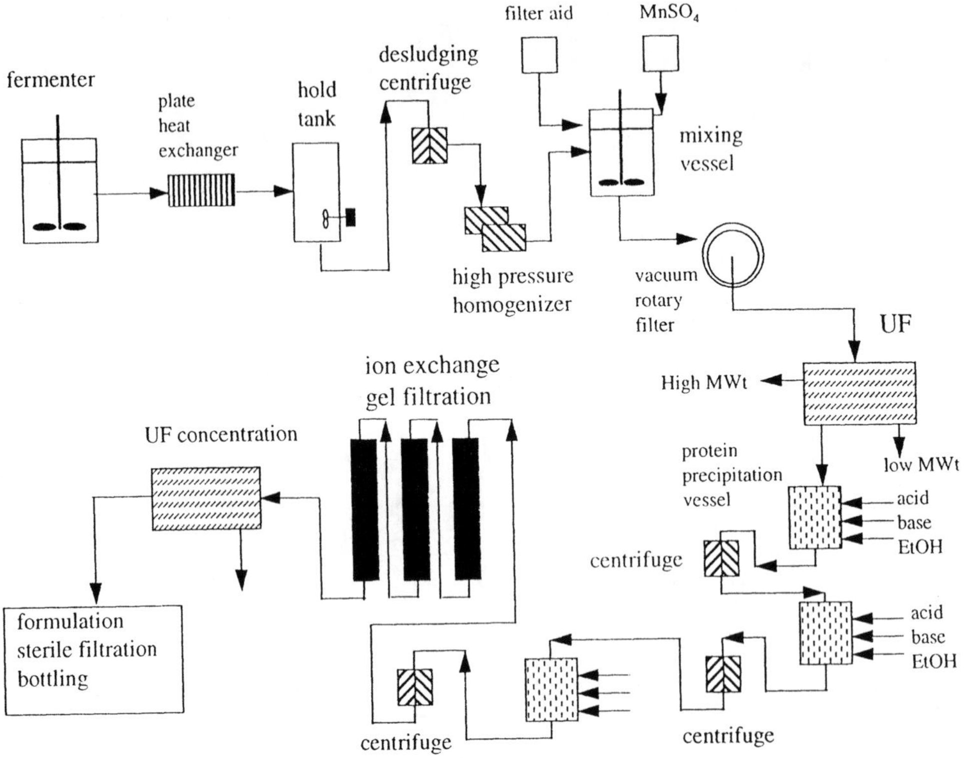

Figure 8.8. *Flow scheme for the production and purification of the target intracellular protein from recombinant yeast.*

The recovery train is illustrated in Figure 8.8. The fermenter broth is cooled to 5°C by a plate heat exchanger and held in a holding vessel. The cells are concentrated in a high-speed disk centrifuge and then homogenized in high-pressure homogenizers to disrupt the yeast cells and release the protein. Nucleic acids are precipitated by addition of $MnSO_4$ (0.05M) in a mixing vessel. Cell debris and precipitated nucleic acids are removed by addition of filter aid and passage of the slurry through a vacuum rotary drum filter. The filtrate contains a dilute (~1 wt%) solution of intracellular proteins. This solution is then fractionated by ultrafiltration. Very large molecular weight proteins and cellular debris are removed in this step. The UF step allows passage of materials of molecular weight below ~100,000, and retains higher molecular weight material. A second UF step is employed to remove low molecular weight materials (30 kD cut-off). This concentrates the protein solution to around 5 wt% and the solution is then precipitated to remove undesired proteins. A three-stage selective precipitation is used to separate the target protein from other proteins of comparable molecular weight. By varying the pH and using ethanol, these steps can be quite selective. Ion exchange chromatography is then employed to purify the target material. Based on the flowsheet and assumptions about the fermentation and recovery, the equipment can now be sized. For this example the MS index is taken as 900 (1992).

(a) Fermentation area

The steps involved in determining the size of the fermentation equipment are outlined below.

1. required annual target protein production = 5000 kg ÷ 0.4 recovery factor = 12,500 kg/yr
2. total protein to be produced = 12,500 kg ÷ 0.1 fraction target protein = 125,000 kg/yr
3. total yeast to be produced (DCW) = 125,000 kg ÷ 0.4 frac yeast DCW = 312,500 kg/yr
4. total volume of broth = 312,500 kg ÷ 0.04 kg DCW/liter = 7,812,500 liters
5. fermenter cycles per year = 350 days x 24 hours ÷ 30 hour cycle = 280 fermenter cycles/yr
6. assume 1 fermenter
   working volume = 7,812,500 liters ÷ 280 = 27,902 liters
   actual volume = 27,902 ÷ 0.7 (working/actual volume) = 39,860 liters
   ≅ 40,000 liters
7. seed fermenter volume = 40,000 liters x 0.05 volume ratio = 2,000 liters (assume 2 required)
8. pre-seed fermenter volume = 2,000 liters x 0.05 volume ratio = 100 liters (assume 2 required)
9. inoculum development fermenters = 100 liters x 0.05 = 5 liters
10. fermenter costs[20]
   production fermenter cost ($) = (MSI/800) 91,228 $V^{0.35}$ = 373,251
   seed fermenter cost ($) = (MSI/800) 91,228 $V^{0.35}$ = 130,810
   pre-seed fermenters ($) = (MSI/800) 90,000 $V^{0.21}$ = 62,430
   inoculum development($) = (MSI/800) 90,000 $V^{0.21}$ = 44,525

(b) Cell recovery

The cells are harvested over a 4 hour period and chilled via a plate heat exchanger from 30°C to 5°C. The area of the exchanger is obtained from an energy balance, assuming an overall heat transfer coefficient of 150 BTU-$hr^{-1}ft^{-2}$-$°F^{-1}$. Chilled brine is available at 40°F. The log mean temperature difference, ΔT_{LM}, is based on a 10°C permissible increase in the chilled brine temperature. Hence ΔT_{LM} is 7.7°F.

(20) See Appendix B for selected equipment costs.

$Q = mc_p\Delta T$

Q = 27,902 liters (1/3.79 gal/liter) (8.345 lb/gal) (1 BTU/lb-°F) (30-5)(9/5 °F)÷(4 hours)

= 691,155 BTU/hr

$A = Q \div (U \times \Delta T_{LM})$

$\cong 600\ ft^2$

Cost of the heat exchanger is

cost (\$) = 105 (MSI/300) $(ft^2)^{0.62}$ = 105 (MSI/300) $(600)^{0.62}$

= 16,625

Chilled broth is fed to a holding tank, which has capacity to store the fermentation batch. The volume is 40,000 liters, and the cost is obtained from the correlation

cost (\$) = (MSI/800) 3,000 $(40\ meter^3)^{1.0}$

= 135,000

(c) Cell harvesting and disruption

Assuming that the cells are fed from the holding tank to a centrifuge over an 8 hour period, the flow to the continuous centrifuge will be 27,902 ÷ 8 = 3,488 liters/hr. This requirement can be met by two 2,000 liter/hr continuous disk centrifuges. We modify the capacity for yeast by using a hydraulic-to-functional capacity ratio of 2:1. Thus the actual capacity will be 4,000 liters/hr. The cost of each centrifuge is obtained from the correlation below, and the bmf factor is 2.6.

cost (\$) = 65 (MSI/800) $(liters/hr)^{0.73}$ = 65 (MSI/800) $(4,000)^{0.73}$

= 31,116

The slurry from the centrifuges is ~50% (wet weight) and passes twice through a high-pressure homogenizer; each pass is of 4-hours duration.

1. dry weight of cells = 40 gm/l DCW
2. weight of cells from centrifuge = 50% wet weight = 500 gm/l wet weight = 100 gm/l DCW (assuming cells contain 20% dry matter and 80% water)
3. cell cream flow rate = 27,902 liters x (40 ÷ 100) ÷ 4 hours = 2,790 liters/hr

Thus two 1,500 liter/hr homogenizers are required, at a cost given by

f.o.b. cost (\$) = 949 (MSI/800) $(liters/hr)^{0.50}$ = 949 (MSI/800) $(1,500)^{0.5}$

= 41,349

The bmf for these homogenizers is 2.06.

(d) Nucleic acid and cell debris removal

The volume of homogenizate is 11,160 liters. The solution is made up to 0.05 M $MgSO_4$ by addition of a solution of $MgSO_4$ from a holding tank, with a volume of 5,000 liters. The mixing tank is 20,000 liters and the filter-aid slurry tank is 2,000 liters. The cost of these tanks can be determined by the correlations for agitated holding tanks with cooling coils:

f.o.b. cost (\$'s) = (MSI/800) x 1,500 x $V^{1.0}$

After precipitation of nucleic acids, the solution is filtered in a vacuum rotary drum filter. The filtrate will contain about 1 wt% protein content. The area of the drum filter can be estimated from an approximation of the filtration rate for a flocculated protein solution with precoat. Values range from ~50 to 500 liters/m^2-hr (here assumed to be 100 liters/m^2-hr), and we assume that the filtration will be complete in 4 hours. The volume to be treated is 11,160 liters plus the additional volume due to the addition of filter aid and $MnSO_4$. Assume this is 15,000 liters. The area required is thus:

filter area = (15,000¸4 liters/hr) ÷ (100 liters/m^2-hr) = 37.5 m^2 = 404 ft^2

The cost is obtained from the correlation below, using a bmf of 2.4 to estimate the installed cost.

f.o.b. cost (\$) = 1,620 (MSI/300) (0.25 x area in $ft^2)^{0.63}$ = 88,993

(e) Ultrafiltration

We assume that the protein is concentrated 50-fold in the ultrafiltration step, which operates over a four hour period. For removal of high molecular weight debris (cut-off ~ 100,000 Dalton), we assume a transmembrane flux of 25 liters/m^2-hr. Neglecting the retentate volume will introduce a small error, but we can assume that the flow rate required is ~ 12,500 liters ÷ 4 hours = 3,125 liters/hr. The UF area required is thus 3,125 liters/hr ÷ 25 liters/m^2-hr = 125 m^2. The cost of the UF module (bmf 1.42) is

f.o.b. cost (\$) = 1,468 (MSI/800) (area in $m^2)^{0.89}$ = 1,468 (MSI/800) $(125)^{0.89}$ = \$121,375

The low molecular weight material is removed in a second ultrafilter, with a molecular weight cut-off of 30,000 D. The transmembrane flux for a 50-fold concentration is assumed to by 50 liters/m^2-hr, giving an area of 75 m^2. The retentate contains the target protein, concentrated fivefold. The total protein content is about 5 wt %, and the volume is 2,500 liters. The cost of this unit is \$77,034.

(f) Protein precipitation

The approximate volume of permeate from which proteins other than the target protein are to be precipitated is 2,500 liters. Three stages are employed, the first two involve variation of the pH and the third is an ethanol precipitation step. The holding time in each vessel is ~ 1 hour, to allow for growth of the precipitate to a point where it is readily separated from the liquid by a desludging disk centrifuge. Three agitated holding tanks of volume 5,000 liters each are employed. Costs are determined from the correlations employed previously.

One desludging centrifuge is required to remove the protein precipitate. Because precipitation is relatively rapid, the same unit will be employed for each precipitation stage. The liquid flow rate can be calculated over a five hour period at 500 liters/hr, to give a hydraulic flow of 1,000 liters/hr for centrifuge design. The cost of the centrifuge is determined from the correlation

$$\text{cost (\$)} = 65\ (\text{MSI}/800)\ (\text{liters/hr})^{0.73} = 11{,}325$$

Allowance must be made for holding and recovery of the ethanol used in the third precipitation step. Assuming precipitation occurs at ~ 30 vol% ethanol, the amount required is 750 liters per 30-hr fermentation cycle. A holding tank of 5,000 liters and an ethanol distillation capacity of 1,000 liters per day can accommodate these requirements. The target protein is now present in a 3,250 liter 30% ethanol solution, and the protein is then separated by an ion exchange step.

(g) Ion exchange and gel filtration

Assuming that losses of the target protein are small in the final stages, we consider the protein solution to contain 45 kg of the target material in 3,250 liters, i.e. 13.85 gm/liter. During the ion exchange process, the protein will be diluted by the eluent to around 1 gm/liter. To determine the amounts of ion exchange material required, we assume a capacity of 15 gm/100 ml for the cation exchanger.

$$\text{volume of resin} = 45{,}000\ \text{gm} \div 15\ \text{gm}/100\ \text{ml} = 300{,}000\ \text{ml} = 300\ \text{liters}$$

We assume that the first step is a cation exchange column and that the purity of the product is about 25% at this step. Thus 4 x 300 = 1,200 liters of resin are required. The second anion exchange step requires 300 liters of resin. The ion exchange system is purchased as a skid mounted unit, with complete instrumentation and controls, and generally includes holding tanks for eluents. We estimate the installed cost of these units to be $250,000 each. Final purification is accomplished in a gel filtration system. This is similar in cost to each ion exchange unit.

Table 8.19 lists the equipment in the plant, based on the descriptions above. The labor costs for the process are illustrated in Table 8.20. In relatively small plants, labor costs can be a significant fraction of the manufacturing costs. Tables 8.21 and 8.22 provide details on the raw materials and supplies, utilities and a summary of the overall capital cots and manufacturing cost estimates. These were derived by estimating the amounts of raw materials required for a typical fermentation process and the specific details of the recovery process described. Utilities are based on the requirements of the individual pieces of capital equipment, with estimates for general load of the plant (air conditioning, lighting etc.). The manufacturing cost estimate includes amortization of the capital equipment over 10 years at 12%, and amortization of buildings over 30 years at 12%.

This approach is suitable for determining the economic feasibility of the process and for analyzing the effects of process changes on the manufacturing costs. For example, if the target protein can be produced at a higher intracellular concentration, the corresponding reduction in cost can be easily determined. If the protein can be secreted, the costs can be further reduced.

Table 8.19. *Installed Equipment List*

Equipment	Unit Cost	bmf	Cost ($)
Batching Area			
1 15,000 liter glucose syrup storage tank agitated, stainless steel, coils	43,902	1.97	86,487
1 5,000 liter batching tank stainless steel, load cells, agitator	23,992	1.97	47,264
5 1,000 liter blending tanks	28,125	1.97	55,406
Fermentation Area			
2 20 liter inoculum development fermenters	89,050	1.5	133,575
2 100 liter pre-seed fermenters stainless steel, agitators, instrumentation	124,860	1.5	187,290
2 2,000 liter seed fermenters (as above)	261,620	1.5	392,430
1 40,000 liter production fermenter stainless steel, agitator, instrumentation	373,251	1.5	559,877
Recovery			
1 Heat exchanger, 600 sq.ft.	16,625	3.4	56,525
1 40,000 liter chilled broth holding tank	75,295	1.97	148,332
2 Desludging centrifuges, 4,000 liters/h	62,232	2.6	161,803
2 1,500 liter/hr high pressure homogenizers	82,698	2.06	170,358
1 5,000 liter MnSO4 solution/holding tank stainless steel, agitated , insulated	13,632	1.97	26,855
1 20,000 liter mixing tank, stainless steel	29,221	1.97	57,565
1 2,000 liter filter aid slurry tank	8,235	1.97	16,224
1 404 sq.ft. vacuum rotary drum filter (stainless steel)	88,993	2.4	213,583
1 125 sq.m. ultrafiltration unit (100,000 D cutoff)	121,375	1.42	172,353
1 75 sq.m. ultrafiltration unit (30,000 D cutoff)	77,034	1.42	109,388
3 5,000 liter protein precipitation vessels stainless steel, agitated, insulated	40,896	1.97	80,564
1 desludging centrifuge, 1,000 liter/hr	11,325	2.6	29,445
1 5,000 liter ethanol receiving vessel (stainless steel)	13,632	1.97	26,855
1 1,000 liter/day 95% ethanol distillation unit			
2 Ion exchange systems (cation & anion)			500,000
1 Gel filtration system			250,000
1 Sterile filtration & bottling system			100,000
Ancillary Equipment			
75 Pumps and motors , various sizes	150,000	3.75	562,500
1 40 ton chiller (20F) with coolant vessel	70,000	1.42	99,400

Table 8.19 Continued.

1	Centrifugal air compressor, 100 psi, 30 cubic m/min	75,000	3.1	232,500
1	1,500 Kg/hr steam boiler, with water treatment			70,000
1	Cooling tower 125 gpm, 2HP fan	25,875	1.75	45,281
	Laboratory equipment			500,000
	Office equipment			100,000
	Subtotal			***$5,191,859***
	Contingency and fee (40%)			***$2,076,744***
	Total Cost			***$7,268,603***
	Amortized cost for 10 years at 12% interest			***$1,251,400***

Table 8.20. Salaries and Wages

		Number	Wages $/hr	Annual Cost
Salaries				
	Production manager	1		42,000
	Maintenance supervisor	1		38,000
	Shift supervisor	4		140,000
	Secretaries	4		60,000
			Subtotal	***$280,000***
Wages				
	Shipping & receiving	2	8	33,280
	Batching	4	8	66,560
	Fermentation	12	10	249,600
	Recovery	20	10	416,000
	Laboratory (QC)	8	9	149,760
	Maintenance	8	10	166,400
	Subtotal			***$1,081,600***
	Overtime (6%)			***64,896***
	Subtotal			***$1,426,496***
	Fringe benefits (35%)			***499,274***
	TOTAL			***$1,925,770***

Table 8.21. *Raw Materials and Supplies*

	Quantity	Units	Cost per unit ($)	Annual Cost ($)
Fermentation				
95 Dextrose corn syrup	1,275,000	kg	0.30	382,500
Casein hydrolysate	6,875	kg	7.00	48,125
ammonium sulphate	61,875	kg	0.52	32,175
K_2HPO_4	25,780	kg	1.25	32,225
$MgSO_4.2H_2O$	4,300	kg	0.75	3,225
NaCl	8,595	kg	0.10	860
$CaCO_3$	8,595	kg	0.02	189
Antifoam	8,595	kg	0.75	6,446
Miscellaneous (trace elements)				26,500
		Subtotal		***$532,245***
Recovery				
$MnSO_4.H_2O$	111,800	kg	1.5	167,700
Filter aid	82,800	kg	0.33	27,324
95% ethanol	35,000	liters	0.5	17,500
Ultrafilters 100 kD	125	m^2	200	25,000
Ultrafilters 30 kD	75	m^2	200	15,000
Cation resin	1,200	liters	205	246,000
Anion resin	300	liters	225	67,500
Gel media	225	liters	190	42,750
		Subtotal		***$608,774***
Supplies				
Operating	3% salaries & wages			22,963
Laboratory	40% lab. salaries			59,904
Maintenance	100% maintenance salaries			176,384
Other	10% plant overhead			48,144
		Subtotal		***1,448,414***
		Contingency (5%)		***72,421***
		TOTAL		***$1,520,835***

Table 8.22. Utilities

	Annual Consumption	Unit cost	Annual cost
Steam			
sterilization	2,578,125 kg	$15/1000 kg	38,672
solvent recovery	875,450 kg		13,132
contingency (25%)			12,951
			$64,755
Water			
process	12,500,000 liters	$0.50/1,000L	6,250
cooling	22,500,000 liters	$0.50/1,000L	11,250
high purity	8,000,000 liters	$2.00/1,000L	16,000
contingency (25%)			8,375
			$41,875
Electricity			
batching	7,700	$0.10/kW-hr	770
aeration	1,031,250		103,125
agitation	412,500		41,250
centrifuges	35,840		3,584
homogenizers	268,800		26,880
drum filters	134,400		13,440
agitators	67,200		6,720
pumps	140,000		14,000
chillers	336,000		33,600
cooling tower	12,900		1,290
general load (HVAC , lighting etc.)			30,000
		Subtotal	***$487,918***
Waste treatment			
landfill	2,064,500 kg	$0.03/kg	61,935
biological	10,000,000 liters	$0.20/1,000L	2,000
		Subtotal	***$63,935***
	Total Utilities		***$658,483***

Table 8.23. *Summary of Total Costs. Capital Investment and Manufacturing Cost Estimates.*

Capital Investment	***Cost***
Equipment	7,268,603
Buildings (15% of capital)	1,090,290
Land & site development (7%)	508,802
Start-up costs (10%)	726,860
Total fixed capital	*$9,594,556*
Working capital (3 months rate materials, 1 month labor)	*540,690*
Total capital investment	***$10,135,246***

Manufacturing Cost Estimate	
Raw materials and supplies	1,520,835
Salaries and wages	1,925,770
Utilities	658,483
Equipment (amortized 10 yrs, 12%)	1,251,400
Buildings (amortized 30 yrs, 12%)	134,578
Taxes and insurance (4% of fixed capital)	383,782
Plant overhead (25% salaries and wages)	481,442
Total annual cost	***$6,356,290***
Unit cost ($/kg protein) 5,000 kg/year	***$1,271***

8.5.3 Comparison of Alternatives for Phenylalanine Production

The methyl ester of the amino acid L-phenylalanine, together with aspartic acid, form the sweetener Aspartame, which has the structure below.

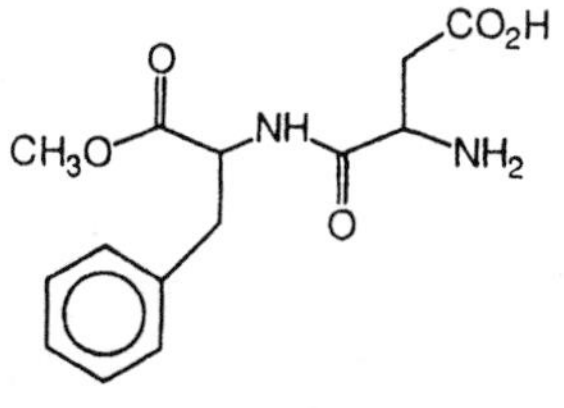

Phenylalanine may be produced by several routes, including batch fermentation based on glucose, batch fermentation using a precursor, and enzymatic synthesis. In determining which process to use in phenylalanine production, the type of cost estimation we have seen for the recombinant protein in the foregoing example would provide a means to assess each approach. The needs for research and development can be determined by examining the sensitivity of the manufacturing and capital costs to process changes. We shall examine and compare some of the alternatives for phenylalanine production.

(i) Batch fermentation with glucose feed.

Two approaches to the development of high-yielding microbial strains can be considered, the traditional route involving regulatory mutants, and the use of genetically-engineered organisms. A number of regulatory mutants have been described in the literature. For example, Choi and Tribe[21] describe an *E. coli* mutant (aro$^-$, phe^r, tyrR, aroR) which in batch fermentation produced 10.9 gm phe/liter, $Y_{P/S}$ of 0.21gm/gm, and a productivity of 0.35 gm phe/liter-hour. In continuous culture the final concentration was reduced to 8.7 gm/liter, but the productivity increased to 0.44 gm/liter-hour. In contrast, amplification of DAPH and chorismate mutase in *E. coli*[22] using the plasmid pBr322 (aroF, tryR, ampR, phe AF) resulted in a strain producing 2 gm/l phe, $Y_{P/S}$ of 0.11 gm/gm and a volumetric productivity of 0.07 gm/liter-hour. Better producers have been produced by combining auxotrophic and regulatory mutants to form a new strain. For example, a tyrosine auxotroph of *Brevibacterium lactofermentum*, PFP$^-$, 5MT$^-$ resistant, produced 25 gm phe/liter[23]. The key variables in the use of this approach are the final phenylalanine concentration that is obtained and the yield of phenylalanine on glucose. The advantages and disadvantages of this fermentation method are summarized below.

Advantages
- well-established technology
- inexpensive raw materials (eg. glucose, salts etc.)
- flexible plant design

Disadvantages
- high capital costs
- complex product recovery (cell removal necessary)
- dilute and contaminated product
- low volumetric productivity
- potential contamination and cell instability problems
- batch technology
- high labor costs
- low utilization of raw materials

(ii) Fed-batch fermentation with precursor addition.

A second approach to improving the productivity is the use of precursor feeding. This is attractive for some fermentations where it is difficult to establish a direct route from glucose to final product. For example, L-tryptophan and L-serine production are tightly regulated, and commercial routes employ anthranilic acid and glycine, respectively, as

(21) Choi Y. and D. Tribe, Biotech. Letters, 4 (4) 223-228 (1982).

(22) Forberg C. and L. Haggstrom, Applied Micro & Biotech 26 136-140 (1987).

(23) Goto, E., Ishiwara, M., Sakurai, S., Enei, H. and K. Takinami, Japanese Patent Appl. Kokai 56-64793 (1981).

precursors. Two potential precursors for phenylalanine are phenylpyruvate and acetamidocinnamic acid. Based on the transamination of phenylpyruvate by glutamate, a concentration of 7.5 gm phe/liter, a yield of 0.76 gm phe per gm precursor, and a volumetric productivity of 0.1 gm/liter-hour were obtained using *Corynebacterium glutamicum*[24]. The duration of this fermentation was 70 hours. Fairly similar results were reported[25] using *Alcaligenus faeslis* with acetamidocinnamic acid as the precursor. Some of the considerations involved in the use of precursors are summarized below.

Advantages
- improved yields on substrates
- natural microbial isolates are used (little strain reversion)
- flexible plant design
- able to use any existing fermentation capacity

Disadvantages
- low volumetric productivities
- expensive precursors
- expensive catalyst is used once
- difficult separation problems
- high capital costs
- high labor costs

Potential Improvements
- genetic amplification of required enzymes
- cell reuse through a maintenance reactor

(iii) Enzymatic Synthesis of Phenylalanine

Several enzymes can be employed for the enzymatic production of phenylalanine. These routes involve the transaminase reaction, or reactions catalyzed by phenylalanine dehydrogenase, phenylalanine ammonia lyase, hydantoinase, and phenylalanine acylase. Which approach is selected depends on the availability of the precursor, and the cost of the enzyme and its half-life under operating conditions. We will review each of these approaches.

(a) Transamination of phenylpyruvic acid

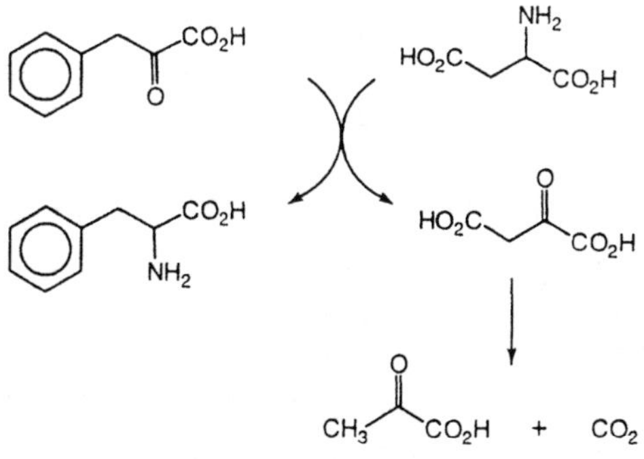

(24) Bulot, E. and C.L. Cooney, Biotech. Letters 7 93-98 (1985).

(25) Nakamichi, K., Nabe, K., Yamada, S., Tosa, T. & I. Chibata, Appl. Micro. Biotech. 19 100-105 (1986).

Reactor	*Phe* (*g/l*)	*Yield* (*g/g*)	*Productivy* (*gm/liter.hour*)	*Half-life* (*days*)	Reference
Immobilized cells (polyazetin)	30	90	1.65	80	Carlton et al. (Bio/technology **4**, 317-320 (1985)
Immobilized cells gelatin entrapped)	30	90	6.0	65	Walter, J.F. (W.R. Grace, ASM Meeting, Washington D.C. 1987)
Free enzyme	31	90	1.7	--	Rozzell, U.S. Patent 4,693,327
Immobilized cells (chitosan)	10	63	1.7	--	Ziehr et al. (Biotech. Bioeng. **29** 482-487 1987)

(b) Phenylalanine dehydrogenase/formate dehydrogenase

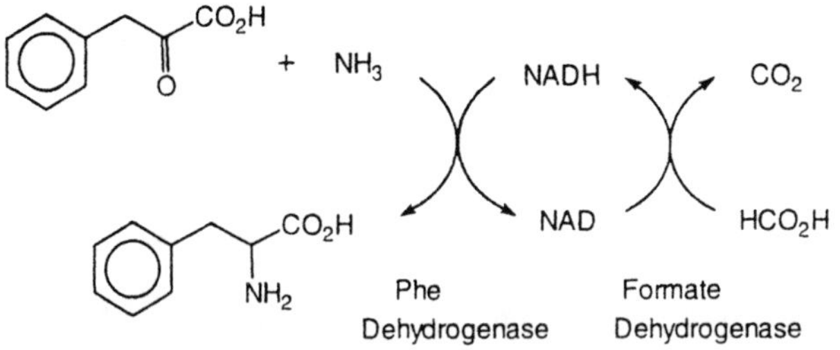

Information from Hummel et al.[26] for enzymes from Brevibacterium sp. in an enzyme membrane reactor with NAD recycling (the NAD being bound to PEG):

phenylalanine concentration	3.0 gm/l
yield	0.93 gm phe/gm phenylpyruvate
volumetric productivity	2.0 gm/liter-hour

The substrate for phenylalanine dehydrogenase in the above example (phenylpyruvic acid) can also be generated from D,L-phenyllactate by oxidation with D,L-hydroxyisocaproate dehydrogenase, which regenerates NADH from the NAD produced by the action of phenylalanine dehydrogenase[27]. The enzymes are present in both *Lactobacillus confusus* and *Rhodococcus*. The relevant production information is as follows:

phenylalanine concentration	5.4 gm/l
yield based on substrate	0.65 gm/gm
productivity	1.2 gm/liter-hour
half life of catalyst	>30 days

(26) Hummel et al., Applied Microbiol. and Biotech. **25** 175-185 (1985).

(27) Schmidt. Vasie-Racki and Wanding, Applied Micro. and Biotech. **26** 42-48 (1987).

(c) Phenylalanine ammonia lyase (PAL).

PAL catalyzes the conversion of phenylalanine to trans-cinnamic acid and ammonia. The reaction can be reversed to favor phenylalanine synthesis by a high concentration of ammonium ions. This route is attractive due its simplicity. The enzyme PAL is present in *Rhodotorula glutinis* and *Endomyces*. Either immobilized cells or cell suspensions can be employed. Typical conversions are

phenylalanine concentration	17-33 g/l
yield	0.71 gm phe/gm cinnamic acid
productivity	1-1.7 gm/liter-hour
cell catalyst half life	40- 80 hours

Higher concentrations and productivities can be obtained using immobilized cells.

(d) Production from D,L phenylalanine hydantoin with hydantoinase.

L-phenylalanine hydantoin can be converted to phenylalanine by the action of the hydantoinase found in *Flavobacerium aminogens*[28]. If a racemic mixture is employed, unconverted D-hydantoin can be racemized and recycled. Production conditions with this organism were reported as

phenylalanine concentration	50 gm/l
yield	1.00 gm phe/gm hydantoin
productivity	20 gm/liter-hour
catalyst half-life	12 days

The reaction is indicated below.

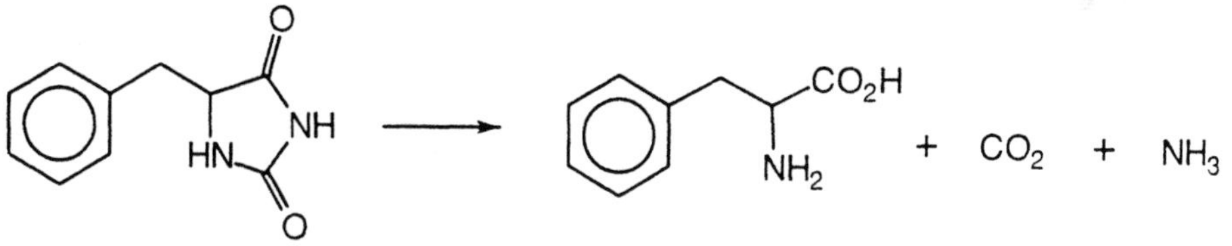

L-Phenylalanine hydantoin

In determining which of the above enzymatic routes might be appropriate, the following advantages and disadvantages of enzymatic synthesis must be considered.

Advantages	*Disadvantages*
• high volumetric productivity	• high raw material costs
• low contamination of the product	• potential of microbial contamination
	• low flexibility of plant design

(28) Yoozcki et al., Proc. Annual Meeting, Agric. Chem. Cos. Japan, p238 (1976).

Advantages	*Disadvantages*
• continuous operation	• relatively few current industrial applications
• low capital investment	• potential need for cofactors in some cases
• ease of automation	• expense of the catalyst
• reuse of catalyst	

In deciding which approach to consider for phenylalanine production, the fermentation route has the advantage that is a well-established commercial technology using inexpensive raw materials. In addition, the plant design is flexible. On the other hand, fermentation involves high labor costs and a fairly complex recovery scheme. Fed-batch processes with precursor feeds require expensive precursors. Enzymatic processes offer the advantages of high volumetric productivities and lower capital requirements. The product is produced in fairly pure form and recovery is straightforward. However, the catalyst, either in the form of suspended or immobilized cells, is typically expensive to produce.

When we consider one of the enzymatic routes to compare with fermentation, the use of phenylalanine ammonia lyase has advantages based on the absence of need for an expensive cofactor and the availability of the substrate at commercial scale. However, the enzyme has a relatively short half-life. The limited life of the cofactors required for the use of dehydrogenases requires development of techniques for their stabilization and retention in the system. The transaminase reaction yields a high product concentration and has a high product yield, but requires an expensive precursor and cofactor. A comparison between a fed-batch fermentation and an immobilized whole cell process based on the transaminase reaction is presented in Tables 8.24 and 8.25, using values relative to the fed-batch fermentation. As can be seen, the enzymatic approach may have an advantage, depending on the cost for which the precursor can be obtained.

Table 8.24. *Process comparison for phenylalanine production. Fed batch fermentation versus transamination. Conversion costs are relative to the fed-batch case. (Data from Walter, J.F., op. cit.)*

	Productivity (gm/l-hr)	$Y_{P/X}$ *(gm P/gm cell)*	*Conversion costs for reaction*	*Conversion costs for recovery*
Fed-batch	1.25	1.0	1.00	1.00
Whole cell immobilization	6.0	115	0.13	0.47

Table 8.25. *Comparison of the economics of fed-batch and enzyme processes. Costs are normalized to the fed batch case, and exclude the cost of the precursor.*

Cost element	*Fed batch fermentation*	*Whole cell immobilization*
Medium	45.2	--
Catalyst	--	1.3
Labor	6.0	3.0
Utilities	19.2	10.3
Equipment	14.0	5.8
Maintenance	10.0	5.5
Total relative cost	*100.0*	*25.9*

Table 8.26. *The economics of large scale fermentations.*

Product	*Selling Price ($/kg)*	*Volumetric productivity (gm/liter-hr)*	*Production costs ($/liter-day)*
Chlorotetracycline	50	0.075	0.09
Lysine	4.4	1.9	0.20
Monosodium glutamate	1.8	3.1	0.13
Itaconic acid	3.2	2.5	0.19
Lactic acid	2.7	2.0	0.13
Penicillin	20	0.5	0.24
Glucose isomerase	15	0.6	0.22
Threonine	16	0.5	0.18
Xanthan gum	10	0.625	0.15
		Average	***0.17***

An important factor in determining which approach to develop is thus the volumetric productivity of the process. If we examine a number of large scale fermentations, illustrated in Table 8.26 below, we see that the average cost of fermentation facilities is approximately $0.17 per liter per day. Thus, to produce phenylalanine by fermentation at a cost of approximately $22/kg, a volumetric productivity of ~ 0.3 gm/liter-hour is required. This is a useful guideline for estimating the costs of a known fermentation process (i.e., where the volumetric productivity is known in advance).

8.6 Appendices.

8.6.1 Selected cost indices

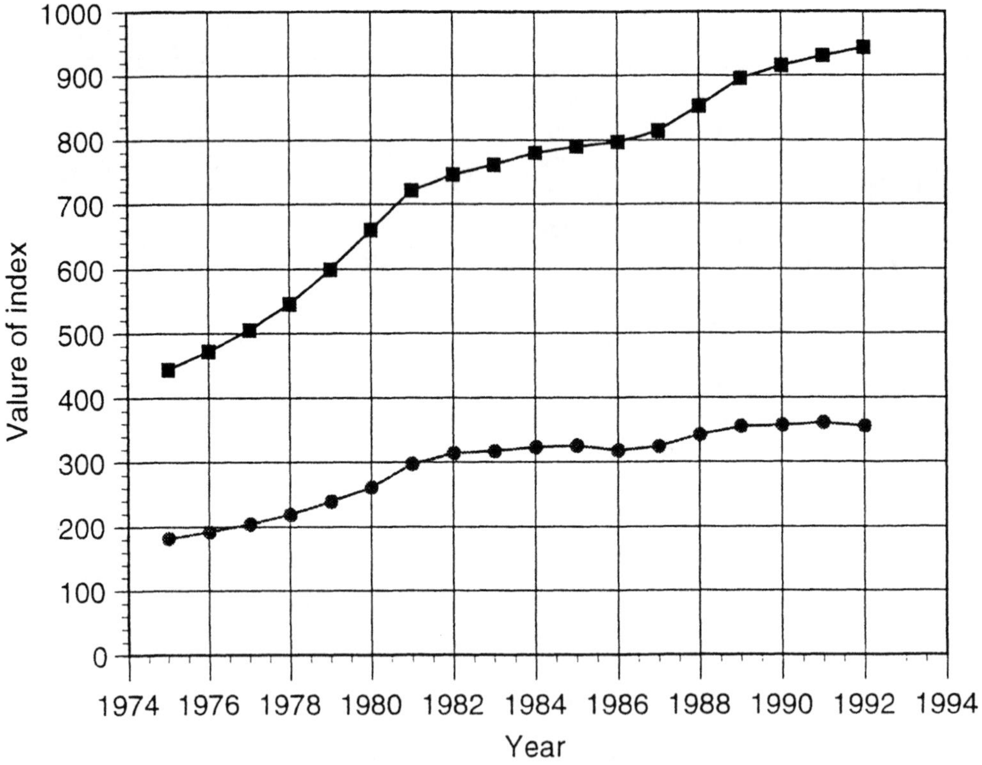

Figure 8.A1. *The Marshall and Swift index (squares) and the Chemical Engineering process plant index (circles) for updating plant equipment costs. The base indices are 100 in 1926 (MSI) and 100 in 1957 for the plant index.*

8.6.2 Selected Equipment Costs and Design Methods[29]

Fermenters

Seed vessels and fermenters can be priced based on volume. Prices include automatic controls of such process variables as pH, temperature, dissolved oxygen, etc. Fermenters are generally 316 stainless steel. Small units (1.0 to 45 m^3 volume) are typically skid mounted and include piping to connect to utilities directly. The direct costs to install such skid-

(29) Cost estimates taken from J. Happel and D.G. Jordan, Chemical Process Economics, Marcel Dekker (1975); and J. Kalk and A. Langlykke, Cost Estimation for Biotechnology Projects, ASM Manual of Industrial Microbiology and Biotechnology (1986).

mounted units is approximately 10-20% of the modular skid-mounted cost. This would include transportation, location on a foundation, and attachment of utilities. Larger fermenters must be field erected, as they cannot be conviently transported. The size of these units is 70 to 250 m^3. For field-erected units, the cost increases with a 0.7 power of size. Field-erected units do not carry a separate installation charge, but indirect costs must be added. For fermenters with additional controls and special requirements the base cost may increase by 50-100%. The bare module factor (bmf) for skid-mounted vessels can be determined as follows: for a vertical process vessel, the bmf is 3.18. With skid-mounted vessels, the price includes all direct labor and materials, so that the bmf becomes 2.12. This is adjusted for stainless steel by considering that the carbon steel total module cost would be (bmf) x (f.o.b. cost). If the installation and indirect costs for stainless steel are the same as those for carbon, and the material cost ratio is 2.25, then

fermenter module cost =f.o.b. cost (SS) + [{f.o.b. cost (SS)}/2.25] x (bmf - 1)
=f.o.b. cost (SS) + [{f.o.b. cost (SS)}/2.25] x (2.12 - 1)
=1.5 x (f.o.b. cost (SS))

Skid-mounted fermenters, basic ("no frills") units, volume V m^3

f.o.b.cost (\$) = (MSI/800) x 90,000 x $V^{0.21}$ for 0.2 to 1.0 m^3
= (MSI/800) x 91,228 x $V^{0.35}$ for 1.0 to 45 m^3

Field-errected fermenters

cost (\$) = (MSI/800) x 33,183 x $V^{0.74}$ for 50 to 250 m^3

Batching and Holding Tanks

Batching and holding tanks for fermentation broth or for other streams being processed may require chilling and generally have some mixing capabilities. They are typically stainless steel (e.g. 304SS). For insulated, agitated stainless steel mixing vessels, the following cost correlation is useful, but should be modified if the agitation requirements or materials of construction change. The bmf for carbon steel tanks is 3.18; for stainless steel this is corrected to 1.97, using the approach outlined for fermentation vessels above. V is in cubic meters.

f.o.b. cost (\$) = (MSI/800) x 8,800 x $V^{0.55}$

Centrifuges

Centrifuges are sized by their hydraulic throughput, but their performance will depend on the particulates being removed from the stream, e.g. bacteria, yeast or solids. The functioning capacity of disk centrifuges would need to be established. To provide a yeast paste, for example, the centrifuge may operate at only 50-100% of its rated hydraulic capacity. For hydraulic capacities between 4,000 and 70,000 liter/hour, the following correlation provides the uninstalled cost for disk centrifuges. The installed cost is estimated using a bmf of 1.6 for a stainless-steel solid-bowl centrifuge; for a dropping-bowl type, the factor is higher at 2.6.

$$\text{cost (\$)} = 65\ (\text{MSI}/800)\ (\text{liters/hr})^{0.73}$$

Ultrafilters

Various types of ultrafilters are available, including flat sheet membranes, spiral-wound and hollow fiber systems. System costs are based on the membrane area required for a given separation. This area is in turn determined by the permeate flux. The flux (e.g liters/hr.m^2) will decrease with time due to fouling of the membranes. Costs are based on the ultrafiltration unit without membranes and membrane costs are added separately. Membranes must generally be replaced annually, and this is often considered as an operating expense, i.e. as a raw material. Typical membrane costs are \$150 to \$250 /m^2. The system cost, excluding membranes, is given below for membrane areas of 10 to 1,000 m^2. The bmf factor for these systems is found from a low installation factor (1.1) and a normal indirect cost factor (1.29). The overall factor is thus 1.42.

$$\text{f.o.b. cost (\$)} = 1{,}468\ (\text{MSI}/800)\ (\text{area in m}^2)^{0.89}$$

Homogenizers

Homogenizers provide a mechanical means for disrupting cells to recover intracellular proteins, such as enzymes or recombinant proteins. Bead mills can be employed on small and pilot scales; at larger scales a high-pressure homogenizer is favored. The cell slurry is pumped at high pressure through a small nozzle. The correlation below is for a 8,000 psi homogenizer, as a function of the feed rate in liters per hour. The bmf for a homogenizer may be taken from the factor for a reciprocating pump, which is 3.38. Making the adjustment for stainless steel, the corrected bmf becomes 2.06.

$$\text{f.o.b. cost (\$)} = 949\ (\text{MSI}/800)\ (\text{liters/hr})^{0.50}$$

Ion Exchangers

Cation and anion exhange are very commonly employed in the separation and recovery of biomolecules. Sizing of ion exhange systems is based on the capacity of the resin and the volume of liquid to be handled. Typical anion and cation exchange resins have a capacity of ~ 0.6 to 1.0 meq/gm; a strong cation exchanger, such as Dowex HCR-S, has a theoretical capacity of 2.0 meq/ml, and might operate at 50% of this value. In the absence of dispersion and mass transfer effects in the column, this sets the total volume of exchanger material required. For protein fractionation, the most common support is cellulose, substituted with diethylaminoethyl(DEAE)- or carboxymethyl(CM)- cellulose. Sephadex ion exchangers have functional groups introduced onto dextran. Cellulose or dextran materials may be in fibrous, bead or microgranular form. Sephadex exchangers have a capacity of 10-50 meq per 100 ml. The capacity available for hemoglobin is thus 5-20 gm per 100 ml; this varies with ionic strength. Flow rates for large-scale systems (~150 liters) are around 30 cm/hr at pressure drops of 0.1 to 0.5 kPa/cm^2.

The geometry of the system can be determined by consideration of the pressure drop through the exchanger bed at the desired flow rate. Ion exchange systems are commonly skid-mounted and the bmf is reduced to account for only the labor and utilities hook-up required. A bmf similar to that for ultrafiltration equipment is suggested: 1.42. The cost of ion exchanger resin is ~$200/liter; a similar value can be assumed for gel filtration material.

Heat Exchangers

Costs for carbon steel heat exchangers are correlated with the surface area. Costs are based on shell and tube, fixed tube sheet, 100 psi exchangers. Installed costs are based on a bmf of 3.4.

f.o.b. cost ($) = 105 (MSI/300) (area in ft^2)$^{0.62}$

Air-cooled heat exchangers can be sized based on 1" O.D., 14 BWG with nine 1/2" diameter aluminum fins per inch of tube length.

Carbon steel tubes (bmf 2.3)

f.o.b. cost ($) = 78 (MSI/300) (area in ft^2)$^{0.75}$

Stainless steel tubes (316SS) (bmf 2.25)

f.o.b. cost ($) = 324 (MSI/300) (area in ft^2)$^{0.75}$

Crystallizers

Continuous forced circulation; carbon steel, tons per day of product (bmf 2.4)

f.o.b. cost ($) = 11,600 (MSI/300) (tons/day)$^{0.55}$

Filters

Plate and frame, cast iron frames, filtration area in ft^2 (bmf 2.68)

f.o.b. cost ($) = 389 (MSI/300) (area in ft^2)$^{0.58}$

Continuous rotary vacuum filter, carbon steel; area in ft^2 (bmf 2.4)

f.o.b. cost ($) = 1,620 (MSI/300) (0.25 x area in ft^2)$^{0.63}$

Rotary vacuum filters are available in various materials of construction and sizes from 3 to 12 ft in diameter. Particulate filter rates range from a low of 10 gal/ft^2-hr to a high of 100 gal/ft^2-hr (400 to 4,000 liters/m^2-hr). For proteins and similar solutes, rates range from 50 to 400 liters/m^2-hr.

Driers

Continuous rotary driers, hot air heat and carbon steel. Costs include motor, drive, fan, dust collector and solids feed (bmf 3.0). Cost does not include air heater. Area is peripheral area of drier in ft^2.

f.o.b. cost ($) = 195 (MSI/300) (area)$^{0.8}$

Cooling water towers

Towers with 15°F cooling range; bmf 1.75

f.o.b. cost ($) = 476 (MSI/300) (gallon per min water)$^{0.6}$

Mechanical Refrigeration Units

For temperatures near 40°F; bmf 1.42

f.o.b. cost ($) = 1,750 (MSI/300) (tons refrigeration)$^{0.55}$

For temperatures near -20°F (bmf 1.42)

f.o.b. cost ($) = 4,550 (MSI/300) (tons refrigeration)$^{0.55}$

Process gas compressors

Centrifugal compressor with electric motor drive; motor horsepower 50 to 500. The HP can be quickly estimated by assuming isothermal compression (bmf is 3.1).

$$HP = 0.0044\, P_1 Q_1 \ln(P_2/P_1)$$

where

P_1 is inlet pressure (psia)
P_2 is outlet gas pressure (psia)
Q_1 is inlet gas flow rate in ft^3/minute
f.o.b. cost ($) = 645 (MSI/300) ($HP^{0.8}$)

8.7 Problems

1. Production of Itaconic Acid

The flow sheet below is that of a proposed plant to produce itaconic acid from glucose. The projected market demand for itaconic acid has been determined by the marketing arm of your company to sharply increase from its current production of several thousand tonnes per year, as itaconic acid will be used with an aromatic diamine in the production of a high volume pyrrolidone.

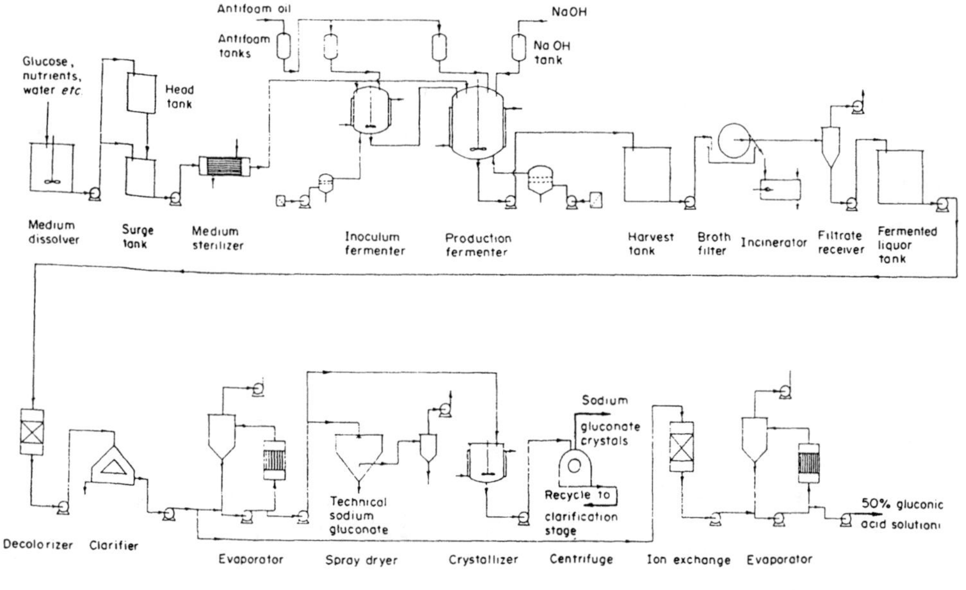

Figure 8.9*. Proposed process flow sheet for the production of itaconic acid from glucose. (From P.E. Milsom and J.L. Meers, "Gluconic and Itaconic Acids", Chap. 35, Comprehensive Biotechnology, Volume 3, eds. H.W. Blanch, S. Drew and D.I.C. Wang (1985)).*

A preliminary cost estimate is desired, for manufacturing and capital costs, based on a 2 million lb/year itaconic acid facility. The plant will use a batch fermentation with a strain of *Aspergillus terreus* with the following characteristics:

yield of itaconic acid from glucose	0.55 gm/gm
total fermentation time	72 hours
initial glucose concentration	150 gm/liter
carbon source	glucose (dextrose)
nitrogen source	$(NH_4)_2SO_4$
inoculum	5% by volume
fermentation temperature	35°C
aeration level	0.5 vvm

The properties of itaconic acid are:

melting point	167°C
solubility at 20°C	7 kg/m^3
solubility at 80°C	60 kg/m^3

The costs of the salts added to the batched medium may be neglected. Develop the process flow sheet for the desired manufacturing level. Provide mass and energy balances around each of the main equipment pieces. Using the information on capital costs of fermentation equipment, sterilizers, etc. given in the appendices, determine the operating costs, the labor and capital costs for the plant. Compare your costs with the current bulk process. What return on investment (ROI) can you achieve?

2. Recovery of Lactic Acid

Several methods for the recovery of lactic acid from a whey-based fermentation are to be considered. The fermentation produces a broth containing 45 gm/liter ammonium lactate, based on the 5% lactose content of whey. The whey was pretreated with proteases to hydrolyze the residual proteins present. The plant produces 5 million kg/yr of lactic acid and a high product purity is desired. The fermentation process is continuous and operates 320 days/yr. Prepare cost estimates for two alternative recovery processes:

(a) Extraction of lactic acid into an organic solvent (isopentyl alcohol). The broth must be first acidified with sulfuric acid so that the lactic acid is present in its neutral form. The dissociation constant of lactic acid is $1.37 \times 10^{-4} M^{-1}$. The distribution coefficient for the undissociated form of lactic acid in the solvent is 3.5. The lactic acid can then be back extracted into a basic aqueous solution of calcium hydroxide and the lactic acid recovered by acidifying the solution with H_2SO_4 and filtering the resulting $CaSO_4$ precipitate. An excess of base to lactic acid of 2 to 1 (molar) has been found to be satisfactory for the back-extraction. The extraction and back-extraction are to be carried out in a series of mixer-settlers.

(b) Recovery of lactic acid directly from filtered broth by electrodialysis. A schematic of the process is provided below. The advantage of this process is that it regenerates ammonium hydroxide that can be recycled to the fermentation for pH control. You may credit this recovery process with this recycle stream. In this process we may also assume that the membranes have a lifetime of 24 months and that the cost of electrical power is 15 cents per Kw-hour. The efficiency of a cell is determined by the number of gram-equivalents of acid or base produced per Faraday (96,500 coulomb) of current input. Assume that this is 0.85. A typical current density is 100 mA/cm^2, and the electrodialysis unit operates at a voltage of 2.0 volts at 100 mA/cm^2. Assume membrane costs are \$350/m^2.

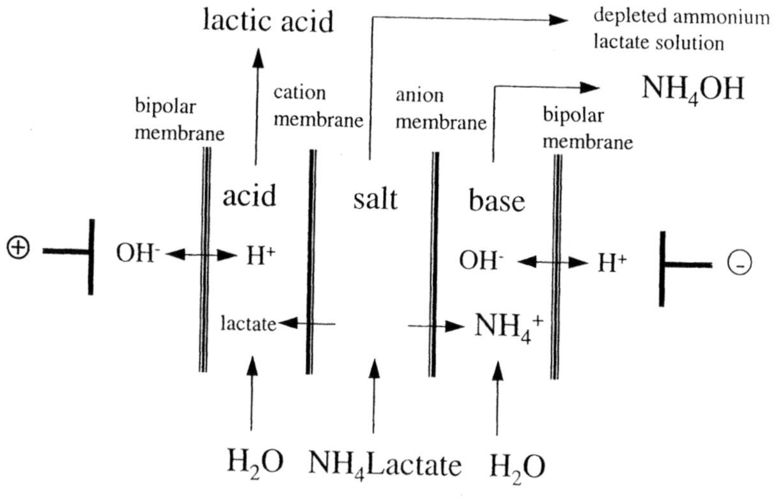

Electrodialysis for the recovery of lactic acid fermentation broth.

Conversion Factors

Length

1 in. = 2.54 cm

100 cm = 1 meter

1 micron (micrometer) = 10^{-6} m = 10^{-4} cm

1 Å (Angstrom) = 10^{-10} m = 10^{-4} micron

Volume

1 liter = 1000 cm^3

1 in^3 = 16.387 cm^3

1 ft^3 = 28.317 liter

1 ft^3 = 7.481 U.S. gal

1 U.S. gal = 3.7854 liter

1 British gal = 1.20094 U.S. gal

1 m^3 = 35.313 ft^3

1 m^3 = 264.17 U.S. gal

Mass

1 lb_m = 453.59 gm

1 kg = 2.2046 lb_m

1 ton (short) = 2000 lb_m

1 ton (long) = 2240 lb_m

1 ton (metric) = 1000 kg

Density

1 gm cm^{-3} = 62.43 lb_m ft^{-3} = 8.345 lb_m U.S. gal^{-1}

1 lb_m ft^{-3} = 16.0185 kg m^{-3}

Force

1 gm-cm sec^{-2} = 10^{-5} kg-m sec^{-2} = 10^{-5} N (Newton) = 7.2330 x 10^{-5} lb_m-ft sec^{-2}

1 lb_f = 4.4482 N

1 gm-cm sec^{-2} = 2.2481 x 10^{-6} lb_f

Pressure

1 Pa (Pascal) = 1 N m^{-2}

1 bar = 10^5 Pa

1 atm = 1.01325 bar

1 psia = 6.8947 x 10^4 gm-cm sec^{-2}

1 dyne cm^{-2} = 2.0886 x 10^{-3} $lb_f ft^{-2}$

1 psia = 1 $lb_f in^{-2}$

1 atm = 14.696 psia

1 atm = 29.921 in Hg at 0°C

1 atm = 33.90 ft H_2O at 4°C

1 $lb_f ft^{-2}$ = 47.88 N m^{-2}

Power

1 watt (W) = 14.34 cal min^{-1} = 1 joule sec^{-1}
1 BTU hr^{-1} = 0.29307 W
1 hp = 0.7457 kW
1 hp = 550 ft lb_f sec^{-1}
1 hp = 0.7068 BTU sec^{-1}

Work, Energy

1 Joule (J) = 1 N m = 1 kg m^2 sec^{-2}
1 J = 10^7 gm-cm^2 sec^{-2} (erg)
1 BTU = 1055.06 J = 252.16 cal
1 cal = 4.184 J
1 BTU = 778.17 ft lb_f
1 hp hr = 0.7457 kW hr
1 ft lb_f = 1.35582 J

Viscosity

1 cp = 10^{-2} gm cm^{-1} sec^{-1}
1 cp = 2.4191 lb_m ft^{-1} hr^{-1}
1 cp = 6.7197 x 10^{-4} lb_m ft^{-1} sec^{-1}
1 Pa sec = 1 N sec m^{-2} = 1000 cp
1 Pa sec = 0.67197 lb_m ft^{-1} sec^{-1}

Thermal conductivity

1 BTU hr^{-1} ft^{-2} $°F^{-1}$ = 4.1365 x 10^{-3} cal sec^{-1} cm^{-1} $°C^{-1}$ = 1.73 W m^{-1} K^{-1}

Heat transfer coefficient

1 BTU hr^{-1} ft^{-2} $°F^{-1}$ = 1.3571 x 10^{-4} cal sec^{-1} cm^{-2} $°C^{-1}$
1 BTU hr^{-1} ft^{-2} $°F^{-1}$ = 5.6783 x 10^{-4} W cm^{-2} $°C^{-1}$

Mass transfer coefficient

k_L	1 gm mol sec^{-1} cm^{-2} mol $frac^{-1}$ = 10 kg mol sec^{-1} m^{-2} mol $frac^{-1}$
	1 lb mol hr^{-1} ft^{-2} mol $frac^{-1}$ = 1.3562 x 10^{-3} kg mol sec^{-1} m^{-2} mol $frac^{-1}$
k_g	1 kg mol sec^{-1} m^{-2} atm^{-1} = 0.98692 x 10^{-5} kg mol sec^{-1} m^{-2} Pa^{-1}
k_La	1 lb mol hr^{-1} ft^{-3} mol $frac^{-1}$ = 4.449 x 10^{-3} kg mol sec^{-1} m^{-3} mol $frac^{-1}$
k_ga	1 lb mol hr^{-1} ft^{-3} atm^{-1} = 1.25996 x 10^{-4} kg mol sec^{-1} m^{-3} atm^{-1}

Useful constants

Avogadro's number N_{avo}

6.0225×10^{23} molecules gm-mol^{-1}

gas law constant R

1.9872 gm cal gm-mol^{-1} K^{-1}

1.9872 BTU lb-mol^{-1} °F

82.057 cm^3 atm gm-mol^{-1} K^{-1}

1545.3 ft lb$_f$ lb-mol^{-1} °R^{-1}

Boltzmann's constant $k_B = R/N_{avo}$

1.3805×10^{-16} erg molecule^{-1} °K^{-1}

1.3805×10^{-23} joule molecule^{-1} °K^{-1}

Planck's constant h

6.6256×10^{-27} erg sec

6.6256×10^{-34} joule sec

Faraday's constant F

9.652×10^{4} coulombs g-equivalent^{-1}

Electronic charge e

4.8029×10^{-10} esu

1.602×10^{-19} coulomb

1 gm-mol ideal gas at 0°C, 760 mm Hg = 22.414 liters

1 lb-mol ideal gas at 0°C, 760 mm Hg = 359.05 ft^3

density of dry air at 0°C, 760 mm Hg = 1.2929 gm/liter = 0.0807 lb$_m$/ft^3

viscosity of air 80°F, 760 mm Hg = 1.24×10^{-5} lb$_m$ ft^{-1} sec^{-1}

viscosity of water at 80°F, 760 mm Hg = 0.578 lb$_m$ ft^{-1} sec^{-1}

Index